Code of Federal Regulations

CODE OF FEDERAL REGULATIONS

T0200322

Title 7
Agriculture

Parts 1760 to 1939

Revised as of January 1, 2020

Containing a codification of documents
of general applicability and future effect

As of January 1, 2020

Published by the Office of the Federal Register
National Archives and Records Administration
as a Special Edition of the Federal Register

Table of Contents

Cite this Code: CFR

*To cite the regulations in
  this volume use title,
  part and section num-
  ber. Thus,* 7 CFR
  1767.10 *refers to title 7,
  part 1767, section 10.*

Explanation

The Code of Federal Regulations is a codification of the general and permanent rules published in the Federal Register by the Executive departments and agencies of the Federal Government. The Code is divided into 50 titles which represent broad areas subject to Federal regulation. Each title is divided into chapters which usually bear the name of the issuing agency. Each chapter is further subdivided into parts covering specific regulatory areas.

Each volume of the Code is revised at least once each calendar year and issued on a quarterly basis approximately as follows:

Title 1 through Title 16..as of January 1
Title 17 through Title 27 ....as of April 1
Title 28 through Title 41 ....as of July 1
Title 42 through Title 50..as of October 1

The appropriate revision date is printed on the cover of each volume.

LEGAL STATUS

The contents of the Federal Register are required to be judicially noticed (44 U.S.C. 1507). The Code of Federal Regulations is prima facie evidence of the text of the original documents (44 U.S.C. 1510).

HOW TO USE THE CODE OF FEDERAL REGULATIONS

The Code of Federal Regulations is kept up to date by the individual issues of the Federal Register. These two publications must be used together to determine the latest version of any given rule.

To determine whether a Code volume has been amended since its revision date (in this case, January 1, 2020), consult the "List of CFR Sections Affected (LSA)," which is issued monthly, and the "Cumulative List of Parts Affected," which appears in the Reader Aids section of the daily Federal Register. These two lists will identify the Federal Register page number of the latest amendment of any given rule.

EFFECTIVE AND EXPIRATION DATES

Each volume of the Code contains amendments published in the Federal Register since the last revision of that volume of the Code. Source citations for the regulations are referred to by volume number and page number of the Federal Register and date of publication. Publication dates and effective dates are usually not the same and care must be exercised by the user in determining the actual effective date. In instances where the effective date is beyond the cut-off date for the Code a note has been inserted to reflect the future effective date. In those instances where a regulation published in the Federal Register states a date certain for expiration, an appropriate note will be inserted following the text.

OMB CONTROL NUMBERS

The Paperwork Reduction Act of 1980 (Pub. L. 96–511) requires Federal agencies to display an OMB control number with their information collection request.

Many agencies have begun publishing numerous OMB control numbers as amendments to existing regulations in the CFR. These OMB numbers are placed as close as possible to the applicable recordkeeping or reporting requirements.

PAST PROVISIONS OF THE CODE

Provisions of the Code that are no longer in force and effect as of the revision date stated on the cover of each volume are not carried. Code users may find the text of provisions in effect on any given date in the past by using the appropriate List of CFR Sections Affected (LSA). For the convenience of the reader, a "List of CFR Sections Affected" is published at the end of each CFR volume. For changes to the Code prior to the LSA listings at the end of the volume, consult previous annual editions of the LSA. For changes to the Code prior to 2001, consult the List of CFR Sections Affected compilations, published for 1949-1963, 1964-1972, 1973-1985, and 1986-2000.

"[RESERVED]" TERMINOLOGY

The term "[Reserved]" is used as a place holder within the Code of Federal Regulations. An agency may add regulatory information at a "[Reserved]" location at any time. Occasionally "[Reserved]" is used editorially to indicate that a portion of the CFR was left vacant and not dropped in error.

INCORPORATION BY REFERENCE

What is incorporation by reference? Incorporation by reference was established by statute and allows Federal agencies to meet the requirement to publish regulations in the Federal Register by referring to materials already published elsewhere. For an incorporation to be valid, the Director of the Federal Register must approve it. The legal effect of incorporation by reference is that the material is treated as if it were published in full in the Federal Register (5 U.S.C. 552(a)). This material, like any other properly issued regulation, has the force of law.

What is a proper incorporation by reference? The Director of the Federal Register will approve an incorporation by reference only when the requirements of 1 CFR part 51 are met. Some of the elements on which approval is based are:

(a) The incorporation will substantially reduce the volume of material published in the Federal Register.

(b) The matter incorporated is in fact available to the extent necessary to afford fairness and uniformity in the administrative process.

(c) The incorporating document is drafted and submitted for publication in accordance with 1 CFR part 51.

What if the material incorporated by reference cannot be found? If you have any problem locating or obtaining a copy of material listed as an approved incorporation by reference, please contact the agency that issued the regulation containing that incorporation. If, after contacting the agency, you find the material is not available, please notify the Director of the Federal Register, National Archives and Records Administration, 8601 Adelphi Road, College Park, MD 20740-6001, or call 202-741-6010.

CFR INDEXES AND TABULAR GUIDES

A subject index to the Code of Federal Regulations is contained in a separate volume, revised annually as of January 1, entitled CFR INDEX AND FINDING AIDS. This volume contains the Parallel Table of Authorities and Rules. A list of CFR titles, chapters, subchapters, and parts and an alphabetical list of agencies publishing in the CFR are also included in this volume.

An index to the text of "Title 3—The President" is carried within that volume.

The Federal Register Index is issued monthly in cumulative form. This index is based on a consolidation of the "Contents" entries in the daily Federal Register.

A List of CFR Sections Affected (LSA) is published monthly, keyed to the revision dates of the 50 CFR titles.

REPUBLICATION OF MATERIAL

There are no restrictions on the republication of material appearing in the Code of Federal Regulations.

INQUIRIES

For a legal interpretation or explanation of any regulation in this volume, contact the issuing agency. The issuing agency's name appears at the top of odd-numbered pages.

For inquiries concerning CFR reference assistance, call 202–741–6000 or write to the Director, Office of the Federal Register, National Archives and Records Administration, 8601 Adelphi Road, College Park, MD 20740-6001 or e-mail *fedreg.info@nara.gov*.

THIS TITLE

Title 7—AGRICULTURE is composed of fifteen volumes. The parts in these volumes are arranged in the following order: Parts 1–26, 27–52, 53–209, 210–299, 300–399, 400–699, 700–899, 900–999, 1000–1199, 1200–1599, 1600–1759, 1760–1939, 1940–1949, 1950–1999, and part 2000 to end. The contents of these volumes represent all current regulations codified under this title of the CFR as of January 1, 2020.

The Food and Nutrition Service current regulations in the volume containing parts 210–299, include the Child Nutrition Programs and the Food Stamp Program. The regulations of the Federal Crop Insurance Corporation are found in the volume containing parts 400–699.

All marketing agreements and orders for fruits, vegetables and nuts appear in the one volume containing parts 900–999. All marketing agreements and orders for milk appear in the volume containing parts 1000–1199.

For this volume, Robert J. Sheehan, III was Chief Editor. The Code of Federal Regulations publication program is under the direction of John Hyrum Martinez, assisted by Stephen J. Frattini.

Title 7—Agriculture

(This book contains parts 1760 to 1939)

SUBTITLE B—REGULATIONS OF THE DEPARTMENT OF AGRICULTURE
(CONTINUED)

Subtitle B—Regulations of the Department of Agriculture (Continued)

CHAPTER XVII—RURAL UTILITIES SERVICE, DEPARTMENT OF AGRICULTURE (CONTINUED)

PARTS 1758–1766 [RESERVED]

PART 1767—ACCOUNTING REQUIREMENTS FOR RUS ELECTRIC BORROWERS

Subpart A—General [Reserved]

AUTHORITY: 7 U.S.C. 901 *et seq.*, 1921 *et seq.*, 6941 *et seq.*

SOURCE: 58 FR 59825, Nov. 10, 1993, unless otherwise noted.

Subpart A—General [Reserved]

Subpart B—Uniform System of Accounts

§ 1767.10 Definitions.

As used in this part:

Accounting borrower is an RUS borrower.

Accounts are the accounts prescribed in this system of accounts.

Actually issued as applied to securities issued or assumed by the utility, are those which have been sold to bona fide purchasers for a valuable consideration, those issued as dividends on stock, and those which have been issued in accordance with contractual requirements direct to trustees of sinking funds.

Actually outstanding as applied to securities issued or assumed by the utility, are those which have been actually issued and are neither retired nor held by or for the utility; provided, however, that securities held by trustees shall be considered as actually outstanding.

Amortization is the gradual extinguishment of an amount in an account by distributing such amount over a fixed period, over the life of the asset or liability to which it applies, or over the period during which it is anticipated the benefit will be realized.

Associated (affiliated) companies are companies or persons that directly, or indirectly through one or more intermediaries, control, or are controlled by, or under common control with, the accounting company.

Book Cost means the amount at which property is recorded in these accounts without deduction of related provisions for accrued depreciation, amortization, or for other purposes.

CFC is the National Rural Utilities Cooperative Finance Corporation.

Continuing property records are company plant records for retirement units and mass property that provide, as either a single record, or in separate records readily obtainable by references made in a single record, the following information:

(1) For each retirement unit:

(i) The name or description of the unit, or both;

(ii) The location of the unit;

7

(iii) The date the unit was placed in service;

(iv) The cost of the unit as set forth in § 1767.16 (b) and (c); and

(v) The plant control account to which the cost of the unit is charged.

(2) For each category of mass property:

(i) A general description of the property and quantity;

(ii) The quantity placed in service by vintage year;

(iii) The average cost as set forth in § 1767.16 (b) and (c); and

(iv) The plant control account to which the costs are charged.

Control (including the terms *controlling, controlled by,* and *under common control with*) is the possession, directly or indirectly, of the power to direct or cause the direction of the management and policies of a company, whether such power is exercised through one or more intermediary companies, or alone, or in conjunction with, or pursuant to an agreement, and whether such power is established through a majority or minority ownership or through voting of securities; common directors, officers, or stockholders; voting trusts; holding trusts; associated companies; contracts; or any other direct or indirect means.

Cost is the amount of money actually paid for property or services. When the consideration given is other than cash in a purchase and sale transaction, as distinguished from a transaction involving the issuance of common stock in a merger or a pooling of interest, the value of such consideration shall be determined on a cash basis.

Cost of removal is the cost of demolishing, dismantling, tearing down or otherwise removing electric plant, including the cost of transportation and handling incidental thereto. It does not include the cost of removal activities associated with asset retirement obligations that are capitalized as part of the tangible long-lived assets that give rise to the obligation. (See § 1767.15(y).

Customer is a consumer or patron.

Debt expense includes all expenses incurred in connection with the issuance and initial sale of evidence of debt, such as fees for drafting mortgages and trust deeds; fees and taxes for issuing or recording evidences of debt; costs of engraving and printing bonds and certificates of indebtedness; fees paid to trustees; specific costs of obtaining governmental authority; fees for legal services; fees and commissions paid underwriters, brokers, and salesmen for marketing such evidences of debt; fees and expenses of listing on exchanges; and other like costs.

Depreciation, as applied to depreciable electric plant, is the loss in service value, not restored by current maintenance, incurred in connection with the consumption or prospective retirement of electric plant in the course of service from causes which are known to be in current operation and against which the utility is not protected by insurance. Among the causes to be given consideration are wear and tear, decay, action of the elements, inadequacy, obsolescence, changes in the art, changes in demand and requirements of public authorities.

Discount, as applied to the securities issued or assumed by the utility, is the excess of the par (stated value of no-par stocks) or face value of the securities plus interest or dividends accrued at the date of the sale over the cash value of the consideration received from their sale.

FASB is the Financial Accounting Standards Board.

Form 7 is the January 2004 revision (or the revision of any other date which may be specified) of such Form 7, Financial and Statistical Report, or any later revision which shall have been at the time prescribed for use by Rural Development.

Form 12 is the December 2002 revision (or the revision of any other date which may be specified) of such Form 12, Operating Report—Financial, or any later revision which shall have been at the time prescribed for use by Rural Development.

G&T is a generation and transmission cooperative.

Investment advances are advances, represented by notes or by book accounts only, with respect to which it is mutually agreed or intended between the creditor and debtor that they shall be settled by the issuance of securities or shall not be subject to current settlement.

Lease, capital is a lease of property used in utility or nonutility operations, which meets one or more of the criteria stated in § 1767.15(s).

Lease, operating is a lease of property used in utility or nonutility operations, which does not meet any of the criteria stated in § 1767.15(s).

Minor items of property are the associated parts or items of which retirement units are composed.

Net salvage value is the salvage value of property retired less the cost of removal.

Nominally issued, as applied to securities issued or assumed by the utility, are those which have been signed, certified, or otherwise executed, and placed with the proper officer for sale and delivery, or pledged, or otherwise placed in some special funds of the utility, but which have not been sold, or issued direct to trustees of sinking funds in accordance with contractual requirements.

Nominally outstanding, as applied to securities issued or assumed by the utility, are those which, after being actually issued, have been reacquired by or for the utility under circumstances which require them to be considered as held alive and not retired, provided, however, that securities held by trustees shall be considered as actually outstanding.

NRECA is the National Rural Electric Cooperative Association.

Original cost, as applied to electric plant, is the cost of such property to the person first devoting it to public service.

Person is an individual, a corporation, a partnership, an association, a joint stock company, a business trust, or any organized group of persons, whether incorporated or not, or any receiver or trustee.

Premium, as applied to securities issued or assumed by the utility, is the excess of the cash value of the consideration received from their sale over the sum of their par (stated value of no-par stocks) or face value and interest or dividends accrued at the date of sale.

Project is a complete unit of improvement or development, consisting of a power house, all water conduits, all dams and appurtenant works and structures (including navigation structures) which are a part of said unit, and all storage, diverting, or forebay reservoirs directly connected therewith, the primary line or lines transmitting power therefrom to the point of junction with the distribution system or with the interconnected primary transmission system, all miscellaneous structures used and useful in connection with said unit or any part thereof, and all water rights, rights of way, ditches, dams, reservoirs, lands, or interest in lands the use and occupancy of which are necessary or appropriate in the maintenance and operation of such unit.

Property retired, as applied to electric plant, is property which has been removed, sold, abandoned, destroyed, or which for any cause has been withdrawn from service.

REA means the Rural Electrification Administration formerly an agency of the United States Department of Agriculture and predecessor agency to RUS with respect to administering certain electric and telephone loan programs.

Regional Market is an organized energy market operated by a public utility, whether directly or through a contractual relationship with another entity.

Regulatory Assets and Liabilities are assets and liabilities that result from rate actions of regulatory agencies. Regulatory assets and liabilities arise from specific revenues, expenses, gains, or losses that would have been included in net income determinations in one period under the general requirements of the Uniform System of Accounts but for it being probable:

(1) That such items will be included in a different period(s) for purposes of developing the rates the utility is authorized to charge for its utility services; or

(2) In the case of regulatory liabilities, that refunds to customers, not provided for in the other accounts, will be required.

Replacing (including replacement) when not otherwise indicated in the context, is the construction or installation of electric plant in place of property retired, together with the removal of the property retired.

Research, Development, and Demonstration (RD&D) includes all expenditures incurred by borrowers either directly or through another person or organization (such as a research institute, industry association, foundation, university, engineering company or similar contractor) in pursuing research, development, and demonstration activities including experiment, design, installation, construction, or operation. This definition includes expenditures for the implementation or development of new and/or existing concepts until technically feasible and commercially feasible operations are verified. Such research, development, and demonstration costs should be reasonably related to the existing or future utility business, broadly defined, of the borrower or in the environment in which it operates or expects to operate. The term includes, but is not limited to, all such costs incidental to the design, development or implementation of an experimental facility, a plant process, a product, a formula, an invention, a system or similar items, and the improvement of already existing items of a like nature; amounts expended in connection with the proposed development and/or proposed delivery of alternate sources of electricity; and the costs of obtaining its own patent, such as attorney's fees expended in making and perfecting a patent application. The term includes preliminary investigations and detailed planning of specific projects for securing for customers non-conventional electric power supplies that rely on technology that has not been verified previously to be feasible. The term does not include expenditures for efficiency surveys; studies of management, management techniques, and organization; or consumer surveys, advertising, promotions, or items of a like nature.

Retirement units are those items of electric plant which, when retired with or without replacement, are accounted for by crediting the book cost thereof to the electric plant accounts in which included.

RUS means the Rural Utilities Service, an agency of the United States Department of Agriculture established pursuant to Section 232 of the Federal Crop Insurance Reform and Department of Agriculture Reorganization Act of 1994 (Pub. L. 103–354, 108 Stat. 3178), successor to REA with respect to administering certain electric and telephone programs. See 7 CFR 1700.1.

RUS Form 7 is the August 1988 revision (or the revision of any other date which may be specified) of such RUS Form 7, Financial and Statistical Report, or any later revision which shall have been at the time prescribed for use by RUS.

RUS Form 12 is the November 1979 revision (or the revision of any other date which may be specified) of such RUS Form 12, Operating Report—Financial, or any later revision which shall have been at the time prescribed for use by RUS.

RUS USoA is the USoA prescribed in this subpart.

Salvage value is the amount received for property retired, less any expenses incurred in connection with the sale or in preparing the property for sale; or, if retained, the amount at which the material recovered is chargeable to materials and supplies, or other appropriate accounts.

Service life is the time between the date electric plant is includible in electric plant in service, or electric plant leased to others, and the date of its retirement. If depreciation is accounted for on a production basis rather than on a time basis, service life should be measured in terms of the appropriate unit of production.

Service value is the difference between original cost and net salvage value of electric plant.

State is a State admitted to the Union, the District of Columbia, and any organized Territory of the United States.

Subsidiary company is a company which is controlled by the utility through ownership of voting stock. (See the definition of control in §1767.10.) A corporate joint venture in which a corporation is owned by a small group of businesses as a separate and specific business or project for the mutual benefit of the members of the group is a subsidiary company for the purposes of this system of accounts.

Utility is an RUS borrower.

Work order is an order authorizing the construction of utility plant. It

10

serves as the basis for the accounts or subaccounts in which costs are recorded.

[58 FR 59825, Nov. 10, 1993, as amended at 59 FR 66440, Dec. 27, 1994; 73 FR 30279, May 27, 2008]

§ 1767.11 Purpose.

(a) The standard form of RUS loan documents for electric borrowers requires that the borrower keep books, records, and accounts in which full and true entries will be made of all of the dealings, business and affairs of the borrower in accordance with the methods and principles of accounting of this part.

(b) This subpart implements these provisions of the RUS loan documents by prescribing the RUS USoA for electric borrowers and by providing accounting methodologies and procedures which are applicable to particular situations.

§ 1767.12 Accounting system requirements.

(a) Each Rural Development electric borrower must maintain and keep its books of accounts and all other books and records that support the entries in such books of accounts in accordance with §§ 1767.13–1767.31.

(b) Each RUS electric borrower shall maintain and keep its books of accounts and all other books and records which support the entries in such books of accounts in accordance with § 1767.41, Accounting Methods and Procedures Required of All RUS Borrowers, herein, which prescribes accounting principles to be applied to specific factual circumstances.

[58 FR 59825, Nov. 10, 1993, as amended at 73 FR 30280, May 27, 2008]

§ 1767.13 Departures from the prescribed RUS Uniform System of Accounts.

(a) No departures are to be made to the prescribed Rural Development USoA without the prior written approval of Rural Development. Requests for departures from the Rural Development USoA shall be addressed, in writing, to the Assistant Administrator, Program Accounting and Regulatory Analysis. (AA–PARA).

(b) RUS borrowers subject to the jurisdiction of a state regulatory authority with jurisdiction over rates and/or accounting for electric utilities will not:

(1) Request approval of such authority to use accounting methodologies and principles that depart from the provisions herein; or

(2) File with such authority, any documents or information, including without limitation, any filings associated with the borrower's rates, based upon accounting methods and principles inconsistent with the provisions of this part.

(c) If any state regulatory authority with jurisdiction over an RUS borrower prescribes accounting methods or principles for the borrower that are inconsistent with the provisions of this part, the borrower must immediately notify the Director, BAD, and provide such documents, information, and reports as RUS may request to evaluate the impact that such accounting methods or principles may have on the interests of RUS.

(1) If RUS determines that the accounting methods and principles do not adversely impact RUS interests, RUS will permit the borrower to use the accounting methods and principles as prescribed by the state regulatory authority to comply with the provisions of the RUS loan documents.

(2) If RUS determines that the accounting methods and principles may adversely impact RUS's interests, RUS may require that, for the purposes of complying with provisions of RUS loan documents, including, without limitation, those provisions relating to financial coverage standards (e.g. "TIER"), the borrower continue to maintain books, records, and accounts in accordance with this subpart.

(i) RUS may, however, approve requests by the borrower to maintain such additional books, records, and accounts as necessary to comply with the requirements of the state regulatory authority.

(ii) Such approval will not waive, modify or amend the requirements of the RUS loan documents or of this subpart.

(d) RUS borrowers will not implement the provisions of Statement of

Financial Accounting Standards (SFAS) No. 71, Accounting for the Effects of Certain Types of Regulation, SFAS No. 90, Regulated Enterprises—Accounting for Abandonments and Disallowances of Plant Costs, SFAS No. 92, Regulated Enterprises—Accounting for Phase-in Plans, without the prior written approval of RUS except as provided for in paragraphs (d)(1) through (d)(5) of this section. Requests for approval shall be addressed, in writing, to the Director, PASD. The specific deferrals set forth in paragraphs (d)(1) through (d)(5) of this section may be implemented without the prior written approval of RUS provided that the deferrals comply with Statement No. 71 and that the RUS borrowers implementing such deferrals continue to meet the requirements set forth in Statement No. 71 for doing so:

(1) The deferral and amortization of prior service pension costs (See § 1767.41, Interpretation No. 606, Pension Costs), remapping expenses (See § 1767.41, Interpretation No. 613, Mapping Costs), and preliminary survey and investigation charges (See § 1767.17, Interpretation No. 111, Engineering Contracts for System Planning);

(2) The deferral of any current period expense or expenses, on a cumulative basis for the fiscal year, only if a borrower would have met each of its financial tests or coverage ratios that it has covenanted with RUS to meet for that fiscal year, had the deferral not been made;

(3) The deferral of any cost that will be fully amortized within the next 12 succeeding months;

(4) The accelerated amortization of any previously deferred expense; and

(5) The deferral of revenues coincident with a moratorium imposed by the National Rural Electric Cooperative Association on its Retirement and Security Program, provided, however, that the deferral is for the sole purpose of offsetting future pension costs.

(e) RUS will consider approval of specific departures from this part upon submission of:

(1) A detailed description of the proposed departure;

(2) The specific accounting journal entries that will be used including the account number and title, and the dollar amounts where appropriate;

(3) The total dollar amount of the departure and the impact on margins during the time period of the departure; and

(4) A resolution from the borrower's Board of Directors authorizing such action; and

(5) Any additional information RUS may deem necessary to adequately evaluate the borrower's request.

(f) RUS will, within 90 days of final receipt of this information, render a decision on the borrower's request for a departure from the prescribed RUS USoA.

(1) If, due to extenuating circumstances, RUS is unable to reach a decision within the required time period, RUS will notify the borrower of the delay within this same 90-day period, and provide a projected decision date.

(2) The requested departure from the prescribed RUS USoA must not be implemented until final approval is granted by RUS.

[58 FR 59825, Nov. 10, 1993, as amended at 60 FR 55429, Nov. 1, 1995; 62 FR 42289, Aug. 6, 1997; 73 FR 30280, May 27, 2008]

§ 1767.14 **Interpretations of the Rural Development uniform system of accounts.**

To maintain uniformity in accounting, borrowers must submit questions concerning interpretations of the Rural Development USoA, in writing, to the AA–PARA, for consideration and decision.

(Approved by the Office of Management and Budget under control number 0572–0002)

[73 FR 30280, May 27, 2008]

§ 1767.15 **General instructions.**

(a) *Records.* (1) Each utility shall keep its books of account, and all other books, records, and memoranda which support the entries in such books of account so as to be able to furnish readily full information as to any item included in any account.

(2) Each entry shall be supported by such detailed information as will permit ready identification, analysis, and verification of all facts relevant thereto.

(3) The books and records referred to herein include not only accounting records in a limited technical sense, but all other records, such as minute books, stock books, reports, correspondence, memoranda, etc., which may be useful in developing the history of or facts regarding any transaction.

(4) No utility shall destroy any such books or records unless the destruction thereof is permitted by the rules and regulations contained in subpart D of this part.

(5) In addition to the prescribed accounts, clearing accounts, temporary or experimental accounts, and subdivisions of any accounts, may be kept, provided the integrity of the prescribed accounts is not impaired.

(6) When the utility chooses to recognize the gain in the year of reacquisition as a taxable gain, Account 411.1, Provision for Deferred Income Taxes—Credit, Utility Operating Income, shall be credited with the amount of the related tax effect, such amount to be allocated to the periods affected in accordance with the provisions of Account 190, Accumulated Deferred Income Taxes.

(7) The arrangement or sequence of the accounts prescribed herein shall not be controlling as to the arrangement or sequence in report forms which may be prescribed by RUS.

(b) *Numbering system.* (1) The account numbering plan used herein consists of a system of three-digit whole numbers as follows:

100–199   Assets and other debits.
200–299   Liabilities and other credits.
300–399   Plant accounts.
400–432, 434–435   Income accounts.
433, 436–439   Retained earnings accounts.
440–459   Revenue accounts.
500–599   Production, transmission, and distribution expenses.
900–949   Customer accounts, customer service and informational, sales, and general and administrative expenses.

(2) In certain instances, numbers have been skipped in order to allow for possible later expansion or to permit better coordination with the numbering system for other utility departments.

(3) The numbers prefixed to account titles are to be considered as parts of the titles.

(i) Each utility, however, may adopt, for its own purposes, a different system of account numbers provided that the numbers herein prescribed shall appear in the descriptive headings of the ledger accounts and in the various sources of original entry.

(ii) If a utility uses a different group of account numbers and it is not practicable to show the prescribed account numbers in the various sources of original entry, such reference to the prescribed account numbers may be omitted from the various sources of original entry.

(iii) Each utility using different account numbers for its own purposes shall keep readily available, a list of such account numbers which it uses and a reconciliation of such account numbers with the account numbers provided herein.

(iv) The utility's records shall be so kept as to permit ready analysis by prescribed accounts (by direct reference to sources of original entry to the extent practicable) and to permit preparation of financial and operating statements directly from such records at the end of each accounting period according to the prescribed accounts.

(c) *Accounting period.* (1) Each utility shall keep its books on a monthly basis so that for each month, all transactions applicable thereto, as nearly as may be ascertained, shall be entered in the books of the utility.

(2) Amounts applicable or assignable to specific utility departments shall be so segregated monthly.

(3) Each utility shall close its books at the end of each fiscal year unless otherwise authorized by RUS.

(d) *Submission of questions.* To maintain uniformity of accounting, utilities shall submit questions of doubtful interpretation to RUS for consideration and decision.

(e) *Item lists.* (1) Lists of "items" appearing in the texts of the accounts or elsewhere herein are for the purpose of more clearly indicating the application of the prescribed accounting.

(2) The lists are intended to be representative, but not exhaustive.

(3) The appearance of an item in a list warrants the inclusion of the item in the account mentioned only when the text of the account also indicates

inclusion inasmuch as the same item frequently appears in more than one list.

(4) The proper entry in each instance must be determined by the texts of the accounts.

(f) *Extraordinary items.* (1) Net income shall reflect all items of profit and loss during the period with the exception of prior period adjustments as described in § 1767.15 (g) and long-term debt as described in § 1767.15 (q).

(2) Those items related to the effects of events and transactions which have occurred during the current period and which are not typical or customary business activities of the company shall be considered extraordinary items.

(3) They will be events and transactions of significant effect which would not be expected to recur frequently and which would not be considered as recurring factors in any evaluation of the ordinary operating processes of business.

(i) In determining significance, items of a similar nature should be considered in the aggregate.

(ii) Dissimilar items should be considered individually; however, if they are few in number, they may be considered in the aggregate.

(iii) To be considered as extraordinary under the above guidelines, an item should be more than approximately 5 percent of income, computed before extraordinary items.

(iv) RUS approval must be obtained to treat an item of less than 5 percent, as extraordinary. (See Accounts 434 and 435.)

(g) *Prior period items.* (1) Items of profit and loss related to the following shall be accounted for as prior period adjustments and excluded from the determination of net income for the current year:

(i) Correction of an error in the financial statements of a prior year

(ii) Adjustments that result from realization of income tax benefits of preacquisition operating loss carryforwards of purchased subsidiaries.

(2) All other items of profit and loss recognized during the year shall be included in the determination of net income for that year.

(h) *Unaudited items.* (1) Whenever a financial statement is required by RUS, if it is known that a transaction has occurred which affects the accounts but the amount involved in the transaction and its effect upon the accounts cannot be determined with absolute accuracy, the amount shall be estimated and such estimated amount included in the proper accounts.

(2) The utility is not required to anticipate minor items which would not appreciably affect the accounts.

(i) *Distribution of pay and expenses of employees.* Charges to electric plant, operating expense, and other accounts for services and expenses of employees engaged in activities chargeable to various accounts, such as construction, maintenance, and operations, shall be based upon the actual time engaged in the respective classes of work, or in case that method is impracticable, upon the basis of a study of the time actually engaged during a representative period.

(j) *Payroll distribution.* (1) Underlying accounting data shall be maintained so that the distribution of the cost of labor charged direct to the various accounts will be readily available.

(2) Such underlying data shall permit a reasonably accurate distribution to be made of the cost of labor charged initially to clearing accounts so that the total labor cost may be classified among construction, cost of removal, electric operating functions (steam generation, nuclear generation, hydraulic generation, transmission, distribution, etc.) and nonutility operations.

(k) *Accounting on an accrual basis.* (1) The utility is required to keep its accounts on the accrual basis.

(i) This requires the inclusion, in its accounts, of all known transactions of appreciable amount which affect the accounts.

(ii) If bills covering such transactions have not been received or rendered, the amounts shall be estimated and appropriate adjustments made when the bills are received.

(2) When payments are made in advance for items such as insurance, rents, taxes, or interest, the amount applicable to future periods shall be charged to Account 165, Prepayments,

and spread over the periods to which applicable, by credits to Account 165, and charges to the accounts appropriate for the expenditure.

(1) *Records for each plant.* (1) Separate records shall be maintained by electric plant accounts of the book cost of each plant owned, including additions by the utility to plant leased from others, and of the cost of operating and maintaining each plant owned or operated.

(2) The term "plant" as used herein includes each generating station and each transmission line or appropriate group of transmission lines.

(m) *Accounting for other departments.* (1) If the utility also operates other utility departments, such as gas or water, it shall keep such accounts for the other departments as may be prescribed by proper authority and in the absence of prescribed accounts, it shall keep such accounts as are proper or necessary to reflect the results of operating each such department.

(2) It is not intended that proprietary and similar accounts which apply to the utility as a whole shall be departmentalized.

(n) *Transactions with associated companies.* (1) Each utility shall keep its accounts and records so as to be able to furnish accurately and expeditiously statements of all transactions with associated companies.

(2) The statements may be required to show the general nature of the transactions, the amounts involved therein and the amounts included in each account prescribed herein with respect to such transactions. Transactions with associated companies shall be recorded in the appropriate accounts for transactions of the same nature. Nothing herein contained, however, shall be construed as restraining the utility from subdividing accounts for the purpose of recording separately transactions with associated companies.

(o) *Contingent assets and liabilities.* (1) Contingent assets represent a possible source of value to the utility contingent upon the fulfillment of conditions regarded as uncertain.

(2) Contingent liabilities include items which may, under certain conditions, become obligations of the utility but which are neither direct nor as-

sumed liabilities at the date of the balance sheet. The utility shall be prepared to give a complete statement of significant contingent assets and liabilities (including cumulative dividends on preference stock) in its audited financial statements; its RUS Form 7, Financial and Statistical Report, or its RUS Form 12, Operating Report—Financial; and at such other times as may be requested by RUS.

(p) *Separate accounts or records for each licensed project.* The accounts or records of each borrower shall be so kept as to show for each project (including pumped storage) under license:

(1) The actual legitimate original cost of the project, including the original cost of the original project, the original cost of additions thereto and betterments thereof, and credits for property retired from service, as determined under RUS's regulations in 7 CFR chapter XVII;

(2) The charges for operation and maintenance of the project property directly assignable to the project;

(3) The credits and debits to the depreciation and amortization accounts, and the balances in such accounts; and

(4) The credits and debits to the operating revenue, income, and retained earnings accounts that can be identified with and directly assigned to the project.

NOTE: The purpose of this instruction is to insure that accounts or records are currently maintained by each borrower from which reports may be made to RUS for use in determining the net investment in each licensed project. The instruction covers only the debit and credit items appearing in the borrower's accounts which may be identified with and assigned directly to any project. In the determination of the net investment, allocations of items affecting the net investment may be required where direct assignment is not practicable.

(q) *Long-term debt: premium, discount and expense, and gain or loss on reacquisition*—(1) *Premium, discount and expense.* (i) A separate premium, discount and expense account shall be maintained for each class and series of long-term debt (including receivers' certificates) issued or assumed by the utility.

(ii) The premium will be recorded in Account 225, Unamortized Premium on Long-Term Debt, the discount will be recorded in Account 226, Unamortized

15

Discount on Long-Term Debt—Debit, and the expense of issuance shall be recorded in Account 181, Unamortized Debt Expense.

(iii) The premium, discount and expense shall be amortized over the life of the respective issues under a plan which will distribute the amounts equitably over the life of the securities.

(A) The amortization shall be charged or credited on a monthly basis with the amounts relating to discount and expense charged to Account 428, Amortization of Debt Discount and Expense.

(B) The amounts relating to premium shall be credited to Account 429, Amortization of Premium on Debt—Credit.

(2) *Reacquisition, without refunding.* (i) When long-term debt is reacquired or redeemed without being converted into another form of long-term debt and when the transaction is not in connection with a refunding operation (primarily redemptions for sinking fund purposes), the difference between the amount paid upon reacquisition and the face value; plus any unamortized premium less any related unamortized debt expense and reacquisition costs; or less any unamortized discount, related debt expense and reacquisition costs applicable to the debt redeemed, retired and cancelled, shall be included in Account 189, Unamortized Loss on Reacquired Debt, or Account 257, Unamortized Gain on Reacquired Debt, as appropriate.

(ii) The utility shall amortize the recorded amounts equally on a monthly basis over the remaining life of the respective security issues (old original debt).

(iii) The amount so amortized shall be charged to Account 428.1, Amortization of Loss on Reacquired Debt, or credited to Account 429.1, Amortization of Gain on Reacquired Debt—Credit, as appropriate.

(3) *Reacquisition, with refunding.* (i) When the redemption of one issue or series of bonds or other long-term obligations is financed by another issue or series before the maturity date of the first issue, the difference between the amount paid upon refunding and the face value; plus any unamortized premium less related debt expense or less any unamortized discount and related

debt expense, applicable to the debt refunded, shall be included in Account 189, Unamortized Loss on Reacquired Debt, or Account 257, Unamortized Gain on Reacquired Debt, as appropriate.

(ii) The utility may elect to account for such amounts as follows:

(A) Write them off immediately when the amounts are insignificant;

(B) Amortize them by equal monthly amounts over the remainder of the original life of the issue retired; or

(C) Amortize them by equal monthly amounts over the life of the new issue.

(iii) Once an election is made, it shall be applied on a consistent basis.

(iv) The amounts in paragraphs (q)(3)(ii)(A), (B), or (C) of this section shall be charged to Account 428.1, Amortization of Loss on Reacquired Debt, or credited to Account 429.1, Amortization of Gain on Reacquired Debt—Credit, as appropriate.

(4) Under methods in paragraphs (q)(3)(ii)(B) and (C) of this section, the increase or reduction in current income taxes resulting from the reacquisition should be apportioned over the remainder of the original life of the issued retired or over the life of the new issue, as appropriate, as directed more specifically in paragraphs (q)(5) and (6) of this section.

(5) When the utility recognizes the loss in the year of reacquisition as a tax deduction, Account 410.1, Provision for Deferred Income Taxes, Utility Operating Income, shall be debited and Account 283, Accumulated Deferred Income Taxes—Other, shall be credited with the amount of the related tax effect, such amount to be allocated to the periods affected in accordance with the provisions of Account 283.

(6) When the utility chooses to recognize the gain in the year of reacquisition as a taxable gain, Account 411.1, Provision for Deferred Income Taxes—Credit, Utility Operating Income, shall be debited with the amount of the related tax effect, such amount to be allocated to the periods affected in accordance with the provisions of Account 190, Accumulated Deferred Income Taxes.

(7) When the utility chooses to use the optional privilege of deferring the

tax on the gain attributable to the re-acquisition of debt by reducing the depreciable basis of utility property for tax purposes, pursuant to Section 108 of the Internal Revenue Code (26 U.S.C. 108), the related tax effects shall be deferred as the income is recognized for accounting purposes, and the deferred amounts shall be amortized over the life of the associated property on a vintage year basis.

(i) Account 410.1, Provision for Deferred Income Taxes, Utility Operating Income, shall be debited, and Account 282, Accumulated Deferred Income Taxes—Other Property, shall be credited with an amount equal to the estimated income tax effect applicable to the portion of the income, attributable to reacquired debt, recognized for accounting purposes during the period.

(ii) Account 282 shall be debited and Account 411.1, Provision for Deferred Income Taxes—Credit, Utility Operating Income, shall be credited with an amount equal to the estimated income tax effects, during the life of the property, attributable to the reduction in the depreciable basis for tax purposes.

(8) The tax effects relating to gain or loss shall be allocated as above to utility operations except in cases where a portion of the debt reacquired is directly applicable to nonutility operations.

(i) In that event, the related portion of the tax effects shall be allocated to nonutility operations.

(ii) Where it can be established that reacquired debt is generally applicable to both utility and nonutility operations, the tax effects shall be allocated between utility and nonutility operations based on the ratio of net investment in utility plant to net investment in nonutility plant.

(9) Premium, discount, or expense on debt shall not be included as an element in the cost of construction or acquisition of property (tangible or intangible), except under the provisions of Account 432, Allowance for Borrowed Funds Used During Construction—Credit.

(10) *Alternate method.* Where a regulatory authority or a group of regulatory authorities having prime rate jurisdiction over the utility specifically disallows the rate principle of amortizing gains or losses on reacquisition of long-term debt without refunding, and does not apply the gain or loss to reduce interest charges in computing the allowed rate of return for rate purposes, the following alternate method may be used to account for gains or losses relating to reacquisition of long-term debt, with or without refunding:

(i) The difference between the amount paid upon reacquisition of any long-term debt and the face value, adjusted for unamortized discount, expenses or premium, as the case may be, applicable to the debt redeemed shall be recognized currently in income and recorded in Account 421, Miscellaneous Nonoperating Income, or Account 426.5, Other Deductions.

(ii) When this alternate method of accounting is used, the utility shall include a footnote to each financial statement, prepared for public use, explaining why this method is being used along with the treatment given for ratemaking purposes.

(r) *Comprehensive interperiod income tax allocation.* (1) Where there are timing differences between the periods in which transactions affect taxable income and the periods in which they enter into the determination of pretax accounting income, the income tax effects of such transactions are to be recognized in the periods in which the differences between book accounting income and taxable income arise and in the periods in which the differences reverse using the deferred tax method.

(2) Comprehensive interperiod tax allocation should be followed whenever transactions enter into the determination of pretax accounting income for the period even though some transactions may affect the determination of taxes payable in a different period.

(3) Utilities are not required to utilize comprehensive interperiod income tax allocation until the deferred income taxes are included as an expense in the rate level by the regulatory authority having rate jurisdiction over the utility.

(4) Where comprehensive interperiod tax allocation accounting is not practiced the utility shall include as a note to each financial statement, prepared for public use, a footnote explanation

setting forth the utility's accounting policies with respect to interperiod tax allocation and describing the treatment for rate making purposes of the tax timing differences by regulatory authorities having rate jurisdiction.

(5) Should the utility be subject to more than one agency having rate jurisdiction, its accounts shall appropriately reflect the ratemaking treatment (deferral or flow through) of each jurisdiction.

(6) Once comprehensive interperiod tax allocation has been initiated either in whole or in part it shall be practiced on a consistent basis and shall not be changed or discontinued without prior RUS approval.

(7) Tax effects deferred currently will be recorded as deferred debits or deferred credits in Accounts 190, Accumulated Deferred Income Taxes; 281, Accumulated Deferred Income Taxes—Accelerated Amortization Property; 282, Accumulated Deferred Income Taxes—Other Property, and 283, Accumulated Deferred Taxes—Other, as appropriate.

(8) The resulting amounts recorded in these accounts shall be disposed of as prescribed in this system of accounts or as otherwise authorized by RUS.

(s) *Criteria for classifying leases.* (1) If, at its inception, a lease meets one or more of the following criteria, the lease shall be classified as a capital lease:

(i) The lease transfers ownership of the property to the lessee by the end of the lease term.

(ii) The lease contains a bargain purchase option.

(iii) The lease term is equal to 75 percent or more of the estimated economic life of the leased property. However, if the beginning of the lease term falls within the last 25 percent of the total estimated economic life of the leased property, including earlier years of use, this criterion shall not be used for purposes of classifying the lease.

(iv) The present value at the beginning of the lease term of the minimum lease payments, excluding that portion of the payments representing executory costs such as insurance, maintenance, and taxes to be paid by the lessor, including any profit thereon, equals or exceed 90 percent of the excess of the fair value of the leased property to the lessor at the inception of the lease over any related investment tax credit retained by the lessor and expected to be realized by lessor.

(A) However, if the beginning of the lease term falls within the last 25 percent of the total estimated economic life of the leased property, including earlier years of use, this criterion shall not be used for purposes of classifying the lease.

(B) The lessee utility shall compute the present value of the minimum lease payments using its incremental borrowing rate, unless it is practicable for the utility to learn the implicit rate computed by the lessor, and the implicit rate computed by the lessor is less than the lessee's incremental borrowing rate. If both of those conditions are met, the lessee shall use the implicit rate.

(2) If, at any time, the lessee and lessor agree to change the provisions of the lease, other than by renewing the lease or extending its term, in a manner that would have resulted in a different classification of the lease under the criteria in paragraph (s)(1) of this section had the changed terms been in effect at the inception of the lease, the revised agreement shall be considered as a new agreement over its term, and the criteria in paragraph (s)(1) of this section shall be applied for purposes of the expiration of the existing lease term, such as the exercise of a lease renewal option other than those already included in the lease term, shall be considered as a new agreement and shall be classified according to the above provision. Changes in estimates (for example, changes in estimates of the economic life or of the residual value of the leased property) or changes in circumstances (for example, default by the lessee) shall not give rise to a new classification of a lease for accounting purposes.

(t) *Accounting for leases.* (1) All leases shall be classified as either capital or operating leases.

(2) The utility shall record a capital lease as an asset in Account 101.1, Property Under Capital Leases, Account 120.6, Nuclear Fuel Under Capital Leases or Account 121, Nonutility Property;

(3) The utility, as a lessee, shall recognize an asset retirement obligation arising from the plant under a capital lease unless the obligation is recorded as an asset and liability under a capital lease. The utility shall record the asset retirement cost by debiting Account 101.1, Property Under Capital Leases, or Account 120.6, Nuclear Fuel Under Capital Leases, or Account 121, Nonutility Property, as appropriate, and crediting the liability for the asset retirement obligation in Account 230, Asset Retirement Obligations. Asset retirement costs recorded in Account 101.1, Account 120.6, or Account 121 shall be amortized by charging rent expense, or Account 518, Nuclear Fuel Expense, or Account 421, Miscellaneous Nonoperating Income, as appropriate, and crediting a separate subaccount of the account in which the asset retirement costs are recorded. Charges for the periodic accretion of the liability in Account 230, Asset Retirement Obligations, shall be recorded by a charge to Account 411.10, Accretion Expense, for electric utility plant, and Account 421, Miscellaneous Nonoperating Income, for nonutility plant and a credit to Account 230, Asset Retirement Obligations.

(4) Rental payments on all leases shall be charged to rent expense, fuel expense, construction work in progress, or other appropriate accounts as they become payable.

(5) For a capital lease, for each period during the lease term, the amounts recorded for the asset and obligation shall be reduced by an amount equal to the portion of each lease payment that would have been allocated to the reduction of the obligation, if the payment had been treated as a payment on an installment obligation (liability) and allocated between interest expense and a reduction of the obligation so as to produce a constant periodic rate of interest on the remaining balance.

(u) *Allowances.* (1) Title IV of the Clean Air Act Amendments of 1990, Pub. L. 101–549, 104 Stat. 2399, 2584 (42 U.S.C. 7407 and 42 U.S.C. 7651), provides for the issuance of allowances as a means to limit the emissions of certain airborne pollutants by various entities, including utilities. Utilities owning allowances, other than those acquired for speculative purposes, shall account for such allowances at cost in Account 158.1, Allowance Inventory, or Account 158.2, Allowances Withheld, as appropriate. Allowances acquired for speculative purposes and identified as such in contemporaneous records at the time of purchase shall be accounted for in Account 124, Other Investments.

(2) When purchased, allowances become eligible for use in different years, and the allocation of the purchase cost cannot be determined by fair value, the purchase cost allocated to allowances of each vintage shall be determined through use of a present-value based measurement. The interest rate used in the present-value measurement shall be the utility's incremental borrowing rate, in the month in which the allowances are acquired, for a loan with a term similar to the period that it will hold the allowances and in an amount equal to the purchase price.

(3) The underlying records supporting Account 158.1 and Account 158.2 shall be maintained in sufficient detail so as to provide the number of allowances and the related cost by vintage year.

(4) Issuances from inventory included in Account 158.1 and Account 158.2 shall be accounted for on a vintage basis using a monthly weighted-average method of cost determination. The cost of eligible allowances not used in the current year shall be transferred to the vintage for the immediately following year.

(5) Account 158.1 shall be credited and Account 509, Allowances, debited so that the cost of the allowances to be remitted for the year is charged to expense monthly based on each month's emissions. This may, in certain circumstances, require allocation of the cost of an allowance between months on a fractional basis.

(6) In any period in which actual emissions exceed the amount allowable based on eligible allowances owned, the utility shall estimate the cost to acquire the additional allowances needed and charge Account 158.1 with the estimated cost. This estimated cost of future allowance acquisitions shall be credited to Account 158.1 and charged to Account 509 in the same accounting period as the related charge to Account 158.1. Should the actual cost of these

allowances differ from the estimated cost, the differences shall be recognized in the then-current period's inventory issuance cost.

(7) Any penalties assessed by the Environmental Protection Agency for the emission of excess pollutants shall be charged to Account 426.3, Penalties.

(8) Gains on dispositions of allowances, other than allowances held for speculative purposes, shall be accounted for as follows. First, if there is uncertainty as to the regulatory treatment, the gain shall be deferred in Account 254, Other Regulatory Liabilities, pending resolution of the uncertainty. Second, if there is certainty as to the existence of a regulatory liability, the gain will be credited to Account 254, with subsequent recognition in income when reductions in charges to customers occur or the liability is otherwise satisfied. Third, all other gains will be credited to Account 411.8, Gains from Disposition of Allowances. Losses on disposition of allowances, other than allowances held for speculative purposes, shall be accounted for as follows. Losses that qualify as regulatory assets shall be charged directly to Account 182.3, Other Regulatory Assets. All other losses shall be charged to Account 411.9, Losses from Disposition of Allowances. (See the definition of regulatory assets and liabilities.) Gains or losses on disposition of allowances held for speculative purposes shall be recognized in Account 421, Miscellaneous Nonoperating Income, or Account 426.5, Other Deductions, as appropriate.

(9) The costs and benefits of exchange-traded allowance futures contracts used to protect the utility from the risk of unfavorable price changes ("hedging transactions") shall be deferred in Account 186, Miscellaneous Deferred Debits, or Account 253, Other Deferred Credits, as appropriate. Such deferred amounts shall be included in Account 158.1, Allowance Inventory, in the month in which the related allowances are acquired, sold or otherwise disposed of. Where the costs or benefits of hedging transactions are not identifiable with specific allowances, the amounts shall be included in Account 158.1 when the futures contract is closed. The costs and benefits of exchange-traded allowance futures contracts entered into as a speculating activity shall be charged or credited to Account 421, Miscellaneous Nonoperating Income, or Account 426.5, Other Deductions, as appropriate.

(v) *Depreciation accounting*—(1) *Method.* Utilities must use a method of depreciation that allocates in a systematic and rational manner the service value of depreciable property over the service life of the property.

(2) *Service lives.* Estimated useful service lives of depreciable property must be supported by engineering, economic, and other depreciation studies.

(3) *Rate.* Utilities must use percentage rates of depreciation that are based on a method of depreciation that allocates in a systematic and rational manner the service value of depreciable property to the service life of the property. Where composite depreciation rates are used, they should be based on the weighted average estimated useful service lives of the depreciable property comprising the composite group.

(w) *Accounting for other comprehensive income.* (1) Utilities shall record items of other comprehensive income in Account 209, Accumulated Other Comprehensive Income. Amounts included in this account shall be maintained by each category of other comprehensive income. Examples of categories of other comprehensive income include foreign currency items, minimum pension liability adjustments, unrealized gains and losses on available-for-sale type securities and cash flow hedge amounts. Supporting records shall be maintained for Account 209 so that the cumulative amount of other comprehensive income for each item included in this account can be readily identified.

(2) When an item of other comprehensive income enters into the determination of net income in the current or subsequent periods, a reclassification adjustment shall be recorded in Account 209 to avoid double counting of that amount.

(3) When it is probable that an item of other comprehensive income will be included in the development of cost-of-service rates in subsequent periods, that amount of unrealized losses or gains will be recorded in Accounts

182.3, Other Regulatory Assets or 254, Other Regulatory Liabilities, as appropriate.

(x) *Accounting for derivative instruments and hedging activities.* (1) Utilities shall recognize derivative instruments as either assets or liabilities in the financial statements and measure those instruments at fair value, except those falling within recognized exceptions. Normal purchases or sales are contracts that provide for the purchase or sale of goods that will be delivered in quantities expected to be used or sold by the utility over a reasonable period in the normal course of business. A derivative instrument is a financial instrument or other contract with all of the following characteristics:

(i) It has one or more underlyings and a notional amount or payment provision. Those terms determine the amount of the settlement or settlements, and, in some cases, whether or not a settlement is required.

(ii) It requires no initial net investment or an initial net investment that is smaller than would be required for other types of contracts that would be expected to have a similar response to changes in market factors.

(iii) Its terms require or permit net settlement, can readily be settled net by a means outside the contract, or provide for delivery of an asset that puts the recipient in a position not substantially different from net settlement.

(2) The accounting for the changes in the fair value of derivative instruments depends upon its intended use and designation. Changes in the fair value of derivative instruments not designated as fair value or cash flow hedges shall be recorded in Account 175, Derivative instrument assets, or Account 244, Derivative Instrument Liabilities, as appropriate, with the gains recorded in Account 421, Miscellaneous Nonoperating Income, and losses recorded in Account 426.5, Other Deductions.

(3) A derivative instrument may be specifically designated as a fair value or cash flow hedge. A hedge is used to manage risk to price, interest rates, or foreign currency transactions. A company shall maintain documentation of the hedge relationship at the inception of the hedge that details the risk management objective and strategy for undertaking the hedge, the nature of the risk being hedged, and how hedge effectiveness will be determined.

(4) If the utility designates the derivative instrument as a fair value hedge against exposure to changes in the fair value of a recognized asset, liability, or a firm commitment, it shall record the change in fair value of the derivative instrument to Account 176, Derivatives in Instrument Assets—Hedges, or Account 245, Derivative Instrument Liabilities—Hedges, as appropriate, with a corresponding adjustment to the subaccount of the item being hedged. The ineffective portion of the hedge transaction shall be reflected in the same income or expense account that will be used when the hedged item enters into the determination of net income. In the case of a fair value hedge of a firm commitment a new asset or liability is created. As a result of the hedge relationship, the new asset or liability will become part of the carrying amount of the item being hedged.

(5) If the utility designates the derivative instrument as a cash flow hedge against exposure to variable cash flows of a probable forecasted transaction, it shall record changes in the fair value of the derivative instrument in Account 176, Derivative Instrument Assets—Hedges, or Account 245, Derivative Instrument Liabilities—Hedges, as appropriate, with a corresponding amount in Account 209, Accumulated Other Comprehensive Income, for the effective portion of the hedge. The ineffective portion of the hedge transaction shall be reflected in the same account or expense account that will be used when the hedged item enters into the determination of net income. Amounts recorded in other comprehensive income shall be reclassified into earning in the same period or periods that the hedged forecasted item enters into the determination of net income.

(y) *Accounting for asset retirement obligations.* (1) An asset retirement obligation represents a liability for the legal obligation associated with the retirement of a tangible long-lived asset that a company is required to settle as a result of an existing or enacted law, statute, ordinance, or written or oral contract or by legal construction of a contract under the doctrine of promissory estoppel. An asset retirement cost represents the amount capitalized when the liability is recognized for the long-lived asset that gives rise to the legal obligation. The amount recognized for the liability and an associated asset retirement cost shall be stated at the fair value of the asset retirement obligation in the period in which the obligation is incurred.

(2) The utility shall initially record a liability for an asset retirement obligation in Account 230, Asset Retirement Obligations, and charge the associated asset retirement costs to electric utility plant (including Accounts 101.1 and 120.6), and nonutility plant, as appropriate, related to the plant that gives rise to the legal obligation. The asset retirement cost shall be depreciated over the useful life of the related asset that gives rise to the obligation. For periods subsequent to the initial recording of the asset retirement obligation, a utility shall recognize the period to period changes of the asset retirement obligation that result from the passage of time due to the accretion of the liability and any subsequent measurement changes to the initial liability for the legal obligation recorded in Account 230, Asset retirement obligations, as follows:

(i) The utility shall record the accretion of the liability by debiting Account 411.10, Accretion Expense, for electric utility plant, Account 413, Expenses of Electric Plant Leased to Others, for electric plant leased to others, and Account 421, Miscellaneous Nonoperating Income, for nonutility plant and crediting Account 230, Asset Retirement Obligations; and

(ii) The utility shall recognize any subsequent measurement changes of the liability initially recorded in Account 230, Asset Retirement Obligation, for each specific asset retirement obligation as an adjustment of that liability in Account 230 with the corresponding adjustment to electric utility plant, electric plant leased to others, and nonutility plant, as appropriate. The utility shall on a timely

basis monitor any measurement changes of the asset retirement obligations.

(3) Gains or losses resulting from the settlement of asset retirement obligations associated with utility plant resulting from the difference between the amount of the liability for the asset retirement obligation included in Account 230, Asset Retirement Obligations, and the actual amount paid to settle the obligation shall be accounted for as follows:

(i) Gains shall be credited to Account 411.6, Gains from Disposition of Utility Plant, and;

(ii) Losses shall be charged to Account 411.7, Losses from Disposition of Utility Plant.

(4) Gains or losses on the settlement of asset retirement obligations associated with nonutility plant resulting from the difference between the amount of the liability for the asset retirement obligation in Account 230, Asset Retirement Obligations, and the amount paid to settle the obligation, shall be accounted for as follows:

(i) Gains shall be credited to Account 421, Miscellaneous Nonoperating Income, and;

(ii) Losses shall be charged to Account 426.5, Other Deductions.

(5) For purposes of analyses a utility shall maintain supporting documentation so as to be able to furnish accurately and expeditiously with respect to each asset retirement obligation the full details of the identity and nature of the legal obligation, the year incurred, the identity of the plant giving rise to the obligation, the full particulars relating to each component and supporting computations related to the measurement of the asset retirement obligation.

[58 FR 59825, Nov. 10, 1993, as amended at 73 FR 30280, May 27, 2008]

§ 1767.16   Electric plant instructions.

(a) *Classification of electric plant at effective date of system of accounts.* (1) The electric plant accounts provided herein are the same as those contained in the prior system of accounts except for inclusion of accounts for nuclear production plant and some changes in classification in the general equipment accounts. Except for these changes, the balances in the various plant accounts, as determined under the prior system of accounts, should be carried forward. Any remaining balance of plant which has not yet been classified, pursuant to the requirements of the prior system, shall be classified in accordance with the following instructions.

(2) The cost to the utility of its unclassified plant shall be ascertained by analysis of the utility's records. Adjustments shall not be made to record in utility plant accounts amounts previously charged to operating expenses or to income deductions in accordance with the USoA in effect at the time or in accordance with the discretion of management as exercised under a USoA, or under accounting practices previously followed.

(3) The detailed electric plant accounts (301 to 399, inclusive) shall be stated on the basis of cost to the utility of plant constructed by it and the original cost, estimated if not known, of plant acquired as an operating unit or system. The difference between the original cost, as above, and the cost to the utility of electric plant after giving effect to any accumulated provision for depreciation or amortization shall be recorded in Account 114, Electric Plant Acquisition Adjustments. The original cost of electric plant shall be determined by analysis of the utility's records or those of the predecessor or vendor companies with respect to electric plant previously acquired as operating units or systems and the difference between the original cost so determined, less accumulated provisions for depreciation and amortization and the cost to the utility with necessary adjustments for retirements from date of acquisition, shall be entered in Account 114, Electric Plant Acquisition Adjustments. Any difference between the cost of electric plant and its book cost, when not properly includible in other accounts, shall be recorded in Account 116, Other Electric Plant Adjustments.

(4) Plant acquired by lease which qualifies as capital lease property under Sec. 1767.15(s), Criteria for Classifying Leases, shall be recorded in Account 101.1, Property Under Capital Leases, or Account 120.6, Nuclear Fuel Under Capital Leases, as appropriate.

(b) *Electric plant to be recorded at cost.* (1) All amounts included in the accounts for electric plant acquired as an operating unit or system, except as otherwise provided in the texts of the intangible plant accounts, shall be stated at the cost incurred by the person who first devoted the property to utility service. All other electric plant shall be included in the accounts at the cost incurred by the utility except for property acquired by lease which qualifies as capital lease property under § 1767.15 (s), Criteria for Classifying Leases, and is recorded in Account 101.1, Property Under Capital Lease, or Account 120.6, Nuclear Fuel Under Capital Leases. Where the term "cost" is used in the detailed plant accounts, it shall have the meaning stated in this paragraph (b).

(2) When the consideration given for property is other than cash, the value of such consideration shall be determined on a cash basis (see, however, the definition of cost in § 1767.10). In the entry recording such transition, the actual consideration shall be described with sufficient particularity to identify it. The utility shall be prepared to furnish RUS the particulars of its determination of the cash value of the consideration if other than cash.

(3) When property is purchased under a plan involving deferred payments, no charge shall be made to the electric plant accounts for interest, insurance, or other expenditures occasioned solely by such form of payment.

(4) The electric plant accounts shall not include the cost or other value of electric plant contributed to the company. Contributions in the form of money or its equivalent toward the construction of electric plant shall be credited to accounts charged with the cost of such construction. Plant constructed from contributions of cash or its equivalent shall be shown as a reduction to gross plant constructed when assembling cost data in work orders for posting to plant ledgers of accounts. The accumulated gross costs of plant accumulated in the work order shall be recorded as a debit in the plant ledger of accounts along with the related amount of contributions concurrently be recorded as a credit.

(c) *Components of construction cost.* The cost of construction properly includible in the electric plant accounts shall include, where applicable, the direct and overhead costs as listed and defined hereunder:

(1) *Contract work* includes amounts paid for work performed under contract by other companies, firms, or individuals, costs incident to the award of such contracts, and the inspection of such work.

(2) *Labor* includes the pay and expenses of employees of the utility engaged on construction work, and related workmen's compensation insurance, payroll taxes, and similar items of expense. It does not include the pay and expenses of employees which are distributed to construction through clearing accounts nor the pay and expenses included in other items hereunder.

(3) *Materials and supplies* includes the purchase price at the point of free delivery plus customs duties, excise taxes, the cost of inspection, loading and transportation, the related stores expenses, and the cost of fabricated materials from the utility's shop. In determining the cost of materials and supplies used for construction, proper allowance shall be made for unused materials and supplies, for materials recovered from temporary structures used in performing the work involved, and for discounts allowed and realized in the purchase of materials and supplies.

NOTE: The cost of individual items of equipment of small value (for example, $500 or less) or of short life, including small portable tools and implements, shall not be charged to utility plant accounts unless the correctness of the accounting therefor is verified by current inventories. The cost shall be charged to the appropriate operating expense or clearing accounts, according to the use of such items, or, if such items are consumed directly in construction work, the cost shall be included as part of the cost of the construction.

(4) *Transportation* includes the cost of transporting employees, materials and supplies, tools, purchased equipment, and other work equipment (when not under own power) to and from points of construction. It includes amounts paid to others as well as the cost of operating the utility's own transportation equipment. (See Item in paragraph (c)(5) of this section.)

(5) *Special machine service* includes the cost of labor (optional), materials and supplies, depreciation, and other expenses incurred in the maintenance, operation and use of special machines, such as steam shovels, pile drivers, derricks, ditchers, scrapers, material unloaders, and other labor saving machines; also expenditures for rental, maintenance and operation of machines of others. It does not include the cost of small tools and other individual items of small value or short life which are included in the cost of materials and supplies. (See Item in paragraph (c)(3) of this section.) When a particular construction job requires the use for an extended period of time of special machines, transportation or other equipment, the net book cost thereof, less the appraised or salvage value at time of release from the job, shall be include in the cost of construction.

(6) *Shop service* includes the proportion of the expense of the utility's shop department assignable to construction work except that the cost of fabricated materials from the utility's shop shall be included in "materials and supplies."

(7) *Protection* includes the cost of protecting the utility's property from fire or other casualties and the cost of preventing damages to others, or to the property of others, including payments for discovery or extinguishment of fires, cost of apprehending and prosecuting incendiaries, witness fees in relation thereto, amounts paid to municipalities and others for fire protection, and other analogous items of expenditures in connection with construction work.

(8) *Injuries and damages* includes expenditures or losses in connection with construction work on account of injuries to persons and damages to the property of others; also the cost of investigation of and defense against actions for such injuries and damages. Insurance recovered or recoverable on account of compensation paid for injuries to persons incident to construction shall be credited to the account or accounts to which such compensation is

charged. Insurance recovered or recoverable on account of property damages incident to construction shall be credited to the account or accounts charged with the cost of the damages.

(9) *Privileges and permits* includes payments for and expenses incurred in securing temporary privileges, permits or rights in connection with construction work, such as for the use of private or public property, streets, or highways, but it does not include rents, or amounts chargeable as franchises and consents for which see Account 302, Franchises and Consents.

(10) *Rents* includes amounts paid for the use of construction quarters and office space occupied by construction forces and amounts properly includible in construction costs for such facilities jointly used.

(11) *Engineers and supervision* includes the portion of the pay and expenses of engineers, surveyors, draftsmen, inspectors, superintendents and their assistants applicable to construction work.

(12) *General administration capitalized* includes the portion of the pay and expenses of the general officers and administrative and general expenses applicable to construction work.

(13) *Engineering services* includes amounts paid to other companies, firms, or individuals engaged by the utility to plan, design, prepare estimates, supervise, inspect, or give general advice and assistance in connection with construction work.

(14) *Insurance* includes premiums paid or amounts provided or reserved as self-insurance for the protection against loss and damages in connection with construction, by fire or other cas-

ualty, injuries or deaths of persons other than employees, damages to property of others, defalcation of employees and agents, and the nonperformance of contractual obligations of others. It does not include workmen's compensation or similar insurance on employees included as "labor" in Item in paragraph (c)(2) of this section.

(15) *Law expenditures* includes the general law expenditures incurred in connection with construction and the court and legal costs directly related thereto, other than law expenses included in "Protection," Item in paragraph (c)(7) of this section, and in Injuries and damages, Item in paragraph (c)(8) of this section.

(16) *Taxes* includes taxes on physical property (including land) during the period of construction and other taxes properly includible in construction costs before the facilities become available for service.

(17) *Allowance for funds used during construction* includes the net cost for the period of construction of borrowed funds used for construction purposes and a reasonable rate on other funds when so used, not to exceed, without prior approval of RUS, allowances computed in accordance with the formula prescribed in Item in paragraph (c)(17)(i) of this section. No allowance for funds used during construction charges shall be included in these accounts upon expenditures for construction projects which have been abandoned.

(i) The formula and elements for the computation of the allowance for funds used during construction shall be:

$$A_i = s\left(\frac{S}{W}\right) + d\left(\frac{D}{D+P+C}\right)\left(1 - \frac{S}{W}\right)$$

$$A_c = \left[1 - \frac{S}{W}\right]\left[P\left(\frac{P}{D+P+C}\right) + c\left(\frac{C}{D+P+C}\right)\right]$$

Where:

A_i = Gross allowance for borrowed funds used during construction rate.

A_c = Allowance for other funds used during construction rate.

S = Average short-term debt.

s = Short-term debt interest rate.
D = Long-term debt.
d = Long-term debt interest rate.
P = Preferred stock.
p = Preferred stock cost rate.
C = Patronage capital assigned.
c = Entity's incremental borrowing rate.
W = Average balance in construction work in progress plus nuclear fuel in process of refinement, conversion, enrichment, and fabrication, less asset retirement costs related to plant under construction.

(ii) The rate shall be determined annually.

(A) The balance for long-term debt, preferred stock, and patronage capital assigned shall be the actual book balances as of the end of the prior year.

(B) The cost rate for long-term debt and preferred stock shall be the weighted average cost.

(C) The cost rate for patronage capital assigned shall be the entity's incremental borrowing rate.

(D) The short-term debt balances and related cost and the average balance for construction work in progress plus nuclear fuel in process of refinement, conversion, enrichment, and fabrication shall be estimated for the current year with appropriate adjustments as actual data becomes available.

NOTE: When only a portion of a plant or project is placed in operation or is completed and ready for service but the construction work as a whole is incomplete, that part of the cost of the property placed in operation or ready for service shall be treated as "Electric Plant in Service," and an allowance for funds used during construction thereon as a charge to construction shall cease. Allowance for funds used during construction on that part of the cost of the plant which is incomplete may continue to be charged to construction until such time as it is placed in operation or is ready for service, except as limited in Item in paragraph (c)(17) of this section.

(18) *Earnings and expenses during construction.* The earnings and expenses during construction shall constitute a component of construction costs.

(i) The earnings shall include revenues received or earned for power produced by generating plants during the construction period and sold or used by the utility.

(A) Where such power is sold to an independent purchaser before intermingling with power generated by other plants, the credit shall consist of the selling price of the energy.

(B) Where the power generated by a plant under construction is delivered to the utility's electric system for distribution and sale, or is delivered to an associated company, or is delivered to and used by the utility for purposes other than distribution and sale (for manufacturing or industrial use, for example), the credit shall be the fair value of the energy so delivered.

(C) Revenue shall also include rentals for lands, buildings, and other property, and miscellaneous receipts not properly includible in other accounts.

(ii) Expenses shall consist of the cost of operating the power plant, and other costs incident to the production and delivery of the power for which construction is credited under paragraph (c)(18)(i) of this section, including the cost of repairs and other expenses of operating and maintaining lands, buildings, and other property, and other miscellaneous and like expenses not properly includible in other accounts.

(19) *Training costs.* (i) When it is necessary that employees be trained to operate or maintain plant facilities that are being constructed and such facilities are not conventional in nature, or are new to the company's operations, these costs may be capitalized as a component of construction cost.

(ii) Once plant is placed in service, the capitalization of training costs shall cease and subsequent training costs shall be expensed. (See §1767.17 (d).)

(20) *Studies.* (i) Studies include the costs of studies such as nuclear operational, safety, or seismic studies, or environmental studies mandated by regulatory bodies relative to plant under construction.

(ii) Studies relative to facilities in service shall be charged to Account 183, Preliminary Survey and Investigation Charges.

(21) Asset retirement. The costs recognized as a result of asset retirement obligations incurred during the construction and testing of utility plant shall constitute a component of construction costs.

(d) *Overhead construction costs.* (1) All overhead construction costs, such as engineering, supervision, general office salaries and expenses, construction engineering and supervision performed by others than the accounting utility, law expenses, insurance, injuries and damages, relief and pensions, taxes and interest, shall be charged to particular jobs or units on the basis of the amounts of such overheads reasonably applicable thereto, to the end that each job or unit shall bear its equitable proportion of such costs and that the entire cost of the unit, both direct and overhead, shall be deducted from the plant accounts as the time the property is retired.

(2) As far as practicable, the determination of payroll charges includible in construction overheads shall be based on time card distributions thereof.

(i) Where this procedure is impractical, special studies shall be made periodically of the time of supervisory employees devoted to construction activities to the end that only such overhead costs as have a definite relation to construction shall be capitalized.

(ii) The addition to direct construction cost of arbitrary percentages or amounts to cover assumed overhead costs is not permitted.

(3) The records supporting the entries for overhead constructions costs shall be so kept as to show:

(i) The total amount of each overhead for each year;

(ii) The nature and amount of each overhead expenditure charged to each construction work order and to each electric plant account; and

(iii) The bases of distribution of such costs.

(e) *Electric plant purchased or sold.* (1) When electric plant constituting an operating unit or system is acquired by purchase, merger, consolidation, liquidation, or otherwise, after the effective date of this system of accounts, the costs of acquisition, including expenses incidental thereto properly includible in electric plant, shall be charged to Account 102, Electric Plant Purchased or Sold.

(2) The accounting for the acquisition shall then be completed as follows:

(i) The original cost of plant, estimated if not known, shall be credited to Account 102, Electric Plant Pur-

chased or Sold, and concurrently charged to the appropriate electric plant in service accounts and to Account 104, Electric Plant Leased to Others; Account 105, Electric Plant Held for Future Use; and Account 107, Construction Work in Progress—Electric, as appropriate.

(ii) The depreciation and amortization applicable to the original cost of the properties purchased shall be charged to Account 102, Electric Plant Purchased or Sold, and concurrently credited to the appropriate account for accumulated provision for depreciation or amortization.

(iii) The cost to the utility of any property includible in Account 121, Nonutility Property, shall be transferred thereto.

(iv) The amount remaining in Account 102, Electric Plant Purchased or Sold, shall then be closed to Account 114, Electric Plant Acquisition Adjustments.

(3) If property acquired in the purchase of an operating unit or system is in such physical condition when acquired that it is necessary to substantially rehabilitate it in order to bring the property up to the standards of the utility, the cost of such work, except replacements, shall be accounted for as a part of the purchase price of the property.

(4) When any property acquired as an operating unit or system includes duplicate or other plant which will be retired by the accounting utility in the reconstruction of the acquired property or its consolidation with previously owned property, the proposed accounting for such property shall be presented to RUS.

(5) In connection with the acquisition of electric plant constituting an operating unit or system, the utility shall procure, if possible, all existing records relating to the property acquired or certified copies thereof, and shall preserve such records in conformity with regulations or practices governing the preservation of records of its own construction.

(6) When electric plant constituting an operating unit or system is sold, conveyed, or transferred to another by sale, merger, consolidation, or otherwise, the book cost of the property sold

or transferred to another shall be credited to the appropriate utility plant accounts, including amounts carried in Account 114, Electric Plant Acquisition Adjustments, and the amounts (estimated if not known) carried with respect thereto in the accounts for accumulated provision for depreciation and amortization and in Account 252, Customer Advances for Construction, shall be charged to such accounts and contra entries made to Account 102, Electric Plant Purchased or Sold. Unless otherwise ordered by RUS, the difference, if any, between:

(i) The net amount of debits and credits, and

(ii) The consideration received for the property (less commissions and other expenses of making the sale) shall be included in Account 421.1, Gain on Disposition of Property, or Account 421.2, Loss on Disposition of Property. (See Account 102, Electric Plant Purchased or Sold.)

NOTE: In cases where existing utilities merge or consolidate because of financial or operating reasons or statutory requirements rather than as a means of transferring title of purchased properties to a new owner, the accounts of the constituent utilities, with the approval of RUS, may be combined. In the event original cost has not been determined, the resulting utility shall proceed to determine such cost as outlined herein.

(f) *Expenditures on leased property.* (1) The cost of substantial initial improvements (including repairs, rearrangements, additions, and betterments) made in the course of preparing for utility service property leased for a period of more than one year, and the cost of subsequent substantial additions, replacements, or betterments to such property, shall be charged to the electric plant account appropriate for the class of property leased.

(i) If the service life of the improvements is terminable by action of the lease, the cost, less net salvage, of the improvements shall be spread over the life of the lease by charges to Account 404, Amortization of Limited-Term Electric Plant.

(ii) If the service life is not terminated by action of the lease but by depreciation proper, the cost of the improvements, less net salvage, shall be accounted for as depreciable plant. The

provisions of (1) are applicable to property leased under either capital leases or operating leases.

(2) If improvements made to property leased for a period of more than one year are of relatively minor cost, or if the lease is for a period of not more than one year, the cost of the improvements shall be charged to the account in which the rent is included, either directly or by amortization thereof.

(g) *Land and land rights.* (1) The accounts for land and land rights shall include the cost of land owned in fee by the utility and rights, interests, and privileges held by the utility in land owned by others, such as leaseholds, easements, water and water power rights, diversion rights, submersion rights, rights-of-way, and other like interests in land.

(i) Do not include in the accounts for land and land rights and rights-of-way costs incurred in connection with first clearing and grading of land and rights-of-way and the damage costs associated with the construction and installation of plant.

(ii) Such costs shall be included in the appropriate plant accounts directly benefited.

(2) Where special assessments for public improvements provide for deferred payments, the full amount of the assessments shall be charged to the appropriate land account and the unpaid balance shall be carried in an appropriate liability account.

(i) Interest on unpaid balances shall be charged to the appropriate interest account.

(ii) If any part of the cost of public improvements is included in the general tax levy, the amount thereof shall be charged to the appropriate tax account.

(3) The net profit from the sale of timber, cord wood, sand, gravel, other resources or other property acquired with the rights-of-way or other lands shall be credited to the appropriate plant accounts to which related. Where land is held for a considerable period of time and timber and other natural resources on the land at the time of purchase increase in value, the net profit (after giving effect to the cost of the

natural resources) from the sale of timber or its products or other natural resources shall be credited to the appropriate utility operating income account when such land has been recorded in Account 105, Electric Plant Held for Future Use, or classified as plant in service, otherwise to Account 421, Miscellaneous Nonoperating Income.

(4) Separate entries shall be made for the acquisition, transfer, or retirement of each parcel of land, and each land right (except rights-of-way for distribution lines), or water right, having a life of more than one year.

(i) A record shall be maintained showing the nature of ownership, full legal description, area, map reference, purpose for which used, city, county, and tax district on which situated, from whom purchased or to whom sold, payment given or received, other costs, contract date and number, date of recording of deed, and book and page of record.

(ii) Entries transferring or retiring land or land rights shall refer to the original entry recording its acquisition.

(5) Any difference between the amount received from the sale of land or land rights, less agents' commissions and other costs incident to the sale, and the book cost of such land or rights, shall be included in Account 411.6, Gains from Disposition of Utility Plant, or 411.7, Losses from Disposition of Utility Plant, when such property has been recorded in Account 105, Electric Plant Held for Future Use, otherwise to Account 421.1, Gain on Disposition of Property, or 421.2, Loss on Disposition of Property, as appropriate, unless a reserve therefor has been authorized and provided. Appropriate adjustments of the accounts shall be made with respect to any structures or improvements located on land sold.

(6) The cost of buildings and other improvements (other than public improvements) shall not be included in the land accounts. If, at the time of acquisition of an interest in land, such interest extends to buildings or other improvements (other than public improvements) which are then devoted to utility operations, the land and improvements shall be separately appraised and a cost allocated to land and buildings or improvements on the basis of the appraisals. If the improvements are removed or wrecked without being used in operations, the cost of removing or wrecking shall be charged and the salvage credited to the account in which the cost of land is recorded.

(7) When the purchase of land for electric operations requires the purchase of more land than needed for such purposes, the charge to the specific land account shall be based upon the cost of the land purchased, less the fair market value of that portion of the land which is not to be used in utility operations. The portion of the cost measured by the fair market value of the land not to be used shall be included in Account 105, Electric Plant Held for Future Use, or Account 121, Nonutility Property, as appropriate.

(8) Provisions shall be made for amortizing amounts carried in the accounts for limited-term interest in land so as to apportion equitably the cost of each interest over the life thereof. (See Account 111, Accumulated Provision for Amortization of Electric Utility Plant, and Account 404, Amortization of Limited-Term Electric Plant.)

(9) The items of cost to be included in the accounts for land and land rights are as follows:

(i) Bulkheads, buried, not requiring maintenance or replacement;

(ii) First cost of acquisition including mortgages and other liens assumed (but not subsequent interest thereon);

(iii) Condemnation proceedings, including court and counsel costs;

(iv) Consents and abutting damages;

(v) Conveyancers' and notaries' fees;

(vi) Fees, commissions, and salaries to brokers, agents, and other in connection with the acquisition of the land or land rights;

(vii) Leases, cost of voiding upon purchase to secure possession of land;

(viii) Removing, relocating, or reconstructing property of others, such as buildings, highways, railroads, bridges, cemeteries, churches, telephone and power lines, etc., in order to acquire quiet possession;

(ix) Retaining walls unless identified with structures;

(x) Special assessments levied by public authorities for public improvements on the basis of benefits for new roads, new bridges, new sewers, new curbing, new pavements, and other public improvements, but not taxes levied to provide for the maintenance of such improvements;

(xi) Surveys in connection with the acquisition, but not amounts paid for topographical surveys and maps where such costs are attributable to structures or plant equipment erected or to be erected or installed on such land;

(xii) Taxes assumed, accrued to date of transfer of title;

(xiii) Title, examining, clearing, insuring, and registering in connection with the acquisition and defending against claims relating to the period prior to the acquisition;

(xiv) Appraisals prior to closing title;

(xv) Cost of dealing with distributees or legatees residing outside of the state or county, such as recording power of attorney, recording will or exemplification of will, recording satisfaction of state tax;

(xvi) Filing satisfaction of mortgage;

(xvii) Documentary stamps;

(xviii) Photographs of property at acquisition;

(xix) Fees and expenses incurred in the acquisition of water rights and grants;

(xx) Cost of fill to extend bulkhead line over land under water, where riparian rights are held, which is not occasioned by the erection of a structure;

(xxi) Sidewalks and curbs constructed by the utility on public property; and

(xxii) Labor and expenses in connection with securing rights of way, where performed by company employees and company agents.

(h) *Structures and improvements.* (1) The accounts for structures and improvements shall include the cost of all buildings and facilities to house, support, or safeguard property or persons, including all fixtures permanently attached to and made a part of buildings and which cannot be removed therefrom without cutting into the walls, ceilings, or floors, or without in some way impairing the buildings, and improvements of a permanent character on or to land.

(2) Also include those costs incurred in connection with the first clearing and grading of land and rights-of-way and the damage costs associated with construction and installation of plant.

(3) The cost of specially provided foundations not intended to outlast the machinery or apparatus for which provided, and the cost of angle irons, and castings installed at the base of an item of equipment, shall be charged to the same account as the cost of the machinery, apparatus, or equipment.

(4) Minor buildings and structures, such as valve towers, patrolmen's towers, telephone stations, fish and wildlife, and recreation facilities which are used directly in connection with or form a part of a reservoir, dam or waterway shall be considered a part of the facility in connection with which constructed or operated and the cost thereof accounted for accordingly.

(5) Where furnaces and boilers are used primarily for furnishing steam for some particular department and only incidentally for furnishing steam for heating a building and operating the equipment therein, the entire cost of such furnaces and boilers shall be charged to the appropriate plant account, and no part to the building account.

(6) Where the structure of a dam forms also the foundation of the power plant building, such foundation shall be considered a part of the dam.

(7) The cost of disposing of materials excavated in connection with construction of structures shall be considered as a part of the cost of such work, except when such material is used for filling, the cost of loading, hauling, and dumping shall be equitably apportioned between the work in connection with which the removal occurs and the work in connection with which the material is used; and when such material is sold, the net amount realized from such sales shall be credited to the work in connection with which the removal occurs. If the amount realized from the sale of excavated materials exceeds the removal costs and the costs in connection with the sale, the excess shall be credited to the land account in which the site is carried.

(8) Lighting or other fixtures temporarily attached to building for purposes

of display or demonstration shall not be included in the cost of the building but in the appropriate equipment account.

(9) The items of cost to be included in the accounts for structures and improvements are as follows:

(i) Architects' plans and specifications including supervision;

(ii) Ash pits (when located within the building);

(iii) Athletic field structures and improvements;.

(iv) Boilers, furnaces, piping, wiring, fixtures, and machinery for heating, lighting, signaling, ventilating, and air conditioning systems, plumbing, vacuum cleaning systems, incinerator and smoke pipe, flues, etc;

(v) Bulkheads, including dredging, riprap fill, piling, decking, concrete, fenders, etc., when exposed and subject to maintenance and replacement;

(vi) Chimneys;

(vii) Coal bins and bunkers;

(viii) Commissions and fees to brokers, agents, architects and others;

(ix) Conduit (not to be removed) with its contents;

(x) Damages to abutting property during construction;

(xi) Docks;

(xii) Door checks and door stops;

(xiii) Drainage and sewerage systems;

(xiv) Elevators, cranes, hoists, etc., and the machinery for operating them;

(xv) Excavation, including shoring, bracing, bridging, refill and disposal of excess excavated material, cofferdams around foundation, pumping water from cofferdams during construction and test borings;

(xvi) Fences and fence curbs (not including protective fences isolating items of equipment, which shall be charged to the appropriate equipment accounts);

(xvii) Fire protection systems when forming a part of a structure;

(xviii) Flagpole;

(xix) Floor covering (permanently attached);

(xx) Foundations and piers for machinery, constructed as a permanent part of a building or other item listed herein;

(xxi) Grading and clearing when directly occasioned by the building of a structure;

(xxii) Intrasite communication system, poles, pole fixtures, wires, and cable;

(xxiii) Landscaping, lawns, shrubbery, etc.;

(xxiv) Leases, voiding upon purchase to secure possession of structures;

(xxv) Leased property, expenditures on;

(xxvi) Lighting fixtures and outside lighting system;

(xxvii) Mailchutes when part of a building;

(xxviii) Marquee, permanently attached to the building;

(xxix) Painting, first cost;

(xxx) Permanent paving, concrete, brick, flagstone, asphalt, etc., within the property lines;

(xxxi) Partitions, including movable;

(xxxii) Permits and privileges;

(xxxiii) Platforms, railings and gratings when constructed as a part of a structure;

(xxxiv) Power boards for services to a building;

(xxxv) Refrigerating systems for general use;

(xxxvi) Retaining walls except when identified with land;

(xxxvii) Roadways, railroads, bridges, and trestles intrasite except railroads provided for in equipment accounts;

(xxxviii) Roofs;

(xxxix) Scales, connected to and forming a part of a structure;

(xl) Screens;

(xli) Sewer systems, for general use;

(xlii) Sidewalks, culverts, curbs and streets constructed by the utility on its property;

(xliii) Sprinkling systems;

(xliv) Sump pumps and pits;

(xlv) Stacks—brick, steel, or concrete, when set on foundation forming part of general foundation and steelwork of a building;

(xlvi) Steel inspection during construction;

(xlvii) Storage facilities constituting a part of a building;

(xlviii) Storm doors and windows;

(xlix) Subways, areaways, and tunnels, directly connected to and forming part of a structure;

(l) Tanks, constructed as part of a building or as a distinct structural unit;

(li) Temporary heating during construction (net cost);

(lii) Temporary water connection during construction (net cost);

(liii) Temporary shanties and other facilities used during construction (net cost);

(liv) Topographical maps;

(lv) Tunnels, intake and discharge, when constructed as part of a structure, including sluice gates, and those constructed to house mains;

(lvi) Vaults constructed as part of a building;

(lvii) Watchmen's sheds and clock systems (net cost when used during construction only);

(lviii) Water basins or reservoirs;

(lix) Water front improvements;

(lx) Water meters and supply system for a building or for general company purposes;

(lxi) Water supply piping, hydrants, and wells;

(lxii) Wharves;

(lxiii) Window shades and ventilators;

(lxiv) Yard drainage system;

(lxv) Yard lighting system; and

(lxvi) Yard surfacing, gravel, concrete, or oil (First cost only).

NOTE: Structures and improvements accounts shall be credited with the cost of coal bunkers, stacks, foundations, subways, and tunnels, the use of which has terminated with the removal of the equipment with which they are associated even though they have not been physically removed.

(i) *Equipment.* (1) The cost of equipment chargeable to the electric plant accounts, unless otherwise indicated in the text of an equipment account, includes the net purchase price thereof, sales taxes, investigation and inspection expenses necessary to such purchase, expenses of transportation when borne by the utility, labor employed, materials, and supplies consumed, and expenses incurred by the utility in unloading and placing the equipment in readiness to operate.

(2) Also include those costs incurred in connection with the first clearing and grading of land and rights-of-way and the damage costs associated with construction and installation of plant.

(3) Exclude from equipment accounts hand and other portable tools, which are likely to be lost or stolen or which have relatively small value (for example, $500 or less) or short life, unless the correctness of the accounting therefor as electric plant is verified by current inventories.

(i) Special tools acquired and included in the purchase price of equipment shall be included in the appropriate plant accounts.

(ii) Portable drills and similar tool equipment when used in connection with the operation and maintenance of a particular plan or department, such as production, transmission, or distribution or in "stores", shall be charged to the plant accounts appropriate for their use.

(4) The equipment accounts shall include angle irons and similar items which are installed at the base of an item of equipment, but piers and foundations which are designed to be as permanent as the buildings which house the equipment, or which are constructed as a part of the building and which cannot be removed without cutting into the walls, ceilings, or floors or, without in some way impairing the building, shall be included in the building accounts.

(5) The equipment accounts shall include the necessary costs of testing or running a plant or parts thereof during an experimental or test period prior to such plant becoming ready for or placed in service.

(i) The utility shall furnish RUS with full particulars of and justification for any test or experimental run extending beyond a period of 120 days for nuclear plant, and a period of 90 days for all other plant.

(ii) Such particulars shall include a detailed operational and downtime log showing days of production, gross kilowatts generated by hourly increments, types, and periods of outages by hours with explanation thereof, beginning with the first date the equipment was either tested or synchronized on the line to the end of the test period.

(6) The cost of efficiency or other tests made subsequent to the date equipment becomes available for service shall be charged to the appropriate expense accounts, except that tests to determine whether equipment meets the specifications and requirements as

to efficiency, or performance guaranteed by manufacturers, made after operations have commenced and within the period specified in the agreement or contract of purchase, may be charged to the appropriate electric plant accounts.

(j) *Additions and retirements of electric plant.* (1) For the purpose of avoiding undue refinement in accounting for additions to and retirements and replacements of electric plant, all property shall be considered as consisting of retirement units and minor items of property.

(2) The addition and retirement of retirement units shall be accounted for as follows:

(i) When a retirement unit is added to electric plant, the cost thereof shall be added to the appropriate electric plant account, except that when units are acquired in the acquisition of any electric plant constituting an operating system, they shall be accounted for as provided in paragraph (e) of this section.

(ii) When a retirement unit is retired from electric plant, with or without replacement, the book cost thereof shall be credited to the electric plant account in which it is included, determined in the manner set forth in Item in paragraph (j)(4) of this section. If the retirement unit is of a depreciable class, the book cost of the unit retired and credited to electric plant shall be charged to the accumulated provision for depreciation applicable to such property. The cost of removal and the salvage shall be charged or credited, as appropriate, to such depreciation account.

(3) The addition and retirement of minor items of property shall be accounted for as follows:

(i) When a minor item of property which did not previously exist is added to plant, the cost thereof shall be accounted for in the same manner as for the addition of a retirement unit, as set forth in Item in paragraph (j)(2)(i) of this section, if a substantial addition results, otherwise the charge shall be to the appropriate maintenance expense account.

(ii) When a minor item of property is retired and not replaced, the book cost thereof shall be credited to the electric plant account in which it is included; and, in the event the minor item is a part of depreciable plant, the account for accumulated provision for depreciation shall be charged with the book cost and cost of removal and credited with the salvage. If, however, the book cost of the minor item retired and not replaced has been or will be accounted for by its inclusion in the retirement unit of which it is a part when such unit is retired, no separate credit to the property account is required when such minor item is retired.

(iii) When a minor item of depreciable property is replaced independently of the retirement unit of which it is a part, the cost of replacement shall be charged to the maintenance account appropriate for the item, except that if the replacement effects a substantial betterment (the primary aim of which is to make the property affected more useful, more efficient, of greater durability, or of greater capacity), the excess cost of the replacement over the estimated cost at current prices of replacing without betterment shall be charged to the appropriate electric plant accounts.

(4) The book cost of electric plant retired shall be the amount at which such property is included in the electric plant accounts, including all components of construction costs. The book cost shall be determined from the utility's records and if this cannot be done, it shall be estimated. When it is impracticable to determine the book cost of each unit, due to the relatively large number or small cost thereof, an appropriate average book cost of the units with due allowance for any differences in size and character, shall be used as the book cost of the units retired.

(5) The book cost of land retired shall be credited to the appropriate land accounts. If the land is sold, the difference between the book cost (less any accumulated provision for depreciation or amortization therefore which has been authorized and provided) and the sale price of the land (less commissions and other expenses of making the sale) shall be recorded in Account 411.6, Gains from Disposition of Utility Plant, or Account 411.7, Losses from Disposition of Utility Plant, when the

property has been recorded in Account 105, Electric Plant Held for Future Use, otherwise to Accounts 421.1, Gain on Disposition of Property, or 421.2, Loss on Disposition of Property, as appropriate. If the land is not used in utility service but is retained by the utility, the book cost shall be charged to Account 105, Electric Plant Held for Future Use, or Account 121, Nonutility Property, as appropriate.

(6) The book cost less net salvage of depreciable electric plant retired shall be charged in its entirety to Account 108, Accumulated Provision for Depreciation of Electric Utility Plant in Service. Any amounts which, by approval or order of RUS, are charged to Account 182.1, Extraordinary Property Losses, shall be credited to Account 108.

(7) The accounting for the retirement of amounts included in Account 302, Franchises and Consents, and Account 303, Miscellaneous Intangible Plant, and the items of limited-term interest in land included in the accounts for land and land rights, shall be as provided for in the text of Account 111, Accumulated Provision for Amortization of Electric Utility Plant in Service; Account 404, Amortization of Limited-Term Electric Plant; and Account 405, Amortization of Other Electric Plant.

(k) *Work order and property record system required.* (1) Each utility shall record all construction and retirements of electric plant by means of work orders or job orders. Separate work orders may be opened for additions to and retirements of electric plant or the retirements may be included with the construction work order, provided, however, that all items relating to the retirements shall be kept separate from those relating to construction and provided, further, that any maintenance costs involved in the work shall likewise be segregated.

(2) Each utility shall keep its work order system so as to show the nature of each addition to or retirement of electric plant, the total cost thereof, the source or sources of costs, and the electric plant account or accounts to which charged or credited. Work orders covering jobs of short duration may be cleared monthly.

(3) Each utility shall maintain records in which, for each plant account, the amounts of the annual additions and retirements are classified so as to show the number and cost of the various record units or retirement units.

(1) *Transfers of property.* When property is transferred from one electric plant account to another, from one utility department to another, such as from electric to gas, from one operating division or area to another, or from Account 101, Electric Plant in Service; Account 104, Electric Plant Leased to Others; Account 105, Electric Plant Held for Future Use, and Account 121, Nonutility Property, the transfer shall be recorded by transferring the original cost thereof from the one account, department, or location to the other. Any related amounts carried in the accounts for accumulated provision for depreciation or amortization shall be transferred in accordance with the segregation of such accounts.

(m) *Common utility plant.* (1) If the utility is engaged in more than one utility service, such as electric, gas, and water, and any of its utility plant is used in common for several utility services or for other purposes to such an extent and in such manner that it is impracticable to segregate it by utility services currently in the accounts, such property, with the approval of RUS, may be designated and classified as "common utility plant."

(2) The book amount of utility plant designated as common plant shall be included in Account 118, Other Utility Plant, and if applicable in part to the electric department, shall be segregated and accounted for in subaccounts as electric plant is accounted for in Accounts 101 to 107, inclusive, and electric plant adjustments in Account 116, Other Electric Plant Adjustments; any amounts classifiable as common plant acquisition adjustments or common plant adjustments shall be subject to disposition as provided in Paragraphs C and B of Accounts 114 and 116, respectively, for amounts classified in those accounts. The original cost of common utility plant in service shall be classified according to the detailed utility plant accounts appropriate for the property.

(3) The utility shall be prepared to show, at any time, and to report to RUS annually, or more frequently, if required, and by utility plant accounts (301 to 399) the book cost of common utility plant, the allocation of such cost to the respective departments using the common utility plant, and the basis of the allocation.

(4) The accumulated provision for depreciation and amortization of the utility shall be segregated so as to show the amount applicable to the property classified as common utility plant.

(5) The expenses of operation, maintenance, rents, depreciation and amortization of common utility plant shall be recorded in the accounts prescribed herein, but designated as common expenses, and the allocation of such expenses to the departments using the common utility plant shall be supported in such manner as to reflect readily the basis of allocation used.

(n) *Transmission and distribution plant.* For the purpose of this system of accounts:

(1) *Transmission system* is all land, conversion structures, and equipment employed at a primary source of supply (i.e. generating station, or point of receipt in the case of purchased power) to change the voltage or frequency of electricity for the purpose of its more efficient or convenient transmission; all land, structures, lines, switching and conversion stations, high tension apparatus, and their control and protective equipment between a generating or receiving point and the entrance to a distribution center or wholesale point; and all lines and equipment whose primary purpose is to augment, integrate or tie together the sources of power supply.

(2) *Distribution system* is all land, structures, conversion equipment, lines, line transformers, and other facilities employed between the primary source of supply (i.e. generating station, or point of receipt in the case of purchased power) and of delivery to customers, which are not includible in transmission system, as defined in Item in paragraph (n)(1) of this section, whether or not such land, structures, and facilities are operated as part of a transmission system or as part of a distribution system.

NOTE: Stations which change electricity from transmission to distribution voltage shall be classified as distribution stations.

(3) Where poles or towers support both transmission and distribution conductors, the poles, towers, anchors, guys, and rights-of-way shall be classified as transmission system. The conductors, cross-arms, braces, grounds, tiewire, and insulators shall be classified as transmission or distribution facilities, according to the purpose for which used.

(4) Where underground conduit contains both transmission and distribution conductors, the underground conduit and right-of-way shall be classified as distribution system. The conductors shall be classified as transmission or distribution facilities according to the purpose for which used.

(5) Land (other than rights-of-way) and structures used jointly for transmission and distribution purposes shall be classified as transmission or distribution according to the major use thereof.

(o) *Hydraulic production plant.* For purpose of this system of accounts hydraulic production plant is all land and land rights, structures and improvements used in connection with hydraulic power generation, reservoirs, dams and waterways, water wheels, turbines, generators, accessory electric equipment, roads, railroads, and bridges and structures and improvements used in connection with fish and wildlife, and recreation.

(p) *Nuclear fuel records required.* Each utility shall keep all the necessary records to support the entries to the various nuclear fuel plant accounts classified under "Assets and Other Debits," Utility Plant Accounts 120.1 through 120.5, inclusive; Account 518, Nuclear Fuel Expense; and Account 157, Nuclear Materials Held for Sale. These records shall be so kept as to readily furnish the basis of the computation of the net nuclear fuel costs.

[58 FR 59825, Nov. 10, 1993, as amended at 73 FR 30281, May 27, 2008]

§ 1767.17 Operating expense instructions.

(a) *Supervision and engineering.* The supervision and engineering includible in the operating expense accounts shall consist of the salary, employee pensions and benefits, social security and other payroll taxes, injuries and damages, and other expenses of superintendents, engineers, clerks, other employees, and consultants engaged in supervising and directing the operation and maintenance of each utility function. Whenever allocations are necessary in order to arrive at the amount to be included in any account, the method and basis of allocation shall be reflected by underlying records.

(1) Labor items:

(i) Special tests to determine efficiency of equipment operation;

(ii) Preparing or reviewing budgets, estimates, and drawings relating to operation or maintenance for departmental approval;

(iii) Preparing instructions for operations and maintenance activities;

(iv) Reviewing and analyzing operating results;

(v) Establishing organizational setup of departments and executing changes therein;

(vi) Formulating and reviewing routines of departments and executing changes therein;

(vii) General training and instruction of employees by supervisors whose pay is chargeable hereto. Specific instructions and training in a particular type of work is chargeable to the appropriate functional account (See paragraph (c)(19) of this section); and

(viii) Secretarial work for supervisory personnel, but not general clerical and stenographic work chargeable to other accounts.

(2) Expense items:

(i) Employee pensions and benefits;

(ii) Social security and other payroll taxes;

(iii) Injuries and damages;

(iv) Consultants' fees and expenses; and

(v) Meals, traveling, and incidental expenses.

(b) *Maintenance.* (1) The cost of maintenance chargeable to the various operating expense and clearing accounts includes labor, employee pensions and benefits, social security and other payroll taxes, injuries and damages, materials, overheads, and other expenses incurred in maintenance work. A list of work operations applicable generally to utility plant is included in this paragraph (b). Other work operations applicable to specific classes of plant are listed in functional maintenance expense accounts.

(2) Materials recovered in connection with the maintenance of property shall be credited to the same account to which the maintenance cost was charged.

(3) If the book cost of any property is carried in Account 102, Electric Plant Purchased or Sold, the cost of maintaining such property shall be charged to the accounts for maintenance of property of the same class and use, the book cost of which is carried in other electric plant in service accounts. Maintenance of property leased from others shall be treated as provided in paragraph (c) of this section.

(4) Items:

(i) Direct field supervision of maintenance;

(ii) Inspecting, testing, and reporting on condition of plant specifically to determine the need for repairs, replacements, rearrangements, and changes and inspecting and testing the adequacy of repairs which have been made;

(iii) Work performed specifically for the purpose of preventing failure, restoring serviceability or maintaining life of plant;

(iv) Rearranging and changing the location of plant not retired;

(v) Repairing for reuse materials recovered from plant;

(vi) Testing for, locating, and clearing trouble;

(vii) Net cost of installing, maintaining, and removing temporary facilities to prevent interruptions in service; and

(viii) Replacing or adding minor items of plant which do not constitute a retirement unit.

(c) *Rents.* (1) The rent expense accounts provided under the several functional groups of expense accounts shall include all rents, including taxes paid by the lessee on leased property, for property used in utility operations, except minor amounts paid for occasional or infrequent use of any property or equipment and all amounts paid for use of equipment that, if owned, would be includible in plant Accounts 391 to 398 inclusive, which shall be treated as an expense item and included in the appropriate function account and rents which are chargeable to clearing accounts, and distributed therefrom to the appropriate account.

(2) If rents cover property used for more than one function such as production and transmission, or by more than one department, the rents shall be apportioned to the appropriate rent expense or clearing accounts of each department on an actual, or if necessary, an estimated basis.

(3) When a portion of property or equipment rented from others for use in connection with utility operations is subleased, the revenue derived from such subleasing shall be credited to the rent revenue account in operating revenues; provided, however, that in case the rent was charged to a clearing account, amounts received from subleasing the property shall be credited to such clearing account.

(4) The cost, when incurred by the lessee, of operating and maintaining leased property, shall be charged to the accounts appropriate for the expense if the property were owned.

(5) The cost incurred by the lessee of additions and replacements to electric plant leased from others shall be account for as provided in § 1767.16 (f).

(d) *Training costs.* (1) When it is necessary that employees be trained to specifically operate or maintain plant facilities that are being constructed, the related costs shall be accounted for as a current operating and maintenance expense.

(2) These expenses shall be charged to the appropriate functional accounts currently as they are incurred.

(3) When the training costs involved relate to facilities which are not conventional in nature, or are new to the company's operations, see § 1767.16 (c)(19), for the accounting.

[58 FR 59825, Nov. 10, 1993, as amended at 62 FR 42290, Aug. 6, 1997]

§ 1767.18　Assets and other debits.

The asset and other debits accounts identified in this section shall be used by all RUS borrowers.

ASSETS AND OTHER DEBITS

Utility Plant

101　Electric Plant in Service
101.1　Property Under Capital Leases
102　Electric Plant Purchased or Sold
103　Experimental Electric Plant Unclassified
104　Electric Plant Leased to Others
105　Electric Plant Held for Future Use
106　Completed Construction not Classified—Electric
107　Construction Work in Progress—Electric
107.1　Construction Work in Progress—Contract
107.2　Construction Work in Progress—Force Account
107.3　Construction Work in Progress—Special Equipment
108　Accumulated Provision for Depreciation of Electric Utility Plant
108.1　Accumulated Provision for Depreciation of Steam Production Plant
108.2　Accumulated Provision for Depreciation of Nuclear Production Plant
108.3　Accumulated Provision for Depreciation of Hydraulic Production Plant
108.4　Accumulated Provision for Depreciation of Other Production Plant
108.5　Accumulated Provision for Depreciation of Transmission Plant
108.6　Accumulated Provision for Depreciation of Distribution Plant
108.7　Accumulated Provision for Depreciation of General Plant
108.8　Retirement Work in Progress
108.9　Accumulated Provision for Depreciation of Asset Retirement
109–110　[Reserved]
111　Accumulated Provision for Amortization of Electric Utility Plant
112–113　[Reserved]
114　Electric Plant Acquisition Adjustments
115　Accumulated Provision for Amortization of Electric Plant Acquisition Adjustments
116　Other Electric Plant Adjustments
118　Other Utility Plant
119　Accumulated Provision for Depreciation and Amortization of Other Utility Plant
120.1　Nuclear Fuel in Process of Refinement, Conversion, Enrichment, and Fabrication
120.2　Nuclear Fuel Materials and Assemblies—Stock Account
120.3　Nuclear Fuel Assemblies in Reactor
120.4　Spent Nuclear Fuel
120.5　Accumulated Provision for Amortization of Nuclear Fuel Assemblies
120.6　Nuclear Fuel Under Capital Leases

Other Property and Investments

121　Nonutility Property
122　Accumulated Provision for Depreciation and Amortization of Nonutility Property

123 Investment in Associated Companies
123.1 Patronage Capital from Associated Co-operatives
123.3 Investment in Associated Organizations—Federal Economic Development Loans
123.4 Investment in Associated Organizations—Non-Federal Economic Development Loans
123.11 Investment in Subsidiary Companies
123.21 Subscriptions to Capital Term Certificates—Supplemental Financing
123.22 Investments in Capital Term Certificates—Supplemental Financing
123.23 Other Investments in Associated Organizations
124 Other Investments
124.1 Other Investments—Federal Economic Development Loans
124.2 Other Investments—Non-Federal Economic Development Loans
125 Sinking Funds
126 Depreciation Fund
128 Other Special Funds

Current and Accrued Assets

131 Cash
131.1 Cash—General
131.2 Cash—Construction Fund—Trustee
131.3 Cash—Installation Loan and Collection Fund
131.4 Transfer of Cash
131.12 Cash—General—Economic Development Loan Funds
131.13 Cash—General—Economic Development Grant Funds
131.14 Cash—General—Economic Development Non-Federal Revolving Funds
132 Interest Special Deposits
133 Dividend Special Deposits
134 Other Special Deposits
135 Working Funds
136 Temporary Cash Investments
141 Notes Receivable
141.1 Accumulated Provision for Uncollectible Notes—Credit
142 Customer Accounts Receivable
142.1 Customer Accounts Receivable—Electric
142.2 Customer Accounts Receivable—Other
143 Other Accounts Receivable
144 Accumulated Provision for Uncollectible Accounts—Credit
144.1 Accumulated Provision for Uncollectible Customer Accounts—Credit
144.2 Accumulated Provision for Uncollectible Merchandising Accounts—Credit
144.3 Accumulated Provision for Uncollectible Accounts, Officers and Employees—Credit
144.4 Accumulated Provision for Other Uncollectible Accounts—Credit 145 Notes Receivable from Associated Companies
145 Notes Receivable from Associated Companies

146 Accounts Receivable from Associated Companies
151 Fuel Stock
152 Fuel Stock Expenses Undistributed
153 Residuals
154 Plant Materials and Operating Supplies
155 Merchandise
156 Other Materials and Supplies
157 Nuclear Materials Held for Sale
158.1 Allowance Inventory
158.2 Allowances Withheld
163 Stores Expense Undistributed
165 Prepayments
165.1 Prepayments—Insurance
165.2 Other Prepayments
171 Interest and Dividends Receivable
172 Rents Receivable
173 Accrued Utility Revenues
174 Miscellaneous Current and Accrued Assets
175 Derivative Instrument Assets
176 Derivative Instrument Assets—Hedges

Deferred Debits

181 Unamortized Debt Expense
182.1 Extraordinary Property Losses
182.2 Unrecovered Plant and Regulatory Study Costs
182.3 Other Regulatory Assets
183 Preliminary Survey and Investigation Charges
184 Clearing Accounts
184.1 Transportation Expense—Clearing
184.2 Clearing Accounts—Other
185 Temporary Facilities
186 Miscellaneous Deferred Debits
187 Deferred Losses from Disposition of Utility Plant
188 Research, Development, and Demonstration Expenditures
189 Unamortized Loss on Reacquired Debt
190 Accumulated Deferred Income Taxes

ASSETS AND OTHER DEBITS

Utility Plant

101 Electric Plant in Service

A. This account shall include the original cost of electric plant, included in Accounts 301 to 399, prescribed herein, owned and used by the utility in its electric utility operations, and having an expectation of life in service of more than one year from date of installation, including such property owned by the utility but held by nominees.

B. (See also Account 106 for unclassified construction costs of completed plant actually in service.)

C. The cost of additions to and betterments of property leased from others, which are includible in this account, shall be recorded in subdivisions separate and distinct from those relating to owned property. (See § 1767.16 (f).)

101.1 Property Under Capital Leases

A. This account shall include the amount recorded under capital leases for plant leased from others and used by the utility in its utility operations.

B. The electric property included in this account shall be classified separately according to the detailed accounts (301 to 399) prescribed for electric plant in service.

C. Records shall be maintained with respect to each capital lease reflection:

(1) Name of lessor, (2) basic details of lease, (3) terminal date, (4) original cost or fair market value of property leased, (5) future minimum lease payments, (6) executory costs, (7) present value of minimum lease payments, (8) the amount representing interest and the interest rate used, and (9) expenses paid. Records shall also be maintained for plant under a lease, to identify the asset retirement obligation and cost originally recognized for each lease and the periodic charges and credits made to the asset retirement obligations and asset retirement costs.

102 Electric Plant Purchased or Sold

A. This account shall be charged with the cost of electric plant acquired as an operating unit or system by purchase, merger, consolidation liquidation, or otherwise, and shall be credited with the selling price of like property transferred to others pending the distribution to appropriate accounts in accordance with § 1767.16 (e).

B. Within 6 months from the date of acquisition or sale of property recorded herein, the borrower shall file with RUS the proposed journal entries to clear from this account the amounts recorded herein.

103 Experimental Electric Plant Unclassified

A. This account shall include the cost of electric plant which was constructed as a research, development, and demonstration plant under the provisions of Paragraph C, Account 107, Construction Work in Progress—Electric, and due to the nature of the plant, it is desirous to operate it for a period of time in an experimental status.

B. Amounts in this account shall be transferred to Account 101, Electric Plant in Service, or Account 121, Nonutility Property, as appropriate when the project is no longer considered as experimental.

C. The depreciation on property in this account shall be charged to Account 403.8, Depreciation Expense, for asset retirement costs, as appropriate, and credited to Account 108, Accumulated Provision for Depreciation of Electric Utility Plant. The amounts herein shall be depreciated over a period which would correspond to the estimated useful life of the relevant project considering the characteristics involved. How-

ever, when projects are transferred to Account 101, Electric Plant in Service, a new depreciation rate based upon the remaining service life and undepreciated amounts, will be established.

D. Records shall be maintained with respect to each unit of experiment so that full details may be obtained as to the cost, depreciation, and the experimental status.

E. Should it be determined that experimental plant recorded in this account will fail to satisfactorily perform its function, the costs thereof shall be accounted for as directed or authorized by RUS.

104 Electric Plant Leased to Others

A. This account shall include the original cost of electric plant owned by the utility, but leased to others as operating units or systems, where the lessee has exclusive possession.

B. The property included in this account shall be classified according to the detailed accounts (301 to 399) prescribed for electric plant in service and this account shall be maintained in such detail as though the property were used by the owner in its utility operations.

105 Electric Plant Held for Future Use

A. This account shall include the original cost of electric plant (except land and land rights) owned and held for future use in electric service under a definite plan for such use, to include: (1) Property acquired (except land and land rights) but never used by the utility in electric service, but held for such service in the future under a definite plan, and (2) property (except land and land rights) previously used by the utility in service but retired from such service and held pending its reuse in the future, under a definite plan, in electric service.

B. This account shall also include the original cost of land and land rights owned and held for future use in electric service under a plan for such use, to include land and land rights: (1) Acquired but never used by the utility in electric service, but held for such service in the future under a plan, and (2) previously held by the utility in service, but retired from such service and held pending its reuse in the future under a plan, in electric service. (See § 1767.16 (g).)

C. In the event that property recorded in this account shall no longer be needed or appropriate for future utility operations, the borrower shall notify RUS of such condition and request approval of journal entries to remove such property from this account.

D. Gains or losses from the sale of land and land rights or other disposition of such property previously recorded in this account and not placed in utility service shall be recorded directly in Accounts 411.6 or 411.7, as appropriate, except when determined to be significant by RUS. Upon such a determination, the amounts shall be transferred to Account 256, Deferred Gains from Disposition of Utility Plant, or Account 187, Deferred Losses from Disposition of Utility Plant, and amortized to Account 411.6, Gains from Disposition of Utility Plant, or Account 411.7, Losses from Disposition of Utility Plant, as appropriate.

E. The property included in this account shall be classified according to the detail accounts (301 to 399) prescribed for electric plant in service and the account shall be maintained in such detail as though the property were in service.

NOTE: Materials and supplies, meters and transformers held in reserve, and normal spare capacity of plant in service shall not be included in this account.

106 Completed Construction not Classified— Electric

At the end of the year or such other date as a balance sheet may be required by RUS, this account shall include the total of the balances of work orders for electric plant which has been completed and placed in service but which work orders have not been classified for transfer to the detailed electric plant accounts.

NOTE: For the purpose of reporting to RUS, the classification of electric plant in service by accounts is required, the utility shall also report the balance in this account tentatively classified as accurately as practicable according to prescribed account classifications. The purpose of this provision is to avoid any significant omissions in reported amounts of electric plant in service.

107 Construction Work in Progress— Electric

A. This account shall include the total of the balances of work orders for electric plant in process of construction.

B. Work orders shall be cleared from this account as soon as practicable, after completion of the job. Further, if a project, such as a hydroelectric project, a steam station, or a transmission line, is designed to consist of two or more units or circuits which may be placed in service at different dates, any expenditures which are common to and which will be used in the operation of the project as a whole shall be included in electric plant in service upon the completion and the readiness for service of the first unit. Any expenditures which are identified exclusively with units of property not yet in service shall be included in this account.

C. Expenditures on research, development, and demonstration projects for construction of utility facilities are to be included in a separate subdivision in this account. Records must be maintained to show separately each project along with complete detail of the nature and purpose of the research, development, and demonstration project together with the related costs.

D. Account 107 shall be subaccounted as follows:

107.1 Construction Work in Progress—Contract

107.2 Construction Work in Progress—Force Account

107.3 Construction Work in Progress—Special Equipment

108 Accumulated Provision for Depreciation of Electric Utility Plant

A. This account shall be credited with the following:

1. Amounts charged to Account 403, Depreciation Expense, or to clearing accounts for current depreciation expense for electric plant in service.

2. Amounts charged to Account 421, Miscellaneous Nonoperating Income, for depreciation expense on property included in Account 105, Electric Plant Held for Future Use. Include, also, the balance of accumulated provision for depreciation on property when transferred to Account 105, Electric Plant Held for Future Use, from other property accounts. Normally, Account 108 will not be used for current depreciation provision because, as provided herein, the service life during which depreciation is computed commences with the date property is includible in electric plant in service; however, if special circumstances indicate the propriety of current accruals for depreciation, such charges shall be made to Account 421, Miscellaneous Nonoperating Income.

3. Amounts charged to Account 413, Expenses of Electric Plant Leased to Others, for electric plant included in Account 104, Electric Plant Leased to Others.

4. Amounts charged to Account 416, Costs and Expenses of Merchandising, Jobbing, and Contract Work, or to clearing accounts for current depreciation expense.

5. Amounts of depreciation applicable to electric properties acquired as operating units or systems. (See § 1767.16 (e).)

6. Amounts charged to Account 182.1, Extraordinary Property Losses, when authorized by RUS.

7. Amounts of depreciation applicable to electric plant donated to the utility.

The utility shall maintain separate subaccounts for depreciation applicable to electric plant in service, electric plant leased to others, and electric plant held for future use.)

B. At the time of retirement of depreciable electric utility plant, this account shall be charged with the book cost of the property retired and the cost of removal and shall be credited with the salvage value and any other amounts recovered, such as insurance.

When retirement, costs of removal and salvage are entered originally in retirement work orders, the net total of such work orders may be included in a separate subaccount hereunder. Upon completion of the work order, the proper distribution to subdivisions of this account shall be made as provided in the following paragraph.

C. Account 108 shall be subaccounted as follows:

108.1 Accumulated Provision for Depreciation of Steam Production Plant

108.2 Accumulated Provision for Depreciation of Nuclear Production Plant

108.3 Accumulated Provision for Depreciation of Hydraulic Production Plant

108.4 Accumulated Provision for Depreciation of Other Production Plant

108.5 Accumulated Provision for Depreciation of Transmission Plant

108.6 Accumulated Provision for Depreciation of Distribution Plant

108.7 Accumulated Provision for Depreciation of General Plant

108.8 Retirement Work in Progress

108.9 Accumulated Provision for Depreciation of Asset Retirement Costs

These subsidiary records shall reflect the current credits and debits to this account in sufficient detail to show separately for each such functional classification: (1) the amount of accrual for depreciation, (2) the book cost of property retired, (3) cost of removal, (4) salvage, and (5) other items, including recoveries from insurance.

D. When transfers of plant are made from one electric plant account to another, or from or to another utility department, of from or to nonutility property accounts, the accounting for depreciation shall be as provided in §1767.16 (1).

E. The utility is restricted in its use of the accumulated provision for depreciation to the purposes set forth above. It shall not transfer any portion of this account to retained earnings or make any other use thereof without authorization by RUS.

109–110 [Reserved]

111 Accumulated Provision for
Amortization of Electric Utility Plant

A. This account shall be credited with the following:

1. Amounts charged to Account 404, Amortization of Limited-Term Electric Plant, for the current amortization of limited-term electric plant investments.

2. Amounts charged to Account 421, Miscellaneous Nonoperating Income, for amortization expense on property included in Account 105, Electric Plant Held for Future Use. Include also the balance of accumulated provision for amortization on property when transferred to Account 105, Electric Plant Held for Future Use, from other property accounts. See also Paragraph A(2), Account 108, Accumulated Provision for Depreciation of Electric Utility Plant.

3. Amounts charged to Account 405, Amortization of Other Electric Plant.

4. Amounts charged to Account 413, Expenses of Electric Plant Leased to Others, for the current amortization of limited-term or other investments subject to amortization included in Account 104, Electric Plant Leased to Others.

5. Amounts charged to Account 425, Miscellaneous Amortization, for the amortization of intangible or other electric plant which does not have a definite or terminable life and is not subject to charges for depreciation expense, with RUS approval.

(The utility shall maintain subaccounts of this account for the amortization applicable to electric plant in service, electric plant leased to others and electric plant held for future use.)

B. When any property to which this account applies is sold, relinquished, or otherwise retired from service, this account shall be charged with the amount previously credited in respect to such property. The book cost of the property so retired less the amount chargeable to this account and less the net proceeds realized at retirement shall be included in Account 421.1, Gain on Disposition of Property, or Account 421.2, Loss on Disposition of Property, as appropriate.

C. For general ledger and balance sheet purposes, this account shall be regarded and treated as a single composite provision for amortization. For purposes of analysis, however, each utility shall maintain subsidiary records in which this account is segregated according to the following functional classification for electric plant: (1) Steam production, (2) Nuclear production, (3) Hydraulic production, (4) Other production, (5) Transmission, (6) Distribution, and (7) General. These subsidiary records shall reflect the current credits and debits to this account in sufficient detail to show separately for each such functional classification: (1) the amount of accrual for amortization, (2) the book cost of property retired, (3) cost of removal, (4) salvage, and (5) other items, including recoveries from insurance.

D. The utility is restricted in its use of the accumulated provision for amortization to the purposes set forth above. It shall not transfer any portion of this account to retained earnings or make any other use thereof without authorization by RUS.

112–113 [Reserved]

114 Electric Plant Acquisition Adjustments

A. This account shall include the difference between the cost to the accounting utility of electric plant acquired as an operating unit or system by purchase, merger, consolidation, liquidation, or otherwise, and the original cost, estimated, if not known, of such property, less the amount or amounts credited by the accounting utility at the time of acquisition to accumulated provisions for depreciation and amortization and contributions in aid of construction with respect to such property.

B. With respect to acquisitions after the effective date of this system of accounts, this account shall be subdivided so as to show the amounts included herein for each property acquisition and to electric plant in service, electric plant held for future use, and electric plant leased to others. (See § 1767.16 (e).)

C. Debit amounts recorded in this account related to plant and land acquisition may be amortized to Account 425, Miscellaneous Amortization, over a period not longer than the estimated remaining life of the properties to which such amounts relate. Amounts related to the acquisition of land only may be amortized to Account 425 over a period of not more than 15 years. Should a utility wish to account for debit amounts in this account in any other manner, it shall petition RUS for authority to do so. Credit amounts recorded in this account shall be accounted for as directed by RUS.

115 Accumulated Provision for Amortization of Electric Plant Acquisition Adjustments

This account shall be credited or debited with amounts which are includible in Account 406, Amortization of Electric Plant Acquisition Adjustments, or Account 425, Miscellaneous Amortization, for the purpose of providing for the extinguishment of amounts in Account 114, Electric Plant Acquisition Adjustments, in instances where the amortization of Account 114 is not being made by direct write-off of the account.

116   Other Electric Plant Adjustments

A. This account shall include the difference between the original cost, estimated if not known, and the book cost of electric plant to the extent that such difference is not properly includible in Account 114, Electric Plant Acquisition Adjustments. (See § 1767.16 (a)(3))

B. Amounts included in this account shall be classified in such manner as to show the origin of each amount and shall be disposed of as RUS may approve or direct.

NOTE: The provisions of this account shall not be construed as approving or authorizing the recording of appreciation of electric plant.

118   Other Utility Plant

This account shall include the balances in accounts for utility plant, other than electric plant, such as gas, or railway.

119   Accumulated Provision for Depreciation and Amortization of Other Utility Plant

This account shall include the accumulated provision for depreciation and amortization applicable to utility property other than electric plant.

120.1 Nuclear Fuel in Process of Refinement, Conversion, Enrichment, and Fabrication

A. This account shall include the original cost to the utility of nuclear fuel materials while in process of refinement, conversion, enrichment, and fabrication into nuclear fuel assemblies and components, including processing, fabrication, and necessary shipping costs. This account shall also include the salvage value of nuclear materials which are actually being reprocessed for use and were transferred from Account 120.5, Accumulated Provision for Amortization of Nuclear Fuel Assemblies. (See § 1767.10 (a)(27).)

B. This account shall be credited and Account 120.2, Nuclear Fuel Materials and Assemblies—Stock Account, shall be debited for the cost of completed fuel assemblies delivered for use in refueling or to be held as spares. In the case of the initial core loading, the transfer shall be made directly to Account 120.3, Nuclear Fuel Assemblies in Reactor, upon the conclusion of the experimental or test period of the plant prior to its becoming available for service.

Items

1. Cost of natural uranium, uranium ores concentrates or other nuclear fuel sources, such as thorium, plutonium, and U–233.

2. Value of recovered nuclear materials being reprocessed for use.

3. Milling process costs.

4. Sampling and weighing, and assaying costs.

5. Purification and conversion process costs.

6. Costs of enrichment by gaseous diffusion or other methods.

7. Costs of fabrication into fuel forms suitable for insertion in the reactor.

8. All shipping costs of materials and components, including shipping of fabricated fuel assemblies to the reactor site.

9. Use charges on leased nuclear materials while in process of refinement, conversion, enrichment, and fabrication.

120.2   Nuclear Fuel Materials and Assemblies—Stock Account

A. This account shall be debited and Account 120.1, Nuclear Fuel in Process of Refinement, Conversion, Enrichment and Fabrication, shall be credited with the cost of fabricated fuel assemblies delivered for use in refueling or to be carried in stock as spares. It shall also include the original cost of fabricated fuel assemblies purchased in completed form. This account shall also include the original cost of partially irradiated fuel assemblies being held in stock for re-insertion in a reactor which had been transferred from Account 120.3, Nuclear Fuel Assemblies in Reactor.

B. When fuel assemblies included in this account are inserted in a reactor, this account shall be credited and Account 120.3, Nuclear Fuel Assemblies in Reactor, debited for the cost of such assemblies.

C. This account shall also include the cost of nuclear materials and byproduct materials being held for future use and not actually in process in Account 120.1, Nuclear Fuel in Process of Refinement, Conversion, Enrichment and Fabrication.

120.3 Nuclear Fuel Assemblies in Reactor

A. This account shall include the cost of nuclear fuel assemblies when inserted in a reactor for the production of electricity. The amounts included herein shall be transferred from Account 120.2, Nuclear Fuel Materials and Assemblies—Stock Account, except for the initial core loading which will be transferred directly from Account 120.1, Nuclear Fuel in Process of Refinement, Conversion, Enrichment and Fabrication.

B. Upon removal of fuel assemblies from a reactor, the original cost of the assemblies removed shall be transferred to Account 120.4, Spent Nuclear Fuel, or Account 120.2, Nuclear Fuel Materials and Assemblies—Stock Account, as appropriate.

120.4 Spent Nuclear Fuel

A. This account shall include the original cost of nuclear fuel assemblies, in the process of cooling, transferred from Account 120.3, Nuclear Fuel Assemblies in Reactor, upon removal from a reactor pending reprocessing.

B. This account shall be credited and Account 120.5, Accumulated Provision for Amortization of Nuclear Fuel Assemblies, debited for fuel assemblies, after the cooling period is over, at the cost recorded in this account.

120.5 Accumulated Provision for Amortization of Nuclear Fuel Assemblies

A. This account shall be credited and Account 518, Nuclear Fuel Expense, shall be debited for the amortization of the net cost of nuclear fuel assemblies used in the production of energy. The net cost of nuclear fuel assemblies subject to amortization shall be the original cost of nuclear fuel assemblies, plus or less the expected net salvage value of uranium, plutonium, and other byproducts.

B. This account shall be credited with the net salvage value of uranium, plutonium, and other nuclear by-products when such items are sold, transferred or otherwise dis-

posed. Account 120.1, Nuclear Fuel in Process of Refinement, Conversion, Enrichment and Fabrication, shall be debited with the net salvage value of nuclear materials to be reprocessed. Account 157, Nuclear Materials Held for Sale, shall be debited for the net salvage value of nuclear materials not to be reprocessed but to be sold or otherwise disposed of and Account 120.2, Nuclear Fuel Materials and Assemblies—Stock Account, will be debited with the net salvage value of nuclear materials that will be held for future use and not actually in process, in Account 120.1, Nuclear Fuel in Process of Refinement, Conversion, Enrichment, and Fabrication.

C. This account shall be debited and Account 120.4, Spent Nuclear Fuel, shall be credited with the cost of fuel assemblies at the end of the cooling period.

120.6 Nuclear Fuel Under Capital Leases

A. This account shall include the amount recorded under capital leases for nuclear fuel leased from others for use by the utility in its utility operations.

B. Records shall be maintained with respect to each capital lease reflecting: (1) name of lessor, (2) basic details of lease, (3) terminal date, (4) original cost or fair market value of nuclear fuel leased, (5) future minimum lease payments, (6) the amount representing interest and the interest rate used, and (7) expenses paid.

Other Property and Investments

121 Nonutility Property

A. This account shall include the book cost of land, structure, and equipment or other tangible or intangible property owned by the utility, but used in utility service and not properly includible in Account 105, Electric Plant Held for Future Use. This account shall also include, where applicable, amounts recorded for asset retirement costs associated with nonutility plant.

B. This account shall also include the amount recorded under capital leases for property leased from others and used by the utility in its nonutility operations. Records shall be maintained with respect to each lease reflecting: (1) name of lessor, (2) basic details of lease, (3) terminal date, (4) original cost or fair market value of property leased, (5) future minimum lease payments, (6) executory costs, (7) present value of minimum lessee payments, (8) the amount representing interest and the interest rate used, and (9) expenses paid.

C. This account shall be subdivided so as to show the amount of property used in operations which are nonutility in character but nevertheless constitute a distinct operating activity of the company (such as operation of an ice department where such activity is not classed as a utility) and the amount of miscellaneous property not used in operations. The records in support of each subaccount shall be maintained so as to show an appropriate classification of the property.

NOTE: The gain from the sale or other disposition of property included in this account which had been previously recorded in Account 105, Electric Plant Held for Future Use, shall be accounted for in accordance with Paragraph C of Account 105.

122 Accumulated Provision for Depreciation and Amortization of Nonutility Property

This account shall include the accumulated provision for depreciation and amortization applicable to nonutility property.

123 Investment in Associated Companies

A. This account shall include the book cost of investments in securities issued or assumed by associated companies and investment advances to such companies, including interest accrued thereon when such interest is not subject to current settlement, provided that the investment does not relate to a subsidiary company. (If the investment relates to a subsidiary company, it shall be included in Account 123.11, Investment in Subsidiary Companies.) Include herein the offsetting entry to the recording of amortization of discount or premium on interest bearing investments. (See Account 419, Interest and Dividend Income.)

B. This account shall be maintained in such manner as to show the investment in securities of, and advances to, each associated company together with full particulars regarding any of such investments that are pledged.

NOTE A: Securities and advances of associated companies owned and pledged shall be included in this account, but such securities, if held in special deposits or in special funds, shall be included in the appropriate deposit or fund account. A complete record of securities pledged shall be maintained.

NOTE B: Securities of associated companies held as temporary cash investments are includible in Account 136, Temporary Cash Investments.

NOTE C: Balances in open accounts with associated companies, which are subject to current settlement, are includible in Account 146, Accounts Receivable from Associated Companies.

NOTE D: The utility may write down the cost of any security in recognition of a decline in the value thereof. Securities shall be written off or written down to a nominal value if there is no reasonable prospect of substantial value. Fluctuations in market value shall not be recorded but a permanent impairment in the value of securities shall be recognized in the accounts. When securities are written off or written down, the amount of the adjustment shall be charged to Account 426.5, Other Deductions, or to an appropriate account for accumulated provisions for loss in value established as a separate subdivision of this account.

C. Account 123 shall be subaccounted as follows:

123.1 Patronage Capital from Associated Cooperatives

123.3 Investment in Associated Organizations—Federal Economic Development Loans

123.4 Investment in Associated Organizations—Non-Federal Economic Development Loans

123.11 Investment in Subsidiary Companies

123.21 Subscriptions to Capital Term Certificates—Supplemental Financing

123.22 Investment in Capital Term Certificates—Supplemental Financing

123.23 Other Investments in Associated Organizations

123.1 Patronage Capital from Associated Cooperatives

This account shall include patronage capital credits allocated to the accounting borrower by G&T cooperatives. It shall also include capital credits, deferred patronage refunds, or like items from other associated cooperatives. The account shall be maintained so as to reflect separately, the allocations of patronage capital and patronage refunds from each organization that makes such allocations to the borrower.

123.3 Investment in Associated Organizations—Federal Economic Development Loans

This account shall include investment advances of Federal funds received from a Rural Economic Development Grant to associated organizations for authorized rural economic development projects.

123.4 Investment in Associated Organizations—Non-Federal Economic Development Loans

This account shall include investment advances of non-Federal funds from the Rural Economic Development Grant revolving fund to associated organizations for authorized rural economic development projects.

123.11 Investment in Subsidiary Companies

A. This account shall include the cost of investments in securities issued or assumed by subsidiary companies and investment advances to such companies, including interest accrued thereon when such interest is not subject to current settlement, plus the equity in undistributed earnings or losses of such subsidiary companies since acquisition. This account shall be credited with any dividends declared by such subsidiaries.

B. This account shall be maintained in such a manner as to show separately for each subsidiary: the cost of such investments in the securities of the subsidiary at the time of acquisition; the amount of equity in the subsidiary's undistributed net earnings or net losses since acquisition; advances or loans to such subsidiary; and full particulars regarding any such investments that are pledged.

123.21 Subscriptions to Capital Term Certificates—Supplemental Financing

This account shall include the total subscriptions to capital term certificates of CFC. When subscriptions are paid, this account shall be credited and Account 123.22, Investments in Capital Term Certificates—Supplemental Financing, debited.

123.22 Investments in Capital Term Certificates—Supplemental Financing

This account shall include paid subscriptions in capital term certificates of CFC or other supplemental lenders.

123.23 Other Investments in Associated Organizations

This account shall include investments in capital stock, securities, membership fees, and investment advances to associated organizations other than provided for elsewhere. This account shall be maintained in such a manner as to show the investment in stock and securities of and advances to each associated organization.

Items

1. Investments in capital stock of associated organizations.
2. Investments in securities issued by associated organizations.
3. Membership fees in associated organizations, including NRECA, and Statewide associations of RUS-financed borrowers.
4. Investment advances to associated organizations.

124 Other Investments

A. This account shall include the book cost of investments in securities issued or assumed by nonassociated companies, investment advances to such companies, and any investments not accounted for elsewhere. This account shall also included unrealized holding gains and losses on trading and available-for-sale types of security investments. Include also the offsetting entry to the recording of amortization of discount or premium on interest bearing investments. (See Account 419, Interest and Dividend Income.)

B. The records shall be maintained in such manner as to show the amount of each investment and the investment advances to each person.

C. Account 124 shall be subaccounted as follows:

124.1 Other Investments—Federal Economic Development Loans
124.2 Other Investments—Non-Federal Economic Development Loans

NOTE A: Securities owned and pledged shall be included in this account, but securities held in special deposits or in special funds shall be included in appropriate deposit or fund accounts. A complete record of securities pledged shall be maintained.

NOTE B: Securities held as temporary cash investments shall not be included in this account.

NOTE C: See Note D of Account 123.

124.1 Other Investments—Federal Economic Development Loans

This account shall include investment advances of Federal funds received from a Rural Economic Development Grant to nonassociated organizations for authorized rural economic development projects.

124.2 Other Investments—Non-Federal Economic Development Loans

This account shall include investment advances of non-Federal funds from the Rural Economic Development Grant revolving fund to nonassociated organizations for authorized rural economic development projects.

125 Sinking Funds

This account shall include the amount of cash and book cost of investments held in sinking funds. This account shall also include unrealized holding gains and losses on trading and available-for-sale types of investments. A separate account, with appropriate title, shall be kept for each sinking fund. Transfers from this account to special deposit accounts, may be as necessary for the purpose of paying matured sinking fund obligations, or obligations called for redemption but not presented, or the interest thereon.

126 Depreciation Fund

This account shall include the amount of cash and the book cost of investments which have been segregated in a special fund for the purpose of identifying such assets with the accumulated provisions for depreciation. This account shall also include unrealized holding gains and losses on trading and available-for-sale types of security investments.

128 Other Special Funds

This account shall include the amount of cash and book cost of investments which have been segregated in special funds for insurance, employee pensions, savings, relief, hospital, and other purposes not provided for elsewhere. This account shall also include unrealized holding gains and losses on trading and available-for-sale types of security investments. A separate account, with appropriate title, shall be kept for each fund.

NOTE: Amounts deposited with a trustee under the terms of an irrevocable trust agreement for pensions or other employee benefits shall not be included in this account.

Current and Accrued Assets

Current and accrued assets are cash, those assets which are readily convertible into cash or are held for current use in operations or construction, current claims against others, payment of which is reasonably assured, and amounts accruing to the utility which are subject to current settlement, except such items for which accounts other than those designated as current and accrued assets are provided. There shall not be included in the category of accounts designated as current and accrued assets any item, the amount or collectibility of which is not reasonably assured, unless an adequate provision for possible loss has been made therefor. Items of current character but of doubtful value may be written down, and for record purposes carried in these accounts at nominal value.

131   Cash

A. This account shall include the amount of current cash funds except working funds.

B. Account 131 shall be subaccounted as follows:

131.1   Cash—General
131.2   Cash—Construction Fund—Trustee
131.3   Cash—Installation Loan and Collection Fund
131.4   Transfer of Cash
131.12 Cash—General—Economic Development Loan Funds
131.13 Cash—General—Economic Development Grant Funds
131.14 Cash—General—Economic Development Non-Federal Revolving Funds

131.1   Cash—General

This account shall include all cash of the organization not provided for elsewhere. Separate subaccounts may be maintained for each bank account in which general cash is maintained. Funds held by others for current obligations shall be recorded in Account 134, Other Special Deposits.

131.2   Cash—Construction Fund—Trustee

This account shall include the cash received from the Rural Utilities Service, CFC, and any other source of supplemental financing for financing the construction, purchase, and operation of electric facilities. RUS construction loan fund advances shall be charged to this account and credited to Account 224.4, RUS Notes Executed—Construction—Debit. CFC and other supplemental lender construction loan fund advances shall be charged to this account and credited to Account 224.13, Supplemental Financing Notes Executed—Debit.

131.3   Cash—Installation Loan and Collection Fund

A. This account shall include the cash advanced on installation loans made subsequent to September 13, 1957. Such advances shall be debited to this account as received and credited to Account 224.10, RUS Notes Executed—Installation—Debit. This account shall also include interest and principal collections received on consumers' loans financed from RUS loans made subsequent to September 13, 1957.

B. Payments shall be made from this account solely for financing consumers' loans for the purpose of wiring of consumers' premises, and the acquisition and installation of electrical and plumbing appliances and equipment by consumers. The cash in this account is also used for the payment of principal and interest on installation loans made by RUS, subsequent to September 13, 1957, in accordance with the terms of the loan agreement.

131.4   Transfer of Cash

This account shall be used in transferring funds from one bank account to another. This account is charged when the check is drawn for the transfer and entered in the check register, and credited when the amount transferred is entered in the cash receipts book. This account is to be used as a clearing account and should not have a balance at the end of an accounting period.

131.12   Cash—General—Economic Development Funds

This account shall include the cash received from the Rural Utilities Service for Rural Economic Development Loans. Economic development loan advances shall be charged to this account and credited to Account 224.17, RUS Notes Executed—Economic Development—Debit.

131.13   Cash—General—Economic Development Grant Funds

This account shall include cash received from the Rural Utilities Service for Rural Economic Development Grants. Economic development grant funds shall be charged to this account and credited to Account 224.18, Other Long-Term Debt—Grant Funds; Account 208, Donated Capital; or Account 421, Miscellaneous Nonoperating Income, as appropriate. This account shall be credited and either Account 123.3, Investment in Associated Organizations—Federal Economic Development Loans, or Account 124.1, Other Investments—Federal Economic Development Loans, shall be debited, as appropriate, with the amount of an economic development revolving fund loan.

131.14 Cash—General—Economic Development Non-Federal Revolving Funds

This account shall include all non-Federal funds comprising the economic development revolving fund. It shall include all funds supplied by the borrower as well as all cash received from the repayment of loans made from the economic development revolving fund. This account shall be credited and either Account 123.4, Investment in Associated Organizations—Non-Federal Economic Development Loans, or Account 124.2, Other Investments—Non-Federal Economic Development Loans, shall be debited, as appropriate, with the amount of an economic development revolving fund loan.

132 Interest Special Deposits

This account shall include special deposits with fiscal agents or others for the payment of interest.

133 Dividend Special Deposits

This account shall include special deposits with fiscal agents or others for the payment of dividends.

134 Other Special Deposits

This account shall include deposits with fiscal agents or others for special purposes other than the payment of interest and dividends. Such special deposits may include cash deposited with Federal, state, or municipal authorities as a guaranty for the fulfillment of obligations; cash deposited with trustees to be held until mortgaged property sold, destroyed, or otherwise disposed of is replaced; and cash realized from the sale of the accounting utility's securities and deposited with trustees to be held until invested in property of the utility. Entries to this account shall specify the purpose for which the deposit is made.

NOTE: Assets available for general corporate purposes shall not be included in this account. Further, deposits for more than one year, which are not offset by current liabilities, shall not be charged to this account but to Account 128, Other Special Funds.

135 Working Funds

This account shall include cash advanced to officers, agents, employees, and others as petty cash or working funds.

136 Temporary Cash Investments

A. This account shall include the book cost of investments, such as demand and time loans, bankers' acceptances, United States Treasury certificates, marketable securities, and other similar investments, acquired for the purpose of temporarily investing cash.

B. This account shall be so maintained as to show separately temporary cash investments in securities of associated companies and of others. Records shall be kept of any pledged investments.

141 Notes Receivable

A. This account shall include the book cost, not includible elsewhere, of all collectible obligations in the form of notes receivable and similar evidences (except interest coupons) of money due on demand or within one year from the date of issue, except, however, notes receivable from associated companies. (See Account 136, Temporary Cash Investments, and Account 145, Notes Receivable from Associated Companies.)

NOTE: The face amount of notes receivable discounted, sold, or transferred without releasing the utility from liability as endorser thereon, shall be credited to a separate subdivision of this account and appropriate disclosure shall be made in the financial statements of any contingent liability arising from such transactions.

B. Account 141 shall be subaccounted as follows:

141.1 Accumulated Provision for Uncollectible Notes—Credit

141.1 Accumulated Provision for Uncollectible Notes—Credit

This account shall be credited with amounts provided for losses on notes receivable which may become uncollectible, and also with collections on notes previously charged hereto. Concurrent charges shall be made to Account 904, Uncollectible Accounts.

142 Customer Accounts Receivable

A. This account shall include amounts due from customers for utility service and for merchandising, jobbing, and contract work. This account shall not include amounts due from associated companies.

B. This account shall be maintained so as to permit ready segregation of the amounts due for merchandising, jobbing, and contract work.

C. Account 142 shall be subaccounted as follows:

142.1 Customer Accounts Receivable—Electric
142.2 Customer Accounts Receivable—Other

142.1 Customer Accounts Receivable—
Electric

This account shall include amounts due from customers for utility service.

142.2 Customer Accounts Receivable—Other

This account shall include amounts due from customers for merchandising, jobbing, and contract work.

143 Other Accounts Receivable

A. This account shall include amounts due the utility upon open accounts, other than amounts due from associated companies and from customers for utility services and merchandising, jobbing and contract work.

B. This account shall be maintained so as to show separately amounts due on subscriptions to capital stock and from officers and employees. The account shall not include amounts advanced to officers or others as working funds. (See Account 135, Working Funds.)

144 Accumulated Provision for
Uncollectible Accounts—Credit

A. This account shall include amounts provided for losses on accounts receivable which may become uncollectible, and also with collections on accounts previously charged hereto. Concurrent charges shall be made to Account 904, Uncollectible Accounts, for amounts applicable to utility operations, and to corresponding accounts for other operations. Records shall be maintained so as to show the write-offs of accounts receivable for each utility department.

B. Account 144 shall be subaccounted as follows:

144.1 Accumulated   Provision   for
Uncollectible Customer Accounts—Credit

144.2 Accumulated   Provision   for
Uncollectible Merchandising Accounts—
Credit

144.3 Accumulated   Provision   for
Uncollectible Accounts, Officers and Employees—Credit

144.4 Accumulated Provision for Other
Uncollectible Accounts—Credit

144.1 Accumulated Provision for
Uncollectible Customer Accounts—Credit

This account shall be credited with amounts provided for losses on accounts receivable which may become uncollectible, and also with collections on accounts previously charged hereto. Concurrent charges shall be made to Account 904, Uncollectible Accounts.

144.2 Accumulated   Provision   for
Uncollectible Merchandising Accounts—
Credit

This account shall be credited with amounts provided for losses on merchandising, jobbing, and contract work which may become uncollectible, and also with collections on accounts previously charged hereto. Concurrent charges shall be made to Account 904, Uncollectible Accounts, for amounts applicable to utility operations, and to corresponding accounts for other operations.

144.3 Accumulated   Provision   for
Uncollectible Accounts, Officers and Employees—Credit

This account shall be credited with amounts provided for losses on accounts receivable from officers and employees which may become uncollectible and also with collections on accounts previously charged hereto. Concurrent charges shall be made to Account 904, Uncollectible Accounts.

144.4 Accumulated Provision for Other
Uncollectible Accounts—Credit

This account shall be credited with amounts provided for losses on accounts receivable which may become uncollectible and for which the recording of this credit has not been provided for elsewhere. This account shall also be credited with collections on accounts previously charged hereto. Concurrent charges shall be made to Account 904, Uncollectible Accounts, for amounts applicable to utility operations and to corresponding accounts for other operations.

145 Notes Receivable from Associated
Companies

This account shall include notes upon which associated companies are liable, and which mature and are expected to be paid in full not later than one year from the date of issue, together with any interest thereon, and debit balances subject to current settlement in open accounts with associated companies. Items which do not bear a specified due date but which have been carried for more than twelve months and items which are not paid within twelve months from due date shall be transferred to Account 123, Investment in Associated Companies.

NOTE: The face amount of notes receivable discounted, sold or transferred without releasing the utility from liability as endorser thereon, shall be credited to a separate subdivision of this account and appropriate disclosure shall be made in the financial statements of any contingent liability arising from such transactions.

146 Accounts Receivable from Associated Companies

This account shall include drafts upon which associated companies are liable, and which mature and are expected to be paid in full not later than one year from the date of issue, together with any interest thereon, and debit balances subject to current settlement in open accounts with associated companies. Items which do not bear a specified due date but which have been carried for more than twelve months and items which are not paid within twelve months from due date shall be transferred to Account 123, Investment in Associated Companies.

NOTE: On the balance sheet, accounts receivable from an associated company may be offset against accounts payable to the same company.

151 Fuel Stock

This account shall include the book cost of fuel on hand.

Items

1. Invoice price of fuel less any cash or other discounts.
2. Freight, switching, demurrage, and other transportation charges, not including, however, any charges for unloading from the shipping medium.
3. Excise taxes, purchasing agents' commissions, insurance, and other expenses directly assignable to cost of fuel.
4. Operating, maintenance and depreciation expenses, and ad valorem taxes on utility-owned transportation equipment used to transport fuel from the point of acquisition to the unloading point.
5. Lease or rental costs of transportation equipment used to transport fuel from the point of acquisition to the unloading point.

152 Fuel Stock Expenses Undistributed

A. This account may include the cost of labor and of supplies used and expenses incurred in unloading fuel from the shipping medium and in the handling thereof prior to its use, if such expenses are sufficiently significant in amount to warrant being treated as a part of the cost of fuel inventory rather than being charged direct to expense as incurred.

B. Amounts included herein shall be charged to expense as the fuel is used to the end that the balance herein shall not exceed the expenses attributable to the inventory of fuel on hand.

Items

Labor:
1. Procuring and handling of fuel.
2. All routine fuel analyses.
3. Unloading from shipping facility and placing in storage.

4. Moving of fuel in storage and transferring from one station to another.
5. Handling from storage or shipping facility to first bunker, hopper, bucket, tank, or holder of boiler house structure.
6. Operation of mechanical equipment such as locomotives, trucks, cars, boats, barges, and cranes.

Supplies and Expenses:
1. Tools, lubricants and other supplies.
2. Operating supplies for mechanical equipment.
3. Transportation and other expenses in moving fuel.
4. Stores expenses applicable to fuel.

153 Residuals

This account shall include the book cost of any residuals produced in the production or manufacturing processes.

154 Plant Materials and Operating Supplies

A. This account shall include the cost of materials purchased primarily for use in the utility business for construction, operation and maintenance purposes. It shall also include the book cost of materials recovered in connection with construction, maintenance, or the retirement of property, such materials being credited to construction, maintenance, or accumulated depreciation provision, respectively, and included herein as follows:

1. Reusable materials consisting of large individual items shall be included in this account at original cost, estimated if not known. The cost of repairing such items shall be charged to the maintenance account appropriate for the previous use.
2. Reusable materials consisting of relatively small items, the identity of which (from the date of original installation to the final abandonment or sale thereof) cannot be ascertained without undue refinement in accounting, shall be included in this account at current prices new for such items. The cost of repairing such items shall be charged to the appropriate expense account as indicated by previous use.
3. Scrap and nonusable materials included in this account shall be carried at the estimated net amount realizable therefrom. The difference between the amounts realized for scrap and nonusable materials sold and the net amount at which the materials were carried in this account, as far as practicable, shall be adjusted to the accounts credited when the materials were charged to this account.

B. Materials and supplies issued shall be credited hereto and charged to the appropriate construction, operating expense, or other account on the basis of a unit price determined by the use of cumulative average,

49

first-in-first-out, or such other method of inventory accounting as conforms with accepted accounting standards consistently applied.

Items

1. Invoice price of materials less cash or other discounts.

2. Freight, switching, or other transportation charges when practicable to include as part of the cost of particular materials to which they relate.

3. Customs duties and excise taxes.

4. Costs of inspection and special tests prior to acceptance.

5. Insurance and other directly assignable charges.

NOTE: Where expenses applicable to materials purchased cannot be directly assigned to particular purchases, they shall be charged to Account 163, Stores Expense Undistributed.

155   Merchandise

This account shall include the book cost of materials and supplies and appliances and equipment held primarily for merchandising, jobbing, and contract work. The principles prescribed in accounting for utility materials and supplies shall be observed with respect to items carried in this account.

156   Other Materials and Supplies

This account shall include the book cost of materials and supplies held primarily for nonutility purposes. The principles prescribed in accounting for utility materials and supplies shall be observed with respect to items carried in this account.

157   Nuclear Materials Held for Sale

This account shall include the net salvage value of uranium, plutonium, and other nuclear materials held by the company for sale or other disposition that are not to be reused by the company in its electric utility operations. This account shall be debited and Account 120.5, Accumulated Provision for Amortization of Nuclear Fuel Assemblies, credited for such net salvage value. Any difference between the amount recorded in this account and the actual amount realized from the sale of materials shall be debited or credited, as appropriate, to Account 518, Nuclear Fuel Expense, at the time of such sale.

158.1   Allowance Inventory

A. This account shall include the cost of allowances owned by the utility and not withheld by the Environmental Protection Agency. See § 1767.15 (u) and Account 158.2, Allowances Withheld.

B. This account shall be credited and Account 509, Allowances, shall be debited concurrent with the monthly emission of sulfur dioxide.

C. Separate subdivisions of this account shall be maintained so as to separately account for those allowances usable in the current year and in each subsequent year. The underlying records of these subdivisions shall be maintained in sufficient detail so as to identify each allowance included; the origin of each allowance; and the acquisition cost, if any, of the allowance.

158.2   Allowances Withheld

A. This account shall include the cost of allowances owned by the utility but withheld by the Environmental Protection Agency. (See § 1767.15 (u).)

B. The inventory cost of the allowances released by the Environmental Protection Agency for use by the utility shall be transferred to Account 158.1, Allowance Inventory.

C. The underlying records of this account shall be maintained in sufficient detail so as to identify each allowance included; the origin of each allowance; and the acquisition cost, if any, of the allowances.

163   Stores Expense Undistributed

A. This account shall include the cost of supervision, labor, and expenses incurred in the operation of general storerooms, including purchasing, storage, handling, and distribution of materials and supplies.

B. This account shall be cleared by adding to the cost of materials and supplies issued, a suitable loading charge which will distribute the expense equitably over stores issues. The balance in the account at the close of the year shall not exceed the amount of stores expenses reasonably attributable to the inventory of materials and supplies, exclusive of fuel, as any amount applicable to fuel costs should be included in Account 152, Fuel Stock Expenses Undistributed.

Items

Labor:

1. Inspecting and testing materials and supplies when not assignable to specific items.

2. Unloading from shipping facility and placing in storage.

3. Supervision of purchasing and stores department to extent assignable to materials handled through stores.

4. Getting materials from stock and in readiness to go out.

5. Inventorying stock received or stock on hand by stores employees but not including inventories by general department employees as part of internal or general audits.

6. Purchasing department activities in checking material needs, investigating

sources of supply, analyzing prices, preparing and placing orders, and related activities to extent applicable to materials handled through stores. (Optional: Purchasing department expenses may be included in administrative and general expenses.)

7. Maintaining stores equipment.

8. Cleaning and tidying storerooms and stores offices.

9. Keeping stock records, including the recording and posting of material receipts and issues and maintaining inventory records of stock.

10. Collecting and handling scrap materials in stores.

Supplies and Expenses:

1. Adjustments of inventories of materials and supplies but not including large differences which can readily be assigned to important classes of materials and equitably distributed among the accounts to which such classes of materials have been charged since the previous inventory.

2. Cash and other discounts not practically assignable to specific materials.

3. Freight and express charges when not assignable to specific items.

4. Heat, light, and power for storerooms and store offices.

5. Brooms, brushes, sweeping compounds and other supplies used in cleaning and tidying storerooms and stores offices.

6. Injuries and damages.

7. Insurance on materials and supplies and on stores equipment.

8. Losses due to breakage, leakage, evaporation, fire or other causes, less credits for amounts received from insurance, transportation companies, or others in compensation of such losses.

9. Postage, printing, stationery, and office supplies.

10. Rent of storage space and facilities.

11. Communication service.

12. Excise and other similar taxes not assignable to specific materials.

13. Transportation expense on inward movement of stores and on transfer between storerooms but not including charges on materials recovered from retirements which shall be accounted for as part of the cost of removal.

NOTE: A physical inventory of each class of materials and supplies shall be made at least every two years.

165 Prepayments

A. This account shall include amounts representing prepayments of insurance, rents, taxes, interest, and miscellaneous items, and shall be kept or supported in such manner as to disclose the amount of each class of prepayment.

B. Account 165 shall be subaccounted as follows:

165.1   Prepayments—Insurance
165.2   Other Prepayments

171 Interest and Dividends Receivable

This account shall include the amount of interest on bonds, mortgages, notes, commercial paper, loans, open accounts, and deposits, the payment of which is reasonably assured, and the amount of dividends declared or guaranteed on stocks owned.

NOTE A: Interest which is not subject to current settlement shall not be included herein but in the account in which the associated principle is recorded.

NOTE B: Interest and dividends receivable from associated companies shall be included in Account 146, Accounts Receivable from Associated Companies.

172 Rents Receivable

This account shall include rents receivable or accrued on property rented or leased by the utility to others.

NOTE: Rents receivable from associated companies shall be included in Account 146, Accounts Receivable from Associated Companies.

173 Accrued Utility Revenues

At the option of the utility, the estimated amount accrued for service rendered, but not billed at the end of any accounting period, may be included herein. If accruals are made for unbilled revenues, accruals shall also be made for unbilled expenses, such as the purchase of energy.

174 Miscellaneous Current and Accrued Assets

This account shall include the book cost of all other current and accrued assets, appropriately designated and supported so as to show the nature of each asset included herein.

175 Derivative Instrument Assets

This account shall include the amounts paid for derivative instruments, and the change in the fair value hedges. Account 421, Miscellaneous Nonoperating Income, shall be credited or debited, as appropriate, with the corresponding amount of the change in the fair value of the derivative instrument.

176 Derivative Instrument Assets—Hedges

A. This account shall include the amounts paid for derivative instruments, and the change in the fair value of derivative instrument assets designated by the utility as cash flow or fair value hedges.

B. When a utility designates a derivative instrument asset as a cash flow hedge it will record the change in the fair value of the derivative instrument in this account with a concurrent charge to Account 209, Accumulated Other Comprehensive Income, with the effective portion of the gain or loss. The ineffective portion of the cash flow hedge shall be charged to the same income or expense account that will be used when the hedged item enters into the determination of net income.

C. When a utility designates a derivative instrument as a fair value hedge it shall record the change in the fair value of the derivative instrument in this account with a concurrent charge to a subaccount of the asset or liability that carries the item being hedged. The ineffective portion of the fair value hedge shall be charged to the same income or expense account that will be used when the hedged item enters into the determination of net income.

Deferred Debits

181 Unamortized Debt Expense

This account shall include expenses related to the issuance or assumption of debt securities. Amounts recorded in this account shall be amortized over the life of each respective issue under a plan which will distribute the amount equitably over the life of the security. The amortization shall be on a monthly basis, and the amounts thereof shall be charged to Account 428, Amortization of Debt Discount and Expense. Any unamortized amounts outstanding at the time that the related debt is prematurely reacquired shall be accounted for as indicated in § 1767.15 (q).

182.1 Extraordinary Property Losses

A. When authorized or directed by RUS, this account shall include extraordinary losses which could not reasonably have been anticipated and which are not covered by insurance or other provisions, such as unforeseen damages to property.

B. Application to RUS for permission to use this account shall be accompanied by a statement giving a complete explanation with respect to the items which it is proposed to include herein, the period over which, and the accounts to which it is proposed to write off the charges, and other pertinent information.

182.2 Unrecovered Plant and Regulatory Study Costs

A. This account shall include: (1) nonrecurring costs of studies and analyses mandated by regulatory bodies related to plants in service, transferred from Account 183, Preliminary Survey and Investigations Charges, and not resulting in construction; and (2) when authorized by RUS, significant unrecovered costs of plant facilities where construction has been cancelled or which have been prematurely retired.

B. This account shall be credited and Account 407, Amortization of Property Losses, Unrecovered Plant and Regulatory Study Costs, shall be debited over the period specified by RUS.

C. Any additional costs incurred, relative to the cancellation or premature retirement, may be included in this account and amortized over the remaining period of the original amortization period. Should any gains or recoveries be realized relative to the cancelled or prematurely retired plant, such amounts shall be used to reduce the unamortized amount of the costs recorded herein.

D. In the event that the recovery of costs included herein is disallowed in the rate proceedings, the disallowed costs shall be charged to Account 426.5, Other Deductions, in the year of such disallowance.

182.3 Other Regulatory Assets

A. This account shall include the amounts of regulatory-created assets, not includable in other accounts, resulting from the rate-making actions of regulatory agencies. (See the definition of regulatory assets and liabilities.)

B. The amounts included in this account are to be established by those charges which would have been included in net income, or accumulated other comprehensive income, determinations in the current period under the general requirements of the Uniform System of Accounts but for it being probable that such items will be included in a different period(s) for purposes of developing the rates that the utility is authorized to charge for its utility services. When specific identification of the particular source of a regulatory asset cannot be made, such as in plant phase-ins, rate moderation plans, or rate levelization plans, Account 407.4, Regulatory Credits, shall be credited. The amounts recorded in this account are generally to be charged, concurrently with the recovery of the amounts in rates, to the same account that would have been charged if included in income when incurred, except all regulatory assets established through the use of Account 407.4 shall be charged to Account 407.3, Regulatory Debits, concurrent with the recovery of the amounts in rates.

C. If rate recovery of all or part of an amount included in this account is disallowed, the disallowed amount shall be charged to Account 426.5, Other Deductions, or Account 435, Extraordinary Deductions, in the year of the disallowance.

D. The records supporting the entries to this account shall be kept so that the utility can furnish full information as to the nature and amount of each regulatory asset included in this account, including justification for inclusion of such amounts in this account.

183 Preliminary Survey and Investigation
Charges

A. This account shall be charged with all expenditures for preliminary surveys, plans, and investigations made for the purpose of determining the feasibility of utility projects under contemplation. If construction results, this account shall be credited and the appropriate utility plant account charged. If the work is abandoned, the charge shall be made to Account 426.5, Other Deductions, or to the appropriate operating expense account.

B. This account shall also include costs of studies and analyses mandated by regulatory bodies related to plant in service. If construction results from such studies, this account shall be credited and the appropriate utility plant account charged with an equitable portion of such study costs directly attributable to new construction. The portion of such study costs not attributable to new construction or the entire cost if construction does not result shall be charged to Account 182.2, Unrecovered Plant and Regulatory Study Costs, or the appropriate operating expense account. The costs of such studies relative to plant under construction shall be included directly in Account 107, Construction Work in Progress—Electric.

C. The records supporting the entries to this account shall be so kept that the utility can furnish complete information as to the nature and the purpose of the survey, plans, or investigations, and the nature and amounts of the sever several charges.

NOTE: The amount of preliminary survey and investigation charges transferred to utility plant shall not exceed the expenditures which may reasonably be determined to contribute directly and immediately and without duplication to utility plant.

184 Clearing Accounts

A. This caption shall include undistributed balances in clearing accounts at the date of the balance sheet. Balances in clearing account shall be substantially cleared not later than the end of the calendar year unless items held therein relate to a future period.

B. Account 184 shall be subaccounted as follows:

184.1 Transportation Expense—Clearing
184.2 Clearing Accounts—Other

185 Temporary Facilities

This account shall include amounts shown by work orders for plant installed for temporary use in utility service for periods of less than one year. Such work orders shall be charged with the cost of temporary facilities and credited with payments received from customers and net salvage realized on removal of the temporary facilities. Any net credit or debit resulting shall be cleared to

Account 451, Miscellaneous Service Revenues.

186 Miscellaneous Deferred Debits

This account shall include all debits not elsewhere provided for, such as miscellaneous work in progress, and unusual or extraordinary expenses, not included in other accounts, which are in process of amortization and items the proper final disposition of which is uncertain.

187 Deferred Losses from Disposition of
Utility Plant

This account shall include losses from the sale or other disposition of property previously recorded in Account 105, Electric Plant Held for Future Use, under the provisions of Paragraphs B, C, and D thereof, where such losses are significant and are to be amortized over a period of 5 years, unless otherwise authorized by RUS. The amortization of the amounts in this account shall be made by debits to Account 411.7, Losses from Disposition of Utility Plant. (See Account 105, Electric Plant Held for Future Use.)

188 Research, Development, and
Demonstration Expenditures

A. This account shall be charged with the cost of all expenditures coming within the meaning of Research, Development, and Demonstration (RD&D) of this USoA (See § 1767.10 (a)(34)) except those expenditures properly chargeable to Account 107, Construction Work in Progress—Electric.

B. Costs that are minor or of a general or recurring nature shall be transferred from this account to the appropriate operating expense function or if such costs are common to the overall operations or cannot be feasibly allocated to the various operating accounts, such costs shall be recorded in Account 930.2, Miscellaneous General Expenses.

C. In certain instances, a company may incur large and significant research, development, and demonstration expenditures which are nonrecurring and which would distort the annual research, development, and demonstration charges for the period. In such a case, the portion of such amounts that cause the distortion may be amortized to the appropriate operating expense account over a period not to exceed 5 years unless otherwise authorized by RUS.

D. The entries in this account must be so maintained as to show separately each project along with complete detail of the nature and purpose of the research, development, and demonstration project together with the related costs.

189 Unamortized Loss on Reacquired Debt

This account shall include the losses on long-term debt reacquired or redeemed. The amounts in this account shall be amortized in accordance with § 1767.15 (q).

190   Accumulated Deferred Income Taxes

A. This account shall be debited and Account 411.1, Provision for Deferred Income Taxes—Credit, Utility Operating Income, or Account 411.2, Provision for Deferred Income Taxes—Credit, Other Income and Deductions, as appropriate, shall be credited with an amount equal to that by which income taxes payable for the year are higher because of the inclusion of certain items in income for tax purposes, which items for general accounting purposes will not be fully reflected in the utility's determination of annual net income until subsequent years.

B. This account shall be credited and Account 410.1, Provision for Deferred Income Taxes, Utility Operating Income, or Account 410.2, Provision for Deferred Income Taxes, Other Income and Deductions, as appropriate, shall be debited with an amount equal to that by which income taxes payable for the year are lower because of prior payment of taxes as provided by Paragraph A above, because of difference in timing for tax purposes of particular items of income or income deductions from that recognized by the utility for general accounting purposes. Such credit to this account and debit to Account 410.1 or Account 410.2 shall, in general, represent the effect on taxes payable in the current year of the smaller amount of book income recognized for tax purposes as compared to the amount recognized in the utility's current accounts with respect to the item or class of items for which deferred tax accounting by the utility was authorized by RUS.

C. Vintage year records with respect to entries to this account, as described above, and the account balance, shall be so maintained as to show the factor of calculation with respect to each annual amount of the item or class of items for which deferred tax accounting by the utility is utilized.

D. The utility is restricted in its use of this account to the purpose set forth above. It shall not make use of the balance in this account or any portion thereof except as provided in the text of this account, without prior approval of RUS. Any remaining deferred tax account balance with respect to an amount for any prior year's tax deferral, the amortization of which or other recognition in the utility's income accounts has been completed, or other disposition made, shall be debited to Account 410.1, Provision for Deferred Income Taxes, Utility Operating Income, or Account 410.2, Provision for Deferred Income Taxes, Other Income and Deductions, as appropriate, or otherwise disposed of as RUS may authorize or direct. (See § 1767.15 (t).)

[58 FR 59825, Nov. 10, 1993, as amended at 59 FR 27436, May 27, 1994; 60 FR 55429, 55430, Nov. 1, 1995; 73 FR 30282, May 27, 2008]

§ 1767.19   Liabilities and other credits.

The liabilities and other credit accounts identified in this section shall be used by all RUS borrowers.

LIABILITIES AND OTHER CREDITS

Margins and Equities

200   Memberships
200.1   Memberships Issued
200.2   Memberships Subscribed But Unissued
201   Patronage Capital
201.1   Patronage Capital Credits
201.2   Patronage Capital Assignable
202–207 [Reserved]
208   Donated Capital
209   Accumulated Other Comprehensive Income
210 [Reserved]
211   Consumers' Contributions for Debt Service
212–214 [Reserved]
215   Appropriated Margins
215.1   Unrealized Gains and Losses—Debt and Equity Securities
216 [Reserved]
216.1   Unappropriated Undistributed Subsidiary Earnings
217   Retired Capital Credits—Gain
218   Capital Gains and Losses
219   Other Margins and Equities
219.1   Operating Margins
219.2   Nonoperating Margins
219.3   Other Margins
219.4   Other Margins and Equities—Prior Periods

Long-Term Debt

221   Bonds
222   Reacquired Bonds
223   Advances from Associated Companies
224   Other Long-Term Debt
224.1   Long-Term Debt—RUS Construction Loan Contract
224.2   RUS Loan Contract—Construction—Debit
224.3   Long-Term Debt—RUS Construction Notes Executed
224.4   RUS Notes Executed—Construction—Debit
224.5   Interest Accrued—Deferred—RUS Construction
224.6   Advance Payments Unapplied—RUS Long-Term Debt—Debit
224.7   Long-Term Debt—Installation Loan Contract
224.8   RUS Loan Contract—Installation—Debit
224.9   Long-Term Debt—Installation Notes Executed

224.10 RUS Notes Executed—Installation—Debit

224.11 Other Long-Term Debt—Subscriptions

224.12 Other Long-Term Debt—Supplemental Financing

224.13 Supplemental Financing Notes Executed—Debit

224.14 Other Long-Term Debt—Miscellaneous

224.15 Notes Executed—Other—Debit

224.16 Long-Term Debt—RUS Economic Development Notes Executed

224.17 RUS Notes Executed—Economic Development—Debit

224.18 Other Long-Term Debt—Grant Funds

225 Unamortized Premium on Long-Term Debt

226 Unamortized Discount on Long-Term Debt—Debit

Other Noncurrent Liabilities

227 Obligations Under Capital Leases—Noncurrent

228.1 Accumulated Provision for Property Insurance

228.2 Accumulated Provision for Injuries and Damages

228.3 Accumulated Provision for Pensions and Benefits

228.4 Accumulated Miscellaneous Operating Provisions

229 Accumulated Provision for Rate Refunds

Current and Accrued Liabilities

231 Notes Payable

232 Accounts Payable

232.1 Accounts Payable—General

232.2 Accounts Payable—RUS Construction

232.3 Accounts Payable—Other

233 Notes Payable to Associated Companies

234 Accounts Payable to Associated Companies

235 Customer Deposits

236 Taxes Accrued

236.1 Accrued Property Taxes

236.2 Accrued U.S. Social Security Tax—Unemployment

236.3 Accrued U.S. Social Security Tax—F.I.C.A.

236.4 Accrued State Social Security Tax—Unemployment

236.5 Accrued State Sales Tax—Consumers

236.6 Accrued Gross Revenue or Gross Receipts Tax

236.7 Accrued Taxes—Other

237 Interest Accrued

238 Patronage Capital and Patronage Refunds Payable

238.1 Patronage Capital Payable

238.2 Patronage Refunds Payable

239 Matured Long-Term Debt

240 Matured Interest

241 Tax Collections Payable

242 Miscellaneous Current and Accrued Liabilities

242.1 Accrued Rentals

242.2 Accrued Payroll

242.3 Accrued Employees' Vacations and Holidays

242.4 Accrued Insurance

242.5 Other Current and Accrued Liabilities

243 Obligations Under Capital Leases—Current

Deferred Credits

251 [Reserved]

252 Customer Advances for Construction

253 Other Deferred Credits

253.1 Other Deferred Credits—Consumers' Energy Prepayments

254 Other Regulatory Liabilities

255 Accumulated Deferred Investment Tax Credits

256 Deferred Gains from Disposition of Utility Plant

257 Unamortized Gain on Reacquired Debt

281 Accumulated Deferred Income Taxes—Accelerated Amortization Property

282 Accumulated Deferred Income Taxes—Other Property

283 Accumulated Deferred Income Taxes—Other

LIABILITIES AND OTHER CREDITS

Margins and Equities

200 Memberships

A. This account shall include the total amount of memberships issued and subscribed.

B. Account 200 shall be subaccounted as follows:

200.1 Memberships Issued

200.2 Memberships Subscribed But Unissued

200.1 Memberships Issued

A. This account shall include the face value of membership certificates outstanding. A detailed record shall be maintained to show for each member, the name, address, date of payment, amount paid, and certificate number.

B. If membership fees are applied against energy bills, this account shall be debited for the full amount of the membership with the offsetting credit to the appropriate accounts receivable, and to accounts payable for any refundable amounts. Any balances that cannot be refunded, due to inability to locate the member or because of bylaw restrictions, shall be credited to Account 208, Donated Capital. If determination of the ultimate disposition of the fees cannot be made immediately, the amount involved should be transferred to Account 253, Other Deferred Credits, until the determination is made.

C. When a transfer fee is collected, the transaction shall be recorded by debiting Account 131.1, Cash—General, and crediting Account 451, Miscellaneous Service Revenues, with the fee collected.

200.2 Memberships Subscribed But Unissued

This account shall include the face value of memberships subscribed for but not issued. When certificates are issued, the amount of the memberships shall be transferred to Account 200.1, Memberships Issued.

201 Patronage Capital

A. This account shall include the total amount of patronage capital assignable and assigned.

B. Account 201 shall be subaccounted as follows:

201.1 Patronage Capital Credits
201.2 Patronage Capital Assignable

201.1 Patronage Capital Credits

A. This account shall include the amounts of patronage capital which have been assigned to individual patrons. A subsidiary record, "patronage capital ledger," shall be maintained, containing an account for each patron who has furnished capital under a capital credits plan.

B. When the return of patrons' capital to individual patrons has been authorized by the board of directors (or trustees), the amounts authorized shall be transferred to Account 238.1, Patronage Capital Payable. (See also Account 217, Retired Capital Credits-Gain.)

201.2 Patronage Capital Assignable

A. This account shall include all amounts transferred from Account 219.1, Operating Margins; Account 219.2, Nonoperating Margins; Account 219.3, Other Margins; and Account 219.4, Other Margins and Equities—Prior Periods, which are assignable to individual patrons' capital accounts.

B. Entries to this account shall be made so as to clearly disclose the nature and source of each transaction. Amounts so assigned shall be transferred to Account 201.1, Patronage Capital Credits.

202–207 [Reserved]

208 Donated Capital

This account shall include credits arising from forfeiture of membership fees and from donations of capital not otherwise provided for. Entries to this account shall be made so as to clearly disclose the nature and source of each transaction.

209 Accumulated Other Comprehensive Income

A. This account shall include revenues, expenses, gains, and losses that are properly includable in other comprehensive income during the period. Examples of other comprehensive income include foreign currency items, minimum pension liability adjustment, unrealized gains and losses on certain investments in debt and equity securities, and cash flow hedges. Records supporting the entries to this account shall be maintained so that the utility can furnish the amount of other comprehensive income for each item included in this account.

B. This account shall also be debited or credited, as appropriate, with amounts of accumulated other comprehensive income that have been included in the determination of net income during the period and in accumulated other comprehensive income in prior periods. Separate records for each category of items shall be maintained to identify the amount of the reclassification adjustments from accumulated other comprehensive income to earning made during the period.

210 [Reserved]

211 Consumers' Contributions for Debt Service

This account shall include the amounts billed to consumers as "amortization charges" for the purpose of servicing long-term debt.

212–214 [Reserved]

215 Appropriated Margins

This account shall include all amounts appropriated as reserves from margins. The account shall be so maintained as to show the amount of each separate reserve and the nature and amounts of the debits and credits thereto.

215.1 Unrealized Gains and Losses—Debt and Equity Securities

This account shall include the unrealized holding gains and losses for available-for-sale securities.

216 [Reserved]

216.1 Unappropriated Undistributed Subsidiary Earnings

This account shall include the balances, either debit or credit, of undistributed retained earnings of subsidiary companies since their acquisition. When dividends are received from subsidiary companies relating to amounts included in this account, this account shall be debited and Account 219.2, Nonoperating Margins, credited.

217 Retired Capital Credits—Gain

A. This account shall include credits resulting from the retirement of patronage capital through settlement of individual patrons' capital credits at less than 100 percent of the capital assigned to the patron. The portion of patronage capital not returned to the patrons, under such settlements, shall be debited to Account 201.1, Patronage Capital Credits, and credited to this account.

B. This account shall also include amounts representing patronage capital authorized to be retired to patrons who cannot be located. Returned checks issued for retirements of patronage capital, after an appropriate waiting period, shall be credited to this account, and a record maintained adequate to enable the cooperative to make payment to the patron if and when a claim has been established by the consumer.

218 Capital Gains and Losses

No entries shall be made to this account without the prior approval of RUS unless it is to distribute past capital gains and losses as capital credits or to eliminate accumulated capital losses in conformance with the bylaws of the cooperative.

219 Other Margins and Equities

A. This account shall include total amount of margins and equities from all sources.

B. Account 219 shall be subaccounted as follows:

219.1 Operating Margins
219.2 Nonoperating Margins
219.3 Other Margins
219.4 Other Margins and Equities—Prior Periods

219.1 Operating Margins

This account shall be debited or credited with the balances arising from transactions, the details of which have been recorded in Accounts 400, 401, 402, 403, 404, 405, 406, 407, 408, 412, 413, 414, 423, 424, 425, 426, 427, 428, and 431. Accounts 400, 401, and 402 are control accounts and, at the option of the borrower may or may not be used. If they are not used, the detailed revenue and expense accounts shall be closed directly to this account.

219.2 Nonoperating Margins

This account shall be debited or credited with the balances arising from transactions, the details of which have been recorded in Accounts 415, 416, 417, 417.1, 418, 419, 419.1, 421, 421.1, 421.2, 422, 434, and 435.

219.3 Other Margins

No entries shall be made to this account unless it is to distribute or eliminate prior balances in conformance with the bylaws of the cooperative.

219.4 Other Margins and Equities—Prior Periods

A. This account shall include significant nonrecurring transactions relating to prior periods. To be significant, the transaction must be of sufficient magnitude to justify redistribution of patronage capital credits already allocated for such prior periods.

B. All entries to this account must receive RUS prior approval.

C. These transactions are limited to items to (1) correct an error in the financial statements of a prior year, and (2) make adjustments that result from realization of income tax benefits of preacquisition operating loss carryforwards. This account shall also include the related income taxes (state and Federal) on items included herein.

D. Amounts in this account shall be transferred at the end of the year to Account 219.1, Operating Margins, or Account 219.2, Nonoperating Margins, as appropriate. Also, at the end of the year, these amounts should be transferred from Account 219.1, or Account 219.2 to Account 201.2, Patronage Capital Assignable, when appropriate.

Long-Term Debt

221 Bonds

This account shall include, in a separate subdivision for each class and series of bonds, the face value of the actually issued and unmatured bonds which have not been retired or cancelled; also the face value of such bonds issued by others, the payment of which has been assumed by the utility.

222 Reacquired Bonds

A. This account shall include the face value of bonds actually issued or assumed by the utility and reacquired by it and not retired or cancelled. The account for reacquired debt shall not include securities which are held by trustees in sinking or other funds.

B. When bonds are reacquired, the difference between face value, adjusted for unamortized discount, expenses or premium, and the amount paid upon reacquisition, shall be included in Account 189, Unamortized Loss on Reacquired Debt, or Account 257, Unamortized Gain on Reacquired Debt, as appropriate. (See § 1767.15 (q).)

223 Advances from Associated Companies

A. This account shall include the face value of notes payable to associated companies and the amount of open book accounts representing advances from associated companies. It does not include notes and open accounts representing indebtedness subject to current settlement which are includible in Account 233, Notes Payable to Associated

Companies, or Account 234, Accounts Payable to Associated Companies.

B. The records supporting the entries to this account shall be so kept that the utility can furnish complete information concerning each note and open account.

224   Other Long-Term Debt

A. This account shall include, until maturity, all long-term debt not otherwise provided for. This covers such items as receivers' certificates, real estate mortgages executed or assumed, assessments for public improvements, notes and unsecured certificates of indebtedness not owned by associated companies, receipts outstanding for long-term debt, and other obligations maturing more than one year from the date of issue or assumption.

B. Account 224 shall be subaccounted as follows:

224.1　Long-Term Debt—RUS Construction Loan Contract
224.2　RUS Loan Contract—Construction—Debit
224.3　Long-Term Debt—RUS Construction Notes Executed
224.4　RUS Notes Executed—Construction—Debit
224.5　Interest Accrued—Deferred—RUS Construction
224.6　Advance Payments Unapplied—RUS Long-Term Debt—Debit
224.7　Long-Term Debt—Installation Loan Contract
224.8　RUS Loan Contract—Installation—Debit
224.9　Long-Term Debt—Installation Notes Executed
224.10　RUS Notes Executed—Installation—Debit
224.11　Other Long-Term Debt—Subscriptions
224.12　Other Long-Term Debt—Supplemental Financing
224.13　Supplemental Lender Notes Executed—Debit
224.14　Other Long-Term Debt—Miscellaneous
224.15　Notes Executed—Other—Debit
224.16　Long-Term Debt—RUS Economic Development Notes Executed
224.17　RUS Notes Executed—Economic Development—Debit
224.18　Other Long-Term Debt—Grant Funds

224.1   Long-Term Debt—RUS Construction Loan Contract

A. This account shall include the contractual obligation to RUS on construction loans covered by loan contract but not by executed notes.

B. This account is to be used at the option of the borrower.

224.2   RUS Loan Contract—Construction—Debit

A. This account shall include the total loans (for construction purposes) which are covered by loan contract but not by executed notes.

B. This account is to be used at the option of the borrower.

224.3   Long-Term Debt—RUS Construction Notes Executed

This account shall include the contractual liability to RUS on construction notes executed. Records shall be maintained to show separately for each class of obligation all details as to the date of obligation, date of maturity, interest date and rate, and securities for the obligation.

224.4   RUS Notes Executed—Construction—Debit

This account shall include the total amount of the unadvanced RUS loans for construction purposes, which are covered by executed notes. When advances are received from the RUS for construction, this account shall be credited and Account 131.2, Cash—Construction Fund—Trustee, debited with the amount of cash advanced.

224.5   Interest Accrued—Deferred—RUS Construction

This account shall include interest on RUS construction obligations deferred by the terms of mortgage notes or extension agreements.

224.6   Advance Payments Unapplied—RUS Long-Term Debt—Debit

A. This account shall include principal payments on mortgage notes paid in advance of the date due and not applied to a specific note. Also, include in this account interest savings which are accrued and added to the advance payment unapplied.

B. At such time as these payments are applied to a specific note or loan balances, this account shall be credited and the long-term debt account debited with the amount so applied.

224.7   Long-Term Debt—Installation Loan Contract

A. This account shall include the contractual obligation to RUS on installation loans covered by loan contract but not covered by executed notes.

B. This account is to be used at the option of the borrower.

224.8   RUS Loan Contract—Installation—Debit

A. This account shall include the total loans for installation purposes which are covered by loan contract but not by executed notes.

B. This account is to be used at the option of the borrower.

224.9 Long-Term Debt—Installation Notes Executed

This account shall include the contractual liability to RUS on installation notes executed.

224.10 RUS Notes Executed—Installation—Debit

This account shall include the total amount of unadvanced loans for installation purposes, which are covered by executed note. When advances are received from RUS, this account shall be credited and Account 131.3, Cash—Installation Loan and Collection Fund, debited with the amount of cash advanced.

224.11 Other Long-Term Debt—Subscriptions

This account shall include the contractual obligation to purchase CFC Capital Term Certificates and any other similar obligation relating to supplemental financing.

224.12 Other Long-Term Debt—Supplemental Financing

This account shall include the contractual liability to CFC or other supplemental lenders for that portion of funds borrowed which mature in more than one year.

224.13 Supplemental Financing Notes Executed—Debit

This account shall include the total amount of the unadvanced loans for construction purposes, which are covered by executed notes to CFC or other supplemental lender. This account shall be debited with the face amount of notes executed. When advances are received from a supplemental lender for construction, this account shall be credited and Account 131.2, Cash—Construction Fund—Trustee, debited with the amount of cash advanced.

224.14 Other Long-Term Debt—Miscellaneous

This account shall include the amount of other long-term debt not provided for elsewhere.

224.15 Notes Executed—Other—Debit

This account shall include the total amount of the unadvanced loans for construction purposes, which are covered by executed notes to others not included in the foregoing accounts. When advances are received from such supplemental lender, this account shall be credited and Account 131.2,

Cash—Construction Fund—Trustee, debited with the amount of cash so advanced.

224.16 Long-Term Debt—RUS Economic Development Notes Executed

This account shall include the contractual liability to RUS on rural economic development notes executed. Records shall be maintained to show separately for each class of obligation all details as to the date of obligation, date of maturity, interest date and rate, and securities for the obligation.

224.17 RUS Notes Executed—Economic Development—Debit

This account shall include the total amount of the unadvanced RUS loans for rural economic development purposes, which are covered by executed notes. When advances are received from the RUS for rural economic development projects, this account shall be credited and Account 131.12, Cash—General—Economic Development Funds, debited with the amount of cash advanced.

224.18 Other Long-Term Debt—Grant Funds

This account shall include the total amount of Rural Development grant funds awarded for rural economic development purposes, which are subject to repayment at the conclusion of the project. (See Sec. 1767.41, Interpretation 626, Rural Economic Development Loan and Grant Program.)

225 Unamortized Premium on Long-Term Debt

A. This account shall include the excess of the cash value of consideration received over the face value upon the issuance or assumption of long-term debt securities.

B. Amounts recorded in this account shall be amortized over the life of each respective issue under a plan which will distribute the amount equitably over the life of the security. The amortization shall be on a monthly basis, with the amounts thereof to be credited to Account 429, Amortization of Premium on Debt—Credit. (See §1767.15 (q).)

226 Unamortized Discount on Long-Term Debt—Debit

A. This account shall include the excess of the face value of long-term debt securities over the cash value of consideration received therefor, related to the issue or assumption of all types and classes of debt.

B. Amounts recorded in this account shall be amortized over the life of the respective issues under a plan which will distribute the amount equitably over the life of the securities. The amortization shall be on a monthly basis, wit the amounts thereof charged to Account 428, Amortization of Debt Discount and Expense. (See §1767.15 (q).)

Other Noncurrent Liabilities

227 Obligations Under Capital Leases— Noncurrent

This account shall include the portion not due within one year, of the obligations recorded for the amounts applicable to leased property recorded as assets in Account 101.1, Property Under Capital Leases; Account 120.6, Nuclear Fuel Under Capital Leases; or Account 121, Nonutility Property.

SPECIAL INSTRUCTIONS

No amounts shall be credited to Accounts 228.1 through 228.4 unless authorized by a regulatory authority or authorities to be collected in the utility's rates.

228.1 Accumulated Provision for Property Insurance

A. This account shall include amounts reserved by the utility for losses through accident, fire, flood, or other hazards to its own property or property leased from others, not covered by insurance. The amounts charged to Account 924, Property Insurance, or other appropriate accounts to cover such risks shall be credited to this account. A schedule of risks covered shall be maintained, giving a description of the property involved, the character of the risks covered and the rates used.

B. Charges shall be made to this account for losses covered, not to exceed the account balance. Details of these charges shall be maintained according to the year the casualty occurred which gave rise to the loss.

228.2 Accumulated Provision for Injuries and Damages

A. This account shall be credited with amounts charged to Account 925, Injuries and Damages, or other appropriate accounts, to meet the probable liability, not covered by insurance, for deaths or injuries to employees and others and for damages to property neither owned nor held under lease by the utility.

B. When liability for any injury or damage is admitted by the utility either voluntarily or because of the decision of a court or other lawful authority, such as a workmen's compensation board, the admitted liability shall be charged to this account and credited to the appropriate current liability account. Details of these charges shall be maintained according to the year the casualty occurred which gave rise to the loss.

NOTE: Recoveries or reimbursements for losses charged to this account shall be credited hereto; the cost of repairs to property of others, if provided for herein, shall be charged to this account.

228.3 Accumulated Provision for Pensions and Benefits

A. This account shall include provisions made by the utility and amounts contributed by employees for pensions, accident and death benefits, savings, relief, hospital, and other provident purposes, where the funds are included in the assets of the utility either in general or in segregated fund accounts.

B. Amounts paid by the utility for the purpose for which this liability is established shall be charged hereto.

C. A separate account shall be kept for each kind of provision included herein.

NOTE: If employee pension or benefit plan funds are not included among the assets of the utility but are held by outside trustees, payments into such funds, or accruals therefor, shall not be included in this account.

228.4 Accumulated Miscellaneous Operating Provisions

A. This account shall include all operating provisions which are not provided for elsewhere.

B. This account shall be maintained in such a manner as to show the amount of each separate provision and the nature and amounts of the debits and credits thereto.

NOTE: This account includes only provisions as may be created for operating purposes and does not include any reservations of income, the credits for which should be recorded in Account 215, Appropriated Margins.

229 Accumulated Provision for Rate Refunds

A. This account shall be credited with amounts charged to Account 449.1, Provision for Rate Refunds, to provide for estimated refunds where the utility is collecting amounts in rates subject to refund.

B. When a refund of any amount recorded in this account is ordered by a regulatory authority, such amount shall be charged hereto and credited to Account 242, Miscellaneous Current and Accrued Liabilities.

C. Records supporting the entries to this account shall be kept so as to identify each amount recorded by the respective rate filing docket number.

Current and Accrued Liabilities

Current and accrued liabilities are those obligations which have either matured or which become due within 1 year from the date thereof; except however, bonds, receivers' certificates, and similar obligations which shall be classified as long-term debt until date of maturity; accrued liabilities, such as income taxes, which shall be classified as accrued liabilities even though payable more than one year from date; compensation awards, which shall be classified as current liabilities regardless of date due; and minor amounts payable in installments which may be classified as current liabilities. If a liability is due more than 1 year from the date of issuance or assumption by the utility, it shall be credited to a long-term debt account appropriate for the transaction; except however, the current liabilities previously mentioned.

230 Asset Retirement Obligations

A. This account shall include the amount of liabilities for the recognition of asset retirement obligations related to electric utility plant and nonutility plant that gives rise to the obligations. This account shall be credited for the amount of the liabilities for asset retirement obligations with amounts charged to the appropriate electric utility plant accounts or nonutility plant account to record the related asset retirement costs.

B. The utility shall charge the accretion expense to Account 411.10, Accretion Expense, for electric utility plant, Account 413, Expenses for Electric Plant Leased to Others, for electric plant leased to others, or Account 421, Miscellaneous Nonoperating Income, for nonutility plant, as appropriate, and credit Account 230, Asset Retirement Obligations.

C. This account shall be debited with amounts paid to settle the asset retirement obligations recorded herein.

D. The utility shall clear from this account any gains or losses resulting from the settlement of asset retirement obligations in accordance with the instruction prescribed in Sec. 1767.15(y).

231 Notes Payable

This account shall include the face value of all notes, drafts, acceptances, or other similar evidences of indebtedness, payable on demand or within a time not exceeding 1 year from the date of issue, to other than associated companies.

232 Accounts Payable

A. This account shall include all amounts payable by the utility within 1 year, which are not provided for in other accounts.

B. Account 232 shall be subaccounted as follows:

232.1 Accounts Payable—General
232.2 Accounts Payable—RUS Construction
232.3 Accounts Payable—Other

233 Notes Payable to Associated Companies

This account shall include amounts owing to associated companies on notes, drafts, acceptances, or other similar evidences of indebtedness payable on demand or not more than 1 year from the date of issue or creation.

NOTE: Notes which are includible in Account 223, Advances from Associated Companies, shall be excluded from this account.

234 Accounts Payable to Associated Companies

This account shall include amounts owing to associated companies on open accounts payable on demand.

NOTE: Accounts which are includible in Account 223, Advances from Associated Companies, shall be excluded from this account.

235 Customer Deposits

This account shall include all amounts deposited with the utility by its customers as security for the payment of bills.

236 Taxes Accrued

A. This account shall be credited with the amount of taxes accrued during the accounting period, corresponding debits being made to the appropriate accounts for tax charges. Such credits may be based upon estimates, but from time to time during the year as the facts become known, the amount of the periodic credits shall be adjusted so as to include, as nearly as can be determined in each year, the taxes applicable thereto. Any amount representing a prepayment of taxes applicable to the period subsequent to the date of the balance sheet, shall be shown under Account 165, Prepayments.

B. If accruals for taxes are found to be insufficient or excessive, correction therefor shall be made through current tax accruals.

C. Accruals for taxes shall be based upon the net amounts payable after credit for any discounts, and shall not include any amounts for interest on tax deficiencies or refunds. Interest received on refunds shall be credited to Account 419, Interest and Dividend Income, and interest paid on deficiencies shall be charged to Account 431, Other Interest Expense.

D. Account 236 shall be subaccounted as follows:

236.1 Accrued Property Taxes
236.2 Accrued U.S. Social Security Tax—Unemployment
236.3 Accrued U.S. Social Security Tax—F.I.C.A.
236.4 Accrued State Social Security Tax—Unemployment
236.5 Accrued State Sales Tax—Consumers
236.6 Accrued Gross Revenue or Gross Receipts Tax
236.7 Accrued Taxes—Other

237 Interest Accrued

This account shall include the amount of interest accrued but not matured on all liabilities of the utility not including, however, interest which is added to the principal of the debt on which incurred. Supporting records shall be maintained so as to show the amount of interest accrued on each obligation.

238 Patronage Capital and Patronage Refunds Payable

A. This account shall include the total amount of patronage capital authorized to be returned and paid to patrons.

B. Account 238 shall be subaccounted as follows:

238.1 Patronage Capital Payable
238.2 Patronage Refunds Payable

238.1 Patronage Capital Payable

This account shall include the amount of patronage capital which has been authorized to be returned to the patron.

238.2 Patronage Refunds Payable

This account shall include the amount of patronage refunds which have been authorized to be paid to patrons.

239 Matured Long-Term Debt

This account shall include the amount of long-term debt (including any obligation for premiums) matured and unpaid, without specific agreement for extension of the time of payment and bonds called for redemption but not presented.

240 Matured Interest

This account shall include the amount of matured interest on long-term debt or other obligations of the utility at the date of the balance sheet unless such interest is added to the principal of the debt on which incurred.

241 Tax Collections Payable

This account shall include the amount of taxes collected by the utility through payroll deductions or otherwise, pending transmittal of such taxes to the proper taxing authority.

NOTE: Do not include liabilities for taxes assessed directly against the utility which are accounted for as part of the utility's own tax expense.

242 Miscellaneous Current and Accrued Liabilities

A. This account shall include the amount of all other current and accrued liabilities not provided for elsewhere appropriately designated and supported so as to show the nature of each liability.

B. Account 242 shall be subaccounted as follows:

242.1 Accrued Rentals
242.2 Accrued Payroll
242.3 Accrued Employees' Vacations and Holidays
242.4 Accrued Insurance
242.5 Other Current and Accrued Liabilities

242.1 Accrued Rentals

This account shall include unpaid joint use pole rentals and other rentals. The records supporting the entries to this account shall be maintained so as to show for each class of rental, the amount accrued, the basis for the accrual, the accounts to which charged, and the amount of rentals paid.

242.2 Accrued Payroll

This account shall include the accrued liability for salaries and wages at the end of an accounting period for which the appropriate expense or other accounts have been charged. This account is to be used whether salaries and wages are paid on a weekly, semimonthly, or monthly basis.

242.3 Accrued Employees' Vacations and Holidays

This account shall include the liability for accrued wages for employees' vacation, holidays, and sick leave.

242.4 Accrued Insurance

A. This account shall most commonly be used in case of workmen's compensation and public liability insurance for recording the excess amounts of earned premium over the advance premiums. Earned premiums are computed each month by applying the insurance rates to the actual payrolls.

B. Until the amount of the advance premiums is exhausted, the earned premium is credited to Account 165, Prepayments. Earned premiums in excess of the advance premiums are credited to this account.

242.5 Other Current and Accrued Liabilities

This account shall include current and accrued liabilities not provided for elsewhere.

243 Obligations Under Capital Leases— Current

This account shall include the portion, due within 1 year, of the obligations recorded for the amounts applicable to leased property recorded as assets in Account 101.1, Property Under Capital Leases; Account 120.6, Nuclear Fuel Under Capital Leases; or Account 121, Nonutility Property.

244 DERIVATIVE INSTRUMENT LIABILITIES

This account shall include the change in the fair value of all derivative instrument liabilities not designated as cash flow or fair value hedges. Account 426, Other Deductions, shall be debited or credited as appropriate with the corresponding amount of the change in the fair value of the derivative instrument.

245 Derivative Instrument Liabilities—
Hedges

A. This account shall include the change in the fair value of derivative instrument liabilities designated by the utility as cash flow or fair value hedges.

B. A utility shall record the change in the fair value of a derivative instrument liability related to a cash flow hedge in this account, with a concurrent charge to Account 209, Accumulated Other Comprehensive Income, with the effective portion of the derivative's gain or loss. The ineffective portion of the cash flow hedge shall be charged to the same income or expense account that will be used when the hedged item enters into the determination of net income.

C. A utility shall record the change in the fair value of a derivative instrument liability related to a fair value hedge in this account, with a concurrent charge to a subaccount of the asset or liability that carries the item being hedged. The ineffective portion or the fair value hedge shall be charged to the same income or expense account that will be used when the hedged item enters into the determination of net income.

Deferred Credits

251 [Reserved]

252 Customer Advances for Construction

This account shall include consumer advances for construction which are to be refunded either wholly or in part. When a customer is refunded the entire amount to which he is entitled, according to the agreement or rule under which the advance was made, the balance, if any, remaining in this account shall be credited to the respective plant accounts.

253 Other Deferred Credits

This account shall include advance billings and receipts and other deferred credit items, not provided for elsewhere, including amounts which cannot be entirely cleared or disposed of until additional information has been received.

253.1 Other Deferred Credits—Consumers'
Energy Prepayments

This account shall include the amount of advance payments made by consumers in connection with electric service.

254 Other Regulatory Liabilities

A. This account shall include the amounts of regulatory liabilities, not includible in other accounts, imposed on the utility by the ratemaking actions of regulatory agencies.

B. The amounts included in this account are to be established by those credits which would have been included in net income, or

accumulated other comprehensive income, determinations in the current period under the general requirements of the Uniform System of Accounts but for it being probable that: (1) Such items will be included in a different period(s) for purposes of developing the rates that the utility is authorized to charge for its utility services; or (2) refunds to customers, not provided for in other accounts, will be required. When specific identification of the particular source of the regulatory liability cannot be made or when the liability arises from revenues collected pursuant to tariffs on file at a regulatory agency, Account 407.3, Regulatory Debits, shall be debited. The amounts recorded in this account generally are to be credited to the same account that would have been credited if included in income when earned except: (1) All regulatory liabilities established through the use of Account 407.3 shall be credited to Account 407.4, Regulatory Credits; and (2) in the case of refunds, a cash account or other appropriate account should be credited when the obligation is satisfied.

C. If it is later determined that the amounts recorded in this account will not be returned to customers through rates or refunds, such amounts shall be credited to Account 421, Miscellaneous Nonoperating Income, or Account 434, Extraordinary Income, as appropriate, in the year such determination is made.

D. The records supporting the entries to this account shall be kept in such a manner that the utility can furnish full information as to the nature and amount of each regulatory liability included in this account, including justification for inclusion of such amounts in this account.

255 Accumulated Deferred Investment Tax
Credits

A. This account shall be credited with all investment tax credits deferred by companies which have elected to follow deferral accounting, partial or full, rather than recognizing, in the income statement, the total benefits of the tax credit as realized. After such election, a company may not transfer amounts from this account, except as authorized herein and in Account 411.4, Investment Tax Credit Adjustments, Utility Operations; Account 411.5, Investment Tax Credit Adjustments, Nonutility Operations; and Account 420, Investment Tax Credits, or with approval of RUS.

B. Where the company's accounting provides that investment tax credits are to be passed on to customers, this account shall be debited and Account 411.4 credited with a proportionate amount determined in relation to the average useful life of electric utility property to which the tax credits relate or such lesser period of time as allowed by a regulatory agency having rate jurisdiction. If, however, the deferral procedure provides that investment tax credits are not to be passed on to customers, the proportionate restorations to income shall be credited to Account 420.

C. Subdivisions of this account, by department, shall be maintained for deferred investment tax credits that are related to nonelectric utility or other operations. Contra entries affecting such account subdivisions shall be appropriately recorded in Account 413, Expenses of Electric Plant Leased to Others; or Account 414, Other Utility Operating Income. Use of deferral or nondeferral accounting procedures adopted for nonelectric utility or other operations are to be followed on a consistent basis.

D. Separate records for electric and nonelectric utility or other operations shall be maintained identifying the properties giving rise to the investment tax credits for each year with the weighted-average service life of such properties and any unused balances of such credits. Such records are not necessary unless the tax credits are deferred.

256   Deferred Gains from Disposition of Utility Plant

This account shall include gains from the sale or other disposition of property previously recorded in Account 105, Electric Plant Held for Future Use, under the provisions of Paragraphs B, C, and D thereof, where such gains are significant and are to be amortized over a period of 5 years, unless otherwise authorized by RUS. The amortization of the amounts in this account shall be made by credits to Account 411.6, Gains from Disposition of Utility Plant. (See Account 105, Electric Plant Held for Future Use.)

257   Unamortized Gain on Reacquired Debt

This account shall include the amounts of discount realized upon reacquisition or redemption of long-term debt. The amounts in this account shall be amortized in accordance with § 1767.15 (q).

SPECIAL INSTRUCTIONS

Accumulated Deferred Income Taxes

Before using the deferred tax accounts provided below, refer to § 1767.15 (r), Comprehensive Interperiod Income Tax Allocation. The text of these accounts are designed primarily to cover deferrals of Federal income taxes. However, they are also to be used when making deferrals of state and local income taxes. Utilities and licensees which, in addition to an electric utility department, have another utility department, gas or water and nonutility property, and which have deferred taxes on income with respect thereto shall separately classify such deferrals in the accounts provided below so as to allow ready identification of items relating to each utility deductions.

281   Accumulated Deferred Income Taxes— Accelerated Amortization Property

A. This account shall include tax deferrals resulting from adoption of the principles of comprehensive interperiod tax allocation described in § 1767.15 (s) that relate to property for which the utility has availed itself of the use of accelerated (5-year) amortization of (1) certified defense facilities as permitted by Section 168 of the Internal Revenue Code, and (2) certified pollution control facilities as permitted by Section 169 of the Internal Revenue Code.

B. This account shall be credited and Account 410.1, Provision for Deferred Income Taxes, Utility Operating Income, or Account 410.2, Provision for Deferred Income Taxes, Other Income and Deductions, as appropriate, shall be debited with tax effects related to property described in Paragraph A above where taxable income is lower than pretax accounting income due to differences between the periods in which revenue and expense transactions affect taxable income and the periods in which they enter into the determination of pretax accounting income.

C. This account shall be debited and Account 411.1, Provision for Deferred Income Taxes—Credit, Utility Operating Income, or Account 411.2, Provision for Deferred Income Taxes-Credit, Other Income and Deductions, as appropriate, shall be credited with taxes related to property described in Paragraph A above where taxable income is higher than pretax accounting income due to differences between the periods in which revenue and expense transactions affect taxable income and the periods in which they enter into the determination of pretax accounting income.

D. The utility is restricted in its use of this account to the purposes set forth above. It shall not transfer the balance in this account or any portion thereof to retained earnings or make any use thereof except as provided in the text of this account without prior approval of RUS. Upon the disposition by sale, exchange, transfer, abandonment, or premature retirement of plant on which there is a related balance therein, this account shall be charged with an amount equal to the related income tax expense, if any, arising from such disposition and Account 411.1, Provision for Deferred Income Taxes— Credit, Utility Operating Income, or Account 411.2, Provision for Deferred Income Taxes— Credit, Other Income and Deductions, as appropriate, shall be credited. When the remaining balance, after consideration of any related income tax expense, is less than $25,000, this account shall be charged and Account 411.1 or Account 411.2, as appropriate, credited with such balance. If after consideration of any related income tax expense, there is a remaining amount of $25,000 or more, RUS shall authorize or direct how such amount shall be accounted for at the time approval for the disposition of accounting is granted. When plant is disposed of by transfer to a wholly owned subsidiary, the related balance in this account shall also be transferred. When the disposition relates to retirement of an item or items under a group method of depreciation where there is no tax effect in the year of retirement, no entries are required in this account if it can be determined that the related balances would be necessary to be retained to offset future group item tax deficiencies.

282 Accumulated Deferred Income Taxes—Other Property

A. This account shall include the tax deferrals resulting from adoption of the principle of comprehensive interperiod income tax allocation described in § 1767.15 (r) which are related to all property other than accelerated amortization property.

B. This account shall be credited and Account 410.1, Provision for Deferred Income Taxes, Utility Operating Income, or Account 410.2, Provision for Deferred Income Taxes, Other Income and Deductions, as appropriate, shall be debited with tax effects related to property described in Paragraph A above where taxable income is lower than pretax accounting income due to differences between the periods in which revenue and expense transactions affect taxable income and the periods in which they enter into the determination of pretax accounting income.

C. This account shall be debited and Account 411.1, Provision for Deferred Income Taxes—Credit, Utility Operating Income, or Account 411.2, Provision for Deferred Income Taxes—Credit, Other Income and Deductions, as appropriate, shall be credited with tax effects related to property described in Paragraph A above where taxable income is higher than pretax accounting income due to differences between the periods in which revenue and expense transactions affect taxable income and the periods in which they enter into the determination of pretax accounting income.

D. The utility is restricted in its use of this account to the purposes set forth above. It shall not transfer the balance in this account or any portion thereof to retained earnings or make any use thereof except as provided in the text of this account without prior approval of RUS. Upon the disposition by sale, exchange, transfer, abandonment, or premature retirement of plant on which there is a related balance herein, this account shall be charged with an amount equal to the related income tax expense, if any, arising from such disposition and Account 411.1, Provision for Deferred Income Taxes—Credit, Utility Operating Income, or Account 411.2, Provision for Deferred Income Taxes—Credit, Other Income and Deductions, shall

be credited. When the remaining balance after consideration of any related tax expenses, is less than $25,000, this account shall be charged and Account 411.1 or Account 411.2, as appropriate, credited with such balance. If after consideration any related income tax expense, there is a remaining amount of $25,00 or more, RUS shall authorize or direct how such amount shall be accounted for at the time approval for the disposition of accounting is granted. When plant is disposed of by transfer to a wholly owned subsidiary, the related balance in this account shall also be transferred. When the disposition relates to retirement of an item or items under a group method of depreciation where there is no tax effect in the year of retirement, no entries are required in this account if it can be determined that the related balance would be necessary to be retained to offset future group item tax deficiencies.

283 Accumulated Deferred Income Taxes—Other

A. This account shall include all credit tax deferrals resulting from the adoption of the principles of comprehensive interperiod income tax allocation described in § 1767.15 (r) other than those deferrals which are includible in Account 281, Accumulated Deferred Income Taxes—Accelerated Amortization Property, and Account 282, Accumulated Deferred Income Taxes—Other Property.

B. This account shall be credited and Account 410.1, Provision for Deferred Income Taxes, Utility Operating Income, or Account 410.2, Provision for Deferred Income Taxes, Other Income and Deductions, as appropriate, shall be debited with tax effects related to items described in Paragraph A above where taxable income is lower than pretax accounting income due to differences between the periods in which revenue and expense transactions affect taxable income and the periods in which they enter into the determination of pretax accounting income.

C. This account shall be debited and Account 411.1, Provision for Deferred Income Taxes—Credit, Utility Operating Income or Account 411.2, Provision for Deferred Income Taxes—Credit, Other Income and Deductions, as appropriate, shall be credited with tax effects related to items described in Paragraph A above where taxable income is higher than pretax accounting income due to differences between the periods in which revenue and expense transactions affect taxable income and the periods in which they enter into the determination of pretax accounting income.

D. Records with respect to entries to this account, as described above, and the account balance, shall be so maintained as to show the factors of calculation with respect to each annual amount of the item or class of items.

E. The utility is restricted in its use of this account to the purposes set forth above. It shall not transfer the balance in the account or any portion thereof to retained earnings or to any other account or make any use thereof except as provided in the text of this account, without prior approval of RUS. Upon the disposition by sale, exchange, transfer, abandonment, or premature retirement of items on which there is a related balance herein, this account shall be charged with an amount equal to the related income tax effect, if any, arising from such disposition and Account 411.1, Provision For Deferred Income Taxes—Credit, Utility Operating Income, or Account 411.2, Provision For Deferred Income Taxes-Credit, Other Income and Deductions, as appropriate, shall be credited. When the remaining balance, after consideration of any related tax expenses, is less than $25,000, this account shall be charged and Account 411.1 or Account 411.2, as appropriate, credited with such balance. If after consideration of any related income tax expense, there is a remaining amount of $25,000 or more, RUS shall authorize or direct how such amount shall be accounted for at the time approval for the disposition of accounting is granted.

When plant is disposed of by transfer to a wholly owned subsidiary, the related balance in this account shall also be transferred. When the disposition relates to retirement of an item or items under a group method of depreciation where there is no tax effect in the year of retirement, no entries are required in this account if it can be determined that the related balance would be necessary to be retained to offset future group item tax deficiencies.

[58 FR 59825, Nov. 10, 1993, as amended at 59 FR 27436, May 27, 1994; 60 FR 55430, Nov. 1, 1995; 73 FR 30283, May 27, 2008]

§ 1767.20 Plant accounts.

The plant accounts identified in this section shall be used by all Rural Development borrowers.

INTANGIBLE PLANT

301   Organization
302   Franchises and Consents
303   Miscellaneous Intangible Plant

PRODUCTION PLANT

Steam Production

310   Land and Land Rights
311   Structures and Improvements
312   Boiler Plant Equipment
313   Engines and Engine Driven Generators
314   Turbogenerator Units
315   Accessory Electric Equipment
316   Miscellaneous Power Plant Equipment

317   Asset Retirement Costs for Steam Production Plant

Nuclear Production

320   Land and Land Rights
321   Structures and Improvements
322   Reactor Plant Equipment
323   Turbogenerator Units
324   Accessory Electric Equipment
325   Miscellaneous Power Plant Equipment
326   Asset Retirement Costs for Nuclear Production Plant

Hydraulic Production

330   Land and Land Rights
331   Structures and Improvements
332   Reservoirs, Dams and Waterways
333   Water Wheels, Turbines and Generators
334   Accessory Electric Equipment
335   Miscellaneous Power Plant Equipment
336   Roads, Railroads and Bridges
337   Asset Retirement Costs for Hydraulic Production Plant

Other Production

340   Land and Land Rights
341   Structures and Improvements
342   Fuel Holders, Producers and Accessories
343   Prime Movers
344   Generators
345   Accessory Electric Equipment
346   Miscellaneous Power Plant Equipment
347   Asset Retirement Costs for Other Production Plant

TRANSMISSION PLANT

350   Land and Land Rights
351   [Reserved]
352   Structures and Improvements
353   Station Equipment
354   Tower and Fixtures
355   Poles and Fixtures
356   Overhead Conductors and Devices
357   Underground Conduit
358   Underground Conductors and Devices
359   Roads and Trails
359.1   Asset Retirement Costs for Transmission Plant

DISTRIBUTION PLANT

360   Land and Land Rights
361   Structures and Improvements
362   Station Equipment
363   Storage Battery Equipment
364   Poles, Towers and Fixtures
365   Overhead Conductors and Devices
366   Underground Conduit
367   Underground Conductors and Devices
368   Line Transformers
369   Services
370   Meters
371   Installations on Customers' Premises
372   Leased Property on Customers' Premises
373   Street Lighting and Signal Systems

374 Asset Retirement Costs for Distribution Plant

REGIONAL TRANSMISSION MARKET OPERATION PLANT

380 Land and Land Rights
381 Structures and Improvements
382 Computer Hardware
383 Computer Software
384 Communication Equipment
385 Miscellaneous Regional Transmission and Market Operation Plant
386 Asset Retirement Costs for Regional Transmission and Market Operation Plant

GENERAL PLANT

389 Land and Land Rights
390 Structures and Improvements
391 Office Furniture and Equipment
392 Transportation Equipment
393 Stores Equipment
394 Tools, Shop and Garage Equipment
395 Laboratory Equipment
396 Power Operated Equipment
397 Communication Equipment
398 Miscellaneous Equipment
399 Other Tangible Property
399.1 Asset Retirement Costs for General Plant

INTANGIBLE PLANT

301 Organization

This account shall include all fees paid to Federal or state governments for the privilege of incorporation and expenditures incident to organizing the corporation, partnership, or other enterprise and putting it into readiness to do business.

Items

1. Cost of obtaining certificates authorizing an enterprise to engage in the public-utility business.
2. Fees and expenses for incorporation.
3. Fees and expenses for mergers or consolidations.
4. Office expenses incident to organizing the utility.
5. Stock and minute books and corporate seal.

NOTE A: This account shall not include any discounts upon securities issued or assumed; nor shall it include any costs incident to negotiating loans, selling bonds or other evidences of debt or expenses in connection with the authorization, issuance, or sale of capital stock.

NOTE B: Exclude from this account and include in the appropriate expense account the cost of preparing and filing papers in connection with the extension of the term of incorporation unless the first organization costs have been written off. When charges are made to this account for expenses incurred in mergers, consolidations, or reorganizations, amounts previously included herein or in similar accounts in the books of the companies concerned shall be excluded from this account.

302 Franchises and Consents

A. This account shall include amounts paid to the Federal Government, to a state or to a political subdivision thereof in consideration for franchises, consents, water power licenses, or certificates, running in perpetuity or for a specified term of more than one year, together with necessary and reasonable expenses incident to procuring such franchises, consents, water power licenses, or certificates of permission and approval, including expenses of organizing and merging separate corporations, where statutes require, solely for the purpose of acquiring franchises.

B. If a franchise, consent, water power license, or certificate is acquired by assignment, the charge to this account in respect thereof shall not exceed the amount paid therefor by the utility to the assignor, nor shall it exceed the amount paid by the original grantee, plus the expense of acquisition to such grantee. Any excess of the amount actually paid by the utility over the amount above specified shall be charged to Account 426.5, Other Deductions.

C. When any franchise has expired, the book cost thereof shall be credited hereto and charged to Account 426.5, Other Deductions, or to Account 111, Accumulated Provision for Amortization of Electric Utility Plant, as appropriate.

D. Records supporting this account shall be kept so as to show separately the book cost of each franchise or consent.

NOTE: Annual or other periodic payments under franchises shall not be included herein but in the appropriate operating expense account.

303 Miscellaneous Intangible Plant

A. This account shall include the cost of patent rights, licenses, privileges, and other intangible property necessary or valuable in the conduct of utility operations and not specifically chargeable to any other account.

B. When any item included in this account is retired or expires, the book cost thereof shall be credited hereto and charged to Account 426.5, Other Deductions, or Account 111, Accumulated Provision for Amortization of Electric Utility Plant, as appropriate.

C. This account shall be maintained in such a manner that the utility can furnish full information with respect to the amounts included herein.

PRODUCTION PLANT

Steam Production

310 Land and Land Rights

This account shall include the cost of land and land rights used in connection with steam-power generation. (See § 1767.16 (g).)

311 Structures and Improvements

This account shall include the cost, in place, of structures and improvements used in connection with steam-power generation. (See § 1767.16 (h).)

NOTE: Include steam production roads and railroads in this account.

312 Boiler Plant Equipment

This account shall include the cost installed of furnaces, boilers, coal and ash handling and coal preparing equipment, steam and feed water piping, boiler apparatus, and accessories used in the production of steam, mercury, or other vapor, to be used primarily for generating electricity.

Items

1. Ash handling equipment, including hoppers, gates, cars, conveyors, hoists, sluicing equipment, including pumps and motors, sluicing water pipe and fittings, sluicing trenches and accessories, except sluices which are a part of a building.

2. Boiler feed system, including feed water heaters, evaporator condensers, heater drain pumps, heater drainers, deaerators, and vent condensers, boiler feed pumps, surge tanks, feed water regulators, feed water measuring equipment, and all associated drives.

3. Boiler plant cranes and hoists and associated drives.

4. Boilers and equipment, including boilers and baffles, economizers, superheaters, soot blowers, foundations and settings, water walls, arches, grates, insulation, blowdown system, drying out of new boilers, also associated motors or other power equipment.

5. Breeching and accessories, including breeching, dampers, soot spouts, hoppers and gates, cinder eliminators, breeching insulation, soot blowers and associated motors.

6. Coal handling and storage equipment, including coal towers, coal lorries, coal cars, locomotives and tracks when devoted principally to the transportation of coal, hoppers, downtakes, unloading and hoisting equipment, skip hoists and conveyors, weighing equipment, magnetic separators, cable ways, and housings and supports for coal handling equipment.

7. Draft equipment, including air preheaters and accessories, induced and forced draft fans, air ducts, combustion control mechanisms, and associated motors or other power equipment.

8. Gas-burning equipment, including holders, burner equipment and piping, and control equipment.

9. Instruments and devices, including all measuring, indicating, and recording equipment for boiler plant service together with mountings and supports.

10. Lighting systems.

11. Oil-burning equipment, including tanks, heaters, pumps with drive, burner equipment and piping, and control equipment.

12. Pulverized fuel equipment, including pulverizers, accessory motors, primary air fans, cyclones and ducts, dryers, pulverized fuel bins, pulverized fuel conveyors and equipment, burners, burner piping, priming equipment, air compressors, and motors.

13. Stacks, including foundations and supports, stack steel and ladders, stack brickwork, stack concrete, stack lining, stack painting (first), when set on separate foundations, independent of substructures or superstructures of building.

14. Station piping, including pipe, valves, fittings, separators, traps, desuperheaters, hangers, excavation, and covering for station piping system, including all steam, condensate, boiler feed and water supply piping, but not condensing water, plumbing, building heating, oil, gas, air piping or piping specifically provided for in Account 313.

15. Stoker or equivalent feeding equipment, including stokers and accessory motors, clinker grinders, fans and motors.

16. Ventilating equipment.

17. Water purification equipment, including softeners and accessories, evaporators and accessories, heat exchanges, filters, tanks for filtered or softened water, pumps, and motors.

18. Water-supply systems, including pumps, motors, strainers, raw-water storage tanks, boiler wash pumps, intake and discharge pipes, and tunnels not a part of a building.

19. Wood fuel equipment, including hoppers, fuel hogs and accessories, elevators and conveyors, bins and gates, spouts, measuring equipment and associated drives.

NOTE: When the system for supplying boiler or condenser water is elaborate, and when it includes a dam, reservoir, canal, pipe line, cooling ponds, or where gas or oil is used as a fuel for producing steam and is supplied through a pipe line system owned by the utility, the cost of such special facilities shall be charged to a subdivision of Account 311, Structures and Improvements.

313 Engines and Engine Driven Generators

This account shall include the cost installed of steam engines, reciprocating or rotary, and their associated auxiliaries; and engine-driven main generators, except turbogenerator units.

Items

1. Air cleaning and cooling apparatus, including blowers, drive equipment, air ducts, not a part of building, louvers, pumps, and hoods.
2. Belting, shafting, pulleys, and reduction gearing.
3. Circulating pumps, including connections between condensers and intake and discharge tunnels.
4. Cooling system, including towers, pumps, tank, and piping.
5. Condensers, including condensate pumps, air and vacuum pumps, ejector unloading valves and vacuum breakers, expansion devices, and screens.
6. Cranes and hoists, including items wholly identified with items listed herein.
7. Engines, reciprocating or rotary.
8. Fire-extinguishing systems.
9. Foundations and settings, especially constructed for and not expected to outlast the apparatus for which provided.
10. Generators-Main, a.c. or d.c., including field rheostats and connections for self-excited units, and excitation systems when identified with the generating unit.
11. Governors.
12. Lighting systems.
13. Lubricating systems, including gauges, filters, tanks, pumps, piping, and motors.
14. Mechanical meters, including gauges, recording instruments, sampling and testing equipment.
15. Piping-main exhaust, including connections between generator and condenser and between condenser and hotwell.
16. Piping-main stream, including connections from main throttle valve to turbine inlet.
17. Platforms, railings, steps, and gratings appurtenant to apparatus listed herein.
18. Pressure oil system, including accumulators, pumps, piping, and motors.
19. Throttle and inlet valve.
20. Tunnels, intake and discharge, for condenser system, when not a part of a structure.
21. Water screens and motors.

314 Turbogenerator Units

This account shall include the cost installed of main turbine-driven units and accessory equipment used in generating electricity by steam.

Items

1. Air leaning and cooling apparatus, including blowers, drive equipment, air ducts not a part of building, louvers, pumps, and hoods.
2. Circulating pumps, including connections between condensers and intake and discharge tunnels.
3. Condensers, including condensate pumps, air and vacuum pumps, ejectors, unloading valves and vacuum breakers, expansion devices, and screens.
4. Generator hydrogen, gas piping, and detrainment equipment.
5. Cooling system, including towers, pumps, tanks, and piping.
6. Cranes and hoists, including items wholly identified with items listed herein.
7. Excitation system, when identified with main generating units.
8. Fire-extinguishing systems.
9. Foundations and settings, especially constructed for and not expected to outlast the apparatus for which provided.
10. Governors.
11. Lighting systems.
12. Lubricating systems, including gauges, filters, water separators, tanks, pumps, piping, and motors.
13. Mechanical meters, including gauges, recording instruments, sampling and testing equipment.
14. Piping-main exhaust, including connections between turbogenerator and condenser and between condenser and hotwell.
15. Piping-main steam, including connections from main throttle valve to turbine inlet.
16. Platforms, railings, steps, and gratings appurtenant to apparatus listed herein.
17. Pressure oil systems, including accumulators, pumps, and piping motors.
18. Steelwork, specially constructed for apparatus listed herein.
19. Throttle and inlet valve.
20. Tunnels, intake and discharge, for condenser system, when not a part of structure, and water screens.
21. Turbogenerators-main, including turbine and generator, field rheostats and electric connections for self-excited units.
22. Water screens and motors.
23. Moisture separator for turbine steam.
24. Turbine lubricating oil (initial charge).

315 Accessory Electric Equipment

This account shall include the cost installed of auxiliary generating apparatus, conversion equipment, and equipment used primarily in connection with the control and switching of electric energy produced by steam power, and the protection of electric circuits and equipment, except electric motors used to drive equipment included in other accounts. Such motors shall be included in the account in which the equipment with which they are associated is included.

Items

1. Auxiliary generators, including boards, compartments, switching equipment, control equipment, and connections to auxiliary power bus.

2. Excitation system, including motor, turbine and dual-drive exciter sets and rheostats, storage batteries and charging equipment, circuit breakers, panels and accessories, knife switches and accessories, surge arresters, instrument shunts, conductors and conduit, special supports for conduit, generator field and exciter switch panels, exciter bus tie panels, generator and exciter rheostats and special housing and protective screens.

3. Generator main connections, including oil circuit breakers and accessories, disconnecting switches and accessories, operating mechanisms and interlocks, current transformers, potential transformers, protective relays, isolated panels and equipment, conductors and conduit, special supports for generator main leads, grounding switch, and special housings and protective screens.

4. Station buses including main, auxiliary, transfer, synchronizing and fault ground buses, including oil circuit breakers and accessories, disconnecting switches and accessories, operating mechanisms and interlocks, reactors and accessories, voltage regulators and accessories, compensators, resistors, starting transformers, current transformers, potential transformers, protective relays, storage batteries and charging equipment, isolated panels and equipment, conductors and conduit, special supports, special housings, concrete pads, general station grounding system, special fire-extinguishing system, and test equipment.

5. Station control system, including station switchboards with panel wiring, panels with instruments and control equipment only, panels with switching equipment mounted or mechanically connected, truck-type boards complete, cubicles, station supervisory control boards, generator and exciter signal stands, temperature recording devices, frequency-control equipment, master clocks, watt-hour meters and synchronoscope in the turbine room, station totalizing wattmeter, boiler-room load indicator equipment, storage batteries, panels and charging sets, instrument transformers for supervisory metering, conductors and conduit, special supports for conduit, switchboards, batteries, special housing for batteries, protective screens, and doors.

NOTE A: Do not include in this account transformers and other equipment used for changing the voltage or frequency of electricity for the purposes of transmission or distribution.

NOTE B: When any item of equipment listed herein is used wholly to furnish power to equipment included in another account, its cost shall be included in such other account.

316 Miscellaneous Power Plant Equipment

This account shall include the cost installed of miscellaneous equipment in and about the steam generating plant devoted to general station use, and which is not properly includible in any of the foregoing steam-power production accounts.

Items

1. Compressed air and vacuum cleaning systems, including tanks, compressors, exhausters, air filters, and piping.

2. Cranes and hoisting equipment, including cranes, cars, crane rails, monorails, and hoists with electric and mechanical connections.

3. Fire-extinguishing equipment for general station use.

4. Foundations and settings specially constructed for and not expected to outlast the apparatus for which provided.

5. Locomotive cranes not includible elsewhere.

6. Locomotives not includible elsewhere.

7. Marine equipment, including boats and barges.

8. Miscellaneous belts, pulleys, and countershafts.

9. Miscellaneous equipment, including atmospheric and weather indicating devices, intrasite communication equipment, laboratory equipment, signal systems, callophones, emergency whistles and sirens, fire alarms, insect-control equipment, and other similar equipment.

10. Railway cars not includible elsewhere.

11. Refrigerating systems, including compressors, pumps, and cooling coils.

12. Station maintenance equipment, including lathes, shapers, planers, drill presses, hydraulic presses, and grinders with motors, shafting, hangers, and pulleys.

13. Ventilating equipment, including items wholly identified with apparatus listed herein.

NOTE: When any item of equipment listed herein is wholly used in connection with equipment included in another account, its cost shall be included in such other account.

317 Asset Retirement Costs for Steam Production Plant

This account shall include asset retirement costs on plant included in the steam production function.

Nuclear Production

320 Land and Land Rights

This account shall include the cost of land and land rights used in connection with nuclear power generation. (See § 1767.16(g).)

321 Structures and Improvements

This account shall include the cost, in place, of structures and improvements used and useful in connection with nuclear power generation. (See § 1767.16 (h).)

NOTE: Include vapor containers and nuclear production roads and railroads in this account.

322 Reactor Plant Equipment

This account shall include the installed cost of reactors, reactor fuel handling and storage equipment, pressurizing equipment, coolant charging equipment, purification and discharging equipment, radioactive waste treatment and disposal equipment, boilers, steam and feed water piping, reactor and boiler apparatus and accessories and other reactor plant equipment used in the production of steam to be used primarily for generating electricity, including auxiliary superheat boilers and associated equipment in systems which change temperatures or pressure of steam from the reactor system.

Items

1. Auxiliary superheat boilers and associated fuel storage handling preparation and burning equipment. (See Account 312, Boiler Plant Equipment, for items, but exclude water supply, water flow lines, and steam lines, as well as other equipment not strictly within the superheat function.)

2. Boiler feed system, including feed water heaters, evaporator condensers, heater drain pumps, heater drainers, deaerators, and vent condensers, boiler feed pumps, surge tanks, feed water regulators, feed water measuring equipment, and all associated drivers.

3. Boilers and heat exchangers.

4. Instruments and devices, including all measuring, indicating, and recording equipment for reactor and boiler plant service together with mountings and supports.

5. Lighting systems.

6. Moderators, such as heavy water, and graphite, initial charge.

7. Reactor coolant; primary and secondary systems, initial charge.

8. Radioactive waste treatment and disposal equipment, including tanks, ion exchangers, incinerators, condensers, chimneys, and diluting fans and pumps.

9. Foundations and settings, especially constructed for and not expected to outlast the apparatus for which provided.

10. Reactor including shielding, control rods and mechanisms.

11. Reactor fuel handling equipment, including manipulating and extraction tools, underwater viewing equipment, seal cutting and welding equipment, fuel transfer equipment, and fuel disassembly machinery.

12. Reactor fuel element failure detection system.

13. Reactor emergency poison container and injection system.

14. Reactor pressuring and pressure relief equipment, including pressurizing tanks and immersion heaters.

15. Reactor coolant or moderator circulation charging, purification, and discharging equipment, including tanks, pumps, heat exchangers, demineralizers, and storage.

16. Station piping, including pipes, valves, fittings, separators, traps, desuperheaters, hangers, excavation, and covering for station piping system, including all-reactor coolant, steam, condensate, boiler feed and water supply piping, but not condensing water, plumbing, building heating, oil, gas, or air piping.

17. Ventilating equipment.

18. Water purification equipment, including softeners, demineralizers and accessories, evaporators and accessories, heat exchangers, filters, tanks for filtered or softened water, pumps, and motors.

19. Water supply systems, including pumps, motors, strainers, raw-water storage tanks, boiler wash pumps, intake and discharge pipes and tunnels not a part of a building.

20. Reactor plant cranes and hoists, and associated drives.

NOTE: When the system for supplying boiler or condenser water is elaborate, as when it includes a dam, reservoir, canal, pipe lines, or cooling ponds, the cost of such special facilities shall be charged to a subdivision of Account 321, Structures and Improvements.

323 Turbogenerator Units

This account shall include the cost installed of main turbine-driven units and accessory equipment used in generating electricity by steam.

Items

1. Air cleaning and cooling apparatus, including blowers, drive equipment, air ducts, not a part of building, louvers, pumps, and hoods.

2. Circulating pumps, including connections between condensers, and intake and discharge tunnels.

3. Condensers, including condensate pumps, air and vacuum pumps, ejectors, unloading valves and vacuum breakers, expansion devices, and screens.

4. Generator hydrogen gas piping system and hydrogen detrainment equipment, and bulk hydrogen gas storage equipment.

5. Cooling system, including towers, pumps, tanks, and piping.

6. Cranes and hoists, including items wholly identified with items listed herein.

7. Excitation system, when identified with main generating units.

8. Fire extinguishing systems.

9. Foundations and settings, especially constructed for and not expected to outlast the apparatus for which provided.

10. Governors.

11. Lighting systems.

12. Lubricating systems, including gauges, filters, water separators, tanks, pumps, piping, and motors.

13. Mechanical meters, including gauges, recording instruments, sampling and testing equipment.

14. Piping-main steam, including connections between turbogenerator and condenser and between condenser and hotwell.

15. Piping-main steam, including connections from main throttle valve to turbine inlet.

16. Platforms, railings, steps, and gratings appurtenant to apparatus listed herein.

17. Pressure oil systems, including accumulators, pumps, piping, and motors.

18. Steelwork, specially constructed for apparatus listed herein.

19. Throttle and inlet valve.

20. Tunnels, intake and discharge, for condenser system, when not a part of structure, and water screens.

21. Turbogenerators-main, including turbine and generator, field rheostats and electric connections for self-excited units.

22. Water screens and motors.

23. Moisture separators for turbine steam.

24. Turbine lubricating oil, initial charge.

324 Accessory Electric Equipment

This account shall include the cost installed of auxiliary generating apparatus, conversion equipment, and equipment used primarily in connection with the control and switching of electric energy produced by nuclear power, and the protection of electric circuits and equipment, except electric motors used to drive equipment included in other accounts. Such motors shall be included in the account in which the equipment with which they are associated is included.

NOTE: Do not include in this account transformers and other equipment used for changing the voltage or frequency of electric energy for the purpose of transmission or distribution.

Items

1. Auxiliary generators, including boards, compartments, switching equipment, control equipment, and connections to auxiliary power bus.

2. Excitation system, including motor, turbine and dual-drive exciter sets and rheostats, storage batteries, and charging equipment, circuit breakers, panels and accessories, knife switches and accessories, surge arresters, instrument shunts, conductors and conduit, special supports for conduit, generator field and exciter switch panels, exciter bus tie panels, generator and exciter rheostats and special housing and protective screens.

3. Generator main connections, including oil circuit breakers and accessories, disconnecting switches and accessories, operating mechanisms and interlocks, current transformers, potential transformers, protective relays, isolated panels and equipment, conductors and conduit, special supports for generator main leads, grounding switch, special housings and protective screens.

4. Station buses, including main, auxiliary, transfer, synchronizing and fault ground buses, including oil circuit breakers and accessories, operating mechanisms and interlocks, reactors and accessories, voltage regulators and accessories, compensators, resistors, starting transformers, current transformers, potential transformers, protective relays, storage batteries and charging equipment, isolated panels and equipment, conductors and conduit, special supports, special housings, concrete pads, general station grounding system, fire-extinguishing system, and test equipment.

5. Station control system, including station switchboards with panel wiring, panels with instruments and control equipment only, panels with switching equipment mounted or mechanically connected, truck-type boards complete, cubicles, station supervisory control boards, generator and exciter signal stands, temperature recording devices, frequency-control equipment, master clocks, watt-hour meters and synchronoscope in the turbine room, station totalizing wattmeter, boiler-room load indicator equipment, storage batteries, panels and charging sets, instrument transformers for supervisory metering, conductors and conduit, special supports for conduit, switchboards, batteries, special housing for batteries, protective screens, and doors.

NOTE: When any item of equipment listed herein is used wholly to furnish power to equipment included in another account, its cost shall be included in such other account.

325 Miscellaneous Power Plant Equipment

This account shall include the cost installed of miscellaneous equipment in and about the nuclear generating plant devoted to general station use, which is not properly includible in any of the foregoing nuclear-power production accounts.

Items

1. Compressed air and vacuum cleaning systems, including tanks, compressors, exhausters, air filters, and piping.

2. Cranes and hoisting equipment, including cranes, cars, crane rails, monorails, and hoists with electric and mechanical connections.

3. Fire-extinguishing equipment for general station and site use.

4. Foundations and settings specially constructed for and not expected to outlast the apparatus for which provided.

5. Locomotive cranes not includible elsewhere.

6. Locomotives not included elsewhere.

7. Marine equipment, including boats and barges.

8. Miscellaneous belts, pulleys, and countershafts.

9. Miscellaneous equipment, including atmospheric and weather recording devices, intrasite communication equipment, laboratory equipment, signal systems, callophones, emergency whistles and sirens, fire alarms, insect-control equipment, and other similar equipment.

10. Railway cars or special shipping containers not includible elsewhere.

11. Refrigerating systems, including compressors, pumps, and cooling coils.

12. Station maintenance equipment, including lathes, shapers, planers, drill presses, hydraulic presses, and grinders with motors, shafting, hangers, and pulleys.

13. Ventilating equipment, including items wholly identified with apparatus listed herein.

14. Station and area radiation monitoring equipment.

NOTE: When any item of equipment listed herein is wholly used in connection with equipment included in another account, its cost shall be included in such other account.

326 Asset Retirement Costs for Nuclear Production Plant

This account shall include asset retirement costs on plant included in the nuclear production function.

Hydraulic Production

330 Land and Land Rights

This account shall include the cost of land and land rights used in connection with hydraulic power generation. (See §1767.16 (g).) It shall also include the cost of land and land rights used in connection with (1) the conservation of fish and wildlife, and (2) recreation. Separate subaccounts shall be maintained for each of the above.

331 Structures and Improvements

This account shall include the cost, in place, of structures and improvements used in connection with hydraulic power generation. (See §1767.16 (h).) It shall also include the cost, in place, of structures and improvements used in connection with (1) the conservation of fish and wildlife, and (2) recreation. Separate subaccounts shall be maintained for each of the above.

332 Reservoirs, Dams, and Waterways

This account shall include the cost in place of facilities used for impounding, collecting, storage, diversion, regulation, and delivery of water used primarily for generating electricity. It shall also include the cost in place of facilities used in connection with (1) the conservation of fish and wildlife, and (2) recreation. Separate subaccounts shall be maintained for each of the above. (See §1767.16 (h)(3).)

Items

1. Bridges and culverts, when not a part of roads or railroads.

2. Clearing and preparing land.

3. Dams, including wasteways, spillways, flash boards, spillway gates with operating and control mechanisms, tunnels, gate houses, and fish ladders.

4. Dikes and embankments.

5. Electric system, including conductors, control system, transformers, and lighting fixtures.

6. Excavation, including shoring, bracing, bridging, refill, and disposal of excess excavated material.

7. Foundations and settings specially constructed for and not expected to outlast the apparatus for which provided.

8. Intakes, including trash racks, rack cleaners, control gates and valves with operating mechanisms, and intake house when not a part of station structure.

9. Platforms, railings, steps, and gratings appurtenant to structures listed herein.

10. Power line wholly identified with items included herein.

11. Retaining walls.

12. Water conductors and accessories, including canals, tunnels, flumes, penstocks, pipe conductors, forebays, tailraces, navigation locks and operating mechanisms, water-hammer and surge tanks, and supporting trestles and structures.

13. Water storage reservoirs, including dams, flashboards, spillway gates and operating mechanisms, inlet and outlet tunnels, regulating valves and valve towers, silt and mud sluicing tunnels with valve or gate towers, and all other structures wholly identified with any of the foregoing items.

333 Water Wheels, Turbines and Generators

This account shall include the cost installed of water wheels and hydraulic turbines (from connection with penstock or flume to tailrace) and generators driven thereby devoted to the production of electricity by water power or for the production of power for industrial or other purposes, if the equipment used for such purposes is a part of the hydraulic power plant works.

Items

1. Exciter water wheels and turbines, including runners, gates, governors, pressure regulators, oil pumps, operating mechanisms, scroll cases, draft tubes, and draft-tube supports.

2. Fire-extinguishing equipment.

3. Foundations and settings, specially constructed for and not expected to outlast the apparatus for which provided.

4. Generator cooling system, including air cooling and washing apparatus, air fans and accessories, and air ducts.

5. Generators-main, a.c. or d.c., including field rheostats and connections for self-excited units and excitation system when identified with the generating unit.

6. Lighting systems.

7. Lubricating systems, including gauges, filters, tanks, pumps, and piping.

8. Main penstock valves and appurtenances, including main valves, control equipment, bypass valves and fittings, and other accessories.

9. Main turbines and water wheels, including runners, gates, governors, pressure regulators, oil pumps, operating mechanisms, scroll cases, draft tubes, and draft-tube supports.

10. Mechanical meters and recording instruments.

11. Miscellaneous water-wheel equipment, including gauges, thermometers, meters, and other instruments.

12. Platforms, railings, steps, and gratings appurtenant to apparatus listed herein.

13. Scroll case filling and drain system, including gates, pipe, valves, and fittings.

14. Water-actuated pressure-regulator system, including tanks and housings, pipes, valves, fittings and insulators, piers and anchorage, and excavation and backfill.

334 Accessory Electric Equipment

This account shall include the cost installed of auxiliary generating apparatus, conversion equipment, and equipment used primarily in connection with the control and switching of electric energy produced by hydraulic power and the protection of electric circuits and equipment, except electric motors used to drive equipment included in other accounts, such motors being included in the account in which the equipment with which they are associated is included.

Items

1. Auxiliary generators, including boards, compartments, switching equipment, control equipment, and connections to auxiliary power bus.

2. Excitation system, including motor, turbine, and dual-drive exciter sets and rheostats, storage batteries and charging equipment, circuit breakers, panels and accessories, knife switches and accessories, surge arresters, instrument shunts, conductors and conduit, special supports for conduit, generator field and exciter switch panels, exciter bus tie panels, generator and exciter rheostats and special housings and protective screens.

3. Generator main connections, including oil circuit breakers and accessories, disconnecting switches and accessories, operating mechanisms and interlocks, current transformers, potential transformers, protective relays, isolated panels and equipment, conductors and conduit, special supports for generator main leads, grounding switch, and special housings and protective screens.

4. Station buses, including main, auxiliary, transfer, synchronizing, and fault ground buses, including oil circuit breakers and accessories, disconnecting switches and accessories, operating mechanisms and interlocks, reactors and accessories, voltage regulators and accessories, compensators, resistors starting transformers, current transformers, potential transformers, protective relays, storage batteries, and charging equipment, isolated panels and equipment, conductors and conduit, special supports, special fire-extinguishing system, and test equipment.

5. Station control system, including station switchboards with panel wiring, panels with instruments and control equipment only, panels with switching equipment mounted for mechanically connected, truck-type boards complete, cubicles, station supervisory control devices, frequency control equipment, master clocks, watt-hour meter, station totalizing watt-meter, storage batteries, panels and charging sets, instrument transformers for supervisory metering, conductors and conduit, special supports for conduit, switchboards, batteries, special housings for batteries, protective screens, and doors.

NOTE A: Do not include in this account transformers and other equipment used for changing the voltage or frequency of electricity for the purpose of transmission or distribution.

NOTE B: When any item of equipment listed herein is used wholly to furnish power to equipment, it shall be included in such equipment account.

335 Miscellaneous Power Plant Equipment

This account shall include the cost installed of miscellaneous equipment in and about the hydroelectric generating plant which is devoted to general station use and is not properly includible in other hydraulic production accounts. It shall also include the cost of equipment used in connection with (1) the conservation of fish and wildlife, and (2) recreation. Separate subaccounts shall be maintained for each of the above.

Items

1. Compressed air and vacuum cleaning systems, including tanks, compressors, exhausters, air filters, and piping.

2. Cranes and hoisting equipment, including cranes, cars, crane rails, monorails, and

hoists with electric and mechanical connections.

3. Fire-extinguishing equipment for general station use.

4. Foundations and settings, specially constructed for and not expected to outlast the apparatus for which provided.

5. Locomotive cranes not includible elsewhere.

6. Locomotives not includible elsewhere.

7. Marine equipment, including boats and barges.

8. Miscellaneous belts, pulleys, and countershafts.

9. Miscellaneous equipment, including atmospheric and weather indicating devices. Intrasite communication equipment, laboratory equipment, insect control equipment, signal systems, callophones, emergency whistles and sirens, fire alarms, and other similar equipment.

10. Railway cars, not includible elsewhere.

11. Refrigerating system, including compressors, pumps, and cooling coils.

12. Station maintenance equipment, including lathes, shapers, planers, drill presses, hydraulic presses, and grinders with motors, shafting, hangers, and pulleys.

13. Ventilating equipment, including items wholly identified with apparatus listed herein.

NOTE: When any item of equipment, listed herein, is used wholly in connection with equipment included in another account, its cost shall be included in such other account.

336 Roads, Railroads, and Bridges

This account shall include the cost of roads, railroads, trails, bridges, and trestles used primarily as production facilities. It also includes those roads necessary to connect the plant with highway transportation systems, except when such roads are dedicated to public use and maintained by public authorities.

Items

1. Bridges, including foundations, piers, girders, trusses, and flooring.

2. Clearing land.

3. Railroads, including grading, ballast, ties, rails, culverts, and hoists.

4. Roads, including grading, surfacing, and culverts.

5. Structures, constructed and maintained in connection with items listed herein.

6. Trails, including grading, surfacing, and culverts.

7. Trestles, including foundations, piers, girders, trusses, and flooring.

NOTE A: Roads intended primarily for connecting employees' houses with the power plant, and roads used primarily in connection with fish and wildlife, and recreation activities, shall not be included herein but in Account 331, Structures and Improvements.

NOTE B: The cost of temporary roads and bridges necessary during the period of construction but abandoned or dedicated to public use upon completion of the plant, shall not be included herein but shall be charged to the accounts appropriate for the construction.

337 Asset Retirement Costs for Hydraulic Production Plant

This account shall include asset retirement costs on plant included in the hydraulic production function.

Other Production

340 Land and Land Rights

This account shall include the cost of land and land rights used in connection with other power generation. (See §1767.16 (g).)

341 Structures and Improvements

This account shall include the cost in place of structures and improvements used in connection with other power generation. (See §1767.16 (h).)

342 Fuel Holders, Producers, and Accessories

This account shall include the cost installed of fuel handling and storage equipment used between the point of fuel delivery to the station and the intake pipe through which fuel is directly drawn to the engine, also the cost of gas producers and accessories devoted to the production of gas for use in prime movers driving main electric generators.

Items

1. Blower and fans.
2. Boilers and pumps.
3. Economizers.
4. Exhauster outfits.
5. Flues and piping.
6. Pipe system.
7. Producers.
8. Regenerators.
9. Scrubbers.
10. Steam injectors.
11. Tanks for storage of oil and gasoline.
12. Vaporizers.

343 Prime Movers

This account shall include the cost installed of Diesel or other prime movers devoted to the generation of electric energy, together with their auxiliaries.

Items

1. Air-filtering system.
2. Belting, shafting, pulleys, and reduction gearing.

3. Cooling system, including towers, pumps, tanks, and piping.

4. Cranes and hoists, including items wholly identified with apparatus listed herein.

5. Engines, Diesel, gasoline, gas, or other internal combustion.

6. Foundations and settings specially constructed for and not expected to outlast the apparatus for which provided.

7. Governors.

8. Ignition system.

9. Inlet valve.

10. Lighting systems.

11. Lubricating systems, including filters, tanks, pumps, and piping.

12. Mechanical meters, including gauges, recording instruments, sampling, and testing equipment.

13. Mufflers.

14. Piping.

15. Starting systems, compressed air, or other, including compressors and drives, tanks, piping, motors, boards and connections, and storage tanks.

16. Steelwork, specially constructed for apparatus listed herein.

17. Waste heat boilers and antifluctuators.

344 Generators

This account shall include the cost installed of Diesel or other power driven main generators.

Items

1. Cranes and hoists, including items wholly identified with such apparatus.

2. Fire-extinguishing equipment.

3. Foundations and settings, specially constructed for and not expected to outlast the apparatus for which provided.

4. Generator cooling system, including air cooling and washing apparatus, air fans and accessories, and air ducts.

5. Generators-main, a.c. or d.c., including field rheostats and connections for self-excited units and excitation system when identified with the generating unit.

6. Lighting systems.

7. Lubricating system, including tanks, filters, strainers, pumps, piping, and coolers.

8. Mechanical meters and recording instruments.

9. Platforms, railings, steps, and gratings appurtenant to apparatus listed herein.

NOTE: If prime movers and generators are so integrated that it is not practical to classify them separately, the entire unit may be included in Account 344, Generators.

345 Accessory Electric Equipment

This account shall include the cost installed of auxiliary generating apparatus, conversion equipment, and equipment used primarily in connection with the control and switching of electric energy produced in other power generating stations, and the protection of electric circuits and equipment, except electric motors used to drive equipment included in other accounts. Such motors shall be included in the account in which the equipment with which it is associated is included.

Items

1. Auxiliary generators, including boards, compartments, switching equipment, control equipment, and connections to auxiliary power bus.

2. Excitation system, including motor, turbine and dual-drive exciter sets and rheostats, storage batteries and charging equipment, circuit breakers, panels and accessories, knife switches and accessories, surge arresters, instrument shunts, conductors and conduit, special supports for conduit, generator field and exciter switch panels, exciter bus tie panels, generator and exciter rheostats and special housings and protective screens.

3. Generator main connections, including oil circuit breakers and accessories, disconnecting switches and accessories, operating mechanisms and interlocks, current transformers, potential transformers, protective relays, isolated panels and equipment, conductors and conduit, special supports for generator main leads, grounding switch, and special housing and protective screens.

4. Station control system, including station switchboards with panel wiring, panels with instruments and control equipment only, panels with switching equipment mounted or mechanically connected, trunk-type boards complete, cubicles, station supervisory control boards, generator and exciter signal stands, temperature-recording devices, frequency control equipment, master clocks, watt-hour meter, station totalizing wattmeter, storage batteries, panels and charging sets, instrument transformers for supervisory metering, conductors and conduit, special supports for conduit, switchboards, batteries, special housing for batteries, protective screens, and doors.

5. Station buses, including main, auxiliary, transfer, synchronizing and fault ground buses, including oil circuit breakers and accessories, disconnecting switches and accessories, operating mechanisms and interlocks, reactors and accessories, voltage regulators and accessories, compensators, resistors, starting transformers, current transformers, potential transformers, protective relays, storage batteries and charging equipment, isolated panels and equipment, conductors and conduit, special supports, special housings, concrete pads, general station ground system, special fire-extinguishing system, and test equipment.

NOTE A: Do not include in this account transformers and other equipment used for changing the voltage or frequency of electric energy for the purpose of transmission or distribution.

NOTE B: When any item of equipment listed herein is used wholly to furnish power to equipment included in another account, its cost shall be included in such other account.

346 Miscellaneous Power Plant Equipment

This account shall include the cost installed of miscellaneous equipment in and about the other power generating plant, devoted to general station use, and not properly includible in any of the foregoing other power production accounts.

Items

1. Compressed air and vacuum cleaning systems, including tanks, compressors, exhausters, air filters, and piping.
2. Cranes and hoisting equipment, including cranes, cars, crane rails, monorails, and hoists with electric and mechanical connections.
3. Fire-extinguishing equipment for general station use.
4. Foundations and settings, specially constructed for and not expected to outlast the apparatus for which provided.
5. Miscellaneous equipment, including atmospheric and weather indicating devices, intrasite communication equipment, laboratory equipment, signal systems, callophones, emergency whistles and sirens, fire alarms, and other similar equipment.
6. Miscellaneous belts, pulleys, and countershafts.
7. Refrigerating systems including compressors, pumps, and cooling coils.
8. Station maintenance equipment, including lathes, shapers, planters, drill presses, hydraulic presses, and grinders with motors, shafting, hangers, or pulleys.
9. Ventilating equipment, including items wholly identified with apparatus listed herein.

NOTE: When any item of equipment, listed herein is used wholly in connection with equipment included in another account, its cost shall be included in such other account.

347 Asset Retirement Costs for Other Production Plant

This account shall include asset retirement costs on plant included in the other production function.

TRANSMISSION PLANT

350 Land and Land Rights

This account shall include the cost of land and land rights used in connection with transmission operations. (See §1767.16 (g).)

351 [Reserved]

352 Structures and Improvements

This account shall include the cost, in place, of structures and improvements used in connection with transmission operations. (See §1767.16 (h).)

353 Station Equipment

This account shall include the cost installed of transforming, conversion, and switching equipment used for the purpose of changing the characteristics of electricity in connection with its transmission or for controlling transmission circuits.

Items

1. Bus compartments, concrete, brick, and sectional steel, including items permanently attached thereto.
2. Conduit, including concrete and iron duct runs not a part of a building.
3. Control equipment, including batteries, battery charging equipment, transformers, remote relay boards, and connections.
4. Conversion equipment, including transformers, indoor and outdoor, frequency changers, motor generator sets, rectifiers, synchronous converters, motors, cooling equipment, and associated connections.
5. Fences.
6. Fixed and synchronous condensers, including transformers, switching equipment, blowers, motors and connections.
7. Foundations and settings, specially constructed for and not expected to outlast the apparatus for which provided.
8. General station equipment, including air compressors, motors, hoists, cranes, test equipment, and ventilating equipment.
9. Platforms, railings, steps, and gratings appurtenant to apparatus listed herein.
10. Primary and secondary voltage connections, including bus runs and supports, insulators, potheads, lightning arresters, cable and wire runs from and to outdoor connections or to manholes and the associated regulators, reactors, resistors, surge arresters, and accessory equipment.
11. Switchboards, including meters, relays, and control wiring.
12. Switching equipment, indoor and outdoor, including oil circuit breakers and operating mechanisms, truck switches, and disconnect switches.
13. Tools and appliances.

354 Towers and Fixtures

This account shall include the cost installed of towers and appurtenant fixtures used for supporting overhead transmission conductors.

Items

1. Anchors, guys, and braces.
2. Brackets.

3. Crossarms, including braces.
4. Excavation, backfill, and disposal of excess excavated material.
5. Foundations.
6. Guards.
7. Insulator pins and suspension bolts.
8. Ladder and steps.
9. Railings.
10. Towers.

355 Poles and Fixtures

This account shall include the cost installed of transmission line poles, wood, steel, concrete, or other material, together with appurtenant fixtures used for supporting overhead transmission conductors.

Items

1. Anchors, head arm and other guys, including guy guards, guy clamps, strain insulators, and pole plates.
2. Brackets.
3. Crossarms and braces.
4. Excavation and backfill, including disposal of excess excavated material.
5. Extension arms.
6. Gaining, roofing, stenciling, and tagging.
7. Insulator pins and suspension belts.
8. Paving.
9. Pole steps.
10. Poles, wood, steel, concrete, or other material.
11. Racks complete with insulators.
12. Reinforcing and stubbing.
13. Settings.
14. Shaving and painting.

356 Overhead Conductors and Devices

This account shall include the cost installed of overhead conductors and devices used for transmission purposes.

Items

1. Circuit breakers.
2. Conductors, including insulated and bare wires and cables.
3. Ground wires and ground clamps.
4. Insulators, including pin, suspension, and other types.
5. Lightning arresters.
6. Switches.
7. Other line devices.

357 Underground Conduit

This account shall include the cost installed of underground conduit and tunnels used for housing transmission cables or wires. (See § 1767.16 (n).)

Items

1. Conduit, concrete, brick or tile, including iron pipe, fiber pipe, Murray duct, and standpipe on pole or tower.
2. Excavation, including shoring, bracing, bridging, backfill, and disposal of excess excavated material.

3. Foundations and settings specially constructed for and not expected to outlast the apparatus for which provided.
4. Lighting systems.
5. Manholes, concrete or brick, including iron or steel, frames and covers, hatchways, gratings, ladders, cable racks and hangers, permanently attached to manholes.
6. Municipal inspection.
7. Pavement disturbed, including cutting and replacing pavement, pavement base and sidewalks.
8. Permits.
9. Protection of street openings.
10. Removal and relocation of subsurface obstructions.
11. Sewer connections, including drains, traps, tide valves, and check valves.
12. Sumps, including pumps.
13. Ventilating equipment.

358 Underground Conductors and Devices

This account shall include the cost installed of underground conductors and devices used for transmission purposes.

Items

1. Armored conductors, buried, including insulators, insulating materials, splices, potheads, and trenching.
2. Armored conductors, submarine, including insulators, insulating materials, splices in terminal chambers, and potheads.
3. Cables in standpipe, including pothead and connection from terminal chamber of manhole to insulators on pole.
4. Circuit breakers.
5. Fireproofing, in connection with any items listed herein.
6. Hollow-core oil-filled cable, including straight or stop joints, pressure tanks, auxiliary air tanks, feeding tanks, terminals, potheads and connections, and ventilating equipment.
7. Lead and fabric covered conductors, including insulators, compound filled, oil filled, or vacuum splices, and potheads.
8. Lightning arresters.
9. Municipal inspection.
10. Permits.
11. Protection of street openings.
12. Racking of cables.
13. Switches.
14. Other line devices.

359 Roads and Trails

This account shall include the cost of roads, trails, and bridges used primarily as transmission facilities.

Items

1. Bridges, including foundation piers, girders, trusses, and flooring.
2. Clearing land.
3. Roads, including grading, surfacing, and culverts.

4. Structures, constructed and maintained in connection with items included herein.

5. Trails, including grading, surfacing, and culverts.

NOTE: The cost of temporary roads, and bridges necessary during the period of construction but abandoned or dedicated to public use upon completion of the plant, shall be charged to the accounts appropriate for the construction.

359.1 Asset Retirement Costs for Transmission Plant

This account shall include asset retirement costs on plant included in the transmission plant function.

DISTRIBUTION PLANT

360 Land and Land Rights

This account shall include the cost of land and land rights used in connection with distribution operations. (See §1767.16 (g).)

NOTE: Do not include the cost of permits to erect poles, or towers or to trim trees in this account. (See Account 364, Poles, Towers and Fixtures, and Account 365, Overhead Conductors and Devices.)

361 Structures and Improvements

This account shall include the cost, in place, of structures and improvements used in connection with distribution operations. (See §1767.16 (h).)

362 Station Equipment

This account shall include the cost installed of station equipment, including transformer banks, which are used for the purpose of changing the characteristics of electricity in connection with its distribution.

Items

1. Bus compartments, concrete, brick and sectional steel, including items permanently attached thereto.

2. Conduit, including concrete and iron duct runs not part of building.

3. Control equipment, including batteries, battery charging equipment, transformers, remote relay boards, and connections.

4. Conversion equipment, indoor and outdoor, frequency changers, motor generator sets, rectifiers, synchronous converters, motors, cooling equipment, and associated connections.

5. Fences.

6. Fixed and synchronous condensers, including transformers, switching equipment, blowers, motors, and connections.

7. Foundations and settings, specially constructed for and not expected to outlast the apparatus for which provided.

8. General station equipment, including air compressors, motors, hoists, cranes, test equipment, and ventilating equipment.

9. Platforms, railings, steps, and gratings appurtenant to apparatus listed herein.

10. Primary and secondary voltage connections, including bus runs and supports, insulators, potheads, lightning arresters, cable and wire runs from and to outdoor connections or to manholes and the associated regulators, reactors, resistors, surge arresters, and accessory equipment.

11. Switchboards, including meters, relays, and control wiring.

12. Switching equipment, indoor and outdoor, including oil circuit breakers and operating mechanisms, truck switches, disconnect switches.

NOTE: The cost of rectifiers, series transformers, and other special station equipment devoted exclusively to street lighting service shall not be included in this account, but in Account 373, Street Lighting and Signal Systems.

363 Storage Battery Equipment is account shall include the cost installed of storage battery equipment used for the purpose of supplying electricity to meet emergency or peak demands.

Items

1. Batteries, including elements, tanks, and tank insulators.

2. Battery room connections, including cable or bus runs and connections.

3. Battery room flooring, when specially laid for supporting batteries.

4. Charging equipment, including motor generator sets and other charging equipment and connections, and cable runs from generator or station bus to battery room connections.

5. Miscellaneous equipment, including instruments, and water stills.

6. Switching equipment, including endcell switches and connections, boards and panels, used exclusively for battery control, not part of general station switchboard.

7. Ventilating equipment, including fans and motors, louvers, and ducts not part of building.

NOTE: Storage batteries used for control and general station purposes shall not be included in this account but in the account appropriate for their use.

364 Poles, Towers and Fixtures

This account shall include the cost installed of poles, towers, and appurtenant fixtures used for supporting overhead distribution conductors and service wires.

Items

1. Anchors, head arm, and other guys, including guy guards, guy clamps, strain insulators, and pole plates.
2. Brackets.
3. Crossarms and braces.
4. Excavation and backfill, including disposal of excess excavated material.
5. Extension arms.
6. Foundations.
7. Guards.
8. Insulator pins and suspension bolts.
9. Paving.
10. Permits for construction.
11. Pole steps and ladders.
12. Poles, wood, steel, concrete, or other material.
13. Racks complete with insulators.
14. Railings.
15. Reinforcing and stubbing.
16. Settings.
17. Shaving, painting, gaining, roofing, stenciling, and tagging.
18. Towers.
19. Transformer racks and platforms.

365   Overhead Conductors and Devices

This account shall include the cost installed of overhead conductors and devices used for distribution purposes.

Items

1. Circuit breakers.
2. Conductors, including insulated and bare wires and cables.
3. Ground wires and clamps.
4. Insulators, including pin, suspension, and other types, and tie wire or clamps.
5. Lightning arresters.
6. Railroad and highway crossing guards.
7. Splices.
8. Switches.
9. Tree trimming, initial cost including the cost of permits therefor.
10. Other line devices.
11. Oil circuit reclosers (OCR).
12. Sectionalizers.
13. Labor costs for installation of OCRs and Sectionalizers, first only.

NOTE: The cost of conductors used solely for street lighting or signal systems shall not be included in this account but in Account 373, Street Lighting and Signal Systems.

366   Underground Conduit

This account shall include the cost installed of underground conduit and tunnels used for housing distribution cables or wires.

Items

1. Conduit, concrete, brick and tile, including iron pipe, fiber pipe, Murray duct, and standpipe on pole or tower.

2. Excavation, including shoring, bracing, bridging, backfill, and disposal of excess excavated material.
3. Foundations and settings specially constructed for and not expected to outlast the apparatus for which constructed.
4. Lighting systems.
5. Manholes, concrete or brick, including iron or steel frames and covers, hatchways, gratings, ladders, cable racks, and hangers permanently attached to manholes.
6. Municipal inspection.
7. Pavement disturbed, including cutting and replacing pavement, pavement base, and sidewalks.
8. Permits.
9. Protection of street openings.
10. Removal and relocation of subsurface obstructions.
11. Sewer connections, including drains, traps, tide valves, and check valves.
12. Sumps, including pumps.
13. Ventilating equipment.

NOTE: The cost of underground conduit used solely for street lighting or signal systems shall be included in Account 373, Street Lighting and Signal Systems.

367   Underground Conductors and Devices

This account shall include the cost installed of underground conductors and devices used for distribution purposes.

Items

1. Armored conductors, buried, including insulators, insulating materials, splices, potheads, and trenching.
2. Armored conductors, submarine, including insulators, insulating materials, splices in terminal chamber, and potheads.
3. Cables in standpipe, including pothead and connection from terminal chamber or manhole to insulators on pole.
4. Circuit breakers.
5. Fireproofing, in connection with any items listed herein.
6. Hollow-core oil-filled cable, including straight or stop joints, pressure tanks, auxiliary air tanks, feeding tanks, terminals, potheads and connections.
7. Lead and fabric covered conductors, including insulators, compound-filled, oil-filled or vacuum splices, and potheads.
8. Lightning arresters.
9. Municipal inspection.
10. Permits.
11. Protection of street openings.
12. Racking of cables.
13. Switches.
14. Other line devices.

NOTE: The cost of underground conductors and devices used solely for street lighting or signal systems shall be included in Account 373, Street Lighting and Signal Systems.

368 Line Transformers

A. This account shall include the cost installed of overhead and underground distribution line transformers and pole-type and underground voltage regulators owned by the utility, for use in transforming electricity to the voltage at which it is to be used by the customer, whether actually in service or held in reserve.

B. When a transformer is permanently retired from service, the original installed cost thereof shall be credited to this account.

C. The records covering line transformers shall be so kept that the utility can furnish the number of transformers of various capacities in service and those in reserve, and the location and the use of each transfer.

Items

1. Installation, labor of (first installation only).
2. Transformer cut-out boxes.
3. Transformer lightning arresters.
4. Transformers, line and network.
5. Capacitors.
6. Network protectors.
7. Voltage regulators.

NOTE: The cost of removing and resetting line transformers shall not be charged to this account but to Account 583, Overhead Line Expenses, or Account 584, Underground Line Expenses, as appropriate. The cost of line transformers used solely for street lighting or signal systems shall be included in Account 373, Street Lighting and Signal Systems.

369 Services

This account shall include the cost installed of overhead and underground conductors leading from a point where wires leave the last pole of the overhead system or the distribution box or manhole, or the top of the pole of the distribution line, to the point of connection with the customer's outlet or wiring. Conduit used for underground service conductors shall be included herein.

Items

1. Brackets.
2. Cables and wires.
3. Conduit.
4. Insulators.
5. Municipal inspection.
6. Overhead to underground, including conduit or standpipe and conductor from last splice on pole to connection with customer's wiring.
7. Pavement disturbed, including cutting and replacing pavement, pavement base, and sidewalks.
8. Permits.
9. Protection of street openings.
10. Service switch.
11. Suspension wire.

370 Meters

A. This account shall include the cost installed of meters or devices and appurtenances thereto, for use in measuring the electricity delivered to its users, whether actually in service or held in reserve.

B. When a meter is permanently retired from service, the installed cost included herein shall be credited to this account.

C. The records covering meters shall be so kept that the utility can furnish information as to the number of meters of various capacities in service and in reserve as well as the location of each meter owned.

Items

1. Alternate current, watt-hour meters.
2. Current limiting devices.
3. Demand indicators.
4. Demand meters.
5. Direct current watt-hour meters.
6. Graphic demand meters.
7. Installation, labor of (first installation only).
8. Instrument transformers.
9. Maximum demand meters.
10. Meter badges and their attachments.
11. Meter boards and boxes.
12. Meter fittings, connections, and shelves (first set).
13. Meter switches and cut-outs.
14. Prepayment meters.
15. Protective devices.
16. Testing new meters.

NOTE A: This account shall not include meters for recording output of a generating station, or substation meters. It includes only those meters used to record energy delivered to customers.

NOTE B: The cost of removing and resetting meters shall be charged to Account 586, Meter Expenses.

371 Installations on Customers' Premises

This account shall include the cost installed of equipment on the customer's side of a meter when the utility incurs such cost and when the utility retains title to and assumes full responsibility for maintenance and replacement of such property. This account shall not include leased equipment. (See Account 372, Leased Property on Customers' Premises.)

Items

1. Cable vaults.
2. Commercial lamp equipment.
3. Foundations and settings specially provided for equipment included herein.
4. Frequency changer sets.
5. Motor generator sets.
6. Motors.
7. Switchboard panels, high or low tension.
8. Wire and cable connections to incoming cables.

NOTE: Do not include in this account any costs incurred in connection with merchandising, jobbing, or contract work activities.

372 Leased Property on Customers' Premises

This account shall include the cost of electric motors, transformers, and other equipment on customers' premises (including municipal corporations), leased or loaned to customers, but not including property held for sale.

NOTE A: The cost of setting and connecting such appliances or equipment on the premises of customers and the cost of resetting or removal shall not be charged to this account but to operating expenses, Account 587, Customer Installations Expenses.

NOTE B: Do not include in this account any costs incurred in connection with merchandising, jobbing, or contract work activities.

373 Street Lighting and Signal Systems

This account shall include the cost installed of equipment used wholly for public street and highway lighting or traffic, fire alarm, police, and other signal systems.

Items

1. Armored conductors, buried or submarine, including insulators, insulating materials, splices, and trenching.
2. Automatic control equipment.
3. Conductors, overhead or underground, including lead or fabric covered, parkway cables, including splices, and insulators.
4. Lamps, arc, incandescent, or other types, including glassware, suspension fixtures, and brackets.
5. Municipal inspection.
6. Ornamental lamp posts.
7. Pavement disturbed, including cutting and replacing pavement, pavement base, and sidewalks.
8. Permits.
9. Posts and standards.
10. Protection of street openings.
11. Relays or time clocks.
12. Series contactors.
13. Switches.
14. Transformers, pole or underground.

374 Asset Retirement Costs for Distribution Plant

This account shall include asset retirement costs on plant included in the distribution plant function.

REGIONAL TRANSMISSION AND MARKET OPERATION PLANT

380 Land and Land Rights

This account shall include the cost of land and land rights used in connection with regional transmission and market operations.

381 Structures and Improvements

This account shall include the cost in place of structures and improvement used for regional transmission and market operations.

382 Computer Hardware

This account shall include the cost of computer hardware and miscellaneous information technology equipment to provide scheduling, system control and dispatching, system planning, standards development, market monitoring, and market administration activities. Records shall be maintained identifying to the maximum extent practicable computer hardware owned and used for:
(1) Scheduling, system control and dispatching, (2) System planning and standards development, and (3) Market monitoring and market administration activities.

Items

1. Personal computers
2. Servers
3. Workstations
4. Energy Management System (EMS) hardware
5. Supervisory Control and Data Acquisition (SCADA) system hardware
6. Peripheral equipment
7. Networking components

383 Computer Software

This account shall include the cost of off-the-shelf and in-house developed software purchased and used to provide scheduling, system control and dispatching, system planning, standards development, market monitoring, and market administration activities. Records shall be maintained identifying to the maximum extent practicable the cost of software used for:
(1) Scheduling, system control and dispatching,
(2) System planning and standards development, and
(3) Market monitoring and market administration activities.

Items

1. Software licenses
2. User interface software
3. Modeling software
4. Database software
5. Tracking and monitoring software
6. Energy Management System (EMS) software
7. Supervisory Control and Data Acquisition (SCADA) system software
8. Evaluation and assessment system software
9. Operating, planning and transaction scheduling software
10. Reliability applications
11. Market application software

384 Communication Equipment

This account shall include the cost of communication equipment owned and used to acquire or share data and information used to control and dispatch the system.

Items

1. Fiber optic cable
2. Remote terminal units
3. Microwave towers
4. Global Positioning System (GPS) equipment
5. Servers
6. Workstations
7. Telephones

385 Miscellaneous Regional Transmission and Market Operation Plant

This account shall include the cost of regional transmission and market operation plant and equipment not provided for elsewhere.

386 Asset Retirement Costs for Regional Transmission and Market Operation Plant

This account shall include asset retirement costs on regional transmission and market operations plant and equipment.

GENERAL PLANT

389 Land and Land Rights

This account shall include the cost of land and land rights used for utility purposes, the cost of which is not properly includible in other land and land rights accounts. (See §1767.16 (g).)

390 Structures and Improvements

This account shall include the cost, in place, of structures and improvements used for utility purposes, the cost of which is not properly includible in other structures and improvements accounts. (See §1767.16 (h).)

391 Office Furniture and Equipment

This account shall include the cost of office furniture and equipment owned by the utility and devoted to utility service, and not permanently attached to buildings, except the cost of such furniture and equipment which the utility elects to assign to other plant accounts on a functional basis.

Items

1. Bookcases and shelves.
2. Desks, chairs, and desk equipment.
3. Drafting-room equipment.
4. Filing, storage, and other cabinets.
5. Floor covering.
6. Library and library equipment.
7. Mechanical office equipment, such as accounting machines, and typewriters.
8. Safes.
9. Tables.

392 Transportation Equipment

This account shall include the cost of transportation vehicles used for utility purposes.

Items

1. Airplanes.
2. Automobiles.
3. Bicycles.
4. Electrical vehicles.
5. Motor trucks.
6. Motorcycles.
7. Repair cars or trucks.
8. Tractors and trailers.
9. Other transportation vehicles.

393 Stores Equipment

This account shall include the cost of equipment used for the receiving, shipping, handling, and storage of materials and supplies.

Items

1. Chain falls.
2. Counters.
3. Cranes (portable).
4. Elevating and stacking equipment (portable).
5. Hoists.
6. Lockers.
7. Scales.
8. Shelving.
9. Storage bins.
10. Trucks, hand and power driven.
11. Wheelbarrows.

394 Tools, Shop and Garage Equipment

This account shall include the cost of tools, implements, and equipment used in construction, repair work, general shops and garages and not specifically provided for or includible in other accounts.

Items

1. Air compressors.
2. Anvils.
3. Automobile repair shop equipment.
4. Battery charging equipment.
5. Belts, shafts and countershafts.
6. Boilers.
7. Cable pulling equipment.
8. Concrete mixers.
9. Drill presses.
10. Derricks.
11. Electric equipment.
12. Engines.
13. Forges.
14. Furnaces.
15. Foundations and settings specially constructed for and not expected to outlast the equipment for which provided.
16. Gas producers.
17. Gasoline pumps, oil pumps, and storage tanks.
18. Greasing tools and equipment.

19. Hoists.
20. Ladders.
21. Lathes.
22. Machine tools.
23. Motor-driven tools.
24. Motors.
25. Pipe threading and cutting tools.
26. Pneumatic tools.
27. Pumps.
28. Riveters.
29. Smithing equipment.
30. Tool racks.
31. Vises.
32. Welding apparatus.
33. Work benches.

395 Laboratory Equipment

This account shall include the cost installed of laboratory equipment used for general laboratory purposes and not specifically provided for or includible in other departmental or functional plant accounts.

Items

1. Ammeters.
2. Current batteries.
3. Frequency changers.
4. Galvanometer.
5. Inductometers.
6. Laboratory standard millivolt meters.
7. Laboratory standard volt meters.
8. Meter-testing equipment.
9. Millivolt meters.
10. Motor generator sets.
11. Panels.
12. Phantom loads.
13. Portable graphic ammeters, voltmeters, and wattmeters.
14. Portable loading devices.
15. Potential batteries.
16. Potentiometers.
17. Rotating standards.
18. Standard cell, reactance, resistor, and shunt.
19. Switchboards.
20. Synchronous timers.
21. Testing panels.
22. Testing resistors.
23. Transformers.
24. Voltmeters.
25. Other testing, laboratory, or research equipment not provided for elsewhere.

396 Power Operated Equipment

This account shall include the cost of power operated equipment used in construction or repair work exclusive of equipment includible in other accounts. Include, also, the tools and accessories acquired for use with such equipment and the vehicle on which such equipment is mounted.

Items

1. Air compressors, including driving unit and vehicle.
2. Back filling machines.

3. Boring machines.
4. Bulldozers.
5. Cranes and hoists.
6. Diggers.
7. Engines.
8. Pile drivers.
9. Pipe cleaning machines.
10. Pipe coating or wrapping machines.
11. Tractors-Crawler type.
12. Trenchers.
13. Other power operated equipment.

NOTE: It is intended that this account include only such large units as are generally self-propelled or mounted on movable equipment.

397 Communication Equipment

This account shall include the cost installed of telephone, telegraph, and wireless equipment for general use in connection with utility operations.

Items

1. Antennae.
2. Booths.
3. Cables.
4. Distributing boards.
5. Extension cords.
6. Gongs.
7. Hand sets, manual and dial.
8. Insulators.
9. Intercommunicating sets.
10. Loading coils.
11. Operators' desks.
12. Poles and fixtures used wholly for telephone or telegraph wire.
13. Radio transmitting and receiving sets.
14. Remote control equipment and lines.
15. Sending keys.
16. Storage batteries.
17. Switchboards.
18. Telautograph circuit connections.
19. Telegraph receiving sets.
20. Telephone and telegraph circuits.
21. Testing instruments.
22. Towers.
23. Underground conduit used wholly for telephone or telegraph wires and cable wires.

398 Miscellaneous Equipment

This account shall include the cost of equipment, and apparatus used in the utility operations, which is not includible in other accounts.

Items

1. Hospital and infirmary equipment.
2. Kitchen equipment.
3. Employees' recreation equipment.
4. Radios.
5. Restaurant equipment.
6. Soda fountains.
7. Operators' cottage furnishings.
8. Other miscellaneous equipment.

NOTE: Miscellaneous equipment of the nature indicated above wherever practicable, shall be included in the utility plant accounts on a functional basis.

399 Other Tangible Property

This account shall include the cost of tangible utility plant not provided for elsewhere.

399.1 Asset Retirement Costs for General Plant

This account shall include asset retirement costs on plant included in the general plant function.

[58 FR 59825, Nov. 10, 1993, as amended at 73 FR 30284, May 27, 2008]

§ 1767.21 Operating income.

The operating income accounts identified in this section shall be used by all RUS borrowers.

UTILITY OPERATING INCOME

400 Operating Revenues
401 Operation Expense
402 Maintenance Expense
403 Depreciation Expense
403.1 Depreciation Expense—Steam Production Plant
403.2 Depreciation Expense—Nuclear Production Plant
403.3 Depreciation Expense—Hydraulic Production Plant
403.4 Depreciation Expense—Other Production Plant
403.5 Depreciation Expense—Transmission Plant
403.6 Depreciation Expense—Distribution Plant
403.7 Depreciation Expense—General Plant
403.8 Depreciation Expense-Asset Retirement Costs
403.9 Depreciation Expense-Regional Transmission and Market Operation Plant
404 Amortization of Limited-Term Electric Plant
405 Amortization of Other Electric Plant
406 Amortization of Electric Plant Acquisition Adjustments
407 Amortization of Property Losses, Unrecovered Plant and Regulatory Study Costs
407.3 Regulatory Debits
407.4 Regulatory Credits
408 Taxes Other than Income Taxes
408.1 Taxes—Property
408.2 Taxes—U.S. Social Security—Unemployment
408.3 Taxes—U.S. Social Security—F.I.C.A.
408.4 Taxes—State Social Security—Unemployment
408.5 Taxes—State Sales—Consumers
408.6 Taxes—Gross Revenue or Gross Receipts Tax
408.7 Taxes—Other

409 [Reserved]
409.1 Income Taxes, Utility Operating Income
409.2 Income Taxes, Other Income and Deductions
409.3 Income Taxes, Extraordinary Items
410 [Reserved]
410.1 Provision for Deferred Income Taxes, Utility Operating Income
410.2 Provision for Deferred Income Taxes, Other Income and Deductions
411 [Reserved]
411.1 Provision for Deferred Income Taxes—Credit, Utility Operating Income
411.2 Provision for Deferred Income Taxes—Credit, Other Income and Deductions
411.3 [Reserved]
411.4 Investment Tax Credit Adjustments, Utility Operations
411.5 Investment Tax Credit Adjustments, Nonutility Operations
411.6 Gains from Disposition of Utility Plant
411.7 Losses from Disposition of Utility Plant
411.8 Gains from Disposition of Allowances
411.9 Losses from Disposition of Allowances
411.10 Accretion Expense
412 Revenues from Electric Plant Leased to Others
413 Expenses of Electric Plant Leased to Others
414 Other Utility Operating Income

UTILITY OPERATING INCOME

400 Operating Revenues

There shall be shown under this caption the total amount included in the electric operating revenue accounts provided herein.

401 Operation Expense

There shall be shown under this caption the total amount included in the electric operation expense accounts provided herein. (See note to § 1767.17 (c).)

402 Maintenance Expense

There shall be shown under this caption the total amount included in the electric maintenance expense accounts provided herein.

403 Depreciation Expense

A. This account shall include the amount of depreciation expense for all classes of depreciable electric plant in service except such depreciation expense as is chargeable to clearing accounts or to Account 416, Costs and Expenses of Merchandising, Jobbing and Contract Work.

B. The utility shall keep such records of property and property retirements as will reflect the service life of property which has been retired and aid in estimating probable service life by mortality, turnover, or other appropriate methods; and also such records as will reflect the percentage of salvage and costs of removal for property retired from each account, or subdivision thereof, for depreciable electric plant.

NOTE A: Depreciation expense applicable to property included in Account 104, Electric Plant Leased to Others, shall be charged to Account 413, Expenses of Electric Plant Leased to Others.

NOTE B: Depreciation expenses applicable to transportation equipment, shop equipment, tools, work equipment, power operated equipment, and other general equipment may be charged to clearing accounts as necessary in order to obtain a proper distribution of expenses between construction and operation.

NOTE C: Depreciation expense applicable to transportation equipment used for transportation of fuel from the point of acquisition to the unloading point shall be charged to Account 151, Fuel Stock.

C. Account 403 shall be subaccounted as follows:

403.1 Depreciation Expense—Steam Production Plant
403.2 Depreciation Expense—Nuclear Production Plant
403.3 Depreciation Expense—Hydraulic Production Plant
403.4 Depreciation Expense—Other Production Plant
403.5 Depreciation Expense—Transmission Plant
403.6 Depreciation Expense—Distribution Plant
403.7 Depreciation Expense—General Plant
403.8 Depreciation Expense-Asset Retirement Costs
403.9 Depreciation Expense-Regional Transmission and Market Operation Plant

404 Amortization of Limited-Term Electric Plant

This account shall include amortization charges applicable to amounts included in the electric plant accounts for limited-term franchises, licenses, patent rights, limited-term interests in land, and expenditures on leased property where the service life of the improvements is terminable by action of the lease. The charges to this account shall be such as to distribute the book cost of each investment as evenly as may be over the period of its benefit to the utility. (See Account 111, Accumulated Provision for Amortization of Electric Utility Plant.)

405 Amortization of Other Electric Plant

A. When authorized by RUS, this account shall include charges for amortization of intangible or other electric utility plant which does not have a definite or terminable life and which is not subject to charges for depreciation expense.

B. This account shall be supported in such detail as to show the amortization applicable to each investment being amortized, together with the book cost of the investment and the period over which it is being written off.

406 Amortization of Electric Plant Acquisition Adjustments

This account shall be debited or credited, as appropriate, with amounts includible in operating expenses, pursuant to approval or order of RUS, for the purpose of providing for the extinguishment of the amount in Account 114, Electric Plant Acquisition Adjustments.

407 Amortization of Property Losses, Unrecovered Plant and Recovery Study Costs

This account shall be charged with amounts credited to Account 182.1, Extraordinary Property Losses, when RUS has authorized the amount in the latter account to be amortized by charges to electric operations.

407.3 Regulatory Debits

This account shall be debited, when appropriate, with the amounts credited to Account 254, Other Regulatory Liabilities, to record regulatory liabilities imposed on the utility by the ratemaking actions of regulatory agencies. This account shall also be debited, when appropriate, with the amounts credited to Account 182.3, Other Regulatory Assets, concurrent with the recovery of such amounts in rates.

407.4 Regulatory Credits

This account shall be credited, when appropriate, with the amounts debited to Account 182.3, Other Regulatory Assets, to establish regulatory assets. This account shall also be credited, when appropriate, with the amounts debited to Account 254, Other Regulatory Liabilities, concurrent with the return of such amounts to customers through rates.

408 Taxes Other Than Income Taxes

A. This account shall include the amounts of ad valorem, gross revenue, or gross receipts taxes, state unemployment insurance, franchise taxes, Federal excise taxes, social security taxes, and all other taxes assessed by Federal, state, county, municipal, or other local governmental authorities, except income taxes.

B. These accounts shall be charged in each accounting period with the amounts of taxes which are applicable thereto, with concurrent credits to Account 236, Taxes Accrued, or Account 165, Prepayments, as appropriate. When it is not possible to determine the exact amounts of taxes, the amounts shall be estimated and adjustments made in current accruals as the actual tax levies become known.

C. The charges to these accounts shall be made or supported so as to show the amount of each tax and the basis upon which each charge is made. In the case of a utility rendering more than one utility service, taxes of the kind includible in these accounts shall be assigned directly to the utility department the operation of which gave rise to the tax, in so far as practicable. Where the tax is not attributable to a specific utility department, it shall be distributed among the utility departments or nonutility operations on an equitable basis after appropriate study to determine such basis.

NOTE A: Special assessments for street and similar improvements shall be included in the appropriate utility plant or nonutility property account.

NOTE B: Taxes specifically applicable to construction and retirement activities shall be included in the cost of construction or the retirement.

NOTE C: Gasoline and other sales taxes shall be charged as far as practicable to the same account as the materials on which the tax is levied.

NOTE D: Social security and other forms of payroll taxes shall be charged to nonutility operations, the specific functional operations, maintenance, and administrative expense accounts, and to construction and retirement activities on a basis related to payroll either directly or by transfers from this account.

NOTE E: Property taxes applicable to the various utility functions shall be charged to the specific functional operations and administrative expense accounts either directly or by transfers from this account.

NOTE F: Interest on tax refunds or deficiencies shall not be included in these accounts but in Account 419, Interest and Dividend Income, or Account 431, Other Interest Expense, as appropriate.

D. Account 408 shall be subaccounted as follows:

408.1 Taxes—Property
408.2 Taxes—U.S. Social Security—Unemployment
408.3 Taxes—U.S. Social Security—F.I.C.A.
408.4 Taxes—State Social Security—Unemployment
408.5 Taxes—State Sales—Consumers
408.6 Taxes—Gross Revenue or Gross Receipts Tax
408.7 Taxes—Other

409 [Reserved]

SPECIAL INSTRUCTIONS

Accounts 409.1, 409.2, and 409.3

A. These accounts shall include the amount of local, state, and Federal income taxes on income properly accruable during the period covered by the income statement to meet the actual liability for such taxes. Concurrent credits for the tax accruals shall be made to Account 236, Taxes Accrued, and as the exact amounts of taxes become known, the current tax accruals shall be adjusted by charges or credits to these accounts.

B. The accruals for income taxes shall be apportioned among utility departments and to Other Income and Deductions so that, as nearly as practicable, each tax shall be included in the expenses of the utility department or Other Income and Deductions, the income from which gave rise to the tax. The tax effects relating to interest charges shall be allocated between utility and nonutility operations. The basis for this allocation shall be the ratio of net investment in utility plant to net investment in nonutility plant.

NOTE A: Taxes assumed by the utility on interest shall be charged to Account 431, Other Interest Expense.

NOTE B: Interest on tax refunds or deficiencies shall not be included in these accounts but in Account 419, Interest and Dividend Income, or Account 431, Other Interest Expense, as appropriate.

409.1 Income Taxes, Utility Operating Income

This account shall include the amount of those local, state, and Federal income taxes which relate to utility operating income. This account shall be maintained so as to allow ready identification of tax effects (both positive and negative) relating to Utility Operating Income (by department), Utility Plant Leased to Others, and Other Utility Operating Income.

409.2 Income Taxes, Other Income and Deductions

This account shall include the amount of those local, state, and Federal income taxes (both positive and negative), which relate to Other Income and Deductions.

409.3 Income Taxes, Extraordinary Items

This account shall include the amount of those local, state, and Federal income taxes (both positive and negative), which relate to Extraordinary Items.

410 [Reserved]

SPECIAL INSTRUCTIONS

Accounts 410.1, 410.2, 411.1, and 411.2

A. Accounts 410.1 and 410.2 shall be debited, and Accumulated Deferred Income Taxes, shall be credited, with amounts equal to any current deferrals of taxes on income or any allocations of deferred taxes originating in prior periods, as provided by the texts of Accounts 190, 281, 282, and 283. There shall not be netted against entries required to be made to these accounts any credit amounts appropriately includible in Account 411.1 or Account 411.2.

B. Accounts 411.1 or 411.2 shall be credited, and Accumulated Deferred Income Taxes, shall be debited, with amounts equal to any allocations of deferred taxes originating in prior periods or any current deferrals of taxes on income, as provided by the texts of Accounts 190, 281, 282, and 283. There shall not be netted against entries required to be made to these accounts any debit amounts appropriately includible in Account 410.1 or Account 410.2.

410.1 Provision for Deferred Income Taxes, Utility Operating Income

This account shall include the amounts of those deferrals of taxes and allocations of deferred taxes which relate to Utility Operating Income (by department).

410.2 Provision for Deferred Income Taxes, Other Income and Deductions

This account shall include the amounts of those deferrals of taxes and allocations of deferred taxes which relate to Other Income and Deductions.

411 [Reserved]

411.1 Provision for Deferred Income Taxes— Credit, Utility Operating Income

This account shall include the amounts of those allocations of deferred taxes and deferrals of taxes, credit, which relate to Utility Operating Income (by department).

411.2 Provision for Deferred Income Taxes— Credit, Other Income and Deductions

This account shall include the amounts of those allocations of deferred taxes and deferrals of taxes, credit, which relate to Other Income and Deductions.

411.3 [Reserved]

SPECIAL INSTRUCTIONS

Accounts 411.4 and 411.5

A. Account 411.4 shall be debited with the amounts of investment tax credits related to electric utility property that are credited to Account 255, Accumulated Deferred Investment Tax Credits, by companies which do not apply the entire amount of the benefits of the investment credit as a reduction of the overall income tax expense in the year in which such credit is realized. (See Account 255).

B. Account 411.4 shall be credited with the amounts debited to Account 255 for proportionate amounts of tax credit deferrals allocated over the average useful life of electric utility property to which the tax credits relate or such lesser period of time as may be adopted and consistently followed by the company.

C. Account 411.5 shall be debited and credited as directed in paragraphs A and B, for investment tax credits related to nonutility property.

411.4 Investment Tax Credit Adjustments, Utility Operations

This account shall include the amount of those investment tax credit adjustments related to property used in Utility Operations (by department).

411.5 Investment Tax Credit Adjustments, Nonutility Operations

This account shall include the amount of those investment tax credit adjustments related to property used in Nonutility Operations.

411.6 Gains from Disposition of Utility Plant

A. This account shall include, as approved by RUS, amounts relating to gains from the disposition of future use utility plant including amounts which were previously recorded in and transferred from Account 105, Electric Plant Held for Future Use, under the Provisions of Paragraphs B, C, and D thereof. Income taxes relating to gains recorded in this account shall be recorded in Account 409.1, Income Taxes, Utility Operating Income.

B. The utility shall record in this account gains resulting from the settlement of asset retirement obligations related to utility plant in accordance with the accounting prescribed in Sec. 1767.15(y).

411.7 Losses from Disposition of Utility Plant

A. This account shall include, as approved by RUS, amounts relating to losses from the disposition of future use utility plant including amounts which were previously recorded in and transferred from Account 105, Electric Plant Held for Future Use, under the provisions of Paragraphs B, C, and D thereof. Income taxes relating to losses recorded in this account shall be recorded in Account 409.1, Income Taxes, Utility Operating Income.

B. The utility shall record in this account losses resulting from the settlement of asset retirement obligations related to utility plant in accordance with the accounting prescribed in Sec. 1767.15(y).

411.8 Gains from Disposition of Allowances

This account shall be credited with the gain on the sale, exchange, or other disposition of allowances in accordance with § 1767.15 (u)(8). Income taxes relating to gains recorded in this account shall be recorded in

Account 409.1, Income Taxes, Utility Operating Income.

411.9 Losses from Disposition of Allowances

This account shall be debited with the loss on the sale, exchange, or other disposition of allowances in accordance with § 1767.15 (u)(8). Income taxes relating to losses recorded in this account shall be recorded in Account 409.1, Income Taxes, Utility Operating Income.

411.10 Accretion Expense

This account shall be charged for accretion expense on the liabilities associated with asset retirement obligations included in Account 230, Asset Retirement Obligations, relating to electric utility plant.

412 Revenues from Electric Plant Leased to Others

This account shall include revenues from electric property constituting a distinct operating unit or system leased by the utility to others, and which property is properly includible in Account 104, Electric Plant Leased to Others.

NOTE: Related taxes shall be recorded in Account 408, Taxes Other Than Income Taxes, or Account 409.1, Income Taxes, Utility Operating Income, as appropriate.

413 Expenses of Electric Plant Leased to Others

A. This account shall include expenses from electric property constituting a distinct operating unit or system leased by the utility to others, and which property is properly includible in Account 104, Electric Plant Leased to Others.

B. The detail of expenses shall be kept or supported so as to show separately the following:

1. Operation.
2. Maintenance.
3. Depreciation.
4. Amortization.

NOTE: Related taxes shall be recorded in Account 408, Taxes Other Than Income Taxes, or Account 409.1, Income Taxes, Utility Operating Income, as appropriate.

414 Other Utility Operating Income

A. This account shall include the revenues received and expenses incurred in connection with the operations of utility plant, the book cost of which is included in Account 118, Other Utility Plant.

B. The expenses shall include every element of cost incurred in such operations, including depreciation, rents, and insurance.

NOTE: Related taxes shall be recorded in Account 408, Taxes Other Than Income

Taxes, or Account 409.1, Income Taxes, Utility Operating Income, as appropriate.

[58 FR 59825, Nov. 10, 1993, as amended at 62 FR 42290, Aug. 6, 1997; 73 FR 30285, May 27, 2008]

§ 1767.22 Other income and deductions.

The other income and deductions accounts identified in this section shall be used by all RUS borrowers.

OTHER INCOME AND DEDUCTIONS

415 Revenues from Merchandising, Jobbing, and Contract Work
416 Costs and Expenses of Merchandising, Jobbing, and Contract Work
417 Revenues from Nonutility Operations
417.1 Expenses of Nonutility Operations
418 Nonoperating Rental Income
418.1 Equity in Earnings of Subsidiary Companies
419 Interest and Dividend Income
419.1 Allowance for Funds Used During Construction
420 Investment Tax Credits
421 Miscellaneous Nonoperating Income
421.1 Gain on Disposition of Property
421.2 Loss on Disposition of Property
422 Nonoperating Taxes
423 Generation and Transmission Cooperative Capital Credits
424 Other Capital Credits and Patronage Capital Allocations
425 Miscellaneous Amortization
426 [Reserved]
426.1 Donations
426.2 Life Insurance
426.3 Penalties
426.4 Expenditures for Certain Civic, Political, and Related Activities
426.5 Other Deductions

OTHER INCOME AND DEDUCTIONS

415 Revenues from Merchandising, Jobbing and Contract Work

A. This account shall include all revenues derived from the sale of merchandise and jobbing or contract work, including any profit or commission accruing to the utility on jobbing work performed by it as agent under contracts whereby it does jobbing work for another for a stipulated profit or commission. Interest related income from installment sales shall be recorded in Account 419, Interest and Dividend Income.

B. Records in support of this account shall be so kept as to permit ready summarization of revenues by such major items as are feasible.

NOTE: The classification of revenues of merchandising, jobbing, and contract work as nonoperating, and thus included in this account, is for accounting purposes. It does not preclude consideration of justification to the contrary for ratemaking or other purposes.

Items

1. Revenues from sale of merchandise and from jobbing and contract work.
2. Discounts and allowances made in settlement of bills for merchandise and jobbing work.

416   Costs and Expenses of Merchandising, Jobbing and Contract Work

A. This account shall include all expenses derived from the sale of merchandise and jobbing or contract work.

B. Records in support of this account shall be so kept as to permit ready summarization of costs and expenses by such major items as are feasible.

NOTE: The classification of costs and expenses of merchandising, jobbing, and contract work as nonoperating, and thus included in this account, is for accounting purposes. It does not preclude consideration of justification to the contrary for ratemaking or other purposes.

Items

Labor:

1. Canvassing and demonstrating appliances in homes and other places for the purpose of selling appliances.
2. Demonstrating and selling activities in sales rooms.
3. Installing appliances on customer premises where such work is done only for purchasers of appliances from the utility.
4. Installing wire, piping, or other property work, on a jobbing or contract basis.
5. Preparing advertising materials for appliance sales purposes.
6. Receiving and handling customer orders for merchandise or for jobbing services.
7. Cleaning and tidying sales rooms.
8. Maintaining display counters and other equipment used in merchandising.
9. Arranging merchandise in sales rooms and decorating display windows.
10. Reconditioning repossessed appliances.
11. Bookkeeping and other clerical work in connection with merchandise and jobbing activities.
12. Supervising merchandise and jobbing operations.
13. Advertising in newspapers, periodicals, radio, and television.
14. Cost of merchandise sold and of materials used in jobbing work.
15. Stores expenses on merchandise and jobbing stocks.
16. Fees and expenses of advertising and commercial artists' agencies.
17. Printing booklets, dodgers, and other advertising data.

18. Premiums given as inducement to buy appliances.
19. Light, heat, and power.
20. Depreciation on equipment used primarily for merchandise and jobbing operations.
21. Rent of sales rooms or of equipment.
22. Transportation expense in delivery and pick-up of appliances by utility's facilities or by others.
23. Stationery and office supplies and expenses.
24. Losses from uncollectible merchandise and jobbing accounts.

417   Revenues from Nonutility Operations

This account shall include revenues applicable to operations which are nonutility in character but nevertheless constitute a distinct operating activity of the enterprise as a whole, such as the operation of an ice department where applicable statutes do not define such operation as a utility, or the operation of a servicing organization for furnishing supervision, management, engineering, and similar services to others.

NOTE: Related taxes shall be recorded in Account 408, Taxes Other Than Income Taxes, or Account 409.2, Income Taxes, Other Income and Deductions, as appropriate.

417.1   Expenses of Nonutility Operations

A. This account shall include expenses applicable to operations which are nonutility in character but nevertheless constitute a distinct operating activity of the enterprise as a whole, such as the operation of an ice department where applicable statutes do not define such operation as a utility, or the operation of a servicing organization for furnishing supervision, management, engineering, and similar services to others.

B. The expenses shall include all elements of costs incurred in such operations, and the accounts shall be maintained so as to permit ready summarization as follows:

1. Operation.
2. Maintenance.
3. Rents.
4. Depreciation.
5. Amortization.

NOTE: Related taxes shall be recorded in Account 408, Taxes Other Than Income Taxes, or Account 409.2, Income Taxes, Other Income and Deductions, as appropriate.

418   Nonoperating Rental Income

A. This account shall include all rent revenues and related expenses of land, buildings, or other property included in Account 121, Nonutility Property, which is not used in operations covered by Account 417 or Account 417.1.

B. The expenses shall include all elements of costs incurred in the ownership and rental of property and the accounts shall be maintained so as to permit ready summarization as follows:

1. Operation.
2. Maintenance.
3. Rents.
4. Depreciation.
5. Amortization.

NOTE: Related taxes shall be recorded in Account 408, Taxes Other Than Income Taxes, or Account 409.2, Income Taxes, Other Income and Deductions, as appropriate.

418.1 Equity in Earnings of Subsidiary Companies

This account shall include the utility's equity in the earnings or losses of subsidiary companies for the year.

419 Interest and Dividend Income

A. This account shall include interest revenues on securities, loans, notes, advances, special deposits, tax refunds, and all other interest-bearing assets, and dividends on stocks of other companies, whether the securities on which the interest and dividends are received are carried as investments or included in sinking or other special fund accounts.

NOTE A: Related taxes shall be recorded in Account 408, Taxes Other Than Income Taxes, or Account 409.2, Income Taxes, Other Income and Deductions, as appropriate.

NOTE B: Interest accrued, the payment of which is not reasonably assured, dividends receivable which have not been declared or guaranteed, and interest or dividends upon reacquired securities issued or assumed by the utility shall not be credited to this account.

419.1 Allowance for Funds Used During Construction

This account shall include concurrent credits for allowance for funds other than borrowed funds used for construction purposes during the period of construction, based upon a reasonable rate. (See § 1767.16 (c)(17).)

420 Investment Tax Credits

This account shall be credited as follows with investment tax credit amounts not passed on to customers:

1. By amounts equal to debits to Account 411.4, Investment Tax Credit Adjustments, Utility Operations, and Account 411.5, Investment Tax Credit Adjustments, Nonutility Operations, for investment tax credits used in calculating income taxes for the year when the company's accounting provides for non-deferral of all or a portion of such credits.

2. By amounts equal to debits to Account 255, Accumulated Deferred Investment Tax Credits, for proportionate amounts of tax credit deferrals allocated over the average useful life of the property to which the tax credits relate, or such lesser period of time as may be adopted and consistently used by the company.

421 Miscellaneous Nonoperating Income

This account shall include all revenue and expense items, except taxes properly includible in the income account, not provided for elsewhere. Related taxes shall be recorded in Account 408, Taxes Other Than Income Taxes, or Account 409.2, Income Taxes, Other Income and Deductions, as appropriate.

Items

1. Profit on sale of timber. (See § 1767.16 (g)(3).)
2. Profits from operations of others realized by the utility under contracts.
3. Gains on disposition of investments. Also, gains on reacquisition and resale or retirement of the utility's debt securities when the gain is not amortized or used by a jurisdictional regulatory agency to reduce embedded debt cost in establishing rates. (See § 1767.15 (q).)
4. This account shall include the accretion expense on the liability for an asset retirement obligation included in Account 230, Asset Retirement Obligations, related to nonutility plant.
5. This account shall include the depreciation expense for asset retirement costs related to nonutility plant.
6. The utility shall record in this account gains resulting from the settlement of asset retirement obligations related to nonutility plant in accordance with the accounting prescribed in § 1767.15(y).

421.1 Gain on Disposition of Property

This account shall be credited with the gain on the sale, conveyance, exchange, or transfer of utility or other property to another. Amounts relating to gains on land and land rights held for future use recorded in Account 105, Electric Plant Held for Future Use, will be accounted for as prescribed in Paragraphs B, C, and D thereof. (See § 1767.16 (e)(6), (g)(5), and (j)(5).) Income taxes on gains recorded in this account shall be recorded in Account 409.2, Income Taxes, Other Income and Deductions.

421.2 Loss on Disposition of Property

This account shall be charged with the loss on the sale, conveyance, exchange, or transfer of utility or other property to another. Amounts relating to losses on land and land rights held for future use recorded in Account 105, Electric Plant Held for Future Use, will be accounted for as prescribed in Paragraphs B, C, and D thereof. (See § 1767.16

(e)(6), (g)(5), and (j)(5).) The reduction in income taxes relating to losses recorded in this account shall be recorded in Account 409.2, Income Taxes, Other Income and Deductions.

422 Nonoperating Taxes

This account shall be charged with taxes relating to nonoperating income.

423 Generation and Transmission Cooperative Capital Credits

This account shall be credited with the annual capital furnished the power supply cooperative through payment of power bills. The amount of capital furnished the power supply cooperative should be recorded in the applicable year even though, in most cases, the power supplier's notice of the allocation will not have been received until after the close of the year to which it relates.

424 Other Capital Credits and Patronage Capital Allocations

This account shall be credited with the capital furnished in connection with patronage of cooperative or mutual-type service organization such as CFC and other financing cooperatives, and insurance, oil product, telephone, and data processing cooperatives. This account should be credited in the year in which the notice of the capital credit or patronage capital allocation is received.

425 Miscellaneous Amortization

This account shall include amortization charges not includible in other accounts which are properly deductible in determining the income of the utility before interest charges. Charges includible herein, if significant in amount, must be in accordance with an orderly and systematic amortization program.

Items

1. Amortization of utility plant acquisition adjustments, or of intangibles included in utility plant in service when not authorized to be included in utility operating expenses by RUS.
2. Other miscellaneous amortization charges allowed to be included in this account by RUS.

426 [Reserved]

SPECIAL INSTRUCTIONS

Accounts 426.1, 426.2, 426.3, 426.4, and 426.5

These accounts shall include miscellaneous expense items which are nonoperating in nature but which are properly deductible before determining total income before interest charges.

NOTE: The classification of expenses as nonoperating and their inclusion in these accounts is for accounting purposes. It does not preclude RUS consideration of proof to the contrary for ratemaking or other purposes.

426.1 Donations

This account shall include all payments or donations for charitable, social, or community welfare purposes.

426.2 Life Insurance

This account shall include all payments for life insurance of officers and employees where the company is the beneficiary (net premiums less the increase in the cash surrender value of policies.)

426.3 Penalties

This account shall include payments by the company for penalties or fines for violation of any regulatory statutes by the company or its officials.

426.4 Expenditures for Certain Civic, Political, and Related Activities

This account shall include expenditures for the purpose of influencing public opinion with respect to the election or appointment of public officials, referenda, legislation, or ordinances (either with respect to the possible adoption of new referenda, legislation or ordinances or repeal or modification of existing referenda, legislation or ordinances) or approval, modification, or revocation of franchises; or for the purpose of influencing the decisions of public officials, but shall not include such expenditures which are directly related to appearances before regulatory or other governmental bodies in connection with the reporting utility's existing or proposed operations.

426.5 Other Deductions

This account shall include other miscellaneous expenses which are nonoperating in nature, but which are properly deductible before determining total income before interest charges.

Items

1. Loss relating to investments in securities written-off or written-down.
2. Loss on sale of investments.
3. Loss on reacquisition, resale, or retirement of the utility's debt securities, when the loss is not amortized and used by a jurisdictional regulatory agency to increase embedded debt cost in establishing rates. (See § 1767.15 (q).)
4. Preliminary survey and investigation expenses related to abandoned projects, when not written-off to the appropriate operating expense account.
5. Costs of preliminary abandonment costs recorded in Account 182.1, Extraordinary

Property Losses, and Account 182.2, Uncovered Plant and Regulatory Study Costs, not allowed to be amortized to Account 407, Amortization of Property Losses, Uncovered Plant and Regulatory Study Costs.

6. The utility shall record in this account losses resulting from the settlement of asset retirement obligations related to nonutility plant in accordance with the accounting prescribed in § 1767.15(y).

[58 FR 59825, Nov. 10, 1993, as amended at 73 FR 30285, May 27, 2008]

§ 1767.23 Interest charges.

The interest charges accounts identified in this section shall be used by all RUS borrowers.

INTEREST CHARGES

427 Interest on Long-Term Debt
427.3 Interest Charged to Construction—Credit
428 Amortization of Debt Discount and Expense
428.1 Amortization of Loss on Reacquired Debt
429 Amortization of Premium on Debt—Credit
429.1 Amortization of Gain on Reacquired Debt—Credit
430 Interest on Debt to Associated Companies
431 Other Interest Expense
432 Allowance for Borrowed Funds Used During Construction—Credit

INTEREST CHARGES

427 Interest on Long-Term Debt

A. This account shall include the amount of interest on outstanding long-term debt issued or assumed by the utility, the liability for which included in Account 221, Bonds, or Account 224, Other Long-Term Debt.

B. This account shall be so kept or supported as to show the interest accruals on each class and series of long-term debt.

NOTE: This account shall not include interest on nominally issued or nominally outstanding long-term debt, including securities assumed.

427.3 Interest Charged to Construction—Credit

This account shall include concurrent credits for interest charged to construction based upon the net cost for the period of construction of borrowed funds used for construction purposes.

428 Amortization of Debt Discount and Expense

A. This account shall include the amortization of unamortized debt discount and expense on outstanding long-term debt. Amounts charged to this account shall be credited concurrently to Account 181, Unamortized Debt Expense, and Account 226, Unamortized Discount on Long-Term Debt—Debit.

B. This account shall be so kept or supported as to show the debt discount and expense on each class and series of long-term debt.

428.1 Amortization of Loss on Reacquired Debt

A. This account shall include the amortization of the losses on reacquisition of debt. Amounts charged to this account shall be credited concurrently to Account 189, Unamortized Loss on Reacquired Debt.

B. This account shall be maintained so as to allow ready identification of the loss amortized applicable to each class and series of long-term debt reacquired. (See § 1767.15 (q).)

429 Amortization of Premium on Debt—Credit

A. This account shall include the amortization of unamortized net premium on outstanding long-term debt. Amounts credited to this account shall be charged concurrently to Account 225, Unamortized Premium on Long-Term Debt.

B. This account shall be so kept or supported as to show the premium on each class and series of long-term debt.

429.1 Amortization of Gain on Reacquired Debt—Credit

A. This account shall include the amortization of the gains realized from reacquisition of debt. Amounts credited to this account shall be charged concurrently to Account 257, Unamortized Gain on Reacquired Debt.

B. This account shall be maintained so as to allow ready identification of the amortized gains applicable to each class and series of long-term debt reacquired. (See § 1767.15 (q).)

430 Interest on Debt to Associated Companies

A. This account shall include the interest accrued on amounts included in Account 223, Advances from Associated Companies, and on all other obligations to associated companies.

B. The records supporting the entries to this account shall be so kept as to show to whom the interest is to be paid, the period covered by the accrual, the rate of interest, and the principal amount of the advances or other obligations on which the interest is accrued.

431   Other Interest Expense

This account shall include all interest charges not provided for elsewhere.

Items

1. Interest on notes payable on demand or maturing one year or less from date and on open accounts, except notes and accounts with associated companies.
2. Interest on customers' deposits.
3. Interest on claims and judgments, tax assessments, and assessments for public improvements past due.
4. Income and other taxes levied upon bondholders of the utility and assumed by it.

432   Allowance for Borrowed Funds Used During Construction—Credit   .

This account shall include concurrent credits for allowance for borrowed funds used during construction, not to exceed amounts computed in accordance with the formula prescribed in § 1767.16(c)(17).

NOTE: This account shall not be recorded in Account 427.3, Interest Charged to Construction—Credit.

[58 FR 59825, Nov. 10, 1993, as amended at 73 FR 30285, May 27, 2008]

§ 1767.24   Extraordinary items.

The extraordinary items accounts identified in this section shall be used by all RUS borrowers.

EXTRAORDINARY ITEMS

434   Extraordinary Income
435   Extraordinary Deductions
435.1   Cumulative Effect on Prior Years of a Change in Accounting Principle

EXTRAORDINARY ITEMS

434   Extraordinary Income

This account shall be credited with nontypical, noncustomary, infrequently recurring gains which would significantly distort the current year's income computed before extraordinary items, if reported other than as extraordinary items. Income tax relating to the amounts recorded in this account shall be recorded in Account 409.3, Income Taxes, Extraordinary Items. (See § 1767.15 (g).)

435   Extraordinary Deductions

This account shall be debited with nontypical, noncustomary, infrequently recurring losses which would significantly distort the current year's income computed before extraordinary items, if reported other than as extraordinary items. Income tax relating to the amounts recorded in this account shall be recorded in Account 409.3, Income Taxes, Extraordinary Items. (See § 1767.15 (f).)

435.1   Cumulative Effect on Prior Years of a Change in Accounting Principle

This account shall include the cumulative effect on margins of prior periods as a result of a change in accounting principle from one that is no longer generally accepted to one that is generally accepted.

§ 1767.25   Retained earnings.

The retained earnings accounts identified in this section shall be used by all RUS borrowers.

RETAINED EARNINGS

433–439 [Reserved]

RETAINED EARNINGS

433–439 [Reserved]

§ 1767.26   Operating revenue.

The operating revenue accounts identified in this section shall be used by all RUS borrowers.

OPERATING REVENUE

Sales of Electricity

440   Residential Sales
440.1   Residential Sales—Excluding Seasonal
440.2   Residential Sales—Seasonal
441   Irrigation Sales
442   Commercial and Industrial Sales
442.1   Commercial and Industrial Sales—1000 kVA or Less
442.2   Commercial and Industrial Sales—Over 1000 kVA
444   Public Street and Highway Lighting
445   Other Sales to Public Authorities
446   Sales to Railroads and Railways
447   Sales for Resale
447.1   Sales for Resale—RUS Borrowers
447.2   Sales for Resale—Other
448   Interdepartmental Sales
449.1   Provision for Rate Refunds

Other Operating Revenues

450   Forfeited Discounts
451   Miscellaneous Service Revenues
453   Sales of Water and Water Power
454   Rent from Electric Property
455   Interdepartmental Rents
456   Other Electric Revenues
456.1   Revenues from Transmission of Electricity of Others
457.1   Regional Transmission Service Revenues
457.2   Miscellaneous Revenue

OPERATING REVENUE

Sales of Electricity

440 Residential Sales

A. This account shall include the net billing for electricity supplied for residential or domestic purposes.

NOTE: When electricity supplied through a single meter is used for both residential and commercial purposes, the total revenue shall be included in this account, or Account 442, Commercial and Industrial Sales, according to the rate schedule that is applied. If the same rate schedules apply to residential and commercial and industrial service, classification shall be made according to principal use.

B. Account 440 shall be subaccounted as follows:

440.1 Residential Sales—Excluding Seasonal
440.2 Residential Sales—Seasonal

440.1 Residential Sales—Excluding Seasonal

A. This account shall include the net billing for electricity supplied for residential and domestic purposes.

B. This account shall also include net billings for single phase service to schools, churches, lodges, and other public buildings.

C. Records shall be maintained so that the quantity of electricity sold and the revenue received under each rate schedule shall be readily available.

NOTE: Net billings for multiphase service to schools, churches, lodges, and other public buildings shall be included in the appropriate subaccount of Account 442, Commercial and Industrial Sales.

440.2 Residential Sales—Seasonal

This account shall include the net billings for electricity supplied for residential and domestic purposes to seasonal consumers.

441 Irrigation Sales

This account shall include the net billings for electricity supplied for irrigation pumping. It need not be used unless such service is provided under a special irrigation rate.

442 Commercial and Industrial Sales

A. This account shall include the net billing for electricity supplied to customers for commercial and industrial purposes.

NOTE A: If the utility classifies large commercial and industrial customers and related revenues on a lesser basis than 1000 kilowatts of demand, or segregates industrial customers and related revenues according to a recognized definition of an industrial customer, such classifications are acceptable in lieu of those otherwise required by the text

of this account on the basis of 1000 kilowatts of demand.

NOTE B: When electricity supplied through a single meter is used for both commercial and residential purposes, the total revenue shall be included in this account, or Account 440, Residential Sales, according to the rate schedule that is applied. If the same rate schedules apply to residential and commercial and industrial service, classification shall be made according to principal use.

B. Account 442 shall be subaccounted as follows:

442.1 Commercial and Industrial Sales—1000 kVA or Less
442.2 Commercial and Industrial Sales—Over 1000 kVA

442.1 Commercial and Industrial Sales—1000 kVA or Less

A. This account shall include the net billing for electricity supplied to consumers for commercial and industrial purposes requiring transformer capacity of 1000 kVA or less.

B. Records shall be maintained so that the quantity of electricity sold and the revenue received under each rate schedule shall be readily available.

NOTE: When electricity supplied through a single meter is used for both commercial and residential purposes, the total revenue shall be included in this account or in Account 440, Residential Sales, based upon primary use.

442.2 Commercial and Industrial Sales—Over 1000 kVA

A. This account shall include the net billing for electricity supplied to consumers for commercial and industrial purposes requiring transformer capacity in excess of 1000 kVA.

B. Records shall be maintained so that the quantity of electricity sold and the revenue received under each rate schedule shall be readily available.

444 Public Street and Highway Lighting

A. This account shall include the net billing for electricity supplied and services rendered for the purposes of lighting streets, highways, parks, and other public places or for traffic or signal system service, for municipalities or other divisions or agencies of state of Federal Governments.

B. Records shall be maintained so that the quantity of electricity sold and the revenue received from each customer shall be readily available. In addition, the records shall be maintained so as to show the revenues from (1) contracts which include both electricity and services, and (2) contracts which include sales of electricity only.

445 Other Sales to Public Authorities

A. This account shall include the net billing for electricity supplied to municipalities or divisions or agencies of Federal or state governments, under special contracts or agreements or service classifications applicable only to public authorities, except such revenues as are includible in Account 444 and Account 447.

B. Records shall be maintained so as to show the quantity of electricity sold and the revenues received from each customer.

446 Sales to Railroads and Railways

A. This account shall include the net billing for electricity supplied to railroads and interurban and street railways, for general railroad use, including the propulsion of cars or locomotives, where such electricity is supplied under separate and distinct rate schedules.

B. Records shall be maintained so that the quantity of electricity sold and the revenue received from each customer shall be readily available.

NOTE: Revenues from incidental use of electricity furnished under a contract for propulsion of cars or locomotives shall be included herein.

447 Sales for Resale

A. This account shall include the net billing for electricity supplied to other electric utilities or to public authorities for resale purposes.

NOTE: Revenues from electricity supplied to other utilities for use by them and not for distribution, shall be included in Account 442, Commercial and Industrial Sales, unless supplied under the same contracts as and not readily separable from revenues includible in this account.

B. Account 447 shall be subaccounted as follows:

447.1 Sales for Resale—RUS Borrowers
447.2 Sales for Resale—Other

447.1 Sales for Resale—RUS Borrowers

A. This account shall include the net billing for electricity supplied to RUS borrowers for resale.

B. Records shall be maintained so as to show the quantity of electricity sold and the revenue received from each customer.

NOTE: Revenues from electricity supplied to other utilities for use by them and not for distribution, shall be included in Account 442, Commercial and Industrial Sales, unless supplied under the same contract as and not readily separable from revenues includible in this account.

447.2 Sales for Resale—Other

A. This account shall include the net billing for electricity supplied for resale to utilities not financed by RUS.

B. Records shall be maintained so as to show the quantity of electricity sold and the revenue received from each customer.

NOTE: Revenues from electricity supplied to other utilities for use by them and not for distribution, shall be included in Account 442, Commercial and Industrial Sales, unless supplied under the same contract as and not readily separable from revenues includible in this account.

448 Interdepartmental Sales

A. This account shall include amounts charged by the electric department at tariff or other specified rates for electricity supplied by it to other utility departments.

B. Records shall be maintained so that the quantity of electricity supplied each other department and the charges therefor shall be readily available.

449.1 Provision for Rate Refunds

A. This account shall be charged with provisions for the estimated pretax effects on net income of the portions of amounts being collected subject to refund which are estimated to be required to be refunded. Such provisions shall be credited to Account 229, Accumulated Provision for Rate Refunds.

B. This account shall also be charged with amounts refunded when such amounts had not been previously accrued.

C. Income tax effects relating to the amounts recorded in this account shall be recorded in Account 410.1, Provision for Deferred Income Taxes, Utility Operating Income, or Account 411.1, Provision for Deferred Income Taxes—Credit, Utility Operating Income, as appropriate.

Other Operating Revenues

450 Forfeited Discounts

This account shall include the amount of discounts forfeited or additional charges imposed because of the failure of customers to pay their electric bills on or before a specified date.

451 Miscellaneous Service Revenues

This account shall include revenues for all miscellaneous services and charges billed to customers which are not specifically provided for in other accounts.

Items

1. Fees for changing, connecting, or disconnecting service.

2. Profit on maintenance of appliances, wiring, piping, or other installations on customers' premises.

3. Net credit or debit (cost less net salvage and less payment from customers) on closing of work orders for plant installed for temporary service of less than one year. (See Account 185, Temporary Facilities.)

4. Recovery of expenses in connection with current diversion cases (billing for the electricity consumed shall be included in the appropriate electric revenue account).

453 Sales of Water and Water Power

A. This account shall include revenues derived from the sale of water for irrigation, domestic, industrial, or other uses or for the development by others of water power or for headwater benefits; also, revenues derived from furnishing water power for mechanical purposes when the investment in the property used in supplying such water or water power is carried as electric plant in service.

B. The records for this account shall be kept in such manner as to permit an analysis of the rates charged and the purposes for which the water was used.

454 Rent from Electric Property

A. This account shall include rents received for the use by others of land, buildings, and other property devoted to electric operations by the utility.

B. When property owned by the utility is operated jointly with others under a definite arrangement for apportioning the actual expenses among the parties to the arrangement, any amount received by the utility for interest or return or in reimbursement of taxes or depreciation on the property shall be credited to this account.

NOTE: Do not include in this account rents from property constituting an operating unit or system. (See Account 412, Revenues from Electric Plant Leased to Others.)

455 Interdepartmental Rents

This account shall include rents credited to the electric department on account of rental charges made against other departments (gas, water, etc.) of the utility. In the case of property operated under a definite arrangement to allocate the costs among the departments using the property, any reimbursement to the electric department for interest or return and depreciation and taxes shall be credited to this account.

456 Other Electric Revenues

This account shall include revenues derived from electric operations not includible in any of the foregoing accounts. It shall also include, in a separate subaccount, revenues received from operation of fish and wildlife and recreation facilities whether operated by the company or by contract concessionaires, such as revenues from leases or rentals of land for cottages, homes, or campsites.

Items

1. Commission on sale or distribution of electricity of others when sold under rates filed by such others.

2. Compensation for minor or incidental services provided for others such as customer billing, and engineering.

3. Profit or loss on the sale of material and supplies not ordinarily purchased for resale and not handled through merchandising and jobbing accounts.

4. Sale of steam, but not including sales made by a steamheating department or transfers of steam under joint facility operations.

5. Include in a separate subaccount, revenues in payment for rights and/or benefits received from others which are realized through research, development, and demonstration ventures. In the event the amounts received are so large as to distort revenues for the year in which received (5 percent of net income before application of the benefit), the amounts shall be credited to Account 253, Other Deferred Credits, and amortized by credits to this account over a period not to exceed 5 years.

456.1 Revenues From Transmission of Electricity of Others

This account shall include revenues from transmission of electricity of others over transmission facilities of the utility.

457.1 Regional Transmission Service Revenues

This account shall include revenues derived from providing scheduling, system control and dispatching services. Include also in this account reimbursements for system planning, standards development, and market monitoring and market compliance activities. Records shall be maintained so as to show: (1) The services supplied and revenues received from each customer and (2) the amounts billed by tariff or specified rates.

457.2 Miscellaneous Revenues

This account shall include revenues and reimbursements for costs incurred by regional transmission service providers not provided for elsewhere. Records shall be maintained so as to show: (1) The services supplied and revenues received from each customer, and (2) the amounts billed by tariff or specified rates.

[58 FR 59825, Nov. 10, 1993, as amended at 73 FR 30285, May 27, 2008]

§ 1767.27 Operation and maintenance expenses.

The operation and maintenance expense accounts identified in this section shall be used by all RUS borrowers.

OPERATION AND MAINTENANCE EXPENSE ACCOUNTS

POWER PRODUCTION EXPENSES

Steam Power Generation

(Operation)

500 Operation Supervision and Engineering
501 Fuel
502 Steam Expenses
503 Steam from Other Sources
504 Steam Transferred—Credit
505 Electric Expenses
506 Miscellaneous Steam Power Expenses
507 Rents
509 Allowances

(Maintenance)

510 Maintenance Supervision and Engineering
511 Maintenance of Structures
512 Maintenance of Boiler Plant
513 Maintenance of Electric Plant
514 Maintenance of Miscellaneous Steam Plant

Nuclear Power Generation

(Operation)

517 Operation Supervision and Engineering
518 Nuclear Fuel Expense
519 Coolants and Water
520 Steam Expenses
521 Steam from Other Sources
522 Steam Transferred—Credit
523 Electric Expenses
524 Miscellaneous Nuclear Power Expenses
525 Rents

(Maintenance)

528 Maintenance Supervision and Engineering
529 Maintenance of Structures
530 Maintenance of Reactor Plant Equipment
531 Maintenance of Electric Plant
532 Maintenance of Miscellaneous Nuclear Plant

Hydraulic Power Generation

(Operation)

535 Operation Supervision and Engineering
536 Water for Power
537 Hydraulic Expenses
538 Electric Expenses
539 Miscellaneous Hydraulic Power Generation Expenses

540 Rents

(Maintenance)

541 Maintenance Supervision and Engineering
542 Maintenance of Structures
543 Maintenance of Reservoirs, Dams, and Waterways
544 Maintenance of Electric Plant
545 Maintenance of Miscellaneous Hydraulic Plant

Other Power Generation

(Operation)

546 Operation Supervision and Engineering
547 Fuel
548 Generation Expenses
549 Miscellaneous Other Power Generation Expenses
550 Rents

(Maintenance)

551 Maintenance Supervision and Engineering
552 Maintenance of Structures
553 Maintenance of Generating and Electric Equipment
554 Maintenance of Miscellaneous Other Power Generation Plant

OTHER POWER SUPPLY EXPENSES

555 Purchased Power
556 System Control and Load Dispatching
557 Other Expenses

TRANSMISSION EXPENSES

(Operation)

560 Operation Supervision and Engineering
561.1 Load Dispatch-Reliability
561.2 Load Dispatch-Monitor and Operate Transmission System
561.3 Load Dispatch-Transmission Service and Scheduling
561.4 Scheduling, System Control and Dispatching Services
561.5 Reliability, Planning and Standards Development
561.6 Transmission Service Studies
561.7 Generation Interconnection Studies
561.8 Reliability Planning and Standards Development Services
561 Load Dispatching
562 Station Expenses
563 Overhead Line Expenses
564 Underground Line Expenses
565 Transmission of Electricity by Others
566 Miscellaneous Transmission Expenses
567 Rents

(Maintenance)

568 Maintenance Supervision and Engineering
569 Maintenance of Structures
569.1 Maintenance of Computer Hardware

569.2 Maintenance of Computer Software
569.3 Maintenance of Communication Equipment
569.4 Maintenance of Miscellaneous Regional Transmission Plant
570 Maintenance of Station Equipment
571 Maintenance of Overhead Lines
572 Maintenance of Underground Lines
573 Maintenance of Miscellaneous Transmission Plant

REGIONAL MARKET EXPENSES

(Operation)

575.1 Operation Supervision
575.2 Day-Ahead and Real-Time Market Administration
575.3 Transmission Rights Market Administration
575.4 Capacity Market Administration
575.5 Ancillary Services Market Administration
575.6 Market Monitoring and Compliance
575.7 Market Administration, Monitoring and Compliance Services
575.8 Rents

(Maintenance)

576.1 Maintenance of Structures and Improvements
576.2 Maintenance of Computer Hardware
576.3 Maintenance of Computer Software
576.4 Maintenance of Communication Equipment
576.5 Maintenance of Miscellaneous Market Operation Plant

DISTRIBUTION EXPENSES

(Operation)

580 Operation Supervision and Engineering
581 Load Dispatching
582 Station Expenses
583 Overhead Line Expenses
584 Underground Line Expenses
585 Street Lighting and Signal System Expenses
586 Meter Expenses
587 Customer Installations Expenses
588 Miscellaneous Distribution Expenses
589 Rents

(Maintenance)

590 Maintenance Supervision and Engineering
591 Maintenance of Structures
592 Maintenance of Station Equipment
593 Maintenance of Overhead Lines
594 Maintenance of Underground Lines
595 Maintenance of Line Transformers
596 Maintenance of Street Lighting and Signal Systems
597 Maintenance of Meters
598 Maintenance of Miscellaneous Distribution Plant

OPERATION AND MAINTENANCE EXPENSE ACCOUNTS

POWER PRODUCTION EXPENSES

Steam Power Generation

(Operation)

500 Operation Supervision and Engineering

This account shall include the cost of labor, employee pensions and benefits, social security and other payroll taxes, injuries and damages, and expenses incurred in the general supervision and direction of the operation of steam power generating stations. Direct supervision of specific activities, such as fuel handling, boiler-room operations, and generator operations shall be charged to the appropriate account. (See §1767.17(a).)

501 Fuel

A. This account shall include the cost of fuel used in the production of steam for the generation of electricity, including expenses in unloading fuel from the shipping media and handling thereof up to the point where the fuel enters the first boiler plant bunker, hopper, bucket, tank, or holder of the boiler-house structure. Records shall be maintained to show the quantity, B.t.u. content and cost of each type of fuel used.

B. The cost of fuel shall be charged initially to Account 151, Fuel Stock, and cleared to this account on the basis of the fuel used. Fuel handling expenses may be charged to this account as incurred or charged initially to Account 152, Fuel Stock Expenses Undistributed. In the latter event, they shall be cleared to this account on the basis of the fuel used. Respective amounts of fuel stock and fuel stock expenses shall be readily available.

Items

Labor:

1. Supervising, purchasing, and handling of fuel.
2. All routine fuel analyses.
3. Unloading from shipping facility and placing in storage.
4. Moving of fuel in storage and transferring fuel from one station to another.
5. Handling from storage or shipping facility to first bunker, hopper, bucket, tank, or holder of boiler-house structure.
6. Operation of mechanical equipment, such as locomotives, trucks, cars, boats, barges, and cranes.

Taxes:

1. Federal and state unemployment.
2. F.I.C.A.
3. Property.

Employee Pensions and Benefits: The portion of employee pensions and benefits specifically identifiable with employees' labor costs charged herein or, in the absence of specific employee identification, the portion of employee pensions and benefits, allocated on the more equitable basis of either direct labor dollars or direct labor hours, applicable to the labor items detailed above, including:

1. Accruals for or payments to pension funds or to insurance companies for pension purposes.

2. Group and life insurance premiums (credit dividends received).

3. Payments for medical and hospital services and expenses of employees when not the result of occupational injuries.

4. Payments for accident, sickness, hospital, and death benefits or insurance.

5. Payments to employees incapacitated for service or on leave of absence beyond periods normally allowed when not the result of occupational injuries or in excess of statutory awards.

6. Expenses in connection with educational and recreational activities for the benefit of employees.

Insurance:

1. Premiums payable to insurance companies for fire, storm, burglary, boiler explosion, lightning, fidelity, riot, and similar insurance.

2. Amounts credited to Account 228.1, Accumulated Provision for Property Insurance, for similar protection.

3. Special costs incurred in procuring insurance.

4. Insurance inspection service.

5. Insurance counsel, brokerage fees, and expenses.

6. Premiums payable to insurance companies for protection against claims from injuries and damages by employees or others, such as public liability, property damages, casualty, employee liability, etc., and amounts credited to Account 228.2, Accumulated Provision for Injuries and Damage, for similar protection.

7. Losses not covered by insurance or reserve accruals on account of injuries or deaths to employees or others and damages to the property of others.

8. Fees and expenses of claim investigators.

9. Payment of awards to claimants for court costs and attorneys' services.

10. Medical and hospital service and expenses for employees as the result of occupational injuries or resulting from claims of others.

11. Compensation payments under workmen's compensation laws.

12. Compensation paid while incapacitated as the result of occupational injuries. (See Account 924, Note A.)

13. Cost of safety, accident prevention, and similar educational activities.

Materials and Expenses:

1. Operating, maintenance, and depreciation expenses and ad valorem taxes on utility-owned transportation equipment used to transport fuel from the point of acquisition to the unloading point.

2. Lease or rental costs of transportation equipment used to transport fuel from the point of acquisition to the unloading point.

3. Cost of fuel including freight, switching, demurrage, and other transportation charges.

4. Excise taxes, insurance, purchasing commissions, and similar items.

5. Stores expenses to extent applicable to fuel.

6. Transportation and other expenses in moving fuel in storage.

7. Tools, lubricants, and other supplies.

8. Operating supplies for mechanical equipment.

9. Residual disposal expenses less any proceeds from sale of residuals.

NOTE: Abnormal fuel handling expenses occasioned by emergency conditions shall be charged to expense as incurred.

502 Steam Expenses

This account shall include the cost of labor, employee pensions and benefits, social security and other payroll taxes, injuries and damages, property insurance, property taxes, materials used, and expenses incurred in production of steam for electric generation. This includes all expenses of handling and preparing fuel beginning at the point where the fuel enters the first boiler plant bunker, hopper, tank, or holder of the boiler-house structure.

Items

Labor:

1. Supervising steam production.

2. Operating fuel conveying, storage, weighing, and processing equipment within boiler plant.

3. Operating boiler and boiler auxiliary equipment.

4. Operating boiler feed water purification and treatment equipment.

5. Operating ash-collecting and disposal equipment located inside the plant.

6. Operating boiler plant electrical equipment.

7. Keeping boiler plant log and records and preparing reports on boiler plant operations.

8. Testing boiler water.

9. Testing, checking, and adjusting meters, gauges, and other instruments and equipment in boiler plant.

10. Cleaning boiler plant equipment when not incidental to maintenance work.

11. Repacking glands and replacing gauge glasses where the work involved is of a minor nature and is performed by regular operating crews. Where the work is of a major character, such as that performed on high-

pressure boilers, the item should be considered as maintenance.

Taxes:

1. Federal and state unemployment.
2. F.I.C.A.
3. Property.

Employee Pensions and Benefits: The portion of employee pensions and benefits specifically identifiable with employees' labor costs charged herein or, in the absence of specific employee identification, the portion of employee pensions and benefits, allocated on the more equitable basis of either direct labor dollars or direct labor hours, applicable to the labor items detailed above, including:

1. Accruals for or payments to pension funds or to insurance companies for pension purposes.
2. Group and life insurance premiums (credit dividends received).
3. Payments for medical and hospital services and expenses of employees when not the result of occupational injuries.
4. Payments for accident, sickness, hospital, and death benefits or insurance.
5. Payments to employees incapacitated for service or on leave of absence beyond periods normally allowed when not the result of occupational injuries or in excess of statutory awards.
6. Expenses in connection with educational and recreational activities for the benefit of employees.

Insurance:

1. Premiums payable to insurance companies for fire, storm, burglary, boiler explosion, lightning, fidelity, riot, and similar insurance.
2. Amounts credited to Account 228.1, Accumulated Provision for Property Insurance, for similar protection.
3. Special costs incurred in procuring insurance.
4. Insurance inspection service.
5. Insurance counsel, brokerage fees, and expenses.
6. Premiums payable to insurance companies for protection against claims from injuries and damages by employees or others, such as public liability, property damages, casualty, employee liability, etc., and amounts credited to Account 228.2, Accumulated Provision for Injuries and Damage, for similar protection.
7. Losses not covered by insurance or reserve accruals on account of injuries or deaths to employees or others and damages to the property of others.
8. Fees and expenses of claim investigators.
9. Payment of awards to claimants for court costs and attorneys' services.
10. Medical and hospital service and expenses for employees as the result of occupational injuries or resulting from claims of others.

11. Compensation payments under workmen's compensation laws.
12. Compensation paid while incapacitated as the result of occupational injuries. (See Account 924, Note A.)
13. Cost of safety, accident prevention, and similar educational activities.

Materials and Expenses:

1. Chemicals and boiler inspection fees.
2. Lubricants.
3. Boiler feed water purchased and pumping supplies.

503 Steam from Other Sources

This account shall include the cost of steam purchased or transferred from another department of the utility or from others under a joint facility operating arrangement for use in prime movers devoted to the production of electricity.

NOTE: The records shall be so kept as to show separately for each company from which stem is purchased, the point of delivery, the quantity, the price, and the total charge. When steam is transferred from another department or from others under a joint operating arrangement, the utility shall be prepared to show full details of the cost of producing such steam, the basis of the charge to electric generation, and the extent and manner of use by each department or party involved.

504 Steam Transferred—Credit

A. This account shall include credits for expenses of producing steam which are charged to others or to other utility departments under a joint operating arrangement. Include also credits for steam expenses chargeable to other electric accounts outside of the steam generation group. Full details of the basis of determination of the cost of steam transferred shall be maintained.

B. If the charges to others or to other departments of the utility include an amount for depreciation, taxes, and return on the joint steam facilities, such portion of the charge shall be credited, in the case of others, to Account 454, Rent from Electric Property, and in the case of other departments of the utility, to Account 455, Interdepartmental Rents.

505 Electric Expenses

This account shall include the cost of labor, employee pensions and benefits, social security and other payroll taxes, injuries and damages, property insurance, property taxes, and materials used, and expenses incurred in operating prime movers, generators, and their auxiliary apparatus, switch gear, and other electric equipment to the points where electricity leaves for conversion for transmission or distribution.

Items

Labor:

1. Supervising electric production.
2. Operating turbines, engines, generators, and exciters.
3. Operating condensers, circulating water systems, and other auxiliary apparatus.
4. Operating generator cooling system.
5. Operating lubrication and oil control system, including oil purification.
6. Operating switchboards, switch gear and electric control, and protective equipment.
7. Keeping electric plant log and records and preparing reports on electric plant operations.
8. Testing, checking, and adjusting meters, gauges, and other instruments, relays, controls, and other equipment in the electric plant.
9. Cleaning electric plant equipment when not incidental to maintenance work.
10. Repacking glands and replacing gauge glasses.

Taxes:

1. Federal and state unemployment.
2. F.I.C.A.
3. Taxes.

Employee Pensions and Benefits: The portion of employee pensions and benefits specifically identifiable with employees' labor costs charged herein or, in the absence of specific employee identification, the portion of employee pensions and benefits, allocated on the more equitable basis of either direct labor dollars or direct labor hours, applicable to the labor items detailed above, including:

1. Accruals for or payments to pension funds or to insurance companies for pension purposes.
2. Group and life insurance premiums (credit dividends received).
3. Payments for medical and hospital services and expenses of employees when not the result of occupational injuries.
4. Payments for accident, sickness, hospital, and death benefits or insurance.
5. Payments to employees incapacitated for service or on leave of absence beyond periods normally allowed when not the result of occupational injuries or in excess of statutory awards.
6. Expenses in connection with educational and recreational activities for the benefit of employees.

Insurance:

1. Premiums payable to insurance companies for fire, storm, burglary, boiler explosion, lightning, fidelity, riot, and similar insurance.
2. Amounts credited to Account 228.1, Accumulated Provision for Property Insurance, for similar protection.
3. Special costs incurred in procuring insurance.

4. Insurance inspection service.
5. Insurance counsel, brokerage fees, and expenses.
6. Premiums payable to insurance companies for protection against claims from injuries and damages by employees or others, such as public liability, property damages, casualty, employee liability, etc., and amounts credited to Account 228.2, Accumulated Provision for Injuries and Damage, for similar protection.
7. Losses not covered by insurance or reserve accruals on account of injuries or deaths to employees or others and damages to the property of others.
8. Fees and expenses of claim investigators.
9. Payment of awards to claimants for court costs and attorneys' services.
10. Medical and hospital service and expenses for employees as the result of occupational injuries or resulting from claims of others.
11. Compensation payments under workmen's compensation laws.
12. Compensation paid while incapacitated as the result of occupational injuries. (See Account 924, Note A.)
13. Cost of safety, accident prevention, and similar educational activities.

Materials and Expenses:

1. Lubricants and control system oils.
2. Generator cooling gases.
3. Circulating water purification supplies.
4. Cooling water purchased.
5. Motor and generator brushes.

506   Miscellaneous Steam Power Expenses

This account shall include the cost of labor, employee pensions and benefits, social security and other payroll taxes, injuries and damages, and materials used and expenses incurred which are not specifically provided for or not readily assignable to other steam generation operation expense accounts.

Items

Labor:

1. General clerical and stenographic work.
2. Guarding and patrolling plant and yard.
3. Building service.
4. Care of grounds including snow removal, and grass cutting.
5. Miscellaneous labor.

Taxes:

1. Federal and state unemployment.
2. F.I.C.A.

Employee Pensions and Benefits: The portion of employee pensions and benefits specifically identifiable with employees' labor costs charged herein or, in the absence of specific employee identification, the portion of employee pensions and benefits, allocated on the more equitable basis of either direct labor dollars or direct labor hours, applicable to the labor items detailed above, including:

1. Accruals for or payments to pension funds or to insurance companies for pension purposes.

2. Group and life insurance premiums (credit dividends received).

3. Payments for medical and hospital services and expenses of employees when not the result of occupational injuries.

4. Payments for accident, sickness, hospital, and death benefits or insurance.

5. Payments to employees incapacitated for service or on leave of absence beyond periods normally allowed when not the result of occupational injuries or in excess of statutory awards.

6. Expenses in connection with educational and recreational activities for the benefit of employees.

Insurance:

1. Premiums payable to insurance companies for protection against claims from injuries and damages by employees or others, such as public liability, property damages, casualty, employee liability, etc., and amounts credited to Account 228.2, Accumulated Provision for Injuries and Damage, for similar protection.

2. Losses not covered by insurance or reserve accruals on account of injuries or deaths to employees or others and damages to the property of others.

3. Fees and expenses of claim investigators.

4. Payment of awards to claimants for court costs and attorneys' services.

5. Medical and hospital service and expenses for employees as the result of occupational injuries or resulting from claims of others.

6. Compensation payments under workmen's compensation laws.

7. Compensation paid while incapacitated as the result of occupational injuries. (See Account 924, Note A.)

8. Cost of safety, accident prevention, and similar educational activities.

Materials and Expenses:

1. General operating supplies, such as tools, gaskets, packing waste, gauge glasses, hose, indicating lamps, record and report forms.

2. First-aid supplies and safety equipment.

3. Employees' service facilities expenses.

4. Building service supplies.

5. Communication service.

6. Miscellaneous office supplies and expenses, printing, and stationery.

7. Transportation expenses.

8. Meals, traveling, and incidental expenses.

9. Research, development, and demonstration expenses.

507 Rents

This account shall include all rents of property of others used, occupied or operated in connection with steam power generation. (See § 1767.17 (c).)

509 Allowances

This account shall include the cost of allowances expensed concurrent with the monthly emission of sulfur dioxide. (See § 1767.15 (u).)

(Maintenance)

510 Maintenance Supervision and Engineering

This account shall include the cost of labor, employee pensions and benefits, social security and other payroll taxes, injuries and damages, and expenses incurred in the general supervision and direction of maintenance of steam generation facilities. Direct field supervision of specific jobs shall be charged to the appropriate maintenance account. (See § 1767.17(a).)

511 Maintenance of Structures

This account shall include the cost of labor, employee pensions and benefits, social security and other payroll taxes, injuries and damages, and materials used and expenses incurred in the maintenance of steam structures, the book cost of which is includible in Account 311, Structures and Improvements. (See § 1767.17(b).)

512 Maintenance of Boiler Plant

A. This account shall include the cost of labor, employee pensions and benefits, social security and other payroll taxes, injuries and damages, and materials used and expenses incurred in the maintenance of steam plant, the book cost of which is includible in Account 312, Boiler Plant Equipment. (See § 1767.17(b).)

B. For the purpose of making charges hereto and to Account 513, Maintenance of Electric Plant, the point at which steam plant is distinguished from electric plant is defined as follows:

1. Inlet flange of throttle valve on prime mover.

2. Flange of all steam extraction lines on prime mover.

3. Hotwell pump outlet on condensate lines.

4. Inlet flange of all turbine-room auxiliaries.

5. Connection to line side of motor starter for all boiler-plant equipment.

513 Maintenance of Electric Plant

This account shall include the cost of labor, employee pensions and benefits, social security and other payroll taxes, injuries and damages, and materials used and expenses incurred in the maintenance of electric plant, the book cost of which is includible in Account 313, Engines and Engine-Driven Generators; Account 314, Turbogenerator Units; and Account 315, Accessory Electric Equipment. (See § 1767.17(b) and Paragraph B of Account 512.)

514 Maintenance of Miscellaneous Steam Plant

This account shall include the cost of labor, employee pensions and benefits, social security and other payroll taxes, injuries and damages, and materials used and expenses incurred in maintenance of miscellaneous steam generation plant, the book cost of which is includible in Account 316, Miscellaneous Power Plant Equipment. (See § 1767.17(b).)

Nuclear Power Generation

(Operation)

517 Operation Supervision and Engineering

This account shall include the cost of labor, employee pensions and benefits, social security and other payroll taxes, injuries and damages, and expenses incurred in the general supervision and direction of the operation of nuclear power generating stations. Direct supervision of specific activities, such as fuel handling, reactor operations, and generator operations shall be charged to the appropriate account. (See § 1767.17(a).)

518 Nuclear Fuel Expense

A. This account shall be debited and Account 120.5, Accumulated Provision for Amortization of Nuclear Fuel Assemblies, credited for the amortization of the net cost of nuclear fuel assemblies used in the production of energy. The net cost of nuclear fuel assemblies subject to amortization shall be the cost of nuclear fuel assemblies plus or less the expected net salvage of uranium, plutonium, and other byproducts and unburned fuel. The utility shall adopt the necessary procedures to assure that charges to this account are distributed according to the thermal energy produced in such periods.

B. This account shall also include the costs involved when fuel is leased.

C. This account shall also include the cost of other fuels, used for ancillary steam facilities, including superheat.

D. This account shall be debited or credited as appropriate for significant changes in the amounts estimated as the net salvage value of uranium, plutonium, and other byproducts contained in Account 157, Nuclear Materials Held for Sale, and the amount realized upon the final disposition of the materials. Significant declines in the estimated realizable value of items carried in Account 157 may be recognized at the time of market price declines by charging this account and crediting Account 157. When the declining change occurs while the fuel is recorded in Account 120.3, Nuclear Fuel Assemblies in Reactor, the effect shall be amortized over the remaining life of the fuel.

519 Coolants and Water

This account shall include the cost of labor, employee pensions and benefits, social security and other payroll taxes, injuries and damages, property insurance, property taxes, and materials used and expenses incurred for heat transfer materials and water used for steam and cooling purposes.

Items

Labor:

1. Operation of water supply facilities.
2. Handling of coolants and heat transfer materials.

Taxes:

1. Federal and state unemployment.
2. F.I.C.A.
3. Taxes.

Employee Pensions and Benefits: The portion of employee pensions and benefits specifically identifiable with employees' labor costs charged herein or, in the absence of specific employee identification, the portion of employee pensions and benefits, allocated on the more equitable basis of either direct labor dollars or direct labor hours, applicable to the labor items detailed above, including:

1. Accruals for or payments to pension funds or to insurance companies for pension purposes.
2. Group and life insurance premiums (credit dividends received).
3. Payments for medical and hospital services and expenses of employees when not the result of occupational injuries.
4. Payments for accident, sickness, hospital, and death benefits or insurance.
5. Payments to employees incapacitated for service or on leave of absence beyond periods normally allowed when not the result of occupational injuries or in excess of statutory awards.
6. Expenses in connection with educational and recreational activities for the benefit of employees.

Insurance:

1. Premiums payable to insurance companies for fire, storm, burglary, boiler explosion, lightning, fidelity, riot, and similar insurance.
2. Amounts credited to Account 228.1, Accumulated Provision for Property Insurance, for similar protection.
3. Special costs incurred in procuring insurance.
4. Insurance inspection service.

5. Insurance counsel, brokerage fees, and expenses.

6. Premiums payable to insurance companies for protection against claims from injuries and damages by employees or others, such as public liability, property damages, casualty, employee liability, etc., and amounts credited to Account 228.2, Accumulated Provision for Injuries and Damage, for similar protection.

7. Losses not covered by insurance or reserve accruals on account of injuries or deaths to employees or others and damages to the property of others.

8. Fees and expenses of claim investigators.

9. Payment of awards to claimants for court costs and attorneys' services.

10. Medical and hospital service and expenses for employees as the result of occupational injuries or resulting from claims of others.

11. Compensation payments under workmen's compensation laws.

12. Compensation paid while incapacitated as the result of occupational injuries. (See Account 924, Note A.)

13. Cost of safety, accident prevention, and similar educational activities.

Materials and Expenses:

1. Chemicals.

2. Additions to or refining of fluids used in reactor systems.

3. Lubricants.

4. Pumping supplies and expenses.

5. Miscellaneous supplies and expenses.

6. Purchased water.

NOTE: Do not include in this account water for general station use or the initial charge for coolants, heat transfer, or moderator fluids, chemicals, or other supplies capitalized.

520 Steam Expenses

This account shall include the cost of labor, employee pensions and benefits, social security and other payroll taxes, injuries and damages, property insurance, property taxes, and materials used and expenses incurred in production of steam through nuclear processes, and similar expenses for operation of any auxiliary superheat facilities.

Items

Labor:

1. Supervising steam production.

2. Fuel handling including removal, insertion, disassembly, and preparation for cooling operations and shipment.

3. Testing instruments and gauges.

4. Health, safety, monitoring, and decontamination activities.

5. Waste disposal.

6. Operating steam boilers and auxiliary steam, superheat facilities.

Taxes:

1. Federal and state unemployment.

2. F.I.C.A.

3. Property.

Employee Pensions and Benefits: The portion of employee pensions and benefits specifically identifiable with employees' labor costs charged herein or, in the absence of specific employee identification, the portion of employee pensions and benefits, allocated on the more equitable basis of either direct labor dollars or direct labor hours, applicable to the labor items detailed above, including:

1. Accruals for or payments to pension funds or to insurance companies for pension purposes.

2. Group and life insurance premiums (credit dividends received).

3. Payments for medical and hospital services and expenses of employees when not the result of occupational injuries.

4. Payments for accident, sickness, hospital, and death benefits or insurance.

5. Payments to employees incapacitated for service or on leave of absence beyond periods normally allowed when not the result of occupational injuries or in excess of statutory awards.

6. Expenses in connection with educational and recreational activities for the benefit of employees.

Insurance:

1. Premiums payable to insurance companies for fire, storm, burglary, boiler explosion, lightning, fidelity, riot, and similar insurance.

2. Amounts credited to Account 228.1, Accumulated Provision for Property Insurance, for similar protection.

3. Special costs incurred in procuring insurance.

4. Insurance inspection service.

5. Insurance counsel, brokerage fees, and expenses.

6. Premiums payable to insurance companies for protection against claims from injuries and damages by employees or others, such as public liability, property damages, casualty, employee liability, etc., and amounts credited to Account 228.2, Accumulated Provision for Injuries and Damage, for similar protection.

7. Losses not covered by insurance or reserve accruals on account of injuries or deaths to employees or others and damages to the property of others.

8. Fees and expenses of claim investigators.

9. Payment of awards to claimants for court costs and attorneys' services.

10. Medical and hospital service and expenses for employees as the result of occupational injuries or resulting from claims of others.

11. Compensation payments under workmen's compensation laws.

12. Compensation paid while incapacitated as the result of occupational injuries. (See Account 924, Note A.)

13. Cost of safety, accident prevention, and similar educational activities.

Materials and Expenses:

1. Chemical supplies.
2. Charts and logs.
3. Health, safety, monitoring, and decontamination supplies.
4. Boiler inspection fees.
5. Lubricants.

521   Steam from Other Sources

This account shall include the cost of steam purchased or transferred from another department of the utility or from others under a joint facility operating arrangement for use in prime movers devoted to the production of electricity.

NOTE: The records shall be so kept as to show separately for each company from which steam is purchased, the point of delivery, the quantity, the price, and the total charge. When steam is transferred from another operating department, the utility shall be prepared to show full details of the cost of producing such steam, the basis of the charges to electric generation, and the extent and manner of use by each department involved.

522   Steam Transferred—Credit

A. This account shall include credits for expenses of producing steam which are charged to others or to other utility departments under a joint operating arrangement. Include also credits for steam expenses chargeable to other electric accounts outside of the steam generation group. Full details of the basis of determination of the cost of steam transferred shall be maintained.

B. If the charges to others or to other departments of the utility include an amount for depreciation, taxes, and return on the joint steam facilities, such portion of the charge shall be credited in the case of others, to Account 454, Rent from Electric Property, and in the case of other departments of the utility, to Account 455, Interdepartmental Rents.

523   Electric Expenses

This account shall include the cost of labor, employee pensions and benefits, social security and other payroll taxes, injuries and damages, property insurance, property taxes, materials used, and expenses incurred in operating turbogenerators, steam turbines and their auxiliary apparatus, switch gear, and other electric equipment to the points where electricity leaves for conversion for transmission or distribution.

Items

Labor:

1. Supervising electric production.
2. Operating turbines, engines, generators, and exciters.
3. Operating condensers, circulating water systems, and other auxiliary apparatus.
4. Operating generator cooling system.
5. Operating lubrication and oil control system, including oil purification.
6. Operating switchboards, switch gear, and electric control and protective equipment.
7. Keeping plant log and records and preparing reports on electric plant operations.
8. Testing, checking and adjusting meters, gauges, and other instruments, relays, controls, and other equipment in the electric plant.
9. Cleaning electric plant equipment when not incidental to maintenance.
10. Repacking glands and replacing gauge glasses.

Taxes:

1. Federal and state unemployment.
2. F.I.C.A.
3. Property.

Employee Pensions and Benefits: The portion of employee pensions and benefits specifically identifiable with employees' labor costs charged herein or, in the absence of specific employee identification, the portion of employee pensions and benefits, allocated on the more equitable basis of either direct labor dollars or direct labor hours, applicable to the labor items detailed above, including:

1. Accruals for or payments to pension funds or to insurance companies for pension purposes.
2. Group and life insurance premiums (credit dividends received).
3. Payments for medical and hospital services and expenses of employees when not the result of occupational injuries.
4. Payments for accident, sickness, hospital, and death benefits or insurance.
5. Payments to employees incapacitated for service or on leave of absence beyond periods normally allowed when not the result of occupational injuries or in excess of statutory awards.
6. Expenses in connection with educational and recreational activities for the benefit of employees.

Insurance:

1. Premiums payable to insurance companies for fire, storm, burglary, boiler explosion, lightning, fidelity, riot, and similar insurance.
2. Amounts credited to Account 228.1, Accumulated Provision for Property Insurance, for similar protection.
3. Special costs incurred in procuring insurance.
4. Insurance inspection service.

5. Insurance counsel, brokerage fees, and expenses.

6. Premiums payable to insurance companies for protection against claims from injuries and damages by employees or others, such as public liability, property damages, casualty, employee liability, etc., and amounts credited to Account 228.2, Accumulated Provision for Injuries and Damage, for similar protection.

7. Losses not covered by insurance or reserve accruals on account of injuries or deaths to employees or others and damages to the property of others.

8. Fees and expenses of claim investigators.

9. Payment of awards to claimants for court costs and attorneys' services.

10. Medical and hospital service and expenses for employees as the result of occupational injuries or resulting from claims of others.

11. Compensation payments under workmen's compensation laws.

12. Compensation paid while incapacitated as the result of occupational injuries. (See Account 924, Note A.)

13. Cost of safety, accident prevention, and similar educational activities.

Materials and Expenses:

1. Lubricants and control system oils.
2. Generator cooling gases.
3. Log sheets and charts.
4. Motor and generator brushes.

524 Miscellaneous Nuclear Power Expenses

This account shall include the cost of labor, employee pensions and benefits, social security and other payroll taxes, injuries and damages, materials used, and expenses incurred which are not specifically provided for or are not readily assignable to other nuclear generation operation accounts.

Items

Labor:

1. General clerical and stenographic work.
2. Plant security.
3. Building service.
4. Care of grounds, including snow removal, and grass cutting
5. Miscellaneous labor.

Taxes:

1. Federal and state unemployment.
2. F.I.C.A.

Employee Pensions and Benefits: The portion of employee pensions and benefits specifically identifiable with employees' labor costs charged herein or, in the absence of specific employee identification, the portion of employee pensions and benefits, allocated on the more equitable basis of either direct labor dollars or direct labor hours, applicable to the labor items detailed above, including:

1. Accruals for or payments to pension funds or to insurance companies for pension purposes.

2. Group and life insurance premiums (credit dividends received).

3. Payments for medical and hospital services and expenses of employees when not the result of occupational injuries.

4. Payments for accident, sickness, hospital, and death benefits or insurance.

5. Payments to employees incapacitated for service or on leave of absence beyond periods normally allowed when not the result of occupational injuries or in excess of statutory awards.

6. Expenses in connection with educational and recreational activities for the benefit of employees.

Insurance:

1. Premiums payable to insurance companies for protection against claims from injuries and damages by employees or others, such as public liability, property damages, casualty, employee liability, etc., and amounts credited to Account 228.2, Accumulated Provision for Injuries and Damage, for similar protection.

2. Losses not covered by insurance or reserve accruals on account of injuries or deaths to employees or others and damages to the property of others.

3. Fees and expenses of claim investigators.

4. Payment of awards to claimants for court costs and attorneys' services.

5. Medical and hospital service and expenses for employees as the result of occupational injuries or resulting from claims of others.

6. Compensation payments under workmen's compensation laws.

7. Compensation paid while incapacitated as the result of occupational injuries. (See Account 924, Note A.)

8. Cost of safety, accident prevention, and similar educational activities.

Materials and Expenses:

1. General operating supplies, such as tools, gaskets, hose, indicating lamps, records and reports forms.

2. First-aid supplies and safety equipment.

3. Employees' service facilities expenses.

4. Building service supplies.

5. Communication service.

6. Miscellaneous office supplies and expenses, printing and stationery.

7. Transportation expenses.

8. Meals, traveling, and incidental expenses.

9. Research, development, and demonstration expenses.

525 Rents

This account shall include all rents of property of others used, occupied, or operated in connection with nuclear generation. (See § 1767.17 (c).)

(Maintenance)

528 Maintenance Supervision and Engineering

This account shall include the cost of labor, employee pensions and benefits, social security and other payroll taxes, injuries and damages, and expenses incurred in the general supervision and direction of maintenance of nuclear generation facilities. Direct field supervision of specific jobs shall be charged to the appropriate maintenance account. (See § 1767.17(a).)

529 Maintenance of Structures

This account shall include the cost of labor, employee pensions and benefits, social security and other payroll taxes, injuries and damages, materials used, and expenses incurred in the maintenance of structures, the book cost of which is includible in Account 321, Structures and Improvements. (See § 1767.17(b).)

530 Maintenance of Reactor Plant Equipment

This account shall include the cost of labor, employee pensions and benefits, social security and other payroll taxes, injuries and damages, materials used, and expenses incurred in the maintenance of reactor plant, the book cost of which is includible in Account 322, Reactor Plant Equipment. (See § 1767.17(b).)

531 Maintenance of Electric Plant

This account shall include the cost of labor, employee pensions and benefits, social security and other payroll taxes, injuries and damages, materials used, and expenses incurred in the maintenance of electric plant, the book cost of which is includible in Account 323, Turbogenerator Units, and Account 324, Accessory Electric Equipment. (See § 1767.17(b).)

532 Maintenance of Miscellaneous Nuclear Plant

This account shall include the cost of labor, employee pensions and benefits, social security and other payroll taxes, injuries and damages, materials used, and expenses incurred in maintenance of miscellaneous nuclear generating plant, the book cost of which is includible in Account 325, Miscellaneous Power Plant Equipment. (See § 1767.17(b).)

Hydraulic Power Generation

(Operation)

535 Operation Supervision and Engineering

This account shall include the cost of labor, employee pensions and benefits, social security and other payroll taxes, injuries and damages, and expenses incurred in the general supervision and direction of the operation of hydraulic power generating stations. Direct supervision of specific activities, such as hydraulic operation, and generator operation shall be charged to the appropriate account. (See § 1767.17(a).)

536 Water for Power

This account shall include the cost of water used for hydraulic power generation.

Items

1. Cost of water purchased from others, including water tolls paid reservoir companies.
2. Periodic payments for licenses or permits from any governmental agency for water rights, or payments based on the use of the water.
3. Periodic payments for riparian rights.
4. Periodic payments for headwater benefits or for detriments to others.
5. Cloud seeding.

537 Hydraulic Expenses

This account shall include the cost of labor, employee pensions and benefits, social security and other payroll taxes, injuries and damages, property insurance, property taxes, materials used, and expenses incurred in operating hydraulic works including reservoirs, dams, and waterways, and in activities directly relating to the hydroelectric development outside the generating station. It shall also include the cost of labor, materials used, and other expenses incurred in connection with the operation of (1) fish and wildlife, and (2) recreation facilities. Separate subaccounts shall be maintained for each of the above.

Items

Labor:

1. Supervising hydraulic operation.
2. Removing debris and ice from trash racks, reservoirs, and waterways.
3. Patrolling reservoirs and waterways.
4. Operating intakes, spillways, sluiceways, and outlet works.
5. Operating bubbler, heater, or other deicing systems.
6. Ice and log jam work.
7. Operating navigation facilities.
8. Operations relating to conservation of game, fish, and forests.
9. Insect control activities.

Taxes:

1. Federal and state unemployment.
2. F.I.C.A.
3. Property.

Employee Pensions and Benefits: The portion of employee pensions and benefits specifically identifiable with employees' labor costs charged herein or, in the absence of specific employee identification, the portion of employee pensions and benefits, allocated on the more equitable basis of either direct labor dollars or direct labor hours, applicable to the labor items detailed above, including:

1. Accruals for or payments to pension funds or to insurance companies for pension purposes.

2. Group and life insurance premiums (credit dividends received).

3. Payments for medical and hospital services and expenses of employees when not the result of occupational injuries.

4. Payments for accident, sickness, hospital, and death benefits or insurance.

5. Payments to employees incapacitated for service or on leave of absence beyond periods normally allowed when not the result of occupational injuries or in excess of statutory awards.

6. Expenses in connection with educational and recreational activities for the benefit of employees.

Insurance:

1. Premiums payable to insurance companies for fire, storm, burglary, boiler explosion, lightning, fidelity, riot, and similar insurance.

2. Amounts credited to Account 228.1, Accumulated Provision for Property Insurance, for similar protection.

3. Special costs incurred in procuring insurance.

4. Insurance inspection service.

5. Insurance counsel, brokerage fees, and expenses.

6. Premiums payable to insurance companies for protection against claims from injuries and damages by employees or others, such as public liability, property damages, casualty, employee liability, etc., and amounts credited to Account 228.2, Accumulated Provision for Injuries and Damage, for similar protection.

7. Losses not covered by insurance or reserve accruals on account of injuries or deaths to employees or others and damages to the property of others.

8. Fees and expenses of claim investigators.

9. Payment of awards to claimants for court costs and attorneys' services.

10. Medical and hospital service and expenses for employees as the result of occupational injuries or resulting from claims of others.

11. Compensation payments under workmen's compensation laws.

12. Compensation paid while incapacitated as the result of occupational injuries. (See Account 924, Note A.)

13. Cost of safety, accident prevention, and similar educational activities.

Materials and Expenses:

1. Insect control materials.

2. Lubricants, packing, and other supplies used in the operation of hydraulic equipment.

3. Transportation expense.

538 Electric Expenses

This account shall include the cost of labor, employee pensions and benefits, social security and other payroll taxes, injuries and damages, property insurance, property taxes, materials used, and expenses incurred in operating prime movers, generators, and their auxiliary apparatus, switchgear, and other electric equipment, to the point where electricity leaves for conversion for transmission or distribution.

Items

Labor:

1. Supervising electric production.

2. Operating prime movers, generators, and auxiliary equipment.

3. Operating generator cooling system.

4. Operating lubrication and oil control systems, including oil purification.

5. Operating switchboards, switchgear, and electric control and protection equipment.

6. Keeping plant log and records and preparing reports on plant operations.

7. Testing, checking and adjusting meters, gauges, and other instruments, relays, controls, and other equipment in the plant.

8. Cleaning plant equipment when not incidental to maintenance work.

9. Repacking glands.

Taxes:

1. Federal and state unemployment.

2. F.I.C.A.

3. Property.

Employee Pensions and Benefits: The portion of employee pensions and benefits specifically identifiable with employees' labor costs charged herein or, in the absence of specific employee identification, the portion of employee pensions and benefits, allocated on the more equitable basis of either direct labor dollars or direct labor hours, applicable to the labor items detailed above, including:

1. Accruals for or payments to pension funds or to insurance companies for pension purposes.

2. Group and life insurance premiums (credit dividends received).

3. Payments for medical and hospital services and expenses of employees when not the result of occupational injuries.

4. Payments for accident, sickness, hospital, and death benefits or insurance.

5. Payments to employees incapacitated for service or on leave of absence beyond periods normally allowed when not the result of occupational injuries or in excess of statutory awards.

6. Expenses in connection with educational and recreational activities for the benefit of employees.

Insurance:

1. Premiums payable to insurance companies for fire, storm, burglary, boiler explosion, lightning, fidelity, riot, and similar insurance.

2. Amounts credited to Account 228.1, Accumulated Provision for Property Insurance, for similar protection.

3. Special costs incurred in procuring insurance.

4. Insurance inspection service.

5. Insurance counsel, brokerage fees, and expenses.

6. Premiums payable to insurance companies for protection against claims from injuries and damages by employees or others, such as public liability, property damages, casualty, employee liability, etc., and amounts credited to Account 228.2, Accumulated Provision for Injuries and Damage, for similar protection.

7. Losses not covered by insurance or reserve accruals on account of injuries or deaths to employees or others and damages to the property of others.

8. Fees and expenses of claim investigators.

9. Payment of awards to claimants for court costs and attorneys' services.

10. Medical and hospital service and expenses for employees as the result of occupational injuries or resulting from claims of others.

11. Compensation payments under workmen's compensation laws.

12. Compensation paid while incapacitated as the result of occupational injuries. (See Account 924, Note A.)

13. Cost of safety, accident prevention, and similar educational activities.

Materials and Expenses:

1. Lubricants and control system oils.

2. Motor and generator brushes.

539 Miscellaneous Hydraulic Power Generation Expenses

This account shall include the cost of labor, employee pensions and benefits, social security and other payroll taxes, injuries and damages, materials used, and expenses incurred which are not specifically provided for or are not readily assignable to other hydraulic generation operation expense accounts.

Items

Labor:

1. General clerical and stenographic work.

2. Guarding and patrolling plant and yard.

3. Building service.

4. Care of grounds including snow removal, and grass cutting.

5. Snow removal from roads and bridges.

6. Miscellaneous labor.

Taxes:

1. Federal and state unemployment.

2. F.I.C.A.

Employee Pensions and Benefits: The portion of employee pensions and benefits specifically identifiable with employees' labor costs charged herein or, in the absence of specific employee identification, the portion of employee pensions and benefits, allocated on the more equitable basis of either direct labor dollars or direct labor hours, applicable to the labor items detailed above, including:

1. Accruals for or payments to pension funds or to insurance companies for pension purposes.

2. Group and life insurance premiums (credit dividends received).

3. Payments for medical and hospital services and expenses of employees when not the result of occupational injuries.

4. Payments for accident, sickness, hospital, and death benefits or insurance.

5. Payments to employees incapacitated for service or on leave of absence beyond periods normally allowed when not the result of occupational injuries or in excess of statutory awards.

6. Expenses in connection with educational and recreational activities for the benefit of employees.

Insurance:

1. Premiums payable to insurance companies for protection against claims from injuries and damages by employees or others, such as public liability, property damages, casualty, employee liability, etc., and amounts credited to Account 228.2, Accumulated Provision for Injuries and Damage, for similar protection.

2. Losses not covered by insurance or reserve accruals on account of injuries or deaths to employees or others and damages to the property of others.

3. Fees and expenses of claim investigators.

4. Payment of awards to claimants for court costs and attorneys' services.

5. Medical and hospital service and expenses for employees as the result of occupational injuries or resulting from claims of others.

6. Compensation payments under workmen's compensation laws.

7. Compensation paid while incapacitated as the result of occupational injuries. (See Account 924, Note A.)

8. Cost of safety, accident prevention, and similar educational activities.

Materials and Expenses:

1. General operating supplies, such as tools, gaskets, packing, waste, hose, indicating lamps, record and report forms.

2. First-aid supplies and safety equipment.

3. Employees' service facilities expenses.

4. Building service supplies.
5. Communication service.
6. Office supplies, printing and stationery.
7. Transportation expenses.
8. Fuel.
9. Meals, traveling, and incidental expenses.
10. Research, development, and demonstration expenses.

540 Rents

This account shall include all rents of property of others used, occupied, or operated in connection with hydraulic power generation, including amounts payable to the United States for the occupancy of public lands and reservations for reservoirs, dams, flumes, forebays, penstocks, and power houses but not including transmission right-of-way. (See §1767.17 (c).)

(Maintenance)

541 Maintenance Supervision and Engineering

This account shall include the cost of labor, employee pensions and benefits, social security and other payroll taxes, injuries and damages, and expenses incurred in the general supervision and direction of the maintenance of hydraulic power generating stations. Direct field supervision of specific jobs shall be charged to the appropriate maintenance account. (See §1767.17(a).)

542 Maintenance of Structures

This account shall include the cost of labor, employee pensions and benefits, social security and other payroll taxes, injuries and damages, materials used, and expenses incurred in maintenance of hydraulic structures, the book cost of which is includible in Account 331, Structures and Improvements. (See §1767.17 (b).) However, the cost of labor, materials used, and expenses incurred in the maintenance of fish and wildlife and recreation facilities, the book cost of which is includible in Account 331, Structures and Improvements, shall be charged to Account 545, Maintenance of Miscellaneous Hydraulic Plant.

543 Maintenance of Reservoirs, Dams, and Waterways

This account shall include the cost of labor, employee pensions and benefits, social security and other payroll taxes, injuries and damages, materials used, and expenses incurred in maintenance of plant includible in Account 332, Reservoirs, Dams, and Waterways. (See §1767.17(b).) However, the cost of labor, materials used, and expenses incurred in the maintenance of fish and wildlife and recreation facilities, the book cost of which is includible in Account 332, Reservoirs, Dams, and Waterways, shall be charged to Account 545, Maintenance of Miscellaneous Hydraulic Plant.

544 Maintenance of Electric Plant

This account shall include the cost of labor, employee pensions and benefits, social security and other payroll taxes, injuries and damages, materials used, and expenses incurred in maintenance of plant includible in Account 333, Water Wheels, Turbines and Generators, and Account 334, Accessory Electric Equipment, (See §1767.17(b).)

545 Maintenance of Miscellaneous Hydraulic Plant

This account shall include the cost of labor, employee pensions and benefits, social security and other payroll taxes, injuries and damages, materials used, and expenses incurred in maintenance of plant, the book cost of which is includible in Account 335, Miscellaneous Power Plant Equipment, and Account 336, Roads Railroads and Bridges. (See §1767.17(b).) It shall also include the cost of labor, materials used, and other expenses incurred in the maintenance of (1) fish and wildlife, and (2) recreation facilities. Separate subaccounts shall be maintained for each of the above.

Other Power Generation

(Operation)

546 Operation Supervision and Engineering

This account shall include the cost of labor, employee pensions and benefits, social security and other payroll taxes, injuries and damages, and expenses incurred in the general supervision and direction of the operation of other power generating stations. Direct supervision of specific activities, such as fuel handling and engine and generator operation shall be charged to the appropriate account. (See §1767.17(a).)

547 Fuel

This account shall include the cost delivered at the station (See Account 151, Fuel Stock) of all fuel, such as gas, oil, kerosene, and gasoline used in other power generation.

548 Generation Expenses

This account shall include the cost of labor, employee pensions and benefits, social security and other payroll taxes, injuries and damages, property insurance, property taxes, materials used, and expenses incurred in operating prime movers, generators, and electric equipment in other power generating stations, to the point where electricity leaves for conversion for transmission or distribution.

Items

Labor:

1. Supervising other power generation operation.

2. Operating prime movers, generators, and auxiliary apparatus and switching and other electric equipment.

3. Keeping plant log and records and preparing reports on plant operations.

4. Testing, checking, cleaning, oiling, and adjusting equipment.

Taxes:

1. Federal and state unemployment.
2. F.I.C.A.
3. Property.

Employee Pensions and Benefits: The portion of employee pensions and benefits specifically identifiable with employees' labor costs charged herein or, in the absence of specific employee identification, the portion of employee pensions and benefits, allocated on the more equitable basis of either direct labor dollars or direct labor hours, applicable to the labor items detailed above, including:

1. Accruals for or payments to pension funds or to insurance companies for pension purposes.

2. Group and life insurance premiums (credit dividends received).

3. Payments for medical and hospital services and expenses of employees when not the result of occupational injuries.

4. Payments for accident, sickness, hospital, and death benefits or insurance.

5. Payments to employees incapacitated for service or on leave of absence beyond periods normally allowed when not the result of occupational injuries or in excess of statutory awards.

6. Expenses in connection with educational and recreational activities for the benefit of employees.

Insurance:

1. Premiums payable to insurance companies for fire, storm, burglary, boiler explosion, lightning, fidelity, riot, and similar insurance.

2. Amounts credited to Account 228.1, Accumulated Provision for Property Insurance, for similar protection.

3. Special costs incurred in procuring insurance.

4. Insurance inspection service.

5. Insurance counsel, brokerage fees, and expenses.

6. Premiums payable to insurance companies for protection against claims from injuries and damages by employees or others, such as public liability, property damages, casualty, employee liability, etc., and amounts credited to Account 228.2, Accumulated Provision for Injuries and Damage, for similar protection.

7. Losses not covered by insurance or reserve accruals on account of injuries or deaths to employees or others and damages to the property of others.

8. Fees and expenses of claim investigators.

9. Payment of awards to claimants for court costs and attorneys' services.

10. Medical and hospital service and expenses for employees as the result of occupational injuries or resulting from claims of others.

11. Compensation payments under workmen's compensation laws.

12. Compensation paid while incapacitated as the result of occupational injuries. (See Account 924, Note A.)

13. Cost of safety, accident prevention, and similar educational activities.

Materials and Expenses:

1. Dynamo, motor, and generator brushes.
2. Lubricants and control system oils.
3. Water for cooling engines and generators.

549   Miscellaneous Other Power Generation Expenses

This account shall include the cost of labor, employee pensions and benefits, social security and other payroll taxes, injuries and damages, materials used, and expenses incurred in the operation of other power generating stations which are not specifically provided for or are not readily assignable to other generation expense accounts.

Items

Labor:

1. General clerical and stenographic work.
2. Guarding and patrolling plant and yard.
3. Building service.
4. Care of grounds, including snow removal, and grass cutting.
5. Miscellaneous labor.

Taxes:

1. Federal and state unemployment.
2. F.I.C.A.

Employee Pensions and Benefits: The portion of employee pensions and benefits specifically identifiable with employees' labor costs charged herein or, in the absence of specific employee identification, the portion of employee pensions and benefits, allocated on the more equitable basis of either direct labor dollars or direct labor hours, applicable to the labor items detailed above, including:

1. Accruals for or payments to pension funds or to insurance companies for pension purposes.

2. Group and life insurance premiums (credit dividends received).

3. Payments for medical and hospital services and expenses of employees when not the result of occupational injuries.

4. Payments for accident, sickness, hospital, and death benefits or insurance.

5. Payments to employees incapacitated for service or on leave of absence beyond periods normally allowed when not the result

of occupational injuries or in excess of statutory awards.

6. Expenses in connection with educational and recreational activities for the benefit of employees.

Insurance:

1. Premiums payable to insurance companies for protection against claims from injuries and damages by employees or others, such as public liability, property damages, casualty, employee liability, etc., and amounts credited to Account 228.2, Accumulated Provision for Injuries and Damage, for similar protection.

2. Losses not covered by insurance or reserve accruals on account of injuries or deaths to employees or others and damages to the property of others.

3. Fees and expenses of claim investigators.

4. Payment of awards to claimants for court costs and attorneys' services.

5. Medical and hospital service and expenses for employees as the result of occupational injuries or resulting from claims of others.

6. Compensation payments under workmen's compensation laws.

7. Compensation paid while incapacitated as the result of occupational injuries. (See Account 924, Note A.)

8. Cost of safety, accident prevention, and similar educational activities.

Materials and Expenses:

1. Building service supplies.

2. First-aid supplies and safety equipment.

3. Communication service.

4. Employees' service facilities expenses.

5. Office supplies, printing and stationery.

6. Transportation expense.

7. Meals, traveling, and incidental expenses.

8. Fuel for heating.

9. Water for fire protection or general use.

10. Miscellaneous supplies, such as hand tools, drills, saw blades, and files.

11. Research, development, and demonstration expenses.

550 Rents

This account shall include all rents of property of others used, occupied, or operated in connection with other power generation. (See §1767.17 (c).)

(Maintenance)

551 Maintenance Supervision and Engineering

This account shall include the cost of labor, employee pensions and benefits, social security and other payroll taxes, injuries and damages, and expenses incurred in the general supervision and direction of the maintenance of other power generating stations. Direct field supervision of specific jobs shall be charged to the appropriate maintenance account. (See §1767.17(a).)

552 Maintenance of Structures

This account shall include the cost of labor, employee pensions and benefits, social security and other payroll taxes, injuries and damages, materials used, and expenses incurred in maintenance of facilities used and expenses incurred in maintenance of facilities used in other power generation, the book cost of which is includible in Account 341, Structures and Improvements, and Account 342, Fuel Holders, Producers and Accessories. (See §1767.17(b).)

553 Maintenance of Generating and Electric Equipment

This account shall include the cost of labor, employee pensions and benefits, social security and other payroll taxes, injuries and damages, materials used, and expenses incurred in maintenance of plant, the book cost of which is includible in Account 343, Prime Movers; Account 344, Generators; and Account 345, Accessory Electric Equipment. (See §1767.17(b).)

554 Maintenance of Miscellaneous Other Power Generation Plant

This account shall include the cost of labor, employee pensions and benefits, social security and other payroll taxes, injuries and damages, materials used, and expenses incurred in maintenance of other power generation plant, the book cost of which is includible in Account 346, Miscellaneous Power Plant Equipment. (See §1767.17(b).)

OTHER POWER SUPPLY EXPENSES

555 Purchased Power

A. This account shall include the cost at point of receipt by the utility of electricity purchased for resale. It shall also include, net settlements for exchange of electricity or power, such as economy energy, off-peak energy for on-peak energy, and spinning reserve capacity. In addition, the account shall include the net settlements for transactions under pooling or interconnection agreements wherein there is a balancing of debits and credits for energy, or capacity. Distinct purchases and sales shall not be recorded as exchanges and net amounts only recorded merely because debit and credit amounts are combined in the voucher settlement.

B. The records supporting this account shall show, by months, the demands and demand charges, kilowatt-hours and prices thereof under each purchase contract and the charges and credits under each exchange or power pooling contract.

NOTE: The records supporting this account shall provide information pertaining to the purchase of power from renewable energy sources.

556 System Control and Load Dispatching

This account shall include the cost of labor, employee pensions and benefits, social security and other payroll taxes, injuries and damages, property insurance, property taxes, and expenses incurred in load dispatching activities for system control. Utilities having an interconnected electric system or operating under a central authority which controls the production and dispatching of electricity may apportion these costs to this account and transmission expense Account 561.1 through 561.4, and Account 581, Load Dispatching—Distribution.

Items

Labor:

1. Allocating loads to plants and interconnections with others.
2. Directing switching.
3. Arranging and controlling clearances for construction, maintenance, test, and emergency purposes.
4. Controlling system voltages.
5. Recording loadings, and water conditions.
6. Preparing operating reports and data for billing and budget purposes.
7. Obtaining reports on the weather and special events.

Taxes:

1. Federal and state unemployment.
2. F.I.C.A.
3. Property.

Employee Pensions and Benefits: The portion of employee pensions and benefits specifically identifiable with employees' labor costs charged herein or, in the absence of specific employee identification, the portion of employee pensions and benefits, allocated on the more equitable basis of either direct labor dollars or direct labor hours, applicable to the labor items detailed above, including:

1. Accruals for or payments to pension funds or to insurance companies for pension purposes.
2. Group and life insurance premiums (credit dividends received).
3. Payments for medical and hospital services and expenses of employees when not the result of occupational injuries.
4. Payments for accident, sickness, hospital, and death benefits or insurance.
5. Payments to employees incapacitated for service or on leave of absence beyond periods normally allowed when not the result of occupational injuries or in excess of statutory awards.
6. Expenses in connection with educational and recreational activities for the benefit of employees.

Insurance:

1. Premiums payable to insurance companies for fire, storm, burglary, boiler explosion, lightning, fidelity, riot, and similar insurance.
2. Amounts credited to Account 228.1, Accumulated Provision for Property Insurance, for similar protection.
3. Special costs incurred in procuring insurance.
4. Insurance inspection service.
5. Insurance counsel, brokerage fees, and expenses.
6. Premiums payable to insurance companies for protection against claims from injuries and damages by employees or others, such as public liability, property damages, casualty, employee liability, etc., and amounts credited to Account 228.2, Accumulated Provision for Injuries and Damage, for similar protection.
7. Losses not covered by insurance or reserve accruals on account of injuries or deaths to employees or others and damages to the property of others.
8. Fees and expenses of claim investigators.
9. Payment of awards to claimants for court costs and attorneys' services.
10. Medical and hospital service and expenses for employees as the result of occupational injuries or resulting from claims of others.
11. Compensation payments under workmen's compensation laws.
12. Compensation paid while incapacitated as the result of occupational injuries. (See Account 924, Note A.)
13. Cost of safety, accident prevention, and similar educational activities.

Expenses:

1. Communication service provided for system control purposes.
2. System record and report forms.
3. Meals, traveling, and incidental expenses.
4. Obtaining weather and special events reports.

557 Other Expenses

A. This account shall be charged with any production expenses including expenses incurred directly in connection with the purchase of electricity, which are not specifically provided for in other production expense accounts. Charges to this account shall be supported so that a description of each type of charge will be readily available.

B. Recoveries from insurance companies, under use and occupancy provisions of policies, of amounts in reimbursement of excessive or added productions costs for which the insurance company is liable under the terms of the policy shall be credited to this account.

TRANSMISSION EXPENSES

(Operation)

560   Operation Supervision and Engineering

This account shall include the cost of labor, employee pensions and benefits, social security and other payroll taxes, injuries and damages, and expenses incurred in the general supervision and direction of the operation of the transmission system as a whole. Direct supervision of specific activities, such as station operation and line operation shall be charged to the appropriate account. (See § 1767.17(a).)

561.1   Load Dispatch—Reliability

This account shall include the cost of labor, employee pensions and benefits, social security and other payroll taxes, injuries and damages, property insurance, property taxes, materials used, and expenses incurred by a regional transmission service provider or other transmission provider to manage the reliability coordination function as specified by the North American Electric Reliability Council (NERC) and individual reliability organizations. These activities shall include performing current and next day reliability analysis. This account shall include the costs incurred to calculate load forecasts, and performing contingency analysis.

561.2   Load Dispatch—Monitor and Operate Transmission System

This account shall include the cost of labor, employee pensions and benefits, social security and other payroll taxes, injuries and damages, property insurance, property taxes, materials used, and expenses incurred by a regional transmission service provider or other transmission provider to monitor, assess and operate the power system and individual transmission facilities in real-time to maintain safe and reliable operation of the transmission system. This account shall also include the expense incurred to manage transmission facilities to maintain system reliability and to monitor real-time flows and direct actions according to regional plans and tariffs if necessary.

Items

1. Receive and analyze outage requests
2. Reschedule outage plans
3. Monitor solution quality field data values, providing model updates to NERC and coordinating network model changes across all systems
4. Conduct operating training related to NERC Certification
5. Monitor generation resources and communicate expected dispatch actions
6. Ensure ancillary service requirements are met
7. Directing switching

8. Controlling system voltages
9. Obtaining reports on the weather and special events
10. Preparing operating reports and data for billing and budget purposes

561.3   Load Dispatch—Transmission Service and Scheduling

This account shall include the cost of labor, employee pensions and benefits, social security and other payroll taxes, injuries and damages, property insurance, property taxes, materials used, and expenses incurred by a regional transmission service provider or other transmission provider to process hourly, daily, weekly and monthly transmission service requests using an automated system such as an Open Access Same-Time Information System (OASIS). It shall include the expenses incurred to operate the automated transmission service request system and to monitor the status of all scheduled energy transactions.

561.4   Scheduling, System Control and Dispatching Services

This account shall include the costs billed to the transmission owner, load serving entity or generator for scheduling, system control and dispatching service. Include in this account service billings for system control to maintain the reliability of the transmission area in accordance with reliability standards, maintaining defined voltage profiles, and monitoring operations of the transmission facilities.

561.5   Reliability, Planning and Standards Development

This account shall include the cost of labor, employee pensions and benefits, social security and other payroll taxes, injuries and damages, property insurance, property taxes, materials used, and expenses incurred for the system planning of the interconnected bulk electric transmission system within a planning authority area.

Items

1. Developing and maintaining transmission system models to evaluate transmission system performance.
2. Maintaining and applying methodologies and tools for the analysis and simulation of the transmission systems for the assessment and development of transmission expansion plans.
3. Assessing, developing and documenting transmission expansion plans.
4. Maintaining transmission system models (steady-state, dynamics, and short circuit).
5. Collecting transmission information and transmission facility characteristics and ratings.

6. Notifying participants of any planned transmission changes that may impact their facilities.

7. Developing and reporting on transmission expansion plans for assessment and compliance with reliability standards.

8. Developing reliability standards for the planning and operation of the interconnected bulk electric transmission systems that serve the United States, Canada and Mexico.

9. Developing criteria and certification procedures for reliability authorities, transmission operators and others.

10. Outside services employed.

NOTE: The cost of supervision, customer records and collection expenses, administrative and general salaries, regulatory commission expenses, general advertising, and rents shall be charged to the customer accounts, service, administrative and general expense accounts contained in the Uniform System of Accounts.

561.6   Transmission Service Studies

This account shall include the cost of labor, employee pensions and benefits, social security and other payroll taxes, injuries and damages, property insurance, property taxes, materials used, and expenses incurred to conduct generation interconnection studies for proposed interconnections with the transmission system. Detailed records shall be maintained for each study undertaken and all reimbursements received for conducting such a study.

561.7   Generation Interconnection Studies

This account shall include the cost of labor, employee pensions and benefits, social security and other payroll taxes, injuries and damages, property insurance, property taxes, materials used, and expenses incurred to conduct generation interconnection studies for proposed interconnections with the transmission system. Detailed records shall be maintained for each study undertaken and all reimbursements received for conducting such a study.

561.8   Reliability Planning and Standards Development Services

This account shall include the costs billed to the transmission owner, load serving entity, or generator for system planning of the interconnected bulk electric transmission service provider for system reliability and resource panning to develop long-term strategies to meet customer demand and energy requirements. This account shall also include fees and expenses for outside services incurred by the regional transmission service provider and billed to the load serving entity, transmission owner or generator.

562   Station Expenses

This account shall include the cost of labor, employee pensions and benefits, social security and other payroll taxes, injuries and damages, property insurance, property taxes, materials used, and expenses incurred in operating transmission substations and switching stations. If transmission station equipment is located in or adjacent to a generating station, the expenses applicable to transmission station operations shall nevertheless be charged to this account.

Items

Labor:

1. Supervising station operation.

2. Adjusting station equipment where such adjustment primarily affects performance, such as regulating the flow of cooling water, adjusting current in fields of a machine or changing voltage of regulators, changing station transformer taps.

3. Inspecting, testing, and calibrating station equipment for the purpose of checking its performance.

4. Keeping station log and records and preparing records on station operation.

5. Operating switching and other station equipment.

6. Standing watch, guarding, and patrolling station and station yard.

7. Sweeping, mopping, and tidying station.

8. Care of grounds, including snow removal, and grass cutting.

Taxes:

1. Federal and state unemployment.

2. F.I.C.A.

3. Property.

Employee Pensions and Benefits: The portion of employee pensions and benefits specifically identifiable with employees' labor costs charged herein or, in the absence of specific employee identification, the portion of employee pensions and benefits, allocated on the more equitable basis of either direct labor dollars or direct labor hours, applicable to the labor items detailed above, including:

1. Accruals for or payments to pension funds or to insurance companies for pension purposes.

2. Group and life insurance premiums (credit dividends received).

3. Payments for medical and hospital services and expenses of employees when not the result of occupational injuries.

4. Payments for accident, sickness, hospital, and death benefits or insurance.

5. Payments to employees incapacitated for service or on leave of absence beyond periods normally allowed when not the result of occupational injuries or in excess of statutory awards.

6. Expenses in connection with educational and recreational activities for the benefit of employees.

Insurance:

1. Premiums payable to insurance companies for fire, storm, burglary, boiler explosion, lightning, fidelity, riot, and similar insurance.

2. Amounts credited to Account 228.1, Accumulated Provision for Property Insurance, for similar protection.

3. Special costs incurred in procuring insurance.

4. Insurance inspection service.

5. Insurance counsel, brokerage fees, and expenses.

6. Premiums payable to insurance companies for protection against claims from injuries and damages by employees or others, such as public liability, property damages, casualty, employee liability, etc., and amounts credited to Account 228.2, Accumulated Provision for Injuries and Damage, for similar protection.

7. Losses not covered by insurance or reserve accruals on account of injuries or deaths to employees or others and damages to the property of others.

8. Fees and expenses of claim investigators.

9. Payment of awards to claimants for court costs and attorneys' services.

10. Medical and hospital service and expenses for employees as the result of occupational injuries or resulting from claims of others.

11. Compensation payments under workmen's compensation laws.

12. Compensation paid while incapacitated as the result of occupational injuries. (See Account 924, Note A.)

13. Cost of safety, accident prevention, and similar educational activities.

Materials and Expenses:

1. Building service expenses.

2. Operating supplies, such as lubricants, commutator brushes, water, and rubber goods.

3. Station meter and instrument supplies, such as ink and charts.

4. Station record and report forms.

5. Tool expense.

6. Transportation expenses.

7. Meals, traveling, and incidental expenses.

563   Overhead Line Expenses

564   Underground Line Expenses

A. These accounts shall include the cost of labor, employee pensions and benefits, social security and other payroll taxes, injuries and damages, property insurance, property taxes, materials used, and expenses incurred in the operation of transmission lines.

B. If the expenses are not substantial for both overhead and underground lines, these accounts may be combined.

Items

Labor:

1. Supervising line operation.

2. Inspecting and testing lightning arresters, circuit breakers, switches, and grounds.

3. Load tests of circuits.

4. Routine line patrolling.

5. Routine voltage surveys made to determine the condition or efficiency of transmission system.

6. Transferring loads, switching and reconnecting circuits and equipment for operating purposes. (Switching for construction or maintenance purposes is not includible in this account.)

7. Routine inspection and cleaning of manholes, conduit, network, and transformer vaults.

8. Electrolysis surveys.

9. Inspecting and adjusting line-testing equipment, such as voltmeters, ammeters, and wattmeters.

10. Regulation and addition of oil or gas in high-voltage cable systems.

Taxes:

1. Federal and state unemployment.

2. F.I.C.A.

3. Property.

Employee Pensions and Benefits: The portion of employee pensions and benefits specifically identifiable with employees' labor costs charged herein or, in the absence of specific employee identification, the portion of employee pensions and benefits, allocated on the more equitable basis of either direct labor dollars or direct labor hours, applicable to the labor items detailed above, including:

1. Accruals for or payments to pension funds or to insurance companies for pension purposes.

2. Group and life insurance premiums (credit dividends received).

3. Payments for medical and hospital services and expenses of employees when not the result of occupational injuries.

4. Payments for accident, sickness, hospital, and death benefits or insurance.

5. Payments to employees incapacitated for service or on leave of absence beyond periods normally allowed when not the result of occupational injuries or in excess of statutory awards.

6. Expenses in connection with educational and recreational activities for the benefit of employees.

Insurance:

1. Premiums payable to insurance companies for fire, storm, burglary, boiler explosion, lightning, fidelity, riot, and similar insurance.

2. Amounts credited to Account 228.1, Accumulated Provision for Property Insurance, for similar protection.

3. Special costs incurred in procuring insurance.

4. Insurance inspection service.

5. Insurance counsel, brokerage fees, and expenses.

6. Premiums payable to insurance companies for protection against claims from injuries and damages by employees or others, such as public liability, property damages, casualty, employee liability, etc., and amounts credited to Account 228.2, Accumulated Provision for Injuries and Damage, for similar protection.

7. Losses not covered by insurance or reserve accruals on account of injuries or deaths to employees or others and damages to the property of others.

8. Fees and expenses of claim investigators.

9. Payment of awards to claimants for court costs and attorneys' services.

10. Medical and hospital service and expenses for employees as the result of occupational injuries or resulting from claims of others.

11. Compensation payments under workmen's compensation laws.

12. Compensation paid while incapacitated as the result of occupational injuries. (See Account 924, Note A.)

13. Cost of safety, accident prevention, and similar educational activities.

Materials and Expenses:

1. Transportation expenses.

2. Meals, traveling, and incidental expenses.

3. Tool expenses.

4. Operating supplies, such as instrument charts, and rubber goods.

565 Transmission of Electricity by Others

This account shall include amounts payable to others for the transmission of the utility's electricity over transmission facilities owned by others.

566 Miscellaneous Transmission Expenses

This account shall include the cost of labor, employee pensions and benefits, social security and other payroll taxes, injuries and damage, materials used, and expenses incurred in transmission map and record work, transmission office expenses, and other transmission expenses not provided for elsewhere.

Items

Labor:

1. General records of physical characteristics of lines and stations, such as capacities.

2. Ground resistance records.

3. Janitor work at transmission office buildings, including care of grounds, snow removal, and grass cutting.

4. Joint pole maps and records.

5. Line load and voltage records.

6. Preparing maps and prints.

7. General clerical and stenographic work.

8. Miscellaneous labor.

Taxes:

1. Federal and state unemployment.

2. F.I.C.A.

Employee Pensions and Benefits: The portion of employee pensions and benefits specifically identifiable with employees' labor costs charged herein or, in the absence of specific employee identification, the portion of employee pensions and benefits, allocated on the more equitable basis of either direct labor dollars or direct labor hours, applicable to the labor items detailed above, including:

1. Accruals for or payments to pension funds or to insurance companies for pension purposes.

2. Group and life insurance premiums (credit dividends received).

3. Payments for medical and hospital services and expenses of employees when not the result of occupational injuries.

4. Payments for accident, sickness, hospital, and death benefits or insurance.

5. Payments to employees incapacitated for service or on leave of absence beyond periods normally allowed when not the result of occupational injuries or in excess of statutory awards.

6. Expenses in connection with educational and recreational activities for the benefit of employees.

Insurance:

1. Premiums payable to insurance companies for protection against claims from injuries and damages by employees or others, such as public liability, property damages, casualty, employee liability, etc., and amounts credited to Account 228.2, Accumulated Provision for Injuries and Damage, for similar protection.

2. Losses not covered by insurance or reserve accruals on account of injuries or deaths to employees or others and damages to the property of others.

3. Fees and expenses of claim investigators.

4. Payment of awards to claimants for court costs and attorneys' services.

5. Medical and hospital service and expenses for employees as the result of occupational injuries or resulting from claims of others.

6. Compensation payments under workmen's compensation laws.

7. Compensation paid while incapacitated as the result of occupational injuries. (See Account 924, Note A.)

8. Cost of safety, accident prevention, and similar educational activities.

Materials and Expenses:

1. Communication service.
2. Building service supplies.
3. Map and record supplies.
4. Transmission office supplies and expenses, printing and stationery.
5. First-aid supplies.
6. Research, development, and demonstration expenses.

567 Rents

This account shall include rents of property of others used, occupied, or operated in connection with the transmission system, including payments to the United States and others for use of public or private lands and reservations for transmission line rights-of-way. (See §1767.17 (c).)

(Maintenance)

568 Maintenance Supervision and Engineering

This account shall include the cost of labor, employee pensions and benefits, social security and other payroll taxes, injuries and damages, and expenses incurred in the general supervision and direction of maintenance of the transmission system. Direct field supervision of specific jobs shall be charged to the appropriate maintenance account. (See §1767.17(a).)

569 Maintenance of Structures

This account shall include the cost of labor, employee pensions and benefits, social security and other payroll taxes, injuries and damages, materials used, and expenses incurred in the maintenance of structures, the book cost of which is includible in Account 352, Structures and Improvements. (See §1767.17(b).)

569.1 Maintenance of Computer Hardware

This account shall include the cost of labor, employee pensions and benefits, social security and other payroll taxes, injuries and damages, materials used and expenses incurred in the maintenance of computer hardware serving the transmission function.

569.2 Maintenance of Computer Software

This account shall include the cost of labor, employee pensions and benefits, social security and other payroll taxes, injuries and damages, materials used and expenses incurred for annual computer software license renewals, annual software update services and the cost of ongoing support for software products serving the transmission function.

Items

1. Telephone Support
2. Onsite support
3. Software updates and minor revisions

569.3 Maintenance of Communication Equipment

This account shall include the cost of labor, employee pensions and benefits, social security and other payroll taxes, injuries and damages, materials used and expenses incurred in the maintenance of communication equipment serving the transmission function.

569.4 Maintenance of Miscellaneous Regional Transmission Plant

This account shall include the cost of labor, employee pensions and benefits, social security and other payroll taxes, injuries and damages, materials used and expenses incurred in the maintenance of miscellaneous regional transmission plant serving the transmission function.

570 Maintenance of Station Equipment

This account shall include the cost of labor, employee pensions and benefits, social security and other payroll taxes, injuries and damages, materials used, and expenses incurred in maintenance of station equipment, the book cost of which is includible in Account 353, Station Equipment. (See §1767.17(b).)

571 Maintenance of Overhead Lines

This account shall include the cost of labor, employee pensions and benefits, social security and other payroll taxes, injuries and damages, materials used, and expenses incurred in maintenance of transmission plant, the book cost of which is includible in Accounts 354, Towers and Fixtures; 355, Poles and Fixtures; 356, Overhead Conductors and Devices; and 359, Roads and Trails. (See §1767.17(b).)

Items

1. Work of the following character on poles, towers, and fixtures:
   a. Installing or removing additional clamps or strain insulators on guys in place.
   b. Moving line or guy pole in relocation of the same pole or section of line.
   c. Painting poles, towers, crossarms, or pole extensions.
   d. Readjusting and changing position of guys or braces.
   e. Realigning and straightening poles, crossarms braces, and other pole fixtures.
   f. Reconditioning reclaimed pole fixtures.
   g. Relocating crossarms, racks, brackets, and other fixtures on poles.
   h. Repairing or realigning pins, racks, or brackets.
   i. Repairing pole supported platform.
   j. Repairs by others to jointly owned poles.
   k. Shaving, cutting rot, or testing poles or crossarms in use or salvaged for reuse.
   l. Stubbing poles already in service.

m. Supporting fixtures and conductors and transferring them to new poles during pole replacements.

n. Maintenance of pole signs, stencils, and tags.

2. Work of the following character on overhead conductors and devices:

a. Overhauling and repairing line cutouts, line switches, and line breakers.

b. Cleaning insulators and bushings.

c. Refusing cutouts.

d. Repairing line oil circuit breakers and associated relays and control wiring.

e. Repairing grounds.

f. Resagging, retyping, or rearranging position or spacing of conductors.

g. Standing by phones, going to calls, cutting faulty lines clear, or similar activities at times of emergencies.

h. Sampling, testing, changing, purifying, and replenishing insulating oil.

i. Repairing line testing equipment.

j. Transferring loads, switching and reconnecting circuits and equipment for maintenance purposes.

k. Trimming trees and clearing brush.

l. Chemical treatment of right of way areas when occurring subsequent to construction of line.

3. Work of the following character on roads and trails:

a. Repairing roadways and bridges.

b. Trimming trees and brush to maintain previous roadway clearance.

c. Snow removal from roads and trails.

d. Maintenance work on publicly owned roads and trails when done by utility at its expense.

Taxes:

1. Federal and state unemployment.

2. F.I.C.A.

Employee Pensions and Benefits: The portion of employee pensions and benefits specifically identifiable with employees' labor costs charged herein or, in the absence of specific employee identification, the portion of employee pensions and benefits, allocated on the more equitable basis of either direct labor dollars or direct labor hours, applicable to the labor items detailed above, including:

1. Accruals for or payments to pension funds or to insurance companies for pension purposes.

2. Group and life insurance premiums (credit dividends received).

3. Payments for medical and hospital services and expenses of employees when not the result of occupational injuries.

4. Payments for accident, sickness, hospital, and death benefits or insurance.

5. Payments to employees incapacitated for service or on leave of absence beyond periods normally allowed when not the result of occupational injuries or in excess of statutory awards.

6. Expenses in connection with educational and recreational activities for the benefit of employees.

Insurance:

1. Premiums payable to insurance companies for protection against claims from injuries and damages by employees or others, such as public liability, property damages, casualty, employee liability, etc., and amounts credited to Account 228.2, Accumulated Provision for Injuries and Damage, for similar protection.

2. Losses not covered by insurance or reserve accruals on account of injuries or deaths to employees or others and damages to the property of others.

3. Fees and expenses of claim investigators.

4. Payment of awards to claimants for court costs and attorneys' services.

5. Medical and hospital services and expenses for employees as the result of occupational injuries or resulting from claims of others.

6. Compensation payments under workmen's compensation laws.

7. Compensation paid while incapacitated as the result of occupational injuries. (See Account 924, Note A.)

8. Cost of safety, accident prevention, and similar educational activities.

572   Maintenance of Underground Lines

This account shall include the cost of labor, employee pensions and benefits, social security and other payroll taxes, injuries and damages, materials used, and expenses incurred in maintenance of transmission plant, the book cost of which is includible in Accounts 357, Underground Conduit, and Account 358, Underground Conductors and Devices. (See § 1767.17(b).)

Items

1. Work of the following character on underground conduit:

a. Cleaning ducts, manholes, and sewer connections.

b. Minor alterations of handholes, manholes, or vaults.

c. Refastening, repairing, or moving racks, ladders, hangers in manholes, or vaults.

d. Plugging and shelving or replugging ducts.

e. Repairs to sewers and drains, walls and floors, rings and covers.

2. Work of the following character on underground conductors and devices:

a. Repairing oil circuit breakers, switches, cutouts, and control wiring.

b. Repairing grounds.

c. Retraining and reconnecting cables in manholes, including transfer of cables from one duct to another.

d. Repairing conductors and splices.

e. Repairing or moving junction boxes and potheads.

f. Refireproofing of cables and repairing supports.

g. Repairing electrolysis preventive devices for cables.

h. Repairing cable bonding systems.

i. Sampling, testing, changing, purifying, and replenishing insulating oil.

j. Transferring loads, switching and reconnecting circuits, and equipment for maintenance purposes.

k. Repairing line testing equipment.

l. Repairs to oil or gas equipment in high-voltage cable system and replacement of oil or gas.

Taxes:

1. Federal and state unemployment.
2. F.I.C.A.

Employee Pensions and Benefits: The portion of employee pensions and benefits specifically identifiable with employees' labor costs charged herein or, in the absence of specific employee identification, the portion of employee pensions and benefits, allocated on the more equitable basis of either direct labor dollars or direct labor hours, applicable to the labor items detailed above, including:

1. Accruals for or payments to pension funds or to insurance companies for pension purposes.

2. Group and life insurance premiums (credit dividends received).

3. Payments for medical and hospital services and expenses of employees when not the result of occupational injuries.

4. Payments for accident, sickness, hospital, and death benefits or insurance.

5. Payments to employees incapacitated for service or on leave of absence beyond periods normally allowed when not the result of occupational injuries or in excess of statutory awards.

6. Expenses in connection with educational and recreational activities for the benefit of employees.

Insurance:

1. Premiums payable to insurance companies for protection against claims from injuries and damages by employees or others, such as public liability, property damages, casualty, employee liability, etc., and amounts credited to Account 228.2, Accumulated Provision for Injuries and Damage, for similar protection.

2. Losses not covered by insurance or reserve accruals on account of injuries or deaths to employees or others and damages to the property of others.

3. Fees and expenses of claim investigators.

4. Payment of awards to claimants for court costs and attorneys' services.

5. Medical and hospital service and expenses for employees as the result of occupational injuries or resulting from claims of others.

6. Compensation payments under workmen's compensation laws.

7. Compensation paid while incapacitated as the result of occupational injuries. (See Account 924, Note A.)

8. Cost of safety, accident prevention, and similar educational activities.

573 Maintenance of Miscellaneous Transmission Plant

This account shall include the cost of labor, employee pensions and benefits, social security and other payroll taxes, injuries and damages, materials used, and expenses incurred in maintenance of owned or leased plant which is assignable to transmission operations and is not provided for elsewhere. (See § 1767.17(b).)

Regional Market Expenses

(Operational)

575.1 Operation Supervision

This account shall include the cost of labor, employee pensions and benefits, social security and other payroll taxes, injuries and damages, and expenses incurred in the general supervision and direction of the regional energy markets.

575.2 Day-Ahead and Real-Time Market Administration

This account shall include the cost of labor, employee pensions and benefits, social security and other payroll taxes, injuries and damages, and expenses incurred to facilitate the Day-Ahead and Real-Time markets. This account shall also include the costs incurred to manage the real-time deployment of resources to meet generation needs and to provide capacity adequacy verification. Include in this account the costs incurred to maintain related sections of the tariff, market rules, operating procedures, and standards and coordinating with neighboring areas.

Items

1. Consultant fees and expenses
2. System record and report forms
3. Meals, traveling and incidental expenses

NOTE: The cost of supervision, customer records and collection expenses, administrative and general salaries, regulatory commission expenses, general advertising, and rents shall be charged to the customer accounts, service, administrative and general expense accounts contained in the Uniform System of Accounts.

575.3 Transmission Rights Market Administration

This account shall include the cost of labor, employee pensions and benefits, social security and other payroll taxes, injuries and damages, and expenses incurred to manage the allocation and auction of transmission rights.

575.4　Capacity Market Administration

This account shall include the cost of labor, employee pensions and benefits, social security and other payroll taxes, injuries and damages, and expenses incurred to manage the allocation of capacity rights.

575.5　Ancillary Services Market Administration

This account shall include the cost of labor, employee pensions and benefits, social security and other payroll taxes, injuries and damages, and expenses incurred to manage all other ancillary services market functions

575.6　Market Monitoring and Compliance

This account shall include the cost of labor, employee pensions and benefits, social security and other payroll taxes, injuries and damages, and expenses incurred to review market data and operational decisions for compliance with market rules. It shall also include the costs incurred to interface with external market monitors.

575.7　Market Administration, Monitoring and Compliance Services

This account shall include the cost billed to the transmission owner, load serving entity or generator for market administration, monitoring and compliance services.

575.8　Rents

This account shall include all rents of property of others used, occupied, or operated in connection with market administration and monitoring. (See Sec. 1767.17(c).) (Maintenance)

576.1　Maintenance of Structures and Improvements

This account shall include the cost of labor, employee pensions and benefits, social security and other payroll taxes, injuries and damages, and expenses incurred in the maintenance of structures used in market administration and monitoring. (See Sec. 1767.17(b).)

576.2　Maintenance of Computer Hardware

This account shall include the cost of labor, employee pensions and benefits, social security and other payroll taxes, injuries and damages, and expenses incurred in the maintenance of computer hardware used in market administration and monitoring.

576.3　Maintenance of Computer Software

This account shall include the cost of labor, employee pensions and benefits, social security and other payroll taxes, injuries and damages, and expenses incurred for annual computer software license renewals, annual software update services and the cost of ongoing support for software products used in market administration and monitoring.

Items

1. Telephone support
2. Onsite support
3. Software updates and minor revisions

576.4　Maintenance of Communication Equipment

This account shall include the cost of labor, employee pensions and benefits, social security and other payroll taxes, injuries and damages, and expenses incurred in the maintenance of communication equipment used in market administration and monitoring.

576.5　Maintenance of Miscellaneous Market Operation Plant

This account shall include the cost of labor, employee pensions and benefits, social security and other payroll taxes, injuries and damages, and expenses incurred in the maintenance of miscellaneous market operation plant used in market administration and monitoring.

DISTRIBUTION EXPENSES

(Operation)

580　Operation Supervision and Engineering

This account shall include the cost of labor, employee pensions and benefits, social security and other payroll taxes, injuries and damages, and expenses incurred in the general supervision and direction of the operation of the distribution system. Direct supervision of specific activities, such as station operation, line operation, and meter department operation shall be charged to the appropriate account. (See § 1767.17(a).)

581　Load Dispatching

This account (the keeping of which is optional with the utility) shall include the cost of labor, employee pensions and benefits, social security and other payroll taxes, injuries and damages, property insurance, property taxes, materials used, and expenses incurred in load dispatching operations pertaining to the distribution of electricity.

Items

Labor:

1. Direct switching.
2. Arranging and controlling clearances for construction, maintenance, test, and emergency purposes.
3. Controlling system voltages.
4. Preparing operating reports.
5. Obtaining reports on the weather and special events.

Taxes:

1. Federal and state unemployment.
2. F.I.C.A.
3. Property.

Employee Pensions and Benefits: The portion of employee pensions and benefits specifically identifiable with employees' labor costs charged herein or, in the absence of specific employee identification, the portion of employee pensions and benefits, allocated on the more equitable basis of either direct labor dollars or direct labor hours, applicable to the labor items detailed above, including:

1. Accruals for or payments to pension funds or to insurance companies for pension purposes.
2. Group and life insurance premiums (credit dividends received).
3. Payments for medical and hospital services and expenses of employees when not the result of occupational injuries.
4. Payments for accident, sickness, hospital, and death benefits or insurance.
5. Payments to employees incapacitated for service or on leave of absence beyond periods normally allowed when not the result of occupational injuries or in excess of statutory awards.
6. Expenses in connection with educational and recreational activities for the benefit of employees.

Insurance:

1. Premiums payable to insurance companies for fire, storm, burglary, boiler explosion, lightning, fidelity, riot, and similar insurance.
2. Amounts credited to Account 228.1, Accumulated Provision for Property Insurance, for similar protection.
3. Special costs incurred in procuring insurance.
4. Insurance inspection service.
5. Insurance counsel, brokerage fees, and expenses.
6. Premiums payable to insurance companies for protection against claims from injuries and damages by employees or others, such as public liability, property damages, casualty, employee liability, etc., and amounts credited to Account 228.2, Accumulated Provision for Injuries and Damage, for similar protection.
7. Losses not covered by insurance or reserve accruals on account of injuries or deaths to employees or others or damages to the property of others.
8. Fees and expenses of claim investigators.
9. Payment of awards to claimants for court costs and attorneys' services.
10. Medical and hospital service and expenses for employees as the result of occupational injuries or resulting from claims of others.
11. Compensation payments under workmen's compensation laws.

12. Compensation paid while incapacitated as the result of occupational injuries. (See Account 924, Note A.)
13. Cost of safety, accident prevention, and similar educational activities.

Expenses:

1. Communication service provided for system control purposes.
2. System record and report forms.
3. Meals, traveling, and incidental expenses.

582 Station Expenses

This account shall include the cost of labor, employee pensions and benefits, social security and other payroll taxes, injuries and damages, property insurance, property taxes, materials used, and expenses incurred in the operation of distribution substations.

Items

Labor:

1. Supervising station operation.
2. Adjusting station equipment where such adjustment primarily affects performance, such as regulating the flow of cooling water, adjusting current in fields of a machine, changing voltage of regulators, or changing station transformer taps.
3. Keeping station log and records and preparing reports on station operation.
4. Inspecting, testing, and calibrating station equipment for the purpose of checking its performance.
5. Operating switching and other station equipment.
6. Standing watch, guarding, and patrolling station and station yard.
7. Sweeping, mopping, and tidying station.
8. Care of grounds, including snow removal, and grass cutting.

Taxes:

1. Federal and state unemployment.
2. F.I.C.A.
3. Property.

Employee Pensions and Benefits: The portion of employee pensions and benefits specifically identifiable with employees' labor costs charged herein or, in the absence of specific employee identification, the portion of employee pensions and benefits, allocated on the more equitable basis of either direct labor dollars or direct labor hours, applicable to the labor items detailed above, including:

1. Accruals for or payments to pension funds or to insurance companies for pension purposes.
2. Group and life insurance premiums (credit dividends received).
3. Payments for medical and hospital services and expenses of employees when not the result of occupational injuries.
4. Payments for accident, sickness, hospital, and death benefits or insurance.

5. Payments to employees incapacitated for service or on leave of absence beyond periods normally allowed when not the result of occupational injuries or in excess of statutory awards.

6. Expenses in connection with educational and recreational activities for the benefit of employees.

Insurance:

1. Premiums payable to insurance companies for fire, storm, burglary, boiler explosion, lightning, fidelity, riot, and similar insurance.

2. Amounts credited to Account 228.1, Accumulated Provision for Property Insurance, for similar protection.

3. Special costs incurred in procuring insurance.

4. Insurance inspection service.

5. Insurance counsel, brokerage fees, and expenses.

6. Premiums payable to insurance companies for protection against claims from injuries and damages by employees or others, such as public liability, property damages, casualty, employee liability, etc., and amounts credited to Account 228.2, Accumulated Provision for Injuries and Damage, for similar protection.

7. Losses not covered by insurance or reserve accruals on account of injuries or deaths to employees or others and damages to the property of others.

8. Fees and expenses of claim investigators.

9. Payment of awards to claimants for court costs and attorneys' services.

10. Medical and hospital service and expenses for employees as the result of occupational injuries or resulting from claims of others.

11. Compensation payments under workmen's compensation laws.

12. Compensation paid while incapacitated as the result of occupational injuries. (See Account 924, Note A.)

13. Cost of safety, accident prevention, and similar educational activities.

Materials and Expenses:

1. Building service expenses.

2. Operating, supplies, such as lubricants, commutator brushes, water, and rubber goods.

3. Station meter and instrument supplies, such as ink and charts.

4. Station record and report forms.

5. Tool expense.

6. Transportation expense.

7. Meals, traveling, and incidental expenses.

NOTE: If the utility owns storage battery equipment used for supplying electricity to customers in periods of emergency, the cost of operating labor and of supplies, such as acid, gloves, hydrometers, thermometers, soda, automatic cell fillers, and acid proof shoes shall be included in this account. If significant in amount, a separate subdivision shall be maintained for such expenses.

583   Overhead Line Expenses

584   Underground Line Expenses

These accounts shall include, respectively, the cost of labor, employee pensions and benefits, social security and other payroll taxes, injuries and damages, property insurance, property taxes, materials used, and expenses incurred in the operation of overhead and underground distribution lines.

Items

Labor:

1. Supervising line operation.

2. Changing line transformer taps.

3. Inspecting and testing lightning arresters, line circuit breakers, switches, and grounds.

4. Inspecting and testing line transformers for the purpose of determining load, temperature, or operation performance.

5. Patrolling lines.

6. Load tests and voltage surveys of feeders, circuits, and line transformers.

7. Removing line transformers and voltage regulators with or without replacement.

8. Installing line transformers or voltage regulators with or without change in capacity provided that the cost of first installation of these items is included in Account 368, Line Transformers.

9. Voltage surveys, either routine or upon request of customers, including voltage tests at customer's main switch.

10. Transferring loads, switching and reconnecting circuits and equipment for operation purpose.

11. Electrolysis surveys.

12. Inspecting and adjusting line testing equipment.

Taxes:

1. Federal and State unemployment.

2. F.I.C.A,

3. Property.

Employee Pensions and Benefits: The portion of employee pensions and benefits specifically identifiable with employees' labor costs charged herein or, in the absence of specific employee identification, the portion of employee pensions and benefits, allocated on the more equitable basis of either direct labor dollars or direct labor hours, applicable to the labor items detailed above, including:

1. Accruals for or payments to pension funds or to insurance companies for pension purposes.

2. Group and life insurance premiums (credit dividends received).

3. Payments for medical and hospital services and expenses of employees when not the result of occupational injuries.

4. Payments for accident, sickness, hospital, and death benefits or insurance.

5. Payments to employees incapacitated for service or on leave of absence beyond periods normally allowed when not the result of occupational injuries or in excess of statutory awards.

6. Expenses in connection with educational and recreational activities for the benefit of employees.

Insurance:

1. Premiums payable to insurance companies for fire, storm, burglary, boiler explosion, lightning, fidelity, riot, and similar insurance.

2. Amounts credited to Account 228.1, Accumulated Provision for Property Insurance, for similar protection.

3. Special costs incurred in procuring insurance.

4. Insurance inspection service.

5. Insurance counsel, brokerage fees, and expenses.

6. Premiums payable to insurance companies for protection against claims from injuries and damages by employees or others, such as public liability, property damages, casualty, employee liability, etc., and amounts credited to Account 228.2, Accumulated Provision for Injuries and Damage, for similar protection.

7. Losses not covered by insurance or reserve accruals on account of injuries or deaths to employees or others and damages to the property of others.

8. Fees and expenses of claim investigators.

9. Payment of awards to claimants for court costs and attorneys' services.

10. Medical and hospital service and expenses for employees as the result of occupational injuries or resulting from claims of others.

11. Compensation payments under workmen's compensation laws.

12. Compensation paid while incapacitated as the result of occupational injuries. (See Account 924, Note A.)

13. Cost of safety, accident prevention, and similar educational activities.

Materials and Expenses:

1. Tool expense.

2. Transportation expense.

3. Meals, traveling, and incidental expenses.

4. Operating supplies, such as instrument charts, and rubber goods.

585 Street Lighting and Signal System Expenses

This account shall include the cost of labor, employee pensions and benefits, social security and other payroll taxes, injuries and damages, property insurance, property taxes, materials used, and expenses incurred in: (1) The operation of street lighting and signal system plant which is owned or leased by the utility; and (2) the operation and maintenance of such plant owned by customers where such work is done regularly as a part of the street lighting and signal system service.

Items

Labor:

1. Supervising street lighting and signal systems operation.

2. Replacing lamps and incidental cleaning of glassware and fixtures in connection therewith.

3. Routine patrolling for lamp outages, extraneous nuisances, or encroachments.

4. Testing lines and equipment including voltage and current measurement.

5. Winding and inspection of time switch and other controls.

Taxes:

1. Federal and state unemployment.

2. F.I.C.A.

3. Property.

Employee Pensions and Benefits: The portion of employee pensions and benefits specifically identifiable with employees' labor costs charged herein or, in the absence of specific employee identification, the portion of employee pensions and benefits, allocated on the more equitable basis of either direct labor dollars or direct labor hours, applicable to the labor items detailed above, including:

1. Accruals for or payments to pension funds or to insurance companies for pension purposes.

2. Group and life insurance premiums (credit dividends received).

3. Payments for medical and hospital services and expenses of employees when not the result of occupational injuries.

4. Payments for accident, sickness, hospital, and death benefits or insurance.

5. Payments to employees incapacitated for service or on leave of absence beyond periods normally allowed when not the result of occupational injuries or in excess of statutory awards.

6. Expenses in connection with educational and recreational activities for the benefit of employees.

Insurance:

1. Premiums payable to insurance companies for fire, storm, burglary, boiler explosion, lightning, fidelity, riot, and similar insurance.

2. Amounts credited to Account 228.1, Accumulated Provision for Property Insurance, for similar protection.

3. Special costs incurred in procuring insurance.

4. Insurance inspection service.

5. Insurance counsel, brokerage fees, and expenses.

6. Premiums payable to insurance companies for protection against claims from injuries and damages by employees or others, such as public liability, property damages, casualty, employee liability, etc., and amounts credited to Account 228.2, Accumulated Provision for Injuries and Damage, for similar protection.

7. Losses not covered by insurance or reserve accruals on account of injuries or deaths to employees or others and damages to the property of others.

8. Fees and expenses of claim investigators.

9. Payment of awards to claimants for court costs and attorneys' services.

10. Medical and hospital service and expenses for employees as the result of occupational injuries or resulting from claims of others.

11. Compensation payments under workmen's compensation laws.

12. Compensation paid while incapacitated as the result of occupational injuries. (See Account 924, Note A.)

13. Cost of safety, accident prevention, and similar educational activities.

Materials and Expenses:

1. Street lamp renewals.
2. Transportation and tool expense.
3. Meals, traveling, and incidental expenses.

586   Meter Expenses

This account shall include the cost of labor, employee pensions and benefits, social security and other payroll taxes, injuries and damages, property insurance, property taxes, materials used, and expenses incurred in the operation of customer meters and associated equipment.

Items

Labor:

1. Supervising meter operation.
2. Clerical work on meter history and associated equipment record cards, test cards, and reports.
3. Disconnecting and reconnecting, removing and reinstalling, sealing and unsealing meters and other metering equipment in connection with initiating or terminating services including the cost of obtaining meter readings, if incidental to such operation.
4. Consolidating meter installations due to elimination of separate meters for different rates of service.
5. Changing or relocating meters, instrument transformers, time switches, and other metering equipment.
6. Resetting time controls, checking operation of demand meters and other metering equipment, when done as an independent operation.
7. Inspecting and adjusting meter testing equipment.

8. Inspecting and testing meters, instrument transformers, time switches, and other metering equipment on premises or in shops excluding inspecting and testing incidental to maintenance.

Taxes:

1. Federal and state unemployment.
2. F.I.C.A.
3. Property.

Employee Pensions and Benefits: The portion of employee pensions and benefits specifically identifiable with employees' labor costs charged herein or, in the absence of specific employee identification, the portion of employee pensions and benefits, allocated on the more equitable basis of either direct labor dollars or direct labor hours, applicable to the labor items detailed above, including:

1. Accruals for or payments to pension funds or to insurance companies for pension purposes.
2. Group and life insurance premiums (credit dividends received).
3. Payments for medical and hospital services and expenses of employees when not the result of occupational injuries.
4. Payments for accident, sickness, hospital, and death benefits or insurance.
5. Payments to employees incapacitated for service or on leave of absence beyond periods normally allowed when not the result of occupational injuries or in excess of statutory awards.
6. Expenses in connection with educational and recreational activities for the benefit of employees.

Insurance:

1. Premiums payable to insurance companies for fire, storm, burglary, boiler explosion, lightning, fidelity, riot, and similar insurance.
2. Amounts credited to Account 228.1, Accumulated Provision for Property Insurance, for similar protection.
3. Special costs incurred in procuring insurance.
4. Insurance inspection service.
5. Insurance counsel, brokerage fees, and expenses.
6. Premiums payable to insurance companies for protection against claims from injuries and damages by employees or others, such as public liability, property damages, casualty, employee liability, etc., and amounts credited to Account 228.2, Accumulated Provision for Injuries and Damage, for similar protection.
7. Losses not covered by insurance or reserve accruals on account of injuries or deaths to employees or others and damages to the property of others.
8. Fees and expenses of claim investigators.
9. Payment of awards to claimants for court costs and attorneys' services.

10. Medical and hospital service and expenses for employees as the result of occupational injuries or resulting from claims of others.

11. Compensation payments under workmen's compensation laws.

12. Compensation paid while incapacitated as the result of occupational injuries. (See Account 924, Note A.)

13. Cost of safety, accident prevention, and similar educational activities.

Materials and Expenses

1. Meter seals and miscellaneous meter supplies.

2. Transportation expenses.

3. Meals, traveling, and incidental expenses.

4. Tool expenses.

NOTE: The cost of the first setting and testing of a meter is chargeable to utility plant, Account 370, Meters.

587 Customer Installations Expenses

This account shall include the cost of labor, employee pensions and benefits, social security and other payroll taxes, injuries and damages, property insurance, property taxes, materials used, and expenses incurred in work on customer installations in inspecting premises and in rendering services to customers of the nature of those indicated by the list of items hereunder.

Items

Labor:

1. Supervising customer installations work.

2. Inspecting premises, including the check of wiring for code compliance.

3. Investigating, locating, and clearing grounds on customers' wiring.

4. Investigating service complaints, including load tests of motors and lighting and power circuits on customers' premises; field investigations of complaints on bills or of voltage.

5. Installing, removing, renewing, and changing lamps and fuses.

6. Radio, television, and similar interference work including erection of new aerials on customers' premises and patrolling of lines, testing of lightning arresters, inspection of pole hardware, and examination on or off premises of customers' appliances, wiring, or equipment to locate cause of interference.

7. Installing, connecting, reinstalling, or removing leased property on customers' premises.

8. Testing, adjusting, and repairing customers' fixtures and appliances in the shop or on premises.

9. Cost of changing customers' equipment due to changes in service characteristics.

10. Investigation of current diversion including setting and removal of check meters and securing special readings thereon; special calls by employees in connection with discovery and settlement of current diversion; changes in customer wiring; and any other labor cost identifiable as caused by current diversion.

Taxes:

1. Federal and state unemployment.

2. F.I.C.A.

3. Property.

Employee Pensions and Benefits: The portion of employee pensions and benefits specifically identifiable with employees' labor costs charged herein or, in the absence of specific employee identification, the portion of employee pensions and benefits, allocated on the more equitable basis of either direct labor dollars or direct labor hours, applicable to the labor items detailed above, including:

1. Accruals for or payments to pension funds or to insurance companies for pension purposes.

2. Group and life insurance premiums (credit dividends received).

3. Payments for medical and hospital services and expenses of employees when not the result of occupational injuries.

4. Payments for accident, sickness, hospital, and death benefits or insurance.

5. Payments to employees incapacitated for service or on leave of absence beyond periods normally allowed when not the result of occupational injuries or in excess of statutory awards.

6. Expenses in connection with educational and recreational activities for the benefit of employees.

Insurance:

1. Premiums payable to insurance companies for fire, storm, burglary, boiler explosion, lightning, fidelity, riot, and similar insurance.

2. Amounts credited to Account 228.1, Accumulated Provision for Property Insurance, for similar protection.

3. Special costs incurred in procuring insurance.

4. Insurance inspection service.

5. Insurance counsel, brokerage fees, and expenses.

6. Premiums payable to insurance companies for protection against claims from injuries and damages by employees or others, such as public liability, property damages, casualty, employee liability, etc., and amounts credited to Account 228.2, Accumulated Provision for Injuries and Damage, for similar protection.

7. Losses not covered by insurance or reserve accruals on account of injuries or deaths to employees or others and damages to the property of others.

8. Fees and expenses of claim investigators.

9. Payment of awards to claimants for court costs and attorneys' services.

10. Medical and hospital service and expenses for employees as the result of occupational injuries or resulting from claims of others.

11. Compensation payments under workmen's compensation laws.

12. Compensation paid while incapacitated as the result of occupational injuries. (See Account 924, Note A.)

13. Cost of safety, accident prevention, and similar educational activities.

Materials and Expenses:

1. Lamp and fuse renewals.

2. Materials used in servicing customers' fixtures, appliances, and equipment.

3. Power, light, heat, telephone, and other expenses of the appliance repair department.

4. Tool expense.

5. Transportation expense, including pickup and delivery charges.

6. Meals, traveling, and incidental expenses.

7. Rewards paid for discovery of current diversion.

NOTE A: Amounts billed customers for any work, the cost of which is charged to this account, shall be credited to this account. Any excess over costs resulting therefrom, shall be transferred to Account 451, Miscellaneous Service Revenues.

NOTE B: Do not include in this account expenses incurred in connection with merchandising, jobbing, and contract work.

588 Miscellaneous Distribution Expenses

This account shall include the cost of labor, employee pensions and benefits, social security and other payroll taxes, injuries and damages, materials used, and expenses incurred in distribution system operation not provided for elsewhere.

Items

Labor:

1. General records of physical characteristics of lines and substations, such as capacities.

2. Ground resistance records.

3. Joint pole maps and records.

4. Distribution system voltage and load records.

5. Preparing maps and prints.

6. Service interruption and trouble records.

7. General clerical and stenographic work except that chargeable to Account 586, Meter Expenses.

Taxes:

1. Federal and state unemployment.

2. F.I.C.A.

Employee Pensions and Benefits: The portion of employee pensions and benefits specifically identifiable with employees' labor costs charged herein or, in the absence of

specific employee identification, the portion of employee pensions and benefits, allocated on the more equitable basis of either direct labor dollars or direct labor hours, applicable to the labor items detailed above, including:

1. Accruals for or payments to pension funds or to insurance companies for pension purposes.

2. Group and life insurance premiums (credit dividends received).

3. Payments for medical and hospital services and expenses of employees when not the result of occupational injuries.

4. Payments for accident, sickness, hospital, and death benefits or insurance.

5. Payments to employees incapacitated for service or on leave of absence beyond periods normally allowed when not the result of occupational injuries or in excess of statutory awards.

6. Expenses in connection with educational and recreational activities for the benefit of employees.

Insurance:

1. Premiums payable to insurance companies for protection against claims from injuries and damages by employees or others such as public liability, property damages, casualty, employee liability, etc., and amounts credited to Account 228.2, Accumulated Provision for Injuries and Damage, for similar protection.

2. Losses not covered by insurance or reserve accruals on account of injuries or deaths to employees or others and damages to the property of others.

3. Fees and expenses of claim investigators.

4. Payment of awards to claimants for court costs and attorneys' services.

5. Medical and hospital service and expenses for employees as the result of occupational injuries or resulting from claims of others.

6. Compensation payments under workmen's compensation laws.

7. Compensation paid while incapacitated as the result of occupational injuries. (See Account 924, Note A.)

8. Cost of safety, accident prevention, and similar educational activities.

Expenses:

1. Operating records covering poles, transformers, manholes, cables, and other distribution facilities. Exclude meter records chargeable to Account 586, Meter Expenses, and station records chargeable to Account 582, Station Expenses, and stores records chargeable to Account 163, Stores Expense Undistributed.

2. Janitor work at distribution office buildings including snow removal and grass cutting.

3. Communication service.

4. Building service expenses.

5. Miscellaneous office supplies and expenses, printing and stationery, maps and records, and first-aid supplies.

6. Research, development, and demonstration expenses.

589 Rents

This account shall include rents of property of others used, occupied, or operated in connection with the distribution system, including payments to the United States and others for the use and occupancy of public lands and reservations for distribution line rights of way. (See §1767.17 (c).)

(Maintenance)

590 Maintenance Supervision and Engineering

This account shall include the cost of labor, employee pensions and benefits, social security and other payroll taxes, injuries and damages, and expenses incurred in the general supervision and direction of maintenance of the distribution system. Direct field supervision of specific jobs shall be charged to the appropriate maintenance account. (See §1767.17(a).)

591 Maintenance of Structures

This account shall include the cost of labor, employee pensions and benefits, social security and other payroll taxes, injuries and damages, materials used, and expenses incurred in maintenance of structures, the book cost of which is includible in Account 361, Structures and Improvements. (See §1767.17(b).)

592 Maintenance of Station Equipment

This account shall include the cost of labor, employee pensions and benefits, social security and other payroll taxes, injuries and damages, materials used, and expenses incurred in maintenance of plant, the book cost of which is includible in Account 362, Station Equipment, and Account 363, Storage Battery Equipment. (See §1767.17(b).)

593 Maintenance of Overhead Lines

This account shall include the cost of labor, employee pensions and benefits, social security and other payroll taxes, injuries and damages, materials used, and expenses incurred in the maintenance of overhead distribution line facilities, the book cost of which is includible in Account 364, Poles, Towers and Fixtures; Account 365, Overhead Conductors and Devices; and Account 369, Services. (See §1767.17(b).)

Items

1. Work of the following character on poles, towers, and fixtures:

a. Installing additional clamps or removing clamps or strain insulators on guys in place.

b. Moving line or guy pole in relocation of pole or section of line.

c. Painting poles, towers, crossarms, or pole extensions.

d. Readjusting and changing position of guys or braces.

e. Realigning and straightening poles, crossarms, braces, pins, racks, brackets, and other pole fixtures.

f. Reconditioning reclaimed pole fixtures.

g. Relocating crossarms, racks, brackets, and other fixtures on poles.

h. Repairing pole supported platform.

i. Repairs by others to jointly owned poles.

j. Shaving, cutting rot, or treating poles or crossarms in use or salvaged for reuse.

k. Stubbing poles already in service.

l. Supporting conductors, transformers, and other fixtures and transferring them to new poles during pole replacements.

m. Maintaining pole signs, stencils, and tags.

2. Work of the following character on overhead conductors and devices:

a. Overhauling and repairing line cutouts, line switches, line breakers, and capacitor installations.

b. Cleaning insulators and bushings.

c. Refusing line cutouts.

d. Repairing line oil circuit breakers and associated relays and control wiring.

e. Repairing grounds.

f. Resagging, retying, or rearranging position or spacing of conductors.

g. Standing by phones, going to calls, cutting faulty lines clear, or similar activities at times of emergency.

h. Sampling, testing, changing, purifying, and replenishing insulating oil.

i. Transferring loads, switching, and reconnecting circuits and equipment for maintenance purposes.

j. Repairing line testing equipment.

k. Trimming trees and clearing brush.

l. Chemical treatment of right-of-way area when occurring subsequent to construction of line.

3. Work of the following character on overhead services:

a. Moving position of service either on pole or on customers' premises.

b. Pulling slack in service wire.

c. Retying service wire.

d. Refastening or tightening service bracket.

Taxes:

1. Federal and state unemployment.

2. F.I.C.A.

Employee Pensions and Benefits: The portion of employee pensions and benefits specifically identifiable with employees' labor costs charged herein or, in the absence of specific employee identification, the portion of employee pensions and benefits, allocated on the more equitable basis of either direct labor dollars or direct labor hours, applicable to the labor items detailed above, including:

1. Accruals for or payments to pension funds or to insurance companies for pension purposes.

2. Group and life insurance premiums (credit dividends received).

3. Payments for medical and hospital services and expenses of employees when not the result of occupational injuries.

4. Payments for accident, sickness, hospital, and death benefits or insurance.

5. Payments to employees incapacitated for service or on leave of absence beyond periods normally allowed when not the result of occupational injuries or in excess of statutory awards.

6. Expenses in connection with educational and recreational activities for the benefit of employees.

Insurance:

1. Premiums payable to insurance companies for protection against claims from injuries and damages by employees or others, such as public liability, property damages, casualty, employee liability, etc., and amounts credited to Account 228.2, Accumulated Provision for Injuries and Damage, for similar protection.

2. Losses not covered by insurance or reserve accruals on account of injuries or deaths to employees or others and damages to the property of others.

3. Fees and expenses of claim investigators.

4. Payment of awards to claimants for court costs and attorneys' services.

5. Medical and hospital service and expenses for employees as the result of occupational injuries or resulting from claims of others.

6. Compensation payments under workmen's compensation laws.

7. Compensation paid while incapacitated as the result of occupational injuries. (See Account 924, Note A.)

8. Cost of safety, accident prevention, and similar educational activities.

594 Maintenance of Underground Lines

This account shall include the cost of labor, employee pensions and benefits, social security and other payroll taxes, injuries and damages, materials used, and expenses incurred in the maintenance of underground distribution line facilities, the book cost of which is includible in Account 366, Underground Conduit; Account 367, Underground Conductors and Devices; and Account 369, Services. (See § 1767.17(b).)

Items

1. Work of the following character on underground conduit:

a. Cleaning ducts, manholes, and sewer connections.

b. Moving or changing position of conduit or pipe.

c. Minor alterations of handholes, manholes, or vaults.

d. Refastening, repairing, or moving racks, ladders, or hangers in manholes or vaults.

e. Plugging and shelving ducts.

f. Repairs to sewers, drains, walls, and floors, rings, and covers.

2. Work of the following character on underground conductors and devices:

a. Repairing circuit breakers, switches, cutouts, network protectors, and associated relays and control wiring.

b. Repairing grounds.

c. Retraining and reconnecting cables in manholes including transfer of cables from one duct to another.

d. Repairing conductors and splices.

e. Repairing or moving junction boxes and potheads.

f. Refireproofing cables and repairing supports.

g. Repairing electrolysis preventive devices for cables.

h. Repairing cable bonding systems.

i. Sampling, testing, changing, purifying, and replenishing insulating oil.

j. Transferring loads, switching and reconnecting circuits and equipment for maintenance purposes.

k. Repairing line testing equipment.

l. Repairing oil or gas equipment in high voltage cable systems and replacement of oil or gas.

3. Work of the following character on underground services:

a. Cleaning ducts.

b. Repairing any underground service plant.

Taxes:

1. Federal and state unemployment.

2. F.I.C.A.

Employee Pensions and Benefits: The portion of employee pensions and benefits specifically identifiable with employees' labor costs charged herein or, in the absence of specific employee identification, the portion of employee pensions and benefits, allocated on the more equitable basis of either direct labor dollars or direct labor hours, applicable to the labor items detailed above, including:

1. Accruals for or payments to pension funds or to insurance companies for pension purposes.

2. Group and life insurance premiums (credit dividends received).

3. Payments for medical and hospital services and expenses of employees when not the result of occupational injuries.

4. Payments for accident, sickness, hospital, and death benefits or insurance.

5. Payments to employees incapacitated for service or on leave of absence beyond periods normally allowed when not the result of occupational injuries or in excess of statutory awards.

6. Expenses in connection with educational and recreational activities for the benefit of employees.

Insurance:

1. Premiums payable to insurance companies for protection against claims from injuries and damages by employees or others, such as public liability, property damages, casualty, employee liability, etc., and amounts credited to Account 228.2, Accumulated Provision for Injuries and Damage, for similar protection.

2. Losses not covered by insurance or reserve accruals on account of injuries or deaths to employees or others and damages to the property of others.

3. Fees and expenses of claim investigators.

4. Payment of awards to claimants for court costs and attorneys' services.

5. Medical and hospital service and expenses for employees as the result of occupational injuries or resulting from claims of others.

6. Compensation payments under workmen's compensation laws.

7. Compensation paid while incapacitated as the result of occupational injuries. (See Account 924, Note A.)

8. Cost of safety, accident prevention, and similar educational activities.

595 Maintenance of Line Transformers

This account shall include the cost of labor, employee pensions and benefits, social security and other payroll taxes, injuries and damages, materials used, and expenses incurred in maintenance of distribution line transformers, the book cost of which is includible in Account 368, Line Transformers. (See § 1767.17(b).)

596 Maintenance of Street Lighting and Signal Systems

This account shall include the cost of labor, employee pensions and benefits, social security and other payroll taxes, injuries and damages, materials used, and expenses incurred in maintenance of plant, the book cost of which is includible in Account 373, Street Lighting and Signal Systems. (See § 1767.17(b).)

597 Maintenance of Meters

This account shall include the cost of labor, employee pensions and benefits, social security and other payroll taxes, injuries and damages, materials used, and expenses incurred in the maintenance of meters and meter testing equipment, the book cost of which is includible in Account 370, Meters,

and Account 395, Laboratory Equipment, respectively. (See § 1767.17(b).)

598 Maintenance of Miscellaneous Distribution Plant

This account shall include the cost of labor, employee pensions and benefits, social security and other payroll taxes, injuries and damages, materials used, and expenses incurred in maintenance of plant, the book cost of which is includible in Accounts 371, Installations on Customers' Premises, and Account 372, Leased Property on Customers' Premises, and any other plant the maintenance of which is assignable to the distribution function and is not provided for elsewhere. (See § 1767.17(b).)

Items

1. Work of similar nature to that listed in other distribution maintenance accounts.

2. Maintenance of office furniture and equipment used by distribution system department.

Taxes:

1. Federal and state unemployment.

2. F.I.C.A.

Employee Pensions and Benefits: The portion of employee pensions and benefits specifically identifiable with employees' labor costs charged herein or, in the absence of specific employee identification, the portion of employee pensions and benefits, allocated on the more equitable basis of either direct labor dollars or direct labor hours, applicable to the labor items detailed above, including:

1. Accruals for or payments to pension funds or to insurance companies for pension purposes.

2. Group and life insurance premiums (credit dividends received).

3. Payments for medical and hospital services and expenses of employees when not the result of occupational injuries.

4. Payments for accident, sickness, hospital, and death benefits or insurance.

5. Payments to employees incapacitated for service or on leave of absence beyond periods normally allowed when not the result of occupational injuries or in excess of statutory awards.

6. Expenses in connection with educational and recreational activities for the benefit of employees.

Insurance:

1. Premiums payable to insurance companies for protection against claims from injuries and damages by employees or others, such as public liability, property damages, casualty, employee liability, etc., and amounts credited to Account 228.2, Accumulated Provision for Injuries and Damage, for similar protection.

131

2. Losses not covered by insurance or reserve accruals on account of injuries or deaths to employees or others and damages to the property of others.

3. Fees and expenses of claim investigators.

4. Payment of awards to claimants for court costs and attorneys' services.

5. Medical and hospital service and expenses for employees as the result of occupational injuries or resulting from claims of others.

6. Compensation payments under workmen's compensation laws.

7. Compensation paid while incapacitated as the result of occupational injuries. (See Account 924, Note A.)

8. Cost of safety, accident prevention, and similar educational activities.

[58 FR 59825, Nov. 10, 1993, as amended at 62 FR 42291, Aug. 6, 1997; 73 FR 20286, May 27, 2008]

§ 1767.28 Customer accounts expenses.

The customer accounts expense accounts identified in this section shall be used by all RUS borrowers.

CUSTOMER ACCOUNTS EXPENSES

(Operation)

901 Supervision
902 Meter Reading Expenses
903 Customer Records and Collection Expenses
904 Uncollectible Accounts
905 Miscellaneous Customer Accounts Expenses

CUSTOMER ACCOUNTS EXPENSES

(Operation)

901 Supervision

This account shall include the cost of labor, employee pensions and benefits, social security and other payroll taxes, injuries and damages, and expenses incurred in the general direction and supervision of customer accounting and collecting activities. Direct supervision of a specific activity shall be charged to Account 902, Meter Reading Expenses, or Account 903, Customer Records and Collection Expenses, as appropriate. (See § 1767.17(a).)

902 Meter Reading Expenses

This account shall include the cost of labor, employee pensions and benefits, social security and other payroll taxes, injuries and damages, materials used, and expenses incurred in reading customer meters, and determining consumption when performed by employees engaged in reading meters.

Items

Labor:

1. Addressing forms for obtaining meter readings by mail.

2. Changing and collecting meter charts used for billing purposes.

3. Inspecting time clocks and checking seals when performed by meter readers and the work represents a minor activity incidental to regular meter reading routine.

4. Reading meters, including demand meters, and obtaining load information for billing purposes. Exclude and charge to Account 586, Meter Expenses, or to Account 903, Customer Records and Collection Expenses, as applicable, the cost of obtaining meter readings, first and final, if incidental to the operation of removing or resetting, sealing or locking, and disconnecting or reconnecting meters.

5. Computing consumption from meter reader's book or from reports by mail when done by employees engaged in reading meters.

6. Collecting from prepayment meters when incidental to meter reading.

7. Maintaining record of customers' keys.

8. Computing estimated or average consumption when performed by employees engaged in reading meters.

Taxes:

1. Federal and state unemployment.

2. F.I.C.A.

Employee Pensions and Benefits: The portion of employee pensions and benefits specifically identifiable with employees' labor costs charged herein or, in the absence of specific employee identification, the portion of employee pensions and benefits, allocated on the more equitable basis of either direct labor dollars or direct labor hours, applicable to the labor items detailed above, including:

1. Accruals for or payments to pension funds or to insurance companies for pension purposes.

2. Group and life insurance premiums (credit dividends received).

3. Payments for medical and hospital services and expenses of employees when not the result of occupational injuries.

4. Payments for accident, sickness, hospital, and death benefits or insurance.

5. Payments to employees incapacitated for service or on leave of absence beyond periods normally allowed when not the result of occupational injuries or in excess of statutory awards.

6. Expenses in connection with educational and recreational activities for the benefit of employees.

Insurance:

1. Premiums payable to insurance companies for protection against claims from injuries and damages by employees or others,

such as public liability, property damages, casualty, employee liability, etc., and amounts credited to Account 228.2, Accumulated Provision for Injuries and Damage, for similar protection.

2. Losses not covered by insurance or reserve accruals on account of injuries or deaths to employees or others and damages to the property of others.

3. Fees and expenses of claim investigators.

4. Payment of awards to claimants for court costs and attorneys' services.

5. Medical and hospital service and expenses for employees as the result of occupational injuries or resulting from claims of others.

6. Compensation payments under workmen's compensation laws.

7. Compensation paid while incapacitated as the result of occupational injuries. (See Account 924, Note A.)

8. Cost of safety, accident prevention, and similar educational activities.

Materials and Expenses:

1. Badges, lamps, and uniforms.

2. Demand charts, meter books and binders and forms for recording readings, but not the cost of preparation.

3. Postage and supplies used in obtaining meter readings by mail.

4. Transportation, meals, and incidental expenses.

903 Customer Records and Collection Expenses

This account shall include the cost of labor, employee pensions and benefits, social security and other payroll taxes, injuries and damages, materials used, and expenses incurred in work on customer applications, contracts, orders, credit investigations, billing and accounting, collections and complaints.

Items

Labor:

1. Receiving, preparing, recording, and handling routine orders for service, disconnections, transfers or meter tests initiated by the customer, excluding the cost of carrying out such orders, which is chargeable to the account appropriate for the work called for by such orders.

2. Investigations of customers' credit and keeping of records pertaining thereto, including records of uncollectible accounts written off.

3. Receiving, refunding, or applying customer deposits and maintaining customer deposit, line extension, and other miscellaneous records.

4. Checking consumption shown by meter readers' reports where incidental to preparation of billing date.

5. Preparing address plates and addressing bills and delinquent notices.

6. Preparing billing data.

7. Operating billing and bookkeeping machines.

8. Verifying billing records with contracts or rate schedules.

9. Preparing bills for delivery and mailing or delivering bills.

10. Collecting revenues, including collection from prepayment meters, unless incidental to meter-reading operations.

11. Balancing collections, preparing collections for deposit, and preparing cash reports.

12. Posting collections and other credits or charges to customer accounts and extending unpaid balances.

13. Balancing customer accounts and controls.

14. Preparing, mailing, or delivering delinquent notices and preparing reports of delinquent accounts.

15. Final meter reading of delinquent accounts when done by collectors incidental to regular activities.

16. Disconnecting and reconnecting service because of nonpayment bills.

17. Receiving, recording, and handling of inquiries, complaints, and requests for investigations from customers, including preparation of necessary orders, but excluding the cost of carrying out such orders, which is chargeable to the account appropriate for the work called for by such orders.

18. Statistical and tabulating work on customer accounts and revenues, but not including special analyses for sales department, rate department, or other general purposes, unless incidental to regular customer accounting routines.

19. Preparing and periodically rewriting meter reading sheets.

20. Determining consumption and computing estimated or average consumption when performed by employees other than those engaged in reading meters.

Taxes:

1. Federal and state unemployment.

2. F.I.C.A.

Employee Pensions and Benefits: The portion of employee pensions and benefits specifically identifiable with employees' labor costs charged herein or, in the absence of specific employee identification, the portion of employee pensions and benefits, allocated on the more equitable basis of either direct labor dollars or direct labor hours, applicable to the labor items detailed above, including:

1. Accruals for or payments to pension funds or to insurance companies for pension purposes.

2. Group and life insurance premiums (credit dividends received).

3. Payments for medical and hospital services and expenses of employees when not the result of occupational injuries.

4. Payments for accident, sickness, hospital, and death benefits or insurance.

5. Payments to employees incapacitated for service or on leave of absence beyond periods normally allowed when not the result of occupational injuries or in excess of statutory awards.

6. Expenses in connection with educational and recreational activities for the benefit of employees.

Insurance:

1. Premiums payable to insurance companies for protection against claims from injuries and damages by employees or others, such as public liability, property damages, casualty, employee liability, etc., and amounts credited to Account 228.2, Accumulated Provision for Injuries and Damage, for similar protection.

2. Losses not covered by insurance or reserve accruals on account of injuries or deaths to employees or others and damages to the property of others.

3. Fees and expenses of claim investigators.

4. Payment of awards to claimants for court costs and attorneys' services.

5. Medical and hospital service and expenses for employees as the result of occupational injuries or resulting from claims of others.

6. Compensation payments under workmen's compensation laws.

7. Compensation paid while incapacitated as the result of occupational injuries. (See Account 924, Note A.)

8. Cost of safety, accident prevention, and similar educational activities.

Materials and Expenses:

1. Address plates and supplies.
2. Cash overages and shortages.
3. Commissions or fees to others for collecting.
4. Payments to credit organizations for investigations and reports.
5. Postage.
6. Transportation expenses, including transportation of customer bills and meter books under centralized billing procedures.
7. Transportation, meals expenses, and incidental expenses.
8. Bank charges, exchange, and other fees for cashing and depositing customers' checks.
9. Forms for recording orders for services, or removals.
10. Rent of mechanical equipment.

NOTE: The cost of work on meter history and meter location records in chargeable to Account 586, Meter Expenses.

904 Uncollectible Accounts

This amount shall be charged with amounts sufficient to provide for losses from uncollectible utility revenues. Concurrent credits shall be made to Account 144, Accumulated Provision for Uncollectible Accounts—Credit. Losses from uncollectible accounts shall be charged to Account 144.

905 Miscellaneous Customer Accounts Expenses

This account shall include the cost of labor, employee pensions and benefits, social security and other payroll taxes, injuries and damages, property insurance, property taxes, materials used, and expenses incurred not provided for in other accounts.

Items

Labor:

1. General clerical and stenographic work.
2. Miscellaneous labor.

Taxes:

1. Federal and state unemployment.
2. F.I.C.A.
3. Property.

Employee Pensions and Benefits: The portion of employee pensions and benefits specifically identifiable with employees' labor costs charged herein, or, in the absence of specific employee identification, the portion of employee pensions and benefits, allocated on the more equitable basis of either direct labor dollars or direct labor hours, applicable to the labor items detailed above, including:

1. Accruals for or payments to pension funds or to insurance companies for pension purposes.

2. Group and life insurance premiums (credit dividends received).

3. Payments for medical and hospital services and expenses of employees when not the result of occupational injuries.

4. Payments for accident, sickness, hospital, and death benefits or insurance.

5. Payments to employees incapacitated for service or on leave of absence beyond periods normally allowed when not the result of occupational injuries or in excess of statutory awards.

6. Expenses in connection with educational and recreational activities for the benefit of employees.

Insurance:

1. Premiums payable to insurance companies for fire, storm, burglary, boiler explosion, lightning, fidelity, riot, and similar insurance.

2. Amounts credited to Account 228.1, Accumulated Provision for Property Insurance, for similar protection.

3. Special costs incurred in procuring insurance.

4. Insurance inspection service.

5. Insurance counsel, brokerage fees, and expenses.

6. Premiums payable to insurance companies for protection against claims from injuries and damages by employees or others,

such as public liability, property damages, casualty, employee liability, etc., and amounts credited to Account 228.2, Accumulated Provision for Injuries and Damage, for similar protection.

7. Losses not covered by insurance or reserve accruals on account of injuries or deaths to employees or others and damages to the property of others.

8. Fees and expenses of claim investigators.

9. Payment of awards to claimants for court costs and attorneys' services.

10. Medical and hospital service and expenses for employees as the result of occupational injuries or resulting from claims of others.

11. Compensation payments under workmen's compensation laws.

12. Compensation paid while incapacitated as the result of occupational injuries. (See Account 924, Note A.)

13. Cost of safety, accident prevention, and similar educational activities.

Materials and Expenses:

1. Communication service.

2. Miscellaneous office supplies and expenses and stationery and printing other than those specifically provided for in Account 902 and Account 903.

[58 FR 59825, Nov. 10, 1993, as amended at 62 FR 42311, Aug. 6, 1997; 62 FR 43201, Aug. 12, 1997]

§ 1767.29 Customer service and informational expenses.

The customer service and informational expense accounts identified in this section shall be used by all RUS borrowers.

CUSTOMER SERVICE AND INFORMATIONAL EXPENSES

(Operation)

907 Supervision
908 Customer Assistance Expenses
909 Informational and Instructional Advertising Expenses
910 Miscellaneous Customer Service and Informational Expenses

CUSTOMER SERVICE AND INFORMATIONAL EXPENSES

(Operation)

907 Supervision

This account shall include the cost of labor, employee pensions and benefits, social security and other payroll taxes, injuries and damages, and expenses incurred in the general direction and supervision of customer service activities, the object of which is to encourage safe, efficient, and economical use of the utility's service. Direct supervision of a specific activity within customer service and informational expense classification shall be charged to the account wherein the costs of such activity are included. (See § 1767.17(a).)

908 Customer Assistance Expenses

This account shall include the cost of labor, employee pensions and benefits, social security and other payroll taxes, injuries and damages, materials used, and expenses incurred in providing instructions or assistance to customers, the object of which is to encourage safe, efficient, and economical use of the utility's service.

Items

Labor:

1. Direct supervision of department.

2. Processing customer inquiries relating to the proper use of electric equipment, the replacement of such equipment, and information related to such equipment.

3. Advice directed to customers as to how they may achieve the most efficient and safest use of electric equipment.

4. Demonstrations, exhibits, lectures, and other programs designed to instruct customers in the safe, economical, or efficient use of electric service, and/or oriented toward conservation of energy.

5. Engineering and technical advice to customers, the object of which is to promote safe, efficient, and economical use of the utility's service.

Taxes:

1. Federal and state unemployment.

2. F.I.C.A.

Employee Pensions and Benefits: The portion of employee pensions and benefits specifically identifiable with employees' labor costs charged herein or, in the absence of specific employee identification, the portion of employee pensions and benefits, allocated on the more equitable basis of either direct labor dollars or direct labor hours, applicable to the labor items detailed above, including:

1. Accruals for or payments to pension funds or to insurance companies for pension purposes.

2. Group and life insurance premiums (credit dividends received).

3. Payments for medical and hospital services and expenses of employees when not the result of occupational injuries.

4. Payments for accident, sickness, hospital, and death benefits or insurance.

5. Payments to employees incapacitated for service or on leave of absence beyond periods normally allowed when not the result of occupational injuries or in excess of statutory awards.

6. Expenses in connection with educational and recreational activities for the benefit of employees.

Insurance:

1. Premiums payable to insurance companies for protection against claims from injuries and damages by employees or others, such as public liability, property damages, casualty, employee liability, etc., and amounts credited to Account 228.2, Accumulated Provision for Injuries and Damage, for similar protection.

2. Losses not covered by insurance or reserve accruals on account of injuries or deaths to employees or others and damages to the property of others.

3. Fees and expenses of claim investigators.

4. Payment of awards to claimants for court costs and attorneys' services.

5. Medical and hospital service and expenses for employees as the result of occupational injuries or resulting from claims of others.

6. Compensation payments under workmen's compensation laws.

7. Compensation paid while incapacitated as the result of occupational injuries. (See Account 924, Note A.)

8. Cost of safety, accident prevention, and similar educational activities.

Materials and Expenses:

1. Supplies and expenses pertaining to demonstrations, exhibits, lectures, and other programs.

2. Loss in value on equipment and appliances used for customer assistance programs.

3. Office supplies and expenses.

4. Transportation, meals, and incidental expenses.

NOTE: Do not include in this account expenses that are provided for elsewhere, such as Accounts 416, Costs and Expenses of Merchandising, Jobbing, and Contract Work; 587, Customer Installations Expenses; and 912, Demonstrating and Selling Expenses.

909 Informational and Instructional Advertising Expenses

This account shall include the cost of labor, employee pensions and benefits, social security and other payroll taxes, injuries and damages, materials used, and expenses incurred in activities which primarily convey information as to what the utility urges or suggests customers should do in utilizing electric service to protect health and safety, to encourage environmental protection, to utilize their electric equipment safely and economically, or to conserve electric energy.

Items

Labor:

1. Direct supervision of information activities.

2. Preparing informational materials for newspapers, periodicals, and billboards and

preparing and conducting informational motion pictures, radio and television programs.

3. Preparing informational booklets and bulletins used in direct mailings.

4. Preparing informational window and other displays.

5. Employing agencies, selecting media, and conducting negotiations in connection with the placement and subject matter of information programs.

Taxes:

1. Federal and state unemployment.

2. F.I.C.A.

Employee Pensions and Benefits: The portion of employee pensions and benefits specifically identifiable with employees' labor costs charged herein or, in the absence of specific employee identification, the portion of employee pensions and benefits, allocated on the more equitable basis of either direct labor dollars or direct labor hours, applicable to the labor items detailed above, including:

1. Accruals for or payments to pension funds or to insurance companies for pension purposes.

2. Group and life insurance premiums (credit dividends received).

3. Payments for medical and hospital services and expenses of employees when not the result of occupational injuries.

4. Payments for accident, sickness, hospital, and death benefits or insurance.

5. Payments to employees incapacitated for service or on leave of absence beyond periods normally allowed when not the result of occupational injuries or in excess of statutory awards.

6. Expenses in connection with educational and recreational activities for the benefit of employees.

Insurance:

1. Premiums payable to insurance companies for protection against claims from injuries and damages by employees or others, such as public liability, property damages, casualty, employee liability, etc., and amounts credited to Account 228.2, Accumulated Provision for Injuries and Damage, for similar protection.

2. Losses not covered by insurance or reserve accruals on account of injuries or deaths to employees or others and damages to the property of others.

3. Fees and expenses of claim investigators.

4. Payment of awards to claimants for court costs and attorneys' services.

5. Medical and hospital service and expenses for employees as the result of occupational injuries or resulting from claims of others.

6. Compensation payments under workmen's compensation laws.

7. Compensation paid while incapacitated as the result of occupational injuries. (See Account 924, Note A.)

8. Cost of safety, accident prevention, and similar educational activities.

Materials and Expenses:

1. Use of newspapers, periodicals, billboards, and radio for informational purposes.

2. Postage on direct mailings to customers exclusive of postage related to billings.

3. Printing of informational booklets, dodgers, and bulletins.

4. Supplies and expenses in preparing informational materials by the utility.

5. Office supplies and expenses.

NOTE A: Exclude from this account and charge to Account 930.2, Miscellaneous General Expenses, the cost of publication of stockholder reports, dividend notices, bond redemption notices, financial statements, and other notices of a general corporate character. Also exclude all expenses of a promotional, institutional, goodwill, or political nature, which are includible in such accounts as 913, Advertising Expenses; 930.1, General Advertising Expenses; and 426.4, Expenditures for Certain Civic, Political and Related Activities.

NOTE B: Entries relating to informational advertising included in this account shall contain or refer to supporting documents which identify the specific advertising message. If references are used, copies of the advertising message shall be readily available.

910 Miscellaneous Customer Service and Informational Expenses

This account shall include the cost of labor, employee pensions and benefits, social security and other payroll taxes, injuries and damages, property insurance, property taxes, materials used, and expenses incurred in connection with customer service and informational activities which are not includible in other customer information expense accounts.

Items

Labor:

1. General clerical and stenographic work not assigned to specific customer service and informational programs.

2. Miscellaneous labor.

Taxes:

1. Federal and state unemployment.

2. F.I.C.A.

3. Property.

Employee Pensions and Benefits: The portion of employee pensions and benefits specifically identifiable with employees' labor costs charged herein or, in the absence of specific employee identification, the portion of employee pensions and benefits, allocated on the more equitable basis of either direct labor dollars or direct labor hours, applicable to the labor items detailed above, including:

1. Accruals for or payments to pension funds or to insurance companies for pension purposes.

2. Group and life insurance premiums (credit dividends received).

3. Payments for medical and hospital services and expenses of employees when not the result of occupational injuries.

4. Payments for accident, sickness, hospital, and death benefits or insurance.

5. Payments to employees incapacitated for service or on leave of absence beyond periods normally allowed when not the result of occupational injuries or in excess of statutory awards.

6. Expenses in connection with educational and recreational activities for the benefit of employees.

Insurance:

1. Premiums payable to insurance companies for fire, storm, burglary, boiler explosion, lightning, fidelity, riot, and similar insurance.

2. Amounts credited to Account 228.1, Accumulated Provision for Property Insurance, for similar protection.

3. Special costs incurred in procuring insurance.

4. Insurance inspection service.

5. Insurance counsel, brokerage fees, and expenses.

6. Premiums payable to insurance companies for protection against claims from injuries and damages by employees or others, such as public liability, property damages, casualty, employee liability, etc., and amounts credited to Account 228.2, Accumulated Provision for Injuries and Damage, for similar protection.

7. Losses not covered by insurance or reserve accruals on account of injuries or deaths to employees or others and damages to the property of others.

8. Fees and expenses of claim investigators.

9. Payment of awards to claimants for court costs and attorneys' services.

10. Medical and hospital service and expenses for employees as the result of occupational injuries or resulting from claims of others.

11. Compensation payments under workmen's compensation laws.

12. Compensation paid while incapacitated as the result of occupational injuries. (See Account 924, Note A.)

13. Cost of safety, accident prevention, and similar educational activities.

Materials and Expenses:

1. Communication service.

2. Printing, postage, and office supplies expenses.

[58 FR 59825, Nov. 10, 1993, as amended at 62 FR 42313, Aug. 6, 1997]

§ 1767.30 Sales expenses.

The sales expense accounts identified in this section shall be used by all RUS borrowers.

SALES EXPENSES

(Operation)

911 Supervision
912 Demonstrating and Selling Expenses
913 Advertising Expenses
916 Miscellaneous Sales Expenses

SALES EXPENSES

(Operation)

911 Supervision

This account shall include the cost of labor, employee pensions and benefits, social security and other payroll taxes, injuries and damages, and expenses incurred in the general direction and supervision of sales activities, except merchandising. Direct supervision of a specific activity, such as demonstrating, selling, or advertising shall be charged to the account wherein the costs of such activity are included. (See § 1767.17(a).)

912 Demonstrating and Selling Expenses

This account shall include the cost of labor, employee pensions and benefits, social security and other payroll taxes, injuries and damages, materials used, and expenses incurred in promotional, demonstrating, and selling activities, except by merchandising, the object of which is to promote or retain the use of utility services by present and prospective customers.

Items

Labor:

1. Demonstrating uses of utility services.
2. Conducting cooking schools, preparing recipes, and related home service activities.
3. Exhibitions, displays, lectures, and other programs designed to promote use of utility services.
4. Experimental and development work in connection with new and improved appliances and equipment, prior to general public acceptance.
5. Solicitation of new customers or of additional business from old customers, including commissions paid employees.
6. Engineering and technical advice to present or prospective customers in connection with promoting or retaining the use of utility services.
7. Special customer canvasses when their primary purpose is the retention of business or the promotion of new business.

Taxes:

1. Federal and state unemployment.
2. F.I.C.A.

Employee Pensions and Benefits: The portion of employee pensions and benefits specifically identifiable with employees' labor costs charged herein or, in the absence of specific employee identification, the portion of employee pensions and benefits, allocated on the more equitable basis of either direct labor dollars or direct labor hours, applicable to the labor items detailed above, including:

1. Accruals for or payments to pension funds or to insurance companies for pension purposes.
2. Group and life insurance premiums (credit dividends received).
3. Payments for medical and hospital services and expenses of employees when not the result of occupational injuries.
4. Payments for accident, sickness, hospital, and death benefits or insurance.
5. Payments to employees incapacitated for service or on leave of absence beyond periods normally allowed when not the result of occupational injuries or in excess of statutory awards.
6. Expenses in connection with educational and recreational activities for the benefit of employees.

Insurance:

1. Premiums payable to insurance companies for protection against claims from injuries and damages by employees or others, such as public liability, property damages, casualty, employee liability, etc., and amounts credited to Account 228.2, Accumulated Provision for Injuries and Damage, for similar protection.
2. Losses not covered by insurance or reserve accruals on account of injuries or deaths to employees or others and damages to the property of others.
3. Fees and expenses of claim investigators.
4. Payment of awards to claimants for court costs and attorneys' services.
5. Medical and hospital service and expenses for employees as the result of occupational injuries or resulting from claims of others.
6. Compensation payments under workmen's compensation laws.
7. Compensation paid while incapacitated as the result of occupational injuries. (See Account 924, Note A.)
8. Cost of safety, accident prevention, and similar educational activities.

Materials and Expenses:

1. Supplies and expenses pertaining to demonstration, experimental, and development activities.
2. Booth and temporary space rental.
3. Loss in value on equipment and appliances used for demonstration purposes.
4. Transportation, meals, and incidental expenses.

913 Advertising Expenses

This account shall include the cost of labor, employee pensions and benefits, social security and other payroll taxes, injuries and damages, materials used, and expenses incurred in advertising designed to promote or retain the use of utility service, except advertising the sale of merchandise by the utility.

Items

Labor:

1. Direct supervision of department.

2. Preparing advertising material for newspapers, periodicals, and billboards, and preparing and conducting motion pictures, radio, and television programs.

3. Preparing booklets and bulletins used in direct mail advertising.

4. Preparing window and other displays.

5. Clerical and stenographic work.

6. Investigating advertising agencies and media and conducting negotiations in connection with the placement and subject matter of sales advertising.

Taxes:

1. Federal and state unemployment.

2. F.I.C.A.

Employee Pensions and Benefits: The portion of employee pensions and benefits specifically identifiable with employees' labor costs charged herein or, in the absence of specific employee identification, the portion of employee pensions and benefits, allocated on the more equitable basis of either direct labor dollars or direct labor hours, applicable to the labor items detailed above, including:

1. Accruals for or payments to pension funds or to insurance companies for pension purposes.

2. Group and life insurance premiums (credit dividends received).

3. Payments for medical and hospital services and expenses of employees when not the result of occupational injuries.

4. Payments for accident, sickness, hospital, and death benefits or insurance.

5. Payments to employees incapacitated for service or on leave of absence beyond periods normally allowed when not the result of occupational injuries or in excess of statutory awards.

6. Expenses in connection with educational and recreational activities for the benefit of employees.

Insurance:

1. Premiums payable to insurance companies for protection against claims from injuries and damages by employees or others, such as public liability, property damages, casualty, employee liability, etc., and amounts credited to Account 228.2, Accumulated Provision for Injuries and Damage, for similar protection.

2. Losses not covered by insurance or reserve accruals on account of injuries or deaths to employees or others and damages to the property of others.

3. Fees and expenses of claim investigators.

4. Payment of awards to claimants for court costs and attorneys' services.

5. Medical and hospital service and expenses for employees as the result of occupational injuries or resulting from claims of others.

6. Compensation payments under workmen's compensation laws.

7. Compensation paid while incapacitated as the result of occupational injuries. (See Account 924, Note A.)

8. Cost of safety, accident prevention, and similar educational activities.

Materials and Expenses:

1. Advertising in newspapers, periodicals, billboards, and radio for sales promotion purposes, but not including institutional or goodwill advertising includible in Account 930.1, General Advertising Expenses.

2. Materials and services given as prizes or otherwise in connection with civic lighting contests, canning, or cooking contests, and bazaars in order to publicize and promote the use of utility services.

3. Fees and expenses of advertising agencies and commercial artists.

4. Novelties for general distribution.

5. Postage on direct mail advertising.

6. Premiums distributed generally, such as recipe books when not offered as inducement to purchase appliances.

7. Printing booklets, dodgers, and bulletins.

8. Supplies and expenses in preparing advertising material.

9. Office supplies and expenses.

NOTE A: The cost of advertisements which set forth the value or advantages of utility service without reference to specific appliances, or, if reference is made to appliances, invites the reader to purchase appliances from his dealer or refer to appliances not carried for sale by the utility, shall be considered sales promotion advertising and charged to this account. However, advertisements which are limited to specific makes of appliances sold by the utility and price and terms, thereof, without referring to the value or advantages of utility service, shall be considered as merchandise advertising and the cost shall be charged to Costs and Expenses of Merchandising, Jobbing and Contract Work, Account 416.

NOTE B: Advertisements which substantially mention or refer to the value or advantages of utility service, together with specific reference to makes of appliance sold by the utility and the price, and terms, thereof, and designed for the joint purpose of increasing the use of utility service and the sales of appliances, shall be considered as a combination advertisement and the costs shall be distributed between this account and Account 416 on the basis of space, time, or other proportional factors.

NOTE C: Exclude from this account and charge to Account 930.2, Miscellaneous General Expenses, the cost of publication of stockholder reports, dividend notices, bond redemption notices, financial statements, and other notices of a general corporate character. Also exclude all institutional or goodwill advertising. (See Account 930.1, General Advertising Expenses.)

916 Miscellaneous Sales Expenses

This account shall include the cost of labor, employee pensions and benefits, social security and other payroll taxes, injuries and damages, property insurance, property taxes, materials used, and expenses incurred in connection with sales activities, except merchandising, which are not includible in other sales expense accounts.

Items

Labor:

1. General clerical and stenographic work not assigned to specific functions.
2. Special analysis of customer accounts and other statistical work for sales purposes not a part of the regular customer accounting and billing routine.
3. Miscellaneous labor.

Taxes:

1. Federal and state unemployment.
2. F.I.C.A.
3. Property.

Employee Pensions and Benefits: The portion of employee pensions and benefits specifically identifiable with employees' labor costs charged herein or, in the absence of specific employee identification, the portion of employee pensions and benefits, allocated on the more equitable basis of either direct labor dollars or direct labor hours, applicable to the labor items detailed above, including:

1. Accruals for or payments to pension funds or to insurance companies for pension purposes.
2. Group and life insurance premiums (credit dividends received).
3. Payments for medical and hospital services and expenses of employees when not the result of occupational injuries.
4. Payments for accident, sickness, hospital, and death benefits or insurance.
5. Payments to employees incapacitated for service or on leave of absence beyond periods normally allowed when not the result of occupational injuries or in excess of statutory awards.
6. Expenses in connection with educational and recreational activities for the benefit of employees.

Insurance:

1. Premiums payable to insurance companies for fire, storm, burglary, boiler explosion, lightning, fidelity, riot, and similar insurance.
2. Amounts credited to Account 228.1, Accumulated Provision for Property Insurance, for similar protection.
3. Special costs incurred in procuring insurance.
4. Insurance inspection service.
5. Insurance counsel, brokerage fees, and expenses.
6. Premiums payable to insurance companies for protection against claims from injuries and damages by employees or others, such as public liability, property damages, casualty, employee liability, etc., and amounts credited to Account 228.2, Accumulated Provision for Injuries and Damage, for similar protection.
7. Losses not covered by insurance or reserve accruals on account of injuries or deaths to employees or others and damages to the property of others.
8. Fees and expenses of claim investigators.
9. Payment of awards to claimants for court costs and attorneys' services.
10. Medical and hospital service and expenses for employees as the result of occupational injuries or resulting from claims of others.
11. Compensation payments under workmen's compensation laws.
12. Compensation paid while incapacitated as the result of occupational injuries. (See Account 924, Note A.)
13. Cost of safety, accident prevention, and similar educational activities.

Materials and Expenses:

1. Communication service.
2. Printing, postage, office supplies, and expenses applicable to sales activities, except those chargeable to Account 913, Advertising Expenses.

[58 FR 59825, Nov. 10, 1993, as amended at 62 FR 42315, Aug. 6, 1997]

§ 1767.31 Administrative and general expenses.

The administrative and general expense accounts identified in this section shall be used by all RUS borrowers.

ADMINISTRATIVE AND GENERAL

(Operation)

920 Administrative and General Salaries
921 Office Supplies and Expenses
922 Administrative Expenses Transferred—Credit
923 Outside Services Employed
924 Property Insurance
925 Injuries and Damages

926 Employee Pensions and Benefits
927 Franchise Requirements
928 Regulatory Commission Expenses
929 Duplicate Charges—Credit
930.1 General Advertising Expenses
930.2 Miscellaneous General Expenses
931 Rents

(Maintenance)

935 Maintenance of General Plant

ADMINISTRATIVE AND GENERAL

(Operation)

920 Administrative and General Salaries

A. This account shall include the compensation (salaries, bonuses, employee pensions and benefits, social security and other payroll taxes, injuries and damages, and other consideration for services, but not including directors' fees) of officers, executives, and other employees of the utility properly chargeable to utility operations and not chargeable directly to a particular operating function.

B. This account may be subdivided in accordance with a classification appropriate to the departmental or other functional organization of the utility.

921 Office Supplies and Expenses

A. This account shall include office supplies and expenses incurred in connection with the general administration of the utility's operations which are assignable to specific administrative or general departments and are not specifically provided for in other accounts. This includes the expenses of the various administrative and general departments, the salaries and wages of which are includible in Account 920.

B. This account may be subdivided in accordance with a classification appropriate to the departmental or other functional organization of the utility.

NOTE: Office expenses which are clearly applicable to any category of operating expenses other than the administrative and general category shall be included in the appropriate account in such category. Further, general expenses which apply to the utility as a whole rather than to a particular administrative function, shall be included in Account 930.2, Miscellaneous General Expenses.

Items

1. Automobile service, including charges through clearing account.
2. Bank messenger and service charges.
3. Books, periodicals, bulletins, and subscriptions to newspapers, newsletters, and tax services.

4. Building service expenses for customer accounts, sales, and administrative and general purposes.
5. Communication service expenses.
6. Cost of individual items of office equipment used by general departments which are of small value or short life.
7. Membership fees and dues in trade, technical, and professional associations paid by a utility for employees. (Company memberships are includible in Account 930.2.)
8. Office supplies and expenses.
9. Payment of court costs, witness fees, and other expenses of legal department.
10. Postage, printing, and stationery.
11. Meals, traveling, and incidental expenses.

922 Administrative Expenses Transferred—
Credit

This account shall be credited with administrative expenses recorded in Account 920 and Account 921 which are transferred to construction costs or to nonutility accounts. (See § 1767.16 (d).)

923 Outside Services Employed

A. This account shall include the fees and expenses of professional consultants and others for general services which are not applicable to a particular operating function or other accounts. It shall include also the pay and expenses of persons engaged for a special or temporary administrative or general purpose in circumstances where the person so engaged is not considered as an employee of the utility.

B. This account shall be so maintained as to permit ready summarization according to the nature of service and the person furnishing the same.

Items

1. Fees, pay, and expenses of accountants and auditors, actuaries, appraisers, attorneys, engineering consultants, management consultants, negotiators, public relations counsel, and tax consultants.
2. Supervision fees and expenses paid under contracts for general management services.

NOTE: Do not include inspection and brokerage fees and commissions chargeable to other accounts or fees and expenses in connection with security issues which are includible in the expenses of issuing securities.

924 Property Insurance

A. This account shall include the cost of insurance or reserve accruals to protect the utility against losses and damages to owned or leased property used in its utility operations. It shall also include the cost of labor, employee pensions and benefits, social security and other payroll taxes, injuries and damages, and the related supplies and expenses incurred in property insurance activities.

B. Recoveries from insurance companies or others for property damages shall be credited to the account charged with the cost of the damage. If the damaged property has been retired, the credit shall be to the appropriate account for accumulated provision for depreciation.

C. Records shall be kept so as to show the amount of coverage for each class of insurance carried, the property covered, and the applicable premiums. Any dividends distributed by mutual insurance companies shall be credited to the accounts to which the insurance premiums were charged.

Items

1. Premiums payable to insurance companies for fire, storm, burglary, boiler explosion, lightning, fidelity, riot, and similar insurance.

2. Amounts credited to Account 228.1, Accumulated Provision for Property Insurance, for similar protection.

3. Special costs incurred in procuring insurance.

4. Insurance inspection service.

5. Insurance counsel, brokerage fees, and expenses.

NOTE A: The cost of insurance or reserve accruals capitalized, shall be charged to construction and retirement either directly or by transfers to construction and retirement work orders from this account.

NOTE B: The cost of insurance or reserve accruals for the following classes of property shall be charged as indicated:

1. Materials, supplies, and stores equipment to Account 163, Stores Expense Undistributed, or appropriate materials account.

2. Transportation and other general equipment to appropriate clearing accounts that may be maintained.

3. Electric plant leased to others to Account 413, Expenses of Electric Plant Leased to Others.

4. Nonutility property to the appropriate nonutility income account.

5. Merchandise and jobbing property to Account 416, Costs and Expenses of Merchandising, Jobbing and Contract Work.

NOTE C: The cost of labor, employee pensions and benefits, social security and other payroll taxes, and the related supplies and expenses of administrative and general employees who are only incidentally engaged in property insurance work may be included in Account 920 and Account 921, as appropriate.

NOTE D: The cost of insurance or reserve accruals applicable to the various utility functions shall be charged to the specific functional operations and the appropropriate miscellaneous administrative expense accounts either directly or by transfers from this account.

925 Injuries and Damages

A. This account shall include the cost of insurance or reserve accruals to protect the utility against injuries and damages claims of employees or others, losses of such character not covered by insurance, and expenses incurred in settlement of injuries and damages claims. It shall also include the cost of labor, employee pensions and benefits, social security and other payroll taxes, injuries and damages, related supplies, and expenses incurred in injuries and damages activities.

B. Reimbursements from insurance companies or others for expenses charged hereto on account of injuries, damages, and insurance dividends or refunds shall be credited to this account.

Items

1. Premiums payable to insurance companies for protection against claims from injuries and damages by employees or others, such as public liability, property damages, casualty, employee liability, etc., and amounts credited to Account 228.2, Accumulated Provision for Injuries and Damage, for similar protection.

2. Losses not covered by insurance or reserve accruals on account of injuries or deaths to employees or others and damages to the property of others.

3. Fees and expenses of claim investigators.

4. Payment of awards to claimants for court costs and attorneys' services.

5. Medical and hospital service and expenses for employees as the result of occupational injuries or resulting from claims of others.

6. Compensation payments under workmen's compensation laws.

7. Compensation paid while incapacitated as the result of occupational injuries. (See Note A.)

8. Cost of safety, accident prevention, and similar educational activities.

NOTE A: Payments to or in behalf of employees for accident or death benefits, hospital expenses, medical expenses, or for salaries while incapacitated for service or on leave of absence beyond periods normally allowed, when not the result of occupational injuries, shall be charged to Account 926, Employee Pensions and Benefits. (See also Note B of Account 926.)

NOTE B: The cost of injuries and damages or reserve accruals capitalized shall be charged to construction and retirement activities either directly or by transfers from this account to the applicable construction and retirement work orders.

NOTE C: The cost of insurance or reserve accruals applicable to the various utility functions shall be charged to the specific functional operations and the appropropriate miscellaneous administrative expense accounts either directly or by transfers from this account.

NOTE D: Exclude herefrom the time and expenses of employees (except those engaged in injuries and damages activities) spent in attendance at safety and accident prevention educational meetings, if occurring during the regular work period.

NOTE E: The cost of labor, employee pensions and benefits, social security and other payroll taxes, and the related supplies and expenses of administrative and general employees who are only incidentally engaged in injuries and damages activities, may be included in Account 920 and Account 921, as appropriate.

926 Employee Pensions and Benefits

A. This account shall include pensions paid to or on behalf of retired employees or accruals to provide for pensions or payments for the purchase of annuities for this purpose, when the utility has definitely, by contract, committed itself to a pension plan under which the pension funds are irrevocably devoted to pension purposes and payments for employee accident, sickness, hospital, and death benefits, or insurance therefor. Include, also, expenses incurred in medical, educational, or recreational activities for the benefit of employees and administrative expenses in connection with employee pensions and benefits.

B. The utility shall maintain a complete record of accruals or payments for pensions and be prepared to furnish full information to RUS of the plan under which it has created or proposes to create a pension fund and a copy of the declaration of trust or resolution under which the pension plan is established.

C. There shall be credited to this account, the portion of pensions and benefits expenses which is applicable to nonutility operations, the specific functional operations, maintenance, and administrative expense accounts, and to construction and retirement activities unless such amounts are distributed directly to the accounts involved and are not included herein in the first instance.

D. Records in support of this account shall be so kept that the total pensions expense, the total benefits expense, the administrative expenses included herein, and the amounts of pensions and benefits expenses transferred to the operations, maintenance, administrative, construction or retirement accounts will be readily available.

Items

1. Payment of pensions to retirees on a nonaccrual basis.

2. Accruals for or payments to pension funds or to insurance companies for pension purposes.

3. Group and life insurance premiums (credit dividends received).

4. Payments for medical and hospital services and expenses of employees when not the result of occupational injuries.

5. Payments for accident, sickness, hospital, and death benefits or insurance.

6. Payments to employees incapacitated for service or on leave of absence beyond periods normally allowed when not the result of occupational injuries or in excess of statutory awards.

7. Expenses in connection with educational and recreational activities for the benefit of employees.

NOTE A: The cost of labor, employee pensions and benefits, social security and other payroll taxes, injuries and damages, and the related supplies and expenses of administrative and general employees who are only incidentally engaged in employee pension and benefit activities may be included in Account 920 and Account 921, as appropriate.

NOTE B: Salaries paid to employees during periods of nonoccupational sickness may be charged to the appropriate labor account rather than to employee benefits.

927 Franchise Requirements

A. This account shall include payments to municipal or other governmental authorities and the cost of materials, supplies, and services furnished such authorities without reimbursement in compliance with franchise, ordinance, or similar requirements; provided, however, that the utility may charge to this account at regular tariff rates, instead of cost, utility service furnished without charge under provisions of franchises.

B. When no direct outlay is involved, concurrent credit for such charges shall be made to Account 929, Duplicate Charges—Credit.

C. The account shall be maintained so as to readily reflect the amounts of cash outlays, utility service supplied without charge, and other items furnished without charge.

NOTE A: Franchise taxes shall not be charged to this account, but to Account 408.1, Taxes Other Than Income Taxes, Utility Operating Income.

NOTE B: Any amount paid as initial consideration for a franchise running for more than one year shall be charged to Account 302, Franchises and Consents.

928 Regulatory Commission Expenses

A. This account shall include all expense (except pay of regular employees only incidentally engaged in such work) properly includible in utility operating expenses, incurred by the utility in connection with formal cases before regulatory commissions or other regulatory bodies or cases in which such a body is a party, including payments made to a regulatory commission for fees assessed against the utility for pay and expenses of such commission, its officers, agents, and employees, and also including payments made to the United States for the administration of the Federal Power Act.

B. Amounts of regulatory commission expenses which, by approval or direction of RUS, are to be spread over future periods shall be charged to Account 182.3, Other Regulatory Assets, and amortized by charges to this account.

C. The utility shall be prepared to show the cost of each formal case.

Items

1. Salaries, fees, retainers, and expenses of counsel, solicitors, attorneys, accountants, engineers, clerks, attendants, witnesses, and others engaged in the prosecution of or defence against petitions or complaints presented to regulatory bodies or in the valuation of property owned or used by the utility in connection with such cases.

2. Office supplies and expenses, payments to public service or other regulatory commissions, stationery and printing, traveling expenses, and other expenses incurred directly in connection with formal cases before regulatory commissions.

NOTE A: Exclude from this account and include in other appropriate operating expense accounts, expenses incurred in the improvement of service, additional inspection, or rendering reports which are made necessary by the rules and regulations, or orders, of regulatory bodies.

NOTE B: Do not include in this account amounts includible in Account 302, Franchises and Consents; Account 181, Unamortized Debt Expense; or Account 214, Capital Stock Expense.

929   Duplicate Charges—Credit

This account shall include concurrent credits for charges which may be made to operating expenses or to other accounts for the use of utility service from its own supply. Include, also, offsetting credits for any other charges made to operating expenses for which there is no direct money outlay.

930.1   General Advertising Expenses

This account shall include the cost of labor, employee pensions and benefits, social security and other payroll taxes, injuries and damages, materials used, and expenses incurred in advertising and related activities, the cost of which by their content and purpose are not provided for elsewhere.

Items

Labor:

1. Supervision.

2. Preparing advertising material for newspapers, periodicals, and billboards and preparing or conducting motion pictures, radio, and television programs.

3. Preparing booklets and bulletins used in direct mail advertising.

4. Preparing window and other displays.

5. Clerical and stenographic work.

6. Investigating and employing advertising agencies, selecting media, and conducting negotiations in connection with the placement and subject matter of advertising.

Taxes:

1. Federal and state unemployment.

2. F.I.C.A.

Employee Pensions and Benefits: The portion of employee pensions and benefits specifically identifiable with employees' labor costs charged herein or, in the absence of specific employee identification, the portion of employee pensions and benefits, allocated on the more equitable basis of either direct labor dollars or direct labor hours, applicable to the labor items detailed above, including:

1. Accruals for or payments to pension funds or to insurance companies for pension purposes.

2. Group and life insurance premiums (credit dividends received).

3. Payments for medical and hospital services and expenses of employees when not the result of occupational injuries.

4. Payments for accident, sickness, hospital, and death benefits or insurance.

5. Payments to employees incapacitated for service or on leave of absence beyond periods normally allowed when not the result of occupational injuries or in excess of statutory awards.

6. Expenses in connection with educational and recreational activities for the benefit of employees.

Insurance:

1. Premiums payable to insurance companies for protection against claims from injuries and damages by employees or others, such as public liability, property damages, casualty, employee liability, etc., and amounts credited to Account 228.2, Accumulated Provision for Injuries and Damage, for similar protection.

2. Losses not covered by insurance or reserve accruals on account of injuries or deaths to employees or others and damages to the property of others.

3. Fees and expenses of claim investigators.

4. Payment of awards to claimants for court costs and attorneys' services.

5. Medical and hospital service and expenses for employees as the result of occupational injuries or resulting from claims of others.

6. Compensation payments under workmen's compensation laws.

7. Compensation paid while incapacitated as the result of occupational injuries. (See Account 924, Note A.)

8. Cost of safety, accident prevention, and similar educational activities.

Materials and Expenses:

1. Advertising in newspapers, periodicals, billboards, and radios.

2. Advertising matter such as posters, bulletins, booklets, and related items.

3. Fees and expenses of advertising agencies and commercial artists.

4. Postage and direct mail advertising.

5. Printing of booklets, dodgers, and bulletins.

6. Supplies and expenses in preparing advertising materials.

7. Office supplies and expenses.

NOTE A: Properly includible in this account is the cost of advertising activities on a local or national basis of a goodwill or institutional nature, which is primarily designed to improve the image of the utility or the industry, including advertisements which inform the public concerning matters affecting the company's operations, such as, the cost of providing service, the company's efforts to improve the quality of service, and the company's efforts to improve and protect the environment. Entries relating to advertising included in this account shall contain or refer to supporting documents which identify the specific advertising message. If references are used, copies of the advertising message shall be readily available.

NOTE B: Exclude from this account and include in Account 426.4, Expenditures for Certain Civic, Political and Related Activities, expenses for advertising activities, which are designed to solicit public support or the support of public officials in matters of a political nature.

930.2 Miscellaneous General Expenses

This account shall include the cost of labor, employee pensions and benefits, social security and other payroll taxes, injuries and damages, property insurance, property taxes, and expenses incurred in connection with the general management of the utility not provided for elsewhere.

Items

Labor:

1. Miscellaneous labor not elsewhere provided for.

Taxes:

1. Federal and state unemployment.

2. F.I.C.A.

3. Property.

Employee Pensions and Benefits: The portion of employee pensions and benefits specifically identifiable with employees' labor costs charged herein or, in the absence of specific employee identification, the portion of employee pensions and benefits, allocated on the more equitable basis of either direct labor dollars or direct labor hours, applicable to the labor items detailed above, including:

1. Accruals for or payments to pension funds or to insurance companies for pension purposes.

2. Group and life insurance premiums (credit dividends received).

3. Payments for medical and hospital services and expenses of employees when not the result of occupational injuries.

4. Payments for accident, sickness, hospital, and death benefits or insurance.

5. Payments to employees incapacitated for service or on leave of absence beyond periods normally allowed when not the result of occupational injuries or in excess of statutory awards.

6. Expenses in connection with educational and recreational activities for the benefit of employees.

Insurance:

1. Premiums payable to insurance companies for fire, storm, burglary, boiler explosion, lightning, fidelity, riot, and similar insurance.

2. Amounts credited to Account 228.1, Accumulated Provision for Property Insurance, for similar protection.

3. Special costs incurred in procuring insurance.

4. Insurance inspection service.

5. Insurance counsel, brokerage fees, and expenses.

6. Premiums payable to insurance companies for protection against claims from injuries and damages by employees or others, such as public liability, property damages, casualty, employee liability, etc., and amounts credited to Account 228.2, Accumulated Provision for Injuries and Damage, for similar protection.

7. Losses not covered by insurance or reserve accruals on account of injuries or deaths to employees or others and damages to the property of others.

8. Fees and expenses of claim investigators.

9. Payment of awards to claimants for court costs and attorneys' services.

10. Medical and hospital service and expenses for employees as the result of occupational injuries or resulting from claims of others.

11. Compensation payments under workmen's compensation laws.

12. Compensation paid while incapacitated as the result of occupational injuries. (See Account 924, Note A.)

13. Cost of safety, accident prevention, and similar educational activities.

Expenses:

1. Industry association dues for company memberships.

145

2. Contributions for conventions and meetings of the industry.

3. Research, development, and demonstration expenses not charged to other operation and maintenance expense accounts on a functional basis.

4. Communication service not chargeable to other accounts.

5. Trustee, registrar, and transfer agent fees and expenses.

6. Stockholders meeting expenses.

7. Dividend and other financial notices.

8. Printing and mailing dividend checks.

9. Directors' fees and expenses.

10. Publishing and distributing annual reports to stockholders.

11. Public notices of financial, operating, and other data required by regulatory statutes, not including, however, notices required in connection with security issues or acquisitions of property.

931   Rents

This account shall include rents properly includible in utility operating expenses for the property of others used, occupied, or operated in connection with the customer accounts, customer service and informational, sales, general, and administrative functions of the utility. (See § 1767.17 (c).)

(Maintenance)

935   Maintenance of General Plant

A. This account shall include the cost assignable to customer accounts, sales, administrative, and general functions of labor, employee pensions and benefits, social security and other payroll taxes, injuries and damages, materials used, and expenses incurred in the maintenance of property, the book cost of which is includible in Account 390, Structures and Improvements; Account 391, Office Furniture and Equipment; Account 397, Communication Equipment; and Account 398, Miscellaneous Equipment. (See § 1767.17(b).)

B. Maintenance expenses on office furniture and equipment used elsewhere than in general, commercial, and sales offices shall be charged to the following accounts:

1. Steam Power Generation, Account 514.

2. Nuclear Power Generation, Account 532.

3. Hydraulic Power Generation, Account 545.

4. Other Power Generation, Account 554.

5. Transmission, Account 573.

6. Distribution, Account 598.

7. Merchandise and Jobbing, Account 416.

8. Garages, Shops, etc., Appropriate clearing account, if used.

NOTE: Maintenance of plant included in other general equipment accounts shall be included herein unless charged to clearing accounts or to the particular functional

maintenance expense account indicated by the use of the equipment.

[58 FR 59825, Nov. 10, 1993, as amended at 62 FR 42317, Aug. 6, 1997]

§§ 1767.32–1767.40   [Reserved]

§ 1767.41   Accounting methods and procedures required of all RUS borrowers.

All RUS borrowers shall maintain and keep their books of accounts and all other books and records which support the entries in such books of accounts in accordance with the accounting principles prescribed in this section. Interpretations Nos. 133, 134, 137, 403, 404, 602, 606, 618, 627, 628, and 629 adopt and implement the provisions of standards issued by the Financial Accounting Standards Board (FASB). Each interpretation includes a synopsis of the requirements of the standard as well as specific accounting requirements and interpretations required by RUS. The synopsis provides general information to assist borrowers in determining whether the standard applies to an individual cooperative's operations. The synopsis is not intended to change the requirements of the FASB standards unless it is set forth in the section entitled RUS Accounting Requirements in each interpretation. If a particular borrower believes a conflict exists between the FASB standard and an RUS interpretation, the borrower shall contact the Director, PASD, to seek resolution of the issue.

Numerical Index

Number	Title
101	Work Order Procedures
102	Line Conversion
103	Sacrificial Anodes and the Replacement of a Neutral
104	Terminal Facilities
105	Pole Top Disconnect Switch
106	Steel Pole Reinforcers
107	Mobile Substations
108	Security Lights
109	Joint Use
110	First Clearing and Grading of Land and Rights of Way
111	Engineering Contracts for System Planning
112	Determination of Availability of Service
113	Temporary Facilities (Services)

Subject Matter Index

101 Work Order Procedures

When a minor item of property is removed from service and not replaced, a retirement work order is not required except in the case of a conductor. The cost of the minor item shall remain in the appropriate plant account until the retirement unit, of which it is a part, is retired. However, as conductor is recorded in feet and is not part of any specific retirement unit, conductor shall be retired even though the amount taken down and not replaced is less than a retirement unit (two spans).

When minor items of plant are removed and not replaced, material salvaged shall be recorded on a material salvage ticket. Items of material recorded on this ticket shall be charged to the materials and supplies account and credited in the miscellaneous columns of the Materials Register to the Accumulated Provision for Depreciation. In this example, it is assumed that the cost of removal is nil. If, however, costs are incurred during the removal of minor items of plant, these costs shall reduce the credit to the Accumulated Provision for Depreciation.

When a staking sheet supporting a single work order reflects a combination of new construction and replacements, or system improvements, the predominant cost shall be the governing factor in determining the amount of cost RUS will finance. To illustrate, assume that a service is to be run to a new home near the end of an existing line. On inspection, the pole from which the service is to be run is found to be in very poor physical condition and must be replaced. In addition, a single span of wire and a service are presently connected to this pole which serve no purpose. The home originally served has been demolished and the existing span, pole, and service were retired. In other words, what started out to be simply the installation of a new service now includes the retirement of a span of wire, a pole, and a service; the replacement of a pole; and the running of a new service. Assuming the replacement of the pole is the costliest part of this project, the construction and retirement activity shall be classified as an ordinary replacement even though the work includes new construction and retirements without replacement.

102 Line Conversion

If it is necessary to move a conductor from one location to another on a pole assembly during the conversion of a line from one phase to another phase, the cost of moving the conductor is capitalizable as a system improvement.

103 Sacrificial Anodes and the Replacement of a Neutral

Many utilities conduct studies to determine whether sacrificial anodes are needed to protect underground cable against corrosion. The following procedures shall be followed to account for sacrificial anodes and the replacement of a neutral:

1. If the study results in the installation of sacrificial anodes, the cost of the study shall be capitalized to Account 367, Underground Conductors and Devices. If the study does not result in the installation of anodes, the cost shall be charged to Account 594, Maintenance of Underground Lines.

2. Costs incurred in the first installation are capitalizable even though anodes are considered minor items of property. However, only the first costs of installation shall be capitalized. All subsequent replacements of anodes shall be expensed.

3. Sacrificial anodes do not constitute a record unit; therefore, the cost of anodes shall be added to the cost of the underground cable unit.

4. Because a neutral is part of an underground cable record unit, and is not, in and of itself, a record unit, the cost to replace a corroded neutral shall be charged to Account 594, Maintenance of Underground Lines.

104 Terminal Facilities

Borrowers are sometimes required to construct terminal facilities in the transmission line of another utility in order to receive power from their power supplier. The document executed between the borrower and the utility is normally referred to as a "License Agreement". The license agreement may stipulate that certain items of the terminal facilities are to be transferred to, and become the property of, the other utility upon completion of the construction. The accounting for this type of transaction shall be as follows:

1. All construction costs incurred shall be charged to a work order. Upon completion of the construction and accumulation of all costs, the cost of the facilities that become the property of another utility shall be transferred from construction work-in-progress to Account 303, Miscellaneous Intangible Plant. The cost of the plant for which the borrower retains title shall be charged to the appropriate plant accounts.

2. The cost of the facilities recorded in Account 303 shall be amortized to Account 405, Amortization of Other Electric Plant, over the contract term or the estimated useful service life of the plant, whichever is shorter. If the related contract or contracts for this power supply are terminated, the unamortized balance shall be expensed, in the current period, in Account 557.

105 Pole Top Disconnect Switch

The installation of pole top service disconnect switches, where title is retained by the utility, shall be capitalized in Account 371, Installations on Customers' Premises. If a switch cabinet is purchased with a current transformer included as an integral part of the cabinet, the entire cost of the switch shall be charged to Account 371. If the current transformer is installed outside of the switch cabinet, the transformer, meter, and meter base, together with the first installation costs, shall be capitalized, upon purchase, in Account 370, Meters.

Payments received from the customer toward construction costs shall be credited to Account 371, Installations on Customers' Premises. Such payments, together with any amount not financed by RUS, shall be entered in column 9 of the RUS Form 219, Inventory of Work Orders. The associated maintenance costs shall be charged to Account 587, Customer Installations Expenses, or to Account 597, Maintenance of Meters, as appropriate.

When pole top disconnect switches are installed and title is held by the customer, the cost of the material shall be charged to Account 456, Other Electric Revenues and the receipts from the sale of line material shall be credited to Account 456. The portion of the receipts for resale material as well as that for installation shall be credited to Account 415, Revenues from Merchandising, Jobbing, and Contract Work. The cost of resale material sold and the cost of installation shall be charged to Account 416, Costs and Expenses of Merchandising, Jobbing and Contract Work.

Future maintenance costs incurred by the cooperative that are not billed to the customer shall be charged to Account 587, Customer Installations Expenses.

106 Steel Pole Reinforcers

The cost associated with the purchase and installation of steel pole reinforcers shall be charged to Account 593, Maintenance of Overhead Lines.

107 Mobile Substations

Mobile substations shall be accounted for in a manner similar to that for a spare and are, therefore, included as part of transmission or distribution station equipment, depending upon the use of the mobile substation. The mobile substation, together with the trailer on which it is permanently mounted, shall be capitalized upon purchase. A general purpose truck or tractor used to relocate a mobile substation and trailer shall be classified as transportation equipment.

The composite depreciation rate used for transmission plant or distribution plant, as appropriate, shall be applied to the mobile substation.

108 Security Lights

Where a pole supports both a secondary wire and a security light, the cost of the pole shall be charged to Account 364, Poles, Towers, and Fixtures, even though the plant investment in security lights is recorded in Account 371, Installations on Customers' Premises.

109 Joint Use

There are many cases in which an electric utility and a communications utility enter into an agreement that provides for joint use of poles. Under the terms of these agreements, either utility may occupy the poles of the other upon payment of a stipulated annual rental. If such joint occupancy necessitates the use of a higher than standard pole, the new pole shall be provided at the expense of the utility having the need for the higher pole.

When an electric utility replaces, at its own expense, a standard pole belonging to the communications utility with a higher pole, the cost of the higher pole, less net salvage (if any) of the pole replaced, shall be charged to the account in which the pole rental is included.

Contributions made to an electric utility by a communications utility for the costs incurred in stubbing joint use electric poles shall be credited to Account 593, Maintenance of Overhead Lines. The cost of pole stubbing on electric plant distribution facilities shall be charged to Account 593.

An investment in outside plant that is held in joint ownership shall be recorded in the appropriate plant accounts at its cost to the utility. For continuing property record purposes, jointly owned property units shall be priced at their cost to the utility and shall be appropriately segregated in the CPRs to indicate joint ownership.

110 First Clearing and Grading of Land and Rights of Way

Utility accounting practice requires the costs associated with the first clearing and grading of land and rights of way and any resulting damage thereto, to be included in the accounts for structures and improvements or equipment to which such costs relate. Since the first clearing, as well as clearing which is "directly occasioned by the building of a structure," is done, not for the purpose of enhancing the value of the land or the rights of way, but for the purpose of constructing plant, these costs are more directly related to the construction of plant than to the purchase of land or rights of way. The accounts shall be charged as follows:

1. For overhead transmission pole lines, Account 356, Overhead Conductors and Devices;

2. For overhead distribution lines, Account 365, Overhead Conductors and Devices; and

3. For underground distribution lines, Account 366, Underground Conduit, for a conduit installation; or Account 367, Underground Conductors and Devices, for a direct burial installation.

111 Engineering Contracts for System Planning

Engineering costs for long-range system plans shall be charged to Account 183, Preliminary Survey and Investigation Charges, as incurred. The cost of engineering services incurred in preparing a long-range system plan represents a legitimate component of the total cost of construction of all system improvements detailed in the plan. The amount of engineering costs to be associated with any specific system improvement is the annual costs incurred up to the time of the allocation (not previously allocated), plus that portion of the initial cost which relates to the particular construction in question. If any major system improvement included in the engineering plan is not constructed, or if the study is superseded by another complete study, the cost of that portion of the original study not resulting in construction shall be charged to Account 182.2, Unrecovered Plant and Regulatory Study Costs, if the costs are to be recovered through future rates. Costs recorded in Account 182.2 shall be amortized to Account 407, Amortization of Property Losses, Unrecovered Plant and Regulatory Study Costs, as the costs are recovered through the rates. Any costs included in Account 182.2 that are disallowed for rate-making purposes shall be charged to Account 426.5, Other Deductions.

The allocation of engineering services to the various construction projects requires the exercise of judgment. In some cases, system improvements are continuous over a period of months or years, thus permitting the engineering cost to be spread monthly as overhead in relation to the direct costs incurred in construction. (If a substantial amount of retirement work is performed in connection with system improvements, a proportionate share of the engineering cost shall be allocated on the basis of direct retirement labor.) If the system improvements detailed in the plan are not performed in a continuous manner, the engineering cost shall be allocated on the basis of the estimated costs of the various larger system improvement projects which result from the long-range plan.

If construction is performed by contract, the engineering cost applicable thereto shall be transferred from Account 183 to Account 107, Construction Work-in-Progress—Electric, and thereby spread to the appropriate plant accounts on the basis of contract costs.

In the case of system improvement construction performed on the basis of work orders, engineering costs shall be transferred to Account 107, Construction Work-in-Progress—Electric, and included in total work order costs as either overhead or special services. If engineering services are not readily identifiable with individual work orders, they shall be capitalized as overhead. If engineering costs for each work order are readily separable from the engineering costs for all other work orders, they shall be capitalized as special services.

In summarizing system improvement work orders on the RUS Form 219, Inventory of Work Orders, the amount of engineering costs previously approved for advance on the long range plan, if any, shall be deducted to determine the balance of loan funds subject to advance by RUS.

112 Determination of Availability of Service

Costs relating to the determination of availability of service, rates, and similar items for individual applicants shall be charged to Account 912, Demonstrating and Selling Expenses. If it is expected that construction will result, the costs incurred to provide service, including staking, shall be charged to Account 107, Construction Work-in-Progress—Electric. If construction does not result, Account 107 shall be credited and Account 426.5, Other Deductions, shall be charged.

113 Temporary Facilities (Services)

Plant installed for temporary use, a period of less than 1.ar, shall be recorded in Account 185, Temporary Facilities, net of any payments received from customers. Upon retirement, this net cost plus cost of removal, less any salvage value, shall be cleared to Account 451, Miscellaneous Service Revenues.

When a temporary service is installed at the site of a building under construction, the location of the permanent service entrance and the load and its characteristics are usually known. The temporary service is of the proper capacity and is so located or has sufficient slack, that it can be relocated to serve the new building as a permanent service. Under these conditions, the service shall be charged to Account 369, Services, when first installed. The cost of moving and attaching the service to the permanent service entrance shall be charged to Account 593, Maintenance of Overhead Lines or Account 594, Maintenance of Underground Lines, as appropriate.

114 Construction Work-in-Progress Damaged or Destroyed by Storm

When installed plant, not yet completed or completed but not yet placed in service, has been damaged or destroyed by storm, the cost of the repair and restoration shall be added to the cost of construction and capitalized if the plant was constructed under force account or work order construction, and the utility paid for the cost of the repairs. If the plant was constructed under contract, the contractor is required to deliver the plant in new condition. Therefore, any repairs required prior to the completion of construction and acceptance by the utility, are ordinarily borne by the contractor.

115 Liquidated Damages

Liquidated damages are amounts paid by or assessed against contractors for the completion of construction after an agreed upon date. Liquidated damages shall be credited to Account 107, Construction Work-in-Progress—Electric. Since these damages accrue during the construction period, they become one of the components of construction cost. Even though a portion of these damages may compensate the utility for costs which are not "identifiable," no portion of the damages shall be credited to revenue or expense.

When a contractor has been paid in full from loan funds or from funds to be reimbursed by loan funds without a deduction for liquidated damages, the amount of liquidated damages received shall be deposited in the Construction Fund. This amount shall be reflected by a decrease in column 5, "Total Expenditures to Date," of the RUS Form 595, Financial Requirement and Expenditure Statement, and as an increase in column 6, "Cash Balance." If liquidated damages are obtained by withholding an equivalent amount from the contractor's payment, the net result will be the same.

116   Nonrefundable Payments for Construction

Nonrefundable payments (contributions) from customers and developers for underground construction shall first be credited to Account 107.2, Construction Work-in-Progress—Force Account. When the constructed plant is unitized and distributed to the individual plant accounts, the contributions shall be credited to those plant accounts which gave rise to the contribution.

When a customer or developer furnishes a trench or other service in connection with buried plant, the cooperative shall debit Account 107.2 with the actual or estimated cost of the service performed, and account for the credit as set forth above.

117   Refunds of Overpayments for Materials and Equipment

Refunds of overpayments for materials and equipment previously purchased are occasionally received as the result of legal action brought against electrical suppliers for price fixing in violation of antitrust laws. Such refunds shall be accounted for as follows:

1. The refund shall first be applied to any litigation costs that were incurred.

2. Refunds for special equipment items shall be accounted for, in detail, on the Summary of Special Equipment Costs and credited against the appropriate plant accounts.

3. Other material or equipment items that were installed through work orders or a materials furnished contract shall be adjusted on an amended work order. The amended work order shall include full details of the refund.

4. Continuing property records shall be adjusted to reflect the above transactions.

5. Amounts approved for advance on the RUS Form 595, Financial Requirement and Expenditure Statement, and on the loan budget records, shall be adjusted. For special equipment items, the adjustment shall be requested in a letter to RUS. For materials installed by work order or contract, the adjustments shall be made through credits shown on the RUS Form 219, Inventory of Work Orders.

6. Refunds for material currently in stock shall be credited to Account 154, Plant Materials and Operating Supplies.

7. If the material was used in maintenance activities or operations, the refund shall be credited to the appropriate maintenance or operations expense account.

8. Refunds for materials or equipment financed from loan funds shall be deposited in the Construction Fund—Trustee Account or remitted to RUS as a special payment on a note. Other refunds shall be deposited in the general funds.

118   Load Control Equipment

The primary purpose of a Load Management System is to optimize load dispatch and to reduce or minimize system peaks in order to reduce purchases of power or to delay or eliminate the need for construction of new plant. A Load Management System may be used on integrated systems, or on generation, transmission, or distribution systems separately. The telemetry equipment used for data acquisition and interpretation may be included at various points on a system, such as generation, transmission, or distribution substation, switchyards or on consumers' premises.

An effective load control program should be coordinated with the G&T and requires full participation of all member distribution systems. The G&T monitors the power load of the total member distribution system to predict the time of the system's peak load. An optimal load control strategy is developed by the G&T and is passed on from the G&T computer system to the load control computer systems of the member distribution cooperatives.

The equipment at the member distribution system level is the type actually being used by an integrated power system to operate a load control program. The equipment used may vary from one integrated power system to another. The selection of equipment used is determined by the information needs of the integrated power system, and the method selected to operate the load control system.

Some equipment performs only SCADA-type functions. This equipment is included with the equipment that performs only load control functions because SCADA-type equipment is an integral part of a load control program. An effective load control strategy requires current information on loads so that member distribution systems can determine the actual loads to be shed and the duration of the load control.

The function and location of the load control equipment are the primary factors in determining the account in which the equipment shall be recorded. The following example depicts a common load control system and the associated accounting. Equipment type may vary, thereby necessitating the use of accounts not prescribed below. In all instances, however, the function and location of the equipment shall dictate the appropriate account classification.

G&T Borrower

1. Coordinating System Equipment

Coordinating System Equipment is the data acquisition, processing and control hardware and software used to coordinate the load control efforts of the member distribution system. Generally, this equipment is dedicated to load control use and is not shared with other electric utility activities.

The purpose of the G&T load control computer system is to reduce or minimize the peak power requirements of the entire member distribution system. This involves load dispatching to control transmission circuits and breakers. The computer system for load control shall, therefore, be recorded in Account 353, Station Equipment, with the associated operating expenses recorded in Account 561, Load Dispatching, and maintenance expenses recorded in Account 570, Maintenance of Station Equipment.

2. *Coordinating System Communications Link*

The G&T load control computer system is usually linked to the load control computer system for each member distribution system by a radio or telephone link that is dedicated to that purpose and is not shared with other communication activities. Under such circumstances, communications equipment shall be classified in Account 353, Station Equipment. If the communications equipment is shared with general use or voice communications equipment, however, the equipment shall be classified in Account 397, Communication Equipment.

3. *Depreciation*

Load control equipment shall be recorded in separate subaccounts of the primary plant accounts detailed above and shall be depreciated based upon the owner's estimate of the equipment's useful service life.

Distribution Borrower

1. *Member System Equipment*

Member system equipment is the data acquisition, processing and control hardware and software used as a subset to the overall load control efforts by the integrated power system.

The member system computer for each distribution member system accepts the control strategy from the G&T coordinating system and develops the tables that determine the control loads that are to be shed and the duration of the load control. The member system computer for each distribution system monitors the usage at each of its delivery points. This usage data is then transmitted to the G&T coordinating system for use in developing load projects and evaluating control strategies for the integrated power system. The member system computer is generally dedicated to load control use and is not shared with other electric utility operations.

The member computer system shall be recorded in Account 362, Station Equipment. The associated operating expenses shall be recorded in Account 581, Load Dispatching, and maintenance expenses shall be recorded in Account 592, Maintenance of Station Equipment.

2. *Substation Remote Controllers*

Substation Remote Controllers are located at the distribution substation. They accept control signals from the member system computer and couple the signal to the portion of the distribution system to which it is connected. Substation Remote Controllers also serve as a receiver of inbound signals from transponders located in the distribution system. They also send data back to the member system computer.

Substation Remote Controllers shall be recorded in Account 362, Station Equipment. The associated operating expenses shall be recorded in Account 582, Station Expenses, and maintenance expenses shall be recorded in Account 592, Maintenance of Station Equipment.

3. *Substation Injection Units*

Substation Injection Units are used only in power line based systems and are located in distribution substations. A major function of the Substation Injection Unit is to receive load control signals from the member system computer and inject them into the power line based system to be transmitted to the Load Control Receivers. Substation Injection Units can also perform control and SCADA functions similar to those performed by Substation Remote Controllers.

Substation Injection Units shall be recorded in Account 362, Station Equipment. The associated operating expenses shall be recorded in Account 582, Station Expenses, and maintenance expenses shall be recorded in Account 592, Maintenance of Station Equipment.

4. *Remote Terminal Units*

Remote Terminal Units perform electric utility SCADA functions in a distribution substation or delivery point. These functions include monitoring equipment for abnormal operating conditions, monitoring analog quantities such as conductor voltage or substation load, and controlling of certain equipment within the substation.

Remote Terminal Units shall be recorded in Account 362, Station Equipment. The associated operating expenses shall be recorded in Account 582, Station Expenses, and maintenance expenses shall be recorded in Account 592, Maintenance of Station Equipment.

5. *Line Device Transponder*

A Line Device Transponder directly controls a piece of distribution apparatus, such as a voltage regulator or a power factor correction capacitor, located on a distribution feeder and not accessible to a Remote Terminal Unit. The Line Device Transponder actuates the control functions and reports back to the member system computer upon completion of the requested action. This transponder is located at the site of the distribution apparatus being controlled.

Line Device Transponders shall be recorded in Account 368, Line Transformers. The associated operating expense shall be recorded in Account 583, Overhead Line Expenses, or Account 584, Underground Line Expenses, as appropriate, and maintenance expenses shall be recorded in Account 595, Maintenance of Line Transformers.

6. *Communications Verification Transponders*

Communication Verification Transponders are used to respond to inquiries from Substation Remote Controllers. In power line based systems, these transponders are used to verify the performance of the communications system. They are also used during adverse system operations to isolate sections of the distribution system that are experiencing an outage.

Communication Verification Transponders shall be recorded in Account 362, Station Equipment. The associated operating expenses shall be recorded in Account 582, Station Expenses, and maintenance expenses shall be recorded in Account 592, Maintenance of Station Equipment.

7. *Load Control Receivers*

The Load Control Receiver, also known as a load control switch, is located at the site of the consumer's load. These receivers directly control the electric supply to an end-use appliance, such as an electric water heater, central air conditioning compressor, or irrigation pump. The amount of time that an appliance will be turned off by the load control receiver is preset. When the member system computer determines that load shedding is necessary, it sends a signal to the communication link which then sends signals directly to the Load Control Receivers. In a power line based system, the signal from the communications link is sent by radio or telephone line to the Substation Injection Units, which then signals the Load Control Receivers to shut down the appliances for the present time. In nonpower line based systems, the signal from the communications link is sent by radio directly to the Load Control Receivers.

Load Control Receivers are located on the consumer's side of the meter. When the member distribution system retains title to the Load Control Receivers and assumes full responsibility for maintenance and replacement of the equipment, it shall be classified in Account 371, Installations on Customer's Premises. Load Control Receivers that are donated or given to consumers shall be charged to Account 908, Customer Assistance Expenses.

Operating and maintenance expenses applicable to Load Control Receivers recorded in Account 371 shall be charged to Account 587, Customer Installations Expenses, and Account 598, Maintenance of Miscellaneous Distribution Plant, respectively. Expenses applicable to Load Control Receivers donated

or given to consumers shall be recorded in Account 908, Customer Assistance Expenses.

Load Control Receivers may be moved on a continual basis from one customer location to another and are, therefore, considered to be special equipment items. When ownership is maintained by the member distribution cooperative, Load Control Receivers shall be accounted for in accordance with the special equipment procedures outlined in Accounting Interpretation No. 119 of this section.

8. *Communication Links*

The communication link in the member distribution systems between the Member System Computer, the Substation Remote Controllers or Substation Injection Units, Remote Terminal Units, Line Device Transponders, Communication Verification Transponders, and Load Control Receivers is usually accomplished by radio, telephone line, or power line based system. The communication links are normally dedicated to the SCADA and load control functions being served. Under such circumstances, communications equipment shall be recorded in Account 362, Station Equipment. If, however, the communication equipment used is shared with general use or voice communications equipment, the equipment shall be charged to Account 397, Communication Equipment.

9. *Depreciation*

Load control equipment shall be recorded in separate subaccounts of the primary plant accounts detailed above and shall be depreciated based upon the manufacturer's estimate of the equipment's useful service life.

119 SPECIAL EQUIPMENT

Special Equipment items are classified as such because they are continually being moved from one location to another due to load changes and maintenance practices. The USoA provides accounting that differs from that used for other types of materials. The cost, new, of special equipment items shall be capitalized at the time of purchase; it shall not be charged to Account 154 as is the case with other materials. The first installation cost, as well as all incidental costs necessary to prepare the equipment for use, shall be capitalized with the material upon purchase. All subsequent costs of removing, resetting, changing, renewing oil, and repairing constitute operations and maintenance expenses. The capitalized cost of special equipment items, including the first installation, shall be removed from the electric plant accounts only when the items are abandoned or retired from the system.

Meters, line-type transformers, oil circuit reclosers, sectionalizers, current and potential transformers, meter sockets, and other metering equipment listed in Account 370, Meters, as well as pole-type and underground voltage regulators in Account 368, Line Transformers, are considered to be special equipment items. Similarly, load control receivers (load control switches) recorded in Account 371, Installations on Customers' Premises, are considered to be items of special equipment. (*See* Interpretation No. 118.) Transformers, voltage regulators, metering equipment, and current and potential transformers for substations are not.

Special equipment items which are classified as nonusable shall be segregated in the warehouse and retired from service. The Summary of Special Equipment Costs shall be retitled Summary of Special Equipment Costs Retired and used for this purpose. A journal entry reflecting this information shall be prepared and posted to the books. Since loan funds for special equipment, including first installation costs, are approved for advance by the Rural Development upon receipt of the borrower's written estimate of funds required, and not on the basis of an Inventory of Work Orders, it is improper to take a credit for any salvage involved in the retirement of special equipment on the Inventory of Work Orders.

Electric borrowers that wish to receive a waiver from the special equipment accounting requirements should submit a letter request to Rural Development. In order to expedite these requests the letter to Rural Development should state that the borrower will adhere to the following requirements to account for special equipment using the work order procedure rather than the special equipment accounting procedures prescribed by Rural Development:

1. New purchases of special equipment items are to be charged to Account 154, Materials and Supplies, upon purchase.

2. Labor, material and overhead costs associated with the initial installation and all subsequent installations of special equipment are recorded on construction work orders and charged to the appropriate plant accounts upon closeout of the construction work order.

3. Labor and overhead costs associated with the removal of special equipment items, whether the items removed are placed in inventory or permanently retired and disposed of, are recorded on retirement work orders and charged or credited to the depreciation reserve account upon closeout of the retirement work order.

4. The special equipment items retired and salvaged for reuse are returned to the materials and supplies account at the average material cost in the materials and supplies account and credited to the depreciation reserve upon closeout of the retirement work order.

In addition to recognition of the requirements noted above, the borrower should indicate how it plans to account for the items of special equipment that have been charged to the plant accounts but not installed (in inventory). Two acceptable methods to account for this equipment are: (1) Leave the equipment in the plant accounts until the inventory is depleted and charge only new purchases to materials and supplies, or (2) credit the plant accounts for the installed cost of the equipment in inventory, charge the equipment cost to materials and supplies, and charge the installation cost to the appropriate operations expense account. Also, under the second method, the borrower must submit a "negative" special equipment summary to Rural Development to return to the balance in reserve for the current loan the installed cost of special equipment in inventory on the date of transition.

120 Meter Sockets and Meters

When a utility furnishes meter sockets, ownership by the utility of the meter socket or base, as well as the meter itself, is established by virtue of them being furnished without cost to the consumer by the cooperative. While no agreement as to ownership between the cooperative and the property owner exists, cooperative ownership is implied by long standing practice and tradition in the electric utility industry.

121 Minimum—Maximum Voltmeters

A minimum—maximum voltmeter is used to record the minimum and maximum voltages at a specific line location over a period of time. It is normally installed on a pole in connection with a 1½ kVA transformer, a meter base and connecting wires, and other small items of materials. Meter bases are ordinarily set for these voltmeters throughout the system, and a lesser number of voltmeters are rotated among them periodically to obtain voltage readings. An average system may have one voltmeter to two installations, with a maximum of 20 or 25 voltmeters for the whole system.

Minimum—maximum voltmeters shall be recorded, through work orders, in Account 370, Meters, when installed. The cost of the transformers shall remain in Account 368, Line Transformers, with the cost of the meter bases remaining in Account 370, Meters. The miscellaneous material used in installing the transformer and the meter base shall be charged to Account 370, Meters.

Maintenance expense shall be charged to either Account 595, Maintenance of Line Transformers, or Account 597, Maintenance of Meters, as appropriate. Costs associated with reading the voltmeters shall be charged to Account 583, Overhead Line Expenses, and the cost of relocating or changing the complete installation or any part thereof, other than retirement of the meter base, shall be charged to Account 583, Overhead Line Expenses, or Account 586, Meter Expenses.

122 Retrofitting Demand Meters

A demand meter measures the amount of electricity used over a period of time in kilowatt-hours (kWh) and indicates the maximum kilowatts (kW) required at any one time by means of a pointer.

Electronic or solid state demand meters have a direct readout which reads kilowatt demand to two decimal places. The use of a direct readout demand meter may result in increased revenues as pointer readings tend to register lower than actual usages.

The process of retrofitting a demand meter replaces the pointer with a direct readout. The cost of such a replacement is usually expensed as a minor item of property; however, since the use of a direct readout results in a substantial betterment, the excess cost of the replacement over the estimated cost, at current prices, of replacing the pointer without the betterment is capitalized.

123 Transformer Conversions

The conversion of an overhead transformer to an underground transformer constitutes a betterment and shall, therefore, be capitalized.

124 Transclosures

Transclosures are enclosures or cabinets in which line transformers are mounted. The cost of transclosures that are purchased separately from the transformer shall be charged to Account 154, Plant Materials and Operating Supplies, when received, and capitalized, upon installation, to Account 368, Line Transformers, as a separate unit of property. If the case and the transformer are inseparable, the unit is considered a transformer and shall be capitalized upon purchase.

125 Retirement Units

Services

A retirement unit shall consist of a complete service rather than the individual wires comprising that service. If each separate wire of a service were treated as a retirement unit, the retirement unit would represent a comparatively small cost. Such a small unit of property would substantially increase the number of retirement work orders. The complete service shall, therefore, be considered a retirement unit.

Minor Items

When minor items of property are added separately from complete retirement units, the costs of these items shall be included in work orders, and by unitizing all costs of completed construction for a month, these minor items shall be spread to the retirement units of which they normally form a part. For example, to convert a two-phase line to a three-phase line requires the addition of a conductor, an insulator and a pole-top pin. A pole-top pin is typically capitalized as a component of the cost of the pole to which it is attached. Assuming this is the only work order for the month, the cost of this pin shall be charged to the conductor, so that its cost is included in the total cost of the project. In actual practice, however, this does not happen as it is normal to have a number of work orders for a given month, which include the setting of poles. In allocating the cost of all construction projects for the month, part of the cost of pole-top pins shall be allocated to poles even though the work orders on which they were capitalized did not include poles.

The retirement and replacement of isolated single retirement units cannot be charged to maintenance; a retirement and construction work order shall be used.

126 Establishment of Continuing Property Records

The costs of installing a system of continuing property records shall be charged to Account 930.2, Miscellaneous General Expenses, and may include:

1. Labor and expenses incurred in developing an inventory of property;

2. Labor and material costs incurred in connection with developing pole records including map preparation and pole cards; and

3. Labor and material costs (ledger sheets, etc.) incurred in connection with the installation of the record system.

127 Continuing Property Records for Buildings

When establishing continuing property records for a building where there is no detailed breakdown of contract costs, it is necessary to estimate the cost of the each component part. It should be noted that the establishment of continuing property records is not required for buildings; however, if CPRs are not maintained, all repairs including the replacement of major component parts shall be expensed in the period incurred.

128 Sale of Property

All proceeds deposited in the Construction Fund account from the sale of property, regardless of materiality, shall be reflected on the RUS Form 595, Financial Requirement and Expenditure Statement. Proceeds from the sale of property shall be reported on the Form 595, by budget purpose, as a reduction in total expenditures to date, column 5; and an increase in the cash balance, column 6.

Proceeds from the sale of property shall not be used to maintain an "Employee Fund." A utility may, pursuant to board policy, use general funds for employee welfare equivalent in amount to proceeds received from the sale of scrap property. If general funds, in an amount equivalent to proceeds received from the sale of scrap property, are used for employee welfare, Account 926, Employee Pensions and Benefits, shall be charged.

129 Gain or Loss on the Sale of an Office Building

A gain on the sale of an office building shall be recorded in Account 421.1, Gain on the Disposition of Property, with a loss recorded in Account 421.2, Loss on the Disposition of Property. If the gain or loss will materially distort current year's net margins, such gain or loss is reportable as an extraordinary item in Account 434, Extraordinary Income, or Account 435, Extraordinary Deductions.

130 Salvage and Obsolete Material

The value of material salvaged from the retirement of units of property reduces the loss on the retirement and shall be so applied. The value assigned to salvage shall be credited to Account 108.8, Retirement Work-in-Progress, which results in reducing net charges to the provision for depreciation when the work order is completed and cleared.

If salvage is sold, any difference between the realized value and the estimated value of the salvaged material shall be charged or credited to the appropriate provision for depreciation.

Salvage resulting from maintenance where no retirement units are involved shall be debited to the materials and supplies account, and credited to the appropriate maintenance account.

Occasionally a utility will have a loss due to obsolescence of materials on hand. If the loss is due to obsolescence of new material, the loss shall be charged to Account 426.5, Other Deductions. If the loss is due to obsolescence of used material, the loss shall be charged to the appropriate subaccount of Account 108, Accumulated Provision for Depreciation.

131 Plant Acquisition Adjustments

Plant acquisition adjustments shall be amortized to the operating expense accounts. These adjustments are recorded in Account 114, Electric Plant Acquisition Adjustments, and amortized to Account 406, Amortization of Electric Plant Acquisition Adjustments, or Account 425, Miscellaneous Amortization, as required by the regulatory commission having jurisdiction. Accounts 406 and 425 shall be closed to operating margins.

132 General Plant

When the unit method of depreciation is used for general plant items, gains and losses on sales, trades or disposals of equipment shall be recorded as such. If the composite method of depreciation is used, gains or losses on the disposal of general plant items shall be recorded in the appropriate depreciation reserve account.

A truck which is used only for transporting power operated equipment mounted thereon shall be charged, together with the installed equipment, to Account 396, Power Operated Equipment. If the same type of truck is used for transporting materials and supplies, tools and work equipment, personnel, or other items, the cost of the truck shall be charged to Account 392, Transportation Equipment.

Depreciation and other expenses relating to power operated equipment shall be accumulated in a subaccount of Account 184, Clearing Accounts, and distributed monthly on an equitable basis to the accounts properly chargeable.

Depreciation expense on vehicles and other work equipment, furniture and office equipment, and other such plant used in the construction of utility plant, is a proper component of construction cost. To avoid a duplicate advance of funds, however, the amount of depreciation on such items that has previously been financed from loan funds shall be deducted from Inventories of Work Orders submitted to RUS. This amount shall be specifically identified, and shown either monthly or annually as a single item in column 9 on the RUS Form 219, Inventory of Work Orders.

133 Plant Abandonments and Disallowances of Plant Costs

In December 1986, the Financial Accounting Standards Board issued Statement of Financial Accounting Standards No. 90, Regulated Enterprises—Accounting for Abandonments (Statement No. 90) and Disallowances of Plant Costs. This section provides an overview of the requirements outlined in Statement No. 90 together with the specific accounts that shall be used to record a plant abandonment or a disallowance of plant costs.

Plant Abandonments

When an abandonment becomes probable, the cost of the abandoned asset shall be removed from Construction Work-in-Progress or Plant-in-Service, as applicable. Before making this transfer, however, a determination must be made as to whether recovery of the allowed cost is likely to be provided with a full return on the investment during the period from the time the abandonment becomes probable, to the time when recovery is completed, or with a partial or no return on the investment. This determination shall be made based upon the facts and circumstances of the specific abandonment, and past practices and current policies of regulatory jurisdiction.

If a full return on the investment is likely to be provided, any disallowance of all or part of the cost of abandoned plant that is both probable and reasonably estimated shall be recognized as a loss in the current year with the carrying basis of the asset reduced by an equal amount. The remaining cost of abandoned plant shall be recorded as a separate new asset.

If partial or no return on the investment is likely to be provided, any disallowance of abandoned plant costs that is both probable and reasonably estimated shall be recognized as a loss. The present value of the future revenues expected to be provided to recover the allowable cost of the abandoned plant and return on the investment, if any, shall be reported as a separate new asset. The discount rate used to compute the present value shall be the borrower's incremental borrowing rate, which is the rate that the borrower would have to pay to borrow an equivalent amount for a period equal to the expected recovery period. In determining the value of expected future revenues, the borrower shall consider the probable time period before the recovery is expected to begin and the probable time period over which recovery is expected to be provided.

The amount of the new asset shall be adjusted from time to time, as necessary, if new information indicates that the estimates used to record the new asset have changed. The carrying value of the new asset, however, shall not be adjusted for changes in the incremental borrowing rate. The amount of any adjustments shall be recorded as a gain or loss.

During the period between the date on which a new asset is recognized and the date on which recovery begins, the carrying amount shall be increased by accruing a carrying charge. The rate used to accrue the carrying charge shall be:

1. If a full return on the investment is likely, a rate equal to the allowed overall cost of capital in the jurisdiction in which recovery is expected to be provided shall be used.

2. If partial or no return is likely, the asset shall be amortized in a manner that will produce a constant return on the unamortized investment in the new asset equal to the rate at which the expected revenues were discounted.

Due to the nonprofit environment in which electric cooperatives operate, full recovery of interest expense on plant related long-term debt equates to full recovery of the rate of return for an investor-owned utility. Therefore, if a cooperative is permitted full recovery of the interest expense incurred on the long-term debt borrowed to finance construction of an abandoned plant, no discounting of the asset is required nor is accrual of the carrying charge permitted.

If, at the time the provisions of Statement No. 90 are first applied, the borrower elects to restate the financial statements, the financial statements for all periods presented shall be restated and the financial statements shall disclose the nature of the restatement and its effect on margins before extraordinary items, net margins, and patronage capital at the beginning of the earliest period presented. If the borrower elects not to restate the financial statements, the effect of applying Statement No. 90 shall be reported as a change in accounting principle and the financial statements shall disclose the nature of the change and the effect of applying Statement No. 90 on margins before extraordinary items and net margins.

The specific accounts that shall be used to record transactions involving plant abandonments are as follows:

1. In the year of the abandonment, the unrecoverable portion of the cost of abandoned plant included in construction work-in-progress shall be recognized as a loss by a charge to Account 426.5, Other Deductions, and a credit to Account 107, Construction Work-in-Progress.

2. The balance of the cost remaining in the construction work-in-progress account shall be credited to Account 107 and charged to Account 182.2, Unrecovered Plant and Regulatory Study Costs.

3. The difference between the charge to Account 182.2 and the present value of expected future revenues for recovery of the new asset, shall be recorded as a credit to Account 182.2 and a debit to Account 426.5. The credit to Account 182.2 shall be segregated from the amount charged to Account 182.2 by the use of a separate subaccount. Statement No. 90 does not require this segregation; however, it is necessary under the USoA to provide for the appropriate segregation of operating and nonoperating income.

4. During the waiting period for recovery of the new asset to begin, carrying charges shall be accrued by a debit to Account 182.2 with a concurrent credit to Account 421, Miscellaneous Nonoperating Income. Debits to Account 182.2 shall be treated as reductions to the credit subaccount of Account 182.2.

5. The borrower shall amortize the amount debited to Account 182.2 by charges to operating income, consistent with the way the amortized amounts are recovered through rates. These charges to income shall be recorded in Account 407, Amortization of Property Losses, Unrecovered Plant and Regulatory Study Costs.

6. As the recoverable amount recorded in Account 182.2 is recovered through rates, the borrower shall accrue income by charges to Account 182.2 and credits to Account 421, Miscellaneous Nonoperating Income. Accruals shall be computed by applying the same rate used to derive the present value of the asset established in Account 182.2, to the

unamortized balance in that account. Accrued amounts charged to Account 182.2 shall be treated as reductions to the credit subaccount withinAccount 182.2.

Prior to implementing the accounting prescribed above, the borrower shall submit the details of each plant abandonment to RUS for approval.

Disallowances of Costs of Recently Completed Plant

When it becomes probable that a portion of the cost of recently completed plant will be disallowed for rate making purposes and a reasonable estimate of the amount of the disallowance can be made, the estimated amount of the probable disallowance shall be deducted from the reported cost of the plant and recognized as a loss. If a portion of the costs is explicitly, but indirectly disallowed, the equivalent amount of the cost shall be deducted from the reported cost of the plant and recognized as a loss. The specific accounts that shall be used to record transactions involving the disallowance of plant costs are as follows:

1. Estimated disallowed plant costs which the borrower records as a credit to Account 101, Electric Plant-in-Service, shall be charged to Account 426.5, Other Deductions.

2. If the loss qualifies as an extraordinary item under the criteria set forth in General Instruction No. 7 of the USoA, the borrower shall record the loss in Account 435, Extraordinary Deductions. To be considered extraordinary, an item shall be more than five percent of income computed before extraordinary items. If a borrower believes that a loss of less than five percent should be treated as an extraordinary item; the borrower shall, with commission approval, record the loss in Account 435 and report the loss as an extraordinary item. If the borrower is not subject to state commission jurisdiction, RUS approval is required.

134 Utility Plant Phase-in Plans

In August 1987, the Financial Accounting Standards Board issued Statement of Financial Accounting Standards No. 92, Regulated Enterprises—Accounting for Phase-in Plans (Statement No. 92). This section provides an overview of the requirements outlined in Statement No. 92.

The term phase-in plan is used to refer to any method of recognition of allowable costs in rates that meets all of the following criteria:

1. The method was adopted by the regulator in connection with a major, newly completed plant of the regulated enterprise or one of its suppliers or a major plant scheduled for completion in the near future.

2. The method defers the rates intended to recover allowable costs beyond the period in which those allowable costs would be charged to expense under generally accepted accounting principles applicable to enterprises in general.

3. The method defers the rates intended to recover allowable costs beyond the period in which those rates would have been ordered under the rate-making methods routinely used prior to 1982 by that regulator for similar allowable costs of that regulated enterprise.

If a phase-in plan is ordered by a regulator in connection with a plant on which no substantial physical construction had been performed before January 1, 1988, none of the allowable costs that are deferred for future recovery by the regulator under the plan for rate-making purposes, shall be capitalized for general-purpose financial reporting purposes (financial reporting).

If a phase-in plan is ordered by a regulator in connection with a plant completed before January 1, 1988, or a plant on which substantial physical construction had been performed before January 1, 1988, the criteria specified below shall be applied to that plan. If the phase-in plan meets all of those criteria, all allowable costs that are deferred for future recovery by the regulator under the plan shall be capitalized for financial reporting purposes as a separate asset (a deferred charge). If any one of those criteria is not met, none of the allowable costs that are deferred for future recovery by the regulator under the plan shall be capitalized for financial reporting. The criteria for determining whether capitalization is appropriate are:

1. The allowable costs in question are deferred pursuant to a formal plan that has been agreed to by the regulator;

2. The plan specifies the timing of recovery of all allowable costs that will be deferred under the plan;

3. All allowable costs deferred under the plan are scheduled for recovery within 10 years of the date when the deferral began; and

4. The percentage increase in rates scheduled under the plan for each future year is no greater than the percentage increase in rates scheduled under the plan for each immediately preceding year. That is, the scheduled percentage increase in year two is no greater than the percentage increase granted in year one, the scheduled percentage increase in year three is no greater than the percentage increase in year two, etc.

By definition, a phase-in plan approved prior to 1982 that contains provisions contrary to those detailed above is not subject to the provisions of Statement No. 92. This exemption, however, only relates to a specific utility and a specific regulator. For example, a utility cannot use a phase-in plan approved by its regulator for a different utility as justification for its phase-in plan exceeding the 10-year limit imposed by Statement No. 92.

A phase-in plan is a method of rate making intended to moderate a sudden increase in rates while providing the regulated enterprise with recovery of its investment and a return on that investment during the recovery period. A disallowance is a rate-making action that prevents the regulated enterprise from recovering either some amount of its investment or some amount of return on its investment. Statement No. 90 specifies the accounting for disallowances of plant costs (see item 133 of this regulation). If a method of rate making that meets the criteria for a phase-in plan includes an indirect disallowance of plant costs, that disallowance shall be accounted for in accordance with Statement No. 90. Cumulative amounts capitalized under phase-in plans shall be reported as a separate asset in the balance sheet. The net amount capitalized in each period or the net amount of previously capitalized allowable costs recovered during each period shall be reported as a separate item of other income or expense in the income statement. Allowable costs capitalized shall not be reported as reductions of other expenses.

The terms of any phase-in plan in effect during the year or ordered for future years shall be disclosed in the financial statements. Statement No. 92 does not permit capitalization for financial reporting of allowable costs deferred for future recovery by the regulator pursuant to a phase-in plan that does not meet the criteria or a phase-in plan related to plant on which substantial physical construction was not completed before January 1, 1988. Nevertheless, the financial statements shall include disclosures of the net amount deferred at the balance sheet date for rate-making purposes, and the net change in deferrals for rate-making purposes during the year for those plans.

If the provisions of Statement No. 92 are applied retroactively, the financial statements of all periods presented shall be restated. In addition, the restated financial statements shall, in the year that Statement No. 92 is first applied, disclose the nature of any restatement and its effect on margins before extraordinary items, net margins, and on patronage capital at the beginning of the earliest period presented. If the financial statements for prior years are not restated, the effects of applying Statement No. 92 to existing phase-in plans shall be reported as a change in accounting principle and the financial statements shall disclose the effect of adopting Statement No. 92 on margins before extraordinary items and net margins.

The application of Statement No. 92 to an existing phase-in plan shall be delayed if both of the following conditions are met:

1. The enterprise has filed a rate application to have the plan amended to meet the criteria of Statement No. 92 or intends to do so as soon as practicable; and

2. It is reasonably possible that the regulator will change the terms of the phase-in plan so that it will meet the criteria of Statement No. 92.

If the above conditions are met, the provisions of Statement No. 92 shall be applied to the existing phase-in plan on the earlier of the date when one of the conditions ceases to be met or the date when the final rate order is received, amending or refusing to amend the phase-in plan. However, if the enterprise delays filing its application for the amendment or the regulator does not process the application in the normal period of time, the application of Statement No. 92 shall not be further delayed.

In applying the criteria of Statement No. 92 to a plan that was in existence prior to the first fiscal year beginning after December 15, 1987, and that was revised to meet that criteria, the 10-year criterion and the requirement concerning the percentage increase shall be measured from the date of the amendment rather than from the date of the first scheduled deferrals under the original plan. All phase-in plans must receive RUS approval prior to implementation.

135 Accounting for Removal or Relocation of Electric Facilities Resulting from the Action of Others

Under arrangements with another party, a borrower agrees, or is obliged, to remove, relocate, rearrange, or otherwise make changes in utility property, other than for the purpose of rendering utility service to the other party, for which the utility is reimbursed for all or a portion of the costs incurred.

Plant Accounting

The relocation of the line shall be accounted for as follows:

1. If all of the assemblies in the line are retired or completely removed and later reinstalled or if the line is constructed in a new location before the old line is removed, construction and retirement work orders shall be prepared except for the costs relating to special equipment items (transformers, oil circuit reclosers, etc.) which shall be charged to operations expense.

2. If a line is moved in its entirety to a new location except for isolated retirement units (such as at the end of the line) or poles not suitable for resetting, the cost of moving the portion of line that is moved intact shall be charged to maintenance expense while the cost related to the change in isolated retirement units or the replacement of poles not suitable for resetting shall be accounted for

through use of construction and retirement work orders.

3. If a line is moved intact without any change in assemblies, the cost shall be charged to maintenance expense.

Reimbursement

If the borrower receives reimbursement for the costs related to the relocation of the line, the reimbursement shall be accounted for by crediting operation and maintenance expenses to the extent of actual expenses occasioned by the plant changes and crediting the remainder to the accumulated provision for depreciation, unless contractual terms definitely characterize residual or specific amounts as applicable to the cost of replacement. In the latter event, appropriate credits shall be entered in the plant accounts.

Reimbursement received from a telephone company for adding a pole or replacing a present pole with a taller pole under joint use contracts falls within this latter category. In this instance, appropriate credits are charged against the plant accounts.

Financing

The total reimbursement, less any portion for operations and maintenance costs, shall be entered in the "Contributions in Aid of Construction" section at the bottom of the Construction Work Order. When the Inventory of Work Orders (RUS Form 219) is prepared, enter only enough of the contribution in column 9 to reduce to zero the amount in column 10, "Loan Funds Subject to Advance by RUS." This entry is made although none of the reimbursement received is recorded in the accounting records as a contribution in aid of construction.

136 Storm Damage

As a result of recent hurricane, flood, and ice storm damage, the Rural Utilities Service (RUS) has received several inquiries concerning the proper accounting for storm damage costs and the associated funds received from the Federal Emergency Management Administration (FEMA).

Storm damage costs should be accounted for under the work order procedure. Units of property destroyed or otherwise removed from service must be reflected on retirement work orders and units of property installed must be shown on construction work orders. To ensure that the accounting for construction and retirement costs is as accurate as possible, an effort should be made to accurately accumulate material, labor, and overhead costs. Even when extreme care has been exercised, however, it may still be necessary to use estimates to develop the appropriate cost figures.

When a storm occurs, a utility typically incurs a large retirement loss, all or a part of which should be charged to the accumu-

lated provision for depreciation. Storm damage costs over and above construction and retirement costs represent maintenance expense. Maintenance costs include the costs of resagging lines, straightening poles, and replacing minor items of property. When extensive damage has occurred, the need to restore the property to an operating condition without delay usually results in excessive costs being incurred. Standard property unit costs may be used as a guide in determining the amount to be capitalized. It should be noted, however, that when standard property unit costs are used, all excess costs are charged to maintenance expense.

Because of the storm's destruction, property is retired prematurely and as a result, extraordinary retirement losses occur. When such extraordinary losses occur, they should be recorded in the year in which the losses are incurred. If the recording of such losses will materially distort the income statement, such losses may be charged to Account 435, Extraordinary Deductions. These costs may be deferred and amortized to future periods only if the provisions of Statement of Financial Accounting Standards No. 71, Accounting for the Effects of Certain Types of Regulation (Statement No. 71), are applied. Under the provisions of Statement No. 71, a utility may defer certain costs, provided such costs are included in the utility's rate base and recovered through future rates. If an RUS borrower elects to apply the provisions of Statement No. 71, RUS approval is required. To obtain RUS approval, a borrower must submit:

a. A detailed description of the plan including the nature of the expense item, the amount of the deferral, the specific time period for rate recovery, and justifying support for the time period selected;

b. The accounting journal entries being used by the cooperative to record the expense deferral and amortization of deferred costs; and

c. A copy of the state Commission order authorizing recovery of the deferred costs through future rates, or in the absence of commission jurisdiction, a resolution from the cooperative's board of directors authorizing such recovery.

To assist in the restoration of the damaged facilities, the Federal government often provides assistance through Federal Emergency Management Agency (FEMA).

Under current FEMA procedures, FEMA provides funds for the restoration of facilities based upon the cost estimates submitted by the entity requesting assistance. If the FEMA grant is for less than 100 percent of the cost estimates, and does not specify offset expenses, thereby providing the borrower with the maximum opportunity to utilize Rural Development Utilities Program loan funds to finance capitalizable costs. When the funds are received, they should be accounted for by first applying the funds received as a credit to maintenance expense and administrative and general costs. Any remaining funds should then be applied as a credit to construction and retirement costs.

ACCOUNTING JOURNAL ENTRIES

Dr. 108.8X, Retirement Work in Progress—Storm Damage $1,015.17
    Cr. 107.4, Construction Work in Progress—Storm Damage $1,015.17
To transfer the removal costs recorded in Column 11 of Retirement Work Order #4401X to Account 108.8X.

Dr. 107.4, Construction Work in Progress—Storm Damage $4,141.55
    Cr. 108.8X, Retirement Work in Progress—Storm Damage $4,141.55
To remove material salvaged in the _____ rebuild from Account 107.4. The original entry debited Account 154, Plant Materials and Operating Supplies, and credited Account 107.4. (See Column 12 of Retirement Work Order #4401X.)

Dr. 108.8X, Retirement Work in Progress—Storm Damage $312,230.41
    Cr. 364, Poles Towers and Fixtures $133,377.55
    Cr. 365, Overhead Conductors and Devices 59,683.08
    Cr. 368, Lines Transformers 19,704.60
    Cr. 369, Services 97,651.23
    Cr. 373, Street Lighting and Signal Systems 1,813.95
To remove the original cost of property destroyed and retired from the classified plant accounts. This retirement is recorded, in detail, on Retirement Work Order #4401X. It is understood that this retirement covers all distribution property retired or destroyed in the _____ area exclusive of substations and special equipment items (meters, meter sockets, current and potential transformers, transformers, voltage regulators, oil circuit reclosers (OCR), and sectionalizers).

Dr. 108.6, Accumulated Provision for Depreciation of Distribution
Plant $309,104.03
    Cr. 108.8X, Retirement Work in Progress—Storm Damage $309,104.03
To record the net loss due to the retirement of distribution lines in the _____ area. (See Retirement Work Order #4401X.)

Dr. 364, Poles, Towers and Fixtures $99,075.40
Dr. 365, Overhead Conductors and Devices 104,142.22
Dr. 368, Line Transformers 25,036.07
Dr. 369, Services 28,865.08
Dr. 373, Street Lighting and Signal Systems 2,101.60
    Cr. 107.4, Construction Work in Progress—Storm Damage $259,220.37
To record, in the proper classified plant accounts, Construction Work Order #4401 covering the _____ rebuild.

This entry includes:
    Material Issued $150,336.49
    Less: Materials Returned 15,631.39

    Net Material Used 134,705.10
    Labor and overhead estimated by using standard record
    unit costs 124,515.27

    Total 259,220.37

Dr. 108.8X, Retirement Work in Progress—Storm Damage 2,384.00
    Cr. 107.4, Construction Work in Progress—Storm Damage $2,384.00
To transfer the removal costs associated with the retirement of old transmission lines ($1,966) and substations ($418) to Account 107.4. This cost is shown in Column 11 of Retirement Work Order #4400X.

Dr. 107.4, Construction Work in Progress—Storm Damage $1,939.74
    Cr. 108.8X, Retirement Work in Progress—Storm Damage $1,939.74
To remove material salvaged from transmission lines ($1,545.74) and substations ($394.00) from Account 107.4. The original entry debited Account 154 and credited Account 107.4. (See Column 12 of Retirement Work Order #4400X.)

Dr. 108.8X, Retirement Work in Progress—Storm Damage $162,172.06
    Cr. 355, Poles and Fixtures $47,738.45
    Cr. 356, Overhead Conductors & Devices 80,304.11
    Cr. 362, Station Equipment 34,129.50
To remove the original cost of transmission lines and substations destroyed and retired from the classified plant accounts. (See Retirement Work Order #4400X.) (New substations were built and separately accounted for on Work Order #4406.)

ACCOUNTING JOURNAL ENTRIES—Continued

Dr. 108.5, Accumulated Provision for Depreciation of Transmission
Plant $128,462.82
Dr. 108.6, Accumulated Provision for Depreciation of Distribution
Plant 34,153.50
    Cr. 108.8X, Retirement Work in Progress—Storm Damage $162,616.32

To record the net loss due to the retirement of transmission lines ($128,462.82) and sub-stations ($34,153.50). (See Retirement Work Order #4400X):

	Sub-stations	Trans-mission plant
Original Cost	$34,129.50	$128,042.56
Add: Cost of Removal	418.00	1,966.00
	34,547.50	130,008.56
Less: Material Salvaged	394.00	1,545.74
Total	34,153.50	128,462.82

Dr. 355, Poles and Fixtures $161,784.05
Dr. 356, Overhead Conductors and Devices 124,704.77
    Cr. 107.4, Construction Work in Progress—Storm Damage $286,488.82

To record, in the proper classified plant accounts, the costs of a 69 kV transmission line (_____) as detailed in Work Order #4400. This work order includes construction costs as follows:

Material Used (Net) $171,665.62
    Labor and overhead estimated by using standard record
    unit costs 114,823.20
    Total 286,488.82

Dr. 107.4, Construction Work in Progress—Storm Damage $329.40
    Cr. 108.8X, Retirement Work in Progress—Storm Damage $329.40

To correct the journal entry for cash received from the sale of scrapped meters and transformers. The original entry credited Account 107.4 at the time of receipt.

Transformers $318.00
Meters 11.40
    Net Materials Used 329.40

Dr. 108.8X, Retirement Work in Progress—Storm Damage $137,671.22
    Cr. 365, Overhead Conductors and Devices $4,557.00
    Cr. 368, Line Transformers 112,815.22
    Cr. 370, Meters 20,299.00

To remove the cost of meters, transformers, and OCRs lost or destroyed from the primary plant accounts. (See Retirement Work Order #4402X.)

737 Transformers $112,815.22
31 OCRs 4,557.00
1,532 Meters 20,299.00
    Total 137,671.22

Dr. 108.6, Accumulated Provision for Depreciation of Distribution
Plant $137,341.82
    Cr. 108.8X, Retirement Work in Progress $137,341.82

To record the net loss due to the retirement of meters, transformers, and OCRs. (See Retirement Work Order #4402X.)

Original Cost $137,671.22
Salvaged Realized 329.40
    Total 137,341.82

Dr. 186, Miscellaneous Deferred Debits $1,319.85
    Cr. 107.4, Construction Work in Progress—Storm Damage $1,319.85

To record the engineering costs associated with future construction work in the _____ area.

Dr. 593, Maintenance of Overhead Lines $607.24
Dr. 595, Maintenance of Line Transformers 19,365.86
Dr. 597, Maintenance of Meters 6,595.56
    Cr. 107.4, Construction Work in Progress—Storm Damage $26,568.66

To charge the costs of repairing damaged meters, transformers, voltage regulators, and OCRs to the appropriate expense accounts. Repair costs were originally charged to Account 107.4.

	593	595	597
Meters	....	....	$6,595.56
Transformers	....	$18,869.95	
Voltage Regulators	....	495.91	
Oil Circuit Reclosers	$607.24		
Total	607.24	19,365.86	6,595.56

Dr. 920, Administrative and General Salaries $32,000.00
Dr. 921, Office Supplies and Expenses 4,421.69
    Cr. 107.4, Construction Work in Progress—Storm Damage $36,421.69

To charge the administrative costs incurred to obtain the FEMA grant to the appropriate expense accounts. Administrative costs were originally charged to Account 107.4.

    Salaries $32,000.00
    Office Supplies 4,421.69

    Total $36,421.69

Dr. 571, Maintenance of Overhead Lines $3,675.60
Dr. 593, Maintenance of Overhead Lines 33,080.40
    Cr. 107.4, Construction Work in Progress Storm Damage $36,756.00

To allocate expenses remaining in Account 107.4 to distribution and transmission maintenance expense. It was estimated that only 10 percent is applicable to transmission.

Dr. 426.5, Other Deductions $275,000.00
Dr. 435, Extraordinary Deductions
Dr. 182.1, Extraordinary Property Losses
    Cr. 108.5, Accumulated Provision for Depreciation of Transmission
    Plant $35,000.00
    Cr. 108.6, Accumulated Provision for Depreciation of Distribution
    Plant 240,000.00

To restore the accumulated provisions for depreciation to their appropriate levels based upon a study of plant currently in service.

NOTE: Account 426.5, Other Deductions, should be used to record the retirement loss as a current period expense. Account 435, Extraordinary Deductions, may be used when the loss will materially distort the income statement. Account 182.1, Extraordinary Property Losses, should be used when such costs are being deferred under the provisions of Statement No. 71. Costs recorded in this account should be amortized to Account 407, Amortization of Property Losses, as the costs are recovered through rates.

Dr. 131.1, Cash—General $1,000,000.00
    Cr. 253, Other Deferred Credits $1,000,000.00

To record the receipt of funds from the Federal Emergency Management Administration (FEMA).

Dr. 253, Other Deferred Credits $1,000,000.00
    Cr. 108.5, Accumulated Provision for Depreciation of Trans-
    mission Plant $74,205.00
    Cr. 108.6, Accumulated Provision for Depreciation of Dis-
    tribution Plant 191,575.00
    Cr. 186, Miscellaneous Deferred Debits 872.00
    Cr. 355, Poles and Fixtures 129,056.00
    Cr. 356, Overhead Conductors and Devices 99,408.00
    Cr. 364, Poles, Towers and Fixtures 78,916.00
    Cr. 365, Overhead Conductors and Devices 82,840.00
    Cr. 368, Line Transformers 20,056.00
    Cr. 369, Services 23,108.00
    Cr. 373, Street Lighting and Signal Systems 1,744.00
    Cr. 426.5, Other Deductions 219,220.00
    Cr. 571, Maintenance of Overhead Lines 2,900.00
    Cr. 593, Maintenance of Overhead Lines 26,600.00

Cr. 595, Maintenance of Line Transformers		15,300.00
Cr. 597, Maintenance of Meters		5,200.00
Cr. 920, Administrative and General Salaries		25,491.00
Cr. 921, Office Supplies and Expenses		3,509.00

To allocate FEMA funds to the proper accounts..

Summary of Costs

Maintenance:

Account 571, Maintenance of Overhead Lines	$3,675.60
Account 593, Maintenance of Overhead Lines	33,687.24
Account 595, Maintenance of Line Transformers	19,365.86
Account 597, Maintenance of Meters	6,595.56
Total Maintenance Costs	63,324.26

Retirement Loss:

Account 108.5, Accumulated Provision for Depreciation of Transmission Plant	93,462.82
Account 108.6, Accumulated Provision for Depreciation of Distribution Plant	240,599.35
Account 426.5, Other Deductions	275,000.00
Total Retirement Loss	609,062.17

Construction:

Account 186, Miscellaneous Deferred Debits	1,319.85
Account 355, Poles and Fixtures	161,784.05
Account 356, Overhead Conductors and Devices	124,704.77
Account 364, Poles, Towers and Fixtures	99,075.40
Account 365, Overhead Conductor and Devices	104,142.22
Account 368, Line Transformers	25,036.07
Account 369, Services	28,865.08
Account 373, Street Lighting and Signal Systems	2,101.60
Total Construction Cost	547,029.04

Administrative:

Account 920, Administrative and General Salaries	$32,000.00
Account 921, Office Supplies and Expenses	4,421.69
Total Administrative Cost	36,421.69

Maintenance	63,324.26
Retirement Loss	609,062.17
Construction	547,029.04
Administrative	36,421.69
Total Costs	1,255,837.16

Distribution of FEMA Funds

Maintenance: 63,324.26 ÷ 1,255,837.16 = .0504 = 5.0%
Retirement: 609,062.17 ÷ 1,255,837.16 = .4850 = 48.5%
Construction: 547,029.04 ÷ 1,255,837.16 = .4356 = 43.6%
Administrative: 36,421.69 ÷ 1,255,837.16 = .0290 = 2.9%

Maintenance: $1,000,000.00 × 5.0% =	$50,000.00
Retirement: $1,000,000.00 × 48.5% =	485,000.00
Construction: $1,000,000.00 × 43.6% =	436,000.00
Administrative: $1,000,000.00 × 2.9% =	29,000.00
Total	1,000,000.00

Distribution of FEMA Funds—Maintenance

Account 571: 3,675.60 ÷ 63,324.26 = .0580 = 5.8%
Account 593: 33,687.24 ÷ 63,324.26 = .5320 = 53.2%
Account 595: 19,365.86 ÷ 63,324.26 = .3058 = 30.6%
Account 597: 6,595.56 ÷ 63,324.26 = .1041 = 10.4%

Account 571: $50,000.00 × 5.8% =	$2,900.00
Account 593: $50,000.00 × 53.2% =	26,600.00

Account 595: $50,000.00 × 30.6% =	15,300.00
Account 597: $50,000.00 × 10.4% =	5,200.00
Total	50,000.00

Distribution of FEMA Funds—Retirement Loss

Account 108.5: 93,462.82 ÷ 609,062.17 = .1535 = 15.3%
Account 108.6: 240,599.35 ÷ 609,062.17 = .3950 = 39.5%
Account 426.5: 275,000.00 ÷ 609,062.17 = .4515 = 45.2%

Account 108.5: $485,000.00 × 15.3% =	$74,205.00
Account 108.6: $485,000.00 × 39.5% =	191,575.00
Account 426.5: $485,000.00 × 45.2% =	219,220.00
Total	485,000.00

Distribution of FEMA Funds—Construction

Account 186: 1,319.85 ÷ 547,029.04 = .0024 = .2%
Account 355: 161,784.05 ÷ 547,029.04 = .2958 = 29.6%
Account 356: 124,704.77 ÷ 547,029.04 = .2280 = 22.8%
Account 364: 99,075.40 ÷ 547,029.04 = .1811 = 18.1%
Account 365: 104,142.22 ÷ 547,029.04 = .1904 = 19.0%
Account 368: 25,036.07 ÷ 547,029.04 = .0457 = 4.6%
Account 369: 28,865.08 ÷ 547,029.04 = .0528 = 5.3%
Account 373: 2,101.67 ÷ 547,029.04 = .0038 = .4%

Account 186: $436,000.00 × .2% =	$872.00
Account 355: $436,000.00 × 29.6% =	129,056.00
Account 356: $436,000.00 × 22.8% =	99,408.00
Account 364: $436,000.00 × 18.1% =	78,916.00
Account 365: $436,000.00 × 19.0% =	82,840.00
Account 368: $436,000.00 × 4.6% =	20,056.00
Account 369: $436,000.00 × 5.3% =	23,108.00
Account 373: $436,000.00 × .4% =	1,744.00
Total	436,000.00

Distribution of FEMA Funds—Administrative

Account 920: 32,000.00 ÷ 36,421.69 = .8786 = 87.9%
Account 921: 4,421.69 ÷ 36,421.69 = .1213 = 12.1%

Account 920: $29,000.00 × 87.9% =	$25,491.00
Account 921: $29,000.00 × 12.1% =	3,509.00
Total	29,000.00

137 Impairment of Long-Lived Assets

Statement of Financial Accounting Standards No. 121, Accounting for the Impairment of Long-Lived Assets and for Long-Lived Assets to be Disposed of (Statement No. 121), requires reporting entities to review all long-lived assets and certain identifiable intangibles that are to be held, used, or disposed of by that entity for impairment whenever events and changes in circumstances indicate that the carrying amount of the asset may not be recoverable. If the sum of the expected future cash flows (undiscounted and without interest charges) is less than the carrying value of the asset, the entity must recognize an impairment loss. The impairment loss is measured as the amount by which the carrying amount of the asset exceeds the fair value of the asset. The impairment loss is reported as a component of income from continuing operations before income taxes for entities presenting an income statement and in the statement of activities of not-for-profit organizations. Statement No. 121 does not apply to assets included in the scope of Statement of Financial Accounting Standards No. 90, Regulated Enterprises—Accounting for Abandonments and Disallowances of Plant Costs.

Assets To Be Held or Used

Entities are required to review long-lived assets and certain identifiable intangibles whenever events or changes in circumstances indicate that the carrying value of the asset may not be recoverable. For example:

1. A significant decrease in the market value of an asset;

2. A significant change in the extent or manner in which an asset is used;

3. A significant physical change in an asset;

4. A significant adverse change in legal factors or in the business climate that could affect the value of an asset;

5. An adverse action or assessment by a regulator;

6. An accumulation of costs significantly in excess of the amount originally expected to acquire or construct an asset; and

7. A current period operating or cash flow loss combined with a history of operating or cash flow losses or a projection or forecast that demonstrates continued losses associated with an asset used for the purpose of producing revenue.

The impairment of the asset is measured by estimating the future cash flows expected to result from the use of the asset and its disposition. Assets are grouped at the lowest level for which there are identifiable cash flows that are largely independent of the cash flows of other groups of assets. Future cash flows are those cash inflows that are expected to be generated by the asset less the cash outflows expected to be necessary to maintain those inflows. If the future cash flows (undiscounted and without interest charges) are less than the carrying value of the asset, an impairment loss must be recognized. If the expected future cash flows are greater than the carrying value of the asset, no impairment loss exists.

The impairment loss is the amount by which the carrying amount (acquisition cost less accumulated depreciation) of the asset exceeds the fair value of the asset. The fair value of the asset is the amount for which the asset could be bought or sold in an arms-length transaction between willing parties. A quoted market price is the best evidence of fair value. If this information is not available, the fair value should be based upon the best information available. Consideration should be given to the price of similar assets and valuation techniques such as the present value of the expected future cash flows discounted at a rate representative of the risk involved, option-pricing models, matrix pricing, option-adjusted spread models, and fundamental analysis. All available information should be considered when using the above pricing techniques.

If an impairment is recognized, the carrying value of the asset is reduced to the lower of its fair value or its carrying value and, if depreciable, depreciated over the remaining useful life. Previously recognized impairment losses cannot be restored. If the asset was acquired in a business combination and there is goodwill resulting from the transaction, the goodwill is included in the asset grouping and reduced or eliminated before any adjustment is made to the carrying value of the asset.

The following financial statement disclosures are required in the period in which the impairment is recognized:

1. A description of the impaired assets and the facts and circumstances surrounding the impairment;

2. The amount of the impairment and how fair value was determined;

3. The caption in the income statement or the statement of activities in which the impairment loss is aggregated if that loss has not been presented as a separate caption or reported parenthetically on the face of the statement; and

4. If applicable, the business segment(s) affected.

Assets To Be Disposed

Statement No. 121 also applies to all long-lived assets and certain identifiable intangibles for which management, having the authority to approve the action, has committed to a plan of disposal except those assets covered by APB No. 30, Reporting the Results of Operations—Reporting the Effects of Disposal of a Segment of a Business, and Extraordinary, Unusual and Infrequently Occurring Events and Transactions. An asset to be disposed of is carried at the lower of its carrying amount (acquisition cost less accumulated depreciation) or its fair value less cost to sell.

The fair value of the asset to be disposed of is computed in the same manner as that for an asset to be held or used by the entity. Selling costs include the incremental direct cost to transact the sale—broker commissions, legal fees, title transfer, and other closing costs that must be incurred before legal title can be transferred. Costs such as insurance, security service, and utilities are generally excluded unless these costs are part of a contractual agreement that obligates the entity to incur such costs in the future. If the asset's fair value is based upon current market price or the current selling price for a similar asset, the fair value is considered a current amount and is not discounted. If, however, the fair value is based upon discounted expected future cash flows and if the sale is to occur beyond one year, the cost to sell must also be discounted. Assets covered by this statement are not depreciated (amortized) while being held for disposal.

Subsequent revisions in estimates of fair value less cost to sell are reported as adjustments to the carrying amount of the asset to be disposed of as long as the carrying amount of the asset does not exceed the original carrying amount.

The following financial statement disclosures are required in the period in which the impairment is recognized:

1. A description of the assets to be disposed of including the facts and circumstances leading to the expected disposal, the expected disposal date, and the carrying amount of those assets;

2. If applicable, the business segment(s) in which the assets to be disposed of are held;

3. The amount, if any, of the impairment loss resulting from the adoption of this statement;

4. The gain or loss, if any, resulting from subsequent revisions in the estimates of fair value less cost to sell;

5. The caption in the income statement or statement of activities in which the gains or losses are aggregated if those gains or losses have not been presented as a separate caption or reported parenthetically on the face of the statement; and

6. The results of operations for assets to be disposed of to the extent that those results are included in the entity's results of operations for the period and can be identified.

Accounting Requirements

All borrowers must adopt the accounting prescribed by Statement No. 121.

Effective Date and Implementation

Statement No. 121 is effective for financial statements for fiscal years beginning after December 15, 1995. Impairment losses resulting from the application of this statement to assets that are held or used by the entity must be reported in the period in which the recognition criteria are first applied and met. Impairment losses attributable to assets to be disposed of must be reported as the cumulative effect of a change in accounting principle as prescribed in Accounting Principles Board Opinion No. 20, Accounting Changes.

Accounting Journal Entries—Implementation Date

If a borrower has impaired assets that are held or used at the implementation date, the following entry should be recorded:

Dr. 426.5, Other Deductions
Cr. 300 Series of Accounts, Plant Accounts
To record the adoption of Statement No. 121 for the impairment of assets that are held or used.

If a borrower has impaired assets to be disposed of at the implementation date, the following entry should be recorded:

Dr. 435.1, Cumulative Effect on Prior Years of a Change in Accounting Principle
Cr. 300 Series—Plant Accounts
To record the adoption of Statement No. 121 for assets that are to be disposed.

Accounting Journal Entries—Subsequent to Implementation Date

If an asset that is either held, used or to be disposed of becomes impaired, the following entry should be recorded:

Dr. 426.5, Other Deductions
Cr. 300 Series—Plant Accounts

To record the impairment of a plant asset.

If a borrower makes a subsequent revision in the estimate of the fair value less the cost to sell of an asset to be disposed of, the following entry should be recorded:

Dr. 300 Series—Plant Accounts
Cr. 421, Miscellaneous Nonoperating Income
To revise the fair value of an asset to be disposed.

138   Automatic Meter Reading Systems— Turtles

Automatic meter reading systems were developed from technology called power line carrier communication systems. One such system, developed by Hunt Technologies, Inc., is called by its brand name, the Turtle system. In addition to its function as an automated reading device, the Turtle can provide outage detection, power failure counts, and other potential applications. The current Turtle system does not have the capability for applications such as collection of load survey or interval data. A Turtle system consists of:

1. A meter reader mounted (retrofitted) inside the meter;

2. A receiver located in each substation; and

3. Monitoring and programming equipment (software and personal computer) usually located in the headquarters building.

The system transmits continuous information one way from the meter to a receiver located in the substation. The receiver constantly monitors every Turtle meter served by the substation. The substation receiver can be sized to monitor up to 3,000 Turtle meter readers at the same time. The data is then transmitted to the headquarters monitoring equipment via telephone line or an equivalent communication system.

The technical literature and other information provided by the manufacturer indicates that this system can only be used for remote meter reading, outage detection, power failure counts, and phase identification. At this time, there is no indication that the system supports other functions such as home security. Therefore, the accounting prescribed for the Turtle meter reading devices and support equipment relates only to electric utility operations.

Accounting Requirements

The function of the equipment is the primary factor in determining the account in which the equipment shall be recorded. The components of the Turtle automatic meter reading system shall be recorded in Account 370, Meters. The cost of the meter reader encoding device and retrofitting the meter with the meter reader unit shall be capitalized to the cost of the existing meter. Any associated operating expenses shall be charged to Account 586, Meter Expenses, with maintenance expenses charged to Account 597, Maintenance of Meters.

Separate continuing property records shall be established for the meters, either fitted or retrofitted with the device; the receiver; the personal computer; and the system software. The meters, receivers, and personal computer shall be depreciated over the manufacturer's estimated useful service life. The system software shall be depreciated over the estimated useful service life of the program not to exceed 5 years.

139   Global Positioning Systems

The Global Positioning System (GPS) is a worldwide radio-navigation system formed from a network of 24 satellites and their ground stations. Utilities are using this advanced technology geographic data collection system to update and modernize their system maps. GPS uses a system of satellites orbiting the earth to establish plant locations with pinpoint accuracy. By triangulating from three satellites and using radio signals to measure distances and locate items, system-wide maps can be created of the utility's service area. A field inventory is then taken of the utility's plant and plotted onto the map. The GPS consists of base station equipment, remote station equipment, the GPS program, and mapping conversion software.

All equipment associated with GPS is dedicated to the mapping effort. The base station is installed at a fixed location and ties satellite measurements into a solid local reference. The remote station is a portable receiver that is taken into the field to determine locations and is moved from site to site. The GPS program is the application software that operates the station equipment and is used by layout technicians to gather information of existing and new facilities in the field. The conversion software is used for converting the GPS and inventory information gathered in the field into a form usable by the mapping program.

Accounting Requirements

The function and location of the equipment are the primary factors in determining the account in which the equipment shall be recorded. The components of the GPS shall be accounted for as follows:

1. *Remote and Base Station Equipment.* The cost of the equipment, both remote and fixed, shall be capitalized in a subaccount of Account 391, Office Furniture and Equipment.

2. *GPS Program and Conversion Software for Mapping.* The cost of GPS program and conversion software shall be capitalized in a subaccount of Account 391, Office Furniture and Equipment.

3. *GPS/GIS Field Inventory of System.* The cost of performing a GPS/GIS survey and field inventory of the existing system, by either a consultant or the utility's own forces, shall be charged to Account 588, Miscellaneous Distribution Expenses.

140   Radio-Based Automatic Meter Reading Systems

Radio-based automatic meter reading technology allows meters equipped with a low-power radio device called an ERT (Encoder, Receiver, Transmitter) to be read from a remote location. The ERT device can either be retrofitted to an existing meter or purchased installed in a new meter. The ERT device "encodes" energy consumption and transmits this information to a radio transceiver equipped handheld computer. The data collected and stored in the handheld computer is then uploaded to a billing computer using specialized software for that purpose.

Accounting Requirements

The function of the equipment is the primary factor in determining the account in which the equipment shall be recorded. The components of the radio-based automatic meter reading system shall be recorded in Account 370, Meters. The cost of the meter reader encoding device and retrofitting the meter with the meter reader unit shall be capitalized to the cost of the existing meter. Any associated operating expenses shall be charged to Account 586, Meter Expenses, with maintenance expenses charged to Account 597, Maintenance of Meters.

Separate continuing property records shall be established for the meters, either fitted or retrofitted with the device; the handheld computer; and the upload software. The meters and handheld computer shall be depreciated over the manufacturer's estimated useful service life. The upload software shall be depreciated over the estimated useful service life of the program not to exceed 5 years.

201   Supplemental Financing

Many borrowers secure additional financing from sources other than RUS. CFC was established to provide a source of supplemental financing. Although the accounting provided in this section refers to CFC, it is applicable to other sources of supplemental financing as well.

1. *Membership Fees*

When a membership fee is paid to CFC, the payment shall be recorded as a debit to Account 123.23, Other Investments in Associated Organizations.

2. *Subscriptions*

The subscription agreement to purchase Capital Term Certificates (CTCs) is a binding obligation to pay an initial subscription in equal annual payments over the first three years and an additional annual subscription payable in the fourth through fifteenth years.

The annual subscriptions to CFC for the fourth through fifteenth years is 2.0 percent of total operating revenues after deducting the cost of power. Using the best data available, each borrower shall estimate the amount of CTCs that are required to be purchased. Estimates are not expected to be precise and adjustments shall be made when future projections indicate a change is needed. When the agreement to purchase CTCs is made, an entry shall be recorded debiting Account 123.21, Subscriptions to Capital Term Certificates—Supplemental Financing, and crediting Account 224.11, Other Long-Term Debit—Subscriptions. When the CTCs are actually purchased, the following entries shall be recorded:

Dr. 224.11, Other Long-Term Debt—Subscriptions
   Cr. 131.1, Cash—General
Dr. 123.22, Investments in Capital Term Certificates—Supplemental Financing
   Cr. 123.21, Subscriptions to Capital Term Certificates—Supplemental Financing

3. *Interest Receipts*

Interest accrues monthly to the holder of CTCs at a rate in accordance with the terms of the CFC Invitation to Subscribe. The accrual of interest and the receipt of interest proceeds shall be recorded as follows:

Dr. 171, Interest and Dividends Receivable
   Cr. 419, Interest and Dividend Income
To record the monthly accrual of interest.
Dr. 131.1, Cash—General
   Cr. 171, Interest and Dividends Receivable
To record the receipt of interest proceeds from the investment in CTCs.

NOTE: Any amounts received in excess of the previous accruals shall be credited to Account 419.

Interest penalties may be charged by CFC for late payments on any subscription from the date that the payment was due to the date that the payment was actually received. Such charges shall be expensed to Account 431, Other Interest Expense.

4. *Notes*

If a note is due more than one year after the date of the note, the appropriate sub-account of Account 224, Other Long-Term Debt, shall be credited. If the note is due less than one year from the date of the note, Account 231, Notes Payable, shall be credited.

When a loan from CFC has been consummated and a note is executed, Account 224.13, Supplemental Financing Notes Executed—Debit, shall be debited; and Account 224.12, Other Long-Term Debt—Supplemental Financing, credited. When a loan from another source has been consummated, Account 224.15, Notes Executed—Other—Debit, shall be debited; and Account 224.14, Other Long-Term Debt—Miscellaneous, credited.

5. *Loan Proceeds*

Cash proceeds from unsecured short-term loans shall be deposited into the General Fund Account. Cash proceeds from all secured loans shall be deposited into the Construction Fund Trustee Account.

From two to seven percent, depending upon the class of borrower and its debt-equity ratio, of each CFC loan is applied to the purchase of Capital Term Certificates. At the time of a borrower's first requisition under the CFC loan, the following entry shall be recorded:

Dr. 131.2, Cash—Construction Fund—Trustee
Dr. 123.22, Investments in Capital Term Certificates—Supplemental Financing
   Cr. 224.13, Supplemental Financing Notes Executed—Debit
To record the requisition of funds from CFC.

6. *Capital Credits*

As a result of borrowing from CFC or other lenders organized on a cooperative basis, a borrower may receive capital credit allocations. These allocations are usually based upon the borrower's participation in the lending program with participation measured by the amount of interest expense and conversion costs incurred.

To account for patronage capital allocations from cooperative lenders, the following journal entries shall be recorded:

Dr. 123.1, Patronage Capital from Associated Cooperatives
   Cr. 424, Other Capital Credits and Patronage Capital Allocations
To record the allocation of capital credits from a cooperative lender.

NOTE: If any portion of the interest expense was capitalized as a component of construction cost, a similar portion of the capital credit allocation shall be credited to construction rather than to Account 424. The portion credited to construction shall be determined by applying the percentage of interest expense charged to construction for that particular lender to the interest expense incurred for that lender.

Dr. 131.1, Cash—General
   Cr. 123.1, Patronage Capital from Associated Cooperatives
To record the cash receipt of patronage capital credits from cooperative lenders.

301   Forfeited Customers' Deposits

Customers may be required to make deposits to guarantee payment of amounts billed for electric service. When a customer discontinues service, the customer's deposit shall first be applied to unpaid energy bills, with the balance remitted by check to the customer. If the check is returned, it shall be voided and the original entry that was made when the check was issued shall be reversed.

Unclaimed balances of customer deposits shall remain in Account 235, Customer Deposits, until the legal liability of the cooperative to make such a refund has elapsed. When there is no further legal liability to refund the deposit and if it does not escheat to the state, it shall be transferred to Account 144, Accumulated Provision for Uncollectible Customer Accounts—Credit, retaining full information of all particulars.

401 Computer Software Costs

Computer software consists of programs and routines (sets of computer instructions) which direct the operation of the computer. Software may refer to generalized routines useful in computer operations or to programs for specific applications such as payroll.

The distinction between generalized software and application software is important. Generalized software provides operating support for individual applications. This would include programs for such tasks as making printouts of machine-readable records, sorting records, organizing and maintaining files, translating programs written in a symbolic language into machine-language instructions, and scheduling jobs through the computer. These programs are generally furnished by the manufacturer.

Application software consists of a set of instructions for performing a particular data processing task. Application programs are generally written by the user installation, but are frequently obtained as prewritten packages from software vendors. Application software includes programs such as payroll, billing, general ledger, as well as engineering or managerial applications.

Costs incurred with the purchase or development of computer software shall be accounted for as follows:

1. Capitalize in a subaccount of Account 391, Office Furniture and Equipment, all costs for generalized software. Depreciate the cost over the service life (or remaining life) of the main hardware (i.e., containing central processor). If the purchase invoice does not break out or assign a cost to the "generalized software," it is appropriate to include the full amount in hardware costs. Capitalize in a separate subaccount of Account 391, all costs for applications software determined to have a service life of over one year. Depreciate the cost over the estimated useful service life of the program. This depreciation period shall not exceed five (5) years. RUS realizes, however, that there may be circumstances that justify a useful life longer than 5 years. When this is the case and it is management's intent to utilize these programs over an extended period, written justification shall be submitted to RUS for approval.

2. Expense in Account 921, Office Supplies and Expenses, in the period incurred, all costs associated with the maintenance, updating, and conversion of files or revision of all software, and all costs for software with a useful life of less than 1 year. Also expense in Account 921, the unamortized cost of all software determined, during the year, to be no longer used by or useful to the cooperative. Such costs that are clearly applicable to any category of operating expenses other than the administrative and general category, however, shall be included in the appropriate account in such category. In accordance with the USoA, no portion of such costs shall be capitalized to construction or retirement activities.

In determining the total cost of purchased or internally developed software, the following items shall be included:

a. Costs incurred for feasibility studies if they result in the purchase or development of software;

b. All costs related to the actual purchase or development of the software. These costs must be specifically identifiable with the software and properly supported by time cards, invoices, or other documents; and

c. All costs incurred in "testing and debugging" the software.

Computer software costs are properly chargeable to Account 107, Construction Work in Progress, provided that the following criteria are met:

1. The computer program is specifically dedicated to performing a construction related activity, and

2. The cost of the software is itemized separate and apart from other hardware and software costs.

The cost of software programs meeting the above requirements and having an estimated useful service life in excess of 1 year shall be recorded in Account 186, Miscellaneous Deferred Debits, and amortized to Account 107, Construction Work in Progress, over the estimated service life of the program not to exceed 5 years.

All costs related to training personnel in the use of software shall be expensed as incurred.

The accounting in this section is not intended to apply to immaterial amounts. When it is deemed that the costs of the recordkeeping necessary to amortize these costs outweigh the benefits to the members, software costs shall be expensed in the year incurred.

For computer costs relating to load control equipment, refer to Item 118 of this section.

402 Legal Expenses

Utilities may incur legal expenses which pertain to construction activities, loan activities, or general services. The proper accounting treatment for legal expenses is as follows:

1. Legal fees incurred in connection with a construction project, including the court costs directly related thereto, which can be identified and supported as such, shall be capitalized in Account 107, Construction Work-in-Progress, as a cost of construction.

2. Legal fees specifically identified and properly supported as resulting from activities designed to obtain long-term debt, shall be deferred in Account 181, Unamortized Debt Expense.

3. Legal fees for all other services and fees which cannot be properly identified will require expensing to either Account 417.1, Expenses of Nonutility Operations, or Account 923, Outside Services Employed, as appropriate.

To properly support the capitalization or deferral of legal fees, the attorney shall provide an itemization of services performed and the corresponding costs. Only those costs specifically identified by the attorney as being related to construction or loan activities shall be capitalized or deferred as described above.

403 Leases

Lease transactions shall be accounted for as either a capital lease or an operating lease depending upon whether or not the lease meets the criteria for classification as a capital lease. The definitions for capital and operating leases and the criteria used to determine which method shall be used are as follows:

Definitions

1. *Capital Lease:* A lease that transfers substantially all of the benefits and risks inherent in the ownership of the property to the lessee, who accounts for the lease as an acquisition of an asset and the incurrence of a liability.

2. *Operating Lease:* An operating lease is a simple rental agreement which does not meet the criteria for a capital lease. Under the terms of an operating lease, the lessee records the rental payments due over the term of the lease as rent expense.

Criteria

A lease agreement shall be classified as a capital lease if one or more of the following criteria is met:

1. Ownership of the property is transferred to the lessee by the end of the lease term;

2. The lease contains a bargain purchase option;

3. The lease term is equal to 75 percent or more of the estimated useful life of the leased property; or

4. The present value of the lease payments at the inception of the lease equals or exceeds 90 percent of the fair market value of the leased property.

A lease agreement qualifying as a capital lease shall be recorded in either Account 101.1, Property Under Capital Leases; Account 120.6, Nuclear Fuel Under Capital Leases; or Account 121, Nonutility Property, as appropriate, at the present value (at the beginning of the lease term) of the minimum lease payments. If, however, this amount exceeds the fair value of the leased property at the inception of the lease, the asset shall be recorded at its fair market value. An offsetting credit shall be recorded in Account 227, Obligations Under Capital Leases—Noncurrent, with the current portion recorded in Account 243, Obligations Under Capital Leases—Current. Assets recorded in Account 101.1 shall be classified separately according to the detailed accounts (301–399) provided for electric plant in service.

Monthly payments made under the lease obligation shall be charged to rent expense, fuel expense, or construction work-in-progress as they become payable. Similarly, the leased asset and the associated obligation shall be reduced by the current amount due.

The following journal entries shall be used by the lessee to record capital lease transactions:

Dr. 101.1, Property Under Capital Leases
    Cr. 243, Obligations Under Capital Leases—Current
    Cr. 227, Obligations Under Capital Leases—Noncurrent
To record the capital lease agreement.
Dr. 550, Rents
    Cr. 232, Accounts Payable
Dr. 243, Obligations Under Capital Leases—Current
    Cr. 101.1, Property Under Capital Leases
To record the monthly rental payment due.
Dr. 232, Accounts Payable
    Cr. 131.1, Cash—General
To record the monthly lease payment.

Operating leases which are simple rental agreements do not require the recording of an asset or a liability. The entries that are required to record an operating lease by the lessee are as follows:

Dr. 550, Rents
    Cr. 232, Accounts Payable
To record the monthly rental payment due.
Dr. 232, Accounts Payable
    Cr. 131.1, Cash—General
To record the monthly lease payment.

For purposes of illustration, the journal entries presented in this interpretation debit Account 550, Rents. However, Account 507, Rents (steam power generation); Account 525, Rents (nuclear power generation); Account 540, Rents (hydraulic power generation); Account 550, Rents (other power production); Account 567, Rents (transmission expense); Account 589, Rents (distribution expense); and Account 931, Rents (general and administrative), should be charged, as appropriate, depending upon the function of the equipment being leased.

404 Consolidated Financial Statements

In October 1987, the Financial Accounting Standards Board issued Statement of Financial Accounting Standards No. 94, Consolidation of All Majority-Owned Subsidiaries (Statement No. 94). For purposes of reporting to RUS, Statement No. 94 shall be applied as follows:

1. An RUS borrower that is a subsidiary of another entity shall prepare and submit to RUS separate financial statements even though this financial information is presented in the parent's consolidated statements.

2. In those cases in which an RUS borrower has a majority-ownership in a subsidiary, the borrower must prepare consolidated financial statements in accordance with the requirements of Statement No. 94. These consolidated statements must also include supplementary schedules presenting a Balance Sheet and Income Statement for each majority-owned subsidiary included in the consolidated statements.

Although Statement No. 94 requires the consolidation of majority-owned subsidiaries, Forms 7 and 12 must be prepared on a basis consistent with the equity method of accounting for investments. For distribution borrowers, this requires that the investment be shown on Form 7 in Part C, Balance Sheet, on line 7, Investments in Subsidiary Companies, or line 9, Investments in Associated Organizations—Other—General Funds, as appropriate. The result of operation is shown in Part A, Statement of Operations, on line 23, Income (Loss) from Equity Investments. For generation and transmission borrowers, the investments should be shown on Form 12, in Section C, Balance Sheet, on Line 7, Investments in Subsidiary Companies, or Line 9, Investments in Associated Organizations—Other—General Funds, as appropriate. The result of operations should be shown in Section A, Statement of Operations, on line 30, Income (Loss) from Equity Investments.

501 Patronage Capital Assignments

Accounting for patronage capital and margins may vary depending upon the individual cooperative's bylaws. The comments contained in this section relate to the application of the standard bylaw provisions.

The entries required, at year's end, to record patronage capital transactions where there is no major merchandising program are as follows:

Dr. 219.1, Operating Margins
Dr. 219.2, Nonoperating margins
  Cr. 201.2, Patronage Capital Assignable

To record the amount of patronage capital assignable.

Dr. 201.2, Patronage Capital Assignable
  Cr. 201.1, Patronage Capital Credits

To record the allocation of patronage capital to the patrons' accounts.

The procedure for determining the amount of patronage capital assignable to the individual patron on a total dollar basis is as follows:

1. Determine the total amount to be assigned for the year (Account 201.2).

2. Determine patronage from electric service, the total of consumers' billings (Accounts 440–447).

3. Determine the percentage factor to be used in calculating patronage capital to be credited to each consumer account. Divide "1" by "2".

4. Determine the amount of capital to be credited to each consumer. Multiply the individual consumer's billings for the year by the percentage factor obtained in "3" above.

The procedure for determining the amount of patronage capital assignable to the individual patron on a dollar basis, less the cost of power, is as follows:

1. Determine the total amount to be assigned for the year.

2. Determine the total amount of revenue received from each classification of customers.

3. Determine the total cost of power for each classification of customers. (For example, use cost per kWh sold).

4. For each classification of customers subtract the amount obtained in "3" from the amount obtained in "2," to obtain the total amount received, less cost of power, by classification of customers.

5. Add the amounts obtained in "4" to obtain the total amount of revenue, less cost of power.

6. Divide the total amount received, less cost of power for each classification of customers (amounts obtained in "4"), by the total amount received, less cost of power for all customers (amount obtained in "5") to obtain the prorata percentage for each classification of customers.

7. Multiply the total amount to be allocated (amount obtained in "1") by the prorata percentage for each classification of customers (obtained in "6") to obtain the amount to be assigned each classification of customers.

8. Divide the amount to be assigned each classification of customers (amount obtained in "7") by the total amount received from the classification of customers (amount obtain in "2") to obtain the percentage factor for each classification of customers.

9. Determine the total amount received from each individual customer.

10. Multiply the total amount received from each individual customer (amount obtained in "9") by the percentage factor for his classification (amount obtained in "8") to obtain the amount of capital to be assigned each individual customer.

After calculating the patronage capital to be credited to each customer, there is usually a small balance remaining. This small balance shall remain in Account 201.2, Patronage Capital Assignable, and shall be added to the amount to be assigned in the following year.

Proper records shall be maintained to support all capital credit transactions. As a minimum, these records shall show, for each patron, the amount of capital credited for each year as well as the amount and date retired for each year.

The process of transferring capital credits from the Patronage Capital Assignable accounts to the Patrons' Capital Credits Assigned accounts or to the Patrons' Capital Credits accounts and the making of entries to individual patron's records constitutes an assignment of capital credits. This holds true for recordkeeping purposes as well as from a legal point of view. This assignment shall be followed by formal notification to patrons within a reasonable period of time.

In the event that a distribution cooperative incurs a net loss, that loss shall not be allocated to its members (patrons). The loss shall be accumulated and offset by future nonoperating margins.

502 Patronage Capital Retirements

As the board of directors has the responsibility for determining whether the financial condition of the cooperative will permit retirement of capital credits and whether the proposed retirement complies with mortgage and bylaw provisions, the authorization for the retirement shall be set forth in the board minutes. The entries to record the general retirement of capital credits shall be as follows:

Dr. 201.1, Patronage Capital Credits
  Cr. 238.1, Patronage Capital Payable
To record the board of directors' authorization to make payments of capital credits.
Dr. 238.1, Patronage Capital Payable
  Cr. 131.1, Cash—General.
To record actual cash payments of capital credits.

NOTE: To provide better control over the payment of patronage capital credits, a special checking account should be established in an amount equal to the authorized general retirement. Special prenumbered checks shall be used for each general retirement of patronage capital.

To strengthen internal control and to facilitate the settlement of estates, the board should adopt a policy specifying exactly how payments of capital credits shall be made to the estates of deceased patrons. Payments made to estates shall be recorded as follows:

Dr. 201.1, Patronage Capital Credits
  Cr. 131.1, Cash—General
To record the payment of capital credits when an estate is settled by refunding 100 cents on the dollar.
Dr. 201.1, Patronage Capital Credits
  Cr. 131.1, Cash—General
  Cr. 217, Retired Capital Credits—Gain
To record the payment of capital credits when an estate is settled for less than the full amount of capital credited to the deceased customer's account.
Dr. 217, Retired Capital Credits—Gain
  Cr. 201.2, Patronage Capital Assignable
To record the reallocation to current patrons of the amount of the discount, if provided for in the bylaws.

If a capital credit check is returned due to an inability to locate the patron, it shall be held pending a recheck of available records to ascertain the correct address of the patron. If it is determined that the patron cannot be located, the check shall be cancelled and the amount of the check debited to Account 131.1, Cash—General, and credited to Account 217, Retired Capital Credits—Gain. If the state, however, has unclaimed property laws to which the amount is subject, the amount shall be credited to Account 253, Other Deferred Credits, until final disposition has been made. A notation shall be made in the records of the former patron to facilitate payment if his or her whereabouts is subsequently determined.

If the records show that a number of former patrons have moved and left no forwarding address, it is not necessary to prepare a capital credit retirement check for these patrons when a general retirement of capital credits is made. When setting funds aside to make a general retirement, however, appropriate amounts shall be included to cover payments due these patrons. The cooperative shall then make a reasonable effort to locate these patrons through publication of their names in the newsletter or local newspaper. If the patrons are not located, the amounts set aside and the credits to their accounts shall be handled in a manner similar to those for whom payment checks are returned.

Under the standard bylaw provisions recommended by RUS, it is not proper to use capital credits that were assigned to former patrons to liquidate their delinquent bills. When the standard bylaws are in effect and collection efforts have failed, the balance of an uncollectible bill, after application of customers deposits and membership fees, shall be charged against the accumulated provision for uncollectible accounts. If the patron has capital credits assigned to him or her, these remain untouched except for a notation to indicate the amount of the unpaid bill. When a general retirement of capital credits is made at some future date, amounts which would otherwise be due the patron may be applied to satisfy the unpaid bill with the balance refunded to him or her.

503 Operating and Nonoperating Margins

Occasionally questions arise concerning the accounting for the balances in Accounts 218, Capital Gains and Losses; 219.3, Other Margins; 219.4, Other Margins and Equities-Prior Periods; 434, Extraordinary Income; and 435, Extraordinary Deductions. The balance in these accounts shall be accounted for as follows:

1. The balance in Account 219.4, Other Margins and Equities—Prior Periods, shall be transferred, at year's end, to Account 219.1 or 219.2, as appropriate. Accounts 219.1 and 219.2 are then closed to Account 201.2, Patronage Capital Assignable, unless otherwise provided for in the bylaws.

2. The balances in Account 434, Extraordinary Income, and Account 435, Extraordinary Deductions, shall be cleared to Account 219.2 at year's end.

3. The balances in Account 219.3, Other Margins, and Account 218, Capital Gains and Losses, shall remain in these accounts unless they are allocated to patrons or used to absorb future losses as provided for in the bylaws of the cooperative.

When a cooperative is engaged in a major merchandising activity, all costs properly chargeable to the merchandising activity shall be allocated as such to offset the associated revenue. Nonoperating margins generated from this source shall be prorated annually on a patronage basis and credited to those patrons accounts from whom such amounts were obtained. Merchandising activities of this nature may require a bylaw provision allowing for the allocation of margins generated by a major merchandising activity separate from other operating or nonoperating margins.

If, at the time of the adoption of the bylaw provisions for the allocation of nonoperating margins, there are prior years' losses resulting in debit balances in Accounts 218, Capital Gains and Losses; 219.1, Operating Margins; 219.2, Nonoperating Margins; or 219.3, Other Margins; the credit balances in Accounts 218, 219.2, or 219.3 resulting from prior years' operations shall be transferred, to the extent necessary, to offset such deficits. If the board determines that amounts shall be allocated to prior years' patrons, the credit balances remaining in these accounts shall be transferred to Account 201.2, Patronage Capital Assignable.

If there are current year's losses resulting in debit balances in either Account 219.1 or 219.2, credit balances in Accounts 219.2, 219.3, and 218 shall be transferred, to the extent necessary, to offset such deficits. Remaining credit balances allocable to patrons shall be transferred to Account 1.2.

504 Patronage Capital from G&T Cooperatives

When a cooperative receives capital credits from a G&T cooperative, the transaction shall be recorded by a debit to Account 123.1, Patronage Capital from Associated Cooperatives, and a credit to Account 423, Generation and Transmission Cooperative Capital Credits. This entry shall be made prior to the closing of the cooperative's books even though, in most cases, the notice of the G&T allocation is not received until after the close of the year to which it relates. If precise information cannot be obtained from the G&T within a reasonable time, capital credits shall be recorded on an estimated basis. The difference between the estimated amount and the actual shall be recognized in the following year unless the difference is material.

A distribution cooperative shall not recognize its proportionate share of losses incurred by the G&T. G&T losses shall be accumulated and offset as provided for in the bylaws. Unlike distribution cooperatives, a G&T has the option to offset accumulated losses with future operating and/or nonoperating margins.

505 Patronage Capital Furnished by Other Cooperative Service Organizations

Utilities may obtain long-term and short-term loans, telephone or data processing services, or may purchase oil, gasoline, materials, insurance, and various items from cooperative or mutual enterprises. These enterprises often make patronage refunds or provide evidence that an amount equal to such a refund has been credited to the utility as an investment of capital. The refund may be in the form of cash in the year following the purchase or it may be deducted from the next invoice. The notice of patronage credited to the borrower's account may indicate that such capital may be retired at some future date upon certain conditions having been met. The following provides the accounting journal entries for these types of transactions:

1. Insurance policy refunds from mutual companies, in cash or as credits against subsequent purchases, shall be credited to the appropriate expense account. If sufficient information is not available to credit the refunds to the appropriate expense accounts, they shall be credited to Account 165, Prepayments, and reduce premiums for the current year.

2. Patronage capital allocations from cooperatives, other than mutual insurance companies, shall be credited, in the year that the allocation notice is received, to Account

424, Other Capital Credits and Patronage Allocations, or to construction work-in-progress, as appropriate. The allocation of patronage capital credits between Account 424 and construction work-in-progress shall be made on an equitable basis. For example, patronage capital allocations received from a cooperative money lender are allocated between Account 424 and construction work-in-progress based upon the ratio of interest charged to construction for that particular lender to total interest expense incurred for that lender. Patronage capital allocations received from a material supplier are allocated based upon the ratio of materials charged to construction to total materials purchased.

3. The face amount of patronage capital certificates received by the cooperative from the purchase of goods or services from cooperative money lenders (CFC), oil dealers, material suppliers, pole treating plants, communications services, and others shall be charged to either Account 123.1, Patronage Capital from Associated Cooperatives, or Account 124, Other Investments, as appropriate. Account 123.1 shall include investments in only those cooperatives, or enterprises, that are directly related to the electric utility industry and controlled by the electric cooperatives. These include statewide cooperatives, power cooperatives, and NRECA. Other investments in oil cooperatives and insurance companies shall be charged to Account 124.

506 Forfeited Membership Fees

The bylaws of each cooperative prescribe certain rules and regulations concerning membership in the cooperative. Among these are provisions for forfeiture of membership fees. Some bylaws provide for application of membership fees against any unpaid accounts at the time of termination of service. Any remaining balance may be refunded to the member. Balances that cannot be refunded to the member due to an inability to locate the member or due to bylaw restriction, shall be credited to Account 208, Donated Capital, provided they do not escheat to the state. If disposition of the fees cannot be determined immediately, the amount involved shall be transferred to Account 253, Other Deferred Credits, until the determination is made.

601 Employee Benefits

The costs of employees' fringe benefits (hospitalization, retirement, holiday, sick and vacation pay, etc.) shall be accumulated in an appropriate clearing account and allocated monthly on the basis of payroll. Vacation costs shall be accrued monthly by appropriate credits to an accrual account. These monthly accruals shall be allocated on the basis of direct payroll costs to construction, retirement, and the applicable operations, maintenance, and administrative expense accounts.

Sick leave costs are not normally accrued unless the employee is entitled to be paid for accumulated sick leave at the termination of employment. Salary payments and the associated employee pensions and benefits and social security and other payroll taxes for an employee who is actually sick shall be charged to the same account or accounts to which his or her salary is normally charged.

602 Compensated Absences

Statement of Financial Accounting Standards No. 43, Accounting for Compensated Absences (Statement No. 43), requires employers to accrue a liability as an employee earns the right to be paid for future absences. Four criteria were established for this accrual:

1. The employer's obligation for payment for future absences is attributable to employees' services already performed.

2. The obligation relates to employee rights which vest or accumulate. Vested rights are considered those for which the employer is obligated to make payment even if the employee terminates. Rights which accumulate are those earned but unused rights to compensated absences which may be carried forward to one or more periods, subsequent to the period in which they are earned.

3. Payment of the compensation is probable.

4. The amount can be reasonably estimated.

A company's liability shall be estimated based upon payments it expects to make as a result of employees' work already performed. If a reasonable estimate cannot be made, the company shall disclose that fact in the financial statements.

Statement No. 43 does not apply to severance or termination pay, postretirement benefits, deferred compensation, stock or stock options, group insurance, or other long-term fringe benefits.

The entries required to account for the accrual of compensated absences are as follows:

Dr. 435.1, Cumulative Effect on Prior Years of a Change in Accounting Principle
Cr. 242.3, Accrued Employees' Vacation and Holidays
To record the liability for benefits earned in prior years.

Dr. 107, Construction Work in Progress
Dr. 108.8, Retirement Work in Progress
Dr. Various Operations, Maintenance, and Administrative Expense Accounts
Cr. 242.3, Accrued Employees Vacation and Holidays
To record the liability for benefits earned in the current period.

603 Employee Retirement and Group Insurance

Some borrowers have group insurance or retirement plans or both for their employees. As a general rule the cost of these programs is borne partially by the cooperative and partially by its employees. The cooperative may pay the full cost in advance and recover the employee's share through payroll deductions. The accounting for these transactions is as follows:

1. The cooperative's advanced payment of premiums on insurance and retirement agreements shall be charged to Account 165, Prepayments, for the employers portion, and Account 143, Other Accounts Receivable, for the employee's portion.

2. The cost of the employer's portion of a retirement and group insurance program shall be charged to construction and retirement activities and the applicable operations, maintenance, and administrative expense accounts based upon a specific identification with employees' labor costs charged therein or, in the absence of specific employee identification, based upon direct labor dollars or direct labor hours depending upon which allocation technique provides the most equitable distribution of costs.

604 Deferred Compensation

Many utilities participate in the NRECA Deferred Compensation Program. Based upon the provisions of the program, the following accounting entries shall be made:

Dr. 186.XX, Miscellaneous Deferred Debits—Deferred Compensation
Cr. 228.3, Accumulated Provision for Pensions and Benefits
To increase the deferred compensation provision by the amount of the annual deposit to NRECA's Deferred Compensation Fund.

Dr. 128, Other Special Funds—Deferred Compensation
Cr. 131.1, Cash—General
To record the annual deposit to NRECA's Deferred Compensation Fund.

Dr. Construction Work in Progress, Retirement Work in Progress, or the Various Operations, Maintenance, and Administrative Expense Accounts, as appropriate.
Cr. 186.XX, Miscellaneous Deferred Debits—Deferred Compensation
To record monthly accrual of deferred compensation.

NOTE: If an employee joins the deferred compensation program during the year, use entry #1 to record the additional deposit to the NRECA Deferred Compensation Fund and increase the monthly accrual in entry #2 to reflect this deposit.

NRECA provides borrowers that participate in the deferred compensation program with an annual account statement disclosing the activity for each Homestead Fund investment including the number of shares owned, interest income, dividend income, capital gains/losses, and the value of the shares owned at statement date. Funds may be invested in the Short-term Bond Fund, the Value Fund, the Short-term Government Securities Fund, and the Daily Income Fund. Depending upon the Homestead Fund selected, invested funds may earn interest and dividend income and may experience unrealized holding gains or losses. Based upon the information provided on the annual statement, the following journal entries shall be recorded to recognize the increase or decrease in the fund assets:

Dr. 128, Other Special Funds—Deferred Compensation
Cr. 419, Interest and Dividend Income
Cr. 421, Miscellaneous Nonoperating Income
To record an increase in the fund value as of December 31, 19xx, resulting from interest and dividend income and from unrecognized holding gains on trading securities.

Dr. Various Operations, Maintenance, and Administrative Expense Accounts
Cr. 228.3, Accumulated Provision for Pensions and Benefits
To record an increase in the liability to the employee resulting from an increase in the investment account.

Dr. 426.5, Other Deductions
Cr. 128, Other Special Funds—Deferred Compensation
To record a decrease in fund value as of December 31, 19xx, resulting from unrecognized holding losses on trading securities.

Dr. 228.3, Accumulated Provision for Pensions and Benefits
Cr. Various Operations, Maintenance, and Administrative Expense Accounts
To record a decrease in the liability to the employee resulting from a decrease in the investment account.

Payments made to participating employees because of retirement or separation for other reasons shall be recorded using the following entries:

Dr. 131.1, Cash—General
Cr. 128, Other Special Funds—Deferred Compensation
To record the receipt of funds from NRECA.
and
Dr. 228.3, Accumulated Provision for Pensions and Benefits
Cr. 131.1, Cash—General
To record payment to employee for deferred compensation.

If the borrower has elected to bear the market risk of the funds which guarantee that the amount of money an employee receives will not be less than the amount of salary deferred, the following entry shall be recorded if total payment(s) from NRECA are less than the amount of salary deferred:

Dr. Various Operations, Maintenance, and Administrative Expense Accounts
Cr. 131.1, Cash—General
To record payment to employee for deferred compensation. Payment was made because amount returned did not equal salary deferred.

Appropriate disclosure of the terms of the program shall be made in the notes to the financial statements.

605 Life Insurance Premium on Life of a Borrower Employee

Some borrowers insure the life of the manager and/or key employees with the borrower being named as the beneficiary. Such arrangements shall be accounted for as follows:

1. Charge Account 426.2, Life Insurance, for the net amount of the premium paid each year on the insurance policy.
2. At the anniversary date of the policy each year, charge Account 124, Other Investments, and credit Account 426.2, Life Insurance, with the amount of the annual increase in the cash surrender value of the policy; provided such increase is less than the net premium paid for that year. If the annual increase in the surrender value exceeds the net premium paid for the same year, only that portion of the surrender value increase equal to the net premium paid shall be credited to Account 426.2. The remainder is to be credited to Account 419, Interest and Dividend Income.
3. Upon retirement of the insured employee and surrender of the insurance policy, charge Account 131.1, Cash—General, and credit Account 124, Other Investments, for the amount received from the insurance company. If it is decided to grant to the retiring insured employee all, or any portion, of the cash received upon surrender of the policy, Account 926, Employee Pensions and Benefits, shall be charged and Account 131.1 credited for the amount paid to the retiring employee.
4. If the insured employee dies within his term of service, charge Account 131.1, Cash—General, for the face amount of the policy paid by the insurance company. Credit Account 124, Other Investments, for the cash surrender value previously charged thereto, and credit the remainder to Account 421, Miscellaneous Nonoperating Income.

606 Pension Costs

With the issuance of Statement of Financial Accounting Standards No. 87, Employers' Accounting for Pensions (Statement No. 87), there have been significant changes in the accounting and reporting requirements relating to pension costs. This section will highlight the accounting and reporting requirements for the major types of pension plans. It should be noted, however, that the definitions and accounting procedures outlined in this section relate to financial accounting and they may differ from those used for tax accounting.

Defined Benefit Pension Plans

A defined benefit pension plan is a plan that defines an amount of pension benefit to be provided, usually as a function of one or more factors such as age, years of service, or compensation. In a defined benefit plan, the employer promises to provide, in addition to current wages, retirement income payments in future years after the employee retires or terminates service. Generally, the amount of benefit to be paid depends upon a number of future events that are incorporated into the plan's benefit formula, after including how long the employee and any survivors live, how many years of service the employee renders, and the employee's compensation in the years immediately before retirement or termination.

Under a defined benefit plan, the determination of pension costs, assets, liabilities, and the disclosures in the financial statements require many calculations and assumptions to be made. This section provides a general overview of the accounting and reporting requirements associated with a defined benefit pension plan. Consult Statement No. 87 for guidance in making the necessary calculations and assumption.

The accounting and reporting requirements related to a defined benefit pension plan are as follows:

1. The following components shall be included in the periodic recognition of net pension cost by an employer sponsoring a defined benefit pension plan:

a. The service cost component recognized in a period shall be determined as the actuarial present value of benefits attributed by the pension plan formula to employee service during that period. The measurement of the service cost component requires use of an attribution method and assumptions.

b. The interest cost component recognized in a period shall be determined as the increase in the projected benefit obligation due to the passage of time. Measuring the projected benefit obligation as a present value requires accrual of an interest cost at rates equal to the assumed discount rates.

c. For a funded plan, the actual return on plan assets, if any, shall be determined based upon the fair value of plan assets at the beginning and the end of the period, adjusted for contributions and benefit payments.

d. Plan amendments (including initiation of a plan) often include provisions that grant increased benefits based upon services rendered in prior period. Because plan amendments are granted with the expectation that the employer will realize economic benefits in future period, Statement No. 87 does not require the cost of providing such retroactive benefits (prior service cost) to be included in net periodic pension cost entirely in the year of the amendment but provides for recognition during the future service periods of those employees active at the date of the amendment who are expected to receive benefits under the plan.

The cost of retroactive benefits (including benefits that are granted to retirees) is the increase in the projected benefit obligation at the date of the amendment. Except as noted below, prior service cost shall be amortized by assigning an equal amount to each future period of service of each employee active at the date of the amendments who is expected to receive benefits under the plan. If all or almost all of the plan's participants are inactive, the cost of retroactive plan amendments affecting benefits of inactive participants shall be amortized based upon the remaining life expectancy of those participants rather than the remaining service period.

To reduce the complexity and detail of the computations required, consistent use of an alternative amortization approach that more rapidly reduces the unrecognized cost of retroactive amendments is acceptable. For example, a straight-line amortization of the cost over the average remaining service period of employees expected to receive benefits under the plan is acceptable. The alternative method used shall be disclosed.

In some situations, a history of regular plan amendments and other evidence may indicate that the period during which the employee expects to realize economic benefits from an amendment granting retroactive benefits is shorter than the entire remaining service period of the active employees. Identification of such situations requires an assessment of the individual circumstances and the substance of the particular plan situation. In those circumstances, the amortization of prior service cost shall be accelerated to reflect the more rapid expiration of the employer's economic benefits and to recognize the cost in the periods benefited.

A plan amendment can reduce rather than increase the projected benefit obligation. Such a reduction shall be used to reduce an existing unrecognized prior service cost, and the excess, if any, shall be amortized on the same basis as the cost of benefit increases.

e. Gains and losses are changes in the amount of either the projected benefit obligation or plan assets resulting from experience different from that assumed and changes in assumptions. Gains and losses include amounts that have been realized. Because gains and losses may reflect refinements in estimates as well as real changes in economic values, and because some gains in one period may be offset by losses in another or vice versa, the recognition of gains and losses as components of net pension cost of the period in which they arise is not required.

The expected return on plan assets shall be determined based upon the expected long-term rate of return on plan assets and the market-related value of plan assets. The market-related value of plan assets shall be either fair value or a calculated value that recognizes changes in fair value in a systematic and rational manner over not more than 5 years. Different ways of calculating market-related value may be used for different classes of assets but the manner of determining market-related value shall be applied consistently from year to year for each asset class.

Asset gains and losses are the differences between the actual return on assets during a period and the expected return on assets for that period. Assets gains and losses include both changes reflected in the market-related value of assets and changes not yet reflected in the market-related value (that is, the difference between the fair value of assets and the market-related value). Asset gains and losses not yet reflected in market-related values are not required to be amortized.

As a minimum, amortization of an unrecognized gain or loss (excluding asset gains and losses not yet reflected in market-related value) shall be included as a component of net pension cost for a year if, as of the beginning of the year, that unrecognized net gain or loss exceeds 10 percent of the greater of the projected benefit obligation or the market-related value of plan assets. If amortization is required, the minimum amortization shall be that excess divided by the average remaining service period of active employees expected to receive benefits under the plan. If all or almost all of a plan's participants are inactive, the average remaining life expectancy of the inactive participants shall be used instead of average remaining service life.

Any systematic method of amortization of gains and losses may be used in lieu of the minimum specified in the previous paragraph provided that the minimum is used in any period in which the minimum is greater (i.e., reduces the net balance by more), the method is applied consistently, the method is applied similarly to both gains and losses, and the method is disclosed.

The gain or loss component of net periodic pension cost shall consist of the difference between the actual return on plan assets and the expected return on plan assets and amortization of the unrecognized net gain or loss from previous periods.

2. A liability (unfunded accrued pension cost) shall be recognized if the net periodic pension cost recognized pursuant to Statement No. 87 exceeds amounts the employer has contributed to the plan. An asset (prepaid pension cost) shall be recognized if the net periodic pension cost is less than the

amounts the employer has contributed to the plan.

If the accumulated benefit obligation exceeds the fair value of plan assets, the employer shall recognize a liability (including unfunded accrued pension cost) that is at least equal to the unfunded accumulated benefit obligation. Recognition of an additional minimum liability is required if an unfunded accumulated benefit obligation exists and an asset has been recognized as a prepaid pension cost, the liability already recognized as unfunded accrued pension cost is less than the unfunded accumulated benefit obligation, or no accrued or prepaid pension cost has been recognized.

If an additional minimum liability is recognized, an equal amount shall be recognized as an intangible asset, provided that the asset does not exceed the amount of unrecognized prior service cost. If an additional liability required to be recognized exceeds unrecognized prior service cost, the excess (which represents a net loss not yet recognized as a net periodic pension cost) shall be reported as a separate component (reduction) of equity.

When a new determination of the amount of additional liability is made to prepare a balance sheet, the related intangible asset and separate component of equity shall be eliminated or adjusted, as necessary.

3. An employer sponsoring a defined benefit pension plan shall disclose the following information:

a. A description of the plan including employee groups covered, type of benefit formula, funding policy, types of assets held and significant nonbenefit liabilities, if any, and the nature and effect of significant matters affecting comparability of information for all period presented.

b. The amount of net periodic pension cost for the period showing separately the service cost component, the interest cost component, the actual return on assets for the period, and the net total of other components.

c. A schedule reconciling the funded status of the plan with amounts reported in the employer's balance sheet, showing separately, the fair value of plan assets, the projected benefit obligation identifying the accumulated benefit obligation and the vested benefit obligation, the amount of unrecognized prior service cost, the amount of unrecognized net gain or loss including asset gains and losses not yet reflected in market-related value), the amount of any remaining unrecognized net obligation or net asset existing at the date of initial application of Statement No. 87, the amount of any additional liability recognized, and the amount of net pension asset or liability recognized in the balance sheet (which is the net result of combining the previous six items).

d. The weighted-average assumed discount rate and rate of compensation increase (if applicable) used to measure the projected benefit obligation and the weighted-average expected long-term rate of return on plan assets.

e. If applicable, the amount and type of securities of the employer and related parties included in plan assets, and the approximate amount of annual benefits of employees and retirees covered by annuity contracts issued by the employer and related parties. Also, if applicable, the alternative amortization periods used.

f. An employer that sponsors two or more separate defined benefit pension plans shall determine net periodic pension cost, liabilities, and assets by separately applying the provisions of Statement No. 87 to each plan. In particular, unless an employer clearly has a right to use the assets of one plan to pay benefits of another, a liability required to be recognized for one plan shall not be reduced or eliminated because another plan has assets in excess of its accumulated benefit obligation or because the employer has prepaid pension cost related to another plan.

The required disclosures may be aggregated for all of an employer's single-employer defined benefit plans, or plans may be disaggregated into groups so as to provide the most useful information. Plans with assets in excess of the accumulated benefit obligation, however, shall not be aggregated with plans that have accumulated benefit obligations that exceed plan assets.

Annuity Contracts

An annuity contract is a contract in which an insurance company unconditionally undertakes a legal obligation to provide specified benefits to specific individuals in return for a fixed consideration or premium. An annuity contract is irrevocable and involves the transfer of significant risk from the employer to the insurance company. Some annuity contracts (participating annuity contracts) provide that the purchaser (either the plan or the employer) may participate in the experience of the insurance company. Under these contracts, the insurance company ordinarily pays dividends to the purchaser. If the substance of a participating contract is such that the employer remains subject to all or most of the risks and rewards associated with the benefit obligation covered and the assets transferred to the insurance company, that contract is not an annuity contract for purposes of Statement No. 87.

To the extent that benefits currently earned are covered by annuity contracts, the cost of these benefits shall be the cost of purchasing the contracts, except as noted below. That is, if all benefits attributed by the plan's benefits formula to service in the current period are covered by nonparticipating annuity contracts, the cost of the contracts determines the service cost component of net pension cost for that period.

181

Benefits provided by the pension benefit formula beyond benefits provided by annuity contracts (for example, benefits related to future compensation levels) shall be accounted for according to the provisions applicable to plans not involving insurance contracts.

Benefits covered by annuity contracts shall be excluded from the projected benefit obligation and the accumulated benefit obligation. Except as noted below, annuity contracts shall be excluded from plan assets.

Some annuity contracts provide that the purchaser (either the plan or the employer) may participate in the experience of the insurance company. Under these contracts, the insurance company ordinarily pays dividends to the purchaser, the effect of which is to reduce the cost of the plan. The purchase price of a participating annuity contract ordinarily is higher than the price of an equivalent contract without participation rights. The cost of the participation right shall be recognized, at the date of purchase, as an asset. In subsequent periods, the participation right shall be measured at its fair value if the contract is such that the fair value is reasonably estimable. Otherwise, the participation right shall be measured at its amortized cost (not in excess of its net realizable value), and the cost shall be amortized systematically over the expected dividend period under the contract.

Other Contracts with Insurance Companies

Insurance contracts that are, in substance, equivalent to the purchase of annuities shall be accounted for as such. Other contracts with insurance companies shall be accounted for as investments and measured at fair value. For some contracts, the best available evidence of fair value may be contract value. If a contract has a determinable cash surrender value or conversion value, that is presumed to be its fair value.

Defined Contribution Plans

A defined contribution pension plan is a plan that provides pension benefits in return for services rendered, provides an individual account for each participant, and has terms that specify how contributions to the individual's accounts are to be determined rather than the amount of pension benefits the individual is to receive. Under a defined contribution plan, the pension benefits a participant will receive depend only upon the amount contributed to the participant's account, the returns earned on investments of those contributions, and forfeitures of other participants' benefits that may be allocated to the participant's account.

To the extent that a plan's defined contributions to an individual's account are to be made for periods in which that individual renders services, the net pension cost for a period shall be the contribution called for in that period. If a plan calls for contributions for periods after an individual retires or terminates, the estimated cost shall be accrued during the employee's service period.

An employer that sponsors one or more defined contribution plans shall disclose the following separately from its defined benefit plan disclosures:

1. A description of the plan(s) including employee groups covered, the basis for determining contributions, and the nature and effect of significant matters affecting comparability of information for all periods presented.

2. The amount of cost recognized during the period.

A pension plan having characteristics of both a defined benefit plan and a defined contribution plan requires careful analysis. If the substance of the plan is to provide a defined benefit, as may be the case with some "target benefit" plans, the accounting and disclosure requirements shall be determined in accordance with the provisions applicable to a defined benefit plan.

Multiemployer Plans

A multiemployer plan is a pension plan to which two or more unrelated employers contribute, usually pursuant to one or more collective-bargaining agreements. A characteristic of multiemployer plans is that assets contributed by one participating employer may be used to provide benefits to employees of other participating employers since assets contributed by an employer are not segregated in a separate account or restricted to provide benefits only to employees of that employer.

An employer participating in a multiemployer plan shall recognize as net pension cost, the required contribution for the period and shall recognize as a liability, any contributions due and unpaid. The required contribution includes both current costs and prior service costs. If an employer elects to fund prior service cost in full at the inception of the plan, the total payment becomes the employer's required contribution, and accordingly, its pension cost for the period.

The following provisions are applicable to RUS borrowers participating in a multiemployer pension plan:

1. An electric utility participating in a multiemployer plan may defer current period pension expenses if the provisions of Statement of Financial Accounting Standards No. 71 (Statement No. 71), Accounting for the Effects of Certain Types of Regulation, are applied.

Under the provisions of Statement No. 71, pension costs may be deferred provided such costs are recovered through future rates.

2. An electric utility instituting an amendment to the NRECA Retirement and Security plan enters into a contractual agreement to pay the costs incurred (prior service pension costs) for the amendment. In such cases, the agreement is noncancelable and payable regardless of continued participation in the plan.

Since the utility is unconditionally committed to making these payments and such payments are not contingent upon the utility's continued participation in the plan, the recognition of that liability is appropriate. The costs associated with this liability shall be expensed, in their entirety, when the liability is recognized.

The accounting journal entries required to record the transactions associated with a multiemployer pension plan are as follows:

Sample 1—Current Pension Expense

The journal entry required to record the normal costs associated with the NRECA Retirement and Security Program is as follows:

Dr. Various Operations, Maintenance, and Administrative Expense Accounts
Dr. 107, Construction Work-in-Progress
Dr. 108.8, Retirement Work-in-Progress
Cr. 131.1, Cash—General
To record the payment of pension costs to NRECA.

NOTE: This entry shall not be recorded during the moratorium.

Sample 2—Prior Service Pension Expense

The journal entries required to record the prior service costs associated with the NRECA Retirement and Security Program are as follows:

1. If the RUS borrower elects to pay the prior service pension costs in full, and there is no deferral of costs under the provision of Statement No. 71, the following entry shall be recorded:

Dr. Various Operations, Maintenance, and Administrative Expense Accounts
Dr. 107, Construction Work-in-Progress
Dr. 108.8, Retirement Work-in-Progress
Cr. 131.1, Cash—General
To record the payment of prior service pension costs to NRECA.

2. If the RUS borrower elects to finance prior service pension costs over a period of years and there is no deferral of costs under the provisions of Statement No. 71, the following entries shall be recorded:

Dr. Various Operations, Maintenance, and Administrative Expense Accounts
Dr. 107, Construction Work-in-Progress
Dr. 108.8, Retirement Work-in-Progress
Cr. 224, Other Long-Term Debt
To record the liability to NRECA for prior service pension costs.

Dr. 224, Other Long-Term Debt
Dr. 427, Interest on Long-Term Debt

Cr. 131.1, Cash—General
To record the annual payment to NRECA for prior service pension costs.

3. If the RUS borrower elects to finance prior service pension costs over a period of years and such costs are being deferred and amortized in accordance with the provisions of Statement No. 71, the following entries shall be recorded:

Dr. 182.3, Other Regulatory Assets
Cr. 224, Other Long-Term Debt
To record the liability to NRECA for prior service pension costs.

Dr. Various Operations, Maintenance, and Administrative Expense Accounts
Dr. 107, Construction Work-in-Progress
Dr. 108.8, Retirement Work-in-Progress
Cr. 182.3, Other Regulatory Assets
To record the amortization of deferred prior service pension costs.

Dr. 224, Other Long-Term Debt
Dr. 427, Interest on Long-Term Debt
Cr. 131.1, Cash—General
To record the annual payment to NRECA for prior service pension costs.

4. If the RUS borrower elects to pay the prior service pension costs in full and such costs are being deferred and amortized in accordance with the provisions of Statement No. 71, the following entries shall be recorded:

Dr. 182.3, Other Regulatory Assets
Cr. 131.1, Cash—General
To record the payment to NRECA for prior service pension costs.

Dr. Various Operations, Maintenance, and Administrative Expense Accounts
Dr. 107, Construction Work-in-Progress
Dr. 108.8, Retirement Work-in-Progress
Cr. 182.3, Other Regulatory Assets
To record the amortization of deferred prior service pension costs.

It should be noted that although the above entries relate specifically to the NRECA Retirement and Security Program, they are applicable to all multiemployer pension plans.

An employer that participates in one or more multiemployer plans shall disclose the following separately from disclosures for a single-employer plan:

1. A description of the multiemployer plan(s) including the employee groups covered, the type of benefits provided (defined benefit or defined contribution), and the nature and effect of significant matters affecting comparability of information for all periods presented.

2. The amount of cost recognized during the period.

Multiple-Employer Plans

A multiple-employer plan is, in substance, aggregations of single-employer plans combined to pool their assets for investment purposes to reduce the cost of plan administration. Under a multiple-employer plan, assets are segregated and specifically identified to an employer. In addition, such plans may have features that allow participating employers to have different benefit formulas. Such plans shall be considered single-employer plans for financial accounting purposes and each employer's accounting shall be based upon its respective interest in the plan.

607 Unproductive Time

Lost time relating to construction, operations and maintenance shall be allocated on the basis of direct payroll costs to the appropriate construction, operations or maintenance accounts in the month incurred. Lost time is defined as time on duty during which productive work is not performed due to inclement weather conditions, material shortages, machine repairs, or other reasons.

If lost time attributable to construction has a material effect on the construction accounts in any one month, these costs shall be deferred and distributed over a reasonable period of time by means of a predetermined percentage based upon direct labor.

608 Training Costs, Attendance at Meetings, Etc.

Utilities engage in many types of training programs. Seminars are conducted for directors, managers, office managers, attorneys, engineers, and others. Bookkeepers and office managers attend accountants' meetings. Safety engineers attend safety schools and subsequently conduct regular safety meetings at the cooperative. Costs incurred for the various types of training activities shall be accounted for as follows:

1. Managers' and directors' expenses to attend the NRECA national and state conventions shall be charged to Account 930.2, Miscellaneous General Expenses.

2. Management or engineering seminar fees, salary time attending such seminars including the associated pensions and benefits expense and payroll taxes, and the related per diem and expenses shall be charged to the functional expense accounts. Salaries paid to employees shall also be charged to the appropriate functional expense account. Fees and expenses for directors' attendance shall be charged to Account 930.2, Miscellaneous General Expenses.

3. When the office manager, bookkeeper, or work order clerk attends a state or regional accounting meeting, their salary time and the associated employee pensions and benefits and social security and other payroll taxes shall be charged to the account to which the employees' time is ordinarily charged.

4. Employees' salary time employee and the associated pensions and benefits and social security and other payroll taxes spent attending regular safety meetings conducted by the cooperative shall be charged to the account to which the employees' time is ordinarily charged.

5. A safety engineer's salary time and the associated employee pensions and benefits and social security and other payroll taxes spent attending a statewide safety school shall be charged to Account 925, Injuries and Damages.

6. The salary time and the associated employee pensions and benefits and social security and other payroll taxes spent by a manager or line foreman conducting weekly safely meetings shall be charged to the appropriate functional expense accounts including Account 590, Maintenance, Supervision and Engineering, and Account 920, Administrative and General Services.

609 Maintenance and Operations

"Operations" is the general term used to describe activities involved in the delivery of electric service, by means of a distribution system, to the end user. It pertains to the use of the utility's electric plant facilities and does not include activities intended to prevent or remedy an impending or actual breakdown of those facilities. These activities are classified as maintenance.

"Maintenance" is the general term used to describe the activities involved in the upkeep and repair, but not the enlargement or improvement, of property owned or leased and operated by the company. It does not include the replacement of retirement units.

610 Financial Forecast

Costs incurred and salaries paid to perform a 10-year financial forecast shall be charged to Account 920, Administrative and General Salaries. Related office supplies and expenses shall be charged to Account 921, Office Supplies and Expenses. When a forecast is performed by an outside consultant, the cost shall be charged to Account 923, Outside Services Employed.

611 Advertising Expense

The cost of advertising and the cost of informing the public about the electric cooperative's activities shall be charged to Account 930.2, Miscellaneous General Expenses.

Most of a cooperative's advertising is instructional in nature and relates the cooperative's history and current activities. This type of advertising activity should not be confused with that directed towards the enactment of a specific law or laws directed toward obtaining a specific decision from a regulatory body. Political advertising of the type defined above shall be charged to Account 426.4, Expenditures for Certain Civic, Political, and Related Activities.

612 Special Power Cost Study

A special power cost study is defined as a study to determine whether sufficient power will be available in the future. If additional power or power sources are needed, the study determines whether generation or purchase will supply the lesser cost. The study also indicates when additional power will be needed. As costs are incurred, they shall be charged to a subaccount of Account 186, Miscellaneous Deferred Debits. Upon completion of the study, the costs shall be charged to Account 557, Other Expenses, or amortized to Account 557 over a period of time not to exceed 5 years.

613 Mapping Costs

The purpose of posting completed work orders to system maps is to improve the operation of the system. These costs shall, therefore, be charged to Account 588, Miscellaneous Distribution Expenses. However, the cost of system mapping in the planning stage of construction is an acceptable overhead cost of the resulting construction.

614 Member Relations Costs

Many electric cooperatives hire employees whose duties concern a mixture of power use and member relations activities. The salaries for these employees shall be charged to Account 930.2, Miscellaneous General Expenses, except as provided below:

1. Account 912, Demonstrating and Selling Expenses, shall be charged with all labor, material, advertising, and other expenses incurred in promotional, demonstrating, and selling activities; the objective of which is to promote or retain the use of utility services by present or prospective customers.

2. Account 930.1, General Advertising Expenses, shall be charged with labor, material, and other expenses incurred in advertising and related activities, the cost of which by their content and purpose, are not provided for elsewhere.

3. Account 416, Costs and Expenses of Merchandising, Jobbing, and Contract Work, shall be charged with all costs specifically related to merchandising activities when the utility is engaged in a major merchandising program.

4. Account 426.4, Expenditures for Certain Civic, Political, and Related Activities, shall be charged with expenditures for the purpose of influencing public opinion with respect to the election or appointment of public officials, referenda, legislation, or ordinances (either with respect to the possible adoption of new referenda, legislation or ordinances or repeal or modification of existing referenda, legislation or ordinances); or approval, modi-fication, or revocation of franchises; or for the purpose of influencing the decisions of public officials. Account 426.4 shall not include expenditures which are directly related to appearances before regulatory or other governmental bodies in connection with the borrower's existing or proposed operations.

615 Statewide Fees

Additional fees collected by a statewide association from its members for construction of a statewide building shall be charged to Account 930.2, Miscellaneous General Expenses. Any amounts that are to be repaid by the state association shall be charged to Account 143, Other Accounts Receivable, or Account 123.23, Other Investments in Associated Organizations, depending upon the terms of the repayment.

616 Power Supply/Distribution Cooperative Borrowings

When a power supply cooperative borrows money from a distribution cooperative as the result of a long-term loan agreement, the money shall be recorded on the books of the power supply cooperative as general funds unless restricted to a specific purpose. If restricted, the funds shall be recorded in Account 128, Other Special Funds. The resulting liability shall be recorded in Account 224, Other Long-Term Debt.

The transaction shall be charged to Account 123.23, Other Investments in Associated Organizations, on the books of the distribution cooperative.

617 Rate Discount Allowed by the Power Cooperative to Distribution Cooperatives Owning Connecting Transmission Lines

A distribution cooperative purchases power from a power cooperative. The distribution cooperative owns and operates the transmission line between the power cooperative's facilities and the distribution facilities. Because of this, power is sold at the standard rate at which the power cooperative sells to other distribution cooperatives who do not own their transmission lines, less a discount. The discount or reduction in rate is based upon the distribution cooperative's expense in operating and maintaining its transmission facilities. The contract between the power cooperative and the distribution cooperative must specifically state that the member shall receive a reduced rate or discount from the seller's rate to other member cooperatives.

Under this type of arrangement, the distribution cooperative shall record the cost of purchased power by charging the net amount to Account 555, Purchased Power.

618 Theft Losses not Covered by Insurance

Utilities may suffer losses as a result of thefts of cash, materials and supplies, equipment, or electric plant-in-service that is not covered by insurance. The charges for nominal uninsured losses shall be recorded in the following accounts:

1. Cash—Account 924, Property Insurance, shall be charged.

2. Plant materials and operating supplies—Account 163, Stores Expense Undistributed, shall be charged.

3. Equipment—Account 163, Stores Expense Undistributed, shall be charged for stores equipment; and Account 184, Transportation Expense—Clearing, for transportation and garage equipment. The appropriate miscellaneous operations or administrative expense account (Account 506, 524, 539, 549, 566, 588, 905, 910, 916, or 930.2, as appropriate) shall be charged for all other equipment.

4. Electric Plant-in-Service—A retirement work order shall be prepared for electric plant constituting a unit of property. The loss due to retirement shall be charged to Account 108.6, Accumulated Provision for Depreciation of Distribution Plant. If the plant does not constitute a retirement unit, the loss shall be charged to the appropriate maintenance expense account.

619 Self Billing

To maintain the books of accounts on an accrual basis, bills for customers who self bill and have not sent in a reading or remittance, shall be estimated. A journal entry shall be made to record the estimated revenue and kWh sold by debiting accounts receivable and crediting the appropriate revenue accounts. The estimated bill shall be posted to the customer's account and identified by an appropriate symbol indicating that it is an estimate. Reconciliation with the general ledger control is made in the usual manner.

620 Purchase Rebates

Some vendors from which electric cooperatives purchase plant materials and supplies and merchandise for resale are making purchase rebates based upon the quantity or dollar volume of purchases. These "quantity discounts" may be in the form of cash or credit memoranda, in the form of prepaid package travel arrangements, or a combination of such methods. The rebate shall be accounted for as a reduction in the cost of the material or appliances upon which it was based.

In some instances, the rebate may be for material or appliances that are no longer in stock or cannot be identified. If the rebate is based upon the purchase of plant materials and operating supplies that are normally charged to Account 154, Plant Materials and Operating Supplies, a credit shall be made to Account 163, Stores Expense Undistributed. If the rebate is based upon appliances and equipment held for merchandising or contract work, the credit shall be spread over the items in Account 155, Merchandise. To avoid materially distorting the cost of the remaining appliances, if a portion of the items upon which the rebate was based are no longer in stock, a portion of the credit shall be prorated to Account 416, Cost and Expenses of Merchandising, Jobbing, and Contract Work, on the basis of the number of items sold to the quantity remaining in stock.

If the rebate is in the form of a travel package or travel arrangements, the value of the rebate shall be estimated and recorded as a reduction of the cost of the material or appliances upon which it was based in a manner similar to that of the cash rebates discussed above. The beneficiary of the travel or travel allowance shall be designated by or in accordance with policy established by the board of directors. The contra charge to the reduction in cost shall be to an appropriate account depending upon the relationship of the recipient to the cooperative. For employees, this shall be Account 926, Employee Pensions and Benefits; for directors or patrons, Account 930.2, Miscellaneous General Expenses.

621 Integrity Fund

The CFC Integrity Fund was established to assist borrowers in their attempts to stop takeover bids by investor-owned utilities. A borrower makes a contribution to the Integrity Fund in the form of cash or patronage capital refunds. CFC retains the contribution for a 5-year period during which time the borrower earns interest on the balance in its account. Each year, the borrower receives a statement indicating (both for the total fund and the individual borrower's share) the amount contributed, interest earned, disbursements made, and the ending balance. The disbursements from the fund are allocated to each contributing borrower's account based upon their individual account balances. At the end of the 5-year period, the balance in the account, if any, is refunded to the contributing borrower.

Since the contributing borrower will receive a refund only if its funds are not totally disbursed, the contribution shall be charged to expense in Account 426.1, Donations. If any part of the contribution is returned at the end of the 5-year period, the refund shall be credited to Account 421, Miscellaneous Nonoperating Income.

622 In-Substance Defeasance

An in-substance defeasance has been defined as the process whereby a debtor irrevocably places cash or other assets in a trust to be used solely for the purpose of satisfying scheduled payments of both principal and interest related to a specific debt obligation. Under the structural arrangements of an in-substance defeasance, the probability that the debtor will be required to make additional future debt payments is remote. In these specific circumstances, debt has been determined to be extinguished even though the debtor has not been legally released from his obligations under the debt instrument.

The trust established in a defeasance transaction is restricted as to the nature of the assets held. The trust must be funded with monetary assets that are essentially risk free as to the amount, timing, and collection of interest and principal. For debt denominated in United States dollars, "risk free" assets are limited to:

1. Direct obligations of the United States government;

2. Obligations guaranteed by the United States government; and

3. Securities that are backed by United States government obligations as collateral under an arrangement by which the interest and principal payments on the collateral, flow immediately through to the holder of the security.

The monetary assets of the trust must provide cash flows sufficient to coincide with the scheduled interest and principal payments on the defeased debt. If the trust is expected to pay the costs associated with the defeasance, such as trustee fees, these costs must be considered in determining the amount of funds required by the trust.

The principles of in-substance defeasance apply only to debt with specific maturities and fixed payment schedules and, as such, do not apply to debt with variable terms in which advance determination of debt service requirements is not possible.

Generally accepted accounting principles (GAAP) address the extinguishment of debt in Accounting Principles Board Opinion No. 26, and Statement of Financial Accounting Standard No. 76, Extinguishment of Debt. In accordance with these two statements, debt which has been defeased remains recorded in the regulated books of account as do the assets placed in the irrevocable trust. They are not, however, recognized as an asset and liability for financial reporting purposes. The transaction, including the total amount of debt outstanding and the total amount of debt that is considered extinguished at the end of the period, must be disclosed in the footnotes to the financial statements as long as the debt remains outstanding.

Debt is frequently extinguished before its scheduled maturity. Debt may be extinguished by the use of the borrower's general funds, or by the reacquisition of another debt issue at a different interest rate or varying terms. As these assets are expected to be revenue producing during those years, both the assets and the revenue they generate may be utilized to meet maturing debt payments. Therefore, in most instances, the dollar value of the assets initially placed in the trust do not equal the dollar value of the outstanding principal balance. The difference represents an "economic " gain or loss to the borrower.

To provide consistency in reporting among all RUS borrowers, any gain or loss that is recognized for financial statement purposes should be reported in accordance with the provisions of General Instruction No. 17 of this part. Therefore, the gain or loss should be amortized (for reporting purposes) in equal monthly amounts over the remaining life of the original debt issue or the remaining life of the new issue. The gain or loss may be reported in the current period only in those instances in which it is immaterial to the financial statements.

The RUS Form 7, Financial and Statistical Report, and the RUS Form 12, Operating Report—Financial, must, however, reflect the actual amounts recorded in the books and records of the borrower.

623 Satellite or Cable Television Services

Many electric borrowers have become involved in either providing satellite or cable television services or obtaining satellite or cable television services for their own use. This section outlines the accounting to be followed when recording transactions involving satellite or cable television services.

1. *Separate Subsidiary*

If a borrower provides satellite or cable television services through a separate subsidiary, the investment in the subsidiary shall be recorded in Account 123.11, Investment in Subsidiary Companies. The net income or loss of the subsidiary shall be debited or credited to Account 123.11, as appropriate, with an offsetting entry to Account 418.1, Equity in Earnings of Subsidiary Companies.

2. *Segment of Current Operations*

If a borrower provides satellite or cable television services as part of its normal operations, the investment in satellite or cable television equipment shall be recorded in Account 121, Nonutility Property. All income associated with these services shall be recorded in Account 417, Revenues from Nonutility Operations, and the associated expenses shall be charged to Account 417.1, Expenses of Nonutility Operations.

3. *Sale and Installation of Satellite or Cable Television Equipment*

If a borrower sells or installs satellite or cable television equipment, the equipment purchased for resale shall be recorded in Account 156, Other Materials and Supplies, until sold. The revenues generated from such

187

sales or installations shall be recorded in Account 415, Revenues from Merchandising, Jobbing, and Contract Work, and the associated expenses shall be charged to Account 416, Costs and Expenses of Merchandising, Jobbing, and Contract Work.

4. *Equipment Purchased for Own Use*

If a borrower purchases satellite or cable television equipment for its own use, the investment in the equipment shall be recorded in Account 397, Communication Equipment.

624 Pollution Control Bonds

The construction and installation of pollution control facilities are often financed by issuing tax exempt municipal securities. The funds generated from the sale of these securities are deposited into an account that is controlled by a designated trustee. The funds under the control of the trustee are usually invested, earning interest, until they are needed.

Interest expense accrued on the pollution control bonds during the construction period shall be capitalized in Account 107, Construction Work-in-Progress. After construction is complete, all subsequent accruals of interest expense shall be charged to Account 427, Interest on Long-Term Debt.

Interest income earned during the construction period shall be recorded as a debit to Account 171, Interest and Dividends Receivable, and a credit to Account 107, Construction Work-in-Progress. Upon notification of receipt of the interest in the trustee account, Account 221.XX, Long-Term Debt—Pollution Control Bonds, shall be debited and Account 171, Interest and Dividends Receivable shall be credited. Upon completion of construction, Account 419, Interest and Dividend Income, shall be credited for the amount of interest income earned during the period.

The entries required to account for the transactions associated with the issuance of pollution control bonds are as follows:

Dr. 221.XX, Long-Term Debt—Pollution Control Bonds—Trustee
  Cr. Account 221.X1, Long-Term Debt—Pollution Control Bonds
To record the sale of pollution control bonds.
Dr. 107, Construction Work-in-Progress
  Cr. 232, Accounts Payable
To record costs incurred in construction of pollution control facilities.
Dr. 131.1, Cash—General Funds
  Cr. 221.XX, Long-Term Debt—Pollution Control Bonds—Trustee
To record the transfer of funds from the trustee.
Dr. 107, Construction Work-in-Progress
  Cr. 221.XX, Long-Term Debt—Pollution Control Bonds—Trustee
To record interest expense on pollution control bonds.
Dr. 171, Interest and Dividends Receivable

  Cr. 107, Construction Work-in-Progress
To record earnings from investments made by the trustee.
Dr. 221.XX, Long-Term Debt—Pollution Control Bonds—Trustee
  Cr. 171, Interest and Dividends Receivable
To record receipt of interest income by the trustee account.
Dr. XXX, Various Plant Accounts
  Cr. 107, Construction Work-in-Progress
To close completed construction to the primary plant accounts.

625 Prepayment of Debt

Many RUS borrowers have decided to redeem (prepay) their issues of long-term debt. As a result of this redemption, the borrower may incur a gain (discount) or a loss (penalty) on the early extinguishment of debt. The accounting for this gain or loss is highlighted in this section.

If debt is redeemed without refunding (paid with general funds), the gain or loss incurred shall be recorded in Account 189, Unamortized Loss on Reacquired Debt, or Account 257, Unamortized Gain on Reacquired Debt, as appropriate. The borrower shall amortize the recorded deferral on a monthly basis over the remaining life of the old debt issue. Amounts so amortized shall be charged to Account 428.1, Amortization of Loss on Reacquired Debt, or credited to Account 429.1, Amortization of Gain on Reacquired Debt—Credit, as appropriate.

If the debt is redeemed with refunding (refinanced), the gain or loss incurred shall be recorded in Account 189 or Account 257, as appropriate. The borrower may elect to account for the deferrals as follows:

1. Write them off immediately when the amounts are insignificant;
2. Amortize them by equal monthly amounts over the remaining life of the old debt issue; or
3. Amortize them by equal monthly amounts over the life of the new debt issue.

Once an election has been made, it shall be applied on a consistent basis. Regardless of the option selected, the amortization shall be charged to either Account 428.1 or 429.1, as appropriate.

Where a regulatory authority having jurisdiction over the borrower specifically disallows the rate principle of amortizing gains or losses on the redemption of long-term debt without refunding, and does not apply the gain or loss to interest charges in computing the borrower's rates, the alternative method may be used to account for gains or losses relating to the redemption of long-term debt with or without refunding. The alternative method requires that gains or losses be recorded in Account 421, Miscellaneous Nonoperating Income, or Account 426.5, Other Deductions, as incurred. When the alternative method is used, the borrower shall include a footnote to the financial statements stating the reason for using this method and its treatment for rate making purposes.

626 Rural Economic Development Loan and Grant Program

On December 21, 1987, Section 313, Cushion of Credits Payments Program, was added to the Rural Electrification Act. Section 313 establishes a Rural Economic Development Subaccount and authorizes the Administrator of the Rural Utilities Service to provide zero interest loans or grants to RE Act borrowers for the purpose of promoting rural economic development and job creation projects.

Subpart B, Rural Economic Development Loan and Grant Program, 7 CFR Part 1703, sets forth the policies and procedures relating to the zero interest loan program and for approving and administering grants. The accounting journal entries required to record the transactions associated with a rural economic development loan are as follows:

Dr. 224.17, RUS Notes Executed—Economic Development—Debit
  Cr. 224.16, Long-Term Debt—RUS Economic Development Notes Executed

To record the contractual obligation to RUS for the Economic Development Notes.

Dr. 131.12, Cash—General—Economic Development Funds
  Cr. 224.17, RUS Notes Executed—Economic Development—Debit

To record the receipt of the economic development loan funds.

Dr. 123, Investment in Associated Organizations or
Dr. 124, Other Investments
  Cr. 131.12, Cash—General—Economic Development Funds

To record the disbursement of Economic development loan funds to the project.

Dr. 131.1, Cash—General Funds
  Cr. 421, Miscellaneous Nonoperating Income

To record payment received from the project for loan servicing charges.

Dr. 171, Interest and Dividends Receivable
  Cr. 419, Interest and Dividend Income

To record the interest earned on the investment of rural economic development loan funds.

Dr. 426.1, Donations or
Dr. 426.5, Other Deductions
  Cr. 131.1, Cash—General Funds

To record the payment of interest earned in excess of $500.00 on the investment of rural economic development loan funds.

NOTE: Interest earned in excess of $500.00 must be used for the rural economic development project for which the loan funds were received or returned to RUS.

Dr. 131.12, Cash—General—Economic Development Funds
  Cr. 123, Investment in Associated Organizations or
  Cr. 124, Other Investments

To record receipt of the repayment, by the project, of economic development loan funds.

Dr. 426.5, Other Deductions
  Cr. 123, Investment in Associated Organizations or
  Cr. 124, Other Investments

To record the default, by a project, of economic development loan funds.

Dr. 224.16, Long-Term Debt—RUS Economic Development Notes Executed
  Cr. 131.12, Cash—General—Economic Development Funds

To record the repayment, to RUS, of the economic development loan funds.

The accounting journal entries required to record the transactions associated with a rural economic development grant are as follows:

Dr. 131.13, Cash—General—Economic Development Grant Funds
  Cr. 224.18, Other Long-Term Debt—Grant Funds;
  Cr. 208, Donated Capital; or
  Cr. 421, Miscellaneous Nonoperating Income

To record grant funds disbursed by RUS. If the grant agreement requires repayment of the funds upon termination of the revolving loan program, Account 224.18 should be credited. If the grant agreement states that there is absolutely no obligation for repayment upon termination of the revolving loan program, the funds should be accounted for as a permanent infusion of capital by crediting Account 208. If, however, the grant agreement is silent as to the final disposition of the grant funds, Account 421 should be credited.

Dr. 123.3, Investment in Associated Organizations—Federal Economic Development Loans
  Cr. 131.13, Cash—General—Economic Development Grant Funds

To record advances of Federal funds to associated organizations for authorized rural economic development projects.

Dr. 124.1, Other Investments—Federal Economic Development Loans
  Cr. 131.13, Cash—General—Economic Development Grant Funds

To record advances of Federal funds to nonassociated organizations for authorized rural economic development projects.

Dr. 171, Interest and Dividends Receivable
  Cr. 419, Interest and Dividend Income

189

To record the accrual of interest on loans made to associated and nonassociated organizations with Federal funds for authorized rural economic development projects.

Dr. 131.14, Cash—General—Economic Development Non-Federal Revolving Funds
Cr. 123.3, Investment in Associated Organizations—Federal Economic Development Loans or
Cr. 124.1, Other Investments—Federal Economic Development Loans

To record repayment of loans made with Federal funds.

Dr. 123.4, Investment in Associated Organizations—Non-Federal Economic Development Loans
Cr. 131.14, Cash—General—Economic Development Non-Federal Revolving Funds

To record advances of non-Federal funds to associated organizations for authorized rural economic development projects.

Dr. 124.2, Other Investments—Non-Federal Economic Development Loans
Cr. 131.14, Cash—General—Economic Development Non-Federal Revolving Funds

To record advances of non-Federal funds to nonassociated organizations for authorized rural economic development projects.

Dr. 171, Interest and Dividends Receivable
Cr. 419, Interest and Dividend Income

To record the accrual of interest on loans made to associated and nonassociated organizations with non-Federal funds for authorized rural economic development projects.

Dr. 131.14, Cash—General—Economic Development Non-Federal Revolving Funds
Cr. 123.4, Investment in Associated Organizations—Non-Federal Economic Development Loans or
Cr. 124.2, Other Investments—Non-Federal Economic Development Loans

To record repayment of loans made with non-Federal funds.

627 Postretirement Benefits

Statement of Financial Accounting Standards No. 106, Employers' Accounting for Postretirement Benefits Other than Pensions (Statement No. 106), requires reporting entities to accrue the expected cost of postretirement benefits during the years the employee provides service to the entity. For purposes of applying the provisions of Statement No. 106, members of the board of directors are considered to be employees of the cooperative. Prior to the issuance of Statement No. 106, most reporting entities accounted for postretirement benefit costs on a "pay-as-you-go" basis; that is, costs were recognized when paid, not when the employee provided service to the entity in exchange for the benefits.

As defined in Statement No. 106, a postretirement benefit plan is a deferred compensation arrangement in which an employer promises to exchange future benefits for an employee's current services. Postretirement benefit plans may be funded or unfunded. Postretirement benefits include, but are not limited to, health care, life insurance, tuition assistance, day care, legal services, and housing subsidies provided outside of a pension plan.

This statement applies to both written plans and to plans whose existence is implied from a practice of paying postretirement benefits. An employer's practice of providing postretirement benefits to selected employees under individual contracts with specified terms determined on an employee-by-employee basis does not, however, constitute a postretirement benefit plan under the provisions of this statement.

Postretirement benefit plans generally fall into three categories: single-employer defined benefit plans, multi-employer plans, and multiple-employer plans.

The accounting requirements set forth in this interpretation focus on single-and multiple-employer plans. The accounting requirements set forth in Statement No. 106 for multiemployer plans or defined contribution plans shall be adopted for borrowers electing those types of plans.

Under the provisions of Statement No. 106, there are two components of the postretirement benefit cost: the current period cost and the transition obligation. The transition obligation is a one-time accrual of the costs resulting from services already provided. Statement No. 106 allows the transition obligation to be deferred and amortized on a straight-line basis over the average remaining service period of the active employees. If the average remaining service life of the employees is less than 20 years, a 20-year amortization period may be used.

Accounting Requirements

All RUS borrowers must adopt the accrual accounting provisions and reporting requirements set forth in Statement No. 106. The transition obligation and accrual of the current period cost must be based upon an actuarial study. This study must be updated to allow the borrower to comply with the measurement date requirements of Statement No. 106; however, the study must, at a minimum, be updated every five years. RUS will not allow electric borrowers to account for postretirement benefits on a "pay-as-you-go" basis.

The deferral and amortization of the transition obligation does not require RUS approval provided that it complies with the provisions of Statement No. 106. If, however, a borrower elects to expense the transition obligation in the current period and subsequently defer this expense in accordance with Statement of Financial Accounting Standards No. 71, Accounting for the Effects of Certain Types of Regulation, the deferral must be approved by RUS. In those states in which the commission will not allow the recovery of the transition obligation through future rates, the transition obligation must be expensed, in its entirety, in the year in which Statement No. 106 is adopted. A portion of the transition obligation may be charged to construction and retirement activities provided such charges are properly supported.

Effective Date and Implementation

For plans outside the United States and for defined benefit plans of employers that (a) are nonpublic enterprises and (b) sponsor defined benefit postretirement plans with no more than 500 plan participants in the aggregate, Statement No. 106 is effective for fiscal years beginning after December 15, 1994. For all other plans, Statement No. 106 is effective for fiscal years beginning after December 15, 1992.

RUS borrowers must comply with the implementation dates set forth in Statement No. 106. At the time of the adoption of Statement No. 106, rates must be in place sufficient to recover the current period expense and any amortization of the transition obligation. A copy of a board resolution or commission order, as appropriate, indicating that the transition obligation and current period expense have been included in the borrower's rates must be submitted to RUS.

Accounting Journal Entries—Transition Obligation

The journal entries required to record the transition obligation are as follows:

1. If the borrower elects to expense the transition obligation in the current period and there is no deferral of costs, the following entry shall be recorded:

Dr. 435.1, Cumulative Effect on Prior Years of a Change in Accounting Principle

or

Dr. 926, Employee Pensions and Benefits
Dr. 107, Construction Work-in-Progress
Dr. 108.8, Retirement Work-in-Progress
Cr. 228.3, Accumulated Provision for Pensions and Benefits
To record the current period recognition of the transition obligation for postretirement benefits. Note: A portion of the transition obligation may be charged to construction and retirement activities provided such charges are properly supported.

2. If the borrower elects to defer and amortize the transition obligation in accordance with the provisions of Statement No. 71, the following entry shall be recorded:

Dr. 182.3, Other Regulatory Assets
Cr. 228.3, Accumulated Provision for Pensions and Benefits

To record the deferral of the transition obligation under the provisions of Statement No. 71.

Dr. Various Operations, Maintenance, and Administrative Expense Accounts
Dr. 107, Construction Work-in-Progress
Dr. 108.8, Retirement Work-in-Progress
Cr. 182.3, Other Regulatory Assets
To record the amortization of postretirement benefits expenses as they are recovered through rates in accordance with Statement No. 71.

3. The deferral and amortization of the transition obligation under the provisions of Statement No. 106 is considered to be an off balance sheet item. If, therefore, the borrower elects to defer and amortize the transition obligation on a straight-line basis over the average remaining service period of the active employees or 20 years in accordance with Statement No. 106, no entry is required. Instead, the transition obligation is recognized as a component of postretirement benefit cost as it is amortized. It should be noted, however, that the amount of the unamortized transition obligation must be disclosed in the notes to the financial statements.

Accounting Journal Entries—Current Period Expense

The current period postretirement expense should be recorded by the following entry:

Dr. Various Operations, Maintenance, and Administrative Expense Accounts
Dr. 107, Construction Work-in-Progress
Dr. 108.8, Retirement Work-in-Progress
Cr. 228.3, Accumulated Provision for Pensions and Benefits
To record current period postretirement benefit expense.
Dr. 228.3X, Accumulated Provision for Pensions and Benefits—Funded
Cr. 131.1, Cash—General
To record cash payments on a "pay-as-you-go" basis for postretirement benefits.

Accounting Journal Entry—Funding

If a borrower elects to voluntarily fund its postretirement benefits obligation in an external, irrevocable trust, the following entry shall be recorded:

Dr. 228.3X, Accumulated Provision for Pensions and Benefits—Funded
Cr. 131.1, Cash—General
To record the funding of postretirement benefits expense into an external, irrevocable trust.

If a borrower elects to voluntarily fund its postretirement benefits obligation in an investment vehicle other than an external, irrevocable trust, the following entry shall be recorded:

Dr. 128, Other Special Funds

Cr. 131.1, Cash—General
To record the funding of postretirement benefits expense into an investment vehicle other than an external, irrevocable trust.

628 Postemployment Benefits

Statement of Financial Accounting Standards No. 112, Employers' Accounting for Postemployment Benefits (Statement No. 112) establishes the standards of financial accounting and reporting for employers who provide benefits to former or inactive employees after employment but before retirement. Inactive employees are those who are not currently rendering service to the employer but who have not been terminated, including employees who are on disability leave, regardless of whether they are expected to return to active service. For purposes of applying the provisions of Statement No. 112, former members of the board of directors are considered to be employees of the cooperative.

Postemployment benefits include benefits provided to former or inactive employees, their beneficiaries, and covered dependents. They include, but are not limited to, salary continuation, supplemental benefits (including workmen's compensation), health care, job training and counseling, and life insurance coverage. Benefits may be provided in cash or in kind and may be paid upon cessation of active employment or over a specified period of time.

The cost of providing postemployment benefits is considered to be a part of the compensation provided to an employee in exchange for current service and should, therefore, be accrued as the employee earns the right to be paid for future postemployment benefits. Applying the criteria set forth in Statement of Financial Accounting Standards No. 43, Accounting for Compensated Absences, a postemployment benefit obligation is accrued when all of the following conditions are met:

1. The employer's obligation for payment for future absences is attributable to employees' services already performed;

2. The obligation relates to employee rights that vest or accumulate. Vested rights are considered those rights for which the employer is obligated to make payment even if the employee terminates. Rights that accumulate are those earned, but unused rights to compensated absences that may be carried forward to one or more periods subsequent to the period in which they are earned;

3. Payment of the compensation is probable; and

4. The amount can be reasonably estimated.

If all of these conditions are not met, the employer must account for its postemployment benefit obligation in accordance with Statement of Financial Accounting Standards No. 5, Accounting for Contingencies (Statement No. 5) when it becomes probable that a liability has been incurred and the amount of that liability can be reasonably estimated.

If an obligation for postemployment benefits is not accrued in accordance with the provisions of Statement No. 5 or Statement No. 43 only because the amount cannot be reasonably estimated, the financial statements should disclose that fact.

Accounting Requirements

All RUS borrowers must adopt the accrual accounting provisions and reporting requirements set forth in Statement No. 112 as of the statement's implementation date. A portion of the cumulative effect may be charged to construction and retirement activities provided such charges are properly supported. If a borrower elects to defer the cumulative effect of implementing Statement No. 112 in accordance with the provisions of Statement of Financial Accounting Standards No. 71, Accounting for the Effects of Certain Types of Regulation, the deferral must be approved by RUS.

Effective Date and Implementation

Statement No. 112 is effective for fiscal years beginning after December 15, 1993. Previously issued financial statements should not be restated.

RUS borrowers must comply with the implementation date set forth in Statement No. 112. At the time of the adoption of Statement No. 112, rates must be in place sufficient to recover the current period expense.

Accounting Journal Entries

The journal entries required to account for postemployment benefits are as follows:

Dr. 435.1, Cumulative Effect on Prior Years of a Change in Accounting Principle
Dr. 107, Construction Work in Progress
Dr. 108.8, Retirement Work in Progress
Cr. 228.3, Accumulated Provision for Pensions and Benefits
To record the cumulative effect of implementing Statement No. 112.

NOTE: A portion of the cumulative effect may be charged to construction and retirement activities provided such charges are properly supported. Account 435.1 is closed to Account 219.2, Nonoperating Margins.

If the borrower elects to defer and amortize the cumulative effect in accordance with the provisions of Statement No. 71, the following entry shall be recorded:

Dr. 182.3, Other Regulatory Assets
Cr. 228.3, Accumulated Provision for Pensions and Benefits
To record the deferral of the cumulative effect of implementing Statement No. 112 in

accordance with the provisions of Statement No. 71.

Dr. Various Operations, Maintenance, and Administrative Expense Accounts
Dr. 107, Construction Work in Progress
Dr. 108.8, Retirement Work in Progress
Cr. 182.3, Other Regulatory Assets

To record the amortization of the cumulative effect of implementing Statement No. 112 as it is recovered through rates in accordance with Statement No. 71.

Dr. Various Operations, Maintenance, and Administrative Expense Accounts
Dr. 107, Construction Work in Progress
Dr. 108.8, Retirement Work in Progress
Cr. 228.3, Accumulated Provision for Pensions and Benefits

To record current period postemployment benefit expense.

NOTE: If postemployment benefits are accrued under the criteria set forth in Statement No. 43, this journal entry is made on a monthly basis. If, however, the accrual is based upon the provisions of Statement No. 5, this is a one-time entry unless the liability is reevaluated and subsequently adjusted.

629 Investments in Debt and Equity Securities

Statement of Financial Accounting Standards No. 115, Accounting for Certain Investments in Debt and Equity Securities (Statement No. 115), establishes the standards of financial accounting and reporting for investments in debt securities and for investments in equity securities that have readily determinable fair values. Statement No. 115 does not apply to investments in equity securities accounted for under the equity method nor to investments in consolidated subsidiaries.

At the time of acquisition, an entity must classify debt and equity securities into one of three categories: held-to-maturity, available-for-sale, or trading. At the balance sheet date, the appropriateness of the classifications must be reassessed.

Investments in debt securities are classified as held-to-maturity and are measured at amortized cost in the balance sheet only if the reporting entity has the positive intent and ability to hold these securities to maturity. Debt securities are not classified as held-to-maturity if the entity has the intent to hold the security only for an indefinite period; for example, if the security would become available for sale in response to changes in market interest rates and related changes in the security's prepayment risk, needs for liquidity, changes in the availability of and the yield on alternative investments, changes in funding sources and terms, and changes in foreign currency risk.

Investments in debt securities that are not classified as held-to-maturity and equity securities that have readily determinable fair values are classified as either trading securities or available-for-sale securities and are measured at fair value in the balance sheet. Trading securities are those securities that are bought and held principally for the purpose of selling them in the near future. Trading generally reflects active and frequent buying and selling and trading securities are generally used with the objective of generating profits on short-term differences in prices. Available-for-sale securities are those investments not classified as either trading securities or held-to-maturity securities.

Statement No. 115 requires unrealized holding gains and losses for trading securities to be included in earnings in the current period. Unrealized holding gains and losses for available-for-sale securities are excluded from earnings; however, they are reported as a net amount in a separate component of shareholders' equity until realized.

For individual securities classified as either available-for sale or held-to-maturity, an entity must determine whether a decline in the security's fair value below the amortized cost is other than temporary. If the decline in fair value is determined to be permanent, that is, it is probable that the entity will not be able to collect all amounts due under the contractual terms of the security, the realized loss is accounted for in earnings of the current period. The new cost basis is not adjusted upward for subsequent recoveries in the fair value. Subsequent increases in the fair value of available-for-sale securities are included in the separate component of equity. Subsequent decreases are also included in the separate component of equity.

All trading securities are reported as current assets in the balance sheet and individual held-to-maturity and available-for-sale securities are classified as either current or noncurrent, as appropriate. Cash flows from the purchase, sale, or maturity of available-for-sale securities and held-to-maturity securities are classified in the statement of cash flows as cash flows from investing activities and reported gross for each security classification.

Accounting Requirements

All RUS borrowers must adopt the accounting, reporting, and disclosure requirements set forth in Statement No. 115 as of the statement's implementation date. Unrealized holding gains or losses for trading securities shall be recorded in either Account 421, Miscellaneous Nonoperating Income, or Account 426.5, Other Deductions, as appropriate. Unrealized holding gains or losses for available-for-sale securities held by the corporate entity are recognized as a component of stockholder's equity in Account 215.1, Unrealized Gains and Losses—Debt and Equity Securities. A contra account of the investment account shall be debited or credited accordingly. Unrealized gains and losses for available-for-sale securities held in a decommissioning fund shall increase or decrease, as appropriate, the reported value of the fund.

Effective Date and Implementation

Statement No. 115 is effective for fiscal years beginning after December 15, 1993. At the beginning of the entity's fiscal year, the entity must classify its debt and equity securities on the basis of the entity's current intent. This statement may not be applied retroactively to prior years' financial statements. For fiscal years beginning prior to December 16, 1993, reporting entities are permitted to apply Statement No. 115 as of the end of a fiscal year for which annual financial statements have not previously been issued.

630 Split Dollar Life Insurance

The National Rural Electric Cooperative Association Split Dollar Life Insurance provides life insurance benefits to cooperative employees. The benefits provided under this policy consist of two components, the face value of the insurance policy and the accumulated cash surrender value. While the employee is the owner of the policy, the employee must sign a collateral assignment giving the cooperative absolute right to the cash surrender value of the policy. Under the terms of this collateral assignment, the employee must reimburse the cooperative for the premiums paid upon the employee's termination of employment or attainment of the age of 62 if the employee wishes to maintain the insurance coverage. If death occurs prior to either of these events, the premiums paid to date by the cooperative are deducted from the death benefits payable to the policy beneficiary.

Accounting Requirements

Financial Accounting Standards Board Technical Bulletin 85–4, Accounting for Purchase of Life Insurance (Bulletin 85–4), states that the amount that could be realized under an insurance contract as of the date of the financial statements should be reported as an asset. The change in the cash surrender or contract value of that asset during the period should be reported as an adjustment to the premiums paid in determining the expense or income to be recognized for the period. The cooperative shall, therefore, record the cash surrender value of the policy as an asset because of its absolute right to receive that value based upon the employee's collateral assignment. Any receivable that may occur as a result of the employee reimbursement for the premiums paid is contingent upon the employee electing to maintain the insurance coverage after termination of employment or reaching the age of 62 and is not

recorded as an asset on the cooperative's records.

Accounting Journal Entries

The journal entries required to account for the NRECA Split Dollar Life Insurance Program are as follows:

Dr. 124, Other Investments
Cr. Various Operations, Maintenance, and Administrative Expense Accounts
To record an increase in the cash surrender value of the insurance contract.

or

Dr. Various Operations, Maintenance, and Administrative Expense Accounts
Cr. 124, Other Investments
To record a decrease in the cash surrender value of the insurance contract.

Dr. Various Operations, Maintenance, and Administrative Expense Accounts
Dr. 107, Construction Work-in-Progress
Dr. 108.8, Retirement Work-in-Progress
Cr. 131.1, Cash—General
To record the premium cost of the insurance contract.

631 Special Early Retirement Plan

The Special Early Retirement Plan (SERP) being offered through the National Rural Electric Cooperative Association (NRECA) constitutes an amendment to its Retirement and Security (R&S) program. The SERP is often chosen as a vehicle through which the cooperative may reduce the size of its workforce or replace more highly paid employees with lower paid entry level employees. If an employee covered by an NRECA retirement plan chose to retire before his/her normal retirement date, that employee would receive an actuarially reduced benefit. However, when a cooperative elects to offer a SERP, no such reduction is required. The cooperative selects the criteria under which an employee will be eligible to participate such as age, years of service, or a combination of age and benefit service requirements. As with other amendments to the R&S program, NRECA calculates the cost of the plan based upon the criteria selected by the cooperative and allows the cooperative to pay the cost immediately or on an installment basis.

Under this plan, the employee receives full retirement benefits in the form of either an immediate lump-sum settlement or annuity payments. It is not unusual for the cooperative to add an incentive to encourage participation such as medical or life insurance, either in whole or in part, until age 65. The actuarial analysis provided by NRECA includes the cost of the SERP and the estimated reduction and/or increase in costs associated with Statement of Financial Accounting Standards No. 106, Employer's Accounting for Postretirement Benefits Other Than Pensions (Statement No. 106).

Statement of Financial Accounting Standards No. 87, Employer's Accounting for Pensions (Statement No. 87)

In accordance with the provisions of Statement No. 87, the costs associated with an amendment to a multiemployer plan are recognized when they become due and payable. Since NRECA calculates the amount due and payable at the time of the amendment, the entire amount due, whether paid immediately or financed through NRECA or any other institution, must be recognized as an expense at that time. This cost may, however, be deferred in accordance with the provisions of Statement of Financial Accounting Standards No. 71, Accounting for the Effects of Certain Types of Regulation (Statement No. 71).

Accounting Journal Entries

The journal entry required to record the additional pension costs associated with the SERP is as follows:

Dr. Various Operations, Maintenance, and Administrative Expense Accounts
Dr. 107, Construction Work-in-Progress
Dr. 108.8, Retirement Work-in-Progress
Cr. 131.1, Cash—General
or
Cr. 224, Other Long-Term Debt
To record the prior service pension costs incurred as a result of adopting the SERP.

If the borrower elects to defer and amortize the cost in accordance with Statement No. 71, the following entries shall be recorded:

Dr. 182.3, Other Regulatory Assets
Cr. 131.1, Cash—General
or
Cr. 224, Other Long-Term Debt
To record, under the provisions of Statement No. 71, the deferral of the prior service pension costs incurred as a result of adopting the SERP.

Dr. Various Operations, Maintenance, and Administrative Expense Accounts
Dr. 107, Construction Work-in-Progress
Dr. 108.8, Retirement Work-in-Progress
Cr. 182.3, Other Regulatory Assets
To record the amortization of deferred prior service pension costs as they are recovered through rates in accordance with Statement No. 71.

Statement No. 106

In the event that net reductions in postretirement benefits result from this plan amendment, the reductions are recognized as follows:

1. The amount of the reduction shall first reduce any existing unrecognized prior service cost;

2. Any remaining reductions shall next reduce any unrecognized transition obligation; and

3. Any remaining reduction shall be recognized in a manner consistent with the accounting for prior service postretirement benefit costs.

In accordance with Statement No. 106, prior service postretirement benefit costs are recognized in equal amounts in each remaining year of service for active plan participants. Because it is an off-balance sheet item, only a memorandum entry is required to reduce the amount of unrecognized prior service cost.

At adoption, Statement No. 106 permitted the recognition of the transition obligation in one of two ways. The transition obligation was recognized over the longer of the average remaining service period of current plan participants or 20 years, or it may have been recognized immediately. If the delayed recognition option was chosen under Statement No. 106, this, too, was an off-balance sheet item that requires only a memorandum entry to reduce the amount of unrecognized transition obligation. However, if the immediate recognition option was chosen, the cooperative either recorded the expense in that year or, with RUS approval, deferred the expense under the provisions of Statement No. 71. If the expense were recorded, in total, in the year of adoption, no unrecognized transition obligation remains to reduce. If, however, the transition obligation was deferred in accordance with Statement No. 71, the journal entry required to effect the reduction in Statement No. 106 expense is as follows:

Dr. 228.3, Accumulated Provision for Pensions and Benefits
Cr. 182.3, Other Regulatory Assets
To record a reduction in the deferred Statement No. 106 transition obligation resulting from the adoption of the SERP.

NOTE: The dollar value of this entry must not exceed the deferral shown on the balance sheet.

If, after the two previous reductions have been made, any net credit remains, it shall be recognized in a manner consistent with prior service costs; that is, as an off balance sheet item that is amortized over the remaining service lives (to full eligibility) of the active plan participants. The annual amortization reduces amounts normally charged to the various operations, maintenance, and administrative expense accounts and Account 228.3 as postretirement benefit expenses.

633 Cushion of Credit

On December 21, 1987, Section 313, Cushion of Credits Payments Program, was added to the Rural Electrification Act. Cushion of

195

credit regulations are located in The Code of Federal Regulations (CFR) 7 CFR part 1785. A cushion of credit payment is a voluntary unscheduled payment by a borrower in excess of amounts due and payable. A cushion of credit account is automatically established by Rural Development for each borrower who makes a payment after October 1, 1987, in excess of amounts then due on a Rural Development note. Payments received in the month in which an installment is due will be applied to the installment due. However, if the regular installment payment is received at a later date in the month, the first payment received will be applied retro-

actively to the cushion of credit account and the second will be applied to the installment due. By law, cushion of credit accounts earn five per cent interest annually, accrued daily and posted quarterly. Although the interest earned will appear as a reduction in the interest billed on the borrower's Rural Development notes and will be separately shown on Form 694, Statement of Interest and Principal Due, interest billed must be adjusted by adding back the interest earned while principal is reduced by the amount of the interest earned before recording the debt payment. Below is an example of the adjustment required:

	As billed	Adjustment	Adjusted
Payment Billed	$1,000	....	$1,000
Principal	800	−$50	750
Interest	*200	50	250

* Includes reduction of $50 for interest earned on cushion of credit account.

Cushion of credit is intended to enable the borrower to deposit funds and have those funds available to make scheduled payments (or installments) only. A borrower may not have more cushion of credit funds, including accrued interest, than their entire Rural Development debt which includes loans made in Rural Electric and Telephone (RET) and Federal Financing Bank (FFB). If a borrower makes less than or no payment when their billing invoice is due, cushion of credit will automatically add to or make their payment systematically for them.

Cushion of credit is not available to use for prepayment of loan accounts before maturity except for the following situations:

1. The total amount of cushion of credit principal with accrued interest equals the borrower's total debt

2. The borrower intends to prepay all remaining debt using a combination of payment with all cushion of credit funds available.

ACCOUNTING REQUIREMENTS

All payments made to a cushion of credit account should be recorded as follows:
Dr. 224.6, Advance Payments Unapplied—Long-Term Debt—Debit
Cr. 131.1, Cash—General
All interest earned on the balance of funds in the account should be recorded as follows:
Dr. 224.6, Advance Payments Unapplied—Long-Term Debt—Debit
Cr. 419, Interest and Dividend Income

REPORTING REQUIREMENTS

Previously, Rural Development required that the balance in the cushion of credit account be reported, on the Form 7, Financial and Statistical Report, as a reduction of the Rural Development long-term debt balance.

On January 15, 2003, Rural Development issued letter guidance permitting a proportionate share of the cushion of credit balance to be reported as a reduction in Current Maturities Long-Term Debt. Additionally, beginning with calendar year 2006 submissions, Form 7 has been revised to include a separate line for cushion of credit balances within the long-term debt section of Part C.

For purposes of the audited financial statements, presentation of the balance of the cushion of credit account as a long-term investment is an acceptable alternative to Rural Development.

[58 FR 59825, Nov. 10, 1993, as amended at 59 FR 27436, May 27, 1994; 60 FR 55430, 55435, Nov. 1, 1995; 62 FR 42319, 42323, 42330, Aug. 6, 1997; 73 FR 30288, May 27, 2008]

§§ 1767.42–1767.45　[Reserved]

Subpart C—Depreciation Rates and Procedures [Reserved]

§§ 1767.46–1767.65　[Reserved]

Subpart D—Preservation of Records

SOURCE: 73 FR 30290, May 27, 2008, unless otherwise noted.

§ 1767.66　Purpose.

This subpart establishes policies and procedures for the effective preservation and efficient maintenance of financial records of Electric borrowers.

§ 1767.67 General.

(a) Rural Development endorses the guidelines as described by the Federal Energy Regulatory Commission's (FERC) "Regulations to Govern the Preservation of Records of Public Utilities and Licensees." The FERC guidelines can be found in 18 CFR part 125.

(b) The regulations prescribed in this part apply to all books of account, contracts, records, memoranda, documents, papers, and correspondence prepared by or on behalf of the borrower as well as those which come into its possession in connection with the acquisition of property by purchase, consolidation, merger, etc.

(c) The regulations prescribed in this part shall not be construed as excusing compliance with any other lawful requirements for the preservation of records.

§ 1767.68 Designation of a supervisory official.

Each borrower shall designate one or more officials to supervise the preservation of its records.

§ 1767.69 Index of records.

(a) Each borrower shall maintain a master index of records. The master index shall identify the records retained, the related retention period, and the locations where the records are maintained. The master index shall be subject to review by Rural Development and Rural Development shall reserve the right to add records, or lengthen retention periods upon finding that retention periods may be insufficient for its purposes.

(b) At each office where records are kept or stored the borrower shall arrange, file, and index the records currently at that site so that they may be readily identified and made available to representatives of Rural Development.

§ 1767.70 Record storage media.

The media used to capture and store the data will play an important part of each Rural Development borrower. Each borrower has the flexibility to select its own storage media. The following are required:

(a) The storage media shall have a life expectancy at least equal to the applicable retention period provided for in the master index of records, unless there is a quality transfer from one media to another with no loss of data. Each transfer of data from one media to another shall be verified for accuracy and documented.

(b) Each borrower shall implement internal control procedures that assure the reliability of, and ready access to, data stored on machine-readable media. The borrower's internal control procedures shall be documented by a responsible supervisory official.

(c) Records shall be indexed and retained in such a manner that they are easily accessible.

(d) The borrower shall have the hardware and software available to locate, identify, and reproduce the records in readable form without loss of clarity.

(e) At the expiration of the retention period, the borrower may use any appropriate method to destroy records.

(f) When any records are lost or destroyed before the expiration of the retention period set forth in the master index, a certified statement shall be added to the master index listing, as far as may be determined, the records lost or destroyed and describing the circumstances of the premature loss or destruction.

§ 1767.71 Periods of retention.

(a) Records of Rural Development borrowers of a kind not listed in the FERC regulations should be governed by those applicable to the closest similar records. Financial requirement and expenditure statements, which are not specifically covered by FERC regulations, are recommended to be kept for one year after the "as of date" of Rural Development's loan fund and accounting review.

(b) Consumer accounts' records should be kept for those years for which patronage capital has not been allocated.

(c) Records supporting construction financed by Rural Development shall be retained until audited and approved by Rural Development.

(d) Records related to plant in service must be retained until the facilities are permanently removed from utility service, all removal and restoration activities are completed, and all costs are retired from the accounting records unless accounting adjustments resulting from reclassification and original costs studies have been approved by Rural Development or other regulatory body having jurisdiction.

(e) Life and mortality study data for depreciation purposes must be retained for 25 years or for 10 years after plant is retired, whichever is longer.

§§ 1767.72–1767.85 [Reserved]

PART 1770—ACCOUNTING RE-QUIREMENTS FOR RUS TELE-COMMUNICATIONS BOR-ROWERS

AUTHORITY: 7 U.S.C. 901 et seq.; 7 U.S.C. 1921 et seq.; Pub. L. 103–354, 108 Stat. 3178 (7 U.S.C. 6941 et seq.).

SOURCE: 55 FR 3388, Feb. 1, 1990, unless otherwise noted.

Subpart A—Preservation of Records

SOURCE: 70 FR 25755, May 16, 2005, unless otherwise noted.

§ 1770.1 General.

(a) This subpart establishes RUS polices and procedures for the preservation of records of telecommunications borrowers.

(b) The regulations prescribed in this part apply to all books of account, contracts, records, memoranda, documents, papers, and correspondence prepared by or on behalf of the borrower as well as those which come into its possession in connection with the acquisition of property by purchase, consolidation, merger, etc.

(c) The regulations prescribed in this part shall not be construed as excusing compliance with any other lawful requirements for the preservation of records.

§ 1770.2 Designation of a supervisory official.

Each borrower shall designate one or more officials to supervise the preservation of its records.

§ 1770.3 Index of records.

(a) Each borrower shall maintain a master index of records. The master index shall identify the records retained, the related retention period, and the locations where the records are maintained. The master index shall be subject to review by RUS and RUS shall reserve the right to add records, or lengthen retention periods upon finding that retention periods may be insufficient for its purposes.

(b) At each office where records are kept or stored the borrower shall arrange, file, and index the records currently at that site so that they may be readily identified and made available to representatives of RUS.

§ 1770.4 Record storage media.

Each RUS borrower has the flexibility to select its own storage media subject to the following conditions:

(a) The storage media must have a life expectancy at least equal to the applicable retention period provided for in the master index of records, unless there is quality transfer from one media to another with no loss of data. Each transfer of data from one media to another must be verified for accuracy and documented.

(b) Each borrower is required to implement internal control procedures that assure the reliability of, and ready access to, data stored on machine-readable media. Internal control procedures must be documented by a responsible supervisory official.

(c) The records shall be indexed and retained in such a manner that they are easily accessible.

(d) The borrower shall have the hardware and software available to locate, identify, and reproduce the records in readable form without loss of clarity.

(e) At the expiration of the retention period, the borrower may use any appropriate method to destroy records.

(f) When any records are lost or destroyed before the expiration of the retention period set forth in the master index, a certified statement shall be added to the master index listing, as far as may be determined, the records lost or destroyed and describing the circumstances of the premature loss or destruction.

§ 1770.5   Periods of retention.

(a) Except as provided for in paragraphs (b), (c), and (d) of this section, record retention shall be consistent with Prudent Utility Practice. Prudent Utility Practice shall mean any of the practices, methods, and acts which, in the exercise of reasonable judgment, in light of the facts, including but not limited to, the practices, methods, and acts engaged in or approved by a significant portion of the telecommunications industry prior thereto, known at the time the decision was made, would have been expected to accomplish the desired result consistent with cost effectiveness, reliability, safety, and expeditiousness. It is recognized that Prudent Utility Practice is not intended to be limited to optimum practice, method, or act to the exclusion of all others, but rather is a spectrum of possible practices, methods, or acts which could have been expected to accomplish the desired result at the lowest reasonable cost consistent with

cost effectiveness, reliability, safety, and expedition.

(b) Records supporting construction financed by RUS shall be retained until audited and approved by RUS.

(c) Records related to plant in service must be retained until the facilities are permanently removed from utility service, all removal and restoration activities are completed, and all costs are retired from the accounting records unless accounting adjustments resulting from reclassification and original costs studies have been approved by RUS or other regulatory body having jurisdiction.

(d) Life and mortality study data for depreciation purposes must be retained for 25 years or for 10 years after plant is retired whichever is longer.

§§ 1770.6–1770.9   [Reserved]

Subpart B—Uniform System of Accounts

§ 1770.10   General.

This subpart implements provisions of the standard RUS loan documents with respect to the accounting system accounts to be maintained by telecommunications borrowers of the Rural Utilities Service.

§ 1770.11   Accounting system requirements.

(a) Each RUS borrower subject to the jurisdiction of the Federal Communications Commission (FCC) or a State regulatory body shall maintain its accounts and records in accordance with the rules and regulations prescribed by that regulatory body.

(b) Each RUS borrower not subject to regulatory control as specified in §1770.11(a) shall maintain its accounts and records in accordance with the FCC Uniform System of Accounts as set forth in part 32 of the Commission's Rules and Regulations.

(1) RUS borrowers maintaining the accounts prescribed in 47 CFR part 32 for Class A companies as of June 15, 2005, shall continue to do so. RUS suspends implementation of the reduced number of Class A and B accounts, until the Federal-State Joint Conference has reviewed them.

(2) New borrowers under the RUS telecommunications program shall maintain the accounts prescribed in 47 CFR part 32 for Class A companies.

(3) RUS borrowers maintaining the accounts prescribed for Class B companies may adopt the Class A accounts if they desire more detailed and sophisticated accounting records.

[55 FR 3388, Feb. 1, 1990, as amended at 70 FR 25756, May 16, 2005]

§ 1770.12 Supplementary accounts.

(a) All borrowers shall maintain the supplementary accounts set forth in § 1770.15. These accounts conform in number and title with accounts prescribed in the FCC Uniform System of Accounts. In those instances in which a State regulatory body having jurisdiction over an RUS borrower has prescribed a system of accounts differing from that of the FCC, the account titles prescribed by RUS in § 1770.15 shall remain unchanged; however, the supplementary account numbers shall be changed to conform with the State's accounting system.

(b) In addition to the accounts set forth in § 1770.15, cooperative or other nonprofit borrowers shall maintain the supplementary accounts set forth in § 1770.16.

(c) Borrowers are permitted to deviate from the specific subaccount numbers detailed in §§ 1770.15 and 1770.16 provided that the primary account numbers and account descriptions conform with those prescribed.

(Approved by the Office of Management and Budget under control number 0572–0003)

§ 1770.13 Accounting requirements.

(a) Each borrower shall maintain its books of accounts on the accrual basis of accounting. All transactions shall be recorded in the period in which they occur and reconciled monthly. The books of accounts shall be closed at the end of each fiscal year and financial statements shall be prepared for the period and audited in accordance with the provisions of 7 CFR part 1773, RUS Policy on Audits of Electric and Telephone Borrowers.

(b) All books of accounts, records, and memoranda shall be maintained in such a manner as to fully support the journal entries to which they relate. The books and records referred to herein shall include records of a nontechnical nature such as minute books, stock and membership records, reports, correspondence, and memoranda.

(c) Interpretations of Federal or State requirements shall be referred to the applicable commission exercising jurisdiction over the borrower.

(d) Interpretations of RUS accounting requirements shall be referred to the Assistant Administrator, Program Accounting and Regulatory Analysis, Rural Utilities Service.

[55 FR 3388, Feb. 1, 1990, as amended at 70 FR 25756, May 16, 2005]

§ 1770.14 Continuing property records.

Each borrower shall maintain continuing property records which detail the date of placement, location, description of property, and the original cost of the property record units. The continuing property record and other underlying records of construction costs shall be maintained so that upon retirement of one or more retirement units or of minor items without replacement when not included in the costs of retirement units, the actual cost of the plant retired can be determined.

§ 1770.15 Supplementary accounts required of all borrowers.

Accounts prescribed in the Stockholders' Equity and Patronage Capital section shall be maintained by stock companies and cooperatives as appropriate.

Class of company		Account title
Account No.		
A	B	
		Current Assets
1130.1	1120.11	Cash—General Fund.

Class of company		Account title
Account No.		
A	B	
1130.2	1120.12	Cash—Construction Fund Trustee.
1130.3	1120.13	Cash—Transfer of Funds.
	1120.21	Special Cash Deposits.
1150.1	1120.31	*Petty Cash Fund*
		This account shall include funds in the custody of employees or agents for making minor disbursements. The fund shall be operated on an imprest basis. Expenditures shall be supported by receipts, and reimbursements to the fund shall be for the exact amount of such expenditures and shall be charged to the various accounts to which the expenditures are allocable. At all times, the total of the cash on hand and the unreimbursed expenditures shall equal the amount of the fund.
1150.2	1120.32	Change Fund.
		Supplies
1220.1	1220.1	Materials and Supplies.
1220.2	1220.2	Property Held for Sale or Lease.
1220.3	1220.3	Exempt Materials—Clearing.
		Prepayments
	1280.1	Prepaid Rents.
	1280.2	Prepaid Taxes.
	1280.3	Prepaid Insurance.
	1280.4	Prepaid Directory Expenses.
	1280.5	Other Prepayments.
		Investments
1402.1	1402.1	Investments in Nonaffiliated Companies—Class B RTB Stock.
1402.11	1402.11	Investments in Nonaffiliated Companies—Class B RTB Stock—Cr.
1402.2	1402.2	Investments in Nonaffiliated Companies—Class C RTB Stock.
1402.3	1402.3	Other Investments in Nonaffiliated Companies.
		Property, Plant, and Equipment
2001.1	2001.1	Telecommunications Plant in Service—Classified.
2001.2	2001.2	Telecommunications Plant in Service—Unclassified.
2003.1	2003.1	Telecommunications Plant Under Construction—Contract
2003.2	2003.2	Telecommunications Plant Under Construction—Force Account
2003.3	2003.3	Telecommunications Plant Under Construction—Work Orders
		Telecommunications Plant in Service
	2210.11	Central Office Switching—Analog.
	2210.21	Central Office Switching—Digital.
	2210.31	Central Office Switching—Electro-Mechanical—Step-by-Step.
	2210.32	Central Office Switching—Electro-Mechanical—Crossbar.
	2210.33	Central Office Switching—Electro-Mechanical—Other.
2212.1	2212.1	Digital Electronic Switching—Circuit.
2212.2	2212.2	Digital Electronic Switching—Packet.
	2230.11	Central Office Transmission—Radio Systems—Satellite and Earth Station Facilities.
	2230.12	Central Office Transmission—Radio Systems—Other.
	2230.21	Central Office Transmission—Circuit Equipment.
		Depreciation and Amortization
3100x	3100x	Retirement Work in Progress.
		Current Liabilities
2232.1	2232.1	Circuit Equipment—Electronic.
2232.2	2232.2	Circuit Equipment—Optical.
4010.11	4010.11	Accounts Payable to Affiliated Companies.
4010.21	4010.21	Accounts Payable to Nonaffiliated Companies.
4010.22	4010.22	Accounts Payable—Employees' Income Tax Withheld.
4010.23	4010.23	Accounts Payable—FICA Taxes Withheld.
4010.24	4010.24	Accounts Payable—Federal Excise Taxes.
4010.25	4010.25	Accounts Payable—Payroll.
4070.1	4070.1	Income Taxes Accrued—Federal.
4070.2	4070.2	Income Taxes Accrued—State and Local
4080.1	4080.1	Other Taxes Accrued—Property.
4080.2	4080.2	Other Taxes Accrued—Employer's Portion—FICA.
4080.3	4080.3	Other Taxes Accrued—Federal Unemployment.
4080.4	4080.4	Other Taxes Accrued—State Unemployment.
4080.5	4080.5	Other Taxes Accrued—Miscellaneous.
4120.1	4120.1	Unmatured Interest Accrued—RUS Notes.
4120.2	4120.2	Unmatured Interest Accrued—Telephone Bank Notes.
4120.3	4120.3	Unmatured Interest Accrued—Federal Financing Bank Notes.
4120.4	4120.4	Unmatured Interest Accrued—Bank for Cooperatives Notes.
4120.5	4120.5	Unmatured Interest Accrued—Rural Telephone Finance Cooperative Notes.
4120.6	4120.6	Other Accrued Liabilities.

Class of company		
Account No.		Account title
A	B	

		Long-Term Debt
4210.11	4210.11	Funded Debt—Other.
4210.12	4210.12	RUS Notes.
4210.13	4210.13	Telephone Bank Notes.
4210.14	4210.14	Federal Financing Bank Notes.
4210.15	4210.15	Bank for Cooperatives Notes.
4210.16	4210.16	Rural Telephone Finance Cooperative Notes.
4210.17	4210.17	RUS Notes—Deferred Interest.
4210.18	4210.18	RUS Notes—Advance Payments, Dr.
4210.19	4210.19	Funded Debt—Other—Unadvanced, Dr.
4210.20	4210.20	RUS Notes—Unadvanced, Dr.
4210.21	4210.21	Telephone Bank Notes—Unadvanced, Dr.
4210.22	4210.22	Federal Financing Bank Notes—Unadvanced, Dr.
4210.23	4210.23	Bank for Cooperatives Notes—Unadvanced, Dr.
4210.24	4210.24	Rural Telephone Finance Cooperative Notes—Unadvanced, Dr.
		Stockholders' Equity and Patronage Capital
4540.11	4540.11	Capital Stock Subscribed.
4540.12	4540.12	Memberships Subscribed but Unissued.
4540.13	4540.13	Members' Equity Certificates Subscribed but Unissued.
4540.21	4540.21	Memberships Issued.
4540.22	4540.22	Members' Equity Certificates Issued.
4540.23	4540.23	Members' Equity—Other.
4540.31	4540.31	Installments Paid on Capital Stock.
4540.32	4540.32	Installments Paid on Memberships Subscribed.
4540.33	4540.33	Installments Paid on Equity Certificates Subscribed.
4540.41	4540.41	Other Capital—Miscellaneous.
4550.1	4550.1	Operating Margins.
4550.2	4550.2	Nonoperating Margins.
4550.3	4550.3	Other Margins.
4550.4	4550.4	Patronage Capital Assignable.
4550.5	4550.5	Patrons' Capital Credits Assigned.
4550.6	4550.6	Gain on the Retirement of Capital Credits.
		Plant Specific Operations Expense
	6210.11	Analog Electronic Expense.
	6210.21	Digital Electronic Expense.
	6210.31	Electro-Mechanical Expense.
6212.1	6212.1	Digital Electronic Switching Expense—Circuit.
6212.2	6212.2	Digital Electronic Switching Expense—Packet.
	6230.11	Radio Systems Expense.
	6230.21	Circuit Equipment Expense.
6232.1	6232.1	Circuit Equipment Expense—Electronic.
6232.2	6232.2	Circuit Equipment Expense—Optical.
		Plant Nonspecific Operations Expense
	6560.1	Depreciation Expense.
	6560.2	Amortization Expense.
6620.1	6620.1	Services—Wholesale.
6620.2	6620.2	Services—Retail.
		Operating Taxes.
	7200.1	Operating Investment Tax Credits—Net.
	7200.2	Operating Federal Income Taxes.
	7200.3	Operating State and Local Income Taxes.
7240.1	7200.41	Operating Taxes—Property.
7240.2	7200.42	Operating Taxes—Miscellaneous.
	7200.5	Provision for Deferred Operating Income Taxes—Net.
		Nonoperating Income and Expense
	7300.1	Dividend Income.
	7300.2	Interest Income.
	7300.3	Income From Sinking and Other Funds.
	7300.4	Allowance for Funds Used During Construction.
	7300.5	Gains or Losses from the Disposition of Certain Property.
	7300.6	Other Nonoperating Income and Expense.
		Nonoperating Taxes
	7400.1	Nonoperating Investment Tax Credits—Net.
	7400.2	Nonoperating Federal Income Taxes.
	7400.3	Nonoperating State and Local Income Taxes.
	7400.4	Nonoperating Other Taxes.
	7400.5	Provision for Deferred Nonoperating Income Taxes—Net.
		Extraordinary Items

Class of company		Account title
Account No.		
A	B	
	7600.1	Extraordinary Income Credits.
	7600.2	Extraordinary Income Charges.
	7600.3	Current Income Tax Effect of Extraordinary Items—Net.
	7600.4	Provision for Deferred Income Tax Effect of Extraordinary Items—Net.
1130.1	1120.11	*Cash—General Fund*
		This account shall include all unrestricted funds derived from revenues and other sources which are on deposit in banks or other financial institutions and available on demand. It shall also include funds in transit to the depository for which customers and agents have received credit on their accounts. Separate subaccounts should be maintained for each bank account in which general fund cash is deposited.
1130.2	1120.12	*Cash—Construction Fund Trustee*
		This account shall include all loan funds received from RUS, the Rural Telephone Bank, the Federal Financing Bank, the Bank for Cooperatives, the Rural Telephone Finance Cooperative, and all non-loan funds supplied by the borrower under the terms of the loan contract or otherwise required by RUS. The offsetting credit for funds received from RUS shall be to Account 4210.20, RUS Notes—Unadvanced, Dr.; funds received from the Rural Telephone Bank, to Account 4210.21, Telephone Bank Notes—Unadvanced, Dr.; funds received from the Federal Financing Bank, to Account 4210.22, Federal Financing Bank Notes—Unadvanced, Dr.; funds received from the Bank for Cooperatives, to Account 4210.23, Bank for Cooperatives Notes—Unadvanced, Dr.; and funds received from the Rural Telephone Finance Cooperative, to Account 4210.24, Rural Telephone Finance Cooperative Notes—Unadvanced, Dr.
1130.3	1120.13	*Cash—Transfer of Funds*
		This account shall include all transfers of funds from one bank account to another. This account shall be charged with the amount of a check drawn for the transfer, and credited when the amount transferred is entered into the Cash Receipts Book.
	1120.21	*Special Cash Deposits*
		This account shall include all cash on special deposit, other than in sinking and other special funds provided for elsewhere, to pay dividends, interest, and other debts, when such payments are due one year or less from the date of deposit; the amount of cash deposited to insure the performance of contracts to be performed within one year from the date of the deposit; and other cash deposits of a special nature not provided for elsewhere. This account shall include the amount of cash deposited with trustees to be held until mortgaged property sold, destroyed, or otherwise disposed of is replaced, and also cash realized from the sale of the company's securities and deposited with trustees to be held until invested in physical property of the company or for disbursement when the purposes for which the securities were sold are accomplished.
1150.1	1120.31	*Petty Cash Fund*
		This account shall include funds in the custody of employees or agents for making minor disbursements. The fund shall be operated on an inprest basis. Expenditures shall be supported by receipts, and reimbursements to the fund shall be for the exact amount of such expenditures and shall be charged to the various accounts to which the expenditures are allocable. At all times, the total of the cash on hand and the unreimbursed expenditures shall equal the amount of the fund.
1150.2	1120.32	*Change Fund*
		This account shall include funds in the custody of employees or agents for making change. Records shall be kept of the amount held by each person. Disbursements shall not be made from the fund.
1220.1	1220.1	*Materials and Supplies**
		This account shall include the cost of materials and supplies held in stock including plant supplies, motor vehicles supplies, tools, fuel, other supplies and material and articles of the company in process of manufacture for supply stock.
		Transportation charges and sales and use taxes, as far as practicable, shall be included as a part of the cost of the particular material to which they relate. Transportation and sales and use taxes which are not included as part of the cost of particular material shall be equitably apportioned among the accounts to which material is charged.
		As far as practicable, cash and other discounts on material shall be deducted in determining cost of the particular material to which they relate or credited to the account to which the material is charged. When such deduction is not practicable, discounts shall be equitably apportioned among the accounts to which material is charged.
		Material recovered in connection with construction, maintenance or retirement of property shall be charged to this account as follows:
		—Reusable items that, when installed or in service, were retirement units shall be included in this account at the original cost.
		—Reusable minor items that, when installed or in service, were not retirement units shall be included in this account at current prices new.
		—The cost of repairing reusable material shall be charged to the appropriate Plant Specific Operations Expense accounts.

Class of company		Account title
Account No.		
A	B	

		—Scrap and nonusable material included in this account shall be carried at the estimated amount which will be received therefor. The difference between the amounts realized for scrap and nonusable material sold, and the amounts at which it is carried in this account shall be adjusted in the accounts credited when the material was taken up in this account.
		Interest paid on material bills, the payments of which are delayed, shall be charged to Account 7540, Other Interest Deductions.
		Inventories of materials and supplies shall be taken during each calendar year and the adjustments to this account shall be charged or credited to Account 6512, Provisioning Expense.
1220.2	1220.2	*Property Held for Sale or Lease**
		This account shall include the cost of all items purchased for resale or lease. The cost shall include applicable transportation charges, sales and use taxes, and cash and other purchase discounts. Inventory shortages and overages shall be charged and credited, respectively to Account 7991, Other Nonregulated Revenues.
		*These accounts shall not include items which are related to a nonregulated activity unless that activity involves joint or common use of assets and resources in the provision of regulated and nonregulated products and services.
1220.3	1220.3	*Exempt Materials—Clearing*
		This account shall include the cost of materials and supplies designated as exempt material on the carrier's "Exempt Material List". Charges to this account shall be cleared monthly to the primary plant and maintenance accounts in accordance with percentages developed by the individual carriers.
		When there is a substantial amount of exempt material on hand at the end of the year, substantial enough to distort net income or margins, a physical inventory may be taken. The cost of the inventory on hand shall be debited to this account and credited to the appropriate primary plant and maintenance accounts on a pro-rata basis related to the original charges to these accounts. This entry shall be reversed at the first of the year.
	1280.1	*Prepaid Rents*
		This account shall include the amount of rents paid in advance of the period in which it is chargeable to income, except amounts chargeable to telecommunications plant under construction and minor amounts which may be charged directly to the final accounts. As the term expires for which the rents are paid, this account shall be credited monthly and the appropriate account charged.
	1280.2	*Prepaid Taxes*
		This account shall include the balance of all taxes paid in advance of the period in which they are chargeable to income, except amounts chargeable to telecommunications plant under construction and minor amounts which may be charged directly to the final accounts. As the term expires for which the taxes are paid, this account shall be credited monthly and the appropriate account charged.
	1280.3	*Prepaid Insurance*
		This account shall include the amount of insurance premiums paid in advance of the period in which they are chargeable to income, except premiums chargeable to telecommunications plant under construction and minor amounts which may be charged directly to the final accounts. As the term expires for which the premiums are paid, this account shall be credited monthly and the appropriate account charged.
	1280.4	*Prepaid Directory Expenses*
		This account shall include the cost of preparing, printing, binding, and delivering directories and the cost of soliciting advertisements for directories, except minor amounts which may be charged directly to Account 6620, Services. Amounts in this account, shall be cleared to Account 6620 by monthly charges representing that portion of the expenses applicable to each month.
	1280.5	*Other Prepayments*
		This amount shall include prepayments, other than those includable in Accounts 1280.1 through 1280.4 except minor amounts which may be charged directly to the final accounts. As the term expires for which the payments apply, this account shall be credited monthly and the appropriate account charged.
1402.1	1402.1	*Investments in Nonaffiliated Companies—Class B RTB Stock*
		This account shall include the par value of the required purchase of Class B Rural Telephone Bank stock and the par value of the Class B Rural Telephone Bank stock received as a patronage refund. This account shall be debited at the time the refund is received and Account 1402.11, Investments in Nonaffiliated Companies—Class B RTB Stock—Cr., credited.
		This account shall be credited and Account 1402.11 debited when the patronage refund is redeemed.
1402.11	1402.11	*Investments in Nonaffiliated Companies—Class B RTB Stock—Cr.*
		This account shall include the par value of Class B Rural Telephone Bank stock received as a patronage refund. This account shall be credited at the time the refund is received and Account 1402.1, Investments in Nonaffiliated Companies—Class B RTB Stock, debited.
		This account shall be debited and Account 1402.1 credited when the patronage refund is redeemed.
1402.2	1402.2	*Investments in Nonaffiliated Companies—Class C RTB Stock*

Class of company		Account title
Account No.		
A	B	

		This account shall include the par value of the company's investment in Class C Rural Telephone Bank stock. Cash dividends on Class C stock shall be recorded in Account 7310/7300.1, Dividend Income, when declared.
1402.3	1402.3	*Other Investments in Nonaffiliated Companies*
		This account shall include the acquisition cost of the company's investment in securities issued by non-affiliated companies, other than securities held in special funds which shall be charged to Account 1408, Sinking Funds, and also its investment advances to such parties and special deposits of cash for more than one year from the date of deposit. Declines in value of investments shall be charged to Account 4540.41, Other Capital, if temporary and as a current period loss if permanent. Detailed records shall be maintained to reflect unrealized losses for each investment.
2001.1	2001.1	*Telecommunications Plant in Service—Classified*
		This account shall include the original cost of the property capitalized in Accounts 2110 through 2690.
2001.2	2001.2	*Telecommunications Plant in Service—Unclassified*
		This account shall include the original cost of telecommunications property which has been completed and placed in service but which has not been classified pending completion of final inventories of construction, final cost summaries, etc. The balance in this account is subject to depreciation charges.
2003.1	2003.1	*Telecommunications Plant Under Construction—Short Term—Contract*
		This account shall include all costs incurred in the construction of telecommunications plant performed under contract and the cost of software development projects that are not yet ready for their intended use. Included among these costs are contractor payments and charges for engineering, supervision, taxes, insurance, transportation, and other costs incurred in contract construction. This account shall be maintained such that the various items of cost are readily identifiable.
2003.2	2003.2	*Telecommunications Plant Under Construction—Short Term—Force Account*
		This account shall include all costs incurred in the construction of telecommunications plant performed by the borrowers' own employees and the cost of software development projects performed by the borrowers' own employees that are not yet ready for their intended use. Included among these costs are charges for material, labor, engineering, supervision, taxes, insurance, transportation, supply expense, and other costs incurred in the construction. This account shall be maintained such that the various items of cost are readily identified. Specific subaccounts should be maintained to distinguish individual projects.
2003.3	2003.3	*Telecommunications Plant Under Construction—Short Term—Work Orders*
		This account shall include all costs incurred in the construction of telecommunication plant performed under a work order system or line extension contract. This type of construction generally includes service installations, subscriber extensions, and minor plant improvements after the completion of the initial system. Included among these costs are charges for labor, material and supplies, transportation, payroll taxes, insurance, supervision, and other costs incurred in the construction. Subsidiary records shall be maintained to reflect the cost of the individual jobs. These records shall be reconciled periodically with the general ledger control account. Specific subaccounts should be maintained to accumulate costs incurred under line extension contracts.
	2210.11	*Central Office Switching—Analog**
		This account shall include the original cost of stored program control analog circuit-switching and associated equipment. This account shall also include the original cost of remote analog electronic circuit switches.
	2210.21	*Central Office Switching—Digital**
		This account shall include the original cost of stored program control digital switches and their associated equipment. Included in this account is the original cost of digital switches which utilize either dedicated or non-dedicated circuits. This account shall also include the original cost of remote digital electronic switches.
	2210.31	*Central Office Switching—Electro-Mechanical—Step-by-Step**
		This account shall include the original cost of step-by-step and associated circuit-switching equipment.
	2210.32	*Central Office Switching—Electro-Mechanical—Crossbar**
		This account shall include the original cost of crossbar and associated circuit switching equipment. Also included in this account is the original cost of electronic translator system equipment used in switching.
	2210.33	*Central Office Switching—Electro-Mechanical—Other**
		This account shall include the original cost of all other types of non-electronic circuit-switching equipment such as panel systems and their associated circuit-switching equipment. *Switching plant excludes switchboards which perform operator assistance functions and equipment which is an integral part thereof. It does not exclude equipment used solely for the recording of calling telephone numbers in connection with customer dialed charged traffic, dial tandem switches, and special switchboards used in conjunction with private line service; such equipment shall be classified to the particular switch that it serves.

Class of company		Account title
Account No.		
A	B	
	2230.11	*Central Office Transmission—Radio Systems—Satellite and Earth Station Facilities*
		This account shall include the original cost of an ownership interest in satellites (including land-side spares), other spare parts, materials, and supplies. It shall include launch insurance and other satellite launch costs. This account shall also include the original cost of earth stations and spare parts, materials, and supplies therefor.
	2230.12	*Central Office Transmission—Radio Systems—Other*
		This account shall include the original cost of radio equipment used to provide radio communication channels. Radio equipment is that equipment which is used for the generation, amplification, propagation, reception, modulation, and demodulation of radio waves in free space over which communications channels can be provided. This account shall also include the associated carrier and auxiliary equipment and patch bay equipment which is an integral part of the radio equipment. Such equipment may be located in central office buildings, terminal rooms, or repeater stations or may be mounted on towers, masts, or other supports.
	2230.21	*Central Office Transmission—Circuit Equipment*
		This account shall include the original cost of equipment which is used to reduce the number of physical pairs otherwise required to serve a given number of subscribers by utilizing carrier systems, concentration stages or combinations of both. It shall include equipment that provides for simultaneous use of a number of interoffice channels on a single transmission path. This account shall also include the original cost of equipment which is used for the amplification, modulation, regeneration, circuit patching, balancing or control of signals transmitted over interoffice communications transmission channels. This account shall include the original cost of equipment which utilizes the message path to carry signaling information or which utilizes separate channels between switching offices to transmit signaling information independent of the subscribers' communication paths or transmission channels. This account shall also include the original cost of associated material used in the construction of such plant. Circuit equipment may be located in central offices, in manholes, on poles, in cabinets or huts or at other locations.
		This account excludes carrier and auxiliary equipment and patch bay which are recorded in Account 2230.12, Central Office Transmission—Radio Systems—Other
3100x	3100x	*Retirement Work in Progress*
		This account shall be charged with the original cost of property retired from the telecommunications plant accounts. It shall also be charged with all of the costs incurred in removing the retired plant from service. This account shall be credited with the salvage value of materials recovered in the retirement of the telecommunications plant. At such time as the retirement work order is complete, the net income/loss resulting therefrom shall be transferred from this account to the appropriate primary plant depreciation reserve account.
4010.11	4010.11	*Accounts Payable to Affiliated Companies*
		This account shall include all amounts currently due to affiliated companies for recurring trade obligations, and not provided for in other accounts, such as those for traffic settlements, material and supplies, repairs to telecommunications plant, matured rents, and interest payable under monthly settlements on short-term loans, advances, and open accounts.
4010.21	4010.21	*Accounts Payable to Nonaffiliated Companies*
		This account shall include all amounts currently due to nonaffiliated companies for recurring trade obligations, and not provided for in other accounts, such as those for traffic settlements, materials and supplies, repairs to telecommunications plant, matured rents, and interest payable under monthly settlements on short-term loans, advances, and open accounts.
4010.22	4010.22	*Accounts Payable—Employees' Income Tax Withheld*
		This account shall include income taxes payable that have been withheld from employees' salaries.
4010.23	4010.23	*Accounts Payable—FICA Taxes Withheld*
		This account shall include FICA taxes payable that have been withheld from employees' salaries.
4010.24	4010.24	*Accounts Payable—Federal Excise Taxes*
		This account shall include Federal excise taxes payable.
4010.25	4010.25	*Accounts Payable—Payroll*
		This account shall include amounts payable to the company's employees in the form of salaries or wages.
4070.1	4070.1	*Income Taxes Accrued—Federal*
		For Class A companies, this account shall be credited and Accounts 7220, 7420, and 7630, as appropriate, shall be debited for the amount of Federal income taxes accrued during the current operating period.
		For Class B companies, this account shall be credited and Accounts 7220.2, 7400.2, and 7600.3, as appropriate, shall be debited for the amount of Federal income taxes accrued during the current operating period.
4070.2	4070.2	*Income Taxes Accrued—State and Local*

Class of company		Account title
Account No.		
A	B	
		For Class A companies, this account shall be credited and Accounts, 7230, 7430, and 7630, as appropriate, shall be debited for the amount of state and local income taxes accrued during the current operating period. For Class B companies, this account shall be credited and Accounts, 7200.3, 7400.3, and 7600.3, as appropriate, shall be debited for the amount of state and local income taxes accrued during the current operating period.
4080.1	4080.1	*Other Taxes Accrued—Property*
		This account shall be credited and Account 7240.1/7200.41, Operating Taxes—Property, shall be debited for the amount of property taxes accrued during the current operating period.
4080.2	4080.2	*Other Taxes Accrued—Employer's Portion—FICA*
		This account shall be credited and the appropriate construction, depreciation, or expense account shall be debited for the employer's portion of FICA taxes accrued during the current operating period.
4080.3	4080.3	*Other Taxes Accrued—Federal Unemployment*
		This account shall be credited and the appropriate construction, removal, or expense account shall be debited for the amount of Federal unemployment taxes accrued during the current operating period.
4080.4	4080.4	*Other Taxes Accrued—State Unemployment*
		This account shall be credited and the appropriate construction, removal, or expense account shall be debited for the amount of state unemployment taxes accrued during the current operating period.
4080.5	4080.5	*Other Taxes Accrued—Miscellaneous*
		This account shall be credited and Account 7240.2/7200.42, Operating Taxes—Miscellaneous, shall be debited for the amount of all other taxes accrued during the current operating period and not provided for elsewhere such as a gross receipts tax, franchise taxes, and capital stock taxes.
4120.1	4120.1	*Unmatured Interest Accrued—RUS Notes*
		This account shall include the interest accrued as of the balance sheet date but not payable until after that date on RUS mortgage notes. Interest expense incurred during the period of construction of telecommunications plant shall be charged to Account 2004, Telecommunications Plant Under Construction—Long Term, and credited to Account 7340/7300.4, Allowance for Funds Used During Construction.
4120.2	4120.2	*Unmatured Interest Accrued—Telephone Bank Notes*
		This account shall include the interest accrued as of the balance sheet date but not payable until after that date on Rural Telephone Bank mortgage notes. Interest expense incurred during the period of construction of telecommunications plant shall be charged to Account 2004, Telecommunications Plant Under Construction—Long Term, and credited to Account 7340/7300.4, Allowance for Funds Used During Construction.
4120.3	4120.3	*Unmatured Interest Accrued—Federal Financing Bank Notes*
		This account shall include the interest accrued as of the balance sheet date but not payable until after that date on Federal Financing Bank mortgage notes. Interest expense incurred during the period of construction of telecommunications plant shall be charged to Account 2004, Telecommunications Plant Under Construction—Long Term, and credited to Account 7340/7300.4, Allowance for Funds Used During Construction.
4120.4	4120.4	*Unmatured Interest Accrued—Bank for Cooperatives Notes*
		This account shall include the interest accrued as of the balance sheet date but not payable until after that date on Bank for Cooperatives mortgage notes. Interest expense incurred during the period of construction of telecommunications plant shall be charged to Account 2004, Telecommunications Plant Under Construction—Long Term, and credited to Account 7340/7300.4, Allowance for Funds Used During Construction.
4120.5	4120.5	*Unmatured Interest Accrued—Rural Telephone Finance Cooperative Notes*
		This account shall include the interest accrued as of the balance sheet date but not payable until after that date on Rural Telephone Finance Cooperative mortgage notes. Interest expense incurred during the period of construction of telecommunications plant shall be charged to Account 2004, Telecommunications Plant Under Construction—Long Term, and credited to Account 7340/7300.4, Allowance for Funds Used During Construction.
4120.6	4120.6	*Other Accrued Liabilities*
		This account shall include the amount of wages, compensated absences, interest on indebtedness of the company, dividends on capital stock, and rents accrued as of the balance sheet date but not payable until after the date. This account shall not include interest accrued on RUS, Rural Telephone Bank, Bank for Cooperatives, Federal Financing Bank, or Rural Telephone Finance Cooperative debt.
4210.11	4210.11	*Funded Debt—Other*
		This account shall include the total face amount of unmatured debt, maturing more than one year from the date of issue, issued by the company and not retired, and the total face amount of similar unmatured debt of other companies, the payment of which has been assumed by the company, including funded debt the maturity of which has been extended by specific agreement.

Class of company		Account title
Account No.		
A	B	

4210.12	4210.12	This account shall not include unmatured RUS, Rural Telephone Bank, Federal Financing Bank, Bank for Cooperatives, or Rural Telephone Finance Cooperative debt.
		RUS Notes
		This account shall include the total face amount of unmatured RUS mortgage notes. Account 4210.20, RUS Notes—Unadvanced, Dr., shall be charged and this account credited upon execution of the notes.
		If principal installments are not paid at the maturity date, the amount due shall be transferred to Account 4050, Current Maturities—Long-Term Debt.
4210.13	4210.13	*Telephone Bank Notes*
		This account shall include the total face amount of unmatured Rural Telephone Bank mortgage notes. Account 4210.21, Telephone Bank Notes—Unadvanced, Dr., shall be changed and this account credited upon execution of the notes.
		If principal installments are not paid at the maturity date, the amount due shall be transferred to Account 4050, Current Maturities—Long-Term Debt.
4210.14	4210.14	*Federal Financing Bank Notes*
		This account shall include the total face amount of unmatured Federal Financing Bank mortgage notes. Account 4210.22, Federal Financing Bank Notes—Unadvanced, Dr., shall be charged and this account credited upon execution of the notes.
		If principal installments are not paid at the maturity date, the amount due shall be transferred to Account 4050, Current Maturities—Long-Term Debt.
4210.15	4210.15	*Bank for Cooperatives Notes*
		This account shall include the total face amount of unmatured Bank for Cooperatives mortgage notes. Account 4210.23, Bank for Cooperatives Notes—Unadvanced, Dr., shall be charged and this account credited upon execution of the notes.
		If principal installments are not paid at the maturity date, the amount due shall be transferred to Account 4050, Current Maturities—Long-Term Debt.
4210.16	4210.16	*Rural Telephone Finance Cooperative Notes*
		This account shall include the total face amount of unmatured Rural Telephone Finance Cooperative mortgage notes. Account 4210.24, Rural Telephone Finance Cooperative Notes—Unadvanced, Dr., shall be charged and this account credited upon execution of the notes.
		If principal installments are not paid at the maturity date, the amount due shall be transferred to Account 4050, Current Maturities—Long-Term Debt.
4210.17	4210.17	*RUS Notes—Deferred Interest*
		This account shall include interest accrued on RUS mortgage notes, the payment of which has been deferred in accordance with the terms of the notes or extension agreements. The offsetting charge shall be to Account 7510, Interest on Funded Debt, for Class A companies and Account 7500, Interest and Related Items, for Class B companies.
		If interest payments are not made at the due date, this account shall be debited and Account 4010.21, Accounts Payable to Nonaffiliated Companies, credited with the amount of the matured interest.
4210.18	4210.18	*RUS Notes—Advance Payments, Dr.*
		This account shall include all payments on RUS mortgage notes made in advance of the due date and not applied to a specific quarterly payment. As these payments are applied to specific notes, this account shall be credited and the long-term debt and interest liability accounts debited.
4210.19	4210.19	*Funded Debt—Other—Unadvanced, Dr.*
		This account shall include the total face amount of notes executed to others, for which funds have not been received.
		This account shall be credited and Account 1130.1/1120.11, Cash—General Funds, debited when funds are received from the lender.
4210.20	4210.20	*RUS Notes—Unadvanced, Dr.*
		This account shall include the total face amount of RUS mortgage notes for which funds have not been received.
		This account shall be credited and Account 1130.2/1120.12, Cash—Construction Fund Trustee, debited when funds are received from RUS.
4210.21	4210.21	*Telephone Bank Notes—Unadvanced, Dr.*
		This account shall include the total face amount of Rural Telephone Bank mortgage notes for which funds have not been received.
		This account shall be credited and Account 1130.2/1120.12, Cash—Construction Fund Trustee, debited when funds are received from the Rural Telephone Bank.
4210.22	4210.22	*Federal Financing Bank Notes—Unadvanced, Dr.*
		This account shall include the total face amount of Federal Financing Bank mortgage notes for which funds have not been received.
		This account shall be credited and Account 1130.2/1120.12, Cash—Construction Fund Trustee, debited when funds are received from the Federal Financing Bank.
4210.23	4210.23	*Bank for Cooperatives Notes—Unadvanced, Dr.*

Class of company		Account title
Account No.		
A	B	

		This account shall include the total face amount of Bank for Cooperatives mortgage notes for which funds have not been received.
		This account shall be credited and Account 1130.2/1120.12, Cash—Construction Fund Trustee, debited when funds are received from the Bank for Cooperatives.
4210.24	4210.24	*Rural Telephone Finance Cooperative Notes—Unadvanced, Dr.*
		This account shall include the total face amount of Rural Telephone Finance Cooperative mortgage notes for which funds have not been received.
		This account shall be credited and Account 1130.2/1120.12, Cash—Construction Fund Trustee, debited when funds are received from the Rural Telephone Finance Cooperative.
4540.11	4540.11	*Capital Stock Subscribed.*
		This account shall include the par value of capital stock for which legally enforceable subscriptions have been received but for which, at the date of the balance sheet, stock certificates have not been issued.
		This account shall be debited and Account 4510, Capital Stock, credited when a subscriber has paid the subscription in full and stock certificates are issued.
4540.12	4540.12	*Memberships Subscribed but Unissued.*
		This account shall include the face amount of memberships subscribed but not issued. This account shall be credited at the time the subscription is received and Account 1350.2, Subscriptions to Memberships, debited.
		This account shall be debited and Account 4540.21, Memberships Issued, credited when a subscriber has paid the subscription in full and the membership certificates are issued.
4540.13	4540.13	*Members' Equity Certificates Subscribed but Unissued.*
		This account shall include the face amount of members' equity certificates subscribed but not issued. This account shall be credited at the time the subscription is received and Account 1350.3, Subscriptions to Members' Equity Certificates, debited.
		This account shall be debited and Account 4540.22, Members' Equity Certificates Issued, credited when a subscriber has paid the subscription in full and the members' equity certificates are issued.
4540.21	4540.21	*Memberships Issued.*
		This account shall include the face amount of membership certificates outstanding. A subsidiary membership certificate record shall be maintained to reflect the detail of the balance in this account.
4540.22	4540.22	*Member's Equity Certificates Issued.*
		This account shall include the face amount of members' equity certificates outstanding. A subsidiary members' equity certificate record shall be maintained to reflect the detail of the balance in this account.
4540.23	4540.23	*Members' Equity—Other.*
		This account shall include credit amounts arising from donations, forfeitures of membership fees, forgiveness of debts of the cooperative, and member's equities not otherwise provided for.
4540.31	4540.31	*Installments Paid on Capital Stock.*
		This account shall include the amount of installments paid on capital stock on a partial or installment payment plan by subscribers against whom there is no legally enforceable subscription contract, and who are entitled to be reimbursed the principal amount of their payments, with or without interest, in the event they fail to complete payment for the stock and receive certificates therefore.
		This account shall be debited and Account 4510, Capital Stock, credited with the par value of capital stock when the total subscription is received and the stock certificates are issued. Any difference between the purchase price of the subscription and the par value of the stock shall be credited to Account 4520, Additional Paid-In Capital.
		A subsidiary ledger shall be maintained to record for each subscriber, the amount subscribed, payments made, and the balance due. The balance in this account shall be reconciled monthly with the subscription ledger.
4540.32	4540.32	*Installments Paid on Memberships Subscribed.*
		This account shall include the amount of installments paid by prospective members on membership subscriptions against whom there is no legally enforceable subscription contract, and who are entitled to be reimbursed for the principal amount of their payments, with or without interest, in the event they fail to complete payment for the membership and receive certificates therefor.
		This account shall be debited and Account 4540.21, Memberships Issued, credited with the face amount of the membership when the total subscription is received and the membership certificates are issued.
		A subsidiary ledger shall be maintained to record for each subscriber, the amount subscribed, payments made, and the balance due. The balance in this account shall be reconciled monthly with the subscription ledger.
4540.33	4540.33	*Installments Paid on Equity Certificates Subscribed*
		This account shall include the amount of installments paid by prospective members on equity certificate subscriptions against whom there is no legally enforceable subscription contract, and who are entitled to be reimbursed for the principal amount of their payments, with or without interest, in the event they fail to complete payment for the membership and receive equity certificates therefor.

Class of company		Account title
Account No.		
A	B	

		This account shall be debited and Account 4540.22, Members' Equity Certificates Issued, credited with the face amount of the memberships when the total subscription is received and the equity certificates are issued. A subsidiary ledger shall be maintained to record for each subscriber, the amount subscribed, payments made, and the balance due. The balance in this account shall be reconciled monthly with the subscription ledger.
4540.41	4540.41	*Other Capital—Miscellaneous*
		This account shall include amounts which are credits arising from capital recorded upon the reorganization or recapitalization of the company and temporary declines in the value of marketable securities held for investment purposes.
4550.1	4550.1	*Operating Margins*
		This account shall include amounts received or receivable from the furnishing of telecommunications service in excess of costs incurred in the furnishing of such service. If costs exceed revenues, the excess cost of furnishing telecommunications service shall be recorded as a debit to this account.
4550.2	4550.2	*Nonoperating Margins*
		This account shall include margins arising from transactions or activities not related to the furnishing of telecommunications service. Included in this account are receipts from investments, income from investments, income from nonoperating plant, and revenues derived from services performed for others incident to the company's regulated telecommunications operations.
4550.3	4550.3	*Other Margins*
		This account shall include patronage capital credits assigned to the cooperative by other nonprofit organizations prior to January 1, 1970, which were not credited directly to an operating expense account as a reduction in the cost of furnishing telecommunications service. No entries shall be made to this account unless it is to distribute or eliminate prior balances in conformance with the bylaws of the cooperative.
4550.4	4550.4	*Patronage Capital Assignable*
		This account shall include all amounts transferred from operating margins, nonoperating margins, and other margin accounts which are assignable to individual patrons.
4550.5	4550.5	*Patrons' Capital Credits Assigned*
		This account shall include the amounts of patronage capital which have been credited to individual patrons. A subsidiary patronage capital ledger shall be maintained so as to reflect the amount of capital furnished by each patron and the amount of such capital returned to the patron.
4550.6	4550.6	*Gain on the Retirement of Capital Credits*
		This account shall include credits resulting from the retirement of patronage capital through settlement of individual patrons' accounts at less than 100 percent of the capital assigned to the patron. The portion of patronage capital not returned to patrons under such settlements shall be debited to Account 4550.5, Patrons' Capital Credits Assigned, and credited to this account. This account shall also include amounts representing patronage capital authorized to be retired to patrons who cannot be located. Returned checks issued for retirements of patronage capital, after an appropriate waiting period, shall be credited to this account and a record shall be maintained adequate to enable the cooperative to make payment to the patron if and when a claim has been established by the patron.
	6210.11	*Analog Electronic Expense*
		This account shall include expenses associated with analog electronic switching.
	6210.21	*Digital Electronic Expense*
		This account shall include expenses associated with digital electronic switching.
	6210.31	*Electro-Mechanical Expense*
		This account shall include expenses associated with electro-mechanical switching.
	6230.11	*Radio Systems Expense*
		This account shall include expenses associated with radio systems.
	6230.21	*Circuit Equipment Expense*
		This account shall include expenses associated with circuit equipment.
	6560.1	*Depreciation Expense*
		This account shall include the depreciation expense associated with telecommunications plant in service (Accounts 2112 through 2441) and property held for future telecommunications use (Account 2002).
	6560.2	*Amortization Expense*
		This account shall include the amortization expense associated with capital leases and leasehold improvements (Accounts 2681 and 2682), intangibles (Account 2690), and telecommunications plant adjustments (Account 2005).
	7200.1	*Operating Investment Tax Credits—Net*

Class of company		Account title
Account No.		
A	B	

This account shall be charged and Account 4320, Unamortized Operating Investment Tax Credits—Net, shall be credited with investment tax credits generated from qualified expenditures related to regulated operations which the company defers rather than recognizes currently in income.

This account shall be credited and Account 4320 shall be charged ratably with the amortization of each year's investment tax credits included in Account 4320 for investment services for ratemaking purposes. Such amortization shall be determined in relation to the period of time used for computing book depreciation on the property with respect to which the tax credits relate.

| | 7200.2 | *Operating Federal Income Taxes* |

This account shall be charged and Account 4070.1, Income Taxes Accrued—Federal, shall be credited for the amount of Federal income tax expense incurred in the current operating period. This account shall also reflect subsequent adjustments to amounts previously charged.

Taxes should be accrued each month on an estimated basis and adjustments made as later data becomes available.

| | 7200.3 | *Operating State and Local Income Taxes* |

This account shall be charged and Account 4070.2, Income Taxes Accrued—State and Local, shall be credited for the amount of state and local income tax expense incurred in the current operating period. This account shall also reflect subsequent adjustments to amounts previously charged.

Taxes should be accrued each month on an estimated basis and adjustments made as later data becomes available.

| 7240.1 | 7200.41 | *Operating Taxes—Property* |

This account shall be charged and Account 4080.1, Other Taxes Accrued—Property, shall be credited for the amount of property tax expense incurred in the current operating period. This account shall also reflect subsequent adjustments to amounts previously charged.

Taxes should be accrued each month on an estimated basis and adjustments made as later data becomes available.

| 7240.2 | 7200.42 | *Operating Taxes—Miscellaneous* |

This account shall be charged and Account 4080.5, Other Taxes Accrued—Miscellaneous, shall be credited for the amount of all other taxes accrued during the current operating period and not provided for elsewhere such as gross receipts, franchise, and capital stock tax expense incurred in the current operating period. This account shall also reflect subsequent adjustments to amounts previously charged.

Taxes should be accrued each month on an estimated basis and adjustments made as later data becomes available.

| | 7200.5 | *Provision for Deferred Operating Income Taxes-Net* |

This account shall be charged or credited, as appropriate, with contra entries recorded in either Account 4100, Net Current Deferred Operating Income Taxes, or Account 4340, Net Noncurrent Deferred Operating Income Taxes, as appropriate, for income tax expense that has been deferred.

Subsidiary record categories shall be maintained to distinguish between property and nonproperty related deferrals and so that the company may separately report the amounts contained herein that relate to Federal, state, and local income taxes.

| | 7300.1 | *Dividend Income* |

This account shall include dividends on investments in common and preferred stock, which is the property of the company, whether such stock is owned by the company and held in its treasury, or deposited in trust, or otherwise controlled.

This account shall not include dividends or other returns on securities issued or assumed by the company and held by or for it, whether pledged as collateral, or held in its treasury, in special deposits, or in sinking or other funds.

Dividends on stocks of other companies held in sinking or other funds shall be credited to Account 7300.3, Income from Sinking and Other Funds.

Dividends received and receivable from affiliated companies accounted for on the equity method shall be included in Account 1401, Investments in Affiliated Companies, as a reduction of the carrying value of the investments.

| | 7300.2 | *Interest Income* |

This account shall include interest on securities, including notes and other evidences of indebtedness which are the property of the company, whether such securities are owned by the company and held in its treasury, or deposited in trust (except in sinking or other funds) or otherwise controlled. It shall also include interest on bank balances, certificates of deposits, open accounts, and other analogous items. There shall be included in this account for each month, the applicable amount requisite to extinguish, during the interval between the date of acquisition and the date of maturity, the difference between the purchase price and the par value of securities owned, the income from which is includable in this account. Amounts thus credited or charged shall be concurrently included in the accounts in which the securities are carried. Any such difference remaining unextinguished at the sale or upon the maturity and satisfaction of such securities shall be cleared to Account 7300.6. Other Nonoperating Income and Expense.

| | 7300.3 | *Income from Sinking and Other Funds* |

Class of company		Account title
Account No.		
A	B	

This account shall include the income accrued on cash, securities issued by other companies, and other assets (not including securities issued or assumed by the company) held in sinking and other funds.

There shall be included in this account for each month the applicable amount requisite to extinguish, during the interval between the date of acquisition and the date of maturity, the difference between the purchase price and the par value of securities held in sinking or other funds. Amounts thus credited or charged shall be concurrently included in the accounts in which the securities are carried. Any such differences remaining unextinguished upon the maturity and satisfaction of such securities shall be cleared to Account 7300.6, Other Nonoperating Income and Expense.

7300.4 *Allowance for Funds Used During Construction*

This account shall be credited with such amounts as are charged to the telecommunications plant accounts for the purpose or recording an allowance for funds used for construction purposes.

7300.5 *Gains or Losses from the Disposition of Certain Property*

This account shall include gains or losses resulting from the disposition of land or artworks; plant with traffic, and nonoperating telecommunications plant not previously used in the provision of telecommunication services.

7300.6 *Other Nonoperating Income and Expense*

This account shall include all other items of income and gains or losses from activities not specifically provided for elsewhere such as gains or losses realized on the sale of temporary cash investments or marketable equity securities; fees collected in connection with the exchange of coupon bonds for registered bonds; uncollectible amounts previously credited to Accounts 7300.1, 7300.2, 7300.3, 7300.4, 7300.5, and 7300.6, gains or losses from the extinguishment of debt made to satisfy sinking fund requirements; gains or losses of a nonoperating nature arising from the exchange or translation of foreign currency; net unrealized losses on investments in current marketable equity securities; write-downs or write-offs of the book costs of investments in equity securities due to permanent impairment; amortization of goodwill; the company's share of earnings or losses of affiliated companies accounted for on the equity method; and the net balance of the revenue from and the expenses of property, plant, and equipment, the cost of which is includable in Account 2006, Nonoperating Plant.

7400.1 *Nonoperating Investment Tax Credits—Net*

This account shall be charged and Account 4330, Unamortized Nonoperating Investment Tax Credits—Net, shall be credited with nonoperating investment tax credits generated from qualified expenditures related to other operations which the company has elected to defer rather than recognize currently in income.

This account shall be credited and Account 4330, Unamortized Nonoperating Investment Tax Credits—Net, shall be charged with the amortization of each year's investment tax credits included in such accounts relating to amortization of previously deferred investment tax credits of other property or regulated property, the amortization of which does not serve to reduce costs of service (but the unamortized balance does reduce rate base) for ratemaking purposes. Such amortization shall be determined with reference to the period of time used for computing book depreciation on the property with respect to which the tax credits relate.

7400.2 *Nonoperating Federal Income Taxes*

This account shall be charged and Account 4070.1, Income Taxes Accrued—Federal, shall be credited for the amount of nonoperating Federal income taxes for the current period. This account shall also reflect subsequent adjustments to amounts previously charged.

Taxes shall be accrued each month on an estimated basis and adjustments made as later data becomes available. Companies that adopt the flowthrough method of accounting for investment tax credits shall reduce the calculated provision in this account by the entire amount of the credit realized during the year. Tax credits, if normalized, shall be recorded consistent with the accounting for investment tax credits.

No entries shall be made to this account to reflect interperiod tax allocation.

7400.3 *Nonoperating State and Local Income Taxes*

This account shall be charged and Account 4070.2, Income Taxes Accrued—State and Local, shall be credited for the amount of nonoperating state and local income taxes for the current period. This account shall also reflect subsequent adjustments to amounts previously charged.

Taxes shall be accrued each month on an estimated basis and adjustments made as later data becomes available.

No entries shall be made to this account to reflect interperiod tax allocation.

7400.4 *Nonoperating Other Taxes*

This account shall be charged and Account 4080.5, Other Taxes Accrued—Miscellaneous, shall be credited for all nonoperating taxes, other than Federal, state, and local income taxes, and payroll related taxes for the current period. Among the items includable in this account are property, gross receipts, franchise and capital stock taxes. This account shall also reflect subsequent adjustments to amounts previously charged.

7400.5 *Provision for Deferred Nonoperating Income Taxes—Net*

Class of company		Account title
Account No.		
A	B	
		This account shall be charged or credited, as appropriate, with contra entries recorded in either Account 4110, Net Current Deferred Nonoperating Income Taxes, or Account 4350, Net Noncurrent Deferred Nonoperating Income Taxes, as appropriate, for nonoperating tax expenses that have been deferred. Subsidiary record categories shall be maintained to distinguish between property and nonproperty related deferrals and so that the company may separately report the amounts contained herein that relate to Federal, state, and local income taxes.
	7600.1	*Extraordinary Income Credits*
		This account shall be credited with nontypical, noncustomary, and infrequently recurring gains which would significantly distort the current year's income computed before such extraordinary items, if reported other than as extraordinary items. Income tax relating to the amounts recorded in this account shall be recorded in Account 7600.3, Current Income Tax Effect for Extraordinary Items—Net, and Account 7600.4, Provision for Deferred Income Tax Effect of Extraordinary Items—Net.
	7600.2	*Extraordinary Income Charges*
		This account shall be debited with nontypical, noncustomary, and infrequently recurring losses which would significantly distort the current year's income computed before such extraordinary items, if reported other than as extraordinary items. Income tax relating to the amounts recorded in this account shall be recorded in Account 7600.3, Current Income Tax Effect for Extraordinary Items—Net, and Account 7600.4, Provision for Deferred Income Tax Effect of Extraordinary Items—Net.
	7600.3	*Current Income Tax Effect of Extraordinary Items—Net*
		This account shall be charged or credited and Account 4070.1, Income Taxes Accrued—Federal, or Account 4070.2, Income Taxes Accrued—State and Local, shall be credited or charged, as appropriate, for all current income tax effects (Federal, state, and local) of items included in Account 7600.1, Extraordinary Income Credits, and Account 7600.2, Extraordinary Income Charges.
	7600.4	*Provision for Deferred Income Tax Effect of Extraordinary Items—Net*
		This account shall be charged or credited, as appropriate, with a contra amount recorded in Account 4350, Net Noncurrent Deferred Nonoperating Income Taxes, or Account 4110, Net Current Deferred Nonoperating Income Taxes, for the income tax effects (Federal, state, and local) of items included in Account 7600.1, Extraordinary Income Credits, and Account 7600.2, Extraordinary Income Charges, that have been deferred.

[55 FR 3388, Feb. 1, 1990; 55 FR 17352, Apr. 24, 1990, as amended at 55 FR 53488, Dec. 31, 1990; 70 FR 25757, May 16, 2005]

§ 1770.16 Supplementary accounts required of nonprofit organizations.

Class of company		Account title
Account No.		
A	B	
		Current Assets
1350.1	1350.1	Subscriptions to Capital Stock.
1350.2	1350.2	Subscriptions to Memberships.
1350.3	1350.3	Subscriptions to Members' Equity Certificates.
1350.4	1350.4	Other Current Assets.
		Current Liabilities
4130.1	4130.1	Patronage Capital Payable.
4130.2	4130.2	Other Current Liabilities—Miscellaneous.
		Long-Term Debt
4270.1	4270.1	Members' Redeemable Equity Certificates Subscribed but Unissued.
4270.2	4270.2	Members' Redeemable Equity Certificates Issued.
4270.3	4270.3	Other Long-Term Debt.
1350.1	1350.1	*Subscriptions to Capital Stock*
		This account shall include the balance due from subscribers upon legally enforceable subscriptions to capital stock.
		The purchase price of subscriptions shall be charged to this account at the time the subscription is received. The par value of the stock subscribed shall be credited to Account 4540.11, Capital Stock Subscribed, and the difference between the purchase price and the par value shall be credited to Account 4520, Additional Paid-In Capital.
1350.2	1350.2	*Subscriptions to Memberships*
		This account shall include the balance due on memberships subscribed. The face amount of memberships subscribed shall be charged to this account at the time the subscription is received. The offsetting credit shall be to Account 4540.12, Memberships Subscribed but Unissued.

Class of company		Account title
Account No.		
A	B	
		A subscription ledger shall be maintained to record for each subscriber, the amount subscribed, payments made, and the balance due. The balance in this account shall be reconciled monthly with the subscription ledger.
1350.3	1350.3	*Subscriptions to Members' Equity Certificates*
		This account shall include the balance due on member's equity certificates subscribed. The face amount of certificates subscribed shall be charged to this account at the time the subscription is received. The offsetting credit shall be to Account 4540.13, Members' Equity Certificates Subscribed but Unissued, or to Account 4270.1, Members' Redeemable Equity Certificates Subscribed but Unissued.
		A subscription ledger shall be maintained to record for each subscriber, the amount subscribed, payments made, and the balance due. The balance in this account shall be reconciled monthly with the subscription ledger. The subscription ledger shall be maintained in such a manner as to separately identify redeemable and nonredeemable certificates.
1350.4	1350.4	*Other Current Assets*
		This account shall include the amount of all current assets which are not includable in Accounts 1120 through 1350.3.
4130.1	4130.1	*Patronage Capital Payable*
		This account shall include the amount of patronage capital which has been authorized to be returned to patrons.
4130.2	4130.2	*Other Current Liabilities—Miscellaneous*
		This account shall include liabilities of current character which are not includable in Accounts 4010 through 4130.1.
4270.1	70.1	*Members' Redeemable Equity Certificates Subscribed but Unissued*
		This account shall include the face amount of members' equity certificates which are redeemable at some specified future date for which subscriptions have been received but for which certificates have not been issued. This account shall be credited at the time the subscription is received and Account 1350.3, Subscriptions to Members' Equity Certificates, debited.
		This account shall be debited and Account 4270.2, Members' Redeemable Equity Certificates Issued, credited when a subscriber has paid the subscription in full and the equity certificates are issued.
4270.2	4270.2	*Members' Redeemable Equity Certificates Issued*
		This account shall include the face amount of outstanding members' equity certificates which are redeemable at some specified future date. A subsidiary members' redeemable equity certificate record shall be maintained to reflect the detail of the balance in this account.
4270.3	4270.3	*Other Long-Term Debt*
		This account shall include long-term debt not provided for elsewhere.

§ 1770.17   Expense matrix.

The expense accounts shall be maintained by the following subsidiary record categories, as appropriate to each account. Such subsidiary record categories shall be reported as required by 47 CFR part 43.

(a) *Salaries and wages.* This subsidiary record category shall include compensation to employees, such as wages, salaries, commissions, bonuses, incentive awards, and termination payments.

(b) *Benefits.* This subsidiary record category shall include payroll related benefits on behalf of employees such as the following:

(1) Pensions;

(2) Savings plan contributions (company portion);

(3) Worker's compensation required by law;

(4) Life, hospital, medical, dental, and vision plan insurance, and

(5) Social Security and other payroll taxes.

(c) *Rents.* (1) This subsidiary record category shall include amounts paid for the use of real and personal operating property. Amounts paid for real property shall be included in Account 6121, Land and Buildings Expense. This category includes payments for operating leases but does not include payments for capital leases.

(2) This subsidiary record category is applicable only to the Plant Specific Operations Expense accounts. Incidental rents, e.g., short-term rental car expense, shall be categorized as Other Expenses (see paragraph (d) of this section) under the account which reflects

the function for which the incidental rent was incurred.

(d) *Other expenses.* This subsidiary record category shall include costs which cannot be classified to the other subsidiary record categories. Included are material and supplies, including provisioning (*note also* Account 6512, Provisioning Expense); contracted services; accident and damage payments, insurance premiums; traveling expenses and other miscellaneous costs.

(e) *Clearances.* This subsidiary record category shall include amounts transferred to Construction accounts (*see* 47 CFR 32.2000(c)(2)(iii)), other Plant Specific Operations Expense accounts and/ or Account 3100, Accumulated Depreciation (cost of removal; *see* 47 CFR 32.2000(g)(1)(iii)), as appropriate, from Accounts 6112, Motor Vehicles Expense, 6114, Tools and Other Work Equipment Expense, 6534, Plant Operations and Administration Expense, and 6535, Engineering Expense. There shall also be transfers to Construction or other Plant Specific Operations Expense accounts, as appropriate, from Account 6512, Provisioning Expense. With respect to these expenses, companies may establish such clearing accounts as they deem necessary to accomplish substantially the same results, provided that within thirty (30) days of the opening of such accounts, companies shall notify the FCC of the nature and purpose thereof. Additional clearing accounts affecting other expense areas may be established with prior approval of the FCC. Should companies elect, the initial incurred subsidiary record category identification may be carried through to the final accounts without FCC approval.

[70 FR 25757, May 16, 2005]

§§1770.18–1770.24   [Reserved]

§1770.25   Unusual items and contingent liabilities.

Extraordinary items, prior period adjustments and contingent liabilities shall be submitted to RUS for review before being recorded in the company's books of account. The materiality of corrections of errors in prior periods shall be measured in relation to the summary account level used for reporting purposes for Class A companies, or in relation to total operating revenues or total operating expenses for Class B companies. For Class A companies, no correction in excess of one percent of the aggregate summary account dollars or one million dollars, whichever is higher, may be recorded in current operating accounts without prior approval. For Class B companies, no correction which exceeds one percent of total operating revenues or one percent of total operating expenses, depending on the nature of the item, may be recorded in current operating accounts without prior approval.

[70 FR 25758, May 16, 2005]

Subpart C—Accounting Interpretations

SOURCE: 61 FR 39847, July 31, 1996, unless otherwise noted.

§1770.26   General.

(a) The standard provisions of the security instruments utilized by the Rural Utilities Service (RUS) and the Rural Telephone Bank (RTB) for all telecommunications borrowers require borrowers to at all times keep and safely preserve, proper books, records, and accounts in which full and true entries will be made of all of the dealings, business, and affairs of the borrower in accordance with the methods and principles of accounting prescribed by the state regulatory body having jurisdiction over the borrower and by the Federal Communications Commission (FCC) in its Uniform System of Accounts for telecommunications companies (47 CFR part 32), as those methods and principles of accounting are supplemented from time to time by RUS.

(b) This subpart implements those standard provisions of the RUS and RTB security instruments by prescribing accounting principles, methodologies, and procedures applicable to all telecommunications borrowers for particular situations.

§1770.27   Definitions.

As used in this part:

Borrower is an RUS telecommunications borrower.

Cushion of Credit Account is a 5 percent interest bearing account established by RUS in which all voluntary payments or overpayments on Rural Electric and Telephone Revolving Funds after October 1, 1987, are deposited.

FCC is the Federal Communications Commission.

Part 32 is 47 CFR part 32, Uniform System of Accounts, issued by the Federal Communications Commission.

RAO is the Responsible Accounting Officer of the Federal Communications Commission.

RE Act is the Rural Electrification Act of 1936, as amended (7 U.S.C. 901 *et seq.*).

RETRF is the Rural Electric and Telephone Revolving Fund.

RTB is the Rural Telephone Bank.

RUS is the Rural Utilities Service, an agency of the United States Department of Agriculture, or its predecessor or successor.

§§ 1770.28–1770.45 [Reserved]

Appendix to Subpart C of Part 1770—Accounting Methods and Procedures Required of All Borrowers

All borrowers shall maintain and keep their books of accounts and all other books and records which support the entries in such books of accounts in accordance with the accounting principles prescribed in this appendix.

Numerical Index

Number and Title

Subject Matter Index   *Number*

101 Postretirement Benefits

A. Statement of Financial Accounting Standards No. 106, Employers' Accounting for Postretirement Benefits Other than Pensions (Statement No. 106), requires reporting entities to accrue the expected cost of postretirement benefits during the years the employee provides service to the entity. For purposes of applying the provisions of Statement No. 106, members of the board of directors are considered to be employees of the cooperative. Prior to the issuance of Statement No. 106, most reporting entities accounted for postretirement benefit costs on a "pay-as-you-go" basis; that is, costs were recognized when paid, not when the employee provided service to the entity in exchange for the benefits. (Statement 106 is available from the Financial Accounting Standards Board, 401 Merritt 7, P.O. Box 5116, Norwalk, CT. 06856–5116.)

B. As defined in Statement No. 106, a postretirement benefit plan is a deferred compensation arrangement in which an employer promises to exchange future benefits for an employee's current services. Postretirement benefit plans may be funded or unfunded. Postretirement benefits include, but are not limited to, health care, life insurance, tuition assistance, day care, legal services, and housing subsidies provided outside of a pension plan.

C. Statement No. 106 applies to both written plans and to plans whose existence is implied from a practice of paying postretirement benefits. An employer's practice of providing postretirement benefits to selected employees under individual contracts with specific terms determined on an employee-by-employee basis does not, however, constitute a postretirement benefit plan under the provisions of this statement.

D. Postretirement benefit plans generally fall into three categories: single-employer defined benefit plans, multiemployer plans, and multiple-employer plans.

E. A single-employer plan is a postretirement benefit plan that is maintained by one employer. The term may also be applied to a plan that is maintained by related parties such as a parent and its subsidiaries. A multiemployer plan is a postretirement benefit plan in which two or more unrelated employers contribute, usually pursuant to one or more collective-bargaining agreements. One characteristic of a multiemployer plan is that the assets contributed by one participating employer may be used to provide benefits to employees of other participating employers since assets contributed by an employer are not segregated in a separate account or restricted to provide benefits only to employees of that employer.

F. A multiple-employer plan is a postretirement benefit plan that is maintained by more than one employer but is not a multiemployer plan. A multiple-employer plan is generally not collectively bargained and is intended to allow participating employers to pool their plan assets for investment purposes and reduce the cost of plan administration. A multiple-employer plan maintains separate accounts for each employer so that contributions provide benefits only for employees of the contributing employer.

G. The accounting requirements set forth in this interpretation focus on single- and multiple-employer plans. The accounting requirements set forth in Statement No. 106 for multiemployer plans or defined contribution plans shall be adopted for borrowers electing those types of plans.

H. Under the provisions of Statement No. 106, there are two components of the postretirement benefit cost: the current period cost and the transition obligation. The transition obligation is a one-time accrual of the costs resulting from services already provided. Statement No. 106 allows the transition obligation to be deferred and amortized on a straight-line basis over the average remaining service period of the active employees. If the average remaining service period of the active employees is less than 20 years, a 20-year amortization period may be used.

I. Accounting Requirements

A. All borrowers shall adopt the accrual accounting provisions and reporting requirements as set forth in Statement No. 106. The transition obligation and accrual of the current period cost must be based upon an actuarial study. This study must be updated to allow the borrower to comply with the measurement date requirements of Statement No. 106; however, the study must, at a minimum, be updated every five years. Borrowers may not account for postretirement benefits on a "pay-as-you-go" basis.

B. Under the provisions of Statement No. 106, an entity may recognize the transition obligation, in its entirety, when Statement No. 106 is first adopted or the entity may elect to delay the recognition of the transition obligation. On December 26, 1991, however, the FCC issued 6 FCC Rcd 7560, which requires telecommunications carriers to recognize the transition obligation on a delayed basis. RUS reviewed this issuance and has determined that borrowers must comply with this ruling and recognize the transition obligation on a delayed basis.

C. The deferral and amortization of the transition obligation on a delayed basis is considered to be an off balance sheet item. As a result, an accounting entry is not required at the time of adoption of Statement No. 106. Instead, the transition obligation is recognized as a component of postretirement benefit cost as it is amortized. The amount of the unamortized transition obligation must be disclosed in the notes to the financial statements.

D. In accordance with the provisions of Responsible Accounting Officer (RAO) Letter 20, released by the FCC on April 24, 1992, Account 4310, Other Long-Term Liabilities, shall be used to record the liability accrued for postretirement benefits. (RAO Letter 20 is available from the Federal Communications Commission, 1919 M Street, NW., Washington, DC 20554.) Borrowers shall credit this account for the net periodic cost of postretirement benefits for the current year and shall debit this account for any fund payments made during the current year.

E. Net periodic postretirement benefit cost includes current period service cost, interest cost, return on plan assets, amortization of prior service cost, gains and losses, and amortization of the transition obligation. If fund payments create a debit balance in the postretirement benefits portion of Account 4310, the debit balance applicable to postretirement benefits shall be reported in Account 1410, Other Noncurrent Assets. Account 1410 shall also be used to record any prepaid postretirement benefit cost.

F. The benefits portion of the expense matrix for the appropriate Part 32 expense accounts shall be used to record the current period service cost component of the current year's net periodic postretirement benefit cost. The interest cost component, return on plan assets, amortization of prior service cost, gains and losses, and amortization of the transition obligation shall be charged to the benefits portion of the expense matrix of Account 6728, Other General and Administrative.

II. Effective Date and Implementation

A. For plans outside the United States and for defined benefit plans of employers that (a) are nonpublic enterprises and (b) sponsor defined benefit postretirement plans with no more than 500 plan participants in the aggregate, Statement No. 106 is effective for fiscal years beginning after December 15, 1994. For all other plans, Statement No. 106 is effective for fiscal years beginning after December 15, 1992.

102 Rural Telephone Bank Stock

A. Capital stock issued by the Rural Telephone Bank consists of Class A, Class B, and Class C stock. Class A stock is issued only to the Administrator of RUS on behalf of the United States in exchange for capital furnished to RTB.

B. Class B stock is issued only to recipients of loans under Section 408 of the Rural Electrification Act (RE Act). Borrowers receiving loan funds pursuant to Section 408(a) (1) or (2) of the RE Act are required to invest 5 percent of the amount of loan funds approved in Class B stock. No dividends are payable on Class B stock. All holders of Class B stock are entitled to patronage refunds in the form of Class B stock under the terms and conditions specified in the bylaws of the RTB.

C. Class C stock is available for purchase by borrowers, corporations, and public bodies eligible to borrow under Section 408 of the RE Act, or by organizations controlled by such borrowers, corporations and public bodies. The payment of dividends is in accordance with the bylaws of the RTB.

Accounting Requirements

A. The purchase of RTB stock required by the RE Act shall be debited to Account 1402.1, Investments in Nonaffiliated Companies—Class B RTB Stock. Patronage refunds in the form of additional shares of RTB Class B Stock shall be debited to Account 1402.1 and credited to Account 1402.11, Investments in Nonaffiliated Companies—Class B RTB Stock—Cr.

B. Purchases of Class C RTB stock shall be debited to Account 1402.2, Investments in Nonaffiliated Companies—Class C RTB Stock. Cash dividends received on Class C RTB stock shall be credited to Account 7310, Dividend Income.

C. Once a borrower has repaid all of its RTB loans, it may request that its Class B stock be converted to Class C stock. When the conversion is made, Account 1402.2 shall be debited and Account 1402.1 shall be credited for the face value of the stock converted. Account 1402.21, Investments in Nonaffiliated Companies—Class C RTB Stock—Cr., shall be credited and Account 1402.11 shall be debited for the face value of the Class B stock that has been received as patronage refunds.

103 Cushion of Credit Investments

A. The RUS Cushion of Credit account is an investment account bearing an interest rate of 5 percent. All voluntary payments or overpayments on Rural Electric and Telephone Revolving Fund (RETRF) loans made after October 1, 1987, are deposited into this account in the appropriate borrower's name.

Accounting Requirements

A. The following journal entries shall be used by RUS borrowers to record the transactions associated with cushion of credit payment:

1. Dr. 4210.18, RUS Notes—Advance Payments, Dr. Cr. 1130.1/1120.11, Cash—General Fund. To record the cushion of credit payment.
2. Dr. 4210.18, RUS Notes—Advance Payments, Dr. Cr. 7320/7300.2, Interest Income. To record interest earned on cushion of credit deposits.
3. Dr. 4210.12, RUS Notes, Cr. 4210.18, RUS Notes—Advance Payments, Dr. To apply cushion of credit payments (and interest) to the RUS note.

104 Rural Economic Development Loan and Grant Program

A. On December 21, 1987, Section 313, Cushion of Credit Payments Program (7 U.S.C. 901 et seq.), was added to the RE Act. Section 313 establishes a Rural Economic Development Subaccount and authorizes the Administrator of the RUS to provide zero interest loans or grants to RE Act borrowers for the purpose of promoting rural economic development and job creation projects. Effective December 5, 1994, this authority was assigned to the Administrator, Rural Business and Cooperative Development Service.

B. 7 CFR part 1703, Subpart B, Rural Economic Development Loan and Grant Program, sets forth the policies and procedures relating to the zero interest loan program and for approving and administering grants.

Accounting Requirements

A. The accounting journal entries required to record the transactions associated with a Rural Economic Development grant are as follows:

1. Dr. 1130.4/1120.14, Cash—General Fund—Economic Development Grant Funds. Cr. 4210.25, RUS Notes—Economic Development Grant; Cr. 4540.41, Other Capital—Miscellaneous; or Cr. 7360/7300.6, Other Nonoperating Income. To record grant

funds disbursed by RUS. If the grant agreement requires repayment of the funds upon termination of the revolving loan program, Account 4210.25 shall be credited. If the grant agreement states that there is absolutely no obligation for repayment upon termination of the revolving loan program, the funds shall be accounted for as a permanent infusion of capital by crediting Account 4540.41. If, however, the grant agreement is silent as to the final disposition of the grant funds, Account 7360/7300.6 shall be credited.

2. Dr. 1401.1, Other Investments in Affiliated Companies—Federal Economic Development Grant Loans or Dr. 1402.4, Other Investments in Nonaffiliated Companies—Federal Economic Development Grant Loans Cr. 1130.4/1120.14, Cash—General Fund—Economic Development Grant Funds. To record a Federal revolving loan to an economic development project.

3. Dr. 1130.1/1120.11, Cash—General Fund. Cr. 7360/7300.6, Other Nonoperating Income. To record payment of loan servicing fees charged to the economic development project.

4. Dr. 1130.5/1120.15, Cash—General Fund—Economic Development Non-Federal Revolving Funds. Cr. 1401.1, Other Investments in Affiliated Companies—Federal Economic Development Grant Loans or Cr. 1402.4, Other Investments in Nonaffiliated Companies—Federal Economic Development Grant Loans. To record the repayment, by the project, of the Federal revolving loan.

5. Dr. 1401.2, Other Investments in Affiliated Companies—Non-Federal Economic Development Grant Loans or Dr. 1402.5, Other Investments in Nonaffiliated Companies—Non-Federal Economic Development Grant Loans. Cr. 1130.5/1120.15, Cash—General Fund—Economic Development Non-Federal Revolving Funds. To record a Non-Federal revolving loan to an economic development project.

6. Dr. 1210, Interest and Dividends Receivable Cr. 7320/7300.2, Interest Income. To record the interest earned on a Non-Federal revolving loan to an economic development project.

7. Dr. 1130.5/1120.15, Cash—General Fund—Economic Development Non-Federal Revolving Funds. CR. 1401.2, Other Investments in Affiliated Companies—Non-Federal Economic Development Grant Loans or Cr. 1402.5, Other Investments in Nonaffiliated Companies—Non-Federal Economic Development Grant Loans. To record the repayment, by the project, of the Non-Federal revolving loan.

B. The accounting journal entries required to record the transactions associated with a Rural Economic Development loan are as follows:

1. Dr. 4210.26, Economic Development Notes—Unadvanced, Fr. Cr. 4210.25, Economic Development Notes. To record the contractual obligation to RUS for the Economic Development Notes.

2. Dr. 1130.6/1120.16, Cash—General Fund—Economic Development Loan Funds Cr. 4210.26, Economic Development Notes—Unadvanced, Dr. To record the receipt of the economic development loan funds.

3. Dr. 1401.3, Other Investments in Affiliated Companies—Federal Econmic Development Loans or Dr. 1402.6, Other Investments in Nonaffilitated Companies—Federal Economic Development Loans. Cr. 1130.6/1120.16, Cash—General Fund—Ecoomice Development Loan Funds. To record the discursement of economci development loand funds to the project.

4. Dr. 1130.1/1120.11, Cash—General Fund. Cr. 7360/7300.6, Other Nonoperating Income. To record payment of loan servicing fees charged to the economic development project.

5. Dr. 1210, Interest and Dividends Receivable Cr. 7320/7300.2, Interest Income. To record the interest earned on the investment of rural economic development loan funds.

6. Dr. 7370, Special Charges. Cr. 1130.1, Cash—General Funds. To record the payment of interest earned in excess of $500 on the investment of rural economic development loan funds. NOTE: Interest earned in excess of $500 must be used for the rural economic development project for which the loan funds were received or returned to RUS.

7. Dr. 1130.6/1120.16, Cash—General Fund—Economic Development Loan Funds. Cr. 1401.3, Other Investments in Affiliated Companies—Federal Economic Development Loans or Cr. 1402.6, Other Investments in Nonaffiliated Companies—Federal Economic Development Loans. To record repayment, by the project, of the economic development loan.

8. Dr. 4210.25, Economic Development Notes. Cr. 1130.6/1120.16, Cash—General Fund—Economic Development Loan Funds. To record the repayment, to RUS, of the economic development loan funds.

105   Satellite and Cable Television Services

A. Many RUS borrowers have become involved in providing either satellite or cable television services to their members and others through subsidiaries, joint ventures, or as segments of their current operations.

Accounting Requirements

A. This section outlines the accounting to be followed when recording transactions involving satellite or cable television services.

1. Separate Subsidiary. If a borrower provides satellite or cable television services through a separate subsidiary, the investment in the subsidiary shall be debited to Account 1401, Investments in Affiliated Companies. The net income or loss of the subsidiary shall be debited or credited to Account 1401, as appropriate, with an offsetting entry to Account 7360, Other Nonoperating Income.

2. *Joint Venture.* i. If a borrower provides satellite or cable television services through a joint venture, the borrower's ownership interest dictates the accounting methodology. If the borrower has less than a 20 percent ownership interest in the joint venture, the investment is accounted for under the cost method of accounting in Account 1402, Investments in Nonaffiliated Companies. Under the cost method, the joint venture's net income or loss is not recorded in the borrower's records. Income is recognized only to the extent of any dividends declared by the joint venture. When a dividend is declared, the borrower shall debit Account 1210, Interest and Dividends Receivable, and credit Account 7310, Dividend Income. When the dividend is received in cash, the borrower shall debit Account 1130.1, Cash—General Fund, and credit Account 1210.

ii. If a borrower has a 20-percent or more ownership interest in the joint venture, the investment is accounted for under the equity method in Account 1401, Investments in Affiliated Companies. The borrower's proportionate share of the joint venture's net income or loss shall be debited or credited to Account 1401, as appropriate, with an offsetting entry to Account 7360, Other Nonoperating Income.

3. *Segment of Current Operations.* i. If a borrower provides satellite or cable television service as a segment of its current operations and there are no shared assets between this activity and the regulated telecommunications activities of the borrower, the investment shall be debited to Account 1406.1, Nonregulated Investments—Permanent Investment. The net income or loss from providing such service shall be debited or credited, as appropriate, to Account 1406.3, Nonregulated Investments—Current Net Income, with an offsetting entry to Account 7990, Nonregulated Net Income.

ii. If a borrower provides satellite or cable television service as a segment of current operations and shares assets between this activity and the regulated telecommunications activities of the borrower, the franchise and application fees shall be debited to a subaccount of Account 2690, Intangibles. The cost of the satellite or cable television equipment shall be debited to a subaccount of Account 2231, Radio Systems. Revenues earned from providing satellite or cable service shall be credited to Account 5280, Nonregulated Operating Revenue, while the associated expenses shall be recorded in a subaccount of the applicable regulated expense accounts.

4. *Sale and Installation of Satellite or Cable Television Equipment.* i. If a borrower sells or installs satellite or cable television equipment as a segment of its current operations and there are no shared assets between this activity and the regulated telecommunications activities of the borrower, the purchase of the equipment shall be debited to Account 1406.1, Nonregulated Investments—Permanent Investment. The net income or loss from providing such services shall be debited or credited, as appropriate, to Account 1406.3, Nonregulated Investments—Current Net Income, with an offsetting entry to Account 7990, Nonregulated Net Income.

ii. If a borrower sells or installs satellite or cable television equipment as a segment of its current operations and shares assets between this activity and the regulated telecommunications activities of the borrower, the purchase of the equipment shall be debited to Account 1220.2, Property Held for Sale or Lease. Revenues received for the sale or installation of the equipment shall be credited to Account 5280, Nonregulated Operating Revenue, while the associated expenses shall be debited to a subaccount of the applicable regulated expense accounts.

106 Consolidated Financial Statements

A. In October 1987, FASB issued Statement of Financial Accounting Standards No. 94, Consolidation of All Majority-Owned Subsidiaries (Statement No. 94). (Statement 94 is available from the Financial Accounting Standards Board, 401 Merritt 7, P.O. Box 5116, Norwalk, CT 06856–5116.) For purposes of reporting to RUS, Statement No. 94 shall be applied as follows:

1. A borrower that is a subsidiary of another entity shall prepare and submit to RUS separate financial statements even though this financial information is presented in the parent's consolidated statements.

2. In those cases in which a borrower has a majority-ownership in a subsidiary, the borrower shall prepare consolidated financial statements in accordance with the requirements of Statement No. 94. These consolidated statements must also include supplementary schedules presenting a Balance Sheet and Income Statement for each majority-owned subsidiary included in the consolidated statements.

B. Although Statement No. 94 requires the consolidation of majority-owned subsidiaries, the RUS Form 479, Financial and Statistical Report for Telecommunications Borrowers, shall be prepared on an unconsolidated basis by all borrowers.

107 Allowance for Funds Used During Construction

A. Statement of Financial Accounting Standard No. 34, Capitalization of Interest Cost, established the standards for capitalizing interest cost as a part of the historical cost of acquiring certain assets. In order to capitalize interest, the asset must require a period of time to complete or to get it ready for its intended use. This standard applies to all entities that construct facilities for their own use and should be applied by RUS Telecommunications borrowers as follows:

1. Only actual interest costs incurred on external borrowings qualify to be capitalized. The interest rate used to calculate the amount of interest to be capitalized is based on the companies external borrowings. If a construction project is associated with specific debt, the interest rate on that debt is used to calculate interest cost to be capitalized. If the project is not associated with a specific debt, a weighted average of the rates of all existing debt shall be applied to expenditures for the project. There is no materiality threshold for adoption of this standard (47 CFR 32.26).

2. If a borrower is involved in a joint construction project, all determinations as to the amount of interest incurred and qualified for capitalization must be based on individual financing arrangements with regard to the Interest During Construction rules.

3. The capitalization period shall end when the asset is substantially complete and ready for its intended use.

Disclosures

A. The following information with respect to interest cost shall be disclosed in the financial statements or related notes:

1. For an accounting period in which no interest cost is capitalized, the amount of interest cost incurred and charged to expense during the period.

2. For an accounting period in which some interest cost is capitalized, the total amount of interest cost incurred during the period and the amount thereof that has been capitalized.

108 Reporting Comprehensive Income

A. In June 1997, the Financial Accounting Standards Board issued Statement of Financial Accounting Standards No. 130, Reporting Comprehensive Income. This statement requires that all items that meet the definition of the components of comprehensive income be reported in the financial statements for the period in which they are recognized. Statement 130 establishes a distinction between *comprehensive income* and *other comprehensive income.*

1. *Comprehensive income* is composed of net income and *other comprehensive income.* The net income is the result of operations resulting from the aggregation of revenues, expenses, gains and losses that are not items that comprise other comprehensive income.

2. *Other comprehensive income* is composed of the following:

(a) Foreign currency items,

(b) Minimum pension liability adjustments, and

(c) Unrealized gains and losses on certain investments in debt and equity securities. Gains or losses on investment securities included in the net income of the current period that also had been included in other comprehensive income as unrealized holding gains or losses in a prior period must be adjusted (called reclassification adjustments) in the presentation of other comprehensive income in the current period.

B. *Comprehensive income* expressed as a formula would be:

Net Income ±items of *other comprehensive income = comprehensive income*

While Statement 130 requires that comprehensive income should be divided into two broad display classifications, net income and other comprehensive income, it does not prescribe a specific format for displaying comprehensive income in the financial statements.

C. RUS Telecommunications borrowers that present a single Statement of Operations and Patronage Capital should present the components of *other comprehensive income* below the total for net income and then present the reconciliation of patronage capital (Retained Earnings). Borrowers that present a separate Statement of Patronage Capital (or Retained Earnings) should display the beginning balance of patronage capital (or retained earnings), net income for the period, other items of comprehensive income and total comprehensive income before the presentation of other items of patronage capital (or retained earnings) for the period.

109 Disclosures about Pensions and Other Postretirement Benefits

A. Statement of Financial Accounting Standards (SFAS) No. 132, Employers' Disclosures about Pensions and Other Postretirement Benefits, issued in February 1998, is effective for fiscal years beginning after December 15, 1998. This statement revises employers' disclosure requirements for pension and other postretirement benefit plans. It does not change the measurement or recognition of those plans. The statement also permits reduced disclosures for nonpublic entities, which are defined as any entity other than one:

1. Whose debt or equity securities trade in a public market either on a domestic or foreign stock exchange or in the over-the-counter market, including securities quoted only locally or regionally,

221

2. That makes a filing with a regulatory agency in preparation for the sale of any class of debt or equity securities in a public market, or

3. That is controlled by an entity covered by 1 or 2 above.

Public Entities and Those Controlled by Public Entities

A. A commercial RUS Telecommunications borrower that meets the definition of a public entity and sponsors one or more defined benefit pension or postretirement benefit plan shall provide the following information on a comparative basis for the statements presented:

1. A reconciliation of beginning and ending balances of the benefit obligation showing separately, if applicable, the effects during the period attributable to each of the following:

(a) Service cost,

(b) Interest cost,

(c) Contributions by plan participants,

(d) Actuarial gains and losses,

(e) Foreign currency exchange rate changes,

(f) Benefits paid,

(g) Plan amendments,

(h) Business combinations,

(i) Divestitures,

(j) Curtailments,

(k) Settlements, and

(l) Special termination benefits.

2. A reconciliation of beginning and ending balances of the fair value of plan assets showing separately, if applicable, the effects during the period attributable to each of the following:

(a) Actual return on plan assets,

(b) Foreign currency exchange rate changes,

(c) Contributions by the employer,

(d) Contributions by plan participants,

(e) Benefits paid,

(f) Business combinations,

(g) Divestitures, and

(h) Settlements.

3. The funded status of the plans, the amounts not recognized in the statement of financial position, and the amounts recognized in the statement of financial position, including:

(a) The amount of any unamortized prior service cost.

(b) The amount of any unrecognized net gain or loss (including asset gains and losses not yet reflected in market-related value).

(c) The amount of any remaining unamortized, unrecognized net obligation or net asset existing at the initial date of application of SFAS No. 87, Employers' Accounting for Pensions, or SFAS No. 106, Employers' Accounting for Postretirement Benefits Other Than Pensions.

(d) The net pension or other postretirement benefit prepaid assets or accrued liabilities.

(e) Any intangible asset and the amount of accumulated other comprehensive income recognized pursuant to paragraph 37 of SFAS No. 87, as amended.

4. The amount of net periodic benefit cost recognized, showing separately:

(a) The service cost component,

(b) The interest cost component,

(c) The expected return on plan assets for the period,

(d) The amortization of the unrecognized transition obligation or transition asset,

(e) The amount of recognized gains and losses, the amount of prior service cost recognized, and

(f) The amount of gain or loss recognized due to a settlement or curtailment.

5. The amount included within other comprehensive income for the period arising from a change in the additional minimum pension liability recognized pursuant to paragraph 37 of SFAS No. 87, as amended.

6. On a weighted-average basis, the following assumptions used in the accounting for the plans:

(a) Assumed discount rate,

(b) Rate of compensation increase (for pay-related plans), and

(c) Expected long-term rate of return on plan assets.

7. The assumed health care cost trend rate(s) for the next year used to measure the expected cost of benefits covered by the plan (gross eligible charges) and a general description of the direction and pattern of change in the assumed trend rates thereafter, together with the ultimate trend rate(s) and when that rate is expected to be achieved.

8. The effect of a one-percentage-point increase and the effect of a one-percentage-point decrease in the assumed health care cost trend rates on (for purposes of this disclosure, all other assumptions shall be held constant, and the effects shall be measured based on the substantive plan that is the basis for the accounting):

(a) The aggregate of the service and interest cost components of net periodic postretirement health care benefit cost, and

(b) The accumulated postretirement benefit obligation for health care benefits.

9. If applicable, the amounts and types of securities of the employer and related parties included in plan assets, the approximate amount of future annual benefits of plan participants covered by insurance contracts issued by the employer or related parties, and any significant transactions between the employer or related parties and the plan during the period.

10. If applicable, any alternative amortization method used to amortize prior service amounts or unrecognized net gains and

losses pursuant to paragraphs 26 and 33 of SFAS No. 87 or paragraphs 53 and 60 of SFAS No. 106.

11. If applicable, any substantive commitment, such as past practice or a history of regular benefit increases, used as the basis for accounting for the benefit obligation.

12. If applicable, the cost of providing special or contractual termination benefits recognized during the period and a description of the nature of the event.

13. An explanation of any significant change in the benefit obligation or plan assets not otherwise apparent in the other disclosures.

B. RUS Telecommunications borrowers that sponsor two or more pension or postretirement plans may aggregate the required disclosures. If the disclosures are aggregated, the aggregate benefit obligation and aggregate fair value of plan assets for plans with benefit obligations in excess of plan assets must be disclosed.

C. RUS Telecommunications borrowers sponsoring defined contribution plans shall disclose the amount of cost recognized for defined contribution pension or other postretirement benefit plans during the period separately from the amount of cost recognized for defined benefit plans. The disclosures shall include a description of the nature and effect of any significant changes during the period affecting comparability, such as a change in the rate of employer contributions, a business combination, or a divestiture.

Nonpublic Entities

A. RUS commercial and cooperative type borrowers that meet the definition of a nonpublic entity, as previously defined, may elect to meet the following reduced disclosure requirements:

1. The benefit obligation.
2. Fair value of plan assets.
3. Funded status of the plan.
4. Employer contributions.
5. Participant contributions.
6. Benefits paid.
7. The amounts recognized in the statement of financial position, including the net pension and other postretirement benefit prepaid assets or accrued liabilities and any intangible asset and the amount of accumulated other comprehensive income recognized pursuant to paragraph 37 of SFAS No. 87, as amended.
8. The amount of net periodic benefit cost recognized and the amount included within other comprehensive income arising from a change in the minimum pension liability recognized pursuant to paragraph 37 of SFAS No. 87, as amended.
9. On a weighted-average basis, the following assumptions used in the accounting for the plans: Assumed discount rate, rate of compensation increase (for pay-related

plans), and expected long-term rate of return on plan assets.

10. The assumed health care cost trend rate(s) for the next year used to measure the expected cost of benefits covered by the plan (gross eligible charges) and a general description of the direction and pattern of change in the assumed trend rates thereafter, together with the ultimate trend rate(s) and when that rate is expected to be achieved.

11. If applicable, the amounts and types of securities of the employer and related parties included in plan assets, the approximate amount of future annual benefits of plan participants covered by insurance contracts issued by the employer or related parties, and any significant transactions between the employer or related parties and the plan during the period.

12. The nature and effect of significant nonroutine events, such as amendments, combinations, divestitures, curtailments, and settlements.

B. The majority of RUS Telecommunications borrowers will fall within the definition of nonpublic entities with exception of those held by publicly traded holding companies.

Multiemployer Plans

A. An RUS Telecommunications borrower shall disclose the amount of contributions to multiemployer plans during the period. The borrower may disclose total contributions to multiemployer plans without disaggregating the amounts attributable to pensions and other postretirement benefits. The disclosures shall include a description of the nature and effect of any changes affecting comparability, such as a change in the rate of employer contributions, a business combination, or a divestiture.

B. In some cases, withdrawal from a multiemployer plan results in an obligation to the plan for a portion of the plan's unfunded accumulated postretirement benefit obligation. If it is either probable or reasonably possible that (a) an employer would withdraw from the plan under circumstances that would give rise to an obligation or (b) an employer's contribution to the fund would be increased during the remainder of the contract period to make up a shortfall in the funds necessary to maintain the negotiated level of benefit coverage, the employer shall apply the provisions of SFAS No. 5, Accounting for Contingencies.

DISCLOSURE MATRIX

	Public entities	Nonpublic entities
Change in benefit obligation: Benefit obligation beginning of year	X	
Service Cost	X	

Disclosure Matrix—Continued

	Public entities	Nonpublic entities
Interest Cost	X	
Actuarial Gain	X	
Plan Amendments	X	
Benefits Paid	X	
Benefit obligation at end of year	X	X
Change in plan assets:		
Fair value of plan assets beginning of year	X	
Actual return on plan assets	X	
Employer Contribution	X	X
Contributions by plan participants	X	X
Benefits Paid	X	X
Fair value of plan assets at end of year	X	X
Funded status:		
Unrecognized net actuarial loss (gain)	X	X
Unamortized prior service cost	X	X
Unrecognized transition obligation	X	X
Prepaid (Accrued) benefit cost	X	X
Weighted-average assumptions as of December 31:		
Discount rate	X	X
Expected return on plan assets	X	X
Rate of compensation increase	X	X
Components of net periodic benefit cost:		
Service cost	X	
Interest cost	X	
Expected return on plan assets	X	
Amortization of prior service cost	X	X
Amortization of transition obligation	X	X
Recognized net actuarial loss	X	X
Net periodic benefit cost	X	X

[61 FR 39847, July 31, 1996, as amended at 70 FR 25758, May 16, 2005]

PART 1773—POLICY ON AUDITS OF RUS BORROWERS AND GRANTEES

Subpart A—General Provisions

Sec.

Subpart B—RUS Audit Requirements

Subpart C—RUS Requirements for the Submission and Review of the Reporting Package

Subpart D—RUS Reporting Requirements

Subpart E—RUS Audit Requirements and Documentation

AUTHORITY: 7 U.S.C. 901 *et seq.*, 7 U.S.C. 1921 *et seq.*, 7 U.S.C. 6941 *et seq.*

SOURCE: 83 FR 19907, May 7, 2018, unless otherwise noted.

Subpart A—General Provisions

§ 1773.1 General.

(a) This part implements the standards for audits required by the loan and grant agreements of Rural Utilities Service (RUS) electric and telecommunications borrowers and grantees. The provisions require auditees to prepare and furnish to RUS, at least once during each 12-month period, a full and complete report of its financial condition, operations, and cash flows, in form and substance satisfactory to RUS, audited and certified by an independent auditor, satisfactory to RUS, and accompanied by a report of such audit, in form and substance satisfactory to RUS.

(b) This part is based on the requirements of GAGAS in effect at the time of the audit and applicable RUS regulations and subpart F (Audit Requirements) of 2 CFR part 200 (Uniform Administrative Requirements, Cost Principles, and Audit Requirements for Federal Awards) (2 CFR 200.500–200.521).

(c) This part further sets forth the criteria for selecting auditors satisfactory to RUS and certain audit procedures and audit documentation that must be performed and prepared before an audit report will be accepted by RUS.

(d) Failure to provide an audit in compliance with this part is a serious violation of the RUS Security Agreement. RUS relies on audited financial statements in order to assess and monitor the financial condition of its borrowers and grantees and to fulfill its fiduciary responsibilities.

(e) RUS reserves the right to suspend its acceptance of audits performed by auditors who, in the opinion of RUS, are not meeting the requirements of this part or with unresolved disputes or issues until such time that the matter can be resolved to RUS' satisfaction.

§1773.2 Definitions.

As used in this part:

2 CFR part 200, subpart F means 2 CFR part 200, Uniform Administrative Requirements, Cost Principles and Audit Requirements for Federal Awards, subpart F, Audit Requirements, as adopted by USDA in 2 CFR part 400.

AA–PARA means RUS Assistant Administrator, Program Accounting and Regulatory Analysis.

Administrator means the Administrator of RUS.

Affiliated company means a company that directly or indirectly through one or more intermediaries, control or are controlled by, or are under common control with, the auditee.

AICPA means the American Institute of Certified Public Accountants.

ASC means the Accounting Standards Codification issued by the Financial Accounting Standards Board.

Audit means an examination of financial statements by an independent auditor for the purpose of expressing an opinion on the fairness with which those statements present financial position, results of operations, and changes in cash flows in accordance with accounting principles generally accepted in the United States of America (GAAP) and for determining whether the auditee has complied with applicable laws, regulations, and provisions of loan or grant contracts and grant agreements that could have a material effect on the financial statements.

Audit date means the "as of" date established by the auditee.

Audit documentation has the same meaning as defined in the AICPA's professional auditing standards.

Auditee means an RUS borrower and/or grantee that is required to submit an annual audit as a condition of the award.

Auditor means government auditors as well as certified public accounting firms that perform audits using generally accepted government auditing standards (GAGAS).

BCAS means Broadband Collection and Analysis System (or successor system).

Borrower means an entity that has an outstanding RUS or Federal Financing Bank (FFB) loan or loan guarantee.

CPA means a Certified Public Accountant.

DCS means the Data Collection System (or successor system).

FASB means Financial Accounting Standards Board.

FFB means the Federal Financing Bank, a body corporate and instrumentality of the United States of America under the general supervision of the Secretary of the Department of the Treasury.

Fraud has the same meaning as defined in the AICPA's professional auditing standards.

GAAP has the same meaning as defined in accounting standards issued by the Government Accounting Standards Board (GASB) and the Financial Accounting Standards Board (FASB).

GAGAS means generally accepted government auditing standards as set forth in Government Auditing Standards, issued by the Comptroller General of the United States, Government Accountability Office.

GAO means the United States Government Accountability Office.

GASB means Government Accounting Standards Board.

Governance board means the auditee's board of directors, managing members, or other official body charged with governance.

Grantee means an entity that has a continuing responsibility under a grant agreement with RUS.

Illegal act has the same meaning as defined by the Public Company Accounting Oversight Board.

Material weakness has the same meaning as defined in the AICPA's professional auditing standards.

OIG means the Office of the Inspector General, United States Department of Agriculture.

OMB means The Office of Management and Budget.

Regulatory asset means an asset resulting from an action of a regulator as defined by FASB.

Regulatory liability means a liability imposed on a regulated enterprise by an action of a regulator as defined by FASB.

Related party has the same meaning as defined by FASB.

Reporting package means:

(1) The auditor's report on the financial statements;

(2) The report on internal control over financial reporting and on compliance and other matters;

(3) The report on compliance with aspects of contractual agreements and regulatory requirements;

(4) The schedule of findings and recommendations; and

(5) All supplemental schedules and information required by this part.

RUS means the Rural Utilities Service, an agency of the United States Department of Agriculture.

RUS Bulletin 1773-1, Policy on Audits of RUS Borrowers and Grantees, is a publication prepared by RUS that contains the RUS regulation 7 CFR part 1773 and exhibits of sample audit reports, financial statements, reports on internal control over financial reporting and on compliance and other matters, report on compliance with aspects of contractual agreements and regulatory requirements, and schedule of findings and recommendations used in preparing audits of RUS borrowers and grantees. This bulletin is available on the internet at *https://www.rd.usda.gov/publications/regulations-guidelines/bulletins/program-accounting*.

RUS security agreement means a loan agreement, grant agreement, mortgage, security agreement, or other form of agreement that governs the terms and conditions of, or provides security for, loan and/or grant funds provided by RUS to the auditee.

Significant deficiency has the same meaning as defined in the AICPA's professional auditing standards.

Single Audit Act means Single Audit Act of 1984 (31 U.S.C. 7501 *et seq.*) as implemented by 2 CFR part 200, subpart F.

State means any state or territory of the United States, or the District of Columbia.

Uniform System of Accounts means, for telecommunications borrowers, Bulletin 1770B-1, Accounting Requirements for RUS Telecommunications Borrowers (*https://www.rd.usda.gov/files/UTP_Bulletins_1770B-1.pdf*), and for electric borrowers, as contained in 7 CFR part 1767, Accounting Requirements for RUS Electric Borrowers, subpart B—Uniform System of Accounts, Bulletin 1767B-1, (*https://www.rd.usda.gov/files/UPA_Bulletin_1767B-1.pdf*).

Subpart B—RUS Audit Requirements

§ 1773.3 Annual audit.

(a) Each auditee must have its financial statements audited annually by an auditor selected by the auditee and approved by RUS as set forth in § 1773.4. All auditees must submit audited financial statements on a comparative basis covering two consecutive 12 month periods, unless the entity has not been in existence for two consecutive 12-month audit periods. Consolidated statements of the parent are not an acceptable replacement for an audit of the auditee.

(b) Each auditee must establish an annual audit date within 12 months of the date of the first advance and must prepare annual financial statements for the audit date established. Each auditee must notify the AA–PARA of the audit date at least 90 days prior to the selected audit date.

(c) Auditees must furnish a reporting package to RUS within 120 days of the audit date. (See §1773.21). Until all loans made or guaranteed by RUS are repaid and unliquidated obligations rescinded, auditees that are borrowers must continue to provide annual audited financial statements. Auditees that are grantees must furnish annual audited financial statements in the year of the first advance and until all funds have been advanced or rescinded, and all financial compliance requirements have been fully satisfied.

(d) In addition to the requirements of this part, certain auditees may be subject to the Single Audit Act. An auditee that is defined as a Non-Federal Entity as defined in 2 CFR 200.69 means a state, local government, Indian tribe, institution of higher education (IHE), or nonprofit organization that carries out a Federal award as a recipient or subrecipient and is required to meet the requirements of this part as follows:

(1) Borrowers and/or grantees expending the threshold established for the Single Audit Act (currently $750,000) or more in Federal awards during the year must have an audit performed in accordance with the Single Audit Act. See 2 CFR 200.502, Basis For Determining Federal Awards Expended, for guidance in determining annual expenditures. The audited financial statements must be submitted to RUS and to the Federal Audit Clearinghouse.

(2) For auditees expending less than the threshold for expenditure in Federal awards during the year, RUS reserves its right under 2 CFR 200.503, Relationship to other audit requirements, to arrange for an audit performed in accordance with this part.

(3) Within 30 days of the audit date, auditees must notify the AA–PARA, in writing, of the total Federal awards expended during the year and must state whether the audit will be performed in accordance with the Single Audit Act, or this part.

(i) An auditee electing to comply with this part must select an auditor that meets the qualifications set forth in §1773.5.

(ii) If an audit is performed in accordance with the Single Audit Act, the auditor's reporting on the financial statements that meet the requirements of the Single Audit Act, will be sufficient to satisfy the auditee's obligations under this part.

(e) Subpart F of 2 CFR part 200 does not apply to audits of RUS electric and telecommunications cooperatives and for-profit telecommunications borrowers unless the borrower has contractually agreed with another Federal agency (e.g. Federal Emergency Management Agency) to provide a financial audit performed in accordance with 2 CFR part 200, subpart F. In no circumstance will an auditee be required to submit separate audits performed in accordance with this part and 2 CFR part 200, subpart F.

§1773.4 Auditee's responsibilities.

(a) *Selection of a qualified auditor.* The auditee's governance board is responsible for the selection of a qualified auditor that meets the requirements set forth in §1773.5. When selecting an auditor, the auditee should consider, among other matters:

(1) The qualifications of auditors available to do the work;

(2) The auditor's experience in performing audits of utilities, related industries, or in the case of grantees, experience in auditing entities comparable to the grantee; and

(3) The auditor's ability to complete the audit and submit the reporting package within 90 days of the audit date.

(b) *Board approval of selection.* The board's approval of an auditor must be recorded by a board resolution that states:

(1) The auditor represents that it meets RUS qualifications to perform an audit; and

(2) The auditee and auditor will enter into an audit engagement in accordance with §1773.6.

(c) *Notification of selection.* When the initial selection or subsequent change of an auditor has been made, the auditee must notify the AA–PARA, in writing, at least 90 days prior to the audit date.

(1) Within 30 days of the date of receipt of such notice, RUS will notify

the auditee, in writing, if the selection or change in auditor is not satisfactory.

(2) Notification to RUS that the same auditor has been selected for succeeding audits of the auditee's financial statements is not required; however, the procedures outlined in this part must be followed for each new auditor selected, even though such auditor may previously have been approved by RUS to audit records of other RUS auditees. Changes in the name of an auditor are considered to be a change in the auditor.

(d) *Audit engagement letter.* The auditee must enter into an audit agreement with the auditor that complies with § 1773.6 prior to the initiation of the audit.

(e) *Debarment certification.* The auditee must obtain, from the selected auditor, a lower tier covered transaction certification (Form AD–1048, Certification Regarding Debarment, Suspension, Ineligibility and Voluntary Exclusion—Lower Tier Covered Transactions), as required by Executive Orders 12549 and 12689, Debarment and Suspension, and any rules or regulations issued thereunder.

(f) *Peer review report.* The auditee must obtain, from the selected auditor, a copy of the auditor's current approved peer review report.

(g) *Preparation of schedules.* The auditee must prepare any schedules that are required by the auditor to perform the audit, including a schedule of deferred debits and deferred credits and a detailed schedule of investments in subsidiary and affiliated companies accounted for on the cost, equity, or consolidated basis. The detailed schedule of investments can be included in the notes to the financial statements or as a separate schedule as long as all information required is adequately disclosed. Samples of these schedules can be found in Appendices A–D, of RUS Bulletin 1773–1.

(1) The schedule of deferred debits and deferred credits must include a description of the deferral and a notation as to whether the deferral has received written approval from RUS. If a determination is made that prior written approval is not required, cite the specific authority for the deferral.

(2) The schedule of investments must include investments in subsidiary and affiliated companies, corporations, limited liability corporations and partnerships, joint ventures, etc. accounted for on either the cost, equity or on a consolidated basis. For all investments, the auditee must list the name of the entity, ownership percentage, and the principal business in which the entity is engaged. For investments recorded on the cost basis, the auditee must include the original investment, advances, dividends declared or paid in the current and prior years and the net investment. For investments recorded on the equity or consolidated basis, the auditee must include the ownership percentage, original investment, advances, dividends declared or paid in the current and prior years, and current and prior years' earnings and losses, including accumulated losses in excess of the original investment.

(h) *Scope limitations.* The auditee will not limit the scope of the audit to the extent that the auditor is unable to provide an unqualified opinion that the financial statements are presented fairly in conformity with GAAP due to the scope limitation.

(i) *Submission of reporting package.* The auditee must submit to RUS the required reporting package as set forth in § 1773.21.

(1) A reporting package that fails to meet the requirements detailed in this part will be returned to the auditee with a written explanation of noncompliance.

(2) The auditee must, within 30 days of the date of the letter or email detailing the noncompliance, submit a corrected reporting package to RUS.

(3) If a corrected reporting package is not received within 30 days of the date of the letter or email detailing the noncompliance, RUS will take appropriate action, depending on the severity of the noncompliance.

(j) *Submission of a plan of corrective action.* If the auditor's report contains findings and recommendations but does not include the auditee's response, the auditee must submit written responses to RUS within 180 days of the audit date. The written responses must address:

(1) The corrective action already taken or planned, or the reason the auditee believes no action is necessary; and

(2) The status of corrective action taken on previously reported findings and recommendations.

§ 1773.5 Qualifications of the auditor.

Auditors that meet the qualifications criteria of this section and enter into an audit engagement with the auditee that complies with § 1773.6, will be considered satisfactory to RUS.

(a) *Licensing.* Auditors that audit the financial statements of an RUS auditee must be licensed to perform attestation engagements in the United States of America. Auditors do not have to be licensed by the state in which the auditee is located; however, auditors must abide by the rules and regulations of professional conduct promulgated by the accountancy board of the state in which the auditee is located.

(b) *Independence.* Auditors must be independent as determined by the standards for independence in the AICPA Code of Professional Conduct and in GAGAS in effect at the time of the audit.

(c) *Peer review requirement.* Auditors must be enrolled and participating in a peer review program, and must have undergone a satisfactory peer review of their accounting and audit practice. The peer review must be in effect at the date of the audit report opinion.

(1) *Peer review reports.* RUS reserves the right to request peer review reports from selected auditors.

(2) *Peer review requirements for new auditors.* New auditing firms must meet the requirements of their state board of accountancy with regard to enrolling in a peer review program, timing of the first peer review, and any other peer review requirements.

§ 1773.6 Auditor communication.

(a) GAGAS and AICPA standards require that the auditor communicate with the auditee the auditor's understanding of the services to be performed and document that understanding through a written communication to those charged with governance. To be acceptable to RUS, the auditor's communication must take the form of an audit engagement letter prepared by the auditor and must be formally accepted by the governance board or an audit committee representing the governance board. In addition to the requirements of the AICPA's professional auditing standards and GAGAS, the engagement letter must also include the following:

(1) The nature of planned work and level of assurance to be provided related to internal control over financial reporting and compliance with laws, regulation, and provision of contracts or grant agreements;

(2) That the auditee and auditor acknowledge that the audit is being performed and that the reporting package is being issued to enable the auditee to comply with the provisions of RUS's security instrument which requires compliance with this part;

(3) That the auditor acknowledges the mandatory reporting requirements for fraud, illegal acts, or noncompliance with provisions of laws, regulations, contracts, and grant agreements in § 1773.9. Acceptance of the engagement letter by the auditee is required, thus granting the auditor permission to directly notify the appropriate officials which may include but is not limited to the governance board, RUS, and OIG;

(4) That the auditor acknowledges that it is required under § 1773.7 to contact RUS if the auditor is unable to resolve scope limitations imposed by the auditee, or if such limitations in scope violate this part. Acceptance of the engagement letter by the auditee is required, thus granting the auditor permission to directly notify the AA-PARA as needed;

(5) That the auditee and auditor acknowledge that RUS will consider the auditee to be in violation of its RUS Security Agreement and this part if the auditee fails to have an audit performed and documented in compliance with GAGAS and this part;

(6) That the auditor represents that it meets the requirements under this part to perform the audit;

(7) That the auditor will perform the audit and will prepare the reporting package in accordance with the requirements of this part;

(8) That the auditor will document the audit work performed in accordance with GAGAS, and the requirements of this part; and

(9) That the auditor will make all audit documentation, including the reporting package available to RUS or its representatives (including but not limited to OIG and GAO), upon request, and will permit the photocopying of all such audit documentation.

(b) A copy of the audit engagement letter must be available at the auditee's office for inspection by RUS personnel. One copy of the current audit engagement letter must be maintained in the auditor's audit documentation.

§ 1773.7 Audit standards.

(a) The audit of the financial statements must be performed in accordance with GAGAS and this part in effect at the audit date unless the auditee is directed otherwise, in writing, by RUS.

(b) The audit of the financial statements must include such tests of the accounting records and such other auditing procedures that are sufficient to enable the auditor to express an opinion on the financial statements and to issue the required reporting package.

(c)(1) The auditee will not limit the scope of the audit to the extent that the auditor is unable to meet RUS audit requirements without prior written approval of the AA–PARA.

(2) If the auditor determines during the audit that an unqualified opinion cannot be issued due to a scope limitation imposed by the auditee, the auditor should use professional judgment to determine what levels of the auditee's management and/or those charged with governance should be informed.

(3) After informing the auditee's management and/or those charged with governance, if the scope limitation is not adequately resolved, the auditor should immediately contact the AA–PARA.

§ 1773.8 Audit date.

(a) The annual audit must be performed as of the end of the same calendar month each year unless prior approval to change the audit date is obtained, in writing, from RUS.

(1) An auditee may request a change in the audit date by writing to the AA–PARA at least 60 days prior to the currently approved audit date, providing justification for the change.

(2) The time period between the prior audit date and the newly requested audit date must be no longer than twenty-three months.

(3) Comparative financial statements must be prepared and audited for the 12 months ending as of the new audit date and for the 12 months immediately preceding that period.

§ 1773.9 Disclosure of fraud, and noncompliance with provisions of laws, regulations, contracts, and loan and grant agreements.

(a) In accordance with GAGAS, the auditor is responsible for planning and performing the audit to provide reasonable assurance about whether the financial statements are free of material misstatement due to error or fraud. The auditor must also plan the audit to provide reasonable assurance of detecting material misstatements resulting from violations of provisions of laws, regulations, contracts or loan and grant agreements that could have a direct and material effect on the financial statements.

(b) If specific information comes to the auditor's attention that provides evidence concerning the existence of possible violations of provisions of laws, regulations, contracts or loan and grant agreements that could have a material indirect effect on the financial statements, the auditor should apply audit procedures specifically directed to ascertaining whether a violation of provisions of laws, regulations, contract or grant agreements has occurred.

(c) Pursuant to the terms of its audit engagement letter with the auditee, the auditor must immediately report, in writing, all instances of fraud, illegal acts, and all indications or instances of noncompliance with laws, whether material or not, to:

(1) The president of the auditee's governance board;

(2) AA–PARA; and

(3) OIG, as follows:

(i) For all audits performed in accordance with §1773.3(d) (audits conducted in accordance with 2 CFR part 200 "Uniform Administrative Requirements, Cost Principles, and Audit Requirements for Federal Awards"), report to the USDA–OIG-Audit, National Single Audit Coordinator for USDA, 401 W. Peachtree St NW, Room 2328, Atlanta, GA 30308,

(ii) For all other audits conducted in accordance with §1773.3 report to the appropriate office based on location. See *https://www.usda.gov/oig/national.htm* to determine the correct reporting location.

§1773.10 Access to audit documentation.

Pursuant to the terms of this part and the audit engagement letter, the auditor must make all audit documentation available to RUS, or its designated representative, upon request and must permit RUS, or its designated representative, to photocopy all audit documentation.

§§1773.11–1773.19 [Reserved]

Subpart C—RUS Requirements for the Submission and Review of the Reporting Package

§1773.20 The auditor's submission of the reporting package.

(a) *Time limit.* Within 90 days of the audit date, the auditor must deliver the reporting package to the auditee's governance board. At a minimum, copies should be provided for each member of the governance board and the manager. The auditor must also provide an electronic copy of the audit which meets the requirements of §1773.21 for subsequent transmittal to RUS.

(b) *Communication with the governance board.* In addition to providing sufficient copies of the reporting package for each member of the auditee's governance board, RUS requires that the auditor report all audit findings to the auditee's governance board. RUS recommends that audit findings also be communicated orally unless oral communication would not be adequate. If the information is communicated orally, the auditor must document the communication by appropriate memoranda or notations in the audit documentation. If the auditor communicates in writing, a copy of the written communication must be included in the auditor's audit documentation.

(c) *Matters to be communicated.* Matters communicated to those charged with governance must include, but are not limited to the matters to be communicated as prescribed in the AICPA's professional standards AU–C Section 260, "The Auditor's Communication with Those Charged with Governance".

§1773.21 Auditee's review and submission of the reporting package.

(a) The auditee's governance board should note and record receipt of the reporting package and any action taken in response to the reporting package in the minutes of the board meeting at which such reporting package is presented.

(b) The auditee must furnish RUS with an electronic copy of the reporting package within 120 days of the audit date as provided for in §1773.3.

(c) The auditee must furnish AA-PARA with a copy of its plan for corrective action, if any, within 180 days of the audit date.

(d) The auditee must include in the reporting package a copy of each special report, summary of recommendations or similar communications, if any, received from the auditor as a result of the audit.

(e) All required submissions to RUS described in paragraphs (b) through (d) of this section should be furnished electronically. The electronic copy must be provided in a Portable Document Format (PDF). Auditees with a designation from 0001 through 0199 in the Electric program and 500 through 699 in the Telecommunications programs shall upload the reporting package to the DCS or its successor system. Borrowers and/or grantees with a designation from 1100 through 1199, 1300 through 1399, and 1400 through 1499 in the Broadband program shall upload the reporting package to the BCAS or its successor system. All other borrowers and/or grantees may upload their reporting package through DCS or its successor system. Specific instructions for submission are available from the

Technical Accounting and Auditing Staff.

§§ 1773.22–1773.29 [Reserved]

Subpart D—RUS Reporting Requirements

§ 1773.30 [Reserved]

§ 1773.31 Auditor's report on the financial statements.

The auditor must prepare a written report on comparative balance sheets, statements of revenue and patronage capital (or statement of operations customary to the type of entity reporting) and statements of cash flows. The report must include the manual or printed signature of the auditor, cover all statements presented, and refer to the separate report on internal controls over financial reporting and on compliance and other matters and the report on compliance with aspects of contractual agreements and regulatory requirements issued in conjunction with the auditor's report on the financial statements. The auditor's report on the financial statements should also state that the report on internal controls over financial reporting and on compliance and other matters is an integral part of a GAGAS audit, and in considering the results of the audit, that this report should be read along with the auditor's report on the financial statements.

§ 1773.32 Report on internal control over financial reporting and on compliance and other matters.

(a) As required by GAGAS, the auditor must prepare a written report describing the scope of the auditor's testing of internal control over financial reporting and of compliance with provisions of laws, regulations, contracts, and loan and grant agreements, and that the tests provided sufficient, appropriate evidence to support opinions on the effectiveness of internal control and on compliance with provisions of laws, regulations, contracts, and loan and grant agreements. This report must include the manual or printed signature of the auditor and must include the following items as appropriate:

(1) Significant deficiencies and material weaknesses in internal control;

(2) Instances of fraud and noncompliance with provisions of laws or regulations that have a material effect on the audit and any other instances that warrant the attention of those charged with governance;

(3) Noncompliance with provisions of contracts or grant agreements that have a material effect on the audit; and

(4) Abuse that has a material effect on the audit.

(b) When the auditor detects instances of noncompliance or abuse that have an effect on the financial statements that are less than material but warrant the attention of those charged with governance, they should communicate those findings in writing to those charged with governance in a separate communication. If the auditor has issued a separate communication detailing immaterial instances of noncompliance or abuse, the report on internal controls over financial reporting and on compliance and other matters must be modified to include a statement such as:

"We noted certain immaterial instances of noncompliance [and/or abuse], which we have reported to the management of (auditee's name) in a separate letter dated (month, day, 20XX)."

(c) If the auditor has issued a separate letter to management to communicate other matters involving the design and operation of the internal control over financial reporting, the report on internal controls over financial reporting and on compliance and other matters must be modified to include a statement such as:

"However, we noted other matters involving the internal control over financial reporting that we have reported to the management of (auditee's name) in a separate letter dated (month, day, 20XX)."

(d) The report must contain the status of known but uncorrected deficiencies from prior audits that affect the current audit objective.

§ 1773.33 Report on compliance with aspects of contractual agreements and regulatory requirements.

The auditor must prepare a report on compliance with aspects of contractual agreements and regulatory requirements that includes, at a minimum, comments on:

(a) *Audit procedures.* State whether the audit has been performed in accordance with this part;

(b) *Special reports.* State whether any special reports, summaries of recommendations, or similar communications were furnished to the auditee's management during the course of the audit or during interim audit work, and provide a description of the information furnished;

(c) *Accounting and records.* Comment on whether, during the course of the audit, anything came to the auditor's attention to indicate that the auditee did not maintain adequate and effective accounting procedures and records and utilize adequate and fair methods for accumulating and recording labor, material, and overhead costs, and for distributing these costs to construction, retirement, and maintenance or other expense accounts. Where appropriate, comment on whether anything came to the auditor's attention to indicate that the auditee did not:

(1) Establish continuing property records (CPRs) that are updated on a current basis, at least annually, and are reconciled with the controlling general ledger plant accounts;

(2) Promptly clear construction clearing accounts of costs of completed construction to the proper classified plant accounts and accrue depreciation on such completed construction from the date the plant was placed in service;

(3) Currently and systematically record and properly price retirements of plant;

(4) Properly account for the accumulated provision for depreciation accounts associated with retirements of plant or properly disclose any unusual charges or credits to such accounts; and

(5) Obtain RUS approval for the sale, lease or transfer of capital assets secured under the RUS security agreement when approval is required, and

properly handle any proceeds from the sale or lease of plant, material or scrap in conformance with RUS requirements.

(d) *Materials control.* Comment on whether, during the course of the audit, anything came to the auditor's attention to indicate that the control over materials and supplies was not adequate.

(e) *Compliance with RUS loan and security instrument provisions.* Comment on whether, during the course of the audit, anything came to the auditor's attention to indicate that the following provisions of RUS' loan and security instruments have not been complied with:

(1) For electric auditees, provisions related to:

(i) The requirements for an auditee to obtain written approval of mortgagees to enter into any contract for the management, operation, or maintenance of the auditee's system if the contract covers all or substantially all of the electric system. For purposes of this part, the following contracts shall be deemed as requiring RUS approval:

(A) Management contracts in which the auditee has contracted to have another auditee or other entity manage its affairs;

(B) Operations and maintenance contracts in which the auditee has contracted to have another auditee or other entity operate and/or maintain all or substantially all of the physical plant facilities of the auditee.

(C) Operations and maintenance contracts in which the auditee has contracted to operate and maintain the physical plant facilities of another auditee or other utility system;

(ii) The requirement for an auditee to prepare and furnish mortgagees annual or periodic financial and operating reports on the auditee's financial condition and operations accurately and within the required deadlines. The auditor shall comment on whether, during the course of the audit, anything came to the auditor's attention to indicate that the information represented by the auditee as having submitted to RUS in its most recent December 31 Financial and Operating Report Electric Distribution or Financial and Operating Report Electric

Power Supply was not in agreement with the auditee's audited records. If the auditee represents that an amended report has been filed as of December 31, the comments must relate to the amended report; and

(iii) The requirement for an auditee to use depreciation rates that are within the ranges established by RUS for each primary plant account (See RUS Bulletin 183–1, Depreciation Rates and Procedures *at https://www.rd.usda.gov/files/UPA_Bulletin_183-1.pdf*), or with the requirements of the state regulatory body having jurisdiction over the auditee's depreciation rates in computing monthly accruals.

(2) For telecommunications auditees, provisions related to:

(i) The requirement for an auditee to obtain written approval of the mortgagees to enter into any contract, agreement or lease between the auditee and an affiliate other than as allowed under 7 CFR part 1744, subpart E; and

(ii) The requirement for an auditee to prepare and furnish mortgagees annual or periodic financial and operating reports on the auditee's financial condition and operations accurately and within the required deadlines. The auditor shall comment on whether, during the course of the audit, anything came to the auditor's attention to indicate that the information represented by the auditee as having been submitted to RUS in its most recent December 31 Operating Report for Telecommunications Borrowers was not in agreement with the auditee's audited records. If the auditee represents that an amended report has been filed as of December 31, the comments must be related to the amended report.

(3) For Broadband auditees, provisions relating to the requirement for an auditee to prepare and furnish mortgagee quarterly or periodic financial and operating reports on the auditee's financial condition and operations accurately and within the required deadlines. The auditor shall comment on whether, during the course of the audit, anything came to the auditor's attention to indicate that the information represented by the auditee as having been submitted to RUS in its most recent BCAS filing was not in agreement with the auditee's audited

records. If the auditee represents that an amended report has been filed, the comments must be related to the amended report.

(4) For grantees, provisions related to:

(i) Recipients of Broadband Initiatives Program loans and grants, the requirement for the recipient to prepare and furnish RUS quarterly and annual financial and operating reports on the financial condition and operations of the auditee accurately and within the required deadlines. The auditor shall comment on whether, during the course of the audit, anything came to the auditor's attention to indicate that the information represented by the auditee as having been submitted to RUS in its most recent BCAS filing was not in agreement with the audited records of the auditee. If the auditee represents that an amended report has been filed, the comments must relate to the amended report. The auditor must state whether the Annual Compliance Certificate required by the RUS Security Agreement has been filed in a timely manner with RUS.

(ii) Recipients of all other grant programs within the electric and telecommunications programs, the requirements to prepare and furnish RUS with any required financial reporting accurately and within required deadlines, as appropriate for that specific program. The auditor shall comment on whether, during the course of the audit, anything came to the auditor's attention to indicate that the information represented by the grantee as having been submitted to RUS in its most recent filing was not in agreement with the audited records of the grantee. If the grantee represents that an amended report has been filed, the comments must relate to the amended report.

(f) *Related party transactions.* Comment on whether, during the course of the audit, anything came to the attention of the auditor to indicate that all material related party transactions have not been disclosed in the notes to the financial statements in accordance with ASC 850, entitled "Related Party Disclosures".

(g) *Deferred debits and deferred credits.* For electric auditees, comment on whether, during the course of the audit anything came to the attention of the auditor to indicate that the auditee provided detailed schedule of deferred debits and deferred credits, including, but not limited to, margin stabilization plans, revenue deferral plans, and expense deferrals is not accurately presented. This schedule must be included as supplemental information or within the notes to the financial statements; and

(h) *Investments*. For electric and telecommunications auditees, comment on whether, during the course of the audit, anything came to the auditor's attention to indicate that the auditee provided detailed schedule of investments is not accurately presented. This schedule must be included as supplemental information or within the notes to the financial statements. The auditor must state that the audit did not disclose any investments in subsidiary or affiliated companies.

§1773.34 Schedule of findings and recommendations.

The auditor must prepare a schedule of findings and recommendations to be included with the audited financial statements. The schedule of findings and recommendations shall be developed and presented utilizing the elements of a finding discussed in GAGAS and shall include recommendations for remediation. If the schedule does not include responses from management, as well as any planned corrective actions, those items must be submitted directly to the AA–PARA by management in accordance with §1773.4(j).

§§1773.35–1773.37 [Reserved]

Subpart E—RUS Audit Requirements and Documentation

§1773.38 Scope of engagement.

The audit requirements set forth in §1773.39 through 1773.45 must be met annually by the auditor during the audit of the RUS auditee's financial statements. The auditor must exercise professional judgment in determining whether any auditing procedures in addition to those mandated by GAGAS or this part should be performed on the auditee's financial records in order to afford a reasonable basis for rendering the auditor's report on the financial statements, report on internal controls over financial reporting and on compliance and other matters, report on compliance with aspects of contractual agreements and regulatory requirements, and schedule of findings and recommendations.

§1773.39 Utility plant and accumulated depreciation.

(a) *General*. The audit of these accounts shall include tests of additions, replacements, retirements, and changes. The auditor's audit documentation shall support that the auditor:

(1) Examined direct labor and material transactions to determine whether the auditee's accounting records reflect a complete accumulation of costs;

(2) Examined indirect costs and overhead charges to determine if they conform to the Uniform System of Accounts or the Federal Acquisitions Regulations as required under the RUS Security Agreement;

(3) Reviewed the costs of completed construction and retirement projects to determine if they were cleared promptly from the work in progress accounts to the classified plant in service accounts and the related depreciation accounts;

(4) Examined direct purchases of special equipment and general plant;

(5) Determined the degree of accuracy and control of costing retirements, including tests of salvage and removal costs;

(6) Reviewed the auditee's work order procedures; and

(7) Reviewed depreciation rates for adequate support, and compared them to RUS guidelines to determine that they were in compliance.

(b) *Construction work in progress*. (1) The audit documentation shall include a summary of open work orders reconciled to the general ledger and note on the summary any unusual or atypical projects.

(2) The auditor's audit documentation shall support that the auditor:

(i) Reviewed equipment purchases charged to work orders, including payments and receiving reports;

(ii) Reviewed contracts showing the scope of the work, the nature of the contract, the contract amount, and scheduled payments and reviewed supporting documents to determine that services contracted for were in fact rendered;

(iii) Reviewed time cards and pay rates for a sample of employees who allocate their time to work orders;

(iv) Reviewed the nature of material and supplies issued to the project, traced amounts and quantities to supporting documents, and reviewed the reasonableness of clearing rates for assignment of stores expense to the work order;

(v) Reviewed the accuracy of the computation of overheads applied to the work order; and

(vi) Reviewed other costs charged to the work order for support and propriety.

(3) The auditor's audit documentation shall support that the auditor:

(i) Scheduled payments to contractors and traced to verify payments and supporting invoices;

(ii) Traced contract costs to final closeout documents, to the general ledger, and to the continuing property records; and

(iii) Verified the costs of owner furnished materials, if applicable.

(4) The auditor shall review the auditee's procedures for unitization and classification of work order and contract costs. The auditor's audit documentation shall support that the auditor:

(i) Reviewed the tabulation of record units for construction from the work order staking sheets to the tabulation of record units, to the unitization sheets, and to the continuing property records;

(ii) Reviewed the procedures for unitizing and distributing costs of completed construction to the plant accounts;

(iii) Verified that standard costs were being used;

(iv) Evaluated the basis for development of standard costs; and

(v) Determined that costs of completed construction were cleared promptly from work in progress accounts.

(c) *Continuing property records.* The auditor's audit documentation shall support that the auditor:

(1) Determined whether the subsidiary plant records agree with the controlling general ledger plant accounts;

(2) Noted differences in the audit documentation; and

(3) Commented, in the report on compliance with aspects of contractual agreements and regulatory requirements, on any discrepancies.

(d) *Retirement work-in-progress.* The auditor's audit documentation shall support that the auditor:

(1) Determined that plant retirements are currently and systematically recorded and priced on the basis of the continuing property records, and determined that costs of removal have been properly accounted for;

(2) Explained the method used in computing the cost of units of plant retired if continuing property records have not been established and determined whether costs appeared reasonable; and

(3) Determined the manner in which net losses due to retirements were accounted for and traced clearing entries to the depreciation reserve, the plant accounts, and the continuing property records.

(e) *Provision for accumulated depreciation.* The auditor's audit documentation shall support that the auditor:

(1) Verified the depreciation accruals for the period, including the depreciation base;

(2) Reviewed the basis of the depreciation rates, any change in rates and the reason for the change, and, if appropriate, determined whether the rates are in compliance with RUS requirements or with the requirements of the state regulatory body having jurisdiction over the auditee's depreciation rates;

(3) Reviewed salvage and removal costs; and

(4) Searched for unrecorded retirements.

(f) *Other reserves.* The auditor's audit documentation shall include an account analysis for all other material plant reserves, such as the reserve for the amortization of plant acquisition adjustments. The auditor's audit documentation shall support that appropriate tests of transactions were performed.

(g) *Narrative.* The auditor shall include in the audit documentation a comprehensive narrative on the scope of work performed, observations made, and conclusions reached. Matters covered in this narrative shall include:

(1) The nature of construction and other additions;

(2) The control over, and the accuracy of pricing retirements;

(3) The accuracy of distributing costs to classified utility plant accounts;

(4) An evaluation of the method of:

(i) Capitalizing the direct loadings on labor and material costs;

(ii) Distributing transportation costs and other expense clearing accounts; and

(iii) Capitalizing overhead costs;

(5) The tests of depreciation;

(6) A review of agreements such as those relating to acquisitions, property sales, and leases which affect the plant accounts; and

(7) Notations, if applicable, of RUS approval of property sales and the propriety of the disposition of the proceeds.

§1773.40 Regulatory assets.

The auditor's audit documentation shall support that the auditor tested whether all regulatory assets comply with the requirements of ASC 980. For Electric auditees only, the auditor's audit documentation shall support that all regulatory assets have received RUS approval.

§1773.41 Extraordinary retirement losses.

The auditor's audit documentation shall support that the auditor tested retirement losses, including any required approval by a regulatory commission with jurisdiction in the matter, or RUS, in the absence of commission jurisdiction.

§1773.42 Clearing accounts.

The auditor's audit documentation shall support that the auditor tested all clearing accounts and that transactions selected for testing were reviewed for proper allocation between expense and capital accounts.

§1773.43 Capital and equity accounts.

(a) *Capital stock.* For privately owned companies, the audit documentation shall include analyses of all stock transactions during the audit period. The auditor's audit documentation shall support that the auditor:

(1) Reviewed the subsidiary records and reconciled them to the general ledger control account;

(2) Reviewed authorizations and issuances or redemptions of capital stock for proper approvals by the governance board, stockholders, regulatory commissions and RUS, as required;

(3) Determined that transactions were made in accordance with the appropriate provisions of the articles of incorporation, bylaws, and RUS loan documents; and

(4) Determined that transactions were recorded in accordance with the Uniform System of Accounts.

(b) *Memberships.* For cooperative organizations, the audit documentation shall include an analysis of the membership transactions during the audit period. The auditor's audit documentation shall support that the auditor:

(1) Reviewed the subsidiary records and reconciled them to the general ledger control account; and

(2) Determined that transactions were made in accordance with the appropriate provisions of the articles of incorporation, bylaws, and RUS loan documents.

(c) *Patronage capital, retained earnings, margins, and other equities.* The audit documentation shall include an analysis of the patronage capital, retained earnings, margins and other equities, and any related reserve accounts. The auditor's audit documentation shall support that the auditor:

(1) Determined that the transactions were made in accordance with the appropriate provisions of the articles of incorporation, bylaws, RUS loan documents, Uniform System of Accounts, or orders of regulatory commissions;

(2) Traced payments to underlying support; and

(3) Determined whether, under the terms of the RUS security instrument, restrictions of retained earnings or margins are required and, if so, whether they have been properly recorded.

§ 1773.44 Long-term debt.

The auditor's audit documentation shall support that the auditor:

(a) Confirmed RUS, FFB, and RTB debt to the appropriate confirmation schedule (RUS Form 690, Confirmation Schedule Obligation to the FFB; Form 614, Confirmation Schedule—Long-term Obligation to RUS; or, Confirmation Schedule for RTB Debt);

(b) Confirmed other long-term debt directly with the lender;

(c) Examined notes executed or cancelled during the audit period; and

(d) Tested accrued interest computations.

§ 1773.45 Regulatory liabilities.

The auditor's audit documentation shall support that all regulatory liabilities comply with the requirements of ASC 980. For electric auditees only, the auditor's audit documentation shall document whether all regulatory liabilities have received RUS approval.

§§ 1773.46–1773.48 [Reserved]

§ 1773.49 OMB Control Number.

The information collection requirements in this part are approved by the Office of Management and Budget (OMB) and assigned the OMB Control Number 0572–0095.

PART 1774—SPECIAL EVALUATION ASSISTANCE FOR RURAL COMMUNITIES AND HOUSEHOLDS PROGRAM (SEARCH)

Subpart A—General Provisions

Subpart B—Grant Application Processing

AUTHORITY: 7 U.S.C. 1926(a)(2)(C).

SOURCE: 75 FR 35963, June 24, 2010, unless otherwise noted.

Subpart A—General Provisions

§ 1774.1 General.

The purpose of the Special Evaluation Assistance for Rural Communities and Household (SEARCH) Grant program is to provide financial assistance to the neediest, eligible communities, who lack financial resources to pay for feasibility studies, design assistance and technical assistance. This subpart sets forth the general policies and procedures for making and processing predevelopment planning SEARCH grants for water and waste projects.

§ 1774.2 Definitions.

The following definitions apply to subparts A and B of this part.

Agency. The Rural Utilities Service of the United States Department of Agriculture (USDA) within the Rural Development mission area of the Under Secretary for Rural Development. The Processing Official will administer this water and waste program on behalf of the Rural Utilities Service.

Approval official. The Agency official at the State level who has been delegated the authority to approve grants.

ConAct. Consolidated Farm and Rural Development Act (7 U.S.C. 1926(a)(2)).

Design assistance. Preliminary design and engineering analysis necessary for an application for funding. Design assistance does not include financial assistance for development of plans, specifications, or bidding documents.

DUNS Number. Data Universal Numbering System number obtained from Dun and Bradstreet and used when applying for Federal grants or cooperative agreements. A DUNS number may be obtained at no cost, by calling 1–866–705–5711.

Eligible entity. Entity that meets eligibility requirements to obtain a loan, loan guarantee or grant under Paragraphs 1, 2 or 24 of Section 306(a) of the ConAct (codified at 7 U.S.C. Section 1926(a)(1)(2) and (24)).

Feasibility study. Documentation associated with an objective analysis of project-related technical engineering or environmental impact analyses required to support applications for funding water or waste disposal projects through USDA, Rural Utilities Service or other agencies.

Financially distressed area. An area is considered financially distressed if the median household income of the area to be served is either below the poverty line or below. 80 percent of the statewide non-metropolitan median household income according to the 5-year income data from the American Community Survey (ACS) or, if needed, other Census Bureau data. If there is reason to believe that the ACS or other Census Bureau data does not accurately represent the median household income of the area to be served, the reasons will be documented and the borrower may furnish, or RD may obtain, additional information regarding such median household income data. Information must consist of reliable data from local, regional, State or Federal sources or from a survey conducted by a reliable impartial source.

Grantee. The applicant receiving financial assistance directly from the RUS to carry out the project or program under this program.

Poverty line. The level of income for a family of four, as defined in section 673(2) of the Community Services Block Grant Act (42 U.S.C. 9902(2)).

Processing Official. The Agency official designated by the approval official as having the authority to accept and process applications for water and waste disposal assistance.

Rural area. For the purposes of this SEARCH program, any communities in a city, town, or unincorporated area with populations of 2,500 or fewer inhabitants, according to the most recent decennial Census of the United States (decennial Census).

State. Any of the 50 States, the District of Columbia, the Commonwealth of Puerto Rico, the Territory of Guam, the Commonwealth of the Northern Mariana Islands, the Republic of the Marshall Islands, the Federated States of Micronesia, the Republic of Palau, and the U.S. Virgin Islands.

Technical Assistance. Supervision, oversight, or training by an organization for the development of an application for financial assistance.

[75 FR 35963, June 24, 2010, as amended at 80 FR 9862, Feb. 24, 2015; 84 FR 3669, Feb. 13, 2019]

§1774.3 Availability of forms and regulations.

Information about the forms, instructions, regulations, bulletins, OMB Circulars, Treasury Circulars, standards, documents and publications cited in this part is available from any UDSA/Rural Development Office or the United States Department of Agriculture, Washington, DC 20250–1500 and at *http://www.grants.gov.*

§1774.4 Allocation of funds.

The Secretary may use not more than four percent of the total amount of funds made available for a fiscal year for water and waste disposal activities for SEARCH grants.

§§1774.5–1774.6 [Reserved]

§1774.7 Environmental requirements.

Grants made under this part must comply with the environmental review requirements in accordance with 7 CFR part 1970.

§1774.8 Other Federal Statutes.

Other Federal statutes and regulations are applicable to grants awarded under this part. These include but are not limited to:

(a) 7 CFR part 1, subpart A—USDA implementation of Freedom of Information Act.

(b) 7 CFR part 3—USDA implementation of OMB Circular No. A–129 regarding debt collection.

(c) 7 CFR part 15, subpart A—USDA implementation of Title VI of the Civil Rights Act of 1964, as amended.

(d) 7 CFR part 1970.

(e) 7 CFR part 1901, subpart E—Civil Rights Compliance Requirements.

(f) 2 CFR part 200, as adopted by USDA through 2 CFR part 400, Uniform

Administrative Requirements, Cost Principles, and Audit Requirements for Federal.

(g) 2 CFR part 415—General Program Administrative Requirements.

(h) 2 CFR part 180, as adopted by USDA through 2 CFR part 417, Nonprocurement Debarment and Suspension, implementation of Executive Order 12549 on debarment and suspension.

(i) 2 CFR part 418, New Restrictions on Lobbying, prohibiting the use of appropriated funds to influence Congress or a Federal agency in connection with the making of any Federal grant and other Federal contracting and financial transactions.

(j) 2 CFR part 421, Requirements for Drug-Free Workplace (Financial Assistance), implementing the Drug-Free Workplace Act of 1988 (41 U.S.C 8102).

(k) 7 CFR part 15b, USDA implementation of section 504 of the Rehabilitation Act of 1973 (29 U.S.C. 794), as amended, prohibiting discrimination on the basis of physical or mental handicap in Federally assisted programs.

[75 FR 35963, June 24, 2010, as amended at 79 FR 76005, Dec. 19, 2014; 81 FR 7696, Feb. 16, 2016; 81 FR 11028, Mar. 2, 2016]

§ 1774.9 [Reserved]

Subpart B—Grant Application Processing

§ 1774.10 Applications.

(a) To file an application, an organization must provide their DUNS number. An organization may obtain a DUNS number from Dun and Bradstreet by calling (1–866–705–5711). To file a complete application, the following information should be submitted:

(1) Standard Form 424, "Application for Federal Assistance (For Non-Construction)."

(2) Standard Form 424A & B, "Budget Information—Non-Construction Programs."

(3) Supporting documentation necessary to make an eligibility determination such as financial statements, audits, organizational documents, or existing debt instruments. The Processing Official will advise applicants regarding the required documents. Ap-

plicants that are indebted to RUS will not need to submit documents already on file with the Processing Official as long as such documents are current and valid.

(4) Project narrative detailing the project to be financed with the SEARCH grant funds. The narrative will also provide details on the activities or tasks to be accomplished, objectives, timetables for task completion, and anticipated results.

(5) The applicant's Internal Revenue Service Taxpayer Identification Number (TIN).

(6) Other Forms and certifications. Applicants will be required to submit the following items to the Processing Official, upon notification from the Processing Official to proceed with further development of the full application:

(i) Form RD 442–7, "Operating Budget";

(ii) Form RD 400–1, "Equal Opportunity Agreement";

(iii) Form RD 400–4, "Assurance Agreement";

(iv) Form AD–1047, "Certification Regarding Debarment, Suspension and other Responsibility Matters";

(v) Form AD–1049, Certification regarding Drug-Free Workplace Requirements (Grants) Alternative I For Grantees Other Than Individuals;

(vi) Certifications for Contracts, Grants, and Loans (Regarding Lobbying); and

(vii) Certification regarding prohibited tying arrangements. Applicants that provide electric service must provide the Agency a certification that they will not require users of a water or waste facility financed under this part to accept electric service as a condition of receiving assistance.

(b) Applicants are encouraged to contact the State Office or the Processing Official to find out how to file electronically. The application and supporting documentation must be sent or delivered to the Processing Official, unless it is filed electronically.

§ 1774.11 [Reserved]

§ 1774.12 Eligibility.

The following eligibility requirements must be met:

(a) The applicant must be:

(1) A public body, such as a municipality, county, district, authority, or other political subdivision or a State, territory or commonwealth, or

(2) An organization operated on a not-for-profit basis, such as an association, cooperative, or private corporation. The organization must be an association controlled by a local public body or bodies, or have a broadly based ownership by or membership of people of the local community, or

(3) Indian Tribes on Federal and State reservations and other Federally recognized Indian Tribes.

(b) The area to be served must be financially distressed and rural as defined in §1774.2 of this part.

§1774.13 Limitations.

Grant funds may not be used to:

(a) Fund political or lobbying activities.

(b) Pay for work already completed.

(c) Purchase real estate or vehicles, improve or renovate office space, or repair and maintain privately owned property.

(d) Construct or furnish a building.

(e) Intervene in the Federal regulatory or adjudicatory proceedings.

(f) Sue the Federal Government or any other government entities.

(g) Pay for any other costs that are not allowable under 2 CFR part 200, as adopted by USDA through 2 CFR part 400.

(h) Make contributions or donations to others.

(i) Fund projects that duplicate technical assistance given to implement action plans under the National Forest-Dependent Rural Communities Economic Diversification Act of 1990 (7 U.S.C. 6613). Applicants cannot receive both grants made under this part and grants that the Forest Service makes to implement the action plans for five continuous years from the date of grant approval by the Forest Service.

(j) To pay an outstanding judgment obtained by the United States in a Federal Court (other than in the United States Tax Court), which has been recorded. An applicant will be ineligible to receive a loan or grant until the judgment is paid in full or otherwise satisfied.

[75 FR 35963, June 24, 2010, as amended at 79 FR 76005, Dec. 19, 2014]

§1774.14 Eligible grant purposes.

(a) Eligible predevelopment planning costs are feasibility studies, preliminary design assistance, and technical assistance as each is defined in §1774.2. The eligible predevelopment activities funded with these grant funds must be agreed to and accepted by the Agency prior to the disbursement of the SEARCH grant. The predevelopment planning costs must be related to a proposed project that meets the following requirements:

(1) To construct, enlarge, extend, or otherwise improve rural water, sanitary sewage, solid waste disposal, and storm wastewater disposal facilities.

(2) To construct or relocate public buildings, roads, bridges, fences, or utilities, and to make other public improvements necessary for the successful operation or protection of facilities authorized in paragraph (a)(1) of this section.

(3) To relocate private buildings, roads, bridges, fences, or utilities, and other private improvements necessary for the successful operation or protection of facilities authorized in paragraph (a)(1) of this section.

(b) The Secretary, subject to the limitation in §1774.4 of this part, may fund up to 100 percent of the eligible grant costs, not to exceed $30,000.

§1774.15 Selection Criteria.

Projects will be selected based primarily on the funding priorities in 7 CFR 1780.17. The Program Official discretionary points stated in 7 CFR1780.17 (e) can also include consideration of the following criteria:

(a) Systems with limited resources.

(b) Smallest systems with lowest incomes.

(c) Funds availability.

§1774.16 Grant application processing and approval.

(a) Before starting to assemble the full application, the applicant should arrange through the Processing Official an application conference to provide a basis for orderly application assembly.

The processing office will explain program requirements, public information requirements and provide guidance on preparation of items necessary for final determination.

(b) The Processing Official will determine if the application is properly assembled. If not, the applicant will be notified within fifteen Federal working days as to what additional submittal items are needed.

(c) The Processing Official and Approval Official will coordinate their reviews to ensure that the applicant is advised about eligibility and anticipated fund availability within 45 days of the receipt of a completed application.

(d) The Processing Official will submit the following to the Approval Official:

(1) "Water and Waste Project Information Summary";

(2) Form RD 442-3, "Balance Sheet" or a financial statement or audit that includes a balance sheet;

(3) Letter of Conditions;

(4) Form RD 1942-46, "Letter of Intent to Meet Conditions";

(5) Form RD 1940-1, "Request for Obligation of Funds";

§ 1774.17　Grant closing and disbursement.

(a) *Grant closing.* RUS Bulletin 1780-12 "Water or Waste System Grant Agreement" will be completed and executed in accordance with the requirements of grant approval. The grant will be considered closed when RUS Bulletin 1780-12 has been properly executed. Processing officials or Approval officials are authorized to sign the grant agreement on behalf of RUS.

(b) *Grant disbursements.* Agency policy is not to disburse grant funds from the Treasury until they are actually needed by the applicant. If an approved grant includes applicant or other contributions, then these funds will be disbursed before the disbursal of any Agency grant funds.

(c) *Payment for project costs.* Project costs will be monitored by the RUS processing office. Invoices will be approved by the borrower and submitted to the Processing Official for concurrence. The review and acceptance of project costs by the Agency does not attest to the correctness of the amounts, the quantities shown or that the work has been performed under the terms of the agreements or contracts.

(d) *Use of remaining funds.* Funds remaining after all costs incident to the basic project have been paid or provided for will not include applicant contributions if SEARCH grants funds are financing less than 100 percent of the project. Funds remaining may be considered in direct proportion to the amounts obtained from each source. Remaining funds will be handled as follows:

(1) Remaining funds may be used for eligible grant purposes as described in 1774.14 of this subpart, or

(2) Grant funds not expended will be canceled. Prior to the actual cancellation, the borrower, its attorney and its engineer will be notified of RUS' intent to cancel the remaining funds.

§ 1774.18　Reporting requirements, accounting methods and audits.

All Agency grantees will follow the reporting requirements as outlined in 7 CFR 1780.47.

§ 1774.19　Applications determined ineligible.

If at any time an application is determined ineligible, the processing office will notify the applicant in writing of the reasons. The notification to the applicant will state that an appeal of this decision may be made by the applicant under 7 CFR part 11.

§ 1774.20　Conflict of Interest.

Any processing or servicing activity conducted pursuant to this part involving authorized assistance to Rural Development employees with Water and Environmental Programs responsibility, members of their families, known close relatives, or business or close personal associates, is subject to the provisions of subpart D of part 1900 of this title. Applicants of this assistance are required to identify any known relationship or association with an RUS employee.

§§ 1774.21–1774.23 [Reserved]

§ 1774.24 Exception authority.

The Administrator may, in individual cases, make an exception to any requirement or provision of this part which is not inconsistent with the authorizing statute or other applicable law and is determined to be in the Government's interest. Requests for exceptions must be made in writing by the State Director and supported with documentation to explain the adverse effect on the Government's interest, propose alternative course(s) of action, and show how the adverse affect will be eliminated or minimized if the exception is granted. The exception decision will be documented in writing, signed by the Administrator, and retained in the files.

§§ 1774.25–1774.99 [Reserved]

§ 1774.100 OMB Control Number.

The information collection requirements in this part will not be effective until approved by the Office of Management and Budget (OMB), subject to the submission of a paperwork package to OMB and assigned an OMB Control Number.

PART 1775—TECHNICAL ASSISTANCE GRANTS

Subpart A—General Provisions

AUTHORITY: 5 U.S.C. 301; 7 U.S.C. 1989; 16 U.S.C. 1005.

SOURCE: 70 FR 70878, Jan. 7, 2004, unless otherwise noted.

Subpart A—General Provisions

§ 1775.1 General.

This subpart sets forth the general policies and procedures for the Technical Assistance and Training and the Solid Waste Management Grant Programs. Any processing or servicing activity conducted pursuant to this part involving authorized assistance to Rural Development employees with Water and Environmental Program responsibility, members of their families, known close relatives, or business or close personal associates, is subject to the provisions of subpart D of part 1900 of this title. Applicants for this assistance are required to identify any known relationship or association with an RUS employee.

§ 1775.2 Definitions.

The following definitions apply to subparts A through D of this part.

Association. An entity, including a small city or town, that is eligible for RUS Water and Waste Disposal financial assistance in accordance with 7 CFR 1780.7 (a).

243

Approval official. Any individual with administrative and legal responsibility for Rural Development programs.

DUNS Number. Data Universal Numbering System number obtained from Dun and Bradstreet and used when applying for Federal grants or cooperative agreements. A DUNS number may be obtained at no cost, by calling 1–866–705–5711.

Grant agreement. RUS Guide 1775–1. The agreement outlines the terms and conditions of the grant awards and establishes the guidelines for administering the grant awards.

Grantee. The entity or organization receiving financial assistance directly from the RUS to carry out the project or program under these programs.

Low Income. Median household income (MHI) below 100 percent of the statewide non-metropolitan median household income (SNMHI).

Regional. A multi-State area or any multi-jurisdictional area within a State.

Rural area. Any area not in a city or town with a population in excess of 10,000, according to the most recent decennial Census of the United States. If the applicable population figure cannot be obtained from the most recent decennial Census, RD will determine the applicable population figure based on available population data.

RUS. The Rural Utilities Service, an Agency of the United States Department of Agriculture.

Solid Waste Management. Refers to the operations, maintenance and the recycling of materials disposed of in landfills.

State. Any of the 50 States, the District of Columbia, the Commonwealth of Puerto Rico, the Territory of Guam, the Commonwealth of the Northern Mariana Islands, the Republic of the Marshall Islands, the Federated States of Micronesia, the Republic of Palau, and the U.S. Virgin Islands.

Technical Assistance. Supervision, oversight, or training by an organization for the practical solution of a problem or need of an association as defined in this section.

[70 FR 70878, Jan. 7, 2004, as amended at 80 FR 9862, Feb. 24, 2015]

§ 1775.3 Availability of forms and regulations.

Information about the forms, instructions, regulations, bulletins, OMB Circulars, Treasury Circulars, standards, documents and publications cited in this part is available from any UDSA/Rural Development Office or the Rural Utilities Service, United States Department of Agriculture, Washington, DC 20250–1500.

§ 1775.4 Allocation of funds.

The National Office of the Rural Utilities Service will administer grant funds and will allocate them on a competitive basis.

§ 1775.5 Limitations.

Grant funds may not be used to:

(a) Duplicate current services or replace or substitute support normally provided by other means, such as those performed by an association's consultant in developing a project, including feasibility, design, and cost estimates.

(b) Fund political or lobbying activities.

(c) Purchase real estate or vehicles, improve or renovate office space, or repair and maintain privately owned property.

(d) Pay the costs for construction, improvement, rehabilitation, modification, or operation and maintenance of water, wastewater, and solid waste disposal facilities.

(e) Construct or furnish a building.

(f) Intervene in the Federal regulatory or adjudicatory proceedings.

(g) Sue the Federal Government or any other government entities.

(h) Pay for any other costs that are not allowable under 2 CFR part 200, as adopted by USDA through 2 CFR part 400.

(i) Make contributions or donations to others.

(j) Fund projects that duplicate technical assistance given to implement action plans under the National Forest-Dependent Rural Communities Economic Diversification Act of 1990 (7 U.S.C. 6613). Applicants cannot receive both grants made under this part and grants that the Forest Service makes to implement the action plans for five continuous years from the date of grant approval by the Forest Service.

(1) The Forest Service helps rural communities that are dependent upon national forest resources diversify existing industries and economies. It establishes rural forestry and economic diversification action teams that prepare technical assistance plans for these rural communities to expand their local economies and reduce their dependence on national forest resources. The Forest Service provides assistance to implement the action plans through grants, loans, cooperative agreements, or contracts.

(2) To avoid duplicate assistance, applicants must contact the Forest Service to find out if any geographical areas or local areas in a State have received grants for technical assistance to an economically disadvantaged community. These areas are defined as national forest-dependent communities under 7 U.S.C. 6612. Applicants will provide documentation to the Forest Service and Rural Utilities Service that they have contacted each agency.

(k) To pay an outstanding judgment obtained by the United States in a Federal Court (other than in the United States Tax Court), which has been recorded. An applicant will be ineligible to receive a loan or grant until the judgment is paid in full or otherwise satisfied.

(1) Recruit applications for the RUS's water and waste loan or any other loan or grant program. Grant funds cannot be used to create new business; however, they can be used to assist with application preparation.

[70 FR 70878, Jan. 7, 2004, as amended at 79 FR 76005, Dec. 19, 2014]

§1775.6 Equal opportunity requirements.

The policies and regulations contained in subpart E of part 1901 of this title apply to grants made under this part.

§1775.7 Environmental requirements.

Grants made for the purposes in §§1775.36 and 1775.66 must comply with the environmental review requirements in accordance with 7 CFR part 1970.

[81 FR 11028, Mar. 2, 2016]

§1775.8 Other Federal statutes.

Other Federal statutes and regulations are applicable to grants awarded under this part. These include but are not limited to:

(a) 7 CFR part 1, subpart A—USDA implementation of Freedom of Information Act.

(b) 7 CFR part 3—USDA implementation of OMB Circular No. A–129 regarding debt collection.

(c) 7 CFR part 15, subpart A—USDA implementation of Title VI of the Civil Rights Act of 1964, as amended.

(d) 7 CFR part 1970.

(e) 7 CFR part 1901, subpart E—Civil Rights Compliance Requirements.

(f) 2 CFR part 200, as adopted by USDA through 2 CFR part 400, Uniform Administrative Requirements, Cost Principles, and Audit Requirements for Federal.

(g) 2 CFR part 415—General Program Administrative Requirements.

(h) 2 CFR part 180, as implemented by USDA through 2 CFR part 417, Nonprocurement Debarment and Suspension, implementing Executive Order 12549 on debarment and suspension.

(i) 2 CFR part 418, New Restrictions on Lobbying, prohibiting the use of appropriated funds to influence Congress or a Federal agency in connection with the making of any Federal grant and other Federal contracting and financial transactions.

(j) 2 CFR 421, Requirements for Drug-Free Workplace (Financial Assistance), implementing the Drug-Free Workplace Act of 1988 (41 U.S.C 701).

(k) 2 CFR part 200, subpart F—Audit Requirements.

(l) 29 U.S.C. 794, section 504—Rehabilitation Act of 1973, and 7 CFR part 15B (USDA implementation of statute), prohibiting discrimination based upon physical or mental handicap in Federally assisted programs.

[70 FR 70878, Jan. 7, 2004, as amended at 79 FR 76005, Dec. 19, 2014; 81 FR 7696, Feb. 16, 2016; 81 FR 11028, Mar. 2, 2016]

§1775.9 OMB control number.

The information collection requirements contained in this part have been approved by the Office of Management and Budget and have been assigned OMB control number 0572–0112.

Subpart B—Grant Application Processing

§ 1775.10　Applications.

(a) *Filing period.* Applications may be filed on or after October 1 and must be received by close of business or postmarked by midnight December 31. If an application is received either before October 1 or after December 31, the receiving office will return it to the applicant.

(b) *Where to file.* (1) An applicant will apply to the appropriate State Office of Rural Development if the project will serve a single state.

(2) An applicant will apply to the National Office if the project will serve multiple states. The application must be submitted to the following address: Assistant Administrator, Water and Environmental Programs, Rural Utilities Service, Washington, DC 20250–1570.

(3) Electronic applications will be accepted prior to the filing deadline through the Federal Government's eGrants Web site (Grants.gov) at *http://www.grants.gov.* Applicants should refer to instructions found on the Grants.gov Web site to submit an electronic application. A DUNS number and a Central Contractor Registry (CCR) registration is required prior to electronic submission. The sign-up procedures, required by Grants.gov, may take several business days to complete.

(c) *Application requirements.* To file an application, an organization must provide their DUNS number. An organization may obtain a DUNS number from Dun and Bradstreet by calling (1–866–705–5711). To file a complete application, the following information should be submitted:

(1) Standard Form 424, "Application for Federal Assistance (For Non-Construction)."

(2) Standard Form 424A & B, "Budget Information—Non-Construction Programs."

(3) Form AD–1047, "Certification Regarding Debarment, Suspension, and Other Responsibility Matters—Primary Covered Transaction."

(4) Form AD 1049, "Certification Regarding Drug-Free Workplace Requirements (Grants) Alternative I—For Grantees Other Than Individuals."

(5) Form AD 1048, "Certification Regarding Debarment."

(6) Attachment regarding assistance provided to Rural Development Employees as required by RD Instruction 1900–D.

(7) Form RD 400–4, "Assurance Agreement."

(8) Form RD 400–1, "Equal Opportunity Agreement."

(9) Indirect cost documentation such as cost rate proposals, cost allocation plans, or other election for indirect costs and appropriate certification of indirect costs in accordance with Cost Principles in 2 CFR 200, subpart E, as adopted by USDA through 2 CFR part 400.

(10) Statement of Compliance for Title VI of the Civil Rights Act of 1964.

(11) SF LLL, "Disclosure of Lobbying Activities" (include only if grant is over $100,000).

(12) Certification regarding Forest Service grant.

(d) *Supporting information.* All applications shall be accompanied by:

(1) Evidence of applicant's legal existence and authority in the form of:

(i) Certified copies of current authorizing and organizational documents for new applicants or former grantees where changes were made since the last legal opinion was obtained in conjunction with receipt of an RUS grant, or, certification that no changes have been made in authorizing or organizing documents since receipt of last RUS grant by applicant.

(ii) Current annual corporation report, Certificate of Good Standing, or statement they are not required.

(iii) For public nonprofits, Certificate of Continued Status from local attorney (if applicable).

(iv) Certified list of directors/officers with their respective terms.

(2) Evidence of tax exempt status from the Internal Revenue Service (IRS), if applicable.

(3) Narrative of applicant's experience in providing services similar to those proposed. Provide brief description of successfully completed projects including the need that was identified and objectives accomplished.

(4) Latest financial information to show the applicant's financial capacity

to carry out the proposed work. A current audit report is preferred, however applicants can submit a balance sheet and an income statement in lieu of an audit report.

(5) List of proposed services to be provided.

(6) Estimated breakdown of costs (direct and indirect) including those to be funded by grantee as well as other sources. Sufficient detail should be provided to permit the approval official to determine reasonableness, applicability, and allowability.

(7) Evidence that a Financial Management System is in place or proposed.

(8) Documentation on each of the priority ranking criteria listed in §1775.11 as follows:

(i) List of the associations to be served and the State or States where assistance will be provided. Identify associations by name, or other characteristics such as size, income, location, and provide MHI and population.

(ii) Description of the type of technical assistance and/or training to be provided and the tasks to be contracted.

(iii) Description of how the project will be evaluated and provide clearly stated goals and the method proposed to measure the results that will be obtained.

(iv) Documentation of need for proposed service. Provide detailed explanation of how the proposed services differ from other similar services being provided in the same area.

(v) Personnel on staff or to be contracted to provide the service and their experience with similar projects.

(vi) Statement indicating the number of months it takes to complete the project or service.

(vii) Documentation on cost effectiveness of project. Provide the cost per association to be served or proposed cost of personnel to provide assistance.

(viii) Other factors for consideration, such as emergency situation, training need identified, health or safety problems, geographic distribution, Rural Development Office recommendations, etc.

[70 FR 70878, Jan. 7, 2004, as amended at 79 FR 76005, Dec. 19, 2014]

§1775.11 Priority.

The application and supporting information will be used to determine the applicant's priority for available funds. All applications will be reviewed and scored for funding priority in accordance with RUS Guide 1775–2. Points will be given only for factors that are well documented in the application package and, in the opinion of the RUS, meet the objective outlined under each factor. The following is a listing of the criteria that will be used to select the applications that meet the objectives of the technical assistance program.

(a) Projects proposing to give priority for available services to rural communities having a population less than 5,500 and/or below 2,500.

(b) Projects proposing to give priority for available services to low income communities.

(c) Projects that will provide assistance in a multi-State area.

(d) Points will be awarded for work plans that clearly describe the goals and objectives of the project, how they will be accomplished in targeted communities, and what measurement of accomplishment will be used.

(e) Projects containing needs assessment (i.e. actual issue or problem being addressed) clearly defined and supported by data.

(f) Projects containing evaluation methods that are specific to the activity, clearly defined, measurable, and with projected outcomes.

(g) Applicants proposing to use at least 75 percent of the total grant amount for their own staff, or the staff of an affiliated organization to provide services for a project instead of contracting with an outside organization for the services.

(h) Projects providing technical assistance/training that accomplish the objective within a 12-month or less timeframe.

(i) Projects primarily providing "hands on" technical assistance and training, i.e., on-site assistance as opposed to preparation and distribution of printed material, to communities with existing water and waste systems which are experiencing operation and maintenance or management problems.

(j) Cash or in kind support of project from non-federal sources.

(k) Ability to demonstrate sustainability of project without Federal financial support.

§ 1775.12 Grant processing.

(a) *Single State applications.* (1) Grant applications submitted at the State level will receive a letter acknowledging receipt and confirmation that all information required for a full application was included in the packet. The State will notify the applicant of missing information. The applicant will have 14 business days to respond.

(2) The State Office will review applications for eligibility. Those applicants that are deemed ineligible will be notified. Applicants deemed eligible will be forwarded to the National Office for funding consideration.

(3) The National Office will review all applications received from State Offices. Applications will compete on a priority basis and will be scored and ranked. The applications receiving the highest scores and subject to the availability of funds will be selected for final processing. The National Office will send these applications back to the State Office for processing. The State Office will notify the applicant(s) that they have been selected for funding.

(4) Applicants not selected for funding due to low priority rating shall be notified by the State Office.

(b) *National and multi-State applications.* (1) National and multi-State applications submitted to the National Office will receive a letter acknowledging receipt and confirmation that all information required for a full application was included in the packet. The National Office shall notify the applicant of missing information. The applicant will have 14 business days to respond.

(2) The National Office will review applications for eligibility. Those applications that are deemed ineligible will be notified. Applications deemed eligible will be reviewed and given a rating score. Applications receiving the highest scores will be grouped with those received from State Offices for funding consideration.

(3) The National Office will review all applications received. Applications will compete on a priority basis and will be scored and ranked. The applications receiving the highest scores and subject to the availability of funds will be notified by the National Office that they have been selected for funding. The National Office shall conduct final processing of multi-State and national applications.

(4) Multi-State and National applicants not selected for funding due to low priority rating will be notified by the National Office.

(c) *Low priority applications.* Applications that cannot be funded in the fiscal year received will not be retained for consideration in the following fiscal year and will be handled as outlined in paragraph (a)(4) or (b)(4) of this section.

§ 1775.13 Grant agreement.

Applicants selected for funding will complete a grant agreement, RUS Guide 1775–1, which outlines the terms and conditions of the grant award.

§§ 1775.14–1775.17 [Reserved]

§ 1775.18 Fund disbursement.

Grantees will be reimbursed as follows:

(a) SF–270, "Request for Advance or Reimbursement," will be completed by the grantee and submitted to either the State or National Office not more frequently than monthly.

(b) Upon receipt of a properly completed SF–270, the funds will be requested through the field office terminal system. Ordinarily, payment will be made within 30 days after receipt of a proper request for reimbursement.

(c) Grantees are encouraged to use women- and minority-owned banks (a bank which is owned at least 50 percent by women or minority group members) for the deposit and disbursement of funds.

§ 1775.19 Grant cancellation or major changes.

Any change in the scope of the project, budget adjustments of more than 10 percent of the total budget, or any other significant change in the project must be reported to and approved by the approval official by written amendment to RUS Guide 1775–1.

Any change not approved may be cause for termination of the grant.

§ 1775.20 Reporting.

(a) Grantees shall constantly monitor performance to ensure that time schedules are being met, projected work by time periods is being accomplished, and other performance objectives are being achieved.

(b) SF–425," Federal Financial Report," and a project performance activity report will be required of all grantees on a quarterly basis, due 30 days after the end of each calendar quarter.

(c) A final project performance report will be required with the last SF–425 due 90 days after the end of the last quarter in which the project is completed. The final report may serve as the last quarterly report.

(d) All multi-State grantees are to submit an original of each report to the National Office. Grantees serving only one State are to submit an original of each report to the State Office. The project performance reports should detail, preferably in a narrative format, activities that have transpired for the specific time period and shall include, but not be limited to, the following:

(1) A comparison of actual accomplishments to the objectives established for that period (i.e. number of meetings held, number of people contacted, results of activity);

(2) Analysis of challenges or setbacks that occurred during the grant period;

(3) Copies of fliers, news releases, news articles, announcements and other information used to promote services or projects;

(4) Problems, delays, or adverse conditions which will affect attainment of overall project objectives, prevent meeting time schedules or objectives, or preclude the attainment of particular project work elements during established time periods. This disclosure shall be accompanied by a statement of the action taken or planned to resolve the situation; and

(5) Activities planned for the next reporting period.

[70 FR 70878, Jan. 7, 2004, as amended at 79 FR 76005, Dec. 19, 2014]

§ 1775.21 Audit or financial statements.

The grantee will provide an audit report or financial statements as follows:

(a) Grantees expending $750,000 or more Federal funds per fiscal year will submit an audit conducted in accordance with Subpart F of 2 CFR part 200, as adopted by USDA through 2 CFR part 400. The audit will be submitted with 9 months of the grantee's fiscal year. Additional audits may be required if the project period covers more than one fiscal year.

(b) Grantees expending less than $750,000 will provide annual financial statement covering the grant period, consisting of the organization's statement of income and expense and balance sheet signed by an appropriate official of the organization. Financial statement will be submitted within 90 days after the grantees fiscal year.

[70 FR 70878, Jan. 7, 2004, as amended at 79 FR 76005, Dec. 19, 2014]

§ 1775.22 [Reserved]

§ 1775.23 Grant servicing.

Grants will be serviced in accordance with RUS Guide 1775–1 and subpart E of part 1951 of this title. When grants are terminated for cause, 7 CFR part 11 will be followed.

§ 1775.24 Delegation of authority.

The authority under this part is redelegated to the Assistant Administrator, Water and Environmental Programs, except for the discretionary authority contained in §§ 1775.34 and 1775.68. The Assistant Administrator, Water and Environmental Programs may re-delegate the authority in this part.

§§ 1775.25–1775.30 [Reserved]

Subpart C—Technical Assistance and Training Grants

§ 1775.31 Authorization.

This subpart sets forth additional policies and procedures for making Technical Assistance and Training (TAT) grants authorized under Section 306(a)(14)(A) of the Consolidated Farm and Rural Development Act (CONACT) (7 U.S.C. 1921 *et seq.*, as amended.

§ 1775.32 [Reserved]

§ 1775.33 Objectives.

The objectives of the program are to:

(a) Identify and evaluate solutions to water and waste problems in rural areas.

(b) Assist applicants in preparing applications for water and waste disposal loans/grants.

(c) Assist associations in improving operation and maintenance of existing water and waste facilities in rural areas.

§ 1775.34 Source of funds.

Grants will be made from not less than 1 percent or not more than 3 percent of any appropriations for grants under Section 306(a)(2) of the CONACT. Funds not obligated by September 1 of each fiscal year will be used for water and waste disposal grants made in accordance with part 1780 of this chapter.

§ 1775.35 Eligibility.

(a) Entities eligible for grants must be private nonprofit organizations with tax exempt status, designated by the Internal Revenue Service. A nonprofit organization is defined as any corporation, trust, association, cooperative, or other organization that:

(1) Is operated primarily for scientific, education, service, charitable, or similar purposes in the public interest.

(2) Is not organized primarily for profit.

(3) Uses its net proceeds to maintain, improve, and/or expand its operations.

(b) Entities must be legally established and located within a state as defined in § 1775.2.

(c) Organizations must be incorporated by December 31 of the year the application period occurs to be eligible for funds.

(d) Private businesses, Federal agencies, public bodies, and individuals are ineligible for these grants.

(e) Applicants must also have the proven ability, background, experience (as evidenced by the organization's satisfactory completion of project(s) similar to those proposed), legal authority,

and actual capacity to provide technical assistance and/or training on a regional basis to associations as provided in § 1775.33. To meet the requirement of actual capacity, an applicant must either:

(1) Have the necessary resources to provide technical assistance and/or training to associations in rural areas through its staff, or

(2) Be assisted by an affiliate or member organization which has such background and experience and which agrees, in writing, that it will provide the assistance, or

(3) Contract with a nonaffiliated organization for not more than 49 percent of the grant to provide the proposed assistance.

§ 1775.36 Purpose.

Grants may be made to organizations as defined in § 1775.35 to enable such organizations to assist associations to:

(a) Identify and evaluate solutions to water problems of associations in rural areas relating to source, storage, treatment, and/or distribution.

(b) Identify and evaluate solutions to waste problems of associations in rural areas relating to collection, treatment, and/or disposal.

(c) Prepare water and/or waste disposal loan/grant applications.

(d) Provide technical assistance/ training to association personnel that will improve the management, operation, and maintenance of water and waste facilities.

(e) Pay the expenses associated with providing the technical assistance and/ or training authorized in paragraphs (a) through (d) of this section.

§ 1775.37 Allocation of funds.

At least 10 percent of available funds will be used for funding single State projects based on the priority criteria.

§§ 1775.38–1775.60 [Reserved]

Subpart D—Solid Waste Management Grants

§ 1775.61 Authorization.

This subpart sets forth the policies and procedures for making Solid Waste Management (SWM) grants authorized under section 310B of the CONACT.

§1775.62 [Reserved]

§1775.63 Objectives.

The objectives of the program are to:

(a) Reduce or eliminate pollution of water resources, and

(b) Improve planning and management of solid waste sites.

§1775.64 [Reserved]

§1775.65 Eligibility.

(a) Entities eligible for grants must be either:

(1) Private nonprofit organizations with tax exempt status designated by the Internal Revenue Service. A nonprofit organization is defined as any corporation, trust, association, cooperative, or other organization that:

(i) Is operated primarily for scientific, education, service, charitable, or similar purposes in the public interest.

(ii) Is not organized primarily for profit.

(iii) Uses its net proceeds to maintain, improve, and/or expand its operations.

(2) Public bodies.

(3) Federally acknowledged or State-recognized Native American tribe or group.

(4) Academic institutions.

(b) Entities must be legally established and located within a state as defined in §1775.2.

(c) Organizations must be incorporated by December 31 of the year the application period occurs to be eligible for funds.

(d) Private businesses, Federal agencies, and individuals are ineligible for these grants.

(e) Applicants must also have the proven ability; background; experience, as evidenced by the organization's satisfactory completion of project(s) similar to those proposed; legal authority; and actual capacity to provide technical assistance and/or training on a regional basis to associations as provided in §1775.63. To meet the requirement of actual capacity, an applicant must either:

(1) Have the necessary resources to provide technical assistance and/or training to associations in rural areas through its staff, or

(2) Be assisted by an affiliate or member organization which has such background and experience and which agrees, in writing, that it will provide the assistance, or

(3) Contract with a nonaffiliated organization for not more than 49 percent of the grant to provide the proposed assistance.

§1775.66 Purpose.

Grants may be made to organizations as defined in §1775.65 to enable such organizations to assist associations to:

(a) Provide technical assistance and/or training to reduce the solid waste stream through reduction, recycling, and reuse.

(b) Provide training to enhance operator skills in maintaining and operating active landfills.

(c) Provide technical assistance and/or training for operators of landfills which are closed or will be closed in the near future with the development/implementation of closure plans, future land use plans, safety and maintenance planning, and closure scheduling within permit requirements.

(d) Evaluate current landfill conditions to determine the threats to water resources.

(e) Pay the expenses associated with providing the technical assistance and/or training authorized in paragraphs (a) through (d) of this section.

§1775.67 Allocation of funds.

The maximum amount for a single applicant for a Solid Waste Management project will be 25 percent of available grant funds.

§1775.68 Exception authority.

The Administrator may, in individual cases, make an exception to any requirement or provision of this part which is not inconsistent with the authorizing statute or other applicable law and is determined to be in the Government's interest.

§§ 1775.69–1775.99　[Reserved]

PART 1776—HOUSEHOLD WATER WELL SYSTEM GRANT PROGRAM

Subpart A—General

Sec.
1776.1　Purpose.
1776.2　Uniform Federal Assistance Provisions.
1776.3　Definitions.
1776.4　[Reserved]

Subpart B—HWWS Grants

1776.5　Eligibility to receive a HWWS grant.
1776.6　Funding availability.
1776.7　HWWS grant application process.
1776.8　Methods for submitting applications.
1776.9　Scoring applications.
1776.10　Grant agreement.
1776.11　Revolving loan fund.
1776.12　Use of HWWS grant proceeds.
1776.13　Administrative expenses.

Subpart C—HWWS Loans

1776.14　Eligibility to receive a HWWS loan.
1776.15　Terms of loans.
1776.16　Loan servicing.
1776.17　Revolving loan fund maintenance.

AUTHORITY: 7 U.S.C. 1926e.

SOURCE: 70 FR 28788, May 19, 2005, unless otherwise noted.

Subpart A—General

§ 1776.1　Purpose.

This part sets forth the policies and procedures for making grants to private, non-profit organizations to finance the construction, refurbishing and servicing of individually-owned household water well systems in rural areas for individuals with low or moderate income.

§ 1776.2　Uniform Federal Assistance Provisions.

(a) This program is subject to the general provisions that apply to all grants made by USDA and that are set forth in 2 CFR part 200, Uniform Administrative Requirements, Cost Principles, and Audit Requirements for Federal Awards, as adopted by USDA through 2 CFR part 400, as well as the following:

(1) 2 CFR part 415—General Program Administrative Requirements.

(2) 2 CFR part 180, as adopted by USDA through 2 CFR part 417, Nonprocurement Debarment and Suspension, implementing Executive Order 12549 on debarment and suspension.

(3) 2 CFR part 418, New Restrictions on Lobbying, prohibiting the use of appropriated funds to influence Congress or a Federal agency in connection with the making of any Federal grant and other Federal contracting and financial transactions.

(4) 2 CFR part 421, Requirements for Drug-Free Workplace (Financial Assistance), implementing the Drug-Free Workplace Act of 1988 (41 U.S.C 8102).

(b) [Reserved]

[81 FR 7697, Feb. 16, 2016]

§ 1776.3　Definitions.

Administrative expenses means expenses incurred by a grant recipient that are of the type more particularly described in Section 13 of this part.

Applicant means a private, non-profit organization that applies for a HWWS grant under this part.

Centralized Servicing Center (CSC) means the centralized loan servicing center within the United States Department of Agriculture, Rural Development. CSC provides nationwide services for borrowers that have received financing from Rural Development programs.

Construction means building or assembling a water well system or portion thereof, that is not a water well system or portion thereof being constructed in connection with a new building.

Eligible individual means an individual who is a member of a household the members of which have a combined income (for the most recent 12-month period for which the information is available) that is not more than 100 percent of the median nonmetropolitan household income for the State or territory in which the individual resides, according to 5-year income data from the American Community Survey (ACS) or, if needed, other Census Bureau data. If there is reason to believe that the ACS or other Census Bureau data does not accurately represent the median nonmetropolitan household income for the State or territory in which the individual resides, the reasons will be documented and the applicant may furnish, or RD may obtain, additional information regarding such median household income data. Information must consist of reliable data from local, regional, State or Federal sources or from a survey conducted by a reliable impartial source.

Funding opportunity announcement (FOA) means a publicly available document by which a Federal agency makes know its intentions to award discretionary grants or cooperative agreements, usually as a result of competition for funds. FOA announcements may be known as program announcements, notices of funding availability, solicitations, or other names depending on the agency and type of program. FOA announcements can be found at *www.Grants.gov* in the Search Grants tab and on the funding agency's or program's website.

Grant agreement means the contract between RUS and the grant recipient which sets forth the terms and conditions governing a particular grant awarded under this part.

Grant recipient means an applicant that has been awarded a HWWS grant under this part.

HWWS means household water well system.

HWWS grant means a grant awarded by RUS to a grant recipient under this part.

HWWS loan means a loan made by a grant recipient to a loan recipient using the direct or indirect proceeds of a HWWS grant awarded under this part.

Loan recipient means an eligible individual who has received a HWWS loan.

Refurbishing means to renovate or to restore a water well system or portion thereof to near new condition.

Revolved funds means the cash portion of the revolving loan fund that is not composed of HWWS grant funds, including repayments of revolving HWWS loans, fees, and interest collected on HWWS loans.

Revolving loan fund means the loan fund established by the grant recipient to carry out the purposes of this part, such fund comprising the proceeds of a HWWS grant and other related assets.

Rural area means any area other than a city or town that has a population of greater than 50,000 inhabitants; and the urbanized area contiguous and adjacent to such city or town.

RUS means the Rural Utilities Service, a Federal agency delivering the United States Department of Agriculture's Rural Development Utilities Program.

Servicing means making repairs or performing maintenance on a water well system or portion thereof.

USDA means the United States Department of Agriculture.

[70 FR 28788, May 19, 2005, as amended at 80 FR 9862, Feb. 24, 2015; 83 FR 45033, Sept. 5, 2018]

§1776.4 **[Reserved]**

Subpart B—HWWS Grants

§1776.5 **Eligibility to receive a HWWS grant.**

(a) The applicant must be a private organization.

(b) The applicant must be organized as a non-profit organization.

(c) The applicant must have legal capacity and lawful authority to perform the obligations of a grant recipient under this part. Example 1: If the organization is incorporated as a non-profit corporation, it must have corporate authority under state law and its corporate charter to engage in the practice of making loans to individuals. Example 2: if the organization is an unincorporated association, state law may prevent the organization from entering into binding contracts, such as a grant agreement.

(d) The applicant must have sufficient expertise and experience in lending and in promoting the safe and productive use of individually-owned household water well systems and ground water to assure the likelihood that the objectives of this part can be achieved.

§1776.6 **Funding availability.**

A FOA will be posted to *www.Grants.gov* in fiscal years that funds are available for this program.

253

The FOA will establish the period during which applications for such funds may be submitted for consideration.

[83 FR 45034, Sept. 5, 2018]

§ 1776.7 HWWS Grant application process.

(a) The applicant must complete and submit the following standard forms to RUS to apply for a HWWS grant under this part:

(1) Application for Federal Assistance: *Standard Form 424,*

(2) Budget Information—Non-Construction Programs: *Standard Form 424A,* and

(3) Assurances—Non-Construction Programs: *Standard Form 424B.*

(b) The applicant must submit a written work plan that demonstrates the feasibility of the applicant's lending program to meet the objectives of this part.

(c) The applicant should submit a narrative establishing the basis for any claims that it has substantial expertise in promoting the safe and productive use of individually-owned household water well systems. The Secretary will give priority to an applicant that demonstrates it has substantial experience of this type.

(d) The applicant must submit:

(1) A pro forma balance sheet at start-up and projected balance sheets for at least three additional years,

(2) Financial statements for the last three years, or from inception of the operations of the grant recipient if less than three years, and

(3) Projected cash flow and earnings statements for at least three years, supported by a list of assumptions showing the basis for the projections. The projected earnings statement and balance sheets must include one set of projections specific to the revolving loan fund, and a separate set of projections that detail the proposed applicant organization's total operations.

(e) The applicant may submit such additional information as it elects to support and describe its plan for achieving the objectives of this part.

§ 1776.8 Methods for submitting applications.

(a) Applications may be filed in either paper or electronic format. RUS will not accept applications by fax or e-mail.

(b) Paper applications for HWWS grants may be delivered by the U.S. Postal Service (USPS) or courier delivery services. Applications submitted by mail or courier must be postmarked no later than the filing deadline to be considered for the grant period. Applications delivered by mail or courier must be addressed to the attention of the Assistant Administrator, Water and Environmental Programs as follows: ATTN: Assistant Administrator, WEP, Rural Utilities Service, Stop 1548 Room 5145 South, 1400 Independence Ave. SW., Washington, DC 20250-1548.

(c) Electronic applications may be filed through Grants.gov, the official Federal Government Web site at *http://www.grants.gov.* The applicant must be registered with Grants.gov before they can submit a grant applicant. The applicant should refer to instructions found on the Grants.gov Web site for procedures for registering and using this facility. An applicant who is not registered on Grants.gov should allow a sufficient number of business days to complete the process. Applications submitted electronically must be show an electronic date and time stamp on or before the filing deadline to be considered for the grant period.

(d) The methods of submitting applications may be changed from time to reflect changes in addresses and electronic submission procedures. The applicant should refer to the most recent FOA for notice of any such changes. In the event of any discrepancy, the most recent FOA must be followed.

[70 FR 28788, May 19, 2005, as amended as 83 FR 45033, Sept. 5, 2018]

§ 1776.9 Scoring applications.

(a) Applications that are incomplete or ineligible will be returned to the applicant, accompanied by a statement explaining why the application is being returned.

(b) Promptly after an application period closes, all applications that are complete and eligible will be ranked competitively based on the following scoring criteria:

(1) Degree of expertise and experience in promoting the safe and productive use of individually-owned household

water well systems and ground water. Up to 30 points

(2) Degree of expertise and successful experience in making and servicing loans to individuals. Up to 20 points

(3) Percentage of applicant contributions. Points allowed under this paragraph will be based on written evidence of the availability of funds from sources other than the proceeds of a HWWS grant to pay part of the cost of a loan recipient's project. In-kind contributions will not be considered. Funds from other sources as a percentage of the HWWS grant and points corresponding to such percentages are as follows:

(i) 0 to 9 percent—ineligible;

(ii) 10 to 25 percent—5 points;

(iii) 26 to 30 percent—10 points;

(iv) 31 to 50 percent—15 points; and

(v) 51 percent or more—20 points

(4) Extent to which the work plan demonstrates a well thought out, comprehensive approach to accomplishing the objectives of this part, clearly defines who will be served by the project, and appears likely to be sustainable. Up to 20 points

(5) Extent to which the goals and objectives are clearly defined, tied to the work plan, and measurable. Up to 10 points

(6) Lowest ratio of projected administrative expenses to loans advanced. 10 points

(7) Administrator's discretion, considering such factors as creative outreach ideas for marketing HWWS loans to rural residents, the amount of funds requested in relation to the amount of needs demonstrated in the work plan, previous experiences demonstrating excellent utilization of a revolving loan fund grant, and optimizing the use of agency resources. Up to 10 points

(c) All qualifying applications under this part will be scored based on the criteria contained in this section. Awards will be made based on the highest ranking applications and the amount of financial assistance available for HWWS grants. All applicants will be notified in writing of the score each application receives.

§1776.10 Grant agreement.

(a) RUS and the grantee will enter into an agreement setting forth the terms and conditions governing a particular HWWS grant award. RUS will furnish the form of grant agreement. No funds awarded under this part shall be disbursed to the grant recipient before the grant agreement is binding and RUS has received a fully executed counterpart of the grant agreement.

(b) The grantee or RUS may initiate an amendment or modification to the grant agreement to provide for a loan limit up to $11,000. No change in the grant agreement requested by the grant recipient will be effective unless approved in writing by RUS.

[73 FR 68294, Nov. 18, 2008]

§1776.11 Revolving loan fund.

The grant recipient shall establish and maintain a revolving loan fund for the purposes set forth in §1776.12. All loans made to loan recipients shall be drawn from the revolving loan fund. The loans shall be serviced, and the revolving loan fund shall be maintained, as set forth in §1776.17.

§1776.12 Use of HWWS grant proceeds.

(a) Except as otherwise provided in the next paragraph, HWWS grant proceeds shall be used solely for the purpose of providing loans to eligible individuals for the construction, refurbishing, and servicing of individual household water well systems in rural areas that are or will be owned by the eligible individuals.

(b) A grant recipient may use HWWS grant funds to pay administrative expenses associated with providing the assistance described in the immediately preceding paragraph.

(c) A grant recipient may not use grant funds in any manner inconsistent with the terms of the grant agreement.

§1776.13 Administrative expenses.

(a) Subject to the limitations provided in paragraphs (b), (c) and (d) of this section, the grant recipient may use grant funds to pay administrative expenses associated with providing HWWS loans.

(b) Administrative expenses incurred in any calendar year which exceed 10 percent of the HWWS loans made by the grant recipient during that same period do not qualify for reimbursement.

(c) Administrative expenses incurred prior to the execution of the grant agreement by RUS do not qualify for reimbursement.

(d) Allowability of administrative expense costs shall be determined in accordance with 2 CFR part 200, as adopted by USDA through 2 CFR part 400.

[70 FR 28788, May 19, 2005, as amended at 79 FR 76005, Dec. 19, 2014]

Subpart C—HWWS Loans

§ 1776.14 Eligibility to receive a HWWS loan.

(a) The loan recipient must be an eligible individual.

(b) The loan recipient must either own and occupy the home being improved with the proceeds of the HWWS loan, or be occupying the home as the purchaser under a legally enforceable land purchase contract which is not in default by either the seller or the purchaser.

(c) The home using the water well system being funded from proceeds of the HWWS loan must be located in a rural area.

(d) The water well system being funded from the proceeds of the HWWS loan may not be associated with the construction of a new dwelling.

(e) The water well system being funded from the proceeds of the HWWS loan may not be used to substitute for water service available from collective water systems. Example: Loan recipient wishes to restore an old well which had been abandoned when the dwelling was connected to a water line belonging to a water district.

(f) A loan recipient must not be suspended or debarred from participation in Federal programs.

§ 1776.15 Terms of loans.

(a) HWWS loans under this part—

(1) Shall have an interest rate of 1 percent;

(2) Shall have a term not to exceed 20 years; and

(3) Shall not exceed $11,000 for each household water well system.

(b) The grant recipient must set forth the HWWS loan terms in written documentation signed by the loan recipient.

(c) Grant recipients must develop and use HWWS loan documentation that conforms to the terms of this part, the grant agreement, and the laws of the state or states having jurisdiction.

[70 FR 28788, May 19, 2005, as amended at 73 FR 68295, Nov. 18, 2008]

§ 1776.16 Loan servicing.

(a) If RUS determines that HWWS loans may be serviced by CSC, then the grant recipient will enter into an agreement with the Centralized Servicing Center for servicing all HWWS loans made from the revolving loan fund. All HWWS loan payments will be received by and processed at the Centralized Servicing Center. The grant recipient will be charged a fee for this service, and such fee should be included in the projected financial statements and work plan submitted as part of the grant application. This fee may be reimbursed as an administrative expense as provided in § 1776.13.

(b) If RUS determines that CSC is not able to service HWWS loans, then the grant recipient shall be responsible for servicing, or causing to be serviced, all HWWS loans. Servicing will include preparing loan agreements, processing loan payments, reviewing financial statements and debt reserves balances, and other responsibilities such as enforcement of loan terms. Loan servicing will be in accordance with the work plan RUS approved when the grant was awarded. It will continue as long as any loan made in whole or in part with RUS grant funds is outstanding.

§ 1776.17 Revolving loan fund maintenance.

As long as any part of the HWWS grant remains available for lending, and loans made from the revolving loan fund have an outstanding balance due, the grant recipient must maintain the revolving loan fund for the purposes set forth in § 1776.13.

(a) All HWWS grant funds received by a grant recipient must be deposited into the revolving loan fund.

(b) The grant recipient may transfer additional assets into the revolving loan fund.

(c) All cash and other assets of the revolving loan fund shall be deposited in a separate bank account or accounts.

(d) No cash or other assets of any other fund maintained by the grant recipient shall be commingled with the cash and other assets of the revolving loan fund.

(e) All moneys deposited in such bank account or accounts shall be money of the revolving loan fund.

(f) Loans to loan recipients are advanced from the revolving loan fund.

(g) The revolving loan fund will consist of receivables created by making loans, the grant recipient's security interest in collateral pledged by loan recipients, collections on the receivables, interest, fees, and any other income or assets derived from the operation of the revolving loan fund.

(h) The portion of the revolving loan fund that consists of HWWS grant funds, on a last-in-first-out basis, may be used for only those purposes set forth in this part.

(i) The grant recipient must submit an annual budget of proposed administrative costs for RUS approval. The amount removed from the revolving loan fund for administrative costs in any year must be reasonable; must not exceed the actual cost of operating the revolving loan fund, including loan servicing and providing technical assistance; and must not exceed the amount approved by RUS in the grant recipient's annual budget.

(j) A reasonable amount of revolved funds must be used to create a reserve for bad debts. Reserves should be accumulated over a period of years. The total amount should not exceed maximum expected losses, considering the quality of the grant recipient's portfolio of loans. Unless the grant recipient provides loss and delinquency records that, in the opinion of RUS, justifies different amounts, a reserve for bad debts of 6 percent of outstanding loans must be accumulated over three years and then maintained as set forth in the grant agreement.

(k) Any cash in the revolving loan fund from any source that is not needed for debt service, approved administrative costs, or reasonable reserves must be available for additional loans to loan recipients.

(l) All reserves and other cash in the revolving loan fund not immediately needed for loans to loan recipients or other authorized uses must be deposited in accounts in banks or other financial institutions. Such accounts must be fully covered by Federal deposit insurance or fully collateralized with U.S. Government obligations, and must be interest bearing. Any interest earned thereon remains a part of the revolving loan fund.

PART 1777—SECTION 306C WWD LOANS AND GRANTS

AUTHORITY: 5 U.S.C. 301; 7 U.S.C. 1989; 16 U.S.C. 1005.

SOURCE: 62 FR 33473, June 19, 1997, unless otherwise noted.

§1777.1 General.

(a) This part outlines Rural Utilities Service (RUS) policies and procedures for making Water and Waste Disposal (WWD) loans and grants authorized under section 306C of the Consolidated Farm and Rural Development Act (7 U.S.C. 1926(c)), as amended.

(b) Agency officials will maintain liaison with officials of other Federal, State, regional, and local development agencies to coordinate related programs to achieve rural development objectives.

(c) Agency officials shall cooperate with appropriate State agencies in making loans and/or grants that support State strategies for rural area development.

(d) Funds allocated in accordance with this part will be considered for use by Indian tribes within the State regardless of whether State development strategies include Indian reservations within the State's boundaries. Indians residing on such reservations must have an equal opportunity to participate in this program.

(e) Federal statutes provide for extending the Agency's financial programs without regard to race, color, religion, sex, national origin, marital status, age, or physical/mental handicap (provided the participant possesses the capacity to enter into legal contracts).

§ 1777.2 [Reserved]

§ 1777.3 Objective.

The objective of the Section 306C WWD Loans and Grants program is to provide water and waste disposal facilities and services to low-income rural communities whose residents face significant health risks.

§ 1777.4 Definitions.

Applicant. Entity that receives the Agency loan or grant under this part. The entities can be public bodies such as municipalities, counties, districts, authorities, or other political subdivisions of a State, and organizations operated on a not-for-profit basis such as associations, cooperatives, private corporations, or Indian tribes on Federal and State reservations, and other Federally recognized Indian tribes.

Colonia. Any identifiable community designated in writing by the State or county in which it is located; determined to be a colonia on the basis of objective criteria including lack of potable water supply, lack of adequate sewage systems, and lack of decent, safe, and sanitary housing, inadequate roads and drainage; and existed and was generally recognized as a colonia before October 1, 1989.

Cooperative. A cooperative formed specifically for the purpose of the installation, expansion, improvement, or operation of water supply or waste disposal facilities or systems.

Individual. Recipient of a loan or grant through the applicant to facili-

tate use of the applicant's water and/or waste disposal system.

Rural areas. Includes unincorporated areas and any city or town with a population not in excess of 10,000 inhabitants located in any of the 50 States, the Commonwealth of Puerto Rico, the Western Pacific Territories, Marshall Islands, Federated States of Micronesia, Republic of Palau, and the U.S. Virgin Islands. The population figure is obtained from the most recent decennial Census of the United States. If the applicable population figure cannot be obtained from the most recent decennial Census, RD will determine the applicable population figure based on available population data.

Statewide Nonmetropolitan Median Household Income (SNMHI). Median household income of the State's nonmetropolitan counties and portions of metropolitan counties outside of cities, towns or places of 50,000 or more population.

[62 FR 33473, June 19, 1997, as amended at 69 FR 65519, Nov. 15, 2004; 80 FR 9862, Feb. 24, 2015]

§§ 1777.5-1777.10 [Reserved]

§ 1777.11 Making, processing, and servicing loans and grants.

Unless specifically modified by this part, loans and/or grants will be made, processed, and serviced in accordance with part 1780 of this chapter.

§ 1777.12 Eligibility.

(a) The provisions of paragraphs (a) (1) and (2) of this section do not apply to a rural area recognized as a colonia. Otherwise, the facility financed under this part must provide water and/or waste disposal services to rural areas of a county where, on the date preapplication is received by the Agency, the:

(1) Per capita income of the residents is not more than 70 percent of the most recent national average per capita income, as determined by 5-year income data from the American Community Survey (ACS) or, if needed, other Census Bureau data. If there is reason to believe that the ACS or other Census Bureau data does not accurately represent the per capita income of the

residents, the reasons will be documented and the borrower/applicant may furnish, or RD may obtain, additional information regarding such per capita income data. Information must consist of reliable data from local, regional, State or Federal sources or from a survey conducted by a reliable impartial source, and

(2) Unemployment rate of the residents is not less than 125 percent of the most recent national average unemployment rate, as determined by the Bureau of Labor Statistics.

(b) Residents of the rural area to be served must face significant health risks due to the fact that a significant proportion of the community's residents do not have access to, or are not served by, adequate, affordable, water and/or waste disposal systems. The file should contain documentation to support this determination. The following requirements regarding the documentation must be followed:

(1) The originating documentation must come from an independent third party source that has the experience in specifying the health or sanitary problem that currently exists.

(2) The documentation must state specifically the health or sanitary problems that exist. General statements of problems or support for the project are not acceptable.

(3) Current users of the facility must be experiencing the current health or sanitary problem and not future or possible users.

(4) If no facility exists, documentation must include specific health and sanitary problems associated with individual facilities that currently exist to warrant the health and sanitary determination.

[62 FR 33473, June 19, 1997, as amended at 77 FR 43150, July 24, 2012; 80 FR 9863, Feb. 24, 2015]

§1777.13 Project priority.

Paragraphs (a) through (d) of this section indicate items and conditions which must be considered in selecting applications for further development. When ranking eligible applications for consideration for limited funds, Agency officials must consider the priority items met by each application and the degree to which those priorities are met.

(a) *Applications.* The application and supporting information submitted with it will be used to determine applicant eligibility and the proposed project's priority for available funds. Applicants determined ineligible will be advised of their appeal rights in accordance with 7 CFR part 11.

(b) *State Office review.* All applications will be processed and scored in the area office and then reviewed for funding priority at the State Office using RUS Bulletin 1777-2. Eligible applicants that cannot be funded will be advised that funds are not available and advised of their appeal rights as set forth in 7 CFR part 11.

(c) *National Office.* The National Office will allocate funds on a project-by-project basis as requests are received from the State Office. If the amount of funds requested exceeds the amount of funds available, the total project score will be used to select projects for funding. The RUS Administrator may assign up to 35 additional points which will be considered in the total points for items such as geographic distribution of funds, severity of health risks, etc. Unobligated funds will be pooled by mid-August of each year and made available to all States with eligible colonias applicants on a case-by-case basis.

(d) *Selection priorities.* The priorities described below will be used to rate applications and in selecting projects for funding. Points will be distributed as indicated in paragraphs (d)(1) through (d)(6) of this section and will be used in selecting projects for funding.

(1) *Population.* The proposed project will serve an area with a rural population:

(i) Not in excess of 1,500—30 points.

(ii) More than 1,500 and not in excess of 3,000—20 points.

(iii) More than 3,000 and not in excess of 5,500—10 points.

(2) *Income.* The median household income of population to be served by the proposed project is:

(i) Not in excess of 50 percent of the statewide nonmetropolitan median household income—40 points.

(ii) More than 50 percent and not in excess of 60 percent of the statewide nonmetropolitan median household income—20 points.

(iii) More than 60 percent and not in excess of 70 percent of the statewide nonmetropolitan median household income—10 points.

(3) *Joint financing.* The amount of joint financing committed to the proposed project is:

(i) Twenty percent or more private, local, or State funds except Federal funds channeled through a State agency—10 points.

(ii) Five to 19 percent private, local, or State funds except Federal funds channeled through a State agency—5 points.

(4) *Colonia.* (See definition in § 1777.4). The proposed project will provide water and/or waste disposal services to the residents of a colonia:—50 points. Additional points will be assigned as follows:

(5) *Access and health risks for colonias.*

(i) A colonia that lacks access to both water and waste disposal facilities, resulting in a significant health risk—50 points.

(ii) A colonia that lacks access to either water or waste disposal facilities, resulting in a significant health risk—40 points.

(iii) A colonia that has access to water and waste disposal facilities, but is facing a significant health risk—15 points.

(6) *Discretionary.* In certain cases, and when a written justification is prepared, the State Program Official with loan/grant approval authority may assign up to 15 points for items such as natural disaster, to improve compatibility/coordination between RUS' and other agencies' selection systems, to assist those projects that are the most cost effective, high unemployment rate, severity of health risks, etc.

[77 FR 43151, July 24, 2012]

§§ 1777.14–1777.20 [Reserved]

§ 1777.21 Use of funds.

(a) *Applicant.* Funds may be used to:

(1) Construct, enlarge, extend, or otherwise improve community water and/or waste disposal systems. Otherwise improve would include extending service lines to and/or connecting residence's plumbing to the system.

(2) Make loans and grants to individuals for extending service lines to and/or connecting residences to the applicant's system. The approval official must determine that this is a practical and economical method of connecting individuals to the community water and/or waste disposal system. Loan funds can only be used for loans, and grant funds can only be used for grants.

(3) Make improvements to individual's residence when needed to allow use of the water and/or waste disposal system.

(4) Grants can be made up to 100 percent of eligible project costs.

(b) *Individuals.* Funds may be used to:

(1) Extend service lines to residence.

(2) Connect service lines to residence's plumbing.

(3) Pay reasonable charges or fees for connecting to a community water and/or waste disposal system.

(4) Pay for necessary installation of plumbing and related fixtures within dwellings lacking such facilities. This is limited to one bathtub, sink, commode, kitchen sink, water heater, and outside spigot.

(5) Construction and/or partitioning off a portion of dwelling for a bathroom, not to exceed 4.6 square meters (48 square feet) in size.

(6) Pay reasonable costs for closing abandoned septic tanks and water wells when necessary to protect the health and safety of recipients of a grant in paragraphs (b)(1) or (b)(2) of this section and is required by local or State law.

§§ 1777.22–1777.30 [Reserved]

§ 1777.31 Rates.

(a) Applicant loans will bear interest at the rate of 5 percent per annum.

(b) Individual loans will bear interest at the rate of:

(1) Five percent per annum; or

(2) The Federal Financing Bank rate for loans of a similar term at the time of Agency loan approval, whichever is less.

§§ 1777.32–1777.40 [Reserved]

§ 1777.41 Individual loans and grants.

(a) The amount of loan and grant funds approved by the Agency will be based on the need shown in the application and an implementation plan submitted by the applicant. The implementation plan will include such things as: purpose, how funds will be used, proposed application process, construction requirements, control and disbursement of funds, etc. The implementation plan will be attached to RUS Bulletin 1777–1.

(b) RUS Bulletin 1777–1 is a Memorandum of Agreement which sets forth the procedures and regulations for making and servicing loans and grants made by applicants to individuals. The State Program Official is authorized to enter into a Memorandum of Agreement with any applicant providing loans and/or grants to individuals. The Memorandum of Agreement can be amended to comply with State law and recommendations by the Office of General Counsel. It may also be amended to eliminate references to loans and/or grants if no loan and/or grant is involved. The State Program Official is responsible for:

(1) Ensuring that all provisions of the Agreement are understood.

(2) Determining that the applicant has the ability to make and service loans and/or grants in the manner outlined in the Agreement.

(c) Agency funds remaining after providing individual loans and/or grants will be returned to the Agency. The funds should be disbursed to individuals within 1 year from the date water and/or waste disposal service is available to the individuals. The State Program Official can make an exception to this 1 year requirement if written justification is provided by the applicant.

§ 1777.42   Delegation of authority.

The State Program Official is responsible for the overall implementation of the authorities contained in this part and may redelegate any such authority to appropriate Agency employees.

§ 1777.43   Bulletins.

RUS Bulletin 1780–12 referenced in part 1780 of this chapter and RUS Bulletin 1777–1, 1777–2 and 1777–3 are for use in administering loans and/or grants made under this part. Bulletins, instructions and forms are available from any USDA/Rural Development office or

the Rural Utilities Service, United States Department of Agriculture, Washington, DC 20250–1500.

§§ 1777.44–1777.99   [Reserved]

§ 1777.100   OMB control number.

The reporting and recordkeeping requirements contained in this part have been approved by the Office of Management and Budget and assigned OMB control number 0570–0001. Public reporting burden for this collection of information is estimated to vary from 5 to 30 hours per response with an average of 17.5 hours per response, including the time for reviewing instructions, searching existing data sources, gathering and maintaining the data needed, and completing and reviewing the collection of information. Send comments regarding this burden estimate or any other aspect of this collection of information, including suggestions for reducing this burden, to U.S. Department of Agriculture, Clearance Officer, OIRM, Room 404–W, Washington, DC 20250; and to the Office of Information and Regulatory Affairs, Office of Management and Budget, Washington, DC 20503.

PART 1778—EMERGENCY AND IMMINENT COMMUNITY WATER ASSISTANCE GRANTS

1778.36 [Reserved]
1778.37 Forms, Instructions and Bulletins.
1778.38–1778.99 [Reserved]
1778.100 OMB control number.

AUTHORITY: 5 U.S.C. 301; 7 U.S.C. 1989; 16 U.S.C. 1005.

SOURCE: 68 FR 46078, Aug. 5, 2003, unless otherwise noted.

§ 1778.1 General.

(a) This part outlines policies and procedures for making Emergency Community Water Assistance Grants (ECWAG) authorized under Section 306A of the Consolidated Farm and Rural Development Act, (7 U.S.C. 1926(a)), as amended. Any processing or servicing activity conducted pursuant to this part involving authorized assistance to Agency employees, members of their families, known close relatives, or business or close personal associates, is subject to the provisions of subpart D of part 1900 of this title. Applicants for this assistance are required to identify any known relationship or association with an Agency employee.

(b) Agency officials will maintain liaison with officials of other Federal, State, regional and local development agencies to coordinate related programs to achieve rural development objectives.

(c) Agency officials shall cooperate with appropriate State agencies in making grants that support State strategies for rural area development.

(d) Funds allocated for use in accordance with this part are also to be considered for use by Indian tribes within the State regardless of whether State development strategies include Indian reservations within the State's boundaries. Indians residing on such reservations must have an equal opportunity along with other rural residents to participate in the benefits of this program. This includes equal application of outreach activities of Field Offices.

(e) Federal statutes provide for extending the Agency financial programs without regard to race, color, religion, sex, national origin, marital status, age, or physical/mental handicap (provided the participant possesses the capacity to enter into legal contracts).

§ 1778.2 [Reserved]

§ 1778.3 Objective.

The objective of the ECWAG Program is to assist the residents of rural areas that have experienced a significant decline in quantity or quality of water, or in which such a decline is considered imminent, to obtain or maintain adequate quantities of water that meets the standards set by the Safe Drinking Water Act (42 U.S.C. 300f et seq.) (SDWA).

§ 1778.4 Definitions.

Acute shortage. An acute shortage is a situation in which the system either cannot deliver water at all through its distribution system or can only deliver water on a sporadic basis.

Emergency. Occurrence of an incident such as, but not limited to, a drought; earthquake; flood; tornado; hurricane; disease outbreak; or chemical spill, leakage, or seepage.

Rural areas. Includes any area not in a city or town with a population in excess of 10,000 inhabitants located in any of the fifty States, the Commonwealth of Puerto Rico, the Western Pacific Territories, Marshall Islands, Federated States of Micronesia, Republic of Palau, and the U.S. Virgin Islands. The population figure is obtained from the most recent decennial Census of the United States (decennial Census). If the applicable population figure cannot be obtained from the most recent decennial Census, RD will determine the applicable population figure based on available population data.

Significant decline in quality. A significant decline in quality of potable water occurs when the present community source or delivery system does not meet, as a result of an emergency, the current SDWA requirements. For a private source or delivery system a significant decline in quality occurs when the water is no longer potable as a result of an emergency. As used in this Subpart, the term significant decline in quality may also include a situation where a significant decline is likely to occur within one year from the date of the filing of an application.

Significant decline in quantity. A significant decline in the quantity is caused by a disruption of the potable water supply by an emergency. The disruption in quantity of water prevents the present source or delivery system from supplying potable water needs to rural residents. This would not include a decline in excess water capacity. As used in this Subpart, the term significant decline in quantity may also include a situation where a significant decline is likely to occur within one year from the date of the filing of an application.

Statewide Nonmetropolitan Median Household Income (SNMHI). Median household income of the State's nonmetropolitan counties and portions of metropolitan counties outside of cities, towns or places of 50,000 or more population.

[68 FR 46078, Aug. 5, 2003, as amended at 69 FR 65519, Nov. 15, 2004; 80 FR 9863, Feb. 24, 2015]

§1778.5 [Reserved]

§1778.6 Eligibility.

(a) Grants may be made to public bodies and private nonprofit corporations serving rural areas. Public bodies include counties, cities, townships, incorporated towns and villages, boroughs, authorities, districts, and other political subdivisions of a State. Public bodies also include Indian tribes on Federal and State reservations and other Federally recognized Indian Tribal groups in rural areas.

(b) In the case of grants made to alleviate a significant decline in quantity or quality of water available from the water supplies of rural residents, the applicant must demonstrate that the decline occurred within two years of the date the application was filed with the Agency. This would not apply to grants made for repairs, partial replacement, or significant maintenance on an established water system. In situations involving imminent decline, evidence must be presented to demonstrate that the decline is likely to occur within one year of the date the application is filed with the Agency.

§1778.7 Project priority.

Paragraph (d) of this section indicates items and conditions which must be considered in selecting applications for further development. When ranking eligible applications for consideration for limited funds, Agency officials must consider the priority items met by each application and the degree to which those priorities are met.

(a) *Applications.* The application and supporting information submitted with it will be used to determine the proposed project's priority for available funds.

(b) *State Office review.* All applications will be reviewed and scored for funding priority using RUS Bulletin 1778–1. Eligible applicants that cannot be funded should be advised that funds are not available.

(c) *National Office review.* Each year all funding requests will be reviewed by the National Office beginning 30 days after funds from the annual appropriation are made available to the Agency. Reviews will continue throughout the fiscal year as long as funds are available. Projects selected for funding will be considered based on the priority criteria and available funds. Projects must compete on a national basis for available funds, and the National Office will allocate funds to State offices on a project by project basis.

(d) *Selection priorities.* The priorities described below will be used by the State Program Official to rate applications and by the Assistant Administrator of Water and Environmental Programs to select projects for funding. Points will be distributed as indicated in paragraphs (d)(1) through (d)(5) of this section and will be considered in selecting projects for funding. A copy of RUS Bulletins 1778–1 and 1778–2 used to rate applications, should be placed in the case file for future reference.

(1) *Population.* The proposed project will serve an area with a rural population:

(i) Not in excess of 1,500—30 points.

(ii) More than 1,500 and not in excess of 3,000—20 points.

(iii) More than 3,000 and not in excess of 5,000—15 points.

(iv) Over 5,000—0 points.

(2) *Income.* The median household income of population to be served by the proposed project is:

(i) Not in excess of 70% of the statewide nonmetropolitan median household income—30 points.

(ii) More than 70% and not in excess of 80% of the statewide nonmetropolitan median household income—20 points.

(iii) More than 80% and not in excess of 90% of the statewide nonmetropolitan median household income—10 points.

(iv) Over 90% of the statewide nonmetropolitan median household income—0 points.

(3) *Significant decline.* Points will be assigned for only one of the following paragraphs when the primary purpose of the proposed project is to correct a significant decline that has occurred in the:

(i) Quantity of water available from private individually owned wells or other individual sources of water—30 points; or

(ii) Quantity of water available from an established system's source of water—20 points; or

(iii) Quality of water available from private individually owned wells or other individual sources of water—30 points; or

(iv) Quality of water available from an established system's source of water—20 points.

(4) *Imminent decline.* The proposed project will attempt to avert an imminent decline expected to occur during the one-year period following the filing of an application—10 points.

(NOTE: If points were assigned above for a significant decline, no points will be awarded for imminent decline.)

(5) *Acute shortage.* Grants made in accordance with § 1778.11(b) of this part to assist an established water system remedy an acute shortage of quality water or correct a significant decline in the quantity or quality of water that is available—10 points.

(6) *Discretionary.* In certain cases the Administrator may assign up to 30 points for items such as geographic distribution of funds, rural residents hauling water, severe contamination levels, etc.

§ 1778.8 [Reserved]

§ 1778.9 Uses.

Grant funds may be used for the following purposes:

(a) Waterline extensions from existing systems.

(b) Construction of new waterlines.

(c) Repairs to an existing system.

(d) Significant maintenance to an existing system.

(e) Construction of new wells, reservoirs, transmission lines, treatment plants, and other sources of water.

(f) Equipment replacement.

(g) Connection and/or tap fees.

(h) Pay costs that were incurred within six months of the date an application was filed with the Agency to correct an emergency situation that would have been eligible for funding under this part.

(i) Any other appropriate purpose such as legal fees, engineering fees, recording costs, environmental impact analyses, archaeological surveys, possible salvage or other mitigation measures, planning, establishing or acquiring rights associated with developing sources of, treating, storing, or distributing water.

(j) Assist rural water systems to comply with the requirements of the Federal Water Pollution Control Act (33 U.S.C. 1251 *et seq.*) (FWPCA) or the SDWA when such failure to comply is directly related to a recent decline in quality of potable water. This would not apply to changes in the requirements of FWPCA or SDWA.

(k) Provide potable water to communities through means other than those covered above for not to exceed 120 days when a more permanent solution is not feasible in a shorter time frame.

§ 1778.10 Restrictions.

(a) Grant funds may not be used to:

(1) Assist any city or town with a population in excess of 10,000 inhabitants. The population figure is obtained from the most recent decennial Census. If the applicable population figure cannot be obtained from the most recent decennial Census, RD will determine the applicable population figure based on available population data. Facilities financed by RUS may be located in non-rural areas. However, loan and grant funds may be used to finance only that portion of the facility serving rural areas, regardless of facility location.

(2) Assist a rural area that has a median household income in excess of the statewide nonmetropolitan median household income as determined by 5-year income data from the American Community Survey (ACS) or, if needed, other Census Bureau data. If there is reason to believe that the ACS or other Census Bureau data does not accurately represent the median household income of the rural area, the reasons will be documented and the applicant may furnish, or RD may obtain, additional information regarding such median household income data. Information must consist of reliable data from local, regional, State or Federal sources or from a survey conducted by a reliable impartial source.

(3) Finance facilities which are not modest in size, design, cost, and are not directly related to correcting the potable water quantity or quality problem.

(4) Pay loan or grant finder's fees.

(5) Pay any annual recurring costs that are considered to be operational expenses.

(6) Pay rental for the use of equipment or machinery owned by the rural community.

(7) Purchase existing systems.

(8) Refinance existing indebtedness, except for short-term debt incurred in accordance with §1778.9(h).

(9) Make reimbursement for projects developed with other grant funds.

(10) Finance facilities that are not for public use.

(b) Nothing in paragraph (a)(1) of this section shall preclude rural areas from submitting joint proposals for assistance under this part. Each entity applying for financial assistance under this part to fund their share of a joint project will be considered individually.

[68 FR 46078, Aug. 5, 2003, as amended at 80 FR 9863, Feb. 24, 2015]

§1778.11 Maximum grants.

(a) Grants not to exceed $500,000 may be made to alleviate a significant decline in quantity or quality of water available to a rural area that occurred within two years of filing an application with the Agency, or to attempt to avoid a significant decline that is expected to occur during the twelve month period following the filing of an application.

(b) Grants made for repairs, partial replacement, or significant maintenance on an established system to remedy an acute shortage or significant decline in the quality or quantity of potable water, or an anticipated acute shortage or significant decline, cannot exceed $150,000.

(c) Grants under this part, subject to paragraphs (a) and (b) of this section, shall be made for 100 percent of eligible project costs.

§1778.12 [Reserved]

§1778.13 Set-aside.

(a) At least 70 percent of all grants made under this grant program shall be for projects funded in accordance with §1778.11(a).

(b) At least 50 percent of the funds appropriated for this grant program shall be allocated to rural areas with populations not in excess of 3,000 inhabitants according to the most recent decennical Census.

[68 FR 46078, Aug. 5, 2003, as amended at 80 FR 9863, Feb. 24, 2015]

§1778.14 Other considerations.

(a) *Civil rights compliance requirements.* All grants made under this part are subject to Title VI of the Civil Rights Act of 1964 (42 U.S.C. 2000d *et seq.*) as outlined in subpart E of part 1901 of this title.

(b) *Environmental requirements.* All projects must have appropriate environmental reviews in accordance with RUS requirements.

(c) *Uniform Relocation and Real Property Acquisition Policies Act (42 U.S.C. 4601 et seq.).* All projects must comply with the requirements set forth in 7 CFR part 21.

(d) *Flood and mudslide hazard area precautions.* If the project is located in a flood or mudslide area, then flood or mudslide insurance must be provided as required in subpart A of part 1806 of this title (RD Instruction 426.2).

(e) *Governmentwide debarment and suspension (nonprocurement).* All projects must comply with the requirements of 2 CFR part 180, as adopted by USDA through 2 CFR part 417, Nonprocurement Debarment and Suspension, implementing Executive Order 12549 on debarment and suspension.

(f) *Intergovernmental review.* All projects funded under this part are subject to Executive Order 12372 (3 CFR, 1983 Comp., p. 197), which requires intergovernmental consultation with State and local officials. These requirements are found at 2 CFR part 415, subpart C, "Intergovernmental Review of Department of Agriculture Programs and Activities" and RD Instruction 1970–I, 'Intergovernmental Review,' available in any Agency office or on the Agency's Web site.

(g) *Uniform administrative requirements.* All projects funded under this part are subject to 2 CFR part 200, as adopted by USDA through 2 CFR part 400, Uniform Administrative Requirements, Cost Principles, and Audit Requirements for Federal Awards, and 2 CFR part 415, General Program Administrative Requirements.

(h) *Restrictions on lobbying.* All projects funded under this part are subject to 2 CFR part 418, New Restrictions on Lobbying.

(i) *Requirements for drug-free workplace.* This program is subject to 2 CFR part 421, Requirements for Drug-Free Workplace (Financial Assistance).

[68 FR 46078, Aug. 5, 2003, as amended at 76 FR 80730, Dec. 27, 2011; 79 FR 76006, Dec. 19, 2014; 81 FR 7697, Feb. 16, 2016]

§§ 1778.15–1778.20 [Reserved]

§ 1778.21 Application processing.

(a) The material submitted with the application should include the Preliminary Engineering Report, population and median household income of the area to be served, description of project, and nature of emergency that caused the problem(s) being addressed by the project. The documentation must clearly show that the applicant has had a significant decline in the quantity or quality of potable water or an acute shortage of potable water, or that such a decline or shortage is imminent, and that the proposed project will eliminate or alleviate the problem. For projects to be funded in accordance with § 1778.11 (a), evidence must be furnished that a significant decline in quantity or quality occurred within two years before filing the application with the Agency, or is expected to occur within one year after filing the application.

(b) When favorable action will not be taken on an application, the applicant will be notified in writing by the State Program Official of the reasons why the request was not favorably considered. Notification to the applicant will state that a review of this decision by the Agency may be requested by the applicant in accordance with 7 CFR part 11.

§ 1778.22 Planning development and procurement.

Planning development and procurement for grants made under this part will be in accordance with subpart C of Part 1780 of this chapter. A certification should be obtained from the State agency or the Environmental Protection Agency if the State does not have primacy, stating that the proposed improvements will be in compliance with requirements of the SDWA.

§ 1778.23 Grant closing and disbursement of funds.

(a) Grants will be closed in accordance with § 1780.45 of part 1780 of this chapter.

(b) RUS Bulletin 1780–12, "Water or Waste Grant Agreement," will be executed by all applicants.

(c) The Agency's policy is not to disburse grant funds from the Treasury until they are actually needed by the applicant. Grant funds will be disbursed by using multiple advances.

§§ 1778.24–1778.30 [Reserved]

§ 1778.31 Performing development.

(a) Applicable provisions of subpart C of part 1780 of this chapter will be followed in performing development for grants made under this part.

(b) After filing an application in accordance with § 1778.21 and when immediate action is necessary, the State Program Official may concur in an applicant's request to proceed with construction before funds are obligated provided the RUS environmental requirements are complied with. The applicant must be advised in writing that:

(1) Any authorization to proceed or any concurrence in bid awards, contract concurrence, or other project development activity, is not a commitment by the Agency to provide grant funds under this part.

(2) The Agency is not liable for any debt incurred by the applicant in the event that funds are not provided under this part.

§§ 1778.32–1778.33 [Reserved]

§ 1778.34 Grant servicing.

(a) Grants will be serviced in accordance with § 1951.215 of subpart E of part 1951 of this title and subpart O of part 1951 of this title.

(b) The grantee will provide an audit report in accordance with § 1780.47 of part 1780 of this chapter.

§ 1778.35 Subsequent grants.

Subsequent grants will be processed in accordance with the requirements set forth in this part. The initial and subsequent grants made to complete a previously approved project must comply with the maximum grant requirements set forth in § 1778.11.

§ 1778.36 [Reserved]

§ 1778.37 Forms, Instructions and Bulletins.

Bulletins, instructions and forms referenced are for use in administering grants made under this part and are available from any USDA/Rural Development office or the Rural Utilities Service, United States Department of Agriculture, Washington, DC 20250–1500.

§§ 1778.38–1778.99 [Reserved]

§ 1778.100 OMB control number.

The information collection requirements contained in this part have been approved by the Office of Management and Budget and assigned OMB control number 0572–0110.

PART 1779—WATER AND WASTE DISPOSAL PROGRAMS GUARANTEED LOANS

AUTHORITY: 5 U.S.C. 301, 7 U.S.C. 1989, 16 U.S.C. 1005.

SOURCE: 66 FR 23138, May 8, 2001, unless otherwise noted.

§ 1779.1 General.

(a) This part contains the regulations for Water and Waste Disposal (WW) loans guaranteed by the Agency and applies to lenders, holders, borrowers, and other parties involved in making, guaranteeing, holding, servicing, or liquidating such loans.

(b) The purpose of the WW guaranteed loan program is to provide a loan guarantee for the construction or improvement of water and waste projects serving the financially needy communities in rural areas. This purpose is achieved through bolstering the existing private credit structure through the guarantee of quality loans which will provide lasting benefits.

§ 1779.2 Definitions.

The following general definitions are applicable to the terms used in this part:

Agency. The Rural Utilities Service which is within the Rural Development mission area of the United States Department of Agriculture or its successor agencies with authority delegated by the Secretary of Agriculture to administer the Water and Waste Disposal Programs.

Application. An Agency prescribed form to request an Agency guarantee (available in any Agency office).

Arm's length transaction. The sale, release, or disposition of assets in which the title to the property passes to a ready, willing, and able third party who is not affiliated with, or related to, and has no security, monetary, or stockholder interest in the borrower or transferor at the time of the transaction.

Assignment Guarantee Agreement. The signed agreement among the Agency, the lender, and the holder setting forth the terms and conditions of an assignment of the guaranteed portion of a loan or any part thereof (available in any Agency office).

Borrower. The entity that borrows money from the lender.

Collateral. Property pledged to secure the guaranteed loan.

Conditional Commitment for Guarantee. The Agency's written statement to the lender that the material submitted is approved subject to the completion of all conditions and requirements contained in the commitment (available in any Agency office).

Guaranteed loan. A loan made and serviced by a lender for which the Agency and lender have entered into a Lender's Agreement and for which the Agency has issued a Loan Note Guarantee.

Holder. The person or entity (other than the lender) who holds all or a part of the guaranteed portion of the loan with no servicing responsibilities. When the lender assigns part or all of the guaranteed portion of the loan to an assignee, the assignee becomes a holder when the Assignment Guarantee Agreement is signed by all parties.

Immediate family. Individuals who are closely related by blood or by marriage, or within the same household, such as a spouse, parent, child, brother, sister, aunt, uncle, grandparent, grandchild, niece, or nephew.

In-house expenses. In-house expenses include, but are not limited to, employees' salaries, retainers being paid to lawyers, travel, and overhead.

Insurance. Fire, windstorm, lightning, hail, explosion, riot, civil commotion, aircraft, vehicles, smoke, builder's risk, liability, property damage, flood or mudslide, worker's compensation, fidelity bond, malpractice, or any similar insurance that is available and needed to protect the security or that is required by law.

Joint financing. Two or more lenders (or any combination of lenders and other financial sources) making separate relatively contemporaneous loans or grants to supply the funds required by one borrower. For example, such joint financing may consist of the Agency's financial assistance with the Economic Development Administration, Department of Housing and Urban Development (HUD), or other Federal and State agencies, and private and quasi-public financial institutions.

Lender. The person or organization making and responsible for servicing the loan. The lender is also referred to in this part as the applicant who is requesting a guarantee during the preapplication and application stage of processing.

Lender's Agreement. The signed agreement between the Agency and the lender containing the lender's responsibilities when the Loan Note Guarantee is issued (available in any Agency office).

Loan Note Guarantee. The signed commitment issued by the Agency containing the terms and conditions of the guarantee of an identified loan (available in any Agency office).

Market value. The amount for which property would sell for its highest and best use at a voluntary sale in an arm's length transaction.

Note. An evidence of debt. In those instances where the Agency guarantees a bond issue, "note" shall also be construed to include a bond or other evidence of indebtedness, as appropriate.

Participation. Sale of an interest in a loan in which the lender retains the note, collateral securing the note, and all responsibility for loan servicing and liquidation.

Principals of borrowers. The owners, officers, directors, entities, and supervisors directly involved in the operation and management of the borrower.

Protective advances. Advances made by the lender for the purpose of preserving and protecting the collateral where the debtor has failed to, and will not or cannot, meet obligations to protect or preserve collateral.

Report of loss. An Agency form used by lenders when reporting a loss under an Agency guarantee (available in any Agency office).

Rural and rural area. Any area not in a city or town with a population in excess of 10,000 inhabitants. The population figure is obtained from the most recent decennial Census of the United States (decennial Census). If the applicable population figure cannot be obtained from the most recent decennial Census, RD will determine the applicable population figure based on available population data.

Service area. The area reasonably expected to be served by the project being financed by the guaranteed loan.

State. Any of the 50 States, the Commonwealth of Puerto Rico, the Virgin Islands of the United States, Guam, American Samoa, Commonwealth of the Northern Mariana Islands, Republic of the Marshall Islands, Republic of Palau, and the Federated States of Micronesia.

State Bond Banks and State Bond Pools. An entity authorized by the State to issue State debt instruments and utilize the funds received to finance the construction or improvement of drinking water or waste disposal facilities.

State Director. The Rural Development State Director or the staff member who has been delegated authority to perform action on behalf of the State Director.

Substantive change. Any change in the purpose of the loan or any change in the financial condition of the borrower or the collateral which would jeopardize the performance of the loan.

Transfer and assumption. The conveyance by a debtor to an assuming party of the assets, collateral, and liabilities of the loan in return for the assuming party's binding promise to pay the outstanding debt.

Waste disposal. Sanitary sewer (treatment and collection), solid waste, and storm drainage facilities.

WW. An acronym for Water and Waste Disposal.

[66 FR 23138, May 8, 2001, as amended at 80 FR 9863, Feb. 24, 2015]

§ 1779.3 Full faith and credit.

The Loan Note Guarantee constitutes an obligation supported by the full faith and credit of the United

States and is not contestable except for fraud or misrepresentation (including negligent misrepresentation) of which the lender or holder has actual knowledge, participates in, or condones. A note which provides for the payment of interest on interest shall not be guaranteed and any Loan Note Guarantee or Assignment Guarantee Agreement attached to, or relating to, a note which provides for payment of interest on interest is void. The Loan Note Guarantee will not be enforceable by the lender to the extent any loss is occasioned by violation of usury laws, negligent servicing, or failure to obtain the required security regardless of the time at which the Agency acquires knowledge of the foregoing. Any losses occasioned will not be enforceable by the lender to the extent that loan funds are used for purposes other than those specifically approved by the Agency in its Conditional Commitment for Guarantee. Negligent servicing is defined as the failure to perform those services which a reasonably prudent lender would perform in servicing its own portfolio of loans that are not guaranteed. The term includes not only the concept of a failure to act, but also not acting in a timely manner, acting in a manner contrary to the manner in which a reasonably prudent lender would act up to the time of loan maturity, or until a final loss is paid. The Loan Note Guarantee or Assignment Guarantee Agreement in the hands of a holder shall not cover interest accruing 90 days after the holder has demanded repurchase by the lender, nor shall the Loan Note Guarantee or Assignment Guarantee Agreement in the hands of a holder cover interest accruing 90 days after the lender or Agency has requested the holder to surrender the evidence of debt for repurchase.

§ 1779.4 Conditions of guarantee.

A loan guarantee under this part will be evidenced by a Loan Note Guarantee issued by the Agency. Each lender will also execute a Lender's Agreement.

(a) The entire loan will be secured by the same security with equal lien priority for the guaranteed and non-guaranteed portions of the loan. The non-guaranteed portion of the loan will not be paid first nor given any preference

or priority over the guaranteed portion.

(b) The lender will be responsible for servicing the entire loan and will remain mortgagee or secured party of record notwithstanding the fact that another party may hold a portion of the loan.

(c) When a guaranteed portion of a loan is sold to a holder, the holder shall have all rights of the lender under the Loan Note Guarantee to the extent of the portion purchased. The lender will remain bound by all the obligations under the Loan Note Guarantee, Lender's Agreement, and Agency program regulations. If the Agency makes a payment to a holder, then the lender must reimburse the Agency.

(d) A lender will receive all payments of principal and interest on the account of the entire loan and will promptly remit to each holder a pro rata share, less any lender servicing fee.

(e) The lender may retain all of the unguaranteed portion of the loan or may sell part of the unguaranteed portion of the loan through participation. However, the lender is required to retain 5 percent of the loan amount from the unguaranteed portion in their portfolio.

§§ 1779.5–1779.7 [Reserved]

§ 1779.8 Access to lender's records.

Upon request by the Agency, the lender will permit representatives of the Agency (or other agencies of the U.S. Department of Agriculture authorized by that Department or the U.S. Government) to inspect and make copies of any of the records of the lender pertaining to the guaranteed loans. Such inspection and copying may be made during regular office hours of the lender or at any other time the lender and the Agency agree upon.

§ 1779.9 Environmental review requirements.

Facilities financed under this part must comply with the environmental review requirements in accordance with 7 CFR part 1970. In accordance with Agency guidance documents, the environmental review requirements shall be performed by the applicant simultaneously and concurrently with the project's engineering planning and design. The lender must assist the Agency in ensuring that the borrower complies with the Agency's environmental review requirements and implements any mitigation measure identified in the environmental review document or Conditional Commitment for Guarantee.

[81 FR 11028, Mar. 2, 2016]

§§ 1779.10–1779.11 [Reserved]

§ 1779.12 Inspections.

The lender will notify the Agency of any scheduled field inspections during construction and after issuance of the Loan Note Guarantee. The Agency may attend such field inspections. Any inspections or review conducted by the Agency, including those with the lender, are for the benefit of the Agency only and not for the benefit of other parties in interest. Agency inspections do not relieve any parties in interest of their responsibilities to conduct necessary inspections.

§ 1779.13 Appeals.

Only the borrower, lender, or holder can appeal an Agency decision. In cases where the Agency has denied or reduced the amount of final loss payment to the lender, the adverse decision may be appealed only by the lender. A decision by a lender adverse to the interest of the borrower is not a decision by the Agency, whether or not concurred in by the Agency. Appeals will be handled in accordance with the regulations of the National Appeals Division, U.S. Department of Agriculture, published at 7 CFR part 11.

§§ 1779.14–1779.16 [Reserved]

§ 1779.17 Exception authority.

The Administrator may, in individual cases, make an exception to any requirement or provision of this part which is not inconsistent with the authorizing statute or other applicable law and is determined to be in the Government's interest.

§§ 1779.18–1779.19 [Reserved]

§ 1779.20 Eligibility.

(a) *Availability of credit from other sources.* The Agency must determine that the borrower is unable to obtain the required credit without the loan guarantee from private, commercial, or cooperative sources at reasonable rates and terms for loans for similar purposes and periods of time. The Agency must also determine if an outstanding judgment obtained by the United States in a Federal Court (other than the U.S. Tax Court) has been entered against the borrower or if the borrower has an outstanding delinquent debt with any Federal agency. Such judgment or delinquency shall cause the potential borrower to be ineligible to receive a loan guarantee until the judgment is paid in full or otherwise satisfied or the delinquency is cured.

(b) *Legal authority and responsibility.* (1) Each borrower must have, or will obtain, the legal authority necessary to construct, operate, and maintain the proposed facility and services. They must also have legal authority for obtaining, giving security for, and repaying the proposed loan.

(2) The borrower shall be responsible for operating, maintaining, and managing the facility and services, and providing for the continued availability and use of the facility and services at reasonable rates and terms.

(c) *Applicant.* Eligible entities are:

(1) A public body such as a municipality, county, district, authority, or other political subdivision of a State located in a rural area.

(2) An organization operated on a not-for-profit basis, such as an association, cooperative, or private corporation. The organization must be an association controlled by a local public body or bodies, or have a broadly based ownership by or membership of people of the local community; or

(3) Indian tribes on Federal and State reservations and other federally recognized Indian tribes.

(d) *Facility location.* Facilities must be located in rural areas, except: For utility services such as drinking water, sanitary sewer, solid waste disposal or storm drainage facilities serving both rural and non-rural areas. In such cases, Agency funds may be used to finance only that portion serving rural areas, regardless of facility location.

(e) *Facilities for public use.* All facilities financed under the provisions of this part shall be for public purposes.

(1) Facilities will be installed to serve any user within the service area who desires service and can be feasibly and legally served.

(2) In no case will boundaries for the proposed service area be chosen in such a way that any user or area will be excluded because of race, color, religion, sex, marital status, age, disability, or national origin.

(3) The lender will determine that, when feasible and legally possible, inequities within the proposed project's service area for the same type service proposed will be remedied by the owner on, or before, completion of the project. Inequities are defined as unjustified variations in availability, adequacy, or quality of service. User rate schedules for portions of existing systems or facilities that were developed under different financing, rates, terms, or conditions do not necessarily constitute inequities.

§§ 1779.21–1779.23   [Reserved]

§ 1779.24   Eligible loan purposes.

(a) To construct, enlarge, extend, or otherwise improve rural drinking water, sanitary sewage, solid waste disposal, and storm wastewater disposal facilities.

(b) To construct or relocate public buildings, roads, bridges, fences, or utilities, and to make other public improvements necessary for the successful operation or protection of facilities authorized in paragraph (a) of this section.

(c) To relocate private buildings, roads, bridges, fences, or utilities, and other private improvements necessary for the successful operation or protection of facilities authorized in paragraph (a) of this section.

(d) For payment of other utility connection charges as provided in service contracts between utility systems.

(e) When a necessary part of the project relates to those facilities authorized in paragraphs (a), (b), (c) or (d)

of this section the following may be considered:

(1) Reasonable fees and costs such as: legal, engineering, administrative services, fiscal advisory, recording, environmental analyses and surveys, possible salvage or other mitigation measures, planning, establishing or acquiring rights;

(2) Costs of acquiring interest in land: rights, such as water rights; leases; permits; rights-of-way; and other evidence of land or water control or protection necessary for development of the facility;

(3) Purchasing or renting equipment necessary to install, operate, maintain, extend, or protect facilities;

(4) Cost of additional applicant labor and other expenses necessary to install and extend service;

(5) In unusual cases such as a low-income area, the cost for connecting the user to the main service line;

(6) Interest incurred during construction in conjunction with multiple advances or interest on interim financing;

(7) Initial operating expenses, including interest, for a period ordinarily not exceeding one year when the applicant is unable to pay such expenses;

(8) The purchase of existing facilities when it is necessary either to improve service or prevent the loss of service; and

(9) Refinancing non-Agency debts incurred by, or on behalf of, an applicant when all of the following conditions exist:

(i) The debts being refinanced are a secondary part of the total loan unless the debt being refinanced is an Agency direct loan;

(ii) The debts were incurred for the facility or service being financed or any part thereof; and

(iii) Arrangements cannot be made with the creditors to extend or modify the terms of the debts so that a sound basis will exist for making a loan.

(10) Refinancing Agency debts.

§ 1779.25   Ineligible loan purposes.

Loan funds may not be used to finance:

(a) Facilities which are not modest in size, design, and cost;

(b) Loan or grant finder's fees;

(c) The construction of any new combined storm and sanitary sewer facilities;

(d) Any portion of the cost of a facility which does not serve a rural area;

(e) That portion of project costs normally provided by a business or industrial user, such as wastewater pretreatment;

(f) Rental for the use of equipment or machinery owned by the applicant;

(g) For other purposes not directly related to operating and maintenance of the facility being installed or improved; or

(h) The payment of a judgment which would disqualify an applicant for a loan under §1779.20(a).

§1779.26 [Reserved]

§1779.27 Lenders.

(a) *Eligible lenders.* Eligible lenders may participate in the loan guarantee program. These lenders must be subject to credit examination and supervision by an appropriate agency of the United States or a State that supervises and regulates credit institutions. A lender must have the capability to adequately service loans for which a guarantee is requested. Eligible lenders are:

(1) Any Federal or State chartered bank or savings and loan association;

(2) Any mortgage company that is a part of a bank holding company;

(3) Co-Bank, National Rural Utilities Cooperative Finance Corporation, Farm Credit Bank of the Federal Land Bank, or other Farm Credit System institution with direct lending authority authorized to make loans of the type guaranteed by this part;

(4) An insurance company regulated by a State or National insurance regulatory agency;

(5) State Bond Banks or State Bond Pools; and

(6) Other lenders that possess the legal powers necessary and incidental to making and servicing guaranteed loans involving community development-type projects. Lenders under this category must be approved by the National Office prior to the issuance of the loan guarantee.

(b) *Conflict of interest.* When the lender's officers, stockholders, directors, or partners (including their immediate families) or the borrower, its officers, stockholders, directors, or partners (including their immediate families) own, or have management responsibilities in each other, the lender must disclose such business or ownership relationships. The Agency will determine if such relationships are likely to result in a conflict of interest. This does not preclude lender officials from being on the borrower's board of directors.

§1779.28 Transfer of lenders or borrowers (prior to issuance of Loan Note Guarantee).

(a) Prior to issuance of the loan guarantee, the Agency may approve the transfer of an outstanding Conditional Commitment for Guarantee from the present lender to a new eligible lender: Provided, That:

(1) The former lender states in writing why it does not wish to continue to be the lender for this project;

(2) No substantive changes in ownership or control of the borrower has occurred;

(3) No substantive changes in the borrower's written plan, scope of work, or changes in the purpose or intent of the project has occurred; and

(4) No substantive changes in the loan agreement or Conditional Commitment for Guarantee are required.

(b) The substitute lender must execute a new application for loan and guarantee (available in any Agency office).

(c) If approved, the Agency will issue a letter of amendment to the original Conditional Commitment for Guarantee reflecting the new lender who will acknowledge acceptance of the offer in writing.

(d) Once the Conditional Commitment for Guarantee is issued, the Agency will not approve any substitution of borrowers, including changes in the form of the legal entity, except a change in the legal entity may be requested when the original borrower is replaced with substantially the same individuals or officers with the same interest as originally approved.

§1779.29 Fees and charges by lender.

(a) *Routine charges and fees.* The lender may establish charges and fees for the loan if they do not exceed those

273

charged other borrowers for similar types of transactions. "Similar types of transactions" mean those transactions involving the same type of loan for which a non-guaranteed loan borrower would be assessed charges and fees.

(b) *Late payment fees.* Late payment charges will not be covered by the Loan Note Guarantee. Such charges may not be added to the principal and interest due under any guaranteed note. Late payment charges may be made only if:

(1) They are routinely made by the lender in all types of loan transactions;

(2) Payment has not been received within the customary timeframe allowed by the lender; or

(3) The lender agrees with the borrower, in writing, that the rate or method of calculating the late payment charges will not be changed to increase charges while the Loan Note Guarantee is in effect.

(c) *Guarantee fees.* The guaranteed loan fee will be the applicable guarantee fee rate multiplied by the principal loan amount multiplied by the percent of guarantee. The one-time guarantee fee is paid when the Loan Note Guarantee is issued.

(1) The fee will be paid to the Agency by the lender and is nonreturnable. The lender may pass the fee to the borrower.

(2) The guarantee fee rates are available in any Agency office.

§ 1779.30 Loan guarantee limitations.

(a) The guarantee will be 90 percent of eligible loss.

(b) The lender will retain a minimum of 5 percent of the total loan amount. The retained amount must be from the unguaranteed portion of the loan and cannot be participated to another lender.

§§ 1779.31–1779.32 [Reserved]

§ 1779.33 Interest rates.

(a) *General.* Rates will be negotiated between the lender and the borrower. They may be either fixed or variable rates. Interest rates will be those rates customarily charged borrowers in similar circumstances in the ordinary course of business and are subject to Agency review and approval.

(b) *Variable rate publication.* A variable interest rate must be tied to a base rate published periodically in a recognized national or regional financial publication specifically agreed to by the lender and borrower. Such an agreement must be documented in the borrower or lender loan agreement.

(1) Interest rate caps and incremental adjustment limitations will also be negotiated between the lender and the borrower. Notice of any interest rate change proposed by the lender should allow a sufficient time period for the borrower to obtain any required State or other regulatory approval and to implement any user rate adjustments necessary as a result of the interest rate change. The intervals between interest rate adjustments will be specified in the loan agreement (but not more often than quarterly).

(2) The lender must incorporate within the variable rate note, the provision for adjustment of payments coincident with an interest rate adjustment. This will ensure the outstanding principal balance is properly amortized within the prescribed loan maturity and eliminate the possibility of a balloon payment at the end of the loan.

(c) *Changes.* Any change in the interest rate between the date of issuance of the Conditional Commitment for Guarantee and before the issuance of the Loan Note Guarantee must be approved by the Agency. Approval of such change will be shown as an amendment to the Conditional Commitment for Guarantee.

(d) *Different rates on guaranteed and unguaranteed portion of the loan.* It is permissible to have one interest rate on the guaranteed portion of the loan and another interest rate on the unguaranteed portion of the loan, provided the lender and borrower agree, and:

(1) The rate on the unguaranteed portion does not exceed that currently being charged on loans for similar purposes to borrowers under similar circumstances; and

(2) The rate on the guaranteed portion of the loan will not exceed the rate on the unguaranteed portion. This requirement does not apply when the

unguaranteed rate is variable and the guaranteed portion is fixed.

(e) *Multi-rates*. When multi-rates are used, the lender will provide the Agency with the overall effective interest rate for the entire loan. Multi-rate loans may be either fixed, variable, or a combination of fixed and variable.

§ 1779.34 Terms of loan repayment.

(a) *General*. Principal and interest on the loan will be due and payable as provided in the note except, any interest accrued as the result of the borrower's default on the guaranteed loan over and above that which would have accrued at the note rate on the guaranteed loan will not be guaranteed by the Agency. The lender will structure repayments as established in the loan agreement between the lender and borrower. Ordinarily, such installments will be scheduled for payment as agreed upon by the lender and borrower on terms that reasonably ensure repayment of the loan. However, the first installment to include a repayment of principal may be scheduled for payment after the project is operable and has begun to generate income. Such installment must be due and payable within 3 years from the date of the note and at least annually thereafter. Interest will be due at least annually from the date of the note. Monthly payments will be required except for borrowers with income limited to less frequent intervals.

(b) *Term length*. The maximum time allowable for final maturity for a guaranteed WW loan will be limited to the useful life of the facility, not to exceed 40 years.

(c) *Balloon payments*. The principal balance should be properly amortized within the prescribed loan maturity. Balloon payments at the end of the loan are prohibited.

§§ 1779.35–1779.36 [Reserved]

§ 1779.37 Insurance and fidelity bonds.

The lender must provide evidence that the borrower has adequate insurance and fidelity bond coverage by loan closing or start of construction, whichever occurs first. Adequate coverage must be maintained for the life of the loan and is subject to Agency review and approval.

§§ 1779.38–1779.41 [Reserved]

§ 1779.42 Design and construction requirements.

The lender will provide the Agency with a written certification at the end of construction that all funds were utilized for authorized purposes. The borrower and the lender will authorize designs and plans based upon the preliminary architectural and engineering reports or plans approved by the lender and concurred in by the Agency. The borrower will take into consideration any lender or Agency comments when the facility is being designed.

(a) *Architectural and engineering practices*. All project facilities must be designed utilizing accepted architectural and engineering practices and must conform to applicable Federal, State, and local codes and requirements. The lender must ensure that the planned project will be completed within the available funds and, once completed, will be suitable for the borrower's needs.

(b) *Construction monitoring*. The lender will monitor the progress of construction and undertake the reviews and inspections necessary to ensure that construction proceeds in accordance with the approved plans, specifications, and contract documents and that funds are used for eligible project costs. The lender must expeditiously report any problems in project development to the Agency.

(c) *Equal employment opportunities*. For all construction contracts in excess of $10,000, the contractor must comply with Executive Order 11246 (30 FR 12319, 3 CFR, 1964–1965 Comp., p. 339) entitled "Equal Employment Opportunity" as amended and as supplemented by applicable Department of Labor regulations (41 CFR part 60–1). The borrower and lender are responsible for ensuring that the contractor complies with these requirements.

(d) *Americans with Disabilities Act*. WW loans which involve the construction of, or addition to, facilities that accommodate the public and commercial facilities as defined by the Americans with Disabilities Act (42 U.S.C. 12181 *et seq.*) must comply with that Act. The lender and borrower are responsible for compliance.

(e) *Administrative.* When the Agency reviews the preliminary architectural and engineering reports or plans, they must also consider all applicable Federal laws such as the seismic requirements of Executive Order 12699 (55 FR 835, 3 CFR, 1990 Comp., p. 269), the debarment requirements of 2 CFR part 417, and the Copeland Anti-Kickback Act (18 U.S.C. 874).

[66 FR 23138, May 8, 2001, as amended at 79 FR 76006, Dec. 19, 2014]

§ 1779.43 Other Federal, State, and local requirements.

In addition to the specific requirements of this part and beginning on the date of issuance of the Loan Note Guarantee, proposals for facilities financed in whole or in part with a loan guaranteed by the Agency will be coordinated with all appropriate Federal, State, and local agencies. Borrowers and lenders will be required to comply with any Federal, State, or local laws or regulatory commission rules which are in existence and which affect the project including, but not limited to:

(a) Applicant's authority to design, construct, develop, operate, and maintain the proposed facilities;

(b) Borrowing money, giving security, and raising revenues for repayment;

(c) Land use zoning;

(d) Health, safety, and sanitation standards as well as design and installation standards; and

(e) Protection of the environment and consumer affairs.

§§ 1779.44–1779.46 [Reserved]

§ 1779.47 Economic feasibility requirements.

All projects financed under the provisions of this section must be based on taxes, assessments, revenues, fees, or other sources of revenues in an amount sufficient to provide for facility operation and maintenance, a reasonable reserve, and debt payment. The lender is responsible for determining the credit quality and economic feasibility of the proposed loan and must address all elements of the credit quality in a written financial feasibility analysis which includes adequacy of equity, cash flow, security, history, and management capabilities. Financial feasibility reports must take into consideration any interest rate adjustment which may be instituted under the terms of the note. The lender's financial credit analysis may also serve as the feasibility analysis when sufficient evidence is included to determine economic feasibility as well as financial viability. The borrower's consulting engineer may complete the financial feasibility analysis for WW systems. If the facility is used by businesses and the success or failure of the facility is dependent on individual businesses, then the economic viability of those businesses must be assessed.

(a) *Exceptions.* The Agency loan approval official may exempt the lender from the requirement for an independent financial feasibility report (when requested by the borrower and the lender) provided the approval official determines that the financial feasibility analysis prepared by the borrower fairly represents the financial feasibility of the facility and the financial feasibility analysis contains an accurate projection of the usage, revenues, and expenses of the facility.

(b) *Insufficient information.* When the lender or Agency has insufficient information to determine the borrower's repayment ability, an independent feasibility analysis is required.

§ 1779.48 Collateral.

(a) *Lender responsibility.* The lender is responsible for obtaining and maintaining proper and adequate collateral to protect the interest of the lender, the holder, and the Government.

(b) *Type of collateral.* Collateral must be of such a nature that repayment of the loan is reasonably ensured when considered with the integrity and ability of project management, soundness of the project, and the borrower's prospective earnings. The collateral may include, but is not limited to, the following: General obligation bonds, revenue bonds, pledge of taxes or assessments, assignment of facility revenue, land, easements, rights-of-way, water rights, buildings, machinery, equipment, accounts receivable, contracts, cash, or other accounts or assignments of leases or leasehold interest.

(c) *Separate collateral.* All collateral must secure the entire loan. The lender will not take separate security to secure only the unguaranteed portion of the loan. The lender will not require compensating balances or certificates of deposit as a means of eliminating the lender's exposure on the unguaranteed portion of the loan.

§§ 1779.49–1779.50 [Reserved]

§ 1779.51 Strategic economic and community development.

Applicants with projects that support the implementation of strategic economic development and community development plans are encouraged to review and consider 7 CFR part 1980, subpart K, which contains provisions for providing priority to projects that support the implementation of strategic economic development and community development plans on a Multi-jurisdictional basis.

[81 FR 10456, Mar. 1, 2016]

§ 1779.52 Processing.

(a) *Preapplications.* (1) The preapplication package may be submitted either alone or the necessary information may be submitted simultaneously with the application. The preapplication package will contain:

(i) An Application for Federal Assistance on a form provided by the Agency (available in any Agency office);

(ii) State intergovernmental or other type review comments and recommendations for the borrower's project (clearinghouse comments, if applicable);

(iii) Supporting documentation necessary to make an eligibility determination such as financial statements, audits, copies of organizational documents, or existing debt instruments; and

(iv) Documentation of lender eligibility in accordance with § 1779.27.

(2) If the Agency determines that the project may meet requirements and is likely to be funded, the lender must submit a complete application if it has not previously submitted one.

(b) *Applications.* Contents of application package:

(1) Application for Loan and Guarantee on a form prescribed by the Agency (available in any Agency office);

(2) Proposed loan agreement;

(3) Environmental review documentation in accordance with 7 CFR part 1970.

(4) Preliminary architectural or engineering report (PER);

(5) Cost estimates;

(6) Appraisal reports (as appropriate);

(7) Credit reports (as appropriate);

(8) Financial feasibility analysis and report (as appropriate) if not included in PER; and

(9) Any additional information required.

[66 FR 23138, May 8, 2001, as amended at 81 FR 11028, Mar. 2, 2016]

§ 1779.53 Evaluation of application.

If the Agency determines that the borrower is eligible, the proposed loan is for an eligible purpose, there is reasonable assurance of repayment ability, sufficient collateral and equity exists, the proposed loan complies with all applicable statutes and regulations, the environmental impact analyses is complete, and adequate funds are available, the Agency will provide the lender and the borrower with the Conditional Commitment for Guarantee, listing all conditions for the guarantee. Applicable requirements will include the following:

(a) Approved use of guaranteed loan funds (source and use of funds);

(b) Rates and terms of the loan;

(c) Scheduling of payments;

(d) Number of customers;

(e) Security and lien priority;

(f) Appraisals;

(g) Insurance and bonding;

(h) Financial reporting;

(i) Equal opportunity and nondiscrimination;

(j) Mitigation measures for environmental issues (if necessary);

(k) Americans with Disabilities Act;

(l) By-laws and articles of incorporation changes; and

(m) Other requirements necessary to protect the Government.

§§ 1779.54–1779.58 [Reserved]

§ 1779.59 Review of requirements.

(a) *Lender and borrower.* The lender and borrower must complete and sign the Acceptance of Conditions and return a copy to the Agency as soon as possible. Notwithstanding the preceding sentence, if certain conditions cannot be met, the lender and borrower may propose alternate conditions for Agency consideration.

(b) *Cancellation.* If the lender decides at any time after receiving a Conditional Commitment for Guarantee that it no longer wants a guarantee, the lender must immediately advise the Agency of the cancellation.

(c) *Modifications.* The lender agrees that once the Conditional Commitment for Guarantee is issued and accepted by the lender and borrower, it will not be modified as to the scope of the project, overall facility concept, project purpose, use of proceeds, or other terms and conditions.

§§ 1779.60–1779.62 [Reserved]

§ 1779.63 Conditions precedent to issuance of the Loan Note Guarantee.

The Loan Note Guarantee will not be issued until:

(a) The lender certifies that:

(1) No changes have been made in the lender's loan conditions and requirements since the issuance of the Conditional Commitment for Guarantee except those approved in the interim by the Agency in writing.

(2) All planned property acquisition has been completed and all development has been substantially completed in accordance with plans, specifications, and applicable building codes. No costs have exceeded the amounts approved by the lender and the Agency.

(3) Required insurance is in effect.

(4) The loan has been properly closed and the required security instruments have been obtained on any after-acquired property that cannot be covered initially under State statutory provisions.

(5) The borrower has marketable title to the collateral then owned by the borrower, subject to the instrument securing the loan to be guaranteed and subject to any other exceptions approved, in writing, by the Agency.

(6) When required, the entire amount of the loan for working capital has been disbursed except in cases where the Agency has approved disbursement over an extended time.

(7) All other requirements of the Conditional Commitment for Guarantee have been met.

(8) Lien priorities are consistent with requirements of the Conditional Commitment for Guarantee.

(9) The loan proceeds have been disbursed for purposes and in amounts consistent with the Conditional Commitment for Guarantee and as specified on the application for the guaranteed loan. A copy of a detailed statement by the lender detailing the use of loan funds will be attached to support this certification.

(10) There has been no substantive adverse change in the borrower's financial condition nor any other adverse change in the borrower during the period of time from the Agency's issuance of the Conditional Commitment for Guarantee to issuance of the Loan Note Guarantee. The lender's certification must address all adverse changes of the borrower and the guarantors. For purposes of this paragraph (a)(10), the term borrower includes any parent, affiliate, or subsidiary of the borrower.

(11) All Federal, State, and local design and construction requirements have been met.

(12) The lender understands and will meet the requirements of the Debt Collection Act (31 U.S.C. Chapter 37).

(13) The lender would not make the loan without an Agency guarantee.

(b) The lender has executed and delivered the Lender's Agreement and closing report for the guaranteed loan along with the appropriate guarantee fee.

(c) The lender has advised the Agency of plans to sell or assign any part of the loan as provided in the Lender's Agreement.

(d) Where applicable, the lender must certify that the borrower has obtained:

(1) A legal opinion relative to the title to rights-of-way and easements. Lenders are responsible for ensuring that borrowers have obtained valid,

continuous, and adequate rights-of-way and easements needed for the construction, operation, and maintenance of a facility.

(2) A title opinion or title insurance showing ownership of the land and all mortgages or other lien defects, restrictions, or encumbrances, if any. It is the responsibility of the lender to ensure that the borrower has obtained and recorded such releases, consents, or subordinations to such property rights from holders of outstanding liens or other instruments as may be necessary for the construction, operation, and maintenance of the facility and to provide the required security. For example, when a site is for major structures and the lender and borrower are able to obtain only a right-of-way or easement on such a site rather than a fee simple title, such a title opinion must be requested.

(e) If the Loan Note Guarantee cannot be issued before the Conditional Commitment expires, the lender must submit a written request for an extension of the expiration date. The lender must document and certify to paragraph (a)(1) and (a)(11) of this section specifically identifying any modifications.

(f) Coincident with, or immediately after, loan closing, the lender will contact the Agency and provide those documents and certifications required in this section. For loans to public bodies, lenders may require an opinion from recognized bond counsel regarding the adequacy of the preparation and issuance of the debt instruments. Only when the Agency is satisfied that all conditions for the guarantee have been met will the Loan Note Guarantee be executed.

§ 1779.64 Issuance of Lender's Agreement, Loan Note Guarantee, and Assignment Guarantee Agreement.

(a) *Lender's Agreement.* If the Agency finds that all requirements have been met, the lender and the Agency will execute the Lender's Agreement. The original will be retained by the Agency and a signed duplicate original will be retained by the lender. A separate Lender's Agreement must be executed for each loan to be guaranteed by the Agency.

(b) *Loan Note Guarantee.* (1) Upon receipt of the executed Lender's Agreement and after all requirements have been met, the Agency will execute the Loan Note Guarantee. All originals of the Loan Note Guarantee will be provided to the lender and attached to the note.

(2) If the lender has selected the multi-note system, a Loan Note Guarantee will be prepared and attached to each note the borrower issues. All the notes will be listed on the Loan Note Guarantee. Not more than ten notes will be issued for the guaranteed portion (unless the Agency and borrower agree otherwise) and one note issued for the unguaranteed portion.

(c) *Assignment of Guarantee.* In the event the lender assigns the guaranteed portion of the loan to a holder, the lender, holder, and Agency will execute an Agency prescribed Assignment Guarantee Agreement.

(d) *Failure to meet conditions.* If the Agency determines that it cannot execute the Loan Note Guarantee because all requirements have not been met, the lender will have a reasonable period within which to satisfy the objections. If the lender satisfies the objections within the time allowed, the guarantee will be issued.

(e) *Loan closing report.* The lender will prepare and deliver a guaranteed loan closing report for each loan to be guaranteed and a guarantee fee to the Agency in return for the Loan Note Guarantee.

§ 1779.65 Lender's sale or assignment of the guaranteed portion of loan.

The lender may retain all of the guaranteed loan. The lender must not sell or participate any amount of the guaranteed or non-guaranteed portion of the loan to the borrower or to members of the borrower's immediate families, the borrower's officers, directors, stockholders, other owners, or a subsidiary or affiliate. Disposition of the guaranteed portion of a loan may not be made prior to full disbursement, completion of construction, and acquisition of real estate and equipment without the prior written approval of the Agency. If the lender desires to market all or part of the guaranteed portion of the loan at, or subsequent to, loan closing, the loan must not be in default.

(a) *Assignment.* Any sale or assignment by the lender of the guaranteed portion of the loan must be accomplished in accordance with the conditions in the Lender's Agreement.

(b) *Participation.* The lender may obtain participation in the loan under its normal operating procedures.

(c) *Minimum retention.* The lender is required to hold in its own portfolio or retain a minimum of 5 percent of the total loan amount. This amount must be of the non-guaranteed portion of the loan and cannot be participated to another. The lender may sell the remaining amount of the non-guaranteed portion of the loan only through participation.

§§ 1779.66–1779.68 [Reserved]

§ 1779.69 Loan servicing.

(a) *Lender responsibilities.* The lender is responsible for servicing the entire loan in accordance with the lender's loan agreement. The unguaranteed portion of the loan will not be paid first nor given any preference or priority over the guaranteed portion of the loan. The lender is responsible for taking all servicing actions that a prudent lender would perform in servicing a portfolio of loans that are not guaranteed. This responsibility includes, but is not limited to, the collection of payments; obtaining compliance with the covenants and provisions in the note, loan agreement, security instrument, or any supplemental agreements; obtaining and analyzing financial statements; verifying the payment of taxes and insurance premiums; and maintaining liens on collateral. The lender must notify the Agency of any violation of the loan agreement with the borrower within 30 days of such violation.

(b) *Financial reports.* The lender must obtain the financial statements required by the Loan Agreement. The lender must submit the borrower's annual financial statements to the Agency within 120 days of the end of the borrower's fiscal year. The lender must analyze the financial statements and provide the Agency with a written summary of the lender's analysis and conclusions, including trends, strengths, weaknesses, extraordinary transactions, and other indications of the financial condition of the borrower. Additionally, when applicable, the lender will require an audit in accordance with Office of Management and Budget (OMB) circulars (available in any Agency office). Additionally, when applicable, the lender will require an audit in accordance with subpart F of 2 CFR part 200, as adopted by USDA through 2 CFR part 400.

(c) *Delinquent loans.* The lender will service delinquent loans in accordance with the Lender's Agreement and reasonable and prudent lending standards.

(d) *Loan balances.* The lender must report to the Agency the outstanding principal and interest balance on each guaranteed loan semiannually.

(e) *Collateral inspections.* The lender will inspect the collateral as often as necessary to properly service the loan.

[66 FR 23138, May 8, 2001, as amended at 79 FR 76006, Dec. 19, 2014]

§§ 1779.70–1779.72 [Reserved]

§ 1779.73 Replacement of loss, theft, destruction, mutilation, or defacement of Loan Note Guarantee or Assignment Guarantee Agreement.

(a) *Replacement.* The Agency may issue a replacement Loan Note Guarantee or Assignment Guarantee Agreement which may have been lost, stolen, destroyed, mutilated, or defaced to the lender or holder upon receipt of a certificate of loss and an indemnity bond in accordance with this section.

(b) *Lender responsibilities.* When a Loan Note Guarantee or Assignment Guarantee Agreement is lost, stolen, destroyed, mutilated, or defaced while in the custody of the lender or holder, the lender will coordinate the activities of the party who seeks the replacement documents and will submit the required documents to the Agency for processing. The requirements for replacement are as follows:

(1) A certificate of loss properly notarized which includes:

(i) Legal name and present address of either the lender or the holder who is requesting the replacement forms;

(ii) Legal name and address of the lender of record;

(iii) Capacity of person certifying;

(iv) Full identification of the Loan Note Guarantee or Assignment Guarantee Agreement, including the name of the borrower, Agency case number, date of the Loan Note Guarantee, Assignment Guarantee Agreement, face amount of the evidence of debt purchased, date of evidence of debt, present balance of the loan, percentages of guarantee and, if Assignment Guarantee Agreement, the original named holder and the percentage of the guaranteed portion of the loan assigned to that holder. Any existing parts of the document to be replaced must be attached to the certificate;

(v) A full statement of circumstances of the loss, theft, or destruction of the Loan Note Guarantee or Assignment Guarantee Agreement; and

(vi) The holder shall present evidence demonstrating current ownership of the Loan Note Guarantee and Note or Assignment Guarantee Agreement. If the present holder is not the same as the original holder, a copy of the endorsement of each successive holder in the chain of transfer from the initial holder to present holder must be included. If copies of the endorsement cannot be obtained, best available records of transfer must be presented to the Agency (e.g., order confirmation, canceled checks).

(2) An indemnity bond acceptable to the Agency shall accompany the request for replacement except when the holder is the United States, a Federal Reserve Bank, a Federal Government corporation, a State or Territory, or the District of Columbia.

(3) All indemnity bonds must be issued and payable to the United States of America. The bond shall be in an amount not less than the unpaid principal and interest. The bond shall hold the Government harmless against any claim or demand which might arise or against any damage, loss, costs, or expenses which might be sustained or incurred by reasons of the loss or replacement of the instruments.

§1779.74 [Reserved]

§1779.75 Defaults by borrower.

(a) *Lender notification to Agency.* The lender must notify the Agency when a borrower is 30 days past due on a payment, has not met its responsibilities of providing the required financial statements, or is otherwise in default. The lender will continue to keep the Agency informed on a bimonthly basis until such time as the loan is no longer in default. If a monetary default exceeds 60 days, the lender will arrange a meeting with the borrower to resolve the default. The lender will provide a summary of the meeting and any decisions or actions agreed upon.

(b) *Servicing options.* In considering servicing options, the prospects for providing a permanent cure without adversely affecting the risks to the Agency and the lender must be the paramount objective. Temporary curative actions (such as payment deferments or collateral subordination) must strengthen the loan and be in the best financial interest of the lender and the Agency. Some of these actions may require concurrence of the holder.

(c) *Multi-note.* If the loan was closed with the multi-note option, the lender may need to possess all notes to take some servicing actions. In those situations when the Agency is holder of some of the notes, the Agency may endorse the notes back to the lender, provided a proper receipt is received from the lender which defines the reason for the transfer. Under no circumstances will the Agency endorse the original Loan Note Guarantee to the lender.

§§1779.76–1779.77 [Reserved]

§1779.78 Repurchase of loan.

(a) *Repurchase by lender.* The lender has the option to repurchase the loan from a holder within 30 days of written demand from the holder when the borrower is in default not less than 60 days on payment. The repurchase will be for an amount equal to the unpaid guaranteed portion of principal and accrued interest less the lender's servicing fee. The guarantee does not cover the note interest to the holder on the guaranteed loan accruing after 90 days from the date of the demand letter to the lender. The holder will concurrently send a copy of the demand to the Agency. The lender will accept an assignment without recourse from the holder upon repurchase. The lender is encouraged to repurchase the loan to facilitate the accounting of funds, resolve the problem, and permit the borrower to cure the default, where reasonable. The lender will notify the holder and the Agency of its decision within 30 days of receipt of demand from the holder.

(b) *Agency repurchase.* (1) If the lender does not repurchase as provided in paragraph (a) of this section, the Agency will purchase from the holder the unpaid principal balance of the guaranteed portion together with accrued interest to date of repurchase (less the lender's servicing fee) within 30 days after a specific written demand directed to the Agency. The copy of the demand on the lender is not sufficient. The guarantee will not cover the note interest to the holder on the guaranteed loan accruing after 90 days from the date of the original demand letter. The lender shall not charge the Agency any servicing fees nor are any such fees collectible from the Agency.

(2) The holder's demand to the Agency must include a copy of the written demand made upon the lender. The holder or duly authorized agent must also include evidence of the right to require payment from the Agency. Such evidence will consist of either the original of the Loan Note Guarantee properly endorsed to the Agency or the original of the Assignment Guarantee Agreement properly assigned to the Agency without recourse including all rights, title, and interest in the loan. The Agency will be subrogated to all rights of the holder. The holder must include in the demand the amount due including unpaid principal, unpaid interest to date of demand, and interest subsequently accruing from the date of demand to the proposed payment date. Unless otherwise agreed to by the Agency, such proposed payment will not be later than 30 days from the date of demand.

(3) The lender must promptly provide the Agency with the information necessary for the Agency's determination of the appropriate amount due the holder upon the Agency's notification to the lender of the holder's demand for payment. This information must be certified by an authorized officer of the

lender. Any discrepancy between the amount claimed by the holder and the information submitted by the lender must be resolved before payment will be approved. The Agency will notify both parties and such conflict will suspend the running of the 30-day payment requirement.

(4) Any purchase by the Agency does not change, alter, or modify any of the lender's obligations to the Agency arising from the loan or guarantee nor does it waive any of the Agency's rights against the lender. The Agency may set off against the lender all rights inuring to the Agency as the holder of the instrument against the Agency's obligation to the lender under the Loan Note Guarantee.

(c) *Repurchase for servicing.* When the lender determines that repurchase of the guaranteed portion of the loan is necessary to service the loan, the holder must sell the guaranteed portion to the lender for the unpaid principal and interest balance (less the lender's servicing fee). The guarantee does not cover interest accruing after 90 days from the date the lender's or Agency's letter requesting the holder to tender its guaranteed portion. The lender must not repurchase from the holder for arbitrage purposes to further its own financial gain. Any repurchase must be made only after the lender obtains the Agency written approval. If the lender does not repurchase the portion from the holder, the Agency may, at its option, purchase such guaranteed portion for servicing purposes.

§ 1779.79 [Reserved]

§ 1779.80 Interest rate changes after loan closing.

(a) *General.* Subject to the restrictions below, the borrower, lender, and holder (if any) may collectively effect a permanent reduction in the interest rate on the guaranteed loan at any time during the life of the loan on written agreement by all of the applicable parties. After such a permanent reduction, the Loan Note Guarantee will only cover losses of interest at the reduced interest rate. The Agency must be notified by the lender, in writing, within 10 calendar days of the change. When the Agency is a holder, it will concur only when it is demonstrated that the change is more viable than liquidation and that the Government's financial interests are not adversely affected. Factors which will be considered in making such determination are the Government's cost of borrowing money and the project's enhancement of rural development. The monetary recovery must be greater than the liquidation recovery, and a financial feasibility analysis must show the project's continued viability.

(1) Fixed rates cannot be changed to variable rates to reduce the interest rate to the borrower unless the variable rate has a ceiling which is less than the original fixed rate.

(2) Variable rates can be changed to a lower fixed rate. In a final loss settlement when qualifying rate changes are made with the required written agreements and notification, the interest will be calculated for the periods the given rates were in effect. The lender must maintain records which adequately document the accrued interest claimed.

(3) The lender is responsible for the legal documentation of interest rate changes. However, the lender may not issue a new note.

(b) *Increases.* No increases in interest rates will be permitted under the loan guarantee except the normal fluctuations in approved variable interest rate loans.

§1779.81 Liquidation.

Liquidation will occur when the lender concludes that liquidation of the guaranteed loan is necessary because of default or third party actions that the borrower cannot, or will not, cure or eliminate within a reasonable period of time and the Agency concurs with the lender; or the Agency, at any time, independently concludes that liquidation is necessary. The lender will proceed as expeditiously as possible, including giving any notices or taking any legal actions required by the security instruments.

(a) *General.* If a lender has made a loan guaranteed by the Agency under previous regulations, the lender has the option to liquidate the loan under the provisions of this part or under the provisions of previous regulations. The lender will notify the Agency in writing within 10 days after its decision to liquidate, which regulatory provisions it chooses to use. The lender may not choose some provisions of one regulation and other provisions of the other regulation.

(b) *Acquiring property titles.* If a lender acquires title to property, the Agency may elect to permit the lender the option of calculating the final loss settlement using the net proceeds received at the time of the ultimate disposition of the property. The lender must submit to the Agency a written request to use this option within 15 days of acquiring title and the Agency must agree, in writing, prior to the lender submitting any request for estimated loss payment.

(c) *Liquidation plan.* The lender will (within 30 days after a decision to liquidate) submit to the Agency, in writing, a proposed, detailed liquidation plan. Upon approval by the Agency of the liquidation plan, the lender will commence liquidation. The lender's liquidation plan must include, but is not limited to, the following:

(1) Such proof as the Agency requires to establish the lender's ownership of the guaranteed loan notes and related security instruments, a copy of the payment ledger or other documentation which reflects the outstanding loan balance and accrued interest to date, and the method of computing the interest;

(2) A complete list of collateral;

(3) The recommended liquidation methods for making the maximum collection possible on the indebtedness and the justification for such methods, including the recommended action for acquiring and disposing of all collateral;

(4) Necessary steps for preservation of the collateral;

(5) Copies of the borrower's latest available financial statements;

(6) An itemized list of estimated liquidation expenses expected to be incurred and justification for each expense;

(7) A schedule to periodically report to the Agency on the progress of the liquidation;

(8) Estimated protective advance amounts with justification;

(9) Proposed protective bid amounts on collateral to be sold at auction and a discussion of how the amounts were determined;

(10) If a voluntary conveyance is considered, the proposed amount to be credited to the guaranteed debt;

(11) Legal opinions, as needed; and

(12) If the outstanding balance of principal and interest is less than $250,000, the lender will obtain an estimate of fair market and potential liquidation value of the collateral. If the outstanding balance of principal and interest is $250,000 or more, the lender will obtain an independent appraisal report on all collateral securing the loan which will reflect the fair market value and potential liquidation value. The independent appraiser's fee will be shared equally by the Agency and the lender.

(d) *Partial liquidation plan.* If actions are necessary to immediately preserve and protect the collateral, a partial liquidation plan may be submitted and, when approved, must be followed by a complete liquidation plan prepared by the lender.

(e) *Disposition of collateral.* Disposition of collateral acquired by the lender must be approved, in writing, by the Agency when:

(1) The lender's cost to acquire the collateral of a borrower exceeds the potential recovery value of the security and the lender proposes abandoning the collateral in lieu of liquidation; or

(2) The acquired collateral is to be sold to the borrower, borrower's stockholders or officers, or the lender or lender's stockholders or officers.

(f) *Agency liquidation.* The Agency will liquidate at its option only when it is a holder and there is reason to believe the lender is not likely to initiate liquidation efforts that will result in maximum recovery. When the Agency liquidates, proceeds derived from the sale of the collateral will be applied first to reasonable liquidation expenses and second to the guaranteed portion of the loan.

(g) *Final loss payment.* Final loss payments will be made only after all collateral has been properly accounted for and liquidation expenses are determined to be reasonable and within approved limits. Any estimated loss payments made to the lender will be credited against the final loss on the guaranteed loan. The amount of an estimated loss payment must be credited as a deduction from the principal balance of the loan.

§ 1779.82   [Reserved]

§ 1779.83   Protective advances.

Protective advances can only be added to the loan account for purposes of requirements to preserve the value of the security. Protective advances constitute an indebtedness of the borrower to the lender and must be secured by collateral to the same extent as principal and interest. Protective advances include, but are not limited to, advances made for taxes, annual assessments, ground rent, hazard and flood insurance premiums affecting the collateral (including any other expenses necessary to protect the collateral). Attorney fees are not a protective advance.

(a) *Agency approval.* The Agency must approve, in writing, all protective advances on loans within its loan approval authority which exceed a total cumulative advance amount of $5,000 to the same borrower. Protective advances must be reasonable when associated with the value of the collateral being preserved.

(b) *Preserving collateral.* When considering protective advances, sound judgment must be exercised in determining that the additional funds advanced will actually preserve collateral and recovery is actually enhanced by making the advance.

§ 1779.84   Additional loans or advances.

The lender will not make additional expenditures or new loans to the borrower without first obtaining the written approval of the Agency even though such expenditures or loans will not be guaranteed.

§ 1779.85   Bankruptcy.

(a) *Calculating losses.* Report of Loss form (available in any Agency office) will be used for calculating estimated and final loss determinations.

(b) *Lender responsibility.* The lender is responsible for protecting the guaranteed loan debt and all the collateral securing it in bankruptcy proceedings. These responsibilities include, but are not limited to, the following:

(1) Filing a proof of claim, where necessary, and all necessary papers and pleadings;

(2) Attending and, where necessary, participating in meetings of the creditors and all court proceedings;

(3) Immediately seeking adequate protection of the collateral if it is subject to being used by the trustee in bankruptcy or the debtor in possession;

(4) Where appropriate, seeking involuntary conversion of a pending chapter 11 case to a liquidation proceeding or seeking dismissal of the proceedings; and

(5) Keeping the Agency adequately and regularly informed, in writing, of all aspects of the proceedings.

(c) *Appraisals.* In a chapter 9 or chapter 11 reorganization, the lender must obtain an independent appraisal of the collateral if the Agency believes an independent appraisal is necessary. The Agency and the lender will share the appraisal fee equally.

(d) *Liquidation expenses.* Only expenses authorized by the court of chapter 9 plans or chapter 11 reorganizations, or chapters 11 or 7 liquidation (unless the liquidation is by the lender), may be deducted from the collateral proceeds.

(e) *Repurchase from the holder.* The Agency or the lender, with the approval of the Agency, may initiate the repurchase of the unpaid guaranteed portion of the loan from the holder. If the lender is the holder, an estimated loss payment may be filed at the initiation of a chapter 7 proceeding or after a chapter 9 or chapter 11 proceeding becomes a liquidation proceeding. Any loss payment on loans in bankruptcy must be approved by the Agency.

(f) *Chapter 11 bankruptcy.* If a borrower has filed for protection under chapters 9 or 11 of the United States Code for a reorganization (but not chapter 13) and all or a portion of the debt has been discharged, the lender may request an estimated loss payment of the guaranteed portion of the accrued interest and principal discharged by the court. If the court approves revisions to the chapter 9 plan or chapter 11 reorganization plan, subsequent estimated loss payments may be requested in accordance with the court approved changes. Once the reorganization plan has been satisfactorily completed, the lender is responsible for submitting the documentation necessary for the Agency to review and adjust the estimated loss claim to reflect any actual discharge of principal and interest and to reimburse the lender for any court ordered interest-rate reduction under the terms of the reorganization plan.

(g) *Agency approval of estimated liquidation expenses.* The Agency must approve, in advance and in writing, the lender's estimated liquidation expenses of collateral in a liquidation if the liquidation is performed by the lender. These expenses must be reasonable and customary and not include in-house expenses of the lender.

(h) *Reconciliation.* In the event that the estimated loss payment exceeds the actual loss, the lender will reimburse the Agency the amount in excess of the actual loss plus interest at the note rate from the date of the estimated loss payment.

§§1779.86–1779.87 [Reserved]

§1779.88 Transfers and assumptions.

(a) *General.* For all transfers and assumptions, the lender must concur in the plans for disposition of funds in the transferor's debt service, reserve, and operation and maintenance account. The Agency will approve, in writing, transfers and assumptions of loans to transferees who will continue the original purpose of the guaranteed loan subject to the following applicable provisions:

(1) When the transaction is to a member of the borrower's organization, it will be at an amount which will not result in a loss to the lender.

(2) Transfers to eligible borrowers will receive preference if recovery to the lender from the sale price is not less than it would be if the transfer was to an ineligible borrower.

(3) The present borrower is unable or unwilling to accomplish the objectives

of the guaranteed loan, and the transfer will be to the lender's and Agency's advantage.

(4) The transferee will assume an amount at least equal to either the present market value or the debt, whichever is less.

(b) *Transfers to an eligible borrower.* (1) The total indebtedness may be transferred to an eligible borrower on the same terms.

(2) The total indebtedness may be transferred to another eligible borrower on different terms not to exceed those terms for which an initial guaranteed loan can be made.

(3) Less than the total indebtedness may be transferred to another eligible borrower on the same or different terms and the pro rata share of any eligible loss paid to the lender.

(4) A guaranteed loan for which the transferee is eligible may be made in connection with a transfer subject to the policies and procedures governing the type of loan being made.

(5) If the transferor is to receive a payment for the equity, the total debt must be assumed.

(c) *Ineligible borrower.* Transfers to ineligible borrowers are considered only when needed as a method for servicing problem cases when an eligible transferee is not available. Transfers should not be considered as a means by which members can obtain equity or as a method of providing a source of easy credit for purchasers. Transfers must meet the following requirements:

(1) All transfers to ineligible borrowers will include a one-time nonrefundable transfer fee to the Agency of no more than 1 percent. Transfer fees will be collected, and payments applied, in accordance with paragraph (d) of this section.

(2) For all loans covered by this part, the Agency may approve a transfer of indebtedness to, and assumption of, a loan by a transferee who does not meet the eligibility requirements for the kind of loan being assumed when the ineligible borrower will:

(i) Make a significant down payment, and

(ii) Agree to pay the remaining balance within not more than 15 years. Installments will be at least equal to the amount amortized over a period not greater than the remaining life of the debt being transferred, and the balance will be due the fifteenth year.

(3) Interest rates to ineligible transferees will be the rate specified in the note of the transferor or the rates customarily charged borrowers in similar circumstances in the ordinary course of business and are subject to Agency review and approval. The rates may be either fixed or variable.

(i) Transferees must have the ability to repay as determined by the lender the debt according to the Assumption Agreement and must have the legal authority to enter into the contract. The transferee will submit a current balance sheet to the lender. The lender will obtain and analyze the credit history of the transferee.

(ii) The transferor may receive equity payments only when the full amount of the debt is assumed. However, equity payments will not be made on more favorable terms than those on which the balance of the debt will be paid.

(d) *Transfer fees.* Transfer fees are a one-time nonrefundable cost to be collected by the lender at the time of application or proposal.

(1) The transfer fees will be a standard fee plus the cost of the appraisal.

(2) The lender will collect and submit the fee to the Agency.

(3) The Agency may waive the transfer fee if it determines that such waiver is in the best interest of the Agency.

(e) *Processing transfers and assumptions.* (1) In any transfer and assumption case, the transferor (including any guarantor) may be released from liability by the lender only with prior Agency written concurrence and only when the value of the collateral being transferred is at least equal to the amount of the loan, or part of the loan, being assumed. If the transfer is for less than the entire debt:

(i) The Agency must determine that the transferor and any guarantor have no reasonable debt-paying ability considering their assets and income at the time of transfer, and

(ii) The lender must certify that the transferor has cooperated in good faith, used due diligence to maintain the collateral against loss, and has otherwise fulfilled all of the regulations of this part to the best of the borrower's ability.

(2) The lender will make, in all cases, a complete credit analysis to determine viability of the project (subject to the Agency review and approval) including any requirement for deposit in an escrow account as security to meet the determined equity requirements for the project.

(3) The lender will confirm that the transaction can be properly transferred and the conveyance instruments will be filed, registered, or recorded as appropriate and legally permissible.

(4) The assumption will be made on the lender's form of Assumption Agreement and will contain the Agency case number of the transferor and transferee.

(5) Loan terms cannot be changed by the Assumption Agreement unless previously approved in writing by the Agency with the concurrence of holder and the transferor (including guarantor if it has not been released from personal liability). Any new loan terms cannot exceed those authorized in this part. The lender's request will be supported by:

(i) An explanation of the reasons for the proposed change in the loan terms, and

(ii) Certification that the lien position securing the guaranteed loan will be maintained or improved, and proper hazard insurance will be continued in effect.

(6) In the case of a transfer and assumption, it is the lender's responsibility to see that all such transfers and assumptions will be noted on all originals of the Loan Note Guarantee. The lender will provide the Agency a copy of the Transfer and Assumption Agreement.

(7) If a loss should occur upon a complete transfer of assets and assumption for less than the full amount of the debt and the transferor-debtor (including personal guarantor) is released from personal liability (as provided in paragraph (e)(1)(i) of this section), the lender (if holding the guaranteed portion) may file an estimated Report of Loss to recover their pro rata share of the actual loss at that time. Approved protective advances and accrued interest made during the arrangement of a transfer and assumption, if not assumed by the transferee, will be entered on the estimated Report of Loss.

§1779.89 **Mergers.**

(a) *General.* The Agency may approve mergers or consolidations (herein referred to as "mergers") when the resulting organization will be eligible for an Agency guaranteed loan and assumes all the liabilities and acquires all the assets of the merged borrower. Mergers may be approved when:

(1) The merger is in the best interest of the Government and the merging borrower;

(2) The resulting borrower can meet all required conditions as contained in specific loan note agreements; and

(3) All property can be legally transferred to the resulting borrower.

(b) *Distinguishing mergers from transfers and assumptions.* Mergers occur when one entity combines with another entity in such a way that the first entity ceases to exist as a separate entity while the other continues. In a consolidation, two or more entities combine to form a new, consolidated entity with the original entity ceasing to exist. Such transactions must be distinguished from transfers and assumptions in which a transferor will not necessarily go out of existence, and the transferee will not always take all the transferor's assets nor assume all the transferor's liabilities.

§1779.90 **Disposition of acquired property.**

(a) *General.* When the lender acquires title to the collateral and the final loss claim is not paid until final disposition, the lender must proceed as quickly as possible to develop a plan to fully protect the collateral, and the lender must dispose of the collateral without delay.

(b) *Re-title collateral.* Any collateral accepted by the lender must not be titled in the Agency's name in whole or in part. The Agency's position is that of a guarantor relating to losses, not a lender.

(c) *Collateral preservation.* After acquiring the collateral, the lender must protect the collateral from deterioration (weather, vandalism, etc.). Hazard insurance in an amount necessary to cover the fair market value of the collateral must be maintained.

(d) *Collateral sale.* (1) The lender will prepare and submit to the Agency a plan on the best method of sale, keeping in mind any prospective purchasers. The Agency must approve the plan in writing. If an existing approved liquidation plan addresses the disposition of acquired property, no further review is required unless modification of the plan is needed.

(2) Anytime there is a case when the conversion of collateral to cash can reasonably be expected to result in a negative net recovery amount, abandonment of the collateral should be considered. The Agency must approve abandonment in writing.

§§ 1779.91–1779.93 [Reserved]

§ 1779.94 Determination and payment of loss.

In all liquidation cases, final settlement will be made with the lender after the collateral is liquidated. The Agency will have the right to recover losses paid under the guarantee from any liable party.

(a) *General.* If the lender takes title to collateral, any loss will be based on the collateral value at the time the lender obtains title.

(b) *Loss calculations.* The Report of Loss form (available in any Agency office) will be used for calculations of all estimated and final loss determinations. Estimated loss payments may only be approved after the lender has submitted a liquidation plan approved by the Agency.

(c) *Estimated loss payments.* When the lender is conducting the liquidation and owns any of the guaranteed portion of the loan, it may request an estimated loss payment by submitting an estimate of loss that will occur in connection with liquidation of the loan. An estimated loss payment may be approved after the Agency has approved the liquidation plan.

(1) The lender will prepare and submit a Report of Loss using the appraised value in lieu of amount received from sale of collateral.

(2) The estimated loss payment shall be calculated as of the date of such payment. The total amount of the loss payment remitted by the Agency will be applied by the lender on the guaranteed portion of the loan debt. Such application does not release the borrower from liability. At the time of final loss settlement, the lender may notify the borrower that the loss payment has been so applied.

(3) After liquidation has been completed, a final Report of Loss will be submitted by the lender to the Agency.

(d) *Final report of loss.* In all cases, a final Report of Loss must be submitted to the Agency. Before Agency approval of any final loss report, the lender must account for all funds obtained, disposition of the collateral, all costs incurred, and any other information necessary for the successful completion of liquidation. Upon receipt of the final accounting and Report of Loss, the Agency may conduct an audit and will determine the final loss. The lender will make its records available to, and otherwise assist, the Agency in making any audit it requires of the Report of Loss. The documentation accompanying the Report of Loss must support the loss claimed.

(1) The lender must document and show that all of the collateral has been accounted for and properly liquidated and that liquidation proceeds have been properly accounted for and applied correctly on the loan. The Agency must be satisfied that the lender has accomplished this in the manner contained herein and that the lender has maximized the collections in conducting the liquidation.

(2) The lender must show a breakdown on any protective advance amount as to the payee, purpose of the expenditure, date paid, evidence that the amount expended was proper, and that the amount was actually paid.

(3) The lender must show a breakdown of liquidation expenses as to the payee, purpose of the expenditure, date paid, evidence that the amount expended was proper, and that the amount was actually paid.

(4) Accrued interest should be supported by attachments showing how the amount was accrued by the lender. A copy of the promissory note and ledger will be attached. If the interest rate was a variable rate, the lender must include documentation of

changes in the selected base rate and when the changes in the loan rate became effective.

(e) *Liquidation income.* Any net rental or other income that has been received by the lender from the collateral will be applied on the guaranteed loan debt.

(f) *Liquidation costs.* Certain reasonable liquidation costs will be allowed during the liquidation process. The liquidation costs must be submitted as a part of the liquidation plan. Such costs will be deducted from gross proceeds received from the disposition of collateral unless the costs have been previously determined by the lender (with Agency concurrence) to be protective advances. If changed circumstances after submission of the liquidation plan require a revision of liquidation costs, the lender will obtain the Agency's written concurrence prior to proceeding with the proposed changes. No in-house expenses of the lender will be allowed.

(g) *Protective advance losses.* In those instances where the lender made authorized protective advances, the lender may claim recovery for the guaranteed portion of any loss of monies advanced as well as interest resulting from such protective advances. These claims shall be included in the final Report of Loss.

(h) *Final loss approval.* After the final Report of Loss has been tentatively approved:

(1) If the actual loss is greater than any estimated loss payment, such loss will be paid by the Agency;

(2) If the actual loss is less than any estimated loss payment, the lender will reimburse the Agency;

(3) If the Agency conducted the liquidation, it will provide an accounting to the lender and will pay the lender in accordance with the Loan Note Guarantee.

(i) *Loss limits.* The amount payable by the Agency to the lender cannot exceed the limits contained in the Loan Note Guarantee. If the Agency conducts the liquidation, loss occasioned by accruing interest will be covered by the guarantee only to the date the Agency accepts this responsibility. When the liquidation is conducted by the lender, loss occasioned by accruing interest will be covered to the extent of the

guarantee to the date of final settlement provided the lender proceeds expeditiously with the liquidation plan approved by the Agency.

§ 1779.95 Future recovery.

After a loan has been liquidated and a final loss has been paid by the Agency, any future funds which may be recovered by the lender will be pro-rated between the Agency and the lender in accordance with the guaranteed percentage even if the Loan Note Guarantee has been terminated.

§ 1779.96 Termination of Loan Note Guarantee.

The Loan Note Guarantee under this part will terminate automatically:

(a) Upon full payment of the guaranteed loan; or

(b) Upon full payment of any loss obligation or negotiated loss settlement except for future recovery provisions; or

(c) Upon written request from the lender to the Agency, provided that the lender holds all of the guaranteed portion and the original Loan Note Guarantee is returned to the Agency.

§§ 1779.97–1779.99 [Reserved]

§ 1779.100 OMB control number.

The reporting and recordkeeping requirements contained in this part have been approved by the Office of Management and Budget and have been assigned OMB control number 0572–0122.

PART 1780—WATER AND WASTE LOANS AND GRANTS

Subpart A—General Policies and Requirements

AUTHORITY: 5 U.S.C. 301; 7 U.S.C. 1989; 16 U.S.C. 1005.

SOURCE: 62 FR 33478, June 19, 1997, unless otherwise noted.

Subpart A—General Policies and Requirements

§ 1780.1   General.

(a) This part outlines the policies and procedures for making and processing direct loans and grants for water and waste projects. The Rural Utilities Service (RUS) shall cooperate fully with State and local agencies in making loans and grants to assure maximum support to the State strategy for rural development. Agency officials and their staffs shall maintain coordination and liaison with State agency and substate planning districts.

(b) The income data used in this part to determine median household income must be that which most accurately reflects the income of the service area. The median household income of the service area and the nonmetropolitan median household income of the State will be determined from income data from 5-year income data from the American Community Survey (ACS) or, if needed, other Census Bureau data. If there is reason to believe that the ACS or other Census Bureau data does not accurately represent the median household income within the area to be served, the reasons will be documented and the applicant may furnish, or RD may obtain, additional information regarding such median household income data. Information must consist of reliable data from local, regional, State or Federal sources or from a survey conducted by a reliable impartial source. The nonmetropolitan median household income of the State may only be updated on a national basis by the RUS National Office. This will be done only when median household income data for the same year for all Bureau of the Census areas is available from the Bureau of the Census or other reliable sources. Bureau of the Census areas would include areas such as: Counties, County Subdivisions, Cities, Towns, Townships, Boroughs, and other places.

(c) RUS debt instruments will require an agreement that if at any time it shall appear to the Government that the borrower is able to refinance the amount of the indebtedness to the Government then outstanding, in whole or in part, by obtaining a loan for such purposes from responsible cooperative or private credit sources, at reasonable rates and terms for loans for similar purposes and periods of time, the borrower will, upon request of the Government, apply for and accept such loan in sufficient amount to repay the Government and will take all such actions as may be required in connection with such loan.

(d) Funds allocated for use under this part are also for the use of Indian tribes within the State, regardless of whether State development strategies include Indian reservations within the State's boundaries. Native Americans residing on such reservations must have equal opportunity to participate in the benefits of these programs as compared with other residents of the State. Such tribes might not be subject to State and local laws or jurisdiction. However, any requirements of this part that affect applicant eligibility, the adequacy of RUS's security, or the adequacy of service to users of the facility and all other requirements of this part must be met.

(e) RUS financial programs must be extended without regard to race, color, religion, sex, national origin, marital status, age, or physical or mental handicap.

(f) Any processing or servicing activity conducted pursuant to this part involving authorized assistance to Agency employees, members of their families, known close relatives, or business or close personal associates, is subject to the provisions of subpart D of part 1900 of this title. Applicants for assistance are required to identify any known relationship or association with a RUS employee.

(g) Water and waste facilities will be designed, installed, and operated in accordance with applicable laws which include but are not limited to the Safe Drinking Water Act, Clean Water Act and the Resource Conservation and Recovery Act.

(h) RUS financed facilities will be consistent with any current development plans of State, multijurisdictional areas, counties, or municipalities in which the proposed project is located.

(i) Each RUS financed facility will be in compliance with appropriate State or Federal agency regulations which have control of the appropriation, diversion, storage and use of water and disposal of excess water.

(j) Water and waste applicants must demonstrate that they possess the financial, technical, and managerial capability necessary to consistently comply with pertinent Federal and State laws and requirements. In developing water and waste systems, applicants must consider alternatives of ownership, system design, and the sharing of services.

(k) Applicants should be aware of and comply with other Federal statute requirements including but not limited to:

(1) *Section 504 of the Rehabilitation Act of 1973.* Under section 504 of the Rehabilitation Act of 1973, as amended (29 U.S.C. 794 *et seq.*), no handicapped individual in the United States shall, solely by reason of their handicap, be excluded from participation in, be denied the benefits of, or be subjected to discrimination under any program or activity receiving RUS financial assistance;

(2) *Civil Rights Act of 1964.* All borrowers are subject to, and facilities must be operated in accordance with, title VI of the Civil Rights Act of 1964

(42 U.S.C. 2000d *et seq.*) and subpart E of part 1901 of this title, particularly as it relates to conducting and reporting of compliance reviews. Instruments of conveyance for loans and/or grants subject to the Act must contain the covenant required by § 1901.202(e) of this title;

(3) *The Americans with Disabilities Act (ADA) of 1990.* This Act (42 U.S.C. 12101 *et seq.*) prohibits discrimination on the basis of disability in employment, State and local government services, public transportation, public accommodations, facilities, and telecommunications. Title II of the Act applies to facilities operated by State and local public entities which provides services, programs and activities. Title III of the Act applies to facilities owned, leased, or operated by private entities which accommodate the public; and

(4) *Age Discrimination Act of 1975.* This Act (42 U.S.C. 6101 *et seq.*) provides that no person in the United States shall on the basis of age, be excluded from participation in, be denied the benefits of, or be subjected to discrimination under any program or activity receiving Federal financial assistance.

(1) Applicants for grant assistance will be required to comply with the following requirements as applicable:

(1) 2 CFR part 200, as adopted by USDA through 2 CFR part 400, " Uniform Administrative Requirements, Cost Principles, and Audit Requirements for Federal Awards".

(2) 2 CFR part 415—General Program Administrative Regulations.

(3) 2 CFR part 421-Requirements for Drug-Free Workplace (Financial Assistance).

(m) Applicants for loan or grant assistance will be required to comply with the following requirements as applicable:

(1) 2 CFR part 200, subpart F, "Audit Requirements."

(2) 2 CFR part 180, as adopted by USDA through 2 CFR part 417, Nonprocurement Debarment and Suspension, implementation of Executive Order 12549 and Executive Order 12689 on debarment and suspension.

(3) 2 CFR part 418, New Restrictions on Lobbying.

[62 FR 33478, June 19, 1997, as amended at 79 FR 76006, Dec. 19, 2014; 80 FR 9863, Feb. 24, 2015; 81 FR 7697, Feb. 16, 2016]

§ 1780.2 Purpose.

Provide loan and grant funds for water and waste projects serving the most financially needy communities. Financial assistance should result in reasonable user costs for rural residents, rural businesses, and other rural users.

§ 1780.3 Definitions and grammatical rules of construction.

(a) *Definitions.* For the purposes of this part:

Agency means the Rural Utilities Service and any United States Department of Agriculture (USDA) employee acting on behalf of the Rural Utilities Service in accordance with appropriate delegations of authority.

Agency identified target areas means an identified area in the State strategic plan or other plans developed by the Rural Development State Director.

Approval official means the USDA official at the State level who has been delegated the authority to approve loans or grants.

Equivalent Dwelling Unit (EDU) means the level of service provided to a typical rural residential dwelling.

Parity bonds means bonds which have equal standing with other bonds of the same Issuer.

Poverty line means the level of income for a family of four, as defined in section 673(2) of the Community Services Block Grant Act (42 U.S.C. 9902(2)).

Processing office means the office designated by the State program official to accept and process applications for water and waste disposal assistance.

Project means all activity that an applicant is currently undertaking to be financed in whole or part with RUS assistance.

Protective advances are payments made by a lender for items such as insurance or taxes in order to preserve and protect the security or the lien or priority of the lien securing the loan.

Rural and rural areas means any area not in a city or town with a population in excess of 10,000 inhabitants. The population figure is obtained from the most recent decennial Census of the United States (decennial Census). If the applicable population figure cannot be obtained from the most recent decennial Census, RD will determine the applicable population figure based on available population data.

Rural Development means the mission area of the Under Secretary for Rural Development. Rural Development State and local offices will administer this water and waste program on behalf of the Rural Utilities Service.

RUS means the Rural Utilities Service, an agency of the United States Department of Agriculture established pursuant to section 232 of the Department of Agriculture Reorganization Act of 1994 (Pub. L. 103–354, 108 Stat. 3178), successor to the Farmer's Home Administration and the Rural Development Administration with respect to certain water and waste disposal loan and grant programs.

Service area means the area reasonably expected to be served by the project.

Servicing office means the office designated by the State program official to service water and waste disposal loans and grants.

Similar system cost means the average annual EDU user cost of a system within a community having similar economic conditions and being served by the same type of established system. Similar system cost shall include all charges, taxes, and assessments attributable to the system including debt service, reserves and operation and maintenance costs.

Simplified acquisition threshold means the dollar amount below which an applicant or owner may purchase property or services using small purchase methods as defined further at 2 CFR 200.88.

State program official means the USDA official at the State level who has been delegated the responsibility of administering the water and waste disposal programs under this regulation for a particular State or States.

Statewide nonmetropolitan median household income means the median household income of the State's nonmetropolitan counties and portions of metropolitan counties outside of cities, towns or places of 50,000 or more population.

(b) *Rules of grammatical construction.* Unless the context otherwise indicates, "includes" and "including" are not limiting, and "or" is not exclusive. The terms defined in paragraph (a) of this section include the plural as well as the singular, and the singular as well as the plural.

[62 FR 33478, June 19, 1997, as amended at 69 FR 65519, Nov. 15, 2004; 80 FR 9863, Feb. 24, 2015; 81 FR 7697, Feb. 16, 2016]

§ 1780.4 Availability of forms and regulations.

Information about the availability of forms, instructions, regulations, bulletins, OMB Circulars, Treasury Circulars, standards, documents and publications cited in this part is available from any USDA/Rural Development office or the Rural Utilities Service, United States Department of Agriculture, Washington, DC 20250–1500.

§ 1780.5 [Reserved]

§ 1780.6 Application information.

(a) The Rural Development State Director in each State will determine the office and staff that will be responsible for delivery of the program (processing office) and designate an approving office. Applications will be accepted by the processing office.

(b) The applicant's governing body should designate one person to act as contact person with the Agency during loan and grant processing. Agency personnel should make every effort to involve the applicant's contact person when meeting with the applicant's professional consultants or agents.

§ 1780.7 Eligibility.

Facilities financed by water and waste disposal loans or grants must serve rural areas.

(a) *Eligible applicant.* An applicant must be:

(1) A public body, such as a municipality, county, district, authority, or other political subdivision of a state, territory or commonwealth;

(2) An organization operated on a not-for-profit basis, such as an association, cooperative, or private corporation. The organization must be an association controlled by a local public body or bodies, or have a broadly based ownership by or membership of people of the local community; or

(3) Indian tribes on Federal and State reservations and other Federally recognized Indian tribes.

(b) *Eligible facilities.* Facilities financed by RUS may be located in nonrural areas. However, loan and grant funds may be used to finance only that portion of the facility serving rural areas, regardless of facility location.

(c) *Eligible projects.* (1) Projects must serve a rural area which, if such project is completed, is not likely to decline in population below that for which the project was designed.

(2) Projects must be designed and constructed so that adequate capacity will or can be made available to serve the present population of the area to the extent feasible and to serve the reasonably foreseeable growth needs of the area to the extent practicable.

(3) Projects must be necessary for orderly community development and consistent with a current comprehensive community water, waste disposal, or other current development plan for the rural area.

(d) *Credit elsewhere.* Applicants must certify in writing and the Agency shall determine and document that the applicant is unable to finance the proposed project from their own resources or through commercial credit at reasonable rates and terms.

(e) *Legal authority and responsibility.* Each applicant must have or will obtain the legal authority necessary for owning, constructing, operating, and maintaining the proposed facility or service and for obtaining, giving security for, and repaying the proposed loan. The applicant shall be responsible for operating, maintaining, and managing the facility, and providing for its continued availability and use at reasonable user rates and charges. This responsibility shall be exercised by the applicant even though the facility may be operated, maintained, or managed by a third party under contract or management agreement. Guidance for preparing a management agreement is available from the Agency. Such contracts, management agreements, or leases must not contain options or other provisions for transfer of ownership.

(f) *Economic feasibility.* All projects financed under the provisions of this section must be based on taxes, assessments, income, fees, or other satisfactory sources of revenues in an amount sufficient to provide for facility operation and maintenance, reasonable reserves, and debt payment. If the primary use of the facility is by business and the success or failure of the facility is dependent on the business, then the economic viability of that business must be assessed.

(g) *Federal Debt Collection Act of 1990 (28 U.S.C. 3001 et seq.).* An outstanding judgment obtained by the United States in a Federal Court (other than in the United States Tax Court), which has been recorded, shall cause the applicant to be ineligible to receive a loan or grant until the judgment is paid in full or otherwise satisfied.

[62 FR 33478, June 19, 1997, as amended at 64 FR 29946, June 4, 1999]

§ 1780.8 [Reserved]

§ 1780.9 Eligible loan and grant purposes.

Loan and grant funds may be used only for the following purposes:

(a) To construct, enlarge, extend, or otherwise improve rural water, sanitary sewage, solid waste disposal, and storm wastewater disposal facilities.

(b) To construct or relocate public buildings, roads, bridges, fences, or utilities, and to make other public improvements necessary for the successful operation or protection of facilities authorized in paragraph (a) of this section.

(c) To relocate private buildings, roads, bridges, fences, or utilities, and other private improvements necessary for the successful operation or protection of facilities authorized in paragraph (a) of this section.

(d) For payment of other utility connection charges as provided in service contracts between utility systems.

(e) When a necessary part of the project relates to those facilities authorized in paragraphs (a), (b), (c) or (d) of this section the following may be considered:

(1) Loan or grant funds may be used for:

(i) Reasonable fees and costs such as: legal, engineering, administrative services, fiscal advisory, recording, environmental analyses and surveys, possible salvage or other mitigation measures, planning, establishing or acquiring rights;

(ii) Costs of acquiring interest in land; rights, such as water rights, leases, permits, rights-of-way; and other evidence of land or water control or protection necessary for development of the facility;

(iii) Purchasing or renting equipment necessary to install, operate, maintain, extend, or protect facilities;

(iv) Cost of additional applicant labor and other expenses necessary to install and extend service; and

(v) In unusual cases, the cost for connecting the user to the main service line.

(2) Only loan funds may be used for:

(i) Interest incurred during construction in conjunction with multiple advances or interest on interim financing;

(ii) Initial operating expenses, including interest, for a period ordinarily not exceeding one year when the applicant is unable to pay such expenses;

(iii) The purchase of existing facilities when it is necessary either to improve service or prevent the loss of service;

(iv) Refinancing debts incurred by, or on behalf of, an applicant when all of the following conditions exist:

(A) The debts being refinanced are a secondary part of the total loan;

(B) The debts were incurred for the facility or service being financed or any part thereof; and

(C) Arrangements cannot be made with the creditors to extend or modify the terms of the debts so that a sound basis will exist for making a loan; and

(v) Prepayment of costs for which RUS grant funds were obligated.

(3) Grant funds may be used to restore loan funds used to prepay grant obligated costs.

(f) Construction incurred before loan or grant approval.

(1) Funds may be used to pay obligations for eligible project costs incurred before loan or grant approval if such requests are made in writing by the applicant and the Agency determines that:

(i) Compelling reasons exist for incurring obligations before loan or grant approval;

(ii) The obligations will be incurred for authorized loan or grant purposes; and

(iii) The Agency's authorization to pay such obligations is on the condition that it is not committed to make the loan or grant; it assumes no responsibility for any obligations incurred by the applicant; and the applicant must subsequently meet all loan or grant approval requirements, including environmental and contracting requirements.

(2) If construction is started without Agency approval, post-approval in accordance with this section may be considered, provided the construction meets applicable requirements including those regarding approval and environmental matters.

(g) Water or sewer service may be provided through individual installations or small clusters of users within an applicant's service area. The approval official should consider items such as: quantity and quality of the individual installations that may be developed; cost effectiveness of the individual facility compared with the initial and long term user cost on a central system; health and pollution problems attributable to individual facilities; operational or management problems peculiar to individual installations; and permit and regulatory agency requirements.

(1) Applicants providing service through individual facilities must meet the eligibility requirements in § 1780.7.

(2) The Agency must approve the form of agreement between the applicant and individual users for the installation, operation, maintenance and payment for individual facilities.

(3) If taxes or assessments are not pledged as security, applicants providing service through individual facilities must obtain security necessary

to assure collection of any sum the individual user is obligated to pay the applicant.

(4) Notes representing indebtedness owed the applicant by a user for an individual facility will be scheduled for payment over a period not to exceed the useful life of the individual facility or the RUS loan, whichever is shorter. The interest rate will not exceed the interest rate charged the applicant on the RUS indebtedness.

(5) Applicants providing service through individual or cluster facilities must obtain:

(i) Easements for the installation and ingress to and egress from the facility if determined necessary by RUS; and

(ii) An adequate method for denying service in the event of nonpayment of user fees.

§ 1780.10 Limitations.

(a) Loan and grant funds may not be used to finance:

(1) Facilities which are not modest in size, design, and cost;

(2) Loan or grant finder's fees;

(3) The construction of any new combined storm and sanitary sewer facilities;

(4) Any portion of the cost of a facility which does not serve a rural area;

(5) That portion of project costs normally provided by a business or industrial user, such as wastewater pretreatment, etc.;

(6) Rental for the use of equipment or machinery owned by the applicant;

(7) For other purposes not directly related to operating and maintenance of the facility being installed or improved; and

(8) A judgment which would disqualify an applicant for a loan or grant as provided for in § 1780.7(g).

(b) Grant funds may not be used to:

(1) Reduce EDU costs to a level less than similar system cost;

(2) Pay any costs of a project when the median household income of the service area is more than 100 percent of the nonmetropolitan median household income of the State;

(3) Pay project costs when other loan funding for the project is not at reasonable rates and terms; and

(4) Pay project costs when other funding is a guaranteed loan obtained

in accordance with 7 CFR part 1779 of this title.

(c) Grants may not be made in excess of the following percentages of the RUS eligible project development costs. Facilities previously installed will not be considered in determining the development costs.

(1) 75 percent when the median household income of the service area is below the higher of the poverty line or 80% of the state nonmetropolitan median income and the project is necessary to alleviate a health or sanitary problem.

(2) 45 percent when the median household income of the service area exceeds the 80 percent requirements described in paragraph (c)(1) of this section but is not more than 100 percent of the statewide nonmetropolitan median household income.

(3) Applicants are advised that the percentages contained in paragraphs (c)(1) and (c)(2) of this section are maximum amounts and may be further limited due to availability of funds or the grant determination procedures contained in § 1780.35 (b).

[62 FR 33478, June 19, 1997, as amended at 64 FR 29946, June 4, 1999; 66 FR 23151, May 8, 2001]

§ 1780.11 Service area requirements.

(a) All facilities financed under the provisions of this part shall be for public use. The facilities will be installed so as to serve any potential user within the service area who desires service and can be feasibly and legally served. This does not preclude:

(1) Financing or constructing projects in phases when it is not practical to finance or construct the entire project at one time; and

(2) Financing or constructing facilities where it is not economically feasible to serve the entire area, provided economic feasibility is determined on the basis of the entire system and not by considering the cost of separate extensions to or parts thereof; the applicant publicly announces a plan for extending service to areas not initially receiving service from the system; and potential users located in the areas not to be initially served receive written notice from the applicant that service

will not be provided until such time as it is economically feasible to do so.

(b) Should the Agency determine that inequities exist within the applicants service area for the same type service proposed (i.e., water or waste disposal) such inequities will be remedied by the applicant prior to loan or grant approval or included as part of the project. Inequities are defined as unjustified variations in availability, adequacy or quality of service. User rate schedules for portions of existing systems that were developed under different financing, rates, terms or conditions do not necessarily constitute inequities.

(c) Developers are normally expected to provide utility-type facilities in new or developing areas in compliance with appropriate State statutes. RUS financing will be considered to an eligible applicant only in such cases when failure to complete development would result in an adverse economic condition for the rural area (not the community being developed); the proposal is necessary to the success of a current area development plan; and loan repayment can be assured by:

(1) The applicant already having sufficient assured revenues to repay the loan; or

(2) Developers providing a bond or escrowed security deposit as a guarantee sufficient to meet expenses attributable to the area in question until a sufficient number of the building sites are occupied and connected to the facility to provide enough revenues to meet operating, maintenance, debt service, and reserve requirements. Such guarantees from developers will meet the requirements in §1780.39(c)(4)(ii); or

(3) Developers paying cash for the increased capital cost and any increased operating expenses until the developing area will support the increased costs; or

(4) The full faith and credit of a public body where the debt is evidenced by general obligation bonds; or

(5) The loan is to a public body evidenced by a pledge of tax revenue or assessments; or

(6) The user charges can become a lien upon the property being served and income from such lien can be collected

in sufficient time to be used for its intended purposes.

§1780.12 [Reserved]

§1780.13 Rates and terms.

(a) *General.* (1) Each loan will bear interest at the rate prescribed in RD Instruction 440.1, exhibit B. The interest rates will be set by the Agency for each quarter of the fiscal year. All rates will be adjusted to the nearest one-eighth of one per centum. The rate will be the lower of the rate in effect at the time of loan approval or the rate in effect at the time of loan closing unless the applicant otherwise chooses.

(2) If the interest rate is to be that in effect at loan closing on a loan involving multiple advances of RUS funds using temporary debt instruments, the interest rate charged shall be that in effect on the date when the first temporary debt instrument is issued.

(3) For a loan for a specific project that has been approved, but not closed on or before May 22, 2008, the rate structure in effect at that time will determine the interest rates. For loans approved on or after May 23, 2008, a percentage of the market rate will be used to determine the poverty and intermediate interest rates.

(b) *Poverty rate.* The poverty interest rate will not exceed 5 per centum per annum. Loans approved on or after May 23, 2008, will have the poverty interest rate set at 60 percent of the market rate. All poverty rate loans must comply with the following conditions:

(1) The primary purpose of the loan is to upgrade existing facilities or construct new facilities required to meet applicable health or sanitary standards; and

(2) The median household income of the service area is below the higher of the poverty line, or 80 percent of the Statewide nonmetropolitan median household income.

(c) *Intermediate rate.* The intermediate interest rate will not exceed 7 percent per annum. For a loan for a specific project that has been approved, but not closed on or before May 22, 2008, the intermediate rate is the poverty rate plus one-half of the difference between the poverty rate and the market rate, not to exceed 7 percent per annum. Loans approved on or after May 23, 2008, will have the intermediate interest rate set at 80 percent of the market rate. The intermediate interest rate will apply to loans that do not meet the requirements for the poverty rate and for which the median household income of the service area is not more than 100 percent of the nonmetropolitan median household income of the State.

(d) *Market rate.* The market interest rate will be set using as guidance the average of the Bond Buyer Index (available in any Agency office or the program's Web site) for the four weeks prior to the first Friday of the last month before the beginning of the quarter. The market rate will apply to all loans that do not qualify for a different rate under paragraph (b) or (c) of this section.

(e) *Repayment terms.* The loan repayment period shall not exceed the useful life of the facility, State statute or 40 years from the date of the note or bond, whichever is less. Where RUS grant funds are used in connection with an RUS loan, the loan will be for the maximum term permitted by this part, State statute, or the useful life of the facility, whichever is less, unless there is an exceptional case where circumstances justify making an RUS loan for less than the maximum term permitted. In such cases, the reasons must be fully documented.

(1) Principal payments may be deferred in whole or in part for a period not to exceed 36 months following the date the first interest installment is due. If for any reason it appears necessary to permit a longer period of deferment, the Agency may authorize such deferment. Deferments of principal will not be used to:

(i) Postpone the levying of taxes or assessments;

(ii) Delay collection of the full rates which the borrower has agreed to charge users for its services as soon as those services become available;

(iii) Create reserves for normal operation and maintenance;

(iv) Make any capital improvements except those approved by the Agency which are determined to be essential to the repayment of the loan or to maintain adequate security; and

(v) Make payment on other debt.

(2) *Payment date.* Loan payments will be scheduled to coincide with income availability and be in accordance with State law. If State law only permits principal plus interest (P&I) type bonds, annual or semiannual payments will be used. Insofar as practical monthly payments will be scheduled one full month following the date of loan closing; or semiannual or annual payments will be scheduled six or twelve full months, respectively, following the date of loan closing or any deferment period. Due dates falling on the 29th, 30th or 31st day of the month will be avoided.

(3) In all cases, including those in which RUS is jointly financing with another lender, the RUS payments of principal and interest should approximate amortized installments.

[62 FR 33478, June 19, 1997, as amended at 74 FR 395, Jan. 6, 2009; 82 FR 43671, Sept. 19, 2017]

§ 1780.14 Security.

Loans will be secured by the best security position practicable in a manner which will adequately protect the interest of RUS during the repayment period of the loan. Specific security requirements for each loan will be included in a letter of conditions.

(a) *Public bodies.* Loans to such borrowers, including Federally recognized Indian tribes as appropriate, will be evidenced by notes, bonds, warrants, or other contractual obligations as may be authorized by relevant laws and by borrower's documents, resolutions, and ordinances. Security, in the following order of preference, will consist of:

(1) The full faith and credit of the borrower when the debt is evidenced by general obligation bonds; and/or

(2) Pledges of taxes or assessments; and/or

(3) Pledges of facility revenue and, when it is the customary financial practice in the State, liens will be taken on the interest of the applicant in all land, easements, rights-of-way, water rights, water purchase contracts, water sales contracts, sewage treatment contracts, and similar property rights, including leasehold interests, used or to be used in connection with the facility whether owned at the time

the loan is approved or acquired with loan funds.

(b) *Other-than-public bodies.* Loans to other-than-public body applicants and Federally recognized Indian tribes, as appropriate, will be secured in the following order of preference:

(1) Assignments of borrower income will be taken and perfected by filing, if legally permissible; and

(2) A lien will be taken on the interest of the applicant in all land, easements, rights-of-way, water rights, water purchase contracts, water sales contracts, sewage treatment contracts and similar property rights, including leasehold interest, used, or to be used in connection with the facility whether owned at the time the loan is approved or acquired with loan funds. In unusual circumstances where it is not legally permissible or feasible to obtain a lien on such land (such as land rights obtained from Federal or local government agencies, and from railroads) and the approval official determines that the interest of RUS is otherwise adequately secured, the lien requirement may be omitted as to such land rights. For existing borrowers where the Agency already has a security position on real property, the approval official may determine that the interest of the Government is adequately secured and not require additional liens on such land rights. When the subsequent loan is approved or the acquisition of real property is subject to an outstanding lien indebtedness, the next highest priority lien obtainable will be taken if the approval official determines that the loan is adequately secured.

(c) *Joint financing security.* For projects utilizing joint financing, when adequate security of more than one type is available, the other lender may take one type of security with RUS taking another type. For projects utilizing joint financing with the same security to be shared by RUS and another lender, RUS will obtain at least a parity position with the other lender. A parity position is to ensure that with joint security, in the event of default, each lender will be affected on a proportionate basis. A parity position will conform with the following unless an exception is granted by the approval official:

(1) It is not necessary for loans to have the same repayment terms. Loans made by other lenders involved in joint financing with RUS should be scheduled for repayment on terms similar to those customarily used in the State for financing such facilities.

(2) The use of a trustee or other similar paying agent by the other lender in a joint financing arrangement is acceptable to RUS. A trustee or other similar paying agent will not normally be used for the RUS portion of the funding unless required to comply with State law. The responsibilities and authorities of any trustee or other similar paying agent on projects that include RUS funds must be clearly specified by written agreement and approved by the State program official and the Office of the General Counsel (OGC). RUS must be able to deal directly with the borrower to enforce the provisions of loan and grant agreements and perform necessary servicing actions.

(3) In the event adequate funds are not available to meet regular installments on parity loans, the funds available will be apportioned to the lenders based on the respective current installments of principal and interest due.

(4) Funds obtained from the sale or liquidation of secured property or fixed assets will be apportioned to the lenders on the basis of the pro rata amount outstanding; provided, however, funds obtained from such sale or liquidation for a project that included RUS grant funds will be apportioned as required by the grant agreement.

(5) Protective advances must be charged to the borrower's account and be secured by a lien on the security property. To the extent consistent with State law and customary lending practices in the area, repayment of protective advances made by either lender, for the mutual protection of both lenders, should receive first priority in apportionment of funds between the lenders. To ensure agreement between lenders, efforts should be made to obtain the concurrence of both lenders before one lender makes a protective advance.

§ 1780.15 Other Federal, State, and local requirements.

Proposals for facilities financed in whole or in part with RUS funds will be coordinated with appropriate Federal, State and local agencies. If there are conflicts between this part and State or local laws or regulatory commission regulations, the provisions of this part will control. Applicants will be required to comply with Federal, State, and local laws and any regulatory commission rules and regulations pertaining to:

(a) Organization of the applicant and its authority to own, construct, operate, and maintain the proposed facilities;

(b) Borrowing money, giving security therefore, and raising revenues for the repayment thereof;

(c) Land use zoning; and

(d) Health and sanitation standards and design and installation standards unless an exception is granted by RUS.

§ 1780.16 [Reserved]

§ 1780.17 Selection priorities and process.

When ranking eligible applications for consideration for limited funds, Agency officials must consider the priority items met by each application and the degree to which those priorities are met. Points will be awarded as follows:

(a) *Population priorities.* (1) The proposed project will primarily serve a rural area having a population not in excess of 1,000—25 points;

(2) The proposed project primarily serves a rural area having a population between 1,001 and 2,500—15 points;

(3) The proposed project primarily serves a rural area having a population between 2,501 and 5,500—5 points.

(b) *Health priorities.* The proposed project is:

(1) Needed to alleviate an emergency situation, correct unanticipated diminution or deterioration of a water supply, or to meet Safe Drinking Water Act requirements which pertain to a water system—25 points;

(2) Required to correct inadequacies of a wastewater disposal system, or to meet health standards which pertain to a wastewater disposal system—25 points;

(3) Required to meet administrative orders issued to correct local, State, or Federal solid waste violations—15 points.

(c) *Median household income priorities.* The median household income of the population to be served by the proposed project is:

(1) Less than the poverty line if the poverty line is less than 80% of the statewide nonmetropolitan median household income—30 points;

(2) Less than 80 percent of the statewide nonmetropolitan median household income—20 points;

(3) Equal to or more than the poverty line and between 80% and 100%, inclusive, of the State's nonmetropolitan median household income—15 points.

(d) *Other priorities.* (1) The proposed project will: merge ownership, management, and operation of smaller facilities providing for more efficient management and economical service—15 points;

(2) The proposed project will enlarge, extend, or otherwise modify existing facilities to provide service to additional rural areas—10 points;

(3) Applicant is a public body or Indian tribe—5 points;

(4) Amount of other than RUS funds committed to the project is:

(i) 50% or more—15 points;

(ii) 20% to 49%—10 points;

(iii) 5%—19%—5 points;

(5) Projects that will serve Agency identified target areas—10 points;

(6) Projects that primarily recycle solid waste products thereby limiting the need for solid waste disposal—5 points;

(7) The proposed project will serve an area that has an unreliable quality or supply of drinking water—10 points.

(e) In certain cases the State program official may assign up to 15 points to a project. The points may be awarded to projects in order to improve compatibility and coordination between RUS's and other agencies' selection systems, to ensure effective RUS fund utilization, and to assist those projects that are the most cost effective. A written justification must be prepared and placed in the project file each time these points are assigned.

(f) *Cost overruns.* An application may receive consideration for funding before others at the State or National Office level when it is a subsequent request for a previously approved project which has encountered construction cost overruns. The cost overruns must be due to high bids or unexpected construction problems that cannot be reduced by negotiations, redesign, use of bid alternatives, rebidding or other means. Cost overruns exceeding 20% of the development cost at time of loan or grant approval or where the scope of the original purpose has changed will not be considered under this paragraph.

(g) *National office priorities.* In selecting projects for funding at the National Office level State program official points may or may not be considered. The Administrator may assign up to 15 additional points to account for items such as geographic distribution of funds, the highest priority projects within a state, and emergency conditions caused by economic problems or natural disasters. The Administrator may delegate the authority to assign the 15 points to appropriate National Office staff.

§ 1780.18 Allocation of program funds.

(a) *General.* (1) The purpose of this part is to set forth the methodology and formulas by which the Administrator of the RUS allocates program funds to the States. (The term "State" means any of the States of the United States, the Commonwealth of Puerto Rico, any territory or possession of the United States, or the Western Pacific Areas.)

(2) The formulas in this part are used to allocate program loan and grant funds to Rural Development State offices so that the overall mission of the Agency can be carried out. Considerations used when developing the formulas include enabling legislation, congressional direction, and administration policies. Allocation formulas ensure that program resources are available on an equal basis to all eligible individuals and organizations.

(3) The actual amounts of funds, as computed by the methodology and formulas contained herein, allocated to a State for a funding period, are distributed to each State office. The allocated amounts are available for review in any Rural Development State office.

(b) *Definitions*—(1) *Amount available for allocations.* Funds appropriated or otherwise made available to the Agency for use in authorized programs. On occasion, the allocation of funds to States may not be practical for a particular program due to funding or administrative constraints. In these cases, funds will be controlled by the National Office.

(2) *Basic formula criteria, data source and weight.* Basic formulas are used to calculate a basic State factor as a part of the methodology for allocating funds to the States. The formulas take a number of criteria that reflect the funding needs for a particular program and through a normalization and weighting process for each of the criteria calculate the basic State factor (SF). The data sources used for each criteria are believed to be the most current and reliable information that adequately quantifies the criterion. The weight, expressed as a percentage, gives a relative value to the importance of each of the criteria.

(3) *Basic formula allocation.* The result of multiplying the amount available for allocation less the total of any amounts held in reserve or distributed by base or administrative allocation times the basic State factor for each State. The basic formula allocation (BFA) for an individual State is equal to:

$$BFA = (\text{Amount available for allocation} - NO \text{ reserve} - \text{total base and administrative allocations}) \times SF.$$

(4) *Transition formula.* (i) A formula based on a proportional amount of previous year allocation used to maintain program continuity by preventing large fluctuations in individual State allocations. The transition formula limits allocation shifts to any particular State in the event of changes from year to year of the basic formula, the basic criteria, or the weights given the criteria. The transition formula first checks whether the current year's basic formula allocation is within the transition range (plus or minus 20 percentage points of the proportional amount of the previous year's BFA). The formula follows:

$$\text{Transition Range} = 1.0 + \frac{\text{maximum } 20\%}{100} \times \frac{(\text{Amount available for allocation this year} \times \text{State previous year BFA})}{(\text{Amount available for allocation previous year})}$$

(ii) If the current year's State BFA is not within the transition range in paragraph (b)(4)(i) of this section, the State formula allocation is changed to the amount of the transition range limit closest to the BFA amount. After having performed this transition adjustment for each State, the sum of the funds allocated to all States will differ from the amount of funds available for BFA. This difference, whether a positive or negative amount, is distributed to all States receiving a formula allocation by multiplying the difference by the SF. The end result is the transition formula allocation. The transition range will not exceed 40% (plus or minus 20%), but when a smaller range is used it will be stated in the individual program section.

(5) *Base allocation.* An amount that may be allocated to each State dependent upon the particular program to provide the opportunity for funding at least one typical loan or grant in each Rural Development State office. The amount of the base allocation may be determined by criteria other than that used in the basic formula allocation such as Agency historic data.

(6) *Administrative allocations.* Allocations made by the Administrator in cases where basic formula criteria information is not available. This form of allocation may be used when the Administrator determines the program objectives cannot be adequately met with a formula allocation.

(7) *Reserve.* An amount retained under the National Office control for each loan and grant program to provide flexibility in meeting situations of unexpected or justifiable need occurring during the fiscal year. The Administrator may make distributions from this reserve to any State when it is determined necessary to meet a program need or Agency objective. The Administrator may retain additional amounts to fund authorized demonstration programs.

(8) *Pooling of funds.* A technique used to ensure that available funds are used in an effective, timely and efficient manner. At the time of pooling those funds within a State's allocation for the fiscal year or portion of the fiscal year, depending on the type of pooling, that have not been obligated by the State are placed in the National Office reserve. The Administrator will establish the pooling dates for each affected program.

(i) Mid-year: Mid-year pooling occurs near the midpoint of the fiscal year.

(ii) Year-end: Year-end pooling usually occurs near the first of August.

(iii) Emergency: The Administrator may pool funds at any time that it is determined the conditions upon the initial allocation was based have changed to such a degree that it is necessary to pool funds in order to efficiently carry out the Agency mission.

(9) *Availability of the allocation.* Program funds are made available to the Agency on a quarterly basis.

(10) *Suballocation by the Rural Development State Director.* The State Director may be directed or given the option of suballocating the State allocation to processing offices. When suballocating the State Director may retain a portion of the funds in a State office reserve to provide flexibility in situations of unexpected or justified need. When performing a suballocation the State Director will use the same formula, criteria and weights as used by the National Office.

(c) *Water and waste disposal loans and grants*—(1) *Amount available for allocations.* See paragraph (b)(1) of this section.

(2) *Basic formula criteria, data source and weight.* See paragraph (b)(2) of this section.

(i) The criteria used in the basic formula are:

(A) State's percentage of national rural population will be 50 percent.

(B) State's percentage of national rural population with incomes below the poverty level will be 25 percent.

(C) State's percentage of national nonmetropolitan unemployment will be 25 percent.

(ii) The data sources for each criterion identified in paragraph (c)(2) of this section are specified in paragraphs (c)(2)(ii)(A) through (C) of this section. Each criterion is assigned a specific weight according to its relevance in determining need. The percentage representing each criterion is multiplied by the weight factor and summed to arrive at a State factor (SF). The SF cannot exceed 0.05, as follows:

SF = (criterion in paragraph (b)(2)(i)(A) of this section × 50 percent) + (criterion in paragraph (b)(2)(i)(B) × 25 percent) + (criterion in paragraph (b)(2)(i)(C) of this section × 25 percent)

(A) For the criterion specified in paragraph (b)(2)(i)(A) of this section, the most recent decennial Census data.

(B) For the criterion specified in paragraph (b)(2)(i)(B) of this section, 5-year income data from the American Community Survey (ACS) or, if needed, other Census Bureau data.

(C) For the criterion specified in paragraph (b)(2)(i)(C) of this section, the most recent Bureau of Labor Statistics data.

(3) *Basic formula allocation.* See paragraph (b)(3) of this section. States receiving administrative allocations do not receive formula allocations.

(4) *Transition formula.* See paragraph (b)(4) of this section. The percentage range for the transition formula equals 30 percent (plus or minus 15%).

(5) *Base allocation.* See paragraph (b)(5) of this section. States receiving administrative allocations do not receive base allocations.

(6) *Administrative allocation.* See paragraph (b)(6) of this section. States participating in the formula and base allocation procedures do not receive administrative allocations.

(7) *Reserve.* See paragraph (b)(7) of this section. Any State may request reserve funds by forwarding a request to the National Office. Generally, a request for additional funds will not be honored unless the State has insufficient funds to obligate the loan requested.

(8) *Pooling of funds.* See paragraph (b)(8) of this section. Funds are generally pooled at mid-year and year-end. Pooled funds will be placed in the Na-

tional Office reserve and will be made available administratively.

(9) *Availability of the allocation.* See paragraph (b)(9) of this section. The allocation of funds is made available for States to obligate on an annual basis although the Office of Management and Budget apportions it to the Agency on a quarterly basis.

(10) *Suballocation by the State Director.* See paragraph (b)(10) of this section. The State Director has the option to suballocate funds to processing offices.

[62 FR 33478, June 19, 1997, as amended at 80 FR 9863, Feb. 24, 2015]

§1780.19 Public information.

(a) *Public notice of intent to file an application with the Agency.* Within 60 days of filing an application with the Agency the applicant must publish a notice of intent to apply for a RUS loan or grant. The notice of intent must be published in a newspaper of general circulation in the proposed area to be served.

(b) *General public meeting.* Applicants should inform the general public regarding the development of any proposed project. Any applicant not required to obtain authorization by vote of its membership or by public referendum, to incur the obligations of the proposed loan or grant, must hold at least one public information meeting. The public meeting must be held not later than loan or grant approval. The meeting must give the citizenry an opportunity to become acquainted with the proposed project and to comment on such items as economic and environmental impacts, service area, alternatives to the project, or any other issue identified by Agency. To the extent possible, this meeting should cover items necessary to satisfy all public information meeting requirements for the proposed project. To minimize duplication of public notices and public involvement, the applicant shall, where possible, coordinate and integrate the public involvement activities of the environmental review process into this requirement. The applicant will be required, at least 10 days prior to the meeting, to publish a notice of the meeting in a newspaper of general circulation in the service area, to post a public notice at the applicant's principal office, and to notify the Agency. The applicant will provide the Agency a copy of the published notice and minutes of the public meeting. A public meeting is not normally required for subsequent loans or grants which are needed to complete the financing of a project.

§§ 1780.20–1780.23 [Reserved]

§ 1780.24 Approval authorities.

Appropriate reviews, concurrence, and authorization must be obtained for all loans or grants in excess of the amounts indicated in RUS Staff Instruction 1780–1.

(a) *Redelegation of authority by State Directors.* Unless restricted by memorandum from the RUS Administrator, State Directors can redelegate their approval authorities to State employees by memorandum.

(b) *Restriction of approval authority by the RUS Administrator.* The RUS Administrator can make written restrictions or revocations of the authority given to any approval official.

§ 1780.25 Exception authority.

The Administrator may, in individual cases, make an exception to any requirement or provision of this part which is not inconsistent with the authorizing statute or other applicable law and is determined to be in the Government's interest.

§§ 1780.26–1780.30 [Reserved]

Subpart B—Loan and Grant Application Processing

§ 1780.31 General.

(a) Applicants are encouraged to contact the Agency processing office early in the planning stages of their project. Agency personnel are available to provide general advice and assistance regarding RUS programs, other funding sources, and types of systems or improvements appropriate for the applicants needs. The Agency can also provide access to technical assistance and other information resources for other project development issues such as public information, income surveys, developing rate schedules, system oper-

ation and maintenance, and environmental compliance requirements. Throughout the planning, application processing and construction of the project, Agency personnel will work closely and cooperatively with the applicant and their representatives, other State and Federal agencies and technical assistance providers.

(b) The processing office will handle initial inquiries and provide basic information about the program. They are to provide the application, SF 424.2, "Application for Federal Assistance (For Construction)," assist applicants as needed in completing SF 424.2, and in filing a request for intergovernmental review. Federally recognized Indian tribes are exempt from intergovernmental review. The processing office will explain eligibility requirements and meet with the applicant whenever necessary to discuss application processing.

(c) Applicants can make a written request for an eligibility determination in lieu of filing an SF 424.2 along with the information required by § 1780.33. Applicants seeking only an eligibility determination, should contact the processing office to obtain a list of the items needed to make this determination. An eligibility determination for loan or grant assistance will not give an applicant priority for funding as set forth in § 1780.17.

(d) Applications that are not developed in a reasonable period of time taking into account the size and complexity of the proposed project may be removed from the State's active file. Applicants will be consulted prior to taking such action.

(e) During the earliest discussion with prospective applicants, the Agency will advise prospective applicants on environmental review requirements and evaluation of potential environmental impacts of the proposal. In accordance with 7 CFR part 1970, environmental review requirements shall be performed by the applicant simultaneously and concurrently with the proposal's engineering planning and design.

[62 FR 33478, June 19, 1997, as amended at 63 FR 68655, Dec. 11, 1998; 81 FR 11028, Mar. 2, 2016]

§ 1780.32 Timeframes for application processing.

(a) The processing office will determine if the application is properly assembled. If not, the applicant will be notified within fifteen federal working days as to what additional submittal items are needed.

(b) The processing and approval offices will coordinate their reviews to ensure that the applicant is advised about eligibility and anticipated fund availability within 45 days of the receipt of a completed application.

§ 1780.33 Application requirements.

An initial application consists of the following:

(a) One copy of a completed SF 424.2;

(b) A copy of the State intergovernmental comments or one copy of the filed application for State intergovernmental review; and

(c) Two copies of the preliminary engineering report (PER) for the project.

(1) The PER may be submitted to the processing office prior to the rest of the application material if the applicant desires a preliminary review.

(2) The processing office will forward one copy of the PER with comments and recommendations to the State staff engineer for review upon receipt from the applicant.

(3) The State staff engineer will consult with the applicant's engineer as appropriate to resolve any questions concerning the PER. Written comments will be provided by the State staff engineer to the processing office to meet eligibility determination time lines.

(d) Written certification that other credit is not available.

(e) Supporting documentation necessary to make an eligibility determination such as financial statements, audits, organizational documents, or existing debt instruments. The processing office will advise applicants regarding the required documents. Applicants that are indebted to RUS will not need to submit documents already on file with the processing office.

(f) *Environmental review requirements.* The applicant must comply with the environmental review requirements in accordance with 7 CFR part 1970.

(1) Upon receipt of the Environmental Report, the processing office shall forward one copy of the report with comments and recommendation to the State Environmental Coordinator for review.

(2) The State Environmental Coordinator will consult with the applicant as appropriate to resolve any environmental concerns. Written comments will be provided by the State Environmental Coordinator to the processing office to meet eligibility determination time lines.

(g) The applicant's Internal Revenue Service Taxpayer Identification Number (TIN). The TIN will be used by the Agency to assign a case number which will be the applicant's or transferee's TIN preceded by State and County Code numbers. Only one case number will be assigned to each applicant regardless of the number of loans or grants or number of separate facilities, unless an exception is authorized by the National Office.

(h) Other Forms and certifications. Applicants will be required to submit the following items to the processing office, upon notification from the processing office to proceed with further development of the full application:

(1) Form RD 442–7, "Operating Budget";

(2) Form RD 1910–11, "Application Certification, Federal Collection Policies for Consumer or Commercial Debts";

(3) Form RD 400–1, "Equal Opportunity Agreement";

(4) Form RD 400–4, "Assurance Agreement";

(5) Form AD–1047, "Certification Regarding Debarment, Suspension and other Responsibility Matters";

(6) Form AD–1049, Certification regarding Drug-Free Workplace Requirements (Grants) Alternative I For Grantees Other Than Individuals;

(7) Certifications for Contracts, Grants, and Loans (Regarding Lobbying); and

(8) Certification regarding prohibited tying arrangements. Applicants that provide electric service must provide the Agency a certification that they will not require users of a water or waste facility financed under this part

to accept electric service as a condition of receiving assistance.

[62 FR 33478, June 19, 1997, as amended at 63 FR 68655, Dec. 11, 1998; 81 FR 11028, Mar. 2, 2016]

§ 1780.34 Strategic economic and community development.

Applicants with projects that support the implementation of strategic economic development and community development plans are encouraged to review and consider 7 CFR part 1980, subpart K, which contains provisions for providing priority to projects that support the implementation of strategic economic development and community development plans on a Multi-jurisdictional basis.

[81 FR 10456, Mar. 1, 2016[

§ 1780.35 Processing office review.

Review of the application will usually include the following:

(a) Nondiscrimination. Boundaries for the proposed service area must not be chosen in such a way that any user or area will be excluded because of race, color, religion, sex, marital status, age, handicap, or national origin. This does not preclude construction of the project in phases as noted in § 1780.11 as long as it is not done in a discriminatory manner.

(b) Grant determination. Grants will be determined by the processing office in accordance with the following provisions and will not result in EDU costs below similar system user cost.

(1) Maximum grant. Grants may not exceed the percentages in § 1780.10(c) of the eligible RUS project development costs listed in § 1780.9.

(2) Debt service. Applicants will be considered for grant assistance when the debt service portion of the average annual EDU cost, for users in the applicant's service area, exceeds the following percentages of median household income:

(i) 0.5 percent when the median household income of the service area is equal to or below 80% of the statewide nonmetropolitan median income.

(ii) 1.0 percent when the median household income of the service area exceeds the 0.5 percent requirement but is not more than 100 percent the state-

wide nonmetropolitan household income.

(3) Similar system cost. If the grant determined in paragraph (b)(2) of this section results in an annual EDU cost that is not comparable with similar systems, the Agency will determine a grant amount based on achieving EDU costs that are not below similar system user costs.

(4) Wholesale service. When an applicant provides wholesale sales or services on a contract basis to another system or entity, similar wholesale system cost will be used in determining the amount of grant needed to achieve a reasonable wholesale user cost.

(5) Subsidized cost. When annual cost to the applicant for delivery of service is subsidized by either the state, commonwealth, or territory, and uniform flat user charges regardless of usage are imposed for similar classes of service throughout the service area, the Agency may proceed with a grant in an amount necessary to reduce such delivery cost to a reasonable level.

(c) User charges. The user charges should be reasonable and produce enough revenue to provide for all costs of the facility after the project is complete. The planned revenue should be sufficient to provide for all debt service, debt reserve, operation and maintenance, and, if appropriate, additional revenue for facility replacement of short-lived assets without building a substantial surplus. Ordinarily, the total debt service reserve will be equal to one average annual loan installment which will accumulate at the rate of one-tenth of the total each year.

[62 FR 33478, June 19, 1997, as amended at 64 FR 29946, June 4, 1999]

§ 1780.36 Approving official review.

Projects may be obligated as their applications are completed and approved.

(a) Selection of applications for further processing. The application and supporting information submitted will be used to determine the applications selected for further development and funding. After completing the review, the approval official will normally select those eligible applications with the highest priority scores for further processing. When authorizing the development of an application for funding, the following will be considered:

(1) Funds available in State allocation;

(2) Anticipated allocation of funds for the next fiscal year; and

(3) Time necessary for applicant to complete the application.

(b) *Lower scoring projects.* (1) In cases where preliminary cost estimates indicate that an eligible, high scoring application is unfeasible or would require an amount of funding from RUS that exceeds either 25 percent of a State's current annual allocation or an amount greater than that remaining in the State's allocation, the approval official may instead select the next lower scoring application for further processing provided the high scoring applicant is notified of this action and given an opportunity to revise the proposal and resubmit it.

(2) If it is found that there is no effective way to reduce costs or no other funding sources, the approval official, after consultation with applicant, may submit a request for an additional allocation of funds for the proposed project to the National Office. The request should be submitted during the fiscal year in which obligation is anticipated. Such request will be considered along with all others on hand. A written justification must be prepared and placed in the project file.

§ 1780.37 **Applications determined ineligible.**

If at any time an application is determined ineligible, the processing office will notify the applicant in writing of the reasons. The notification to the applicant will state that an appeal of this decision may be made by the applicant under 7 CFR part 11.

§ 1780.38 **[Reserved]**

§ 1780.39 **Application processing.**

(a) *Processing conference.* Before starting to assemble the full application, the applicant should arrange through the processing office an application conference to provide a basis for orderly application assembly. The processing office will explain program requirements, public information requirements and provide guidance on preparation of items necessary for approval.

(b) *Professional services and contracts related to the facility.* Fees provided for in contracts or agreements shall be reasonable. The Agency shall consider fees to be reasonable if they are not in excess of those ordinarily charged by the profession as a whole for similar work when RUS financing is not involved. Applicants will be responsible for providing the services necessary to plan projects including design of facilities, environmental review and documentation requirements, preparation of cost and income estimates, development of proposals for organization and financing, and overall operation and maintenance of the facility. Applicants should negotiate for procurement of professional services, whereby competitors' qualifications are evaluated and the most qualified competitor is selected, subject to negotiations of fair and reasonable compensation. Contracts or other forms of agreement between the applicant and its professional and technical representatives are required and are subject to RUS concurrence.

(1) *Engineering and architectural services.* (i) Applicants shall publicly announce all requirements for engineering and architectural services, and negotiate contracts for engineering and architectural services on the basis of demonstrated competence and qualifications for the type of professional services required and at a fair and reasonable price.

(ii) When project design services are procured separately, the selection of the engineer or architect shall be done by requesting qualification-based proposals and in accordance with this section.

(iii) Applicants may procure engineering and architectural services in accordance with applicable State statutes or local requirements provided the State Director determines that such procurement meets the intent of this section.

(2) *Other professional services.* Professional services of the following may be

necessary: Attorney, bond counsel, accountant, auditor, appraiser, environmental professionals, and financial advisory or fiscal agent (if desired by applicant). Guidance on entering into an agreement for legal services is available from the Agency.

(3) *Bond counsel.* Unless otherwise provided by subpart D of this part, public bodies are required to obtain the service of recognized bond counsel in the preparation of evidence of indebtedness.

(4) *Contracts for other services.* Contracts or other forms of agreements for other services including management, operation, and maintenance will be developed by the applicant and presented to the Agency for review and concurrence. Guidance on entering into a management agreement is available from the Agency.

(c) *User estimates.* Applicants dependent on users fees for debt payment or operation and maintenance expenses shall base their income and expense forecast on realistic user estimates. For users presently not receiving service, consideration must be given to the following:

(1) An estimated number of maximum users should not be used when setting user fees and rates since it may be several years before all residents will need service by the system. In establishing rates a realistic number of users should be employed.

(2) *New user cash contributions.* The amount of cash contributions required will be set by the applicant and concurred in by the approval official. Contributions should be an amount high enough to indicate sincere interest on the part of the potential user, but not so high as to preclude service to low income families. Contributions ordinarily should be an amount approximating one year's minimum user fee, and shall be paid in full before loan closing or commencement of construction, whichever occurs first. Once economic feasibility is ascertained based on a demonstration of potential user cash contributions, the contribution, membership fee or other fees that may be imposed are not a loan requirement under this section. A new user cash contribution is not required when:

(i) The Agency determines that the potential users as a whole in the applicant's service area cannot make cash contributions; or

(ii) State statutes or local ordinances require mandatory use of the system and the applicant or legal entity having such authority agrees in writing to enforce such statutes, or ordinances.

(3) An enforceable user agreement with a penalty clause is required (RUS Bulletin 1780-9 can be used) except:

(i) For users presently receiving service; or

(ii) Where mandatory use of the system is required.

(4) Individual vacant property owners will not be considered when determining project feasibility unless:

(i) The owner has plans to develop the property in a reasonable period of time and become a user of the facility; and

(ii) The owner agrees in writing to make a monthly payment at least equal to the proportionate share of debt service attributable to the vacant property until the property is developed and the facility is utilized on a regular basis. A bond or escrowed security deposit must be provided to guarantee this monthly payment and to guarantee an amount at least equal to the owner's proportionate share of construction costs. If a bond is provided, it must be executed by a surety company that appears on the Treasury Department's most current list (Circular 570, as amended) and be authorized to transact business in the State where the project is located. The guarantee shall be payable jointly to the borrower and the United States of America.

(5) Applicants must provide a positive program to encourage connection by all users as soon as service is available. The program will be available for review and concurrence by the processing office before loan closing or commencement of construction, whichever occurs first. Such a program shall include:

(i) An aggressive information program to be carried out during the construction period. The applicant should send written notification to all signed users in advance of the date service will be available, stating the date users will be expected to have their connections completed, and the date user charges will begin;

(ii) Positive steps to assure that installation services will be available. These may be provided by the contractor installing the system, local plumbing companies, or local contractors;

(iii) Aggressive action to see that all signed users can finance their connections.

(d) *Interim financing.* For all loans exceeding $500,000, where funds can be borrowed at reasonable interest rates on an interim basis from commercial sources for the construction period, such interim financing may be obtained so as to preclude the necessity for multiple advances of RUS loan funds. However, the approval official may make an exception when interim financing is cost prohibitive or unavailable. Guidance on informing the private lender of RUS's commitment is available from the Agency. When interim commercial financing is used, the application will be processed, including obtaining construction bids, to the stage where the RUS loan would normally be closed, that is immediately prior to the start of construction. The RUS loan should be closed as soon as possible after the disbursal of all interim funds.

(e) *Reserve requirements.* Provision for the accumulation of necessary reserves over a reasonable period of time will be included in the loan documents.

(1) *General obligation or special assessment bonds.* Ordinarily, the requirements for reserves will be considered to have been met if general obligation or other bonds which pledge the full faith and credit of the political subdivision are used, or special assessment bonds are used, and if such bonds provide for the annual collection of sufficient taxes or assessments to cover debt service.

(2) *Other than general obligation or special assessment bonds.* Each borrower will be required to establish and maintain reserves sufficient to assure that loan installments will be paid on time, for emergency maintenance, for extensions to facilities, and for replacement of short-lived assets which have a useful life significantly less than the re-

payment period of the loan. Borrowers issuing bonds or other evidences of debt pledging facility revenues as security will plan their debt reserve to provide for at least one average annual loan installment. The debt reserve will accumulate at the rate of one-tenth of an average annual loan installment each year unless prohibited by state law.

(f) *Membership authorization.* For organizations other than public bodies, the membership will authorize the project and its financing. Form RD 1942–8, "Resolution of Members or Stockholders," may be used for this authorization. The approval official may accept RUS Bulletin 1780–28, "Loan Resolution Security Agreement," without such membership authorization when State statutes and the organization's charter and bylaws do not require such authorization; and

(1) The organization is well established and is operating with a sound financial base; or

(2) The members of the organization have all signed an enforceable user agreement with a penalty clause and have made the required meaningful user cash contribution.

(g) *Insurance.* The purpose of RUS's insurance requirements is to protect the government's financial interest based on the facility financed with loan funds. It is the responsibility of the applicant and not that of RUS to assure that adequate insurance and fidelity or employee dishonesty bond coverage is maintained. The requirements below apply to all types of coverage determined necessary. The approval official may grant exceptions to normal requirements when appropriate justification is provided establishing that it is in the best interest of the applicant and will not adversely affect the government's interest.

(1) Insurance requirements proposed by the applicant will be accepted if the processing office determines that proposed coverage is adequate to protect the government's financial interest. Applicants are encouraged to have their attorney, consulting engineer, and/or insurance provider(s) review proposed types and amounts of coverage, including any deductible provisions.

(2) The use of deductibles may be allowed by RUS providing the applicant has financial resources which would likely be adequate to cover potential claims requiring payment of the deductible.

(3) *Fidelity or employee dishonesty bonds.* Applicants will provide coverage for all persons who have access to funds, including persons working under a contract or management agreement. Coverage may be provided either for all individual positions or persons, or through "blanket" coverage providing protection for all appropriate employees. An exception may be granted by the approval official when funds relating to the facility financed are handled by another entity and it is determined that the entity has adequate coverage or the government's interest would otherwise be adequately protected. The amount of coverage required by RUS will normally approximate the total annual debt service requirements for the RUS loans.

(4) *Property insurance.* Fire and extended coverage will normally be maintained on all structures except as noted below. Ordinarily, RUS should be listed as mortgagee on the policy when RUS has a lien on the property. Normally, major items of equipment or machinery located in the insured structures must also be covered. Exceptions:

(i) Reservoirs, pipelines and other structures if such structures are not normally insured;

(ii) Subsurface lift stations except for the value of electrical and pumping equipment therein.

(5) General liability insurance, including vehicular coverage.

(6) Flood insurance required for facilities located in special flood-and mudslide-prone areas.

(7) *Worker's compensation.* The borrower will carry worker's compensation insurance for employees in accordance with State laws.

(h) [Reserved]

(i) The processing office will assure that appropriate forms and documents listed in RUS Bulletin 1780-6 are complete. Letters of conditions will not be issued unless funds are available.

[62 FR 33478, June 19, 1997, as amended at 63 FR 68655, Dec. 11, 1998; 64 FR 29946, June 4, 1999]

§ 1780.40   [Reserved]

§ 1780.41   Loan or grant approval.

(a) The processing office will submit the following to the approval official:

(1) Form RD 1942-45, "Project Summary";

(2) Form RD 442-7, "Operating Budget";

(3) Form RD 442-3, "Balance Sheet" or a financial statement or audit that includes a balance sheet;

(4) Form RD 442-14, "Association Project Fund Analysis";

(5) "Letter of Conditions";

(6) Form RD 1942-46, "Letter of Intent to Meet Conditions";

(7) Form RD 1940-1, "Request for Obligation of Funds";

(8) Completed environmental review documents including copies of public notices and appropriate proof of publication, if applicable; and

(9) Grant determination, if applicable.

(b) Approval and applicant notification will be accomplished by mailing to the applicant on the obligation date a copy of Form RD 1940-1. The date the applicant is notified is also the date the interest rate at loan approval is established.

[62 FR 33478, June 19, 1997, as amended at 63 FR 68655, Dec. 11, 1998]

§ 1780.42   Transfer of obligations.

An obligation of funds established for an applicant may be transferred to a different (substituted) applicant provided:

(a) The substituted applicant is eligible and has the authority to receive the assistance approved for the original applicant; and

(b) The need, purpose(s) and scope of the project for which RUS funds will be used remain substantially unchanged.

§ 1780.43   [Reserved]

§ 1780.44   Actions prior to loan or grant closing or start of construction, whichever occurs first.

(a) Applicants must provide evidence of adequate insurance and fidelity or employee dishonesty bond coverage.

(b) *Verification of users and other funds.* In connection with a project that involves new users and will be secured by a pledge of user fees or revenues, the processing office will authenticate the number of users. Ordinarily each signed user agreement will be reviewed and checked for evidence of cash contributions. If during the review any indication is received that all signed users may not connect to the system, there will be such additional investigation made as deemed necessary to determine the number of users who will connect to the system.

(c) *Initial compliance review.* An initial compliance review should be completed under subpart E of part 1901 of this title.

(d) *Applicant contribution.* An applicant contributing funds toward the project cost shall deposit these funds in its project account before start of construction. Project costs paid with applicant funds prior to the required deposit time shall be appropriately accounted for.

(e) *Excess RUS loan and grant funds.* If there is a significant reduction in project cost, the applicant's funding needs will be reassessed. Decreases in RUS funds will be based on revised project costs and current number of users, however, other factors including RUS regulations used at the time of loan or grant approval will remain the same. Obligated loan or grant funds not needed to complete the proposed project will be deobligated. Any reduction will be applied to grant funds first. In such cases, applicable forms, the letter of conditions, and other items will be revised.

(f) *Evidence of and disbursement of other funds.* Applicants expecting funds from other sources for use in completing projects being partially financed with RUS funds will present evidence of the commitment of these funds from such other sources. An agreement should be reached with all funding sources on how funds are to be disbursed before the start of construction. RUS funds will not be used to prefinance funds committed to the project from other sources.

(g) *Acquisition of land, easements, water rights, and existing facilities.* Applicants are responsible for acquisition of all property rights necessary for the project and will determine that prices paid are reasonable and fair. RUS may require an appraisal by an independent appraiser or Agency employee.

(1) *Rights-of-way and easements.* Applicants will obtain valid, continuous and adequate rights-of-way and easements needed for the construction, operation, and maintenance of the facility.

(i) The applicant must provide a legal opinion relative to the title to rights-of-way and easements. Form RD 442–22, "Opinion of Counsel Relative to Rights-of-Way," may be used. When a site is for major structures such as a reservoir or pumping station and the applicant is able to obtain only a right-of-way or easement on such a site rather than a fee simple title, the applicant will furnish a title report thereon by the applicant's attorney showing ownership of the land and all mortgages or other lien defects, restrictions, or encumbrances, if any.

(ii) For user connections funded by RUS, applicants will obtain adequate rights to construct and maintain the connection line or other facilities located on the user's property. This right may be obtained through formal easement or user agreements.

(2) *Title for land or existing facilities.* Title to land essential to the successful operation of facilities or title to facilities being purchased, must not contain any restrictions that will adversely affect the suitability, successful operation, security value, or transferability of the facility. Preliminary and final title opinions must be provided by the applicant's attorney. The opinions must be in sufficient detail to assess marketability of the property. Form RD 1927–9, "Preliminary Title Opinion," and Form RD 1927–10, "Final Title Opinion," may be used to provide the required title opinions.

(i) In lieu of receiving title opinions from the applicant's attorney, the applicant may use a title insurance company. If a title insurance company is used, the applicant must provide the Agency a title insurance binder, disclosing all title defects or restrictions, and include a commitment to issue a title insurance policy. The policy should be in an amount at least equal to the market value of the property as improved. The title insurance binder and commitment should be provided to the Agency prior to requesting closing instructions. The Agency will be provided a title insurance policy which will insure RUS's interest in the property without any title defects or restrictions which have not been waived by the Agency.

(ii) The approval official may waive title defects or restrictions, such as utility easements, that do not adversely affect the suitability, successful operation, security value, or transferability of the facility.

(3) *Water rights.* The following will be furnished as applicable:

(i) A statement by the applicant's attorney regarding the nature of the water rights owned or to be acquired by the applicant (such as conveyance of title, appropriation and decree, application and permit, public notice and appropriation and use).

(ii) A copy of a contract with another company or municipality to supply water; or stock certificates in another company which represents the right to receive water.

(4) *Lease agreements.* Where the right of use or control of real property not owned by the applicant is essential to the successful operation of the facility during the life of the loan, such right will be evidenced by written agreements or contracts between the owner of the property and the applicant. Lease agreements shall not contain provisions for restricted use of the site of facility, forfeiture or summary cancellation clauses. Lease agreements shall provide for the right to transfer, encumber, assign and sub-lease without restriction. Lease agreements will ordinarily be written for a term at least equal to the term of the loan. Such lease contracts or agreements will be approved by the approval official with the advice and counsel of OGC, as necessary.

(h) *Obtaining loan closing instructions.* The information required by OGC will be transmitted to OGC with request for closing instructions. Upon receipt of closing instructions, the processing office will discuss with the applicant and its engineer, attorney, and other appropriate representatives, the requirements contained therein and any actions necessary to proceed with closing. State program officials have the

option to work with OGC to obtain waivers for closing instructions in certain cases. Closing instructions are not required for grants.

§ 1780.45 **Loan and grant closing and delivery of funds.**

(a) *Loan closing.* Notes and bonds will be completed on the date of loan closing except for the entry of subsequent RUS multiple advances where applicable. The amount of each note will be in multiples of not less than $100. The amount of each bond will ordinarily be in multiples of not less than $1,000.

(1) Form RD 440-22, "Promissory Note (Association or Organization)," will ordinarily be used for loans to non-public bodies.

(2) RUS Bulletins 1780-27, "Loan Resolution (Public Bodies)," or 1780-28, "Loan Resolution Security Agreement," will be adopted by public and other-than-public bodies. These resolutions supplement other provisions in this part.

(3) Subpart D of this part contains instructions for preparation of notes and bonds evidencing indebtedness of public bodies.

(b) *Loan disbursement.* (1) Multiple advances. Multiple advances will be used only for loans in excess of $100,000. Advances will be made only as needed to cover disbursements required by the borrower over a 30-day period.

(i) Subpart D of this part contains instructions for making multiple advances to public bodies.

(ii) Advances will be requested by the borrower in writing. The request should be in sufficient amounts to pay cost of construction, rights-of-way and land, legal, engineering, interest, and other expenses as needed. The borrower may use Form RD 440-11, "Estimate of Funds Needed for 30 Day Period Commencing XXX," to show the amount of funds needed during the 30-day period.

(2) RUS loan funds obligated for a specific purpose, such as the paying of interest, but not needed at the time of loan closing will remain in the Finance Office until needed unless State statutes require all funds to be delivered to the borrower at the time of closing. Loan funds may be advanced to prepay costs under § 1780.9 (e)(2)(iv). If all funds must be delivered to the borrower at

the time of closing to comply with State statutes, funds not needed at loan closing will be handled as follows:

(i) Deposited in an appropriate borrower account, such as debt service or construction accounts; or

(ii) Deposited in a joint bank account under paragraph (e)(3) of this section.

(c) *Grant closing.* RUS Bulletin 1780–12 "Water or Waste System Grant Agreement" of this part will be completed and executed in accordance with the requirements of grant approval. The grant will be considered closed when RUS Bulletin 1780–12 has been properly executed. Processing or approval officials are authorized to sign the grant agreement on behalf of RUS. For grants that supplement RUS loan funds, the grant should be closed simultaneously with the closing of the loan. However, when grant funds will be disbursed before loan closing, as provided in paragraph (d)(1) of this section, the grant will be closed not later than the delivery date of the first advance of grant funds.

(d) *Grant disbursements.* RUS policy is not to disburse grant funds from the Treasury until they are actually needed by the applicant. Applicant funds will be disbursed before the disbursal of any RUS grant funds. RUS loan funds will be disbursed before the disbursal of any RUS grant funds except when:

(1) Interim financing of the total estimated amount of loan funds needed during construction is arranged; and

(2) All interim funds have been disbursed; and

(3) RUS grant funds are needed before the RUS loan can be closed.

(e) *Use and accountability of funds.* (1) Arrangements will be agreed upon for the prior concurrence by the Agency of the bills or vouchers upon which warrants will be drawn. Form RD 402–2, "Statement of Deposits and Withdrawals," or similar form will be used by the Agency to monitor funds. Periodic reviews of these accounts shall be made by the Agency.

(2) *Pledge of collateral for grants to nonprofit organizations.* Grant funds must be deposited in a bank with Federal Deposit Insurance Corporation (FDIC) insurance coverage. Also, if the balance in the account containing grant funds exceeds the FDIC insur-

ance coverage, the excess amount must be collaterally secured. The pledge of collateral for the excess will be in accordance with Treasury Circular 176.

(3) *Joint RUS/borrower bank account.* RUS funds and any funds furnished by the borrower including contributions to purchase major items of equipment, machinery, and furnishings will be deposited in a joint RUS/borrower bank account if determined necessary by the approval official. When RUS has a Memorandum of Understanding with another agency that provides for the use of joint RUS/borrower accounts, or when RUS is the primary source of funds for a project and has determined that the use of a joint RUS/borrower bank account is necessary, project funds from other sources may also be deposited in the joint bank account. RUS shall not be accountable to the source of the other funds nor shall RUS undertake responsibility to administer the funding program of the other entity. Joint RUS/borrower bank accounts should not be used for funds advanced by an interim lender. When funds exceeds the FDIC insurance coverage, the excess must have a pledge of collateral in accordance with Treasury Circular 176.

(4) *Payment for project costs.* Project costs will be monitored by the RUS processing office. Invoices will be approved by the borrower and their engineer, as appropriate, and submitted to the processing office for concurrence. The review and acceptance of project costs, including construction pay estimates, by RUS does not attest to the correctness of the amounts, the quantities shown or that the work has been performed under the terms of the agreements or contracts.

(f) *Use of remaining funds.* Funds remaining after all costs incident to the basic project have been paid or provided for will not include applicant contributions. Funds remaining, may be considered in direct proportion to the amounts obtained from each source. Remaining funds will be handled as follows:

(1) Remaining funds may be used for eligible loan or grant purposes, provided the use will not result in major

313

changes to the facility(s) and the purpose of the loan and grant remains the same;

(2) RUS loan funds that are not needed will be applied as an extra payment on the RUS indebtedness unless other disposition is required by the bond ordinance, resolution, or State statute; and

(3) Grant funds not expended under paragraph (f)(1) of this section will be canceled. Prior to the actual cancellation, the borrower, its attorney and its engineer will be notified of RUS's intent to cancel the remaining funds. The applicant will be given appropriate appeal rights.

(g) *Post review of loan closing.* In order to determine that the loan has been properly closed the loan docket will be reviewed by OGC. The State program official has the option to consult with OGC to obtain waivers of this review.

[62 FR 33478, June 19, 1997, as amended at 64 FR 29946, June 4, 1999]

§ 1780.46 [Reserved]

§ 1780.47 Borrower accounting methods, management reporting and audits.

(a) Borrowers are required to provide RUS an annual audit or financial statements.

(b) *Method of accounting and preparation of financial statements.* Annual organization-wide financial statements must be prepared on the accrual basis of accounting, in accordance with generally accepted accounting principles (GAAP), unless State statutes or regulatory agencies provide otherwise, or an exception is granted by the Agency. An organization may maintain its accounting records on a basis other than accrual accounting, and make the necessary adjustments so that annual financial statements are presented on the accrual basis.

(c) *Record retention.* Each borrower shall retain all records, books, and supporting material for 3 years after the issuance of the audit or management reports. Upon request, this material will be made available to RUS, Office of the Inspector General (OIG), United States Department of Agriculture (USDA), the Comptroller General, or to their assignees.

(d) *Audits.* All audits are to be performed in accordance with the latest revision of the generally accepted government auditing standards (GAGAS), issued by the Comptroller General of the United States. In addition, the audits are also to be performed in accordance with subpart F of 2 CFR part 200, as adopted by USDA through 2 CFR part 400. The type of audit each borrower is required to submit will be designated by RUS. Further guidance on preparing an acceptable audit can be obtained from RUS. It is not intended that audits required by this part be separate and apart from audits performed in accordance with State and local laws. To the extent feasible, the audit work should be done in conjunction with those audits. Audits must be performed annually except as allowed under the provisions for biennial audits provided in subpart F of 2 CFR part 200. Audits are to be submitted to the processing office as soon as possible after receipt of the auditor's report but no later than nine months after the end of the audit period

(e) *Borrowers exempt from audits.* All borrowers who are exempt from audits, will, within 60 days following the end of each fiscal year, furnish the RUS with annual financial statements, consisting of a verification of the organization's balance sheet and statement of income and expense by an appropriate official of the organization. Forms RD 442-2, "Statement of Budget, Income and Equity," and 442-3 may be used.

(f) *Management reports.* These reports will furnish management with a means of evaluating prior decisions and serve as a basis for planning future operations and financial strategies. In those cases where revenues from multiple sources are pledged as security for an RUS loan, two reports will be required; one for the project being financed by RUS and one combining the entire operation of the borrower. In those cases where RUS loans are secured by general obligation bonds or assessments and the borrower combines revenues from all sources, one management report combining all such revenues is acceptable. The following management data will be submitted by the borrower to the processing office.

These reports at a minimum will include a balance sheet and income and expense statement.

(1) *Quarterly reports.* A quarterly management report will be required for the first year for new borrowers and for all borrowers experiencing financial or management problems for one year from the date problems were noted. If the borrower's account is current at the end of the year, the processing office may waive the required reports.

(2) *Annual management reports.* Prior to the beginning of each fiscal year the following will be submitted to the processing office. (If Form RD 442–2 is used as the annual management report, enter data in column three only of Schedule 1, and complete all of Schedule 2.)

(i) Two copies of the management reports and proposed "Annual Budget".

(ii) Financial information may be reported on Form RD 442–2 which includes Schedule 1, "Statement of Budget, Income and Equity" and Schedule 2, "Projected Cash Flow" or information in similar format.

(iii) A copy of the rate schedule in effect at the time of submission.

(g) *Substitute for management reports.* When RUS loans are secured by the general obligation of the public body or tax assessments which total 100 percent of the debt service requirements, the State program official may authorize an annual audit to substitute for other management reports if the audit is received within nine months after the end of the audit period.

[62 FR 33478, June 19, 1997, as amended at 79 FR 76006, Dec. 19, 2014]

§ 1780.48 **Regional commission grants.**

Grants are sometimes made by regional commissions for projects eligible for RUS assistance. RUS has agreed to administer such funds in a manner similar to administering RUS assistance.

(a) When RUS has funds in the project, no charge will be made for administering regional commission funds.

(b) When RUS has no loan or grant funds in the project, an administrative charge will be made pursuant to the Economy Act of 1932 (31 U.C.S. 1535). A fee of 5 percent of the first $100,000 of a regional commission grant and 1 percent of any amount over $100,000 will be paid to RUS by the commission.

(1) *Appalachian Regional Commission (ARC).* RUS Bulletin 1780–23 will be followed in determining the responsibilities of RUS. The ARC Federal Cochairman and the State program official will provide each other with the necessary notification and certification.

(2) *Other regional commissions.* Title V of the Public Works and Economic Development Act of 1965 (42 U.S.C. 3121 *et seq.*) authorizes other commissions similar to ARC. RUS Bulletin 1780–23 will be used to develop a separate project management agreement between RUS and the commission for each project. The agreement should be prepared by the State program official as soon as notification is received that a commission grant will be made and the amount is confirmed.

(c) Regional commission grants should be obligated as soon as possible in accordance with § 1780.41, except that the announcement procedure referred to in RUS Staff Instruction 1780–2 is not applicable. Regional commission grants will be disbursed from the Finance Office in the same manner as RUS funds.

[62 FR 33478, June 19, 1997, as amended at 64 FR 29946, June 4, 1999]

§§ 1780.49–1780.52 **[Reserved]**

Subpart C—Planning, Designing, Bidding, Contracting, Constructing and Inspections

§ 1780.53 **General.**

This subpart is specifically designed for use by owners including the professional or technical consultants or agents who provide assistance and services such as engineering, environmental, inspection, financial, legal or other services related to planning, designing, bidding, contracting, and constructing water and waste disposal facilities. These procedures do not relieve the owner of the contractual obligations that arise from the procurement of these services. For this subpart, an owner is defined as an applicant, borrower, or grantee.

§ 1780.54 Technical services.

Owners are responsible for providing the engineering, architect and environmental services necessary for planning, designing, bidding, contracting, inspecting, and constructing their facilities. Services may be provided by the owner's "in house" engineer or architect or through contract, subject to Agency concurrence. Engineers and architects must be licensed in the State where the facility is to be constructed.

§ 1780.55 Preliminary engineering reports and environmental review documentation.

Preliminary engineering reports (PERs) must conform to customary professional standards. PER guidelines for water, sanitary sewer, solid waste, and storm sewer are available from the Agency. Environmental review documentation must comply with the environmental review requirements in accordance with 7 CFR part 1970.

[81 FR 11028, Mar. 2, 2016]

§ 1780.56 [Reserved]

§ 1780.57 Design policies.

Facilities financed by the Agency will be designed and constructed in accordance with sound engineering practices, and must meet the requirements of Federal, State and local agencies.

(a) *Environmental review.* Facilities financed by the Agency must undergo an environmental impact analysis in accordance with the National Environmental Policy Act and RUS procedures. Facility planning and design must not only be responsive to the owner's needs but must consider the environmental consequences of the proposed project. Facility design shall incorporate and integrate, where practicable, mitigation measures that avoid or minimize adverse environmental impacts. Environmental reviews serve as a means of assessing environmental impacts of project proposals, rather than justifying decisions already made. Applicants may not take any action on a project proposal that will have an adverse environmental impact or limit the choice of reasonable project alternatives being reviewed prior to the completion of the Agency's environmental review.

(b) *Architectural barriers.* All facilities intended for or accessible to the public or in which physically handicapped persons may be employed must be developed in compliance with the Architectural Barriers Act of 1968 (42 U.S.C. 4151 *et seq.*) as implemented by 41 CFR 101–19.6, section 504 of the Rehabilitation Act of 1973 (42 U.S.C 1474 *et seq.*) as implemented by 7 CFR parts 15 and 15b, and Titles II and III of the Americans with Disabilities Act of 1990 (42 U.S.C. 12101 *et seq.*).

(c) *Energy/environment.* Facility design should consider cost effective energy-efficient and environmentally-sound products and services.

(d) *Fire protection.* Water facilities should have sufficient capacity to provide reasonable fire protection to the extent practicable.

(e) *Growth capacity.* Facilities should have sufficient capacity to provide for reasonable growth to the extent practicable.

(f) *Water conservation.* Owners are encouraged, when economically feasible, to incorporate water conservation practices into a facility's design. For existing water systems, evidence must be provided showing that the distribution system water losses do not exceed reasonable levels.

(g) *Conformity with State drinking water standards.* No funds shall be made available under this part for a water system unless the Agency determines that the water system will make significant progress toward meeting the standards established under title XIV of the Public Health Service Act (commonly known as the 'Safe Drinking Water Act') (42 U.S.C. 300f *et seq.*).

(h) *Conformity with Federal and State water pollution control standards.* No funds shall be made available under this part for a water treatment discharge or waste disposal system unless the Agency determines that the effluent from the system conforms with applicable Federal and State water pollution control standards.

(i) *Combined sewers.* New combined sanitary and storm water sewer facilities will not be financed by the Agency. Extensions to existing combined systems can only be financed when separate systems are impractical.

(j) *Dam safety.* Projects involving any artificial barrier which impounds or diverts water, or the rehabilitation or improvement of such a barrier, must comply with the provisions for dam safety as set forth in the Federal Guidelines for Dam Safety (Government Printing Office stock No. 041–001–00187–5, Superintendent of Documents, Attn: New Orders, P.O. Box 371954, Pittsburgh, PA 15250–7954) as prepared by the Federal Coordinating Council for Science, Engineering and Technology.

(k) *Pipe.* All pipe used shall meet current American Society for Testing Materials (ASTM) or American Water Works Association (AWWA) standards.

(l) *Water system testing.* For new water systems or extensions to existing water systems, leakage shall not exceed limits set by either ASTM or AWWA whichever is the more stringent.

(m) *Metering devices.* Water facilities financed by the Agency will have metering devices for each connection. An exception to this requirement may be granted by the State program official when the owner demonstrates that installation of metering devices would be a significant economic detriment and that environmental considerations would not be adversely affected by not installing such devices. Sanitary sewer projects should incorporate water system metering devices whenever practicable.

(n) *Economical service.* The facility's design must provide the most economical service practicable.

(o) *Seismic safety.* All new structures, fully or partially enclosed, used or intended for sheltering persons or property will be designed with appropriate seismic safety provisions in compliance with the Earthquake Hazards Reduction Act of 1977 (42 U.S.C. 7701 *et seq.*), and Executive Order 12699, Seismic Safety of Federal and Federally Assisted or Regulated New Building Construction (3 CFR, 1990 Comp., p. 269). Designs of components essential for system operation and substantial rehabilitation of structures that are used for sheltering persons or property should incorporate seismic safety provisions to the extent practicable. RUS implementing regulations for seismic safety are in 7 CFR part 1972, subpart C.

[62 FR 33478, June 19, 1997, as amended at 63 FR 68655, Dec. 11, 1998; 64 FR 29946, June 4, 1999]

§§1780.58–1780.60 [Reserved]

§1780.61 Construction contracts.

Contract documents must be sufficiently descriptive and legally binding in order to accomplish the work as economically and expeditiously as possible.

(a) *Standard construction contract documents.* If the construction contract documents utilized are not in the format previously approved by the Agency, OGC's review of the construction contract documents will be obtained prior to their use.

(b) *Contract review and concurrence.* The owner's attorney will review the executed contract documents, including performance and payment bonds, and will certify that they are adequate, and that the persons executing these documents have been properly authorized to do so. The contract documents, engineer's recommendation for award, and bid tabulation sheets will be forwarded to the Agency for concurrence prior to awarding the contract. All contracts will contain a provision that they are not effective until they have been concurred in by the Agency. The State program official or designee is responsible for concurring in construction contracts with the legal advice and guidance of the OGC when necessary.

§1780.62 Utility purchase contracts.

Applicants proposing to purchase water or other utility service from private or public sources shall have written contracts for supply or service which are reviewed and concurred in by the Agency. To the extent practical, the Agency review and concurrence of such contracts should take place prior to their execution by the owner. OGC advice and guidance may be requested. Form RD 442–30, "Water Purchase Contract," may be used when appropriate. If the Agency loan will be repaid from system revenues, the contract will be pledged to the Agency as part of the security for the loan. Such contracts will:

(a) Include a commitment by the supplier to furnish, at a specified point, an adequate quantity of water or other service and provide that, in case of shortages, all of the supplier's users will proportionately share shortages.

(b) Set out the ownership and maintenance responsibilities of the respective parties including the master meter if a meter is installed at the point of delivery.

(c) Specify the initial rates and provide a type of escalator clause which will permit rates for the association to be raised or lowered proportionately as certain specified rates for the supplier's regular customers are raised or lowered. Provisions may be made for altering rates in accordance with the decisions of the appropriate State agency which may have regulatory authority.

(d) Cover period of time which is at least equal to the repayment period of the loan. State program officials may approve contracts for shorter periods of time if the supplier cannot legally contract for such period, or if the owner and supplier find it impossible or impractical to negotiate a contract for the maximum period permissible under State law, provided:

(1) The supplier is subject to regulations of the Federal Energy Regulatory Commission or other Federal or State agency whose jurisdiction can be expected to prevent unwarranted curtailment of supply; or

(2) The contract contains adequate provisions for renewal; or

(3) A determination is made that in the event the contract is terminated, there are or will be other adequate sources available to the owner that can feasibly be developed or purchased.

(e) Set out in detail the amount of connection or demand charges, if any, to be made by the supplier as a condition to making the service available to the owner. However, the payment of such charges from loan funds shall not be approved unless the Agency determines that it is more feasible and economical for the owner to pay such a connection charge than it is for the owner to provide the necessary supply by other means.

(f) Provide for a pledge of the contract to the Agency as part of the security for the loan.

(g) Not contain provisions for:

(1) Construction of facilities which will be owned by the supplier. This does not preclude the use of money paid as a connection charge for construction to be done by the supplier.

(2) Options for the future sale or transfer. This does not preclude an agreement recognizing that the supplier and owner may at some future date agree to a sale of all or a portion of the facility.

(h) If it is impossible to obtain a firm commitment for either an adequate quantity or sharing shortages proportionately, a contract may be executed and concurred in provided adequate evidence is furnished to enable the Agency to make a determination that the supplier has adequate supply and/or treatment facilities to furnish its other users and the applicant for the foreseeable future; and:

(1) The supplier is subject to regulations of the Federal Energy Regulatory Commission or other Federal or State agency whose jurisdiction can be expected to prevent unwarranted curtailment of supply; or

(2) A suitable alternative supply could be arranged within the repayment ability of the borrower if it should become necessary; or

(3) Concurrence in the proposed contract is obtained from the National Office.

§ 1780.63 Sewage treatment and bulk water sales contracts.

Owners entering into agreements with private or public parties to treat sewage or supply bulk water shall have written contracts for such service and all such contracts shall be subject to the Agency concurrence. Section 1780.62 should be used as a guide to prepare such contracts.

§§ 1780.64–1780.66 [Reserved]

§ 1780.67 Performing construction.

Owners are encouraged to accomplish construction through contracts with qualified contractors. Owners may accomplish construction by using their own personnel and equipment provided the owners possess the necessary skills, abilities and resources to perform the work and provided a licensed engineer prepares design drawings and specifications and inspects construction and furnishes inspection reports as required by §1780.76. Inspection services may be provided by individuals as approved by the State staff engineer. Payments for construction will be handled under §1780.76(e).

§1780.68 Owner's contractual responsibility.

This part does not relieve the owner of any responsibilities under its contract. The owner is responsible for the settlement of all contractual and administrative issues arising out of procurement entered into in support of a loan or grant. These include, but are not limited to: source evaluation, protests, disputes, and claims. Matters concerning violation of laws are to be referred to the applicable local, State, or Federal authority.

§1780.69 [Reserved]

§1780.70 Owner's procurement regulations.

Owner's procurement requirements must comply with the following standards:

(a) *Code of conduct.* Owners shall maintain a written code or standards of conduct which shall govern the performance of their officers, employees or agents engaged in the award and administration of contracts supported by Agency funds. No employee, officer or agent of the owner shall participate in the selection, award, or administration of a contract supported by Agency funds if a conflict of interest, real or apparent, would be involved. Examples of such conflicts would arise when: the employee, officer or agent; any member of their immediate family; their partner; or an organization which employs, or is about to employ, any of the above; has a financial or other interest in the firm selected for the award.

(1) The owner's officers, employees or agents shall neither solicit nor accept gratuities, favors or anything of monetary value from contractors, potential contractors, or parties to subagreements.

(2) To the extent permitted by State or local law or regulations, the owner's standards of conduct shall provide for penalties, sanctions, or other disciplinary actions for violations of such standards by the owner's officers, employees, agents, or by contractors or their agents.

(b) *Maximum open and free competition.* All procurement transactions, regardless of whether by sealed bids or by negotiation and without regard to dollar value, shall be conducted in a manner that provides maximum open and free competition. Procurement procedures shall not restrict or eliminate competition. Examples of what are considered to be restrictive of competition include, but are not limited to: placing unreasonable requirements on firms in order for them to qualify to do business; noncompetitive practices between firms; organizational conflicts of interest; and unnecessary experience and bonding requirements. In specifying materials, the owner and its consultant will consider all materials normally suitable for the project commensurate with sound engineering practices and project requirements. The Agency shall consider fully any recommendation made by the owner concerning the technical design and choice of materials to be used for a facility. If the Agency determines that a design or material, other than those that were recommended should be considered by including them in the procurement process as an acceptable design or material in the water or waste disposal facility, the Agency shall provide such owner with a comprehensive justification for such a determination. The justification will be documented in writing.

(c) *Owner's review.* Proposed procurement actions shall be reviewed by the owner's officials to avoid the purchase of unnecessary or duplicate items. Consideration should be given to consolidation or separation of procurement items to obtain a more economical purchase. Where appropriate, an analysis shall be made of lease versus purchase alternatives, and any other appropriate analysis to determine which approach would be the most economical. To foster greater economy and efficiency, owners are encouraged to enter into State and local intergovernmental agreements for procurement or use of common goods and services.

(d) Solicitation of offers, whether by competitive sealed bid or competitive negotiation, shall:

(1) Incorporate a clear and accurate description of the technical requirements for the material, product or service to be procured. When it is impractical or uneconomical to make a clear and accurate description of the technical requirements, a "brand name or equal" description may be used to define the performance or other salient requirements of a procurement. The specific feature of the name brands which must be met by the offeror shall be clearly stated; and

(2) Clearly specify all requirements which offerors must fulfill and all other factors to be used in evaluating bids or proposals.

(e) Affirmative steps should be taken to assure that small, minority, and women businesses are utilized when possible as sources of supplies, equipment, construction and services.

(f) *Contract pricing.* Cost plus a percentage of cost method of contracting shall not be used.

(g) *Unacceptable bidders.* The following will not be allowed to bid on, or negotiate for, a contract or subcontract related to the construction of the project:

(1) An engineer as an individual or firm who has prepared plans and specifications or who will be responsible for monitoring the construction;

(2) Any firm or corporation in which the owner's engineer is an officer, employee, or holds or controls a substantial interest;

(3) The governing body's officers, employees, or agents;

(4) Any member of the immediate family or partners in the entities referred to in paragraphs (g)(1), (g)(2) or (g)(3) of this section; or

(5) An organization which employs, or is about to employ, any person in the entities referred to in paragraphs (g)(1), (g)(2), (g)(3) or (g)(4) of this section.

(h) *Contract award.* Contracts shall be made only with responsible parties possessing the potential ability to perform successfully under the terms and conditions of a proposed procurement. Consideration shall include but not be limited to matters such as integrity, record of past performance, financial and technical resources, and accessibility to other necessary resources. Contracts shall not be made with parties who are suspended or debarred by any Agency of the United States Government.

§ 1780.71   [Reserved]

§ 1780.72   Procurement methods.

Procurement shall be made by one of the following methods and in accordance with requirements of 2 CFR 200.320: Micro-purchases, procurement by small purchase procedures, procurement by sealed bids (formal advertising), procurement by competitive proposals, or procurement by noncompetitive proposals. The sealed bid method is the preferred method for procuring construction.

[81 FR 47689, July 22, 2016]

§ 1780.73   [Reserved]

§ 1780.74   Contracts awarded prior to applications.

Owners awarding construction or other procurement contracts prior to filing an application, must provide evidence that is satisfactory to the Agency that the contract was entered into without intent to circumvent the requirements of Agency regulations.

(a) *Modifications.* The contract shall be modified to conform with the provisions of this part. Where this is not possible, modifications will be made to the extent practicable and, as a minimum, the contract must comply with all State and local laws and regulations as well as statutory requirements and executive orders related to the Agency financing. When all construction is complete and it is impracticable to modify the contracts, the owner must provide the certification required by paragraph (c) of this section.

(b) *Consultant's certification.* Provide a certification by an engineer, licensed in the State where the facility is constructed, that any construction performed complies fully with the plans and specifications.

(c) *Owner's certification.* Provide a certification by the owner that the contractor has complied with applicable statutory and executive requirements related to Agency financing for construction already performed.

§ 1780.75   Contract provisions.

In addition to provisions required for a valid and legally binding contract, any recipient of Agency funds shall include the following contract provisions in all contracts.

(a) *Remedies.* Contracts for more than the Simplified Acquisition Threshold shall contain provisions or conditions which will allow for administrative, contractual, or legal remedies in instances where contractors violate or breach contract terms, and provide for such sanctions and penalties as may be appropriate. A realistic liquidated damage provision should be included in all contracts for construction.

(b) *Termination.* All contracts exceeding $10,000, shall contain suitable provisions for termination by the owner including the manner by which it will be effected and the basis for settlement. In addition, such contracts shall describe conditions under which the contract may be terminated for default as well as conditions where the contract may be terminated because of circumstances beyond the control of the contractor.

(c) *Surety.* In all contracts for construction or facility improvements exceeding the Simplified Acquisition Threshold, the owner shall require bonds or cash deposit in escrow assuring performance and payment each in the amount of 100 percent of the contract cost. The surety will be in the form of performance bonds and payment bonds. For contracts of lesser amounts, the owner may require surety. When a surety is not provided, contractors will furnish evidence of payment in full for all materials, labor, and any other items procured under the contract. Form RD 1924–10, "Release by Claimants," and Form RD 1924–9, "Certificate of Contractor's Release," may be used for this purpose.

Companies providing performance bonds and payment bonds must hold a certificate of authority as an acceptable surety on Federal bonds as listed in Treasury Circular 570 as amended and the surety must be listed as having a license to do business in the State where the facility is located.

(d) *Equal employment opportunity.* All contracts awarded in excess of $10,000 by owners shall contain a provision requiring compliance with Executive Order 11246 (3 CFR, 1966 Comp., p.339), entitled, "Equal Employment Opportunity," as amended by Executive Order 11375 (3 CFR, 1968 Comp., p. 321), and as supplemented by Department of Labor regulations 41 CFR chapter 60.

(e) *Anti-kickback.* All contracts for construction shall include a provision for compliance with the Copeland "Anti-Kickback" Act (18 U.S.C. 874). This Act provides that each contractor shall be prohibited from inducing, by any means, any person employed in the construction, completion, or repair of public work, to give up any part of the compensation to which they are otherwise entitled. The owner shall report suspected or reported violations to the Agency.

(f)–(g) [Reserved]

(h) *Change orders.* The construction contract shall require that all contract change orders be concurred in by the Agency.

(i) *Agency concurrence.* All contracts must contain a provision that they shall not be effective unless and until the State program official or designee concurs in writing.

(j) *Retainage.* All construction contracts shall contain adequate provisions for retainage. No payments will be made that would deplete the retainage nor place in escrow any funds that are required for retainage nor invest the retainage for the benefit of the contractor. The retainage shall not be less than an amount equal to 5 percent of an approved partial payment estimate until the project is substantially complete and accepted by the owner, consulting engineer and Agency. The contract must provide that additional amounts may be retained if the job is not proceeding satisfactorily.

(k) *Clean Air Act (42 U.S.C. 7401–7671q) and the Federal Water Pollution Control*

Act (33 U.S.C. 1251–1388). Contracts and subgrants of amounts in excess of $150,000 must contain a provision that requires the contractor to agree to comply with all applicable standards, orders or regulations issued pursuant to the Clean Air Act (42 U.S.C. 7401–7671q) and the Federal Water Pollution Control Act as amended (33 U.S.C. 1251–1387). Violations must be reported to the Federal awarding agency and the Regional Office of the Environmental Protection Agency (EPA).

(1) *Contract Work Hours and Safety Standards Act (40 U.S.C. 3701–3708).* Where applicable, all contracts awarded by the non-Federal entity in excess of $100,000 that involve the employment of mechanics or laborers must include a provision for compliance with 40 U.S.C. 3702 and 3704, as supplemented by Department of Labor regulations (29 CFR part 5). Under 40 U.S.C. 3702, each contractor must be required to compute the wages of every mechanic and laborer on the basis of a standard work week of 40 hours. Work in excess of the standard work week is permissible provided that the worker is compensated at a rate of not less than one and a half times the basic rate of pay for all hours worked in excess of 40 hours in the work week. The requirements of 40 U.S.C. 3704 are applicable to construction work and provide that no laborer or mechanic must be required to work in surroundings or under working conditions which are unsanitary, hazardous or dangerous. These requirements do not apply to the purchases of supplies or materials or articles ordinarily available on the open market.

(m) *Debarment and suspension.* A contract award (see 2 CFR 180.220) must not be made to parties listed on the governmentwide exclusions in the System for Award Management (SAM), in accordance with the OMB guidelines at 2 CFR part 180, as supplemented by 2 CFR part 417, "Debarment and Suspension." SAM exclusion records contain the names of parties debarred, suspended, or otherwise excluded by agencies, as well as parties declared ineligible under statutory or regulatory authority other than Executive Order 12549.

(n) *Byrd anti-lobbying amendment (31 U.S.C. 1352).* Contractors that apply or bid for an award exceeding $100,000 must file the required certification. Each tier certifies to the tier above that it will not and has not used Federal appropriated funds to pay any person or organization for influencing or attempting to influence an officer or employee of any agency, a member of Congress, officer or employee of Congress, or an employee of a member of Congress in connection with obtaining any Federal contract, grant or any other award covered by 31 U.S.C. 1352. Each tier must also disclose any lobbying with non-Federal funds that takes place in connection with obtaining any Federal award. Such disclosures are forwarded from tier to tier up to the non-Federal award.

(o) *Procurement of recovered materials.* A public body, such as a state government, state agency, municipality, county, district, authority, or other political subdivision of a state, territory or commonwealth, must ensure its contracts include provisions requiring compliance with section 6002 of the Solid Waste Disposal Act, as amended by the Resource Conservation and Recovery Act. The requirements of Section 6002 include procuring only items designated in guidelines of the Environmental Protection Agency (EPA) at 40 CFR part 247 that contain the highest percentage of recovered materials practicable, consistent with maintaining a satisfactory level of competition, where the purchase price of the item exceeds $10,000 or the value of the quantity acquired during the preceding fiscal year exceeded $10,000; procuring solid waste management services in a manner that maximizes energy and resource recovery; and establishing an affirmative procurement program for procurement of recovered materials identified in the EPA guidelines.

[62 FR 33478, June 19, 1997, as amended at 81 FR 7697, Feb. 16, 2016]

§ 1780.76 Contract administration.

Owners shall be responsible for maintaining a contract administration system to monitor the contractors' performance and compliance with the terms, conditions, and specifications of the contracts.

(a) *Preconstruction conference.* Prior to beginning construction, the owner will schedule a preconstruction conference where the consulting engineer will review the planned development with the Agency, owner, resident inspector, attorney, contractor, and other interested parties. The conference will thoroughly cover applicable items included in Form RD 1924–16, "Record of Pre-construction Conference," and the discussions and agreements will be documented.

(b) *Monitoring reports.* The owner is required to monitor construction and provide a report to the Agency giving a full explanation under the following circumstances:

(1) Reasons why approved construction schedules were not met;

(2) Analysis and explanation of cost overruns and how payment is to be made for the same; and

(3) If events occur which have a significant impact upon the project.

(c) *Inspection.* Full-time resident inspection is required for all construction unless a written exception is made by the Agency upon written request of the owner. Unless otherwise agreed, the resident inspector will be provided by the consulting engineer. Prior to the preconstruction conference, the consulting engineer will submit a resume of qualifications of the resident inspector to the owner and to the Agency for acceptance in writing. If the owner provides the resident inspector, it must submit a resume of the inspector's qualifications to the project engineer for comments and the Agency for acceptance in writing prior to the preconstruction conference. The resident inspector will work under the technical supervision of the project engineer and the role and responsibilities will be defined in writing.

(d) *Inspector's daily diary.* The resident inspector will maintain a record of the daily construction progress in the form of a daily diary and daily inspection reports. The daily entries shall be made available to the Agency personnel and will be reviewed during project inspections. The original complete set will be furnished to the owner upon completion of construction. RUS Bulletin 1780–18 is available from the Agency for preparing daily inspection reports or the reports can be provided

in other formats approved by the State staff engineer.

(e) *Payment for Construction.* Form RD 1924–18, "Partial Payment Estimate," or other similar form may be used for construction payments. If Form 1924–18 is not used, prior concurrence by the State staff engineer must be obtained.

(1) Payment of contract retainage will not be made until such retainage is due and payable under the terms of the contact.

(2) Invoices for the payment of construction costs must be approved by the owner, project engineer and concurred in by the Agency.

(3) The review and acceptance of project costs, including construction payment estimates by the Agency shall not attest to the correctness of the amounts, the quantities shown, or that the work has been performed under the terms of agreements or contracts.

(f) *Prefinal inspections.* A prefinal inspection will be made by the owner, resident inspector, project engineer, contractor, representatives of other agencies involved, and Agency representative (preferably the State staff engineer or designee). The inspection results will be recorded by the project engineer and a copy provided to all interested parties.

(g) *Final inspection.* A final inspection will be made by the Agency before final payment is made.

(h) *Changes in development plans.* (1) Changes in development plans shall be reviewed and approved by the Agency provided:

(i) Funds are available to cover any additional costs; and

(ii) The change is for an authorized loan or grant purpose; and

(iii) It will not adversely affect the soundness of the facility operation or the Agency's security; and

(iv) The change is within the scope of the contract,

(2) Changes will be recorded on Form RD 1924–7, "Contract Change Order," or other similar form if approved by the State program official or designee. Regardless of the form, change orders must be approved by the State program official or designee.

(3) Changes should be accomplished only after Agency approval and shall

be authorized only by means of contract change order. The change order will include items such as:

(i) Any changes in labor and material;

(ii) Changes in facility design;

(iii) Any decrease or increase in quantities based on final measurements that are different from those shown in the bidding schedule; and

(iv) Any increase or decrease in the time to complete the project.

(4) All changes shall be recorded on chronologically numbered contract change orders as they occur. Change orders will not be included in payment estimates until approved by all parties.

§§ 1780.77–1780.79 [Reserved]

Subpart D—Information Pertaining to Preparation of Notes or Bonds and Bond Transcript Documents for Public Body Applicants

§ 1780.80 General.

This subpart includes information for use by public body applicants in the preparation and issuance of evidence of debt (bonds, notes, or debt instruments, referred to as bonds in this subpart) and other necessary loan documents.

§ 1780.81 Policies related to use of bond counsel.

The applicant is responsible for preparation of bonds and bond transcript documents. The applicant will obtain the services and opinion of recognized bond counsel experienced in municipal financing with respect to the validity of a bond issue, except for issues of $100,000 or less. With prior approval of the approval official, the applicant may elect not to use bond counsel. Such issues will be closed in accordance with the following:

(a) The applicant must recognize and accept the fact that application processing may require additional legal and administrative time;

(b) It must be established that not using bond counsel will produce significant savings in total legal costs;

(c) The local attorney must be able and experienced in handling this type of legal work;

(d) The applicant must understand that it will likely have to obtain an opinion from bond counsel at its expense should the Agency require refinancing of the debt;

(e) Bonds will be prepared in accordance with this regulation and conform as closely as possible to the preferred methods of preparation stated in § 1780.94; and

(f) Closing instructions must be issued by OGC.

§ 1780.82 [Reserved]

§ 1780.83 Bond transcript documents.

Any questions relating to Agency requirements should be discussed with Agency representatives. Bond counsel or local counsel, as appropriate, must furnish at least two complete sets of the following to the applicant, who will furnish one complete set to the Agency:

(a) Copies of all organizational documents;

(b) Copies of general incumbency certificate;

(c) Certified copies of minutes or excerpts from all meetings of the governing body at which action was taken in connection with the authorizing and issuing of the bonds;

(d) Certified copies of documents evidencing that the applicant has complied fully with all statutory requirements incident to calling and holding a favorable bond election, if one is necessary;

(e) Certified copies of the resolutions, ordinances, or other documents such as the bond authorizing resolutions or ordinances and any resolution establishing rates and regulating use of facility, if such documents are not included in the minutes furnished;

(f) Copies of the official Notice of Sale and the affidavit of publication of the Notice of Sale when State statute requires a public sale;

(g) Specimen bond, with any attached coupons;

(h) Attorney's no-litigation certificate;

(i) Certified copies of resolutions or other documents pertaining to the bond award;

(j) Any additional or supporting documents required by bond counsel;

(k) For loans involving multiple advances of Agency loan funds, a preliminary approving opinion of bond counsel (or local counsel if no bond counsel is involved) if a final unqualified opinion cannot be obtained until all funds are advanced. The preliminary opinion for the entire issue shall be delivered at or before the time of the first advance of funds. It will state that the applicant has the legal authority to issue the bonds, construct, operate and maintain the facility, and repay the loan, subject only to changes occurring during the advance of funds, such as litigation resulting from the failure to advance loan funds, and receipt of closing certificates;

(1) Final unqualified approving opinion of bond counsel, (and preliminary approving opinion, if required) or local counsel if no bond counsel is involved, including an opinion as to whether interest on bonds will be exempt from Federal and State income taxes. With approval of the State program official, a final opinion may be qualified to the extent that litigation is pending relating to Indian claims that may affect title to land or validity of the obligation. It is permissible for such opinion to contain language referring to the last sentence of section 306 (a)(1) or to section 309A (h) of the Consolidated Farm and Rural Development Act (7 U.S.C. 1926 (a)(1) or 1929a (h)).

§§ 1780.84–1780.86 [Reserved]

§ 1780.87 Permanent instruments for Agency loans.

Agency loans will be evidenced by an instrument determined legally sufficient and in accordance with the following order of preference:

(a) *First preference—Form RD 440–22, "Promissory Note"*. Refer to paragraph (b) of this section for methods of various frequency payment calculations.

(b) *Second preference—single instruments with amortized installments*. A single instrument providing for amortized installments which follows Form RD 440–22 as closely as possible. The full amount of the loan must show on the face of the instrument, and there must be provisions for entering the date and amount of each advance on the reverse or an attachment. When principal payments are deferred, the instrument will show that "interest only" is due on interest-only installment dates, rather than specific dollar amounts. The payment period including the "interest only" installment cannot exceed 40 years, the useful life of the facility, or State statute limitations, whichever occurs first. The amortized installment, computed as follows, will be shown as due on installment dates thereafter.

(1) *Monthly payments*. Multiply by twelve the number of years between the due date of the last interest-only installment and the final installment to determine the number of monthly payments. When there are no interest-only installments, multiply by twelve the number of years over which the loan is amortized. Then multiply the loan amount by the amortization factor and round to the next higher dollar.

(2) *Semiannual payments*. Multiply by two the number of years between the due date of the last interest-only installment and the due date of the final installment to determine the correct number of semiannual periods. When there are no interest-only installments, multiply by two the number of years over which the loan is amortized. Then multiply the loan amount by the applicable amortization factor.

(3) *Annual payments*. Subtract the due date of the last interest-only installment from the due date of the final installment to determine the number of annual payments. When there are no interest-only installments, the number of annual payments will equal the number of years over which the loan is amortized. Then multiply the loan amount by the applicable amortization factor and round to the next higher dollar.

(c) *Third preference—single instruments with installments of principal plus interest*. If a single instrument with amortized installments is not legally permissible, use a single instrument providing for installments of principal plus interest accrued on the principal balance. For bonds with semiannual interest and annual principal, the interest is calculated by multiplying the principal balance times the interest rate and dividing this figure by two.

Principal installments are to be scheduled so that total combined interest and principal payments closely approximate amortized payments.

(1) The repayment terms concerning interest only installments described in paragraph (b) of this section apply.

(2) The instrument shall contain in substance provisions indicating:

(i) Principal maturities and due dates;

(ii) Regular payments shall be applied first to interest due through the next principal and interest installment due date and then to principal due in chronological order stipulated in the bond; and

(iii) Payments on delinquent accounts will be applied in the following sequence:

(A) Billed delinquent interest;

(B) Past due interest installments;

(C) Past due principal installments;

(D) Interest installment due; and

(E) Principal installment due.

(d) *Fourth preference—serial bonds with installments of principal plus interest.* If instruments described under the first, second, and third preferences are not legally permissible, use serial bonds with a bond or bonds delivered in the amount of each advance. Bonds will be numbered consecutively and delivered in chronological order. Such bonds will conform to the minimum requirements of §1780.94. Provisions for application of payments will be the same as those set forth in paragraph (c)(2)(ii) of this section.

(e) *Coupon bonds.* Coupon bonds will not be used unless required by State statute. Such bonds will conform to the minimum requirements of §1780.94.

§ 1780.88　[Reserved]

§ 1780.89　Multiple advances of Agency funds using permanent instruments.

Where interim financing from commercial sources is not used, Agency loan proceeds will be disbursed on an "as needed by borrower" basis in amounts not to exceed the amount needed during 30-day periods.

§ 1780.90　Multiple advances of Agency funds using temporary debt instruments.

When none of the instruments described in §1780.87 are legally permissible or practical, a bond anticipation note or similar temporary debt instrument may be used. The debt instrument will provide for multiple advances of Agency funds and will be for the full amount of the Agency loan. The instrument will be prepared by bond counsel, or local counsel if bond counsel is not involved, and approved by the State program official and OGC. At the same time the Agency delivers the last advance, the borrower will deliver the permanent bond instrument and the canceled temporary instrument will be returned to the borrower. The approved debt instrument will show at least the following:

(a) The date from which each advance will bear interest;

(b) The interest rate as determined by §1780.13;

(c) A payment schedule providing for interest on outstanding principal at least annually; and

(d) A maturity date which shall be no earlier than the anticipated issuance date of the permanent instruments and no longer than the 40-year statutory limit.

§§ 1780.91–1780.93　[Reserved]

§ 1780.94　Minimum bond specifications.

The provisions of this section are minimum specifications only and must be followed to the extent legally permissible.

(a) *Type and denominations.* Bond resolutions or ordinances will provide that the instruments be either a bond representing the total amount of the indebtedness or serial bonds in denominations customarily accepted in municipal financing (ordinarily in multiples of not less than $1,000). Single bonds may provide for repayment of principal plus interest or amortized installments. Amortized installments are preferred by the Agency.

(b) *Bond registration.* Bonds will contain provisions permitting registration for both principal and interest. Bonds purchased by the Agency will be registered in the name of "United States of America" and will remain so registered at all times while the bonds are held or insured by the Government. The Agency address for registration purposes will be that of the Finance Office.

(c) *Size and quality.* Size of bonds and coupons should conform to standard practice. Paper must be of sufficient quality to prevent deterioration through ordinary handling over the life of the loan.

(d) *Date of bond.* Bonds will normally be dated as of the day of delivery. However, the borrower may use another date if approved by the Agency. Loan closing is the date of delivery of the bonds or the date of delivery of the first bond when utilizing serial bonds, regardless of the date of delivery of the funds. The date of delivery will be stated in the bond if different from the date of the bond. In all cases, interest will accrue from the date of delivery of the funds.

(e) *Payment date.* Loan payments will be scheduled to coincide with income availability and be in accordance with State law.

(1) If income is available monthly, monthly payments are recommended unless precluded by State law. If income is available quarterly or otherwise more frequently than annually, payments must be scheduled on such basis. However, if State law only permits principal plus interest (P&I) type bonds, annual or semiannual payments will be used.

(2) The payment schedule will be enumerated in the evidence of debt, or if that is not feasible, in a supplemental agreement.

(3) If feasible, the first payment will be scheduled one full month, or other period, as appropriate, from the date of loan closing or any deferment period. Due dates falling on the 29th, 30th, and 31st day of the month will be avoided. When principal payments are deferred, interest-only payments will be scheduled at least annually.

(f) *Extra payments.* Extra payments are derived from the sale of basic chattel or real estate security, refund of unused loan funds, cash proceeds of property insurance and similar actions which reduce the value of basic security. At the option of the borrower, regular facility revenue may also be used as extra payments when regular payments are current. Unless otherwise established in the note or bond, extra payments will be applied as follows:

(1) For loans with amortized debt instruments, extra payments will be applied first to interest accrued to the date of receipt of the payment and second to principal.

(2) For loans with debt instruments with P&I installments, the extra payment will be applied to the final unpaid principal installment.

(3) For borrowers with more than one loan, the extra payment will be applied to the account secured by the lowest priority of lien on the property from which the extra payments was obtained. Any balance will be applied to other Agency loans secured by the property from which the extra payment was obtained.

(4) For assessment bonds, see paragraph (k) of this section.

(g) The place of payments on bonds purchased by the Agency will be determined by the Agency.

(h) *Redemptions.* Bonds will normally contain customary redemption provisions. However, no premium will be charged for early redemption on any bonds held by the Government.

(i) *Additional revenue bonds.* Parity bonds may be issued to complete the project. Otherwise, parity bonds may not be issued unless acceptable documentation is provided establishing that net revenues for the fiscal year following the year in which such bonds are to be issued will be at least 120 percent of the average annual debt serviced requirements on all bonds outstanding, including the newly-issued bonds. For purposes of this section, net revenues are, unless otherwise defined by State statute, gross revenues less essential operation and maintenance expenses. This limitation may be waived or modified by the written consent of bondholders representing 75 percent of the then-outstanding principal indebtedness. Junior and subordinate bonds may be issued in accordance with the loan resolution.

(j) *Precautions.* The following types of provisions in debt instruments should be avoided:

(1) Provisions for the holder to manually post each payment to the instrument.

(2) Provisions for returning the permanent or temporary debt instrument to the borrower in order that it, rather than the Agency, may post the date and amount of each advance or repayment on the instrument.

(3) Provisions that amend covenants contained in RUS Bulletins 1780–27 or 1780–28.

(4) Defeasance provisions in loan or bond resolutions. When a bond issue is defeased, a new issue is sold which supersedes the contractual provisions of the prior issue, including the refinancing requirement and any lien on revenues. Since defeasance in effect precludes the Agency from requiring refinancing before the final maturity date, it represents a violation of the statutory refinancing requirement; therefore, it is disallowed. No loan documents shall include a provision of defeasance.

(k) *Assessment bonds.* When security includes special assessment to be collected over the life of the loan, the instrument should address the method of applying any payments made before they are due. It may be desirable for such payments to be distributed over remaining payments due, rather than to be applied in accordance with normal procedures governing extra payments, so that the account does not become delinquent.

(l) *Multiple debt instruments.* The following will be adhered to when preparing debt instruments:

(1) When more than one loan type is used in financing a project, each type of loan will be evidenced by a separate debt instrument or series of debt instruments;

(2) Loans obligated in different fiscal years and those obligated with different terms in the same fiscal year will be evidenced by separate debt instruments;

(3) Loans obligated for the same loan type in the same fiscal year with the same term may be combined in the same debt instrument;

(4) Loans obligated in the same fiscal year with different interest rates that will be closed at the same interest rate may be combined in the same debt instrument.

[62 FR 33478, June 19, 1997, as amended at 64 FR 29947, June 4, 1999]

§ 1780.95 **Public bidding on bonds.**

Bonds offered for public sale shall be offered in accordance with State law and in such a manner to encourage public bidding. The Agency will not submit a bid at the advertised sale unless required by State law, nor will reference to Agency's rates and terms be included. If no acceptable bid is received, the Agency will negotiate the purchase of the bonds.

§§ 1780.96–1780.100 **[Reserved]**

PART 1781—RESOURCE CONSERVATION AND DEVELOPMENT (RCD) LOANS AND WATERSHED (WS) LOANS AND ADVANCES

1781.25–1781.100 [Reserved]

AUTHORITY: 5 U.S.C. 301; 7 U.S.C. 1989; 16 U.S.C. 1005.

SOURCE: 62 FR 33500, June 19, 1997, unless otherwise noted.

§1781.1 Purpose.

This part prescribes the policies and procedures for making:

(a) Watershed (WS) loans and Watershed (WS) advances for works of improvement in a watershed project; and

(b) Resource Conservation and Development (RCD) loans for measures or projects needed to implement the RCD area plan to achieve objectives in an RCD area.

§1781.2 Policy.

(a) Rural Utilities Service (RUS), is an agency of the United States Department of Agriculture established pursuant to section 232 of the Department of Agriculture Reorganization Act of 1994 (Pub. L. 103–354, 108 Stat. 3178). Natural Resources Conservation Service (NRCS), is an agency of the United States Department of Agriculture established pursuant to section 232 of the Department of Agriculture Reorganization Act of 1994 (Pub. L. 103–354, 108 Stat. 3178), successor to the Soil Conservation Service. RUS will make WS and RCD loans available to sponsoring local public bodies, agencies, and nonprofit organizations to assist them in obtaining the local cost of WS works of improvement and RCD measures. Any processing or servicing activity conducted pursuant to this part involving authorized assistance to RUS employees, members of their families, known close relatives, or business or close personal associates, is subject to the provisions of subpart D of part 1900 of this title. Applicants for this assistance are required to identify any known relationship or association with an RUS employee. RUS will assist the local sponsors and the NRCS in making loans from NRCS construction funds as WS advances when needed for the development of future water supplies or for site preservation.

(b) Rural Development State and local offices will administer these programs on behalf of RUS and will coordinate application processing with the NRCS and other appropriate State and Federal agencies.

[62 FR 33500, June 19, 1997, as amended at 80 FR 9864, Feb. 24, 2015]

§1781.3 Authorities, responsibilities, and delegation of authority.

(a) NRCS provides technical and financial assistance to sponsoring local organizations for developing WS and RCD area plans and for individual RCD measures or projects and watershed works of improvement. The watershed work plan for developing, operating, and maintaining watershed works of improvement must be agreed upon by sponsoring local organizations and NRCS. When approved, it is the basis for extending technical and cost sharing assistance from watershed funds. The RCD area plan is prepared for the development of the RCD area by sponsoring local organizations with assistance from NRCS and other agencies, endorsed by the Governor or by the agency designated by the Governor, and accepted by the Secretary of Agriculture or his delegate. It includes objectives, planned courses of action, and RCD measures or projects to be developed. It is amended as necessary to include continuing activities and needs in the RCD area.

(b) RUS receives and processes applications for WS loans and NRCS WS advances and RCD loans and makes and services such loan and advances. WS loans are made by RUS from either Public Law 534 (78th Cong.) funds authorized in the Flood Control Act of 1944 (33 U.S.C. 701 *et seq.*) or Public Law 566 (83rd Cong.) funds authorized in the Watershed Protection and Flood Prevention Act of 1954 (68 Stat. 666) to cover a part or all of the local cost for a watershed work of improvement.

(c) WS loans and WS advances may be made to project sponsors in watershed project areas for which:

(1) A watershed work plan has been approved administratively or by resolutions adopted by the Committee on Agriculture and Forestry of the Senate and by the Committee on Agriculture of the House of Representatives; and

(2) Federal assistance has been authorized for the installation of works of improvement by the Administrator of NRCS.

(d) RCD loans may be made in areas authorized for RCD program assistance by the Secretary of Agriculture and for which an RCD plan design or area plan has been accepted by the State NRCS Conservationist.

(e) *Delegation of authority.* The Rural Development State Director is authorized to approve WS and RCD loans subject to limitations in RUS Staff Instruction 1780-1 and conditions of this part. The Rural Development State Director is authorized to relegate authority in accordance with this part to the Chief, Community Programs; or other members of the State Office staff.

(f) NRCS is responsible for providing technical and financial assistance to sponsoring local organizations for planning and developing WS and RCD areas. This includes development of WS and RCD plans and WS works of improvement and RCD measures or projects.

(g) RUS is responsible for making and servicing WS loans and advances and RCD loans.

(h) The NRCS-RUS Agreements in RUS Bulletin's 1781 and 1781-2 include further responsibilities and functions of NRCS and RUS in WS and RCD areas.

§ 1781.4 Definitions.

(a) *Watershed (WS) project.* An authorized area in which watershed assistance from NRCS and other U.S. Department of Agriculture (USDA) agencies including WS loans and advances may be provided. Watershed assistance is provided in two types of watershed projects identified by the Public Law under which they are authorized.

(1) *Public Law-534 Watershed.* One of the 11 watersheds authorized by Congress in the Flood Control Act of 1944 (33 U.S.C. 701 *et seq.*), Public Law 78-534 as amended.

(2) *Public Law-566 Watershed.* A small watershed of not more than 250,000 acres authorized in accordance with the Watershed Protection and Flood Prevention Act, August 4, 1954, Public Law 83-566 as amended.

(b) *Resource Conservation and Development (RCD) area.* An area in which RCD program assistance from NRCS and other USDA agencies has been authorized. It usually includes all or part of

more than one county and may be co-terminous with substate planning and development areas. RCD loans are authorized under Section 32 of Title III of the Bankhead-Jones Farm Tenant Act (7 U.S.C. 1011).

(c) *Watershed plan.* A plan agreed upon by sponsoring local organizations and the NRCS for developing, operating, and maintaining watershed works of improvement.

(d) *RC&D measure plan.* A plan document for a land area, directly controlled or under the jurisdiction of the sponsoring public bodies or public nonprofit organization. It involves one of the measure purposes eligible for RC&D cost sharing assistance. The document sets forth what will be done, how, when and by whom, and involves RC&D technical and/or financial assistance.

(e) *RCD area plan.* A plan prepared by sponsoring local organizations with assistance from NRCS and other agencies for the development of the RCD area which has been endorsed by the Governor or his designated agency and accepted by the Secretary of Agriculture or his delegate. It includes objectives, planned courses of action, and RCD measures to be developed. It is amended as necessary to include continuing activities and needs in the RCD area.

(f) *Watershed works of improvement.* Structural, nonstructural, and land treatment measures included in a watershed plan which are to be installed in a watershed project.

(g) *RCD measure or project.* An activity or development indicated in the RCD area plan as being needed to achieve RCD area goals and objectives.

(h) *Cost sharing.* The WS and RCD legislative authorities provide for sharing certain costs of installing WS works of improvement or RCD measures by the Federal Government and by sponsoring local organizations. Federal cost sharing from WS and RCD funds is provided by NRCS for certain WS works of improvement and RCD measures. Information on amounts, purposes, and procedures for cost sharing is available from the NRCS.

(i) *Local cost.* The part of the cost of a WS work of improvement or a RCD measure or project that is to be paid by a sponsoring local organization.

(j) *Public agency or public body.* A State agency or department or instrumentality, county, municipality or other political subdivision or instrumentality of a State or agencies or districts created by or pursuant to State law for making improvements of a public nature or providing public services such as soil and water conservation districts, irrigation districts, drainage districts, flood prevention and control districts, school districts, other special purpose districts, municipal corporations or similar governmental units.

(k) *Non-profit corporation.* Mutual and other irrigation, water users, water supply, drainage, or waste disposal companies or associations, ditch companies, grazing, recreation and forestry associations and similar associations and organizations generally designated as private corporations operating on a non-profit basis. They may be organized and chartered under special law, general nonprofit corporation law, or general profit corporation law, if operated on a nonprofit basis under adequate charter, bylaw, mortgage or supplementary agreement provisions which will assure continued operation in that manner.

(l) *Sponsoring local organization.* A local public agency or body or a local nonprofit corporation having authority under State law to plan, develop, maintain and operate WS works of improvement or RCD measures or projects included in a WS or RCD area plan. The name of the sponsoring local organization must be included in the plan and sponsorship must be evidenced by execution of the plan.

(m) *Watershed loan.* A loan made by RUS from watershed funds to a sponsoring local organization to develop a WS work of improvement.

(n) *RCD loan.* A loan made by RUS from RCD funds to a local sponsoring organization to develop a RCD measure or project. RCD loans are made from RCD funds to enable sponsoring local organizations to provide a part or all of the local share of cost for an RCD measure.

(o) *Watershed advance.* A loan made from NRCS watershed construction funds to develop a future water supply or for the preservation of a site for a work of improvement authorized in a watershed plan.

(p) *Future water supply.* Water storage capacity in a reservoir with related facilities for release or withdrawal of water to meet future needs for municipal or industrial use.

(q) *Preservation of sites.* Acquisition to assure their availability for planned developments. Land, easements, or rights-of-way essential to preserve sites for watershed works of improvement or RCD measures.

(r) *Processing office.* Means the office designated by the Rural Development State Director to accept and process applications for WS and RCD loans and advances.

§1781.5 **Eligibility.**

To be eligible for a WS loan, WS advance, or an RCD loan, the sponsoring local organization must meet the following requirements as applicable. Questions on eligibility will be referred to the Regional Attorney, OGC for legal advise prior to development of a loan docket.

(a) Be named in the WS or RCD plan as a sponsor of the development to be financed.

(b) Be legally organized and established in the WS or RCD area with legal authority, responsibility and capability to develop and operate the facility for which assistance is requested.

(c) Have authority under and comply with Federal, State and local laws on such matters as:

(1) Organizing, installing, operating, and maintaining proposed WS works of improvement or RCD measures or projects.

(2) Borrowing money, giving security, levying taxes, making assessments or raising revenues for operation and maintenance of the facility and repayment of loans.

(3) Land use zoning.

(4) Acquiring necessary property, lands, and rights.

(5) Obtaining approval of construction plans and specifications by appropriate Federal, State, and local agencies and construction facilities.

(6) Health and sanitation standards, water pollution control, and environmental regulations.

(7) Design and installation standards.

(8) Public service commission or similar State public body rules and regulations.

(d) Be financially sound and capable of providing service essential to the rural development needs of the area.

(e) If it is a nonprofit corporation.

(1) Membership should be broadly based and representative of the area benefiting from the facility. Membership on the governing board of the corporation will be limited to those living in the area to be benefited unless for justifiable reasons the Rural Development State Director gives prior approval for other than local residents to serve on the board of directors.

(2) The corporation must propose a facility which will primarily serve or generate other substantial, tangible benefits for farmers and other residents of the area. In the case of a recreational development at least two-thirds of the membership must be farmers and other residing in the area.

(3) Nonprofit corporations will not be formed to serve an area which could be served by a public agency which has adequate authority to provide the needed service unless prior approval of the National Office is obtained.

§ 1781.6 Loan purposes.

(a) *WS and RCD loans.* WS and RCD loans may be used for:

(1) Water development, storage, treatment and conveyance to farms for irrigation and other farm use, including farmstead, livestock, orchard, and crop spraying.

(2) Drainage systems and facilities in farm areas to sustain agricultural production or protect farmers and rural residents from water damage.

(3) Agricultural water management practices for annual streamflow stabilization, recharging ground water reservoirs, and conserving water supplies by management and control of vegetation along waterways and in drainage basins.

(4) Soil conservation and water control facilities such as dikes, terraces, detention reservoirs, stream channels, ditches, and other special land treatment and stabilization measures needed to protect farms and rural residents from water damage, provided such facilities cannot be installed or improved

under, or will not conflict with, other public programs such as those administered by the Corps of Engineers.

(5) Special treatment measures or equipment primarily, though not exclusively, for flood prevention such as:

(i) Facilities and equipment for fire prevention and control.

(ii) Tree planting and establishment of other vegetative cover for stabilizing critical runoff and sediment-producing areas.

(iii) Structural and vegetative measures to stabilize stream channels and gullies.

(iv) Basic farm conservation practices to control runoff, erosion, and sedimentation.

(6) Installing, repairing, and improving water storage facilities, including outlets for immediate and future domestic, municipal and industrial water supply and water quality management, and conveying water to treatment facilities or distribution systems. When payment of loans for such facilities are primarily dependent upon revenues from use of water stored the loan approval official must determine the adequacy of facility for use of the water before a loan is closed.

(7) Public water based recreation and fish and wildlife developer loans will only be made to public bodies for the local share of cost for such developments for which NRCS is providing technical or financial assistance from WS or RCD funds. Loans will not be made for developments larger or more elaborate than that which is included in the WS or RCD plan. Loans may include funds for:

(i) Construction of necessary water resource improvements such as storage capacity in multipurpose and single purpose reservoirs, water level control structures in reservoirs and streams, and stream channel improvements necessary for the development of the facilities. This may include practices for improvement of fish and wildlife habitat and environment and related areas and facilities for proper protection and management of the development.

(ii) Essential developments, improvements, equipment and facilities for access, public health and safety, and efficient operation management and maintenance; such as energy utilities, water supply and waste disposal systems, maintenance buildings, fences, cattle guards, roads and trails, parking, picnicking, camping, beaches, playgrounds, and related shelters and equipment.

(iii) Special areas and structures such as forest and other vegetative cover, marshes, pits, shelters and fish ladders to provide protected natural spawning, breeding, nesting, and feeding for fish and wildlife.

(8) *Soil and water management for agriculture-related pollutant control.* Measures to reduce agriculture-related pollutants that adversely affect the community and the general public. Measures may include, but are not limited to, holding ponds, debris basins, diversions, terraces, and community distribution systems.

(9) Acquiring fee simple title to lands or perpetual easements, or rights-of-way for sites for works of improvement or project measures and related costs for removal, relocation, or replacement of existing improvements including relocation payments for displaced persons, business enterprises and facilities, and other related purposes. Funds for land acquisition will be limited to costs necessary for WS works of improvement or RCD measures. Final construction plans will indicate minimum essential lands and rights-of-way to be acquired. In some cases, sponsoring local organizations may need to acquire lands in excess of actual needs when it is expedient for planned development. If the Rural Development State Director determines that the acquisition of excess land is necessary or expedient for the orderly development of a WS works of improvement, or RCD measure, he may authorize the action subject to the following conditions:

(i) The applicant must agree to sell excess land as soon as practicable and apply the proceeds, together with any income from excess land, on the debt to RUS.

(ii) The applicant must furnish legal evidence of authority to acquire additional land and dispose of it as agreed.

(iii) Evidence must be provided to justify acquisition of additional land.

(iv) Easements for land or water resource protection structures must be perpetual and must not include clauses that terminate the easement with the dissolution or abandonment of the applicant organization. Loan funds will not be used for an easement that deviates in any way from that provided in the standard NRCS form unless modifications of it are approved by both NRCS and RUS.

(10) Acquisition of water supply or water right by purchase or by appropriation under local, State, and Federal laws. The loan may include funds for the purchase of land on which the water supply or water right is presently being used when:

(i) The water supply or water right cannot be purchased without the land; and

(ii) The value of the land is not the major portion of the cost; and

(iii) Any excess land thus acquired will be sold as soon as possible and the proceeds applied on the loan.

(11) Purchase of equipment and machinery necessary for development and operation of planned WS works of improvement or RCD measures or projects including:

(i) *Special-purpose equipment.* Purchase or rent special-purpose equipment to install or maintain any community facility in categories in paragraph (a)(11) of this section or to establish on farms soil and water conservation measures such as terraces, ponds, land leveling for irrigation or drainage, subsoiling, seeding, tree planting, and removal of brush, scattered trees, and stumps, provided:

(A) Such equipment is not otherwise available when needed.

(B) There is sufficient need and local demand to justify ownership or rental.

(C) Rates to be charged include, among other things, an allowance for depreciation, obsolescence, and replacement based upon the recommendations of the equipment manufacturer or the experience of contractors engaged in providing services for similar types of work.

(ii) *Forestry equipment and services.* Purchase or rent basic special-purpose equipment, facilities, certain land or land rights, and supplies needed for furnishing services for the establishment, improvement, protection, and harvesting of timber (not processing) suitable for lumber, pulp, poles or posts;

providing that the forest program and forest practices benefiting from such services are in accordance with approved conservation practices for the development, use, and control of water resources on farms and in forests. Special-purpose equipment may include such items as tractors, bull dozers, plows, planters, trucks, loaders, fire-fighting equipment, and sprayers. Facilities may include such items as ponds and reservoirs, pipelines, buildings for storage of equipment and supplies, nurseries, access roads, fire lanes, and lookout towers. Supplies may include such things as seed, seedlings, fertilizers, fencing, and pesticides. Land or land-rights acquisition will be limited to that necessary for sites for facilities listed above which are directly related to the forestry program. Loans for these purposes may be made only when the equipment, supplies, and facilities to be provided:

(A) Are not readily available when needed.

(B) Will be justified by local need and demand.

(C) Will be available to users at rates sufficient to cover loan amortization, obsolescence, replacement, operation, and cost of supplies.

(D) Will more efficiently serve the group through cooperative effort.

(12) Refinancing debt obligations of the sponsoring local organization that were incurred before application for a WS or RCD loan when that is not the primary purpose of the loan and:

(i) The debt being refinanced was for works of improvement or measures for which loan funds could be used; and

(ii) The debt is a valid obligation of the sponsor; and

(iii) Creditors will not modify payment terms on existing debts, and the organization cannot pay existing debts and a loan from RUS over the same period of time; and

(iv) Long-term debts will not be refinanced unless necessary to provide a sound basis for the loan or WS advance and concurrence is obtained from the National Office.

(13) If repayment is based on revenues, loan funds (not WS advances) can be used for payment of interest installments until the facility is generating enough revenue to make accrued interest payments. Loan funds for interest payments will not exceed the estimated amount that will accrue to the end of the third full calendar year after loan closing without prior approval from the National Office.

(14) Relocation payment to displaced persons, businesses, and farm operations and for relocation assistance advisory services in accordance with the Uniform Relocation Assistance and Real Property Acquisition Policies Act of 1970 (Public Law 91-646, 84 Stat. 1894), the Regulations issued by the Secretary of Agriculture under the Act (7 CFR part 21), and the Memorandum of Understanding Between NRCS and RUS.

(15) Services of engineers, architects, attorneys, auditors, construction foremen, managers, clerks, and others for organizing, planning, surveying, supervising, analyzing, developing, operating, managing, and accounting for activities related to loan processing and closing and development for which the loan is made.

(16) Buildings, fences, roads, utilities, facilities, and relocation:

(i) To construct buildings of modest design essential for the operation and maintenance of the works of improvement or measure.

(ii) To provide support facilities and utilities such as gas, electricity, water, sewer, and waste disposal.

(iii) To build or relocate roads, bridges, utilities, fences, and other improvements when necessary to acquire rights-of-ways or to construct or operate the facility.

(17) Services and fees. To pay costs for services for any purposes listed under this section such as:

(i) Fees or other legal expenses for establishing a water right through appropriation, agreement, permit, or court decree.

(ii) Purchase of water stock or membership in an incorporated water users' association to acquire a water supply.

(iii) Costs of labor, technical or professional services, and fees to be incurred in obtaining the loan and in planning and completing the facilities or services to be financed with loan funds.

(iv) Services such as those listed in paragraph (a)(16) of this section.

(b) *RCD loans.* Purposes for which RCD loans may be made in addition to those included in paragraph (a) of this section are:

(1) *Solid waste management.* Lands, equipment and facilities to collect, transport, and dispose of solid waste in sanitary landfills for which NRCS is providing technical assistance.

(2) *Shifts-in-land use.* Lands for uses such as grazing, forestry, wildlife, natural areas and parks, greenbelts, and other open spaces.

(3) *Purchase existing facilities.* Purchase existing facilities for shift-in-land use, soil and water development, conservation, control and use when it is determined that purchase is necessary to provide efficient service through a facility owned and operated by a public agency (or a nonprofit corporation in a rural area), or the owner is either unwilling or unable to make improvements, enlargement, or extensions needed to provide significant additional or improved service for present users or for a new group of users at reasonable rates.

(c) *NRCS watershed advances.* NRCS watershed advances are loans that may be made from NRCS construction funds for the following purposes included in a watershed work plan agreement:

(1) To pay construction costs including cost of engineering and related services for increasing reservoir capacity (including intake and outlet structures) for a future water supply for municipal, domestic, industrial, or agricultural uses.

(2) To preserve sites for authorized watershed works of improvement by acquiring land, easements, and rights-of-ways or other property rights.

§ 1781.7 Loan and advance limitations and obligations incurred before loan closing.

(a) *WS and RCD loan limitations.* (1) Loans will not be used for:

(i) Land treatment measures on individual farms except as provided in §1781.6(a)(5)(iv).

(ii) Buildings and facilities to be used for lodging, dining or entertainment purposes.

(iii) Building industrial parks or constructing facilities in them, or establishing private industrial or commercial enterprises, or purchasing land to be used primarily for industrial purposes.

(iv) Paying costs allocated to structural measures for flood prevention.

(v) Facilities for the production and harvesting of fish and wildlife such as hatcheries, rearing ponds, and related facilities other than those under natural conditions.

(vi) Facilities primary for treatment and distribution of water or for sewerage, collection and treatment for domestic or industrial use or for municipal or community systems.

(vii) Electric generating, transmission, and distribution facilities, except when provided as part of the minimum basic facilities for recreation and fish and wildlife developments authorized in §1781.6(a)(7).

(viii) Storm and sanitary sewers and solid waste disposal facilities other than authorized in §1781.6(b)(1).

(ix) Payment for a tract of land, easements, or rights-of-ways on which NRCS will share the cost if the amount to be paid with loan funds exceeds the difference between the NRCS share and the value on which the NRCS share is based.

(x) Purchasing tracts of land primarily for later resale to private developers or individuals for agricultural or nonagricultural use.

(xii) Buildings for residential, commercial, or industrial, use.

(xiii) Developments on private property primarily for the benefit of the individual property owner.

(xiv) Payment of that part of the cost of facilities, improvements, and practices that could be earned by participation in agricultural conservation programs unless such cost cannot be covered by purchase orders or assignments to material suppliers or contractors. If a loan is made for such purposes for which practice or cost share payments exceed $500, RUS will obtain an assignment on such payments to be paid on the loan.

(xv) Primarily for water and sewage treatment plants and distribution systems.

(xvi) Drainage facilities primarily for the benefit of other than rural areas.

(xvii) Any single RCD measure that requires a loan of more than $500,000.

(xviii) The total amount of principal outstanding for all WS loans made for one or more watershed works of improvement in a single watershed project, whether made to one or more sponsoring organizations, will not exceed $10,000,000.

(b) *Watershed advance limitations.* (1) A WS advance for future water supply will not be used for acquiring property rights including lands, easements, and rights-of-way; water rights; administration of contracts; storage capacity for immediate municipal use; pipelines from the reservoir to place of use; or for other uses such as irrigation, fish and wildlife, and recreation.

(2) A WS advance for increasing reservoir capacity for future water supply will not exceed 30 percent of the total installation cost of one structure.

(3) A WS advance for site preservation will not exceed that determined necessary by NRCS except to purchase land in excess of actual needs in accordance with the provisions of § 1781.6(a)(7).

(4) Before a project agreement is entered into, there must be satisfactory evidence that the borrower will develop the site to be acquired or will use the future water supply and that revenue will be sufficient to meet all scheduled installments.

(c) *Obligations incurred before loan closing.* (1) WS loans, WS advances, and RCD loans may be used for payment of obligations incurred before loan closing when the Rural Development State Director determines that:

(i) The obligations incurred are necessary for planned developments; and

(ii) The obligations are incurred for authorized loan purposes; and

(iii) Contracts and construction plans meet RUS and NRCS standards; and

(iv) The applicant has legal authority to incur the obligations at the time proposed; and

(v) The Rural Development State Director authorizes such action in a letter to the applicant.

(2) The Rural Development State Director's letter will specifically state that the permission is granted on the condition that RUS is not committed to make a loan and assumes no responsibility for any obligation incurred by the applicant because of the permission

granted and that the loan will be closed subject to compliance with agency regulations including closing instructions of the Regional Attorney Office of the General Counsel.

§ 1781.8 Rates and terms—WS loans and WS advances and RCD loans.

(a) *Interest rates.* The interest rate for WS loans, WS advances and RCD loans will be at a rate not to exceed the current market yield for outstanding municipal obligations with remaining periods to maturity comparable to the average maturity for the loan, adjusted to the nearest 1/8 of 1 percent.

(1) For loans, unless otherwise required by State law, interest will accrue from date of check delivery where Form RD 440–22, "Promissory Note (Association Organization)," is used. Where bonds are used interest will accrue from the applicable dates recorded on the bonds. Where multiple loan disbursements are used interest will accrue from date of check.

(2) Interest on an advance for future water supply will begin as required by State law, when water is first used from the future water storage capacity installed with advance, or ten years from the scheduled date of the completion of the facility, whichever date is the earlier.

(3) Interest on an advance for preservation of sites will begin on the date the advance is closed.

(b) *Length of repayment period.* The repayment period on loans may not exceed the shortest of the following periods:

(1) The statutory limitation on the sponsoring local organization's borrowing authority.

(2) Fifty (50) years for WS loans and WS advances and 30 years for RCD loans from the date when the principal benefits from the WS works of improvement or RCD measure being financed first become available.

(3) The useful life of the WS works of improvement or RCD measure being financed with loan or advance funds.

(c) *Deferred or partial payments.* Deferred or partial payments may be authorized in the following circumstances:

(1) Payments need to be delayed until the receipt of income from taxes or

other revenues is enough to meet a regular installment but not exceed:

(i) The completion date of the facility; or

(ii) The date when benefits from the facility begins; but

(iii) In no case for more than 5 years for other than future water supply.

(2) Payments will depend on the increased returns expected from planned improvements, or from the installation on individual farms of land development or other soil and water improvements essential for obtaining benefits from the improvement to be installed with loan funds.

(3) They will not be used to permit the accelerated payment of other debts, to make capital improvements, or to create operating reserves.

(4) Where prohibited by State statutes; interest payments will not be deferred even though payments on principal may be deferred.

(5) Loans or advances for future water supply will be repaid within the life of the reservoir structure but in no event later than 50 years for WS and 30 years for RCD after the reservoir structure is built. Payments on the principal amount may be deferred one year after the water is first used from the storage capacity installed with the advance or for 10 years from the scheduled completion date of the structures, whichever occurs first.

(i) Interest will begin for a future water supply as required by State law, or when water is first used from the future storage capacity or 10 years from the scheduled date of completion of the facility, whichever occurs first.

(ii) If State law requires that interest be charged and repaid before water is first used or earlier than 10 years from completion date of the structure, interest payments will be scheduled to comply with State law even though payments of principal may be deferred.

(iii) The borrower should be encouraged to begin repayments as soon as practicable after the reservoir is built even though this liberal deferment policy exists.

(iv) WS advances for preservation of sites must be fully repaid before beginning construction of the works of improvement for which such sites were acquired.

(A) Unless a WS advance is to be repaid with a WS loan, installments will be scheduled at the earliest possible date following the date of closing the advance. The date and amount of each such installment will be fixed to coincide with the receipt of income from taxes or other revenues.

(B) Payments for both principal and interest on a WS advance for preservation of sites may be scheduled for payment in one installment to be paid on the date of the closing of a WS loan which includes funds for the repayment of the WS advance.

(C) Interest on a WS advance for preservation of sites will begin on the date the WS advance is closed.

(d) *Payment amortization and application.* (1) A borrower may make prepayments on WS loans, WS advances or RCD loans in any amount at any time.

(2) Payments will be applied first to interest accrued to the date of the receipt of payment, and second to the principal balance. If the regular payments plus any prepayments exceed the cumulative amount due, the excess payments will be applied on the next installment first to interest, then principal. Loan refunds and proceeds from the sale of security property, however, will be applied on the final unpaid installment.

(3) Payments will be scheduled annually beginning one year following the date of loan closing or one year following the end of any approved deferment period, unless another annual due date is required by State statute or upon prior written authorization from the National Office. In those cases where loans are being made under statutes requiring a repayment date other than this, the Rural Development State Director will send a copy of the Regional Attorney's opinion that such is required, to the Finance Office.

(4) When a single obligation instrument is used, amortized installments will be required. When this cannot be done because of state law, serial bonds or a single bond having installments of principal plus interest, stated separately, will be used. In cases where the payment of interest has been deferred, all collections will be applied to interest until such interest has been paid. Also, when a full installment is not

337

paid when due, the payment made will be applied first to accrued interest.

(5) In cases where the indebtedness will be represented by serial bonds or a single bond having installments of principal plus interest, stated separately, annual payments of principal and interest will be scheduled to permit them to be paid in amounts approximately equal to the amounts that would be required for annual amortized installments.

(6) If the borrower will be retiring other debts represented by bonds or notes, the payment on such bonds may be considered in developing the payment schedule for the RUS loan. In some cases, it may be desirable to reduce the amount of payments to RUS in the early years of the loan in order to preclude the necessity for refinancing the outstanding debt. When such payment schedules are proposed, National Office authorization will be obtained prior to loan approval.

(7) *Payment date.* Insofar as loan payments are consistent with income availability, applicable State statutes, and commercial customs in the preparation of bonds or other evidence of indebtedness, they should be scheduled on a monthly basis either in the bond or other evidence of indebtedness or through the use of a supplemental agreement. Such requirements will be accomplished not later than the time of loan closing. When monthly payments are required, such payments will be scheduled beginning one full month following the date of loan closing or the end of any approved deferment period. Subsequent monthly payments will be scheduled each full month thereafter. In those cases where evidence of indebtedness calls for annual or semiannual payments, they will be scheduled beginning six or twelve full months, respectively following the date of loan closing or the end of any approved deferment period. Subsequent payments will be scheduled each sixth or twelfth full month respectively, thereafter. When the evidence of indebtedness is dated the 29th, 30th, or 31st day of a month, the payment date will be scheduled the 28th day of the month.

§ 1781.9 Security, feasibility, evidence of debt, title, insurance and other requirements.

(a) *Security.* WS loans, WS advances, and RCD loans will be secured in accordance with applicable provisions of § 1780.14 of this chapter.

(b) *Feasibility.* All projects financed under the provisions of this part must be based on taxes, assessments, revenues, fees, or other satisfactory sources in an amount that will provide for facility operation and maintenance, a reasonable reserve, and payment of the debt. The Rural Development State Director may obtain needed assistance in determining economic feasibility from officials of NRCS and other appropriate USDA agencies. See § 1780.7(f) of this chapter for applicable economic feasibility requirements and feasibility reports.

(c) *Notes, bonds, and bond transcript documents.* See subpart D of part 1780 of this chapter for applicable requirements and provisions.

(d) *Insurance.* See § 1780.39(g) of this chapter for requirements.

(e) *National flood insurance.* The requirements of the National Flood Insurance Act of 1968 (42 U.S.C. 4001 *et seq.*) as amended by the Flood Disaster Protection Act of 1973 (42 U.S.C. 4003 *et seq.*) will be complied with in accordance with applicable provisions of RD Instruction 1901-L. Also see § 1780.39(g) of this chapter.

(f) *Borrower contracts and bonds.* See subpart C of part 1780 of this chapter for applicable provisions.

(g) *Title requirements.* (1) Title evidence for land, easements, and rights-of-way to be acquired with proceeds of loans or advances will be furnished by the sponsoring local organization in accordance with NRCS policies and procedures.

(2) RUS will specify and approve the form and content of instruments for conveying title to or interest in real estate on which a lien will be taken to secure a WS loan, WS advance, or RCD loan. These should be consistent with the applicable provisions of § 1780.14 of this chapter. The Rural Development State Director will make his decision after consultation with the Regional Attorney and the State Conservationist. He will notify NRCS in writing

of his decision. Thereafter, title clearance will be completed under NRCS regulations except that a marketable title must be obtained on any tract of land, a part of which will be sold as excess land in accordance with §1781.6(a)(9). In addition to the title evidence required by NRCS, applicants will furnish an opinion of legal counsel on all land and interest in land acquired with loan or advance funds.

(h) *Purchasing lands, rights and facilities.* The amounts paid for lands, rights, and facilities with loan funds will be not more than that determined to be reasonable and fair by the loan approval official based upon an appraisal of the current market value made by an Rural Development employee or an independent appraiser.

(i) *Water rights.* Applicants will be required to comply with applicable State and local laws and regulations governing appropriating, diverting, storing and using water, changing the place and manner of use of water, and in disposing of water. All of the rights of any landowner, appropriator, or user of water from any source will be fully honored in all respects as they may be affected by facilities installed with WS loans and advances and RCD loans. If, under the provisions of State law, notice of the proposed diversion or storage of water by the applicant may be filed, the applicant will be required to file such a notice. An applicant must furnish evidence to provide reasonable assurance that its water rights will be or have been properly established, will not interfere with prior vested rights, will likely not be contested or enjoined by other water users or riparian owners, and will be within the provisions of any applicable interstate compact.

§1781.10 [Reserved]

§1781.11 Other considerations.

(a) *Technical assistance.* When pipelines from reservoirs to treatment plants are included in watershed work plans, NRCS will not furnish engineering services for their design or installation. When such pipelines are to be financed by WS or RCD loans, RUS will supervise the activities of the private engineers retained for the purpose. Such RUS supervision will include,

among other things, approval of private engineer's contracts, approval of plans and specifications, authorization of contract awards, spot checks of engineering inspection, and final inspection and acceptance.

(b) *Professional services.* Applicants will be responsible for providing the services necessary to plan projects including design of facilities, preparation of cost and income estimates, development of proposals for organization and financing, and overall operation and maintenance of the facility. Necessary professional services may include such as that of an engineer, architect, attorney, bond counsel, accountant, auditor, and financial advisor or fiscal agent. Form RD 442–19, "Agreement for Engineering Services," may be used when appropriate. RUS Bulletin 1780–7, "Legal Service Agreement" may be used to prepare the agreement for legal services.

(c) *Other services.* Contracts for other services such as management, operation, and maintenance will be developed by the applicant and presented to the RUS official developing the docket for review and approval.

(d) *Fees for services.* Fees provided for in contracts, agreements or services will not be more than those ordinarily charged by the profession for similar work when RUS financing is not involved.

(e) *State pollution control or Environmental Protection Agency standards.* Facilities will be designed, installed and operated to prevent pollution of water in excess of established standards. Effluent disposal will conform with appropriate State and Federal Water Pollution Control Standards.

(f) *Water pollution.* When repayment of a WS loan, WS advance, or RCD loan will be dependent upon income from the use or sale of water, RUS approval will be contingent upon a determination that the proposed use of stored water for recreation or municipal supply might not be permitted by a State health department because the water is being polluted from an upstream or other source.

(g) *Environmental review requirements.* Actions will be taken to comply with the environmental review requirements in accordance with 7 CFR part

1970. When environmental assessments and environmental impact statements have been prepared on WS plans or RCD area plans by NRCS, a separate environmental impact statement or assessment on WS works of improvement or RCD measures for which a WS loan, WS advance, or RCD loan is requested will not be necessary unless the NRCS environmental review fails to meet the requirements of 7 CFR part 1970. If the environmental impact statement or environmental assessment is satisfactory, the Agency should formally adopt the document in accordance with 7 CFR part 1970. If a determination is made that further analysis of the environmental impact is needed, the Agency will make necessary arrangements with the NRCS State Conservationist for such action to be taken before a loan is made.

(h) *National Historic Preservation Act.* All projects will comply with the provisions of the National Historic Preservation Act of 1966 (16 U.S.C. 470 *et seq.*) in accordance with RD Instruction 1901–F.

(i) *Civil Rights Act of 1964.* Recipients of WS loans, WS advances, or RCD loans are subject to Title VI of the Civil Rights Act of 1964 (42 U.S.C. 2000d *et seq.*), which prohibits discrimination because of race, color, or national origin. Borrowers must agree not to discriminate in their operations by signing Form RD 400–4, "Nondiscrimination Agreement," before loan closing. This requirement should be discussed with the applicant as early in the negotiations as possible. Necessary actions will be taken in accordance with RD Instruction 1901–E.

(j) *Appraisals.* When required by the Rural Development State Director, appraisals will be made by an Rural Development official designated or an independent appraiser. Form RD 442–10, "Appraisal Report—Water and Waste Disposal Systems," with appropriate supplements, may be modified as needed for use with the type of facilities being appraised.

(k) *Architectural Barriers Act of 1968.* All facilities financed with RUS loans and grants which are accessible to the public or in which physically handicapped persons may be employed or re-side must be developed in compliance with this act (42 U.S.C. 4151 *et seq.*).

[62 FR 33500, June 19, 1997, as amended at 81 FR 11028, Mar. 2, 2016]

§ 1781.12 Preapplication and application processing.

(a) *WS and RCD loans*—(1) *Preapplications.* (i) The processing office or other person designated by the Rural Development State Director may assist the applicant in completing SF 424.1, "Application for Federal Assistance (For Non-construction)," and will forward one of SF 424.1 to the Rural Development State Director.

(ii) The Rural Development State Director will review SF 424.1 along with other necessary information and will coordinate selection of preapplications to be processed with NRCS. He will consult with NRCS State Conservationist concerning the status of the WS plan or RCD measure plan, the estimated time schedule for construction and cost of the proposed works to be installed with the loan, cost sharing funds to be made available to the applicant, and other pertinent information.

(iii) Form AD–622, "Notice of Preapplication Review Action," will be prepared and signed by the Rural Development State Director within forty-five (45) days from receipt of the preapplication in the processing office stating the results of the review action. An original and one copy of Form AD–622 will be sent to the processing office who will deliver the original to the applicant.

(2) *Applications.* (i) The application includes applicable forms and information indicated in RUS Instruction 1780. When the Rural Development State Director determines that an application will be further processed and Form AD–622 is delivered, he will designate a community program specialist (field), or a member of the community program staff to assist the processing office and the applicant with assembling and processing the application.

(ii) The processing office should arrange needed conferences with the applicant and its legal and engineering consultants, and when necessary, arrange for review of other Rural Development officials, and provide bulletins, forms, instructions and other assistance with assembling and processing the application. A processing checklist and time schedule will be established by using Form RD 1942–40, "Processing Check List (Public Bodies)," or Form RD 1942–39, "Processing Check List (Other than Public Bodies)." The processing office will send a letter and a copy of the processing checklist to the applicant to confirm decisions reached at the conference. The original and a copy of the processing checklist will be kept in the processing office and will be posted current as application processing actions are taken. The copy will be circulated from the processing office to the State Office for use in updating copies of the forms retained, after which it will be returned from the State Office to the processing office.

(3) *Dockets.* WS loan, WS advance, and RCD loan dockets will be developed and assembled in accordance with applicable RUS Instruction 1780.

(b) *Watershed advances.* Applications for WS advances will be developed and processed with NRCS assistance as necessary.

(1) The Rural Development State Director will arrange with the NRCS State Conservationist to be advised when a local sponsoring organization applies to NRCS for a WS advance.

(2) The Rural Development State Director will request the NRCS State Conservationist to provide information justifying the WS advance along with a written recommendation that it be made. This will include:

(i) Economic feasibility of the proposed WS advance.

(ii) Evidence of the legal authority of the sponsoring local organization to incur the obligation and make required payments.

(iii) Any limitations on the issuance of additional bonds or notes which may be imposed by the provisions of bond ordinances or on resolutions which authorize the issuance of any outstanding obligation of the sponsoring local organization.

(iv) The amount of WS advance funds to be provided, purpose for which funds will be used, and date funds will be needed.

(3) When the above information has been made available to the Rural Development State Director, he will send written recommendations concerning further action on the WS advance request to the NRCS State Conservationist including actions to be taken in the preparation of the WS advance docket.

(c) *Combination WS loans and WS advances.* If an applicant requests both a WS loan and WS advance, the application for the WS loan should indicate the amount of the WS advance needed and whether a request for it has been made to NRCS. The Rural Development State Director and the NRCS State Conservationist will coordinate applicable processing actions of such applications. When the Rural Development State Director determines that favorable consideration will be given to an application for a loan or advance, he will provide instructions to the processing office for completing and processing the appropriate docket. Any questions concerning eligibility or other legal matters should be cleared with the Regional Attorney.

(d) *Review of decision.* When it is determined that the preapplication or application cannot be given favorable consideration, the Rural Development State Director will return it to the processing office along with written reasons. When the processing office receives this information, it will notify the applicant in writing of the reasons why the request was not favorably considered. The notification to the applicant will state that the RUS Administrator may be requested to review the decision. This action will be taken in accordance with §1780.37 of this chapter.

(1) Upon receipt of the State Office copy of a review request from the applicant, the Rural Development State Director will furnish a report on the matter to the Administrator.

(2) The Administrator will notify the applicant and the Rural Development State Director in writing of his decision and the reasons therefore.

§1781.13 [Reserved]

§1781.14 Planning, options, and appraisals.

(a) WS and RCD area plans are developed by sponsoring local agencies and

341

organizations with technical assistance from NRCS and other Federal and State agencies. These plans include WS works of improvement and RCD measures to be developed or constructed for which NRCS construction funds may be made available on a cost share basis along with funds provided by the sponsoring local organization, a portion or all of which may be obtained by a WS loan and/or WS advance or a RCD loan.

(b) Current information on the availability of cost share funds and purposes for which they may be used is provided by NRCS. The amount of NRCS cost share funds and the amount of funds to be provided by the sponsoring local organizations will be indicated in each plan. The estimated amount of WS loan, WS advance or RCD loan anticipated by the sponsoring local organization should also be included.

(c) Plans for the development or construction of individual WS works of improvement and RCD measures will normally be developed with NRCS technical assistance. In every case they will be approved by both the NRCS State conservationist and the Rural Development State Director or their designated agent when a WS loan, WS advance or RCD loan is made.

(d) Options and appraisals related to the purchase of real estate for which a WS loan, WS advance, or RCD loan is made must be developed in accordance with NRCS and RUS requirements and approved by RUS. The determination of present market value will be made in accordance with § 1780.44(g) of this chapter.

§ 1781.15 Planning and performing development.

Planning and performing development will be handled in accordance with subpart C of part 1780 of this chapter and guidance from NRCS.

§ 1781.16 [Reserved]

§ 1781.17 Docket preparation and processing.

(a) *Loan dockets.* Dockets for WS loans, WS advances and RCD loans will be prepared in accordance with the applicable provisions of part 1780 of this chapter.

(1) *Time for preparation of docket.* Docket preparation may begin as soon as a preliminary draft of the watershed plan or RCD area plan, together with an estimate of costs and benefits, have been prepared with the assistance of NRCS and approved by the sponsoring local organization applicant. However, the applicant must understand that approval of the WS loan, WS advance, or RCD loan will not be determined until the work plan has been authorized for assistance by NRCS. To the extent practicable, docket preparation may be completed by that time to facilitate the availability of funds when needed.

(2) *Instructions for preparation of docket.* When the Rural Development State Director has determined that plans and other requirements are completed to the extent that preparation of the loan docket may begin, he will send the processing office a memorandum giving complete instructions for docket preparation, with a list of documents to be included in the docket.

(3) *Objectives of the docket.* The docket should include information for use in determining that:

(i) The sponsoring local organization:

(A) Has legal authority to construct and operate the proposed facility, borrow money, give security, incur debt, and generate revenue needed for operation, maintenance, reserves, debt payment, and other cash requirements.

(B) Is a sponsor or cosponsor of the WS plan or RCD work plan and is otherwise eligible for assistance.

(ii) Funds will be used for authorized purposes.

(iii) The source of income to be pledged for debt payment and the security proposed is adequate.

(iv) Actions required for loan closing are administratively satisfactory, legally sufficient and properly documented in accordance with Agency regulations.

(4) *Assembly of the docket.* The docket will be assembled in accordance with paragraph (a)(2) of this section and will include the following:

(i) A copy of the WS works of improvement agreement or RCD measure agreement.

(ii) A copy of the Operation and Maintenance Agreement between NRCS and the WS or RCD sponsoring local organization for the WS works of improvement or the RCD measure.

(iii) A statement from the NRCS State Conservationist concurring in the feasibility of the WS work of improvement or RCD measure and that NRCS is providing financial and/or technical assistance in accordance with applicable WS or RCD authorities.

(5) *Narrative by processing office.* This should be included in or attached to the Project Summary. It should relate project costs to benefits of the WS or RCD loan or WS advance. Minimum and average individual charges, tax levies or assessments should be given where applicable. Where taxes or assessments on land will be levied, acres should be indicated and average cost per acre should be given. Analyses of income from recreational facilities should be based on the best information available from local, State, and Federal agencies concerned with such recreation facilities. Determination of water rates, schedules, and estimated consumption of water should be made by the same methods as for loans for domestic water and irrigation.

(6) *Estimates of right-of-way Costs.* The docket should include, as part of the Project Summary, current estimated costs of easements, rights-of-way, and other land rights which must be acquired. The amount estimated for such purposes in the WS or RCD plan should reflect current conditions.

(b) *Loan processing by State Office*—(1) *Review of the docket.* The processing office will check the docket for accuracy and completeness and forward it to the State Office with their recommendations. The Rural Development State Director will review the docket to determine that:

(i) All documents are accurate and complete.

(ii) The proposed loan complies with WS and RCD program policies and procedures of both RUS and NRCS.

(iii) Security is adequate and the repayment plan is sound.

(iv) Funds requested are for authorized purposes.

(v) Actions are in compliance with requirements of applicable Federal and State laws.

(2) *Letter of conditions.* When the Rural Development State Director determines that the docket is complete and the proposed activity is feasible, he will prepare a proposed letter of conditions under which the application may be further processed. The letter will be delivered to and discussed with the applicant. Upon acceptance of the conditions the applicant will indicate intentions to meet the conditions by a letter of interest and the application will be further processed.

(3) *Legal review.* The complete docket and proposed letter of conditions will be forwarded to the Regional Attorney, OGC for review and preparation of closing instructions. If it is not possible to issue closing instructions at that time, the Regional Attorney, will issue a preliminary legal opinion commenting upon the applicants legal existence, authority to incur debt and give security for the WS loan, WS advance, or RCD loan requested and actions to be taken before closing instructions may be issued.

(4) *Authorization for approval.* When the Rural Development State Director receives closing instructions or a preliminary legal opinion for a WS loan, WS advance, or RCD loan that is not within his approval authority he will send this information along with the docket, the proposed letter of conditions, and a memorandum recommending approval to the National Office. A copy of his memorandum will be sent to the processing office. If the proposed action is within the Rural Development State Director's approval authority he need not submit the material listed in this paragraph (b)(4) to the National Office unless he wants review and comments before approval.

(c) *WS advance processing.* (1) When the Rural Development State Director has concurred with the NRCS State Conservationist in the inclusion of a WS advance in a watershed plan, preparation of the advance docket can be initiated and will be processed in the same manner as for a WS loan. Where both a WS loan and WS advance are planned only one docket will be prepared to include both the WS loan and WS advance.

(2) If the advance appears to be sound and proper, the Rural Development State Director will send a proposed memorandum of concurrence to the

343

NRCS State Conservationist. The memorandum will state that RUS concurs in the execution of a work of improvement agreement for which NRCS will obligate advance funds and that RUS will accept the proposed obligations of the applicant to repay the advance subject to conditions specified in or attached to the memorandum. These conditions will include all appropriate requirements in accordance with paragraph (b)(2) of this section and will specify compliance with closing instructions issued by the Regional Attorney. It will also indicate that preparation of the WS advance docket will be in accordance with paragraph (a) of this section.

(3) The Rural Development State Director and the NRCS State Conservationist will sign the memorandum of concurrence to NRCS when:

(i) It has been determined that funds for the advance will be obligated by NRCS; and

(ii) The WS advance docket, has been approved; and

(iii) Closing instructions have been issued by the Regional Attorney; and

(iv) The Rural Development State Director and NRCS State Conservationist have determined that the applicant can comply with all requirements of the letter of conditions and closing instructions.

§ 1781.18 Feasibility.

(a) Before WS loan, WS advance, or RCD loan is approved, a determination of feasibility will be made by the Rural Development State Director based upon a review of plans developed in cooperation with NRCS personnel. The feasibility determination must have the concurrence of the NRCS State Conservationist before a WS loan, WS advance, or RCD loan is approved.

(b) A written assessment of the project's feasibility will be made by the processing office, Architect/Engineer, and Program Chief in their recommendations or comments on the Project Summary. These should reflect concurrence of the respective NRCS personnel in counterpart positions with whom they cooperate in administering these programs.

§ 1781.19 Approval, closing, and cancellation.

(a) Approval and closing actions will be taken in accordance with the applicable provisions of part 1780 of this chapter and the following requirements have been met:

(1) The WS or RCD plan has been approved for operations by NRCS and the applicant is an official sponsoring or cosponsoring local organization for the plan as evidenced by being included in the list of sponsoring or co-sponsoring local organizations in the plan.

(2) Closing instructions or a preliminary legal opinion has been prepared by the Regional Attorney.

(3) The governing body of the applicant's sponsoring local organization has formally passed and approved the loan resolution.

(4) The Rural Development State Director and NRCS State Conservationist have determined that all planned actions can be carried out as proposed in the project plan and the docket.

(5) The NRCS State Conservationist and Rural Development State Director have mutually agreed on the priority to be given the WS loan or WS advance, or RCD loan. In making this determination, consideration will be given to the relative priority of the WS works of improvement or RCD measures to all other such work in the State and the anticipated availability of Federal and local funds to assure continuity of action and work until the project is completed. When funds are to be provided by NRCS for a WS or RCD loan or a WS advance such funds must be obligated by NRCS before closing.

(6) Public bodies will be required to use bond counsel in accordance with subpart D of part 1780 of this chapter.

(b) When favorable action is not taken on a WS loan, WS advance, or RCD loan, the Rural Development State Director will notify the NRCS State Conservationist and the applicant in writing and, if possible, arrange for a meeting of RUS and NRCS representatives with the applicant to explain the action. WS loans, WS advances, or RCD loans may be canceled before closing.

§1781.20 Disbursement of WS and RCD loan funds and WS advance funds.

(a) WS and RCD loan funds will be disbursed by the processing office in accordance with the applicable provisions of §1780.45 of this chapter and RUS Bulletin 1781-1, paragraph (5). Funds will be made available to the borrower as needed for payment of development or other costs for which the loan is made. The processing office must determine that the payment is for an authorized purpose and is for benefits accrued to the borrower. This will require evidence from NRCS in accordance with the applicable provisions of RUS Bulletin 1781-1, "Memorandum of Understanding Between RUS and NRCS."

(b) WS advance funds may be disbursed in the same manner as WS loan funds if such funds are transferred to RUS by NRCS for disbursement or they may be disbursed by NRCS. When WS advance funds are disbursed by NRCS, payments from advance of funds will be reported to the Rural Development State Director each month to be reported to the Finance Office and charged to the borrower's account. This action will be taken in accordance with the applicable provisions of RUS Bulletin 1781-1 or RUS Bulletin 1781-2 and agreement between the NRCS State Conservationist and Rural Development State Director as follows:

(1) When a future water supply is being developed with NRCS, WS advance funds, the NRCS State Conservationist will send the Rural Development State Director a monthly report of funds disbursed. This will include three (3) copies of Form NRCS-AS-49a and 49b, "Contract Payment Estimate and Construction Progress Report," along with a transmittal Memorandum showing the sequential number (first, second, third, etc.) of the payment, the amount and date of payment, the check number by which the payment was made and the cumulative amount of advance funds disbursed to date. When the works of improvement, for which WS advance funds are used is completed the final report will, in addition to the above, show the date that construction was completed and the total amount of WS advance funds used.

(2) WS advances for construction costs will be set out each month on Form NRCS-49a. The Rural Development State Director should make arrangements with the NRCS State Conservationist to be supplied each month with a copy of Form NRCS 49a when advance funds are included together with an official statement from the NRCS State Administrative Officer giving the date of the check and the exact amount of each advance of funds made under the advance provisions of the project agreement or of any engineering services agreement or other supplementary agreement which further implements the proposal for the advance in the project agreement. The original will be sent immediately to the Finance Office and a copy provided for the processing office file.

(3) When WS advance funds are used to acquire property for site preservation the same reporting procedure as for a future water supply will be used except that Form NRCS-AS-49a and 49b if used, should be adopted to indicate fund use. As payments are made on land on which a mortgage or other security instrument is required, such instruments will be executed in accordance with instructions from the Regional Attorney, OGC.

(4) The Rural Development State Director must send the bond or note evidencing WS advance indebtedness of the borrower to the Finance Office along with reports of payments from advance funds disbursed by NRCS. A copy of the bond or note and copy of each report of payment will be sent to the processing office.

(c) *Actions subsequent to closing of loans or advances.* Actions will be taken in accordance with §1780.44 of this chapter.

§1781.21 Borrower accounting methods, management, reporting, and audits.

These activities will be handled in accordance with the provisions of §1780.47 of this chapter.

§1781.22 Subsequent loans.

Subsequent loans will be processed in accordance with this part.

§ 1781.23 Servicing.

Servicing will be handled in accordance with the provisions of subpart E of part 1951 of this title.

§ 1781.24 State supplements and availability of bulletins, instructions, forms, and memorandums.

(a) State supplements will be issued as needed in accordance with applicable provisions of part 1780 of this chapter.

(b) Bulletins, instructions, forms and memorandums are available from any USDA/Rural Development office or the Rural Utilities Service, United States Department of Agriculture, Washington, DC. 20250–1500.

§§ 1781.25–1781.100 [Reserved]

PART 1782—SERVICING OF WATER AND WASTE PROGRAMS

AUTHORITY: 5 U.S.C. 301; 7 U.S.C. 1981; 16 U.S.C. 1005.

SOURCE: 72 FR 55013, Sept. 28, 2007, unless otherwise noted.

§ 1782.1 Purpose.

This part outlines the Rural Utilities Service's (RUS), an agency delivering the United States Department of Agriculture's (USDA) Rural Development Utilities Programs, hereinafter referred to as Rural Development and/or Agency, policies and procedures for servicing direct and insured Water and Waste Disposal (WWD) loans and grants; Watershed loans and advances; Resource Conservation and Development loans; Technical Assistance and Training grants; Emergency Community Water Assistance grants; Solid Waste Management grants; and section 306C WWD loans and grants.

§ 1782.2 Objectives.

Loan and grant servicing is provided by Rural Development in order to assist recipients in complying with the established objectives and requirements for loans and grants, repaying loans on schedule, acting in accordance with any necessary agreements, and protecting Rural Development's financial interest. Servicing by Rural Development includes, but is not limited to, the review of budgets, management reports, audits, and financial statements; performing operational inspections; providing, arranging, or recommending technical assistance; evaluating environmental impacts of proposed actions by the borrower; and performing civil rights compliance and graduation reviews.

§ 1782.3 Definitions.

The following definitions apply to this part:

Acceleration. A written notice informing the borrower that the total unpaid principal and interest is due and payable immediately.

Adjustment. Satisfaction of a debt, including release of liability, when acceptance by the Agency is conditioned upon completion of payment of the adjusted amount at a specific time or times, with or without the payment of any consideration when the adjustment offer is approved. An adjustment is not a final settlement until all payments under the adjustment agreement have been made.

Administrator. Administrator of the Rural Utilities Service, an agency delivering the United States Department of Agriculture's Utilities Programs.

Agency. The Rural Utilities Service, an Agency delivering the United States Department of Agriculture's Rural Development Utilities Programs, or any employee acting on its behalf in accordance with appropriate delegations of authority.

Assumption of debt. Agreement by one party to legally bind itself to pay the debt incurred by another.

Borrower. Recipient of Agency or predecessor Agency loan assistance.

Cancellation. Final discharge of debt with a release of liability.

Charge-off. Write off of a debt and termination of servicing activity without release of liability. A charge-off is a decision by the Agency to remove debt from Agency receivables, however, future payments may be received.

Compromise. Satisfaction of a debt including a release of liability by accepting a lump-sum payment of less than the total amount owed.

Defeasance. Defeasance is the use of invested proceeds from a new bond issue to repay outstanding bonds in accordance with the repayment schedule of the outstanding bonds. The new issue supersedes the contractual agreements from the prior issue.

Disposition of facility. Relinquishing control of a facility to another entity.

False information. Information, known by the applicant to be incorrect, provided with the intent to obtain benefits which would not have been obtainable based on correct information.

Government. The United States of America, acting through the Agency. USDA, Rural Development and Agency may be used interchangeably throughout this part.

Grantee. Recipient of Agency or predecessor Agency grant assistance, technical assistance, or services.

Letter of Conditions. A written document that describes the conditions which the borrower and/or grantee must meet for funds to be advanced and the loan and/or grant to be closed.

Liquidation. Satisfaction of a debt through the sale of a borrower's assets and discharge of liabilities.

Parity Lien. A lien having an equal lien position to another lender's lien on a borrower's asset.

Reasonable rates and terms. The prevailing commercial rates and terms in the industry that borrowers are expected to pay when borrowing for similar purposes and periods of time.

Rural Development. The mission area of the Under Secretary for Rural Development. Rural Development State and local offices administer the water and waste programs on behalf of the Agency.

Rural Utilities Service (RUS). An Agency of the United States Department of Agriculture's Rural Development mission area established pursuant to section 232 of the Department of Agriculture Reorganization Act of 1994 (Pub. L. 103–354).

Servicing office. The USDA office which maintains the official file of the borrower or grantee and is responsible for the routine servicing of the loan and/or grant account.

Servicing official. USDA official who has been delegated loan and grant approval and servicing authorities subject to any dollar limitations within applicable programs.

Settlement. Compromise, adjustment, cancellation, or charge-off of a debt owed USDA. The term "settlement" is used for convenience in referring to compromise, adjustment, cancellation, or charge-off action, individually or collectively.

Unliquidated obligations. Obligated loan or grant funds that have not been advanced.

USDA. United States Department of Agriculture.

Voluntary conveyance. A method by which title to security is voluntarily transferred to the Government.

§ 1782.4 Availability of forms and regulations.

Information about the availability of forms, regulations, bulletins, and procedures referenced in this chapter are available in any office of Rural Development USDA, Washington, DC 20250–1500 or at the Web site *http://www.usda.gov/rus/water*.

§ 1782.5 Nondiscrimination.

Each instrument of conveyance required for a transfer, assumption, sale of facility, or other servicing action under this subpart will comply with Title VI of the Civil Rights Act of 1964

(Pub. L. 88–352), Title IX of the Education Amendments of 1972 (Pub. L. 92–318), section 504 of the Rehabilitation Act of 1973 (Pub. L. 93–112), and other Federal statutes and regulations issued pursuant thereto that prohibit discrimination on the basis of race, color, national origin, handicap, religion, age, or sex in programs or activities receiving Federal financial assistance. Such provisions apply for as long as the property continues to be used for the same or similar purposes for which the Federal assistance was extended, or for so long as the purchaser owns it, whichever is later.

§ 1782.6 [Reserved]

§ 1782.7 Grants.

Servicing actions relating to Agency grants are governed by the provisions of several regulations and executive orders, including, but not limited to, 2 CFR part 200 as adopted by 2 CFR part 400, and 2 CFR parts 415, 416, 417, and 418 and Executive Order (E.O.) 12803. Grantees remain responsible for property acquired with grant funds in accordance with terms of a grant agreement and applicable regulations.

[79 FR 76006, Dec. 19, 2014]

§ 1782.8 Payments.

Payments will be applied in accordance with the terms of the debt instrument. Information on nontypical payments can be obtained from the Servicing official or office. All new borrowers will use pre-authorized debits as required in their Letter of Conditions.

§ 1782.9 Environmental review requirements.

Servicing actions involving lease or sale of Agency-owned property must comply with the environmental review requirements in accordance with 7 CFR part 1970.

[81 FR 11029, Mar. 2, 2016]

§ 1782.10 Audit requirements.

Audits for loans will be required in accordance with § 1780.47 of this chapter. If the borrower becomes delinquent or is experiencing problems, the servicing official will require an audit or other documentation deemed necessary to resolve the delinquency. The provisions of subpart F of 2 CFR part 200, as adopted by USDA through 2 CFR part 400, address audit requirements for recipients of Federal assistance.

[79 FR 76006, Dec. 19, 2014]

§ 1782.11 Refinancing requirements.

If at any time it appears to the Government that the borrower is able to refinance the amount of the indebtedness then outstanding, in whole or in part, by obtaining a loan for such purposes from responsible cooperative or private credit sources, at reasonable rates and terms, the borrower will, upon request of the Government, apply for and accept such loan in sufficient amount to repay the Government and will take all such actions as may be required in connection with such loan.

§ 1782.12 Sale or exchange of security property.

A cash sale of all or a portion of a borrower's assets or an exchange of security property may be approved subject to the conditions set forth in this section.

(a) *Approval conditions.* Approval may be given when the servicing official determines that:

(1) The consideration is for the full amount of the debt or the present fair market value as determined by an appraisal completed by a qualified Rural Development employee or an independent appraiser as determined appropriate by the approval official;

(2) The sale or exchange will not prevent carrying out the purpose of the loan;

(3) The remaining property is adequate security for the loan and the transaction will not adversely affect the Agency's security position;

(4) If the property to be sold or exchanged will be used for similar purposes that the loan was made, the purchaser will:

(i) Execute Form RD 400–4, "Assurance Agreement." The instrument of conveyance will contain the civil rights covenant referenced in 7 CFR 1901.202(e); and

(ii) Provide the Agency with a written agreement assuming all rights and obligations of the original borrower, and

(5) Proceeds remaining after paying any reasonable and necessary selling expenses are to be used for one or more of the following purposes:

(i) To pay Agency debt, pay on debts secured by a prior lien, and pay on debts secured by a parity or subsequent lien if it is to the Agency's advantage;

(ii) To purchase or acquire property more suited to the borrower's needs, providing the Agency's security position is maintained; and

(iii) To develop or enlarge the facility if necessary to improve the borrower's debt-paying ability, place the operation on a sounder financial basis, or further the loan objectives and purposes.

(b) *Sale of assets financed with Agency grants.* The requirements for the sale or disposition of assets financed with Agency grants are determined by the terms of the grant agreement, 7 CFR parts 3015, 3016, and 3019, and E.O. 12803, as applicable.

(c) *Release from liability.* If a borrower can no longer meet the objectives of the loan, the property may be sold. If the full amount of the borrower's debt is paid or assumed, the State Director may release the borrower from liability.

§ 1782.13 Transfer of security and assumption of loans.

It is the Agency's policy to approve transfers and assumptions to transferees that will continue the original purpose of the loan. Assistant Administrator written concurrence is required when the transfer exceeds the State Director's loan approval authority. The transfer will be approved in accordance with the following requirements:

(a) *General requirements for transferees.* The fulfillment of the following requirements for transfers will be determined by the approval official, in his or her discretion:

(1) The transferees must meet the eligibility requirements of 7 CFR part 1780 and provide the same information required in 7 CFR part 1780, subpart B, for application processing.

(2) The transfer will not be disadvantageous to the Government as determined by the approval official.

(3) If the Agency debt(s) exceeds the present market value of the security as determined by an appraisal, the transferee will assume an amount at least equal to the present market value.

(4) The Agency must concur in plans for disposition of funds in any reserve account, including project construction bank accounts. A reserve account may be considered as a transferable asset.

(5) The transferee will assume all of the borrower's responsibilities regarding loans. The transferee will also agree to accept the original loan conditions plus any conditions set forth by the Agency with regard to the transfer.

(6) A current appraisal will be completed to establish the present market value of the security when the full debt is not being assumed.

(7) There must be no lien, judgement, or similar claims of other parties against the Agency security being transferred unless the transferee is willing to accept such claims. The Agency must also determine that the claims will not prevent the transferee from repaying the Agency debt, meeting all operating and maintenance costs, and maintaining required reserves. The written consent of any other lienholder will be obtained where required.

(8) A letter of conditions establishing requirements to be met in connection with the transfer will be issued, and the transferee will be required to execute Form RD 1942–46, "Letter of Intent to Meet Conditions," prior to closing of the transfer.

(9) The transferee will obtain insurance according to Agency requirements.

(10) The effective date of the transfer is the date the transfer is closed, which is the same date Form RD 1951–15, "Community Programs Assumption Agreement," or other appropriate assumption agreement which is executed and delivered by all necessary parties.

(11) Title to all assets will be conveyed from the transferor to the transferee unless all parties concerned, including the Agency, agree upon other arrangements. All instruments of conveyance will contain the necessary nondiscrimination covenant as referred to in § 1782.5.

(12) If the transfer and assumption is to one or more members of the borrower's organization, there must not be a loss to the Government.

(13) The State Director is authorized to approve transfers to eligible transferees at the same interest rate as on the borrower's note(s) or bond(s). The maturity of the debt instrument for the assumed debt may not exceed the lesser of the repayment period authorized in 7 CFR part 1780 for a "new" loan or the expected life of the facility.

(14) Agency National Office concurrence is required for transfers not in compliance with paragraphs (a)(1) through (13) of this section.

(b) *Loan requirements for eligible transferees.* If a loan is evidenced and secured by a note and lien on real or chattel property, Form RD 1951-15, or other appropriate assumption agreement will be executed by the transferee. If a bond secures a loan, transfer documents will be developed by bond counsel and approved by the Office of the General Counsel (OGC), USDA.

(1) Loans being transferred and assumed may be combined when the security is the same, new terms are being provided, a new debt instrument will be issued, and the loans have the same interest rate and are for the same purpose. If applicable, 7 CFR part 1780 will govern the preparation of any new debt instruments required.

(2) A loan may be made in connection with a transfer if the transferee meets all eligibility and other requirements for the kind of loan being made. Such a loan will be considered as a separate loan and must be evidenced by a separate debt instrument. However, it is permissible to have one authorizing loan resolution or ordinance if permitted by State statutes.

(3) Any development funds remaining in a bank account that are not refunded to the Agency will be transferred to a bank account for the transferee. This will occur simultaneously with the closing of the transfer, and the funds will be used in completing planned development.

(c) *Release from liability.* Transferors may be released from liability when their debt is paid in full or when the debt is settled in accordance with § 1782.20 of this part.

(d) *Transfer of facility financed with Agency grants.* The requirements for the sale or disposition of assets financed with Agency grants are determined by the terms of the grant agreement, 7 CFR parts 3015, 3016, and 3019, and E.O. 12803, as applicable.

§ 1782.14 Protection of service areas—7 U.S.C. 1926(b).

(a) 7 U.S.C. 1926(b) was enacted to protect the service area of Agency borrowers with outstanding loans, or those loans sold in the sale of assets authorized by the "Joint Resolution Making Continuing Appropriations for the Fiscal Year 1987, Pub. L. 99-591, 100 Stat. 3341 (1986)," from loss of users due to actions or activities of other entities in the service area of the Agency financed system. Without this protection, other entities could extend service to users within the service area, and thereby undermine the purpose of the congressionally mandated water and waste loan and grant programs and jeopardize the borrower's ability to repay its Agency debt.

(b) Responsibility for initiating action in response to those actions prohibited by 7 U.S.C. 1926(b) rests with the borrower.

§ 1782.15 Mergers and consolidations.

Mergers and consolidations will be processed the same as a transfer and assumption, although approvals by the Agency will give consideration to the differences under the applicable law regarding the type of transaction under consideration and the unique facts involved in each transaction. Mergers occur when two or more entities combine in such a manner that only one remains in existence. Consolidations occur when two or more entities combine to form a new consolidated entity, and the original entities cease to exist. In both mergers and consolidations, the surviving or emerging entity acquires the assets and assumes the liabilities of the entity or entities that ceased to exist.

§ 1782.16 Defeasance of Agency indebtedness.

Defeasance, or amending outstanding loan instruments and agreements to permit defeasance of Agency debt instruments, is prohibited.

§ 1782.17 Parity lien.

In order for the Agency to agree to a parity lien position, the borrower must submit a written request to the servicing office.

(a) The written request for parity must contain the following items:

(1) An explanation of the purpose of the request for parity; amount of loan for which parity is requested; description of security property; type of security instrument; name and address of financial institution requesting the transaction; and other information determined necessary by the servicing official to evaluate the request.

(2) Current financial statements or an audit, if available or determined necessary by the servicing official.

(3) An annual operating budget which projects income and expenses for a typical year's operation. If construction is involved, the budget must be projected through the first full year of operation following completion of the planned improvements.

(4) A copy of the proposed security instrument.

(5) A certification from the borrower that the Agency debt cannot be refinanced at reasonable rates and terms.

(6) An appraisal, when the primary security is real estate or determined necessary by the servicing official in order to determine the adequacy of loan security or repayment ability.

(7) A certification that any development work will comply with subpart C of part 1780 of this chapter.

(b) Requests for parity must comply with requirements of paragraph (a) of this section, requirements as specified in the bond or loan documents, the requirements as specified in 7 CFR part 1780, subpart D, and as provided in applicable State law.

(c) If the borrower has met all of the requirements in paragraphs (a) and (b) of this section and the proposal is determined to be in the Government's interest, the Agency will then grant approval of the borrower's request for parity. The following factors will be considered in assessing whether the request is in the Government's interest:

(1) The value of the added assets compared with the amount of new debt to be secured;

(2) The value of the assets already pledged under the security documents, and any effects of the proposed transaction on the value of those assets;

(3) The ratio of the total outstanding debt secured under the security documents to the value of all assets pledged as security under the security documents;

(4) The borrower's ability to repay its debt owed to the Government;

(5) The overall financial viability of the borrower;

(6) The borrower's current relationship with the Agency (i.e. no defaults under the loan documents);

(7) Such other factors that may be relevant in individual cases, as determined by the Agency.

§ 1782.18 [Reserved]

§ 1782.19 Third party agreements.

The State Director may authorize third party operation, maintenance, and management of an Agency financed facility. The borrower's attorney must review the contract, management agreement, written lease, or other third party agreement and issue an opinion to the Agency as to their legal sufficiency. The borrower shall retain the legal authority necessary for owning, constructing, operating, and maintaining the facility.

§ 1782.20 Debt Settlement.

Pursuant to 7 U.S.C. 1981, this section prescribes policies for debt settlement of Water and Waste Disposal loans; Watershed loans and advances; Resource Conservation and Development loans; and 306 (c) Water and Waste Facility loans. Within the Omnibus Consolidated Rescissions and Appropriations Act of 1996 (Public Law 104–134) is the Debt Collection Improvement Act of 1996. This law provides that any nontax debt or claim owed to the United States that has been delinquent for a period of 180 days shall be turned over to the Secretary of the Treasury for appropriate action to collect or terminate collection actions on the debt or claim. Debt that is in litigation or foreclosure, with a collection agency or designated Federal debt collection center, or that will be disposed of under an asset sales program, is exempt from transfer to the Secretary.

(a) *General requirements for debt settlement.* (1) The debt or any extension thereof on which settlement is requested must be due and payable. The debt will be due and payable either under the terms of the note or other instrument, or by acceleration, unless the debt is to be cancelled without application under paragraph (e)(2) of this section or charged off under paragraph (f) of this section.

(2) Normally, all security will be disposed of prior to the date of application for debt settlement unless it is necessary to abandon security through the debt settlement process. In such cases, debt settlement may proceed if the servicing official determines that further collection efforts would be ineffective, uneconomical, and not in the best interests of the Government.

(3) Debtors will not be permitted to sell security and use the proceeds as part or all of a compromise/adjustment debt settlement offer.

(4) Requests for debt settlement will consist of Form RD 1956–1 "Application For Settlement of Indebtedness," current financial information, description and estimated market value of collateral, and status of operation (i.e., number of users, compliance with environmental issues, etc.).

(5) Office of General Counsel (OGC) advice on compliance with State or Federal statutes that may affect the debt settlement action must be requested.

(b) *Debts ineligible for settlement.* Debts will not be settled if:

(1) Referral to the Office of Inspector General and/or to OGC is contemplated or pending because of suspected criminal violation,

(2) Civil action to protect the interest of the Government is contemplated or pending,

(3) An investigation for suspected fiscal irregularity is contemplated or pending, or

(4) The debtor requests settlement of a claim that has been referred to or a judgment obtained by the United States Attorney. The settlement offer and any related payment must be submitted directly to the United States Attorney for consideration.

(c) *Types of debt settlement.* Typically, debt settlement will be accomplished through compromise/adjustment, charge-off, or cancellation. Any debt remaining after the security has been liquidated, by sale or transfer, will be cancelled if there are no other assets from which to collect the debt. The servicing official will proceed with advice from OGC and the National Office, as required.

(d) *Compromise and adjustment.* Debts may be compromised or adjusted and security retained by the debtor, provided:

(1) The debtor is unable to pay the indebtedness in full,

(2) The debtor has offered an amount equal to the present fair market value of all security or facility financed, and

(3) The debtor has offered any additional amount that the debtor is able to pay.

(e) *Cancellation.* Non-judgment debts, regardless of the amount, may be cancelled with or without application by the debtor.

(1) *With application by the debtor.* Debts may be cancelled upon application of the debtor, subject to the following conditions:

(i) The servicing official furnishes a favorable recommendation concerning the cancellation;

(ii) There is no known security for the debt and the debtor has no other assets from which the debt could be collected;

(iii) The debtor is unable to pay any part of the debt, and has no reasonable prospect of being able to do so; and

(iv) The debt or any extension thereof is due and payable under the terms of the note or other instrument or due to acceleration by written notice prior to the date of application.

(2) *Without application by debtor.* Debts may be cancelled upon a favorable recommendation of the servicing official in the following instances:

(i) *Debtors discharged in bankruptcy.* If there is no security for the debt, debts discharged in bankruptcy shall be cancelled by the use of Form RD 1956–1. A copy of the Bankruptcy Court's Discharge Order must be attached.

(ii) *Impractical to obtain debtor's signature.* Debts may be cancelled if it is impractical to obtain a signed application and the requirements of paragraphs (e)(1) of this section are met. Form RD 1956–1 will document the specific reason(s) why it was impossible or impracticable to obtain the signature of the debtor. If the debtor refused to sign the application, the reason(s) should be documented.

(f) *Charge-off*—(1) *Judgment debts.* Judgment debts, regardless of the amount, may be charged off without the debtor's signature upon a favorable recommendation of the servicing official provided:

(i) The United States Attorney's file is closed, and

(ii) The requirements of paragraph (e)(2)(ii) of this section, if applicable, have been met, or 2 years have elapsed since any collections were made on the judgment. The debtor must also have no equity in the property subject to the lien or upon which a lien can be obtained.

(2) *Non-judgment debts.* Debts that cannot be settled under other sections of this part may be charged off without the debtor's signature upon a favorable recommendation of the servicing official in the following instances:

(i) When OGC advises in writing that the claim is legally without merit or that evidence necessary to prove the claim in court cannot be provided; or

(ii) When there is no known security for the debt, the debtor has no other assets from which the debt could be collected, and the debtor:

(A) Is unable to pay any part of the debt and has no reasonable prospect of being able to do so; or

(B) Is able to pay part or all of the debt but refuses to do so, and OGC provides an opinion to the effect that the Government cannot enforce collection of a significant amount from assets or income.

§1782.21 [Reserved]

§1782.22 Exception authority.

The Administrator may, in individual cases, make an exception to any requirement or provision of this part which is not inconsistent with the authorizing statute or other applicable law and is determined to be in the Government's interest. Requests for exceptions must be made in writing by the State Director and supported with documentation to explain the adverse effect on the Government's interest, propose alternative course(s) of action, and show how the adverse affect will be eliminated or minimized if the exception is granted. The exception decision will be documented in writing, signed by the Administrator, and retained in the files.

§1782.23 Use of Rural Development loans and grants for other purposes.

(a) If, after making a loan or a grant, the Administrator determines that the circumstances under which the loan or grant was made have sufficiently changed to make the project or activity for which the loan or grant was made available no longer appropriate, the Administrator may allow the borrower or grantee to use property (real and personal) purchased or improved with the loan or grant funds, or proceeds from the sale of property (real and personal) purchased with such funds, for another project or activity that:

(1) Will be carried out in the same area as the original project or activity;

(2) Meets the criteria for a loan or grant described in section 381E(d) of the Consolidated Farm and Rural Development Act (Pub. L. 87–128), as amended; and

(3) Satisfies such additional requirements as are established by the Administrator.

(b) If the new use of the property is under the authority of another USDA Agency Administrator, the other Administrator will be consulted on whether the new use will meet the criteria of the other program. Since the new project or activity must be carried out in the same area as the original project or activity, a new rural area determination will not be necessary.

(c) Borrowers and grantees that wish to use the proceeds for other purposes may make their request through the appropriate Rural Development State Office. Permission to use this option will be exercised on a case-by-case basis on applications submitted through the State Office to the Administrator for consideration. If the proposal is approved, the Administrator will issue a memorandum to the State Director outlining the conditions necessary to complete the transaction.

§§ 1782.24–1782.99 [Reserved]

§ 1782.100 OMB Control Number.

The information collection requirements in this part are approved by the Office of Management and Budget (OMB) and assigned OMB Control Number 0572–0137.

PART 1783—REVOLVING FUNDS FOR FINANCING WATER AND WASTEWATER PROJECTS (REVOLVING FUND PROGRAM)

Subpart A—General

AUTHORITY: 7 U.S.C. 1926 (a)(2)(B).

SOURCE: 69 FR 59772, Oct. 6, 2004, unless otherwise noted.

Subpart A—General

§ 1783.1 What is the purpose of the Revolving Fund Program?

This part sets forth the policies and procedures for making grants to qualified private, non-profit entities to capitalize revolving funds for the purpose of providing financing to eligible entities for pre-development costs associated with proposed water and wastewater projects or with existing water and wastewater systems, and short-term costs incurred for replacement equipment, small-scale extension of services, or other small capital projects that are not part of the regular operations and maintenance activities of existing water and wastewater systems.

§ 1783.2 What Uniform Federal Assistance Provisions apply to the Revolving Fund Program?

(a) This program is subject to the general provisions that apply to all grants made by USDA and that are set forth in 2 CFR part 200, as adopted by USDA through 2 CFR part 400.

(b) This program is subject to the uniform administrative requirements that apply to all grants made by USDA to non-profit organizations and that are set forth in 2 CFR part 415.

(c) 2 CFR part 180, as adopted by USDA through 2 CFR part 417, Nonprocurement Debarment and Suspension, implementing Executive Order 12549 and Executive Order 12689 on debarment and suspension.

(d) This program is subject to 2 CFR part 418, New Restrictions on Lobbying, prohibiting the use of appropriated funds to influence Congress or a Federal agency in connection with the making of any Federal grant and other Federal contracting and financial transactions.

(e) This program is subject to 2 CFR part 421, Requirements for Drug-Free Workplace (Financial Assistance), implementing the Drug-Free Workplace Act of 1988 (41 U.S.C. 8102).

[79 FR 76007, Dec. 19, 2014, as amended at 81 FR 7698, Feb. 16, 2016]

§ 1783.3 What definitions are used in this regulation?

Administrative expenses means expenses incurred by a grant recipient that are of the type more particularly described in § 1783.12.

Applicant means a private, non-profit organization that applies for an RFP grant under this part.

CONACT means the Consolidated Farm and Rural Development Act.

Eligible entity means an entity eligible to obtain a loan, loan guarantee or grant under paragraph 1 or paragraph 2 of section 306(a) the CONACT (codified at 7 U.S.C. 1926(a)(1) and (2)).

Funding opportunity announcement (FOA) means a publicly available document by which a Federal agency makes know its intentions to award discretionary grants or cooperative agreements, usually as a result of competition for funds. FOA announcements may be known as program announcements, notices of funding availability, solicitations, or other names depending on the agency and type of program. FOA announcements can be found at *Grants.gov* in the Search Grants tab and on the funding agency's or program's website.

Grant agreement means the contract between RUS and the grant recipient which sets forth the terms and conditions governing a particular grant awarded under this part.

Grant recipient means a private, non-profit entity that has been awarded a grant under this part.

Loan recipient means an eligible entity that has received an RFP loan.

Revolved funds means the cash portion of the revolving loan fund that is not composed of RFP grant funds, including cash comprising repayments of RFP loans, fees relating to RFP loans and interest collected on RFP loans.

Revolving loan fund means the loan fund established by the grant recipient to carry out the purposes of this part, such fund comprising the proceeds of an RFP grant and other related assets.

RFP means Revolving Fund Program.

RFP grant means a grant from RUS to a grant recipient under this part.

RFP loan means a loan from a grant recipient using the direct or indirect proceeds of an RFP grant awarded under this part.

Rural and rural area means a city, town or unincorporated area that has a population of no more than 10,000 inhabitants. The population figure is obtained from the most recent decennial Census of the United States (decennial Census). If the applicable population figure cannot be obtained from the most recent decennial Census, RD will determine the applicable population figure based on available population data.

RUS means the Rural Utilities Service, a Federal agency delivering the USDA's Rural Development Utilities Program.

USDA means the United States Department of Agriculture.

[69 FR 59772, Oct. 6, 2004, as amended at 80 FR 9864, Feb. 24, 2015; 83 FR 45034, Sept. 5, 2018]

§ 1783.4 [Reserved]

Subpart B—Revolving Loan Program Grants

§ 1783.5 What are the eligibility criteria for grant recipients?

(a) The applicant must be a private entity.

(b) The applicant must be organized as a non-profit entity.

(c) The applicant must have the legal capacity and lawful authority to perform the obligations of a grantee under this part.

Example 1 to paragraph (c): If the organization is incorporated as a non-profit corporation, it must have corporate authority under state law and its corporate charter to engage in the practice of making loans to legal entities.

Example 2 to paragraph (c): If the organization is an unincorporated association, state law may prevent the organization from entering into binding contracts, such as a grant agreement.

(d) The applicant must have sufficient expertise and experience in making and servicing loans to assure the likelihood that the objectives of this part can be achieved.

§ 1783.6 When will applications for grants be accepted?

A FOA will be posted to *www.Grants.gov* in fiscal years that funds are available for this program.

The FOA will establish the period during which applications for such funds may be submitted for consideration.

[83 FR 45034, Sept. 5, 2018]

§ 1783.7　What is the grant application process?

(a) The applicant must complete and submit the following items to RUS to apply for a grant under this part:

(1) Application for Federal Assistance: *Standard Form 424;*

(2) Budget Information—Non-Construction Programs: *Standard Form 424A;*

(3) Assurances—Non-Construction Programs: *Standard Form 424B;*

(4) Evidence of applicant's legal existence and authority in the form of certified copies of organizational documents and a certified list of directors and officers with their respective terms;

(5) Evidence of tax exempt status, and

(6) Most recent annual audit conducted by an independent auditor.

(b) The applicant must submit a written work plan that demonstrates the ability of the applicant to make and service loans to eligible entities under this program and the feasibility of the applicant's lending program to meet the objectives of this part.

(c) The applicant should submit a narrative establishing the basis for any claims that it has substantial expertise in making and servicing loans. The Secretary will give priority to an applicant that demonstrates it has substantial experience of this type.

(d) The applicant may submit such additional information as it elects to support and describe its plan for achieving the objectives of the part.

§ 1783.8　What are the acceptable methods for submitting applications?

(a) Applications for RFP grants may be submitted by U.S. Mail. Applications submitted by mail must be addressed as follows: Rural Utilities Service, U.S. Department of Agriculture, 1400 Independence Avenue, SW., STOP 1548, Washington, DC 20250–1548. The outside of the application should be marked: "Attention: Assistant Administrator, Water and Environmental Programs." Applications submitted by

mail must be postmarked not later than the filing deadline to be considered during the period for which the application was submitted.

(b) In lieu of submitting an application by U.S. Mail, an applicant may file its application electronically by using the Federal Government's eGrants Web site (Grants.gov) at *http://www.grants.gov.* Applicants should refer to instructions found on the Grants.gov Web site for procedures for registering and using this facility. Applicants who have not previously registered on Grants.gov should allow a sufficient number of business days to complete the process necessary to be qualified to apply for Federal Government grants using electronic submissions. Electronic submissions must be filed not later than the filing deadline to be considered during the period for which the application was submitted.

(c) The methods of submitting applications may be changed from time to time to reflect changes in addresses and electronic submission procedures Applicants should refer to the most recent FOA for notice of any such changes. In the event of any discrepancy, the information contained in the FOA must be followed. In the event of any discrepancy, the information contained in the notice must be followed.

[69 FR 59772, Oct. 6, 2004, as amended at 83 FR 45034, Sept. 5, 2018]

§ 1783.9　What are the criteria for scoring applications?

(a) Applications that are incomplete or ineligible will be returned to the applicant, accompanied by a statement explaining why the application is being returned.

(b) Promptly after an application period closes, all applications that are complete and eligible will be ranked competitively based on the following scoring criteria:

(1) Degree of expertise and successful experience in making and servicing commercial loans, with a successful record, for the following number of full years:

(i) At least 1 but less than 3 years— 5 points

(ii) At least 3 but less than 5 years— 10 points

(iii) At least 5 but less than 10 years—20 points

(iv) 10 or more years—30 points

(2) Extent to which the work plan demonstrates a well thought out, comprehensive approach to accomplishing the objectives of this part, clearly defines who will be served by the project, clearly articulates the problem/issues to be addressed, identifies the service area to be covered by the RFP loans, and appears likely to be sustainable. Up to 40 points.

(3) Percentage of applicant contributions. Points allowed under this paragraph will be based on written evidence of the availability of funds from sources other than the proceeds of an RFP grant to pay part of the cost of a loan recipient's project. In-kind contributions will not be considered. Funds from other sources as a percentage of the RFP grant and points corresponding to such percentages are as follows:

(i) Less than 20%—ineligible

(ii) At least 20% but less than 50%—10 points

(iii) 50% or more—20 points

(4) Extent to which the goals and objectives are clearly defined, tied to the work plan, and are measurable. Up to 15 points.

(5) Lowest ratio of projected administrative expenses to loans advanced. Up to 10 points.

(6) The evaluation methods for considering loan applications and making RFP loans are specific to the program, clearly defined, measurable, and are consistent with program outcomes. Up to 20 points.

(7) Administrator's discretion, considering such factors as creative outreach ideas for marketing RFP loans to rural residents; the amount of funds requested in relation to the amount of needs demonstrated in the work plan; previous experiences demonstrating excellent utilization of a revolving loan fund grant; and optimizing the use of agency resources. Up to 10 points.

(c) All qualifying applications under this part will be scored based on the criteria contained in this section. Awards will be made based on the highest ranking applications and the amount of financial assistance available for RFP grants. All applicants will

be notified of the results in writing on form AD–622.

§1783.10 What is the grant agreement?

RUS and the grant recipient will enter into a contract setting forth the terms and conditions governing a particular RFP grant award. RUS will furnish the form of grant agreement. No funds awarded under this part shall be disbursed to the grant recipient before the grant agreement is binding and RUS has received a fully executed counterpart of the grant agreement.

§1783.11 What is the revolving loan fund?

The grant recipient shall establish and maintain a revolving loan fund for the purposes set forth in §1783.12. The revolving loan fund shall be comprised of revolving loan fund grant funds and the grant recipient's contributed funds. All revolving loan fund loans made to loan recipients shall be drawn from the revolving loan fund. All revolving loan fund loans shall be serviced and the revolving loan fund maintained, in accordance with this part and applicable law.

§1783.12 What are eligible uses of grant proceeds?

(a) Grant proceeds shall be used solely for the purpose of establishing the revolving loan fund to provide loans to eligible entities for:

(1) Pre-development costs associated with proposed water and wastewater projects or with existing water and wastewater systems, and

(2) Short-term costs incurred for replacement equipment, small-scale extension of services, or other small capital projects that are not part of the regular operations and maintenance activities of existing water and wastewater systems.

(b) A grant recipient may not use grant funds in any manner inconsistent with the terms of the grant agreement.

§1783.13 What administrative expenses may be funded with grant proceeds?

RFP grant funds may not be used for any purposes not described in §1783.12, including, without limitation, payment or reimbursement of any of the grant recipient's administrative costs or expenses. Administrative expenses may, however, be paid or reimbursed from revolving loan fund assets that are not RFP grant funds, including revolved funds and cash originally contributed by the grant recipient.

Subpart C—Revolving Fund Program Loans

§ 1783.14 What are the eligibility criteria for RFP loan recipients?

(a) A loan recipient must be an eligible entity as defined in § 1783.3.

(b) The loan recipient must be unable to finance the proposed project from their own resources or through commercial credit at reasonable rates and terms.

(c) The loan recipient must have or will obtain the legal authority necessary for owning, constructing, operating and maintaining the proposed service or facility, and for obtaining, giving security for, and repaying the proposed loan.

(d) The project funded by the proceeds of an RFP loan must be located in, or the services provided as the result of such project must benefit, rural areas.

§ 1783.15 What are the terms of RFP loans?

(a) RFP loans under this part—

(1) Shall have an interest rate that is determined by the grant recipient and approved by RUS;

(2) Shall have a terms not to exceed 10 years; and

(3) Shall not exceed the lesser of $100,000 or 75 percent of the total cost of a project. The total outstanding balance for all loans under this program to any one entity shall not exceed $100,000.

(b) The grant recipient must set forth the RFP loan terms in written documentation signed by the loan recipient.

(c) Grant recipients must develop and use RFP loan documentation that conforms to the terms of this part, the grant agreement, and the laws of the state or states having jurisdiction.

§ 1783.16 How will the loans given from the revolving fund be serviced?

The grant recipient shall be responsible for servicing all loans, to include preparing loan agreements, processing loan payments, reviewing financial statements and debt reserves balances, and other responsibilities such as enforcement of loan terms. Loan servicing will be in accordance with the work plan approved by the Agency when the grant is awarded for as long as any loan made in whole or in part with Agency grant funds is outstanding.

PART 1784—RURAL ALASKAN VILLAGE GRANTS

Subpart A—General Provisions

Subpart B—Grant Requirements

Subpart C—Application Processing

Subpart D—Grant Processing

AUTHORITY: 7 U.S.C. 1926d.

SOURCE: 80 FR 52609, Sept. 1, 2015, unless otherwise noted.

Subpart A—General Provisions

§ 1784.1 Purpose.

This part sets forth the policies and procedures that will apply when the Rural Utilities Service (RUS) makes grants under the Rural Alaska Village Grant (RAVG) program (7 U.S.C. 1926d) to rural or native villages in Alaska. The grants will be provided directly to a rural or native village or jointly with either The State of Alaska, Department of Environmental Conservation (DEC) or The Alaska Native Tribal Health Consortium (ANTHC) for the benefit of rural or native villages in Alaska.

§ 1784.2 Definitions.

The following definitions apply to subparts A through E of this part.

ANTHC means the Alaska Native Tribal Health Consortium.

CONACT means the Consolidated Farm and Rural Development Act.

DEC means the State of Alaska, Department of Environmental Conservation.

Dire sanitation conditions means:

(1) Recurring instances of illness reasonably attributed to waterborne communicable disease have been documented or insufficient access to clean water creates a persistent threat of water-washed diseases; or

(2) No community-wide water and sewer system exists and individual residents must haul water to or human waste from their homes and/or use pit privies; or

(3) An appropriate federal agency (such as the Centers for Disease Control and Prevention) or regulatory Agency of the State of Alaska determines that the drinking water and/or sewer system does not meet current regulatory requirements.

Grant recipient means an applicant that has been awarded a Rural Alaskan Village Grant under this part.

IHS means the United States Department of Health and Human Services, Indian Health Service.

Owner means Grant recipient.

RAVG means Rural Alaskan Village Grant, a grant awarded by RUS, DEC, and/or ANTHC to a grant recipient under this part.

Rural or Native Villages in Alaska means a rural community or Native village in Alaska which meets the definition of a village under State statutes and does not have a population in excess of 10,000 inhabitants, according to the U.S. Census American Community Survey.

RD means Rural Development, a federal agency mission area delivering the United States Department of Agriculture's programs to rural communities.

Recipient community means a community that has been awarded a grant under this part.

RUS means the Rural Utilities Service, a federal agency mission area delivering the United States Department of Agriculture's rural utilities programs.

Short-lived assets means repair and replacement items expended each year that are not included in the annual Operational and Maintenance expenses as annual repair and maintenance.

Statewide nonmetropolitan median household income (SNMHI) means the median household income of the State's nonmetropolitan counties and portions of metropolitan counties outside of cities, towns or places of 50,000 or more population.

USDA means the United States Department of Agriculture.

VSW means Village Safe Water Program authorized under the Village Safe Water Act, Alaska Statute Title 46, Chapter 7 (AS 46.07).

§ 1784.3 Objective.

The objective of the RAVG Program is to assist the residents of rural or native villages in Alaska to provide for the development and construction of water and wastewater systems to improve the health and sanitation conditions in those villages through removal of dire sanitation conditions.

§§ 1784.4–1784.7 [Reserved]

Subpart B—Grant Requirements

§ 1784.8 Eligibility.

(a) Grants may be made to the following eligible applicants:

(1) A rural or native village in Alaska; or

(2) DEC on behalf of one or more rural or native village in Alaska; or

(3) ANTHC on behalf of one or more rural or native village in Alaska.

(b) Grants made to DEC or ANTHC may be obligated through a master letter of conditions for more than one rural or native village in Alaska; however, DEC or ANTHC together with each individual rural or native village beneficiary shall execute a grant agreement on a project by project basis. Expenditures for projects will be based on specific scope and be requested on a project by project basis.

(c) For grants proposed to be administered directly by a community, the responsibility to meet the requirements outlined in this part will be met by the community. RUS will be the lead agency on direct administration projects.

(d) The median household income of the rural or native village cannot exceed 110 percent of the statewide nonmetropolitan household income (SNMHI), according to US Census American Community Survey. Alaska census communities considered to be high cost isolated areas or "off the road systems" (*i.e.* communities that cannot be accessed by roads) may utilize up to 150 percent of SNMHI.

(e) For design and construction projects: A dire sanitation condition as defined in § 1784.2 must exist in the village served by the proposed project. For those projects identified under paragraphs (1) and (3) of the dire sanitation definition in § 1784.2, a notice of violation, consent order or other regulatory action from the appropriate regulatory agency must be provided to document the dire sanitation condition. In cases where there is scientific evidence or reports with substantiated evidence of associated health issues, documentation may be accepted from an appropriate federal agency.

(f) In individual cases where a proposed project does not meet the definition of "Dire sanitation condition" in § 1784.2, an applicant may request a special review and eligibility determination from the RUS Administrator in cases where the applicant is able to satisfactorily demonstrate that a water or sewer system is deficient and negatively impacts the health or safety of the community. The decision to review an eligibility determination request and any determinations made subject to this paragraph are not subject to administrative appeal.

(g) In order for an eligible applicant to receive a grant under the Rural Alaska Village Grant program, the State of Alaska shall provide 25 percent in matching funds from non-Federal sources.

(h) In processing grants through DEC and ANTHC, a public meeting must be held to inform the general public regarding the development of any proposed project. Documentation of the public meeting must be received with construction applications.

(1) A notice of intent must be published in a newspaper of general circulation in the proposed area to be served.

(2) For projects where there are no newspapers of general circulation, a posting of the notice in a community building (post office, washeteria, clinic, etc.) frequented by village residents may be used to meet the requirement. This alternative form of notice has been authorized by the RUS Administrator.

§ 1784.9 Grant amount.

Grants will be made for up to 75 percent of the project development and/or construction costs, which does not include project administrative costs. Pursuant to 7 U.S.C. 1926d, the State of Alaska shall provide 25 percent in matching funds from non-Federal sources.

§ 1784.10 Eligible grant purposes.

Grant funds may be used for the following purposes:

(a) To pay reasonable costs associated with providing potable water or waste disposal services to residents of rural or native villages in Alaska. Reasonable costs include construction, planning, pre-development costs (including engineering, design, and rights-of-way establishment), and technical assistance as further defined below:

(1) *Planning.* Grants can be made specifically for planning report costs (including Master Plans, Feasibility Studies, and Detection or Source Studies) associated with the prioritization process.

(2) *Pre-development.* Grants can be made for pre-development costs such as preliminary engineering, environmental, application development, review and establishment of rights-of-way and easement, and full construction design for up to $1,000,000 for each eligible village. Prior to approving additional pre-development costs, a preliminary engineering report (PER) and/or approved PER like document, such as the Cooperative Project Agreement and supplemental documents from ANTHC and an environmental report shall be reviewed and concurred by RUS, DEC, ANTHC, and IHS.

(3) *Training and technical assistance.* Grant funding for technical assistance and training will be available in accordance with Section 306D of the Consolidated Farm and Rural Development Act (7 U.S.C. 1926d) and appropriations current at the time of application. Grants for this purpose will be processed in accordance with 7 CFR part 1775.

(b) To pay reasonable costs associated with the use of a recipient community's equipment during construction. (*i.e.* maintenance, minor repairs, and operational costs). A cost accounting system that is accurate to track expenses must be in place. Use of ANTHC or State of Alaska equipment fleet rental costs will also be eligible. RUS concurrence in the allocation method is required.

(c) *Individual installations.* (1) Individual service installation relates to residential homes only and does not include public facilities or commercial facilities. The only exception to serving a public facility is when the facility is necessary for the successful operation and maintenance of the water or sanitation system (*i.e.* the facility utilized for accepting utility payments and/or holding public meetings for the utility system).

(2) Individual home installations, including wells, septic system, flush tank and haul, in-house plumbing, etc., may be provided. The following guidelines must be followed for individual installations. A certification will be required with the application that provides documentation of the following:

(i) The residents are unable to afford to make the improvements on their own.

(ii) An agreement outlining the installation, operation, and maintenance of facilities must be in place.

(iii) An adequate method for denying service in the event of non-payment of user fees if such fees are required.

(iv) All residents of the community are treated equally.

(v) The improvements provided are reasonable and modest.

(vi) Legal authority (*i.e.* easements) is obtained to construct these improvements.

(vii) Documentation must be provided to RUS indicating the quantity and quality of the individual installations that may be developed; cost effectiveness of the individual facility compared with initial and long term user costs on a central system; health and pollution problems attributable to individual facilities; operational or management problems peculiar to individual installations; and permit of regulatory agency requirements.

§1784.11 Restrictions.

Grant funds may *not* be used to:

(a) Pay any annual recurring costs that are considered to be operational expenses of a facility.

(b) Pay basic/rental fee or depreciation for the use of the recipient community's equipment.

(c) Purchase existing systems.

(d) Pay for items not associated with Rural Utilities Service's approved scope of work. This includes projects developed from other funding sources.

(e) Except as provided in this part, finance any public or commercial facility.

361

§§ 1784.12–1784.15 [Reserved]

Subpart C—Application Processing

§ 1784.16 General.

(a) DEC and ANTHC utilize the National Indian Health Service, Sanitation Deficiency System (SDS) database as a comprehensive source of rural sanitation needs in Alaska. The database provides an inventory of the sanitation deficiencies including water, sewer, and solid waste facilities for existing homes. The sanitation deficiencies data are updated annually by DEC and ANTHC in consultation with the respective rural or native villages. The SDS system is utilized in the RAVG program to help prioritize applications under the Village Safe Water Program.

(b) A prioritized list of projects will be developed each year by RUS, DEC, and ANTHC applying prioritization criteria to the sanitation needs database. Prioritization criteria established by the RUS, DEC, ANTHC, and IHS will be based, at a minimum, on relative health impacts, drinking water and wastewater regulatory requirements, the sanitation conditions in each community and project readiness. The VSW Program process and associated prioritization criteria will be used to prioritize projects and place them on a priority list. The process will be reviewed and approved by RUS, DEC, ANTHC, and IHS. Projects will be funded from the priority list as they meet established planning, design, and construction requirements, subject to available funding.

§ 1784.17 Application for Planning grants.

(a) Entities identified in § 1784.8 of this part may submit a completed Standard Form 424 to apply for funding to establish a Planning report for a rural or Native village.

(b) Funding for planning grants will be allocated annually by RUS, DEC, and ANTHC according to the prioritization list described in § 1784.16(b) of this part.

§ 1784.18 Application for Pre-development grants.

(a) Entities identified in § 1784.8 of this part may submit a completed Standard Form 424, Standard Form 424A, and Standard Form 424B to apply for funding for pre-development costs. Pre-development costs are described in § 1784.10 (a)(1)(iii) of this part.

(b) Funding for pre-development grants will be allocated annually by RUS, DEC, and ANTHC according to the prioritization list described in § 1784.16(b) of this part.

(c) Projects submitted for design only under the pre-development grant, must have RUS approval of a planning or pre-development report prior to consideration for funding.

§ 1784.19 Application for Construction grants.

(a) An application for a construction grant shall include:

(1) Completed Standard Form 424, Standard Form 424C and Standard Form 424D. Current versions of these forms may be found at Grants.gov.

(2) Preliminary Engineering Report, Environmental Report, or approved PER like document, including ANTHC's Cooperative Project Agreement and associated supplemental attachments;

(3) Population and median household income of the area to be served;

(4) Description of the project; and

(5) Approved business plan, including resolution adopting the plan, for the recipient community. The business plan will outline the proposed operation and management costs, rate structures, short-lived asset schedule and associated materials.

(6) Projects submitted for construction must have RUS and ANTHC or DEC approval of a planning or pre-development report prior to consideration for funding.

(b) Funding for construction grants will be allocated annually by RUS, DEC, and ANTHC according to the prioritization list described in § 1784.16(b) of this part.

§ 1784.20 Applications accepted from DEC or ANTHC.

(a) In cases where applications are accepted from DEC or ANTHC, one master application may be submitted covering all rural or native villages to be funded, however, each individual project will be broken out and (for construction grants) each will require its own PER, or PER-like document and Environmental Report.

(b) Each project will be processed individually with individual grant agreements, as appropriate.

(c) Expenditures for projects will be based on specific scope and be requested on a project by project basis.

(d) Funding amounts, as indicated in each grant agreement and letter of conditions, will be for the approved scope of work.

§1784.21 Other forms and certifications.

(a) Referenced bulletins, instructions and forms are for use in administering grants made under this part and are available from any USDA/Rural Development office or the Rural Utilities Service, U. S. Department of Agriculture, Washington, DC 20250–1500.

(b) Applicants will be required to submit the following items to the processing office, upon notification from the processing office to proceed with further development of the full application:

(1) Form RD 400–1, Equal Opportunity Agreement;

(2) Form RD 400–4, Assurance Agreement;

(3) Form AD 1047, Certification Regarding Debarment, Suspension and other Responsibility Matters;

(4) Form AD 1048, Certification regarding Debarment, Suspension, Ineligibility and Voluntary Exclusion—Lower Tier Covered Transactions;

(5) Form AD 1049, Certification regarding Drug-Free Workplace Requirements (Grants) Alternative I for Grantees Other Than Individuals;

(6) RUS Form 266, Compliance Assurance form or written self-certification statement—Civil Rights Compliance;

(7) Standard Form LLL, Disclosure of Lobbying Activities;

(8) RD Instruction 1940–Q, Exhibit A–1, Certifications for Contracts, Grants, and Loans (Regarding Lobbying); and

(9) Certification regarding prohibited tying arrangements. Applicants that provide electric service must provide the Agency a certification that they will not require users of a water or waste facility financed under this part to accept electric service as a condition of receiving assistance.

(c) In the case of grants made to DEC and ANTHC, DEC and ANTHC will certify that the above requirements are included in their agreements with the Villages. The certification and forms listed above must be provided from DEC and ANTHC on an annual basis for utilization in proposed applications.

(d) When favorable action is not taken on an application, the applicant will be notified in writing by the Rural Development State Program Official of the reasons why the request was not favorably considered. Notification to the applicant will state that a review of this decision by the Agency may be requested by the applicant in accordance with 7 CFR part 11.

(e) When favorable action is taken on an application, the applicant will be notified by a letter which establishes conditions that must be understood and agreed to before further consideration may be given to the application. In cases where a master application is submitted by DEC or ANTHC, the letter of conditions will include all projects, and their funding amounts, included in the master application on which favorable action will be taken. The letter of conditions does not constitute loan and/or grant approval, nor does it ensure that funds are or will be available for the project. The grant will be considered approved on the date a signed copy of Form RD 1940–1, Request for Obligation of Funds, is mailed to the applicant.

§1784.22 Other requirements.

Other Federal statutes and regulations are applicable to grants awarded under this part. These include but are not limited to:

(a) 7 CFR part 1, subpart A—USDA implementation of Freedom of Information Act.

(b) 7 CFR part 3—USDA implementation of OMB Circular No. A–129 regarding debt collection.

(c) 7 CFR part 15, subpart A—USDA implementation of Title VI of the Civil Rights Act of 1964, as amended.

(d) 7 CFR part 1970.

(e) 7 CFR part 1901, subpart E—Civil Rights Compliance Requirements.

(f) 2 CFR part 200—Uniform Guidance.

(g) 2 CFR part 215—General Program Administrative Requirements.

(h) 2 CFR part 418—New Restrictions on Lobbying, prohibiting the use of appropriated funds to influence Congress or a Federal agency in connection with the making of any Federal grant and other Federal contracting and financial transactions.

(i) 2 CFR parts 400 and 415—USDA implementation of Uniform Administrative Requirements for Grants and Agreements with Institutions of Higher Education, Hospitals, and Other Nonprofit Organizations.

(j) 2 CFR part 180, as adopted by USDA through 2 CFR 417, Governmentwide Debarment and Suspension (Nonprocurement); 2 CFR part 182, as adopted by USDA through 2 CFR 421, Government-wide Requirements for Drug-Free Workplace (Federal Assistance), implementing Executive Order 12549 on debarment and suspension and the Drug-Free Workplace Act of 1988 (41 U.S.C. 701).

(k) 2 CFR part 200, subpart F—USDA implementation of audit requirements for non-federal organizations.

(l) 29 U.S.C. 794, section 504—Rehabilitation Act of 1973, and 7 CFR part 15B (USDA implementation of statute), prohibiting discrimination based upon physical or mental handicap in federally assisted programs.

(m) Floodplains. The agencies follow the eight-step decision-making process referenced in Section 2(a) of Executive Order 11988, Floodplain Management, when undertaking actions located in floodplains. Pursuant to E. O. 11988, the IHS uses a Class Review process to exclude certain actions from further review under the eight-step process. For all actions that do not qualify for IHS Class Review, the eight-step process shall be completed. All practicable measures to minimize development in floodplains and reduce the risk to human safety, health, and welfare shall be followed, including elevating a new water or wastewater facility at least one foot above the base flood elevation as determined by the Army Corp of Engineers, other qualified survey, or best available data. Since they are considered "critical facilities" as defined by the Federal Emergency Management Agency (FEMA), water and wastewater facilities may be subject to more stringent standards such as relocation out of the floodplain, higher elevation, or other flood proofing measures. If an area has been designated a floodplain by FEMA Flood Insurance Rate Map (FIRM) coverage, flood insurance shall be required for facilities located in flood plains. If an area has no FEMA FIRM coverage the requirement to obtain flood insurance does not apply. If a community is located within a mapped FEMA Flood Insurance Rate Map (FIRM) 100-year floodplain, but is not a participating National Flood Insurance Program (NFIP) community member, then RUS may not fund the project according to 7 CFR 1806 Subpart B.

(n) Project planning, including engineering reports and environmental review documentation, to the maximum extent feasible, must address all water or waste disposal needs for a community in a coordinated manner with other community development projects and take into consideration information presented in available community strategic and comprehensive plans. Any reports or designs completed with funds must be consistent with sound engineering practices and USDA regulations, including 7 CFR part 1970.

[80 FR 52609, Sept. 1, 2015, as amended at 81 FR 11029, Mar. 2, 2016]

§ 1784.23 Lead Agency Environmental Review.

(a) The Agency designated as the lead agency for the purposes of this grant program, will fulfill and agree to be responsible for complying with lead agency requirements for:

(1) National Environmental Policy Act (NEPA) as outlined in 40 CFR 1501.5, Lead agencies;

(2) National Historic Preservation Act (NHPA) Section 106 review process as outlined in 36 CFR part 800.2(a)(2) Lead Federal agency; and

(3) Section 7 of the Endangered Species Act as outlined in 50 CFR 402.07, Designation of lead agency.

(b) All environmental findings and determinations made by the lead agency represent those of the cooperating agencies and will be completed in accordance with the procedures outlined in this section.

(c) RUS will, to the extent possible and in accordance with 40 CFR 1506.2 and 7 CFR part 1970, participate with DEC, IHS, and ANTHC to cooperatively or jointly prepare environmental review documents so that one document will comply with all applicable laws.

(d) For projects administered by DEC and ANTHC, RUS agrees to participate as a cooperating agency in accordance with 40 CFR 1501.6 and 7 CFR part 1970, and relies upon those agencies' procedures for implementing NEPA as further described below.

(e) The lead agency will indicate that RUS is a cooperating agency in all NEPA-related notices published for the proposed action.

(f) A construction grant may not be approved until all environmental findings and determinations have been made according to the following:

(1) *Rural Utilities Service Lead Agency.* If RUS is the lead agency, the environmental review process, including all findings and determinations, will be completed in accordance with 7 CFR part 1970.

(2) *DEC Lead Agency.* In the event DEC is the lead agency, the environmental review process, including all findings and determinations will be completed in accordance with the environmental review process outlined in Appendix A to the June 15, 2011 MOU.

(3) *IHS Lead Agency.* For projects administered by ANTHC, IHS will be the lead agency for the environmental review process, including all findings and determinations. The environmental review process, including all findings and determinations will be completed in accordance with the Department of Health and Human Services policies and procedures in General Administration Manual, Part 30, Council on Environmental Quality regulations at 40 CFR 1500–1508 and with procedures published by IHS in the FEDERAL REGISTER, Vol. 58, No.3, page 569, January 6, 1993. The ANTHC shall notify the funding agencies and the IHS if a change in the project or project scope occurs which could change any previously prepared environmental findings or determinations or could adversely impact the environment. In the event of an unanticipated discovery of a historic property or other environmental resource, the ANTHC shall stop construction activity in the area of the discovery and notify the appropriate authority and the IHS. Mitigation options resulting from unanticipated discoveries, including but not limited to changes in project scope or cancellation of the project will be evaluated by the funding agencies in collaboration with the ANTHC and IHS. If appropriate and necessary, mitigation plans will be negotiated and approved by all parties. When the funding agencies have approved a mitigation plan and IHS has reaffirmed its environmental review process, including all findings and determinations, the ANTHC will be authorized to initiate the agreed to mitigation plan. The IHS shall bear no mitigation costs as it is not a funding agency for projects under this part.

(g) RUS will have an opportunity to review the IHS or DEC environmental review documents, including all findings and determinations to ensure consistency with this part and agency procedures. Where an Environmental Assessment (EA) or Environmental Impact Statement (EIS) is required by the lead agency's environmental policies and procedures, the lead agency will ensure that the scope and content of the EA or EIS satisfies the statutory and regulatory requirements applicable to RUS. Where an EA and EIS is not required under the applicable lead agency's procedures for implementing NEPA, the review by RUS will be limited to ensure that the applicable lead agency's procedures were followed.

(h) The National Historic Preservation Act Section 106 review requirements completed for ANTHC administered projects will be carried out in accordance with the process described in Appendix B of the June 15, 2011 MOU.

[80 FR 52609, Sept. 1, 2015, as amended at 81 FR 11029, Mar. 2, 2016]

§§ 1784.24–1784.25　[Reserved]

Subpart D—Grant Processing

§ 1784.26　Planning, development, and procurement.

(a) If RUS is the lead agency and will provide oversight for the project, a certification should be obtained from the State agency, or the Environmental Protection Agency if the State does not have primacy, stating that the proposed improvements will be in compliance with requirements of the Safe Drinking Water Act and/or Clean Water Act and the applicable requirements of 2 CFR part 200 and 2 CFR part 400.

(b) Applicants that will bid and construct a project in phases, must provide assurance that the full scope of each specific phase of the project will be functional. In the event that the actual cost is anticipated to exceed the funding originally allocated for the project, all potential options will be reviewed and considered, including but not limited to acquiring additional funds or a reduction in project scope. RUS, ANTHC, and VSW will ensure that all items that were funded and within the scope of the project, including all phases, are functional when all funds have been disbursed.

§ 1784.27　Grant closing and disbursement of Funds.

(a) The Water and Waste Grant Agreement for rural and native villages in Alaska, or other approved form(s) will be executed by all applicants. To view all forms and agreements, refer to the USDA RUS Water and Environmental Programs Web site.

(b) Grant funds will be distributed from the Treasury at the time they are actually needed by the applicant using multiple advances. Instructions regarding disbursement of funds can be found in the Letter of Conditions.

(c) If there is a significant reduction in project costs, the applicant's funding needs will be reassessed. Decreases in RUS funds will be based on revised project costs and current number of users. Other factors, including RUS regulations used at the time of grant approval, will continue to be used as published at the time of grant approval. Obligated grant funds not needed to complete the proposed project will be deobligated. In such cases applicable forms, the letter of conditions, and other items will be revised.

§ 1784.28　Grantee accounting methods, management reporting, and audits.

(a) All Agency grantees will follow the reporting requirements as outlined in 7 CFR 1782.

(b) Other reporting requirements are as follows:

(1) During the construction period, for the reporting of expenses incurred for projects under this part, the party responsible for the administration of the project will complete an audit report in accordance with § 1782.10 (which includes GAGAS and 2 CFR part 200 Subpart F "Audit Requirements"). RUS may request a copy of this report.

(2) After the construction period and for the life of the facility, the recipient community will be responsible to meet the requirements outlined in 2 CFR parts 200, 400, 415, 416, and 7 CFR part 1780.47 paragraphs a through d. These requirements must be outlined in funding documents from RUS, ANTHC, and VSW and in agreements with the recipient communities. RUS may request this information for the life of the facility.

(c) The requirements found in 2 CFR parts 200, 400, 415 and 416 shall apply to all grants made under the RAVG program and shall be set forth in the respective grant agreement where required.

§ 1784.29　Grant servicing and accountability.

(a) Grants will be serviced in accordance with 7 CFR part 1782.

(b) RUS reserves the right to request and review project files from grantees at any time.

(c) If at any time an application is determined ineligible, 7 CFR part 11 will be followed.

§ 1784.30　Subsequent grants.

Subsequent grants will be processed in accordance with the requirements set forth in this part. The initial and subsequent grants made to complete a previously approved project must comply with the maximum grant requirements set forth in § 1784.8(f) of this part.

§ 1784.31 Exception authority.

The Administrator may, in individual cases, make an exception to any requirement or provision of this part which is not inconsistent with the authorizing statute or other applicable law and is determined to be in the Government's best interest.

§§ 1784.32–1784.34 [Reserved]

Subpart E—Design, Procurement, Construction, and Inspection

§ 1784.35 General.

This subpart is specifically designed for use by owners including the professional or technical consultants or agents who provide assistance and services such as engineering, environmental, inspection, financial, legal or other services related to planning, designing, bidding, contracting, and constructing water and waste disposal facilities. The selection of engineers for a project design shall be done by a request for proposals by the applicant. These procedures do not relieve the owner of the contractual obligations that arise from the procurement of these services. For this subpart, an owner is defined as the grant recipient.

§ 1784.36 Procurement by applicants eligible under this part

(a) For applicants eligible under § 1784.8(a)(2) and (3), contracting and procurement activities will follow DEC or ANTHC policies, procedures and methods which are based on and shall follow Uniform Administrative Requirements, Cost Principles, and Audit Requirements for Federal Awards (2 CFR part 200). In specifying materials, DEC and ANTHC will consider all materials normally suitable for the project based on sound engineering practices and project requirements.

(b) Contracts for procurement must contain applicable contract provisions listed at Appendix II to 2 CFR part 200.

(c) For grants proposed to be administered directly by applicants eligible under § 1784.8(a)(1), the requirements outlined in 7 CFR part 1780, subpart C will be met by those eligible applicants with the exception of the following requirements:

(1) Preliminary engineering reports and Environmental Reports (§ 1780.55). Refer to the requirements of this subpart and subpart C § 1784.22(n).

(2) Metering devices in § 1780.57(m).

(3) Utility Purchase Contracts in § 1780.62.

(4) Sewage treatment and bulk water sales contracts in § 1780.63.

§ 1784.37 Procurement of recovered materials.

When a grant is made to the DEC, the state and its contractors must comply with section 6002 of the Solid Waste Disposal Act, as amended by the Resource Conservation and Recovery Act. The requirements of Section 6002 include procuring only items designated in guidelines of the Environmental Protection Agency (EPA) at 40 CFR part 247 that contain the highest percentage of recovered materials practicable, consistent with maintaining a satisfactory level of competition, where the purchase price of the item exceeds $10,000 or the value of the quantity acquired during the preceding fiscal year exceeded $10,000; procuring solid waste management services in a manner that maximizes energy and resource recovery; and establishing an affirmative procurement program for procurement of recovered materials identified in the EPA guidelines.

§§ 1784.38–1784.99 [Reserved]

PART 1785—LOAN ACCOUNT COMPUTATIONS, PROCEDURES AND POLICIES FOR ELECTRIC AND TELEPHONE BORROWERS

Subpart A [Reserved]

Subpart B—RUS Cushion of Credit Account Computations and Procedures

AUTHORITY: 7 U.S.C. 901 *et seq.*; Title I, Subtitle D, sec. 1403, Omnibus Budget Reconciliation Act of 1987, Pub. L. 100–203; Pub. L. 103–354, 108 Stat. 3178 (7 U.S.C. 6941 *et seq.*).

Subpart A [Reserved]

Subpart B—RUS Cushion of Credit Account Computations and Procedures

SOURCE: 54 FR 13669, Apr. 5, 1989, unless otherwise noted.

§ 1785.66 General.

This subpart B sets forth policies and procedures on the RUS cushion of credit payments program. The cushion of credit payments program will be maintained only for insured loans evidenced by obligations of the Fund. A subaccount within the Fund is hereby established for purposes of promoting rural economic development. It shall be known as the "Rural Economic Development Subaccount." The assets of the subaccount shall be obtained from crediting (on a monthly basis) a sum determined by multiplying the outstanding cushion of credit payments made after October 1, 1987, by the difference (converted on a monthly basis) between the average weighted interest rate paid on outstanding certificates of beneficial ownership issued by the Fund and the 5 percent rate of interest provided to borrowers on cushion of credit payments, repayment of loans made pursuant to Section 313 of the Act, and other sources as provided by law. This subaccount shall be used to provide grants or zero interest loans to borrowers under the Act for the purpose of promoting rural economic development.

§ 1785.67 Definitions.

Accumulated (deferred) interest means interest allowed to accumulate up to, and including, the basis date of RUS notes covering loans approved before June 5, 1957. The accumulated interest is payable in equal periodic installments over the remaining life of the notes.

Act means the Rural Electrification Act of 1936, as amended (7 U.S.C. 901 *et seq.*).

Advance payment means a voluntary unscheduled payment on an RUS note, made prior to October 2, 1987, credited to the advance payment account of a borrower.

Cushion of Credit Payment means a voluntary unscheduled payment on an RUS note made after October 1, 1987, credited to the cushion of credit account of a borrower.

Current interest means interest payable periodically as it accrues.

Fund means the Rural Electrification and Telephone Revolving Fund established pursuant to the Act.

Interest credit means interest earned on balances in advance payment or cushion of credit accounts. Since the periodic installments are established by the terms of the notes, the interest credits cannot serve to change the total amount of each installment; therefore, an amount equal to the interest credits is added to the principal installment due. On receipt of the full installments, amounts equal to the interest credits (the principal offsets) are added to the respective advance payment accounts.

Prepayment means a voluntary unscheduled payment which the borrower instructs RUS to apply directly and immediately to the principal of an RUS note.

RUS notes means those notes, bonds, or other obligations evidencing indebtedness created by loans made by RUS pursuant to titles I, II, or III of the Act.

Subaccount means the *Rural Economic Development Subaccount* established pursuant to the Act as part of the Fund.

§ 1785.68 Establishing an RUS cushion of credit payment account.

A cushion of credit account shall be automatically established by RUS for each borrower who makes a payment after October 1, 1987, in excess of amounts then due on an RUS note. Such account will bear interest at a rate of 5 percent per annum. All payments on RUS notes which are in excess of required payments and not otherwise designated shall be deposited in the borrowers' respective cushion of credit accounts. Payments received in the month in which an installment is due will be applied to the installment due. However, if the regular installment payment is received at a later date in the month, the first payment received will be applied retroactively to a cushion of credit account and the second will be applied to the installment due.

§ 1785.69 Cushion of credit payment account computations.

(a) *Deposits.* Cushion of credit payments are credited to the borrowers' cushion of credit accounts.

(b) *Interest.* Interest at the rate of 5 percent per annum shall be credited on a quarterly basis to cushion of credit accounts. Interest earned will appear as a reduction in the interest billed on the borrower's RUS notes and will be separately shown on RUS Form 694, "Statement of Interest and Principal Due."

§ 1785.70 Application of RETRF cushion of credit payments.

(a) If a maturing installment on an RUS note or a note which has been guaranteed by RUS is not received by its due date, funds will be withdrawn from the borrower's cushion of credit account and applied as of the installment due date beginning with the oldest of such notes as follows: first, to current interest then due on all notes; second, to the accumulated interest due, if any, on all notes; and third, to the principal then due on all notes. In those instances where a borrower has prior to October 2, 1987, maintained an advance payment account with RUS, its cushion of credit account will be applied in accordance with the provisions of this section prior to using any balance remaining in its advance payment account to pay interest and principal installments on notes. Computations required under this section have been made by RUS as of October 2, 1987; however, on or before May 25, 1989 any borrower may make a one time irrevocable election to have all such computations made as of April 5, 1989, by filing written notice to that effect with Robert D. Ruddy, Director, Fiscal Accounting Division, Rural Utilities Service, Washington, DC 20250–1500.

(b) A borrower may reduce the balance of its cushion of credit account

only if the amount obtained from the reduction is used to make scheduled payments on loans made or guaranteed under the Act.

[54 FR 13669, Apr. 5, 1989; 54 FR 17703, Apr. 25, 1989]

PART 1786—PREPAYMENT OF RUS GUARANTEED AND INSURED LOANS TO ELECTRIC AND TELEPHONE BORROWERS

Subpart A—General [Reserved]

1786.1–1786.24 [Reserved]

Subpart B—Prepayment of RUS Guaranteed Federal Financing Bank Loans Pursuant to Section 306(A) of the RE Act

Subpart C—Special Discounted Prepayments on RUS Direct/Insured Loans

AUTHORITY: 7 U.S.C. 901–950b; Title I, subtitle B, Pub. L. 99–509; Pub. L. 101–624, 104 Stat. 4051; Pub. L. 103–354, 108 Stat. 3178, (7 U.S.C. 6941 et seq.), unless otherwise noted.

SOURCE: 55 FR 1145, Jan. 11, 1990, unless otherwise noted.

Subpart A—General [Reserved]

§§ 1786.1–1786.24　[Reserved]

Subpart B—Prepayment of RUS Guaranteed Federal Financing Bank Loans Pursuant to Section 306(A) of the RE Act

AUTHORITY: 7 U.S.C. 901–950b; Title I, Subtitle B, Pub. L. 99–509; Title I, Pub. L. 100–202; Pub. L. 100–203; Title VI, Pub. L. 100–460; Pub. L. 103–354, 108 Stat. 3178 (7 U.S.C. 6941 et seq.).

SOURCE: 55 FR 1145, Jan. 11, 1990, unless otherwise noted. Redesignated at 55 FR 49250, Nov. 27, 1990.

§ 1786.25　Purpose.

This subpart contains the general regulations of the Rural Utilities Service (RUS) for implementing the provisions of (a) section 306(A) of the Rural Electrification Act of 1936, as amended (RE Act); (b) section 633 of the Rural Development, Agriculture, and Related Agencies Appropriations Act, 1988 (Pub. L. 100–202) (the continuing resolution); and (c) section 637 of the Rural Development, Agriculture, and Related Agencies Appropriations Act, 1989 (Pub. L. 100–460) (the 1989 Appropriations Act) which permit, in certain circumstances, loans made by the Federal Financing Bank (FFB) and guaranteed by the Administrator of RUS to be prepaid by RUS electric and telephone borrowers by paying the outstanding principal balance due on the FFB loan, using a private loan with the existing RUS guarantees or using internally generated funds.

§ 1786.26　Policy.

It is the policy of RUS to facilitate the prepayment of FFB loans in accordance with the provisions of section

306(A) of the RE Act and section 633 of the continuing resolution as modified by section 637 of the 1989 Appropriations Act. Furthermore, consistent with the RE Act, the continuing resolution and the 1989 Appropriations Act, it is the policy of RUS to implement the objectives of the prepayment program in a manner which does not result in an increase in loan guarantee risk or an inappropriate increase in the administrative burden on RUS.

§ 1786.27 Definitions and rules of construction.

(a) *Definitions.* For the purposes of this subpart, the following terms shall have the following meanings:

Administrator means the Administrator of RUS.

Application Category shall have the meaning set forth in § 1786.29(c).

Application period means a period during which RUS is accepting applications to make prepayments pursuant to this subpart, and initially means:

(1) In the case of telephone borrowers, the period commencing on February 12, 1990 and ending on March 12, 1990;

(2) In the case of financially distressed borrowers, the period commencing October 1, 1990 and ending on July 30, 1993; or

(3) In the case of other borrowers, the period to be announced by RUS.

Borrower means any organization which has an outstanding FFB loan guaranteed by RUS under the RE Act.

Business Day means any day other than a Saturday, a Sunday, a legal public holiday under 5 U.S.C. section 6103 for the purposes of statutes relating to pay and leave of employees, or any other day declared to be legal holiday for the purposes of statutes relating to pay and leave of employees by Federal statute or Federal Executive Order.

Continuing Resolution means section 633 of the Rural Development, Agriculture, and Related Agencies Appropriations Act, 1988 (Pub. L. 100–202).

Date Received means the date inscribed on the Notice of Intent to Prepay the Federal Financing Bank, by an authorized official of RUS, as the date the application was received.

Documentation means all or part of the agreements relating to a prepayment under this part, irrespective of whether RUS is a party to each agreement, including all exhibits to such agreements.

Electric Program Applications shall have the meaning specified in § 1786.29(c)(1).

Existing Loan Guarantee means a guarantee of payment issued by RUS to FFB pursuant to the RE Act for an FFB loan made on or before July 2, 1986.

Fees means any fees, costs or charges, incurred in connection with obtaining the private loan used to make the prepayment including without limitation, accounting fees, filing fees, legal fees (including fees and disbursements charged by counsel representing the borrower), printing costs, recording fees, trustee fees, underwriting fees, capital stock purchases or other equity investment requirements of the lender, and other related transaction expenses.

Financially Distressed Borrower means an RUS-financed electric system determined by the Administrator to be either (1) in default or near default on interest or principal payments due on loans made or guaranteed under the RE Act, and is making a good faith effort to increase rates and reduce costs to avoid or mitigate default; or (2) participating in a work out or debt restructuring plan with RUS, either as the borrower being restructured or as a borrower providing assistance as part of the work out or restructuring.

Financially Viable Lender means:

(1) A lender (i) which has a capital and surplus of at least $50 million; (ii) is a beneficiary of an irrevocable letter of credit, in form and substance satisfactory to the Administrator, payable to it in the amount of $50 million; (iii) is the beneficiary of a guarantee, in form and substance satisfactory to the Administrator, in the amount of $50 million from a lending institution with a capital and surplus of at least $50 million; or (iv) has other credit support, in form and substance satisfactory to the Administrator, in the amount of $50 million; or

(2) In the event of a prepayment totalling less than $100 million, a lender (i) which has a capital and surplus of at

least $10 million; (ii) is a beneficiary of an irrevocable letter of credit, in form and substance satisfactory to the Administrator, payable to it in the amount of $10 million; (iii) is the beneficiary of a guarantee, in form and substance satisfactory to the Administrator, in the amount of $10 million from a lending institution with a capital and surplus of at least $10 million; or (iv) has other credit support, in form and substance satisfactory to the Administrator, in the amount of $10 million;

FFB means the Federal Financing Bank, an instrumentality and wholly owned corporation of the United States.

FFB Loan means one or more advances, or a part of one or more advances, made on or before July 2, 1986, by FFB on a promissory note or notes executed by a borrower and guaranteed by RUS pursuant to section 306 of the RE Act (7 U.S.C. 936).

Guarantee means the original endorsement, in the form specified by RUS which is executed by the Administrator and shall be an obligation supported by the full faith and credit of the United States and incontestable except for fraud or misrepresentation of which the holder had actual knowledge at the time it became a holder.

Increase in Loan Guarantee Risk means the change in any of the components of loan guarantee risk associated with the private loan which in the judgment of RUS increases the magnitude or duration of the loan guarantee risk currently assumed by RUS in connection with the existing loan guarantee;

Internally Generated Funds means money belonging to the borrower other than: (1) Proceeds of loans made or guaranteed under the RE Act or (2) funds on deposit in the cash construction trustee account;

Lender means the organization making and servicing the private loan which is to be guaranteed under the provisions of this subpart and used to prepay the FFB loan. The term *lender* does not include the FFB, or any other Government agency.

Loan Guarantee Agreement means the written contract by and among the lender, the borrower, the Adminis-

trator, and such other parties that RUS may require, setting forth the terms and conditions of a guarantee issued pursuant to the provisions of this subpart.

Loan Guarantee Risk means the risk as determined by RUS associated with guaranteeing a loan for a particular borrower. Components of loan guarantee risk include the following:

(1) The outstanding principal balance of a loan;

(2) The dollar weighted average interest rate (stated as an annual percentage rate) on a loan;

(3) The final maturity date of a loan;

(4) The annual principal amortization of the loan; and

(5) Any other factor that as determined by RUS increases the magnitude or duration of the guarantee.

Mortgage means the mortgage and security agreements by and among the borrower and RUS, as from time to time supplemented, amended and restated.

1989 Appropriations Act means the Rural Development, Agriculture, and Related Agencies Appropriations Act, 1989 (Pub. L. 100–460).

Notice of Intent to Prepay the Federal Financing Bank means the notice in the form specified in § 1786.33 hereof.

Prepayment Authority shall have the meaning specified in § 1786.29(a).

Private Loan means a loan or loans to be guaranteed under the provisions of this part and used to prepay an FFB loan.

Pro-rated Percentage shall have the meaning specified in § 1786.30(b)(1).

RE Act means the Rural Electrification Act of 1936 (7 U.S.C. 901–950b), as amended.

REA means the Rural Electrification Administration formerly an agency of the United States Department of Agriculture and predecessor agency to RUS with respect to administering certain electric and telephone loan programs.

RUS means the Rural Utilities Service, an agency of the United States Department of Agriculture established pursuant to Section 232 of the Federal Crop Insurance Reform and Department of Agriculture Reorganization Act of 1994 (Pub. L. 103–354, 108 Stat. 3178), successor to REA with respect to administering certain electric and telephone programs. See 7 CFR 1700.1.

Service or *Servicing* means the following activities:

(1) The billing and collecting of the private loan payments from the borrower;

(2) Notifying the Administrator promptly of any default in the payment of principal and interest on the private loan and submitting a report, as soon as possible thereafter, setting forth the servicer's views as to the reasons for the default, how long the servicer expects the borrower to be in default, and what corrective actions the borrower states it is taking to achieve a current debt service position;

(3) Notifying the Administrator of any known violations or defaults by the borrower under the lending agreement, loan guarantee agreement, the mortgage, or related security instruments, or conditions of which the servicer or the lender is aware which might lead to nonpayment, violation or other default; and

(4) Such other activities as may be specified in the loan guarantee agreement.

Settlement Date means the date the borrower disburses funds to the FFB in order to complete a prepayment pursuant to this subpart, and shall be a date agreed to by RUS, and a date on which both the FFB and the Federal Reserve Bank of New York are open for business.

Standard Electric Program Application shall have the meaning specified in §1786.29(c)(1).

Telephone Borrower means a borrower that provides telephone service as defined in 7 CFR 1735.2(a).

Telephone Program Applications shall have the meaning specified in §1786.29(c)(2).

(b) *Rules of Construction.* Unless the context shall otherwise indicate, the terms defined in §1786.27(a) hereof include the plural as well as the singular, and the singular as well as the plural. The words "herein," "hereof" and "hereunder", and words of similar import, refer to this subpart as a whole.

[55 FR 1145, Jan. 11, 1990, as amended at 55 FR 35426, Aug. 30, 1990. Redesignated at 55 FR 49250, Nov. 27, 1990, as amended at 59 FR 66440, Dec. 27, 1994]

§1786.28 Qualifications.

(a) *Borrowers.* To qualify to prepay an FFB loan pursuant to this subpart, the borrower must:

(1) Demonstrate that the FFB loan was outstanding on July 2, 1986;

(2) Prepay the FFB loan by:

(i) Using a private loan with the existing loan guarantee;

(ii) Using internally generated funds; or

(iii) Using a combination of a private loan with the existing loan guarantee and internally generated funds; and

(3) Certify that any savings resulting from such prepayment will be passed on to its customers, or used to improve the financial strength of the borrower in cases of financial hardship.

(b) *Lenders.* To participate pursuant to this subpart, in a borrower's prepayment of an FFB loan by means of a private loan, the lender must:

(1) Be a private legally organized lender, or a lender established pursuant to the Farm Credit Act of 1971, as amended;

(2)(i) Be subject to credit examination and supervision by either an agency of the United States or a state and be in good standing with its licensing authority and have met the requirements, if any, of licensing, lending and loan servicing in the state where the collateral for the Loan is located;

(ii) Be a financially viable lender; or

(iii) Be a trust administered by an entity meeting the requirements of paragraph (b)(2) (i) or (ii) of this section; and

(3) Have the capability to adequately service the private loan either by using its own resources or by contracting for such resources with a financially viable lender. Under no circumstances may the borrower or an affiliate of the borrower service the private loan. A qualified lender may participate out each private loan to entities other than a Government agency, the borrower, or an affiliate of the borrower, provided that such participation shall be on terms and conditions satisfactory to the Administrator.

(c) *Private Loans.* A borrower who qualifies pursuant to §1786.28(a) may at its option elect to use a private loan to make a prepayment, or a portion of a prepayment, pursuant to this subpart.

Private loans, the proceeds of which are used exclusively to prepay FFB loans, shall be eligible for a guarantee under this subpart. The Administrator shall endorse a guarantee on each note evidencing a qualifying private loan. The private loan shall be structured in a manner which in the judgment of RUS shall not result in an increase in loan guarantee risk and shall comply with the following:

(1) The private loan shall provide for the periodic payment of interest by the borrower not less frequently than annually, at either a variable or fixed rate in a manner which shall not result in an increase in loan guarantee risk. (i.e. The dollar weighted average interest rate on the private loan shall be less than or equal to the dollar weighted average interest rate on the FFB loan being prepaid, so that:

$$C_r = C_o + \frac{\sum_{i=1}^{n}(C_o - A_i)T_i}{(J - n)}$$

Where,

C_r = The revised interest rate cap;
C_o = The original interest rate cap at the time of prepayment;
A_i = The average interest rate actually charged in the ith period;
T_i = Length of the ith period expressed in years;
n = The number of years that have elapsed since the initial prepayment;
J = The initial term of the private loan, at the time of prepayment;
Subject to the constraint that A_i must be less or equal to C_o).

(2) Principal payments on the private loan shall be made either quarterly, semiannually, or annually and shall commence on or before the last day of the calendar year during which the prepayment pursuant to this subpart was made.

(3) With the approval of the Administrator, the lender may refund the private loan with the proceeds of another loan from the same lender, with the existing guarantee and under terms, conditions, and a structure substantially similar to the private loan, on such dates as the lender, the borrower and RUS may agree, provided however, that such a refunding loan shall comply with the provisions of § 1786.28(c) hereof. Additionally, with the approval of the Administrator, the private loan may be prepaid either in whole or in part at any time by the borrower using its general funds.

(4) The private loan and the guaranteed note evidencing the private loan shall not be directly or indirectly part of a transaction the income of which is excluded from gross income for the purposes of Chapter I of the Internal Revenue Code of 1986.

(5) The guaranteed note evidencing the private loan shall not be transferable or assignable except

(i) With the written approval of the Administrator;

(ii) In the event that the guaranteed note evidencing the private loan is held by a trust, to a similar trust, in connection with a refunding loan made by the lender pursuant to § 1786.28(c)(3); or

(iii) As an undivided pro rata interest in a pool of obligations.

(6) The loan documentation shall provide RUS with the right to accelerate the note evidencing the private loan upon the occurrence of any "Event of Default" under the mortgage with the effect that all of the unpaid principal and interest on any such note shall become immediately due and payable to RUS, and RUS shall continue to pay under its guarantee the principal of and interest on such note without taking into account such acceleration. The loan documentation shall also provide RUS with a right, upon the occurrence of such an "Event of Default," to accelerate payment on its guarantee and accelerate payment on the note evidencing the private loan on the earlier of any date the interest rate on the private loan is reset, without premium or penalty; any date the borrower may

374

prepay in accordance with the terms of the private loan, or the tenth anniversary of the date the private loan first bears interest at a fixed interest rate.

(7) The principal of the private loan shall not include amounts attributable to fees associated with the private loan. At the time it submits its application, a borrower may request that the Administrator approve the inclusion of amounts attributable to fees as part of the interest rate on the private loan, if the net effective interest rate including such fees meets the test contained in §1786.28(c)(1). For the purposes of these regulations, such financed fees shall be considered "interest".

(8) Private loans and guaranteed notes evidencing private loans shall otherwise be in form and substance satisfactory to the Administrator.

(d) *Prepayments Without a Guarantee.* Qualifying borrowers may elect to utilize internally generated funds without a guarantee to prepay an FFB loan, or partially prepay an FFB loan, pursuant to this subpart, if

(1) The borrower notifies RUS, of its intent to prepay using internally generated funds in accordance with the application procedures set forth in this subpart; and

(2) The borrower submits a certification to RUS that the prepayment does not, materially adversely affect the financial stability of the borrower and its ability to meet all its obligations, including debt service on all loans made, guaranteed or lien accommodated under the RE Act which will remain outstanding after the date of the prepayment.

(e) *The Use of both a Private Loan and Internally Generated Funds.* Qualifying borrowers may elect to utilize a combination of private loans and internally generated funds without a guarantee, to prepay an FFB loan pursuant to this subpart, if

(1) The private loans comply with the provisions of paragraph (c) of this section, and

(2) The borrower complies with paragraph (d) of this section.

(f) *FFB loans.* A borrower's FFB loans that qualify to be prepaid pursuant to this subpart are:

(1) *Qualifying Borrowers.* In the case of qualifying borrowers other than financially distressed borrowers, FFB advances with long-term maturity dates may be prepaid pursuant to this subpart; and

(2) *Financially distressed borrowers.* FFB loans that are eligible to be prepaid by utilizing the financially distressed borrowers' reserve are advances with long-term maturity dates, and which in the opinion of the Administrator, if prepaid, would result in an economic savings to the financially distressed borrower.

[55 FR 1145, Jan. 11, 1990, as amended at 55 FR 35426, Aug. 30, 1990. Redesignated at 55 FR 49250, Nov. 27, 1990]

§ 1786.29 Prepayment authority, program allocations, categories of prepayment applications and financially distressed borrowers' reserve.

(a) *Prepayment Authority.* So long as the aggregate amount of prepayments made after December 22, 1987, including prepayments made pursuant to §1786.28(d) and §1786.28(e), under section 306(A) of the RE Act, does not exceed $2.5 billion, the approval of the Secretary of the Treasury is not required in order to make a prepayment pursuant to this subpart (such amount of prepayments is hereinafter called prepayment authority).

(b) *Program Allocations.* In accordance with the provisions of section 637 of the 1989 Appropriations Act, $350 million of prepayment authority is allocated to RUS-financed electric systems and $150 million of prepayment authority is allocated to RUS-financed telephone utilities. The amounts of prepayment authority allocated to electric program borrowers and telephone program borrowers shall not be transferred between programs. Borrowers may not sell, assign, or otherwise transfer prepayment authority to another borrower.

(c) *Categories of Prepayment Applications.* Applications received by RUS from borrowers desiring to prepay pursuant to this subpart will be separated into the following two application categories:

(1) *Electric Program Applications.* Electric program applications are applications to make a prepayment pursuant to this subpart from RUS-financed

electric utilities, that qualify in accordance with § 1786.28(a) hereof and which are received by RUS during the application period. Electric program applications will be further subdivided and classified as being either (i) a financially distressed borrower's application, or (ii) a standard electric program application. Applications received from borrowers determined by the Administrator not to be a financially distressed borrower will be classified and processed as a standard electric program application;

(2) *Telephone Program Applications.* Telephone program applications are applications to make a prepayment pursuant to this subpart from RUS-financed telephone utilities that qualify in accordance with § 1786.28(a) hereof and which are received by RUS during the application period;

(d) *Financially distressed borrowers' reserve.* The $350 million of prepayment authority allocated for RUS-financed electric utilities, is initially set aside into a financially distressed borrowers' reserve. This reserve of prepayment authority will be available for prepayments pursuant to this subpart by financially distressed borrowers who apply to make such a prepayment during the application period. In the event that a portion of financially distressed borrowers' reserve remains unsubscribed at the end of the initial application period, the unallocated portion of the financially distressed borrowers' reserve will be allocated to other electric borrowers having submitted applications during an application period to be announced by RUS. Such prepayment applications shall be classified as standard electric program applications.

[55 FR 1145, Jan. 11, 1990, as amended at 55 FR 35427, Aug. 30, 1990. Redesignated at 55 FR 49250, Nov. 27, 1990]

§ 1786.30 Processing procedure.

(a) *Priority of Processing.* The determination of the order or method in which applications or portions of applications will be processed by RUS pursuant to this subpart rests solely within the discretion of the Administrator. RUS expects that a number of prepayment applications will be processed simultaneously. In the event that it becomes necessary to establish priorities of processing, prepayment applications will be processed without regard to the date received, generally in the following order of priority:

(1) Applications from telephone borrowers;

(2) Applications from financially distressed borrowers;

(3) Applications from all other borrowers. When assigning priority to such applications, RUS will consider a number of factors, including without limitation, (i) the number of prepayment applications being processed by the area office; (ii) the novelty or complexity of the proposed transaction; (iii) the method of prepayment; and (iv) the availability of resources. In the event that RUS receives during the initial application period, prepayment applications from such borrowers in an amount less than remaining prepayment authority for each respective program, RUS will establish a new application period and publish a notice to that effect in the FEDERAL REGISTER.

(b) *Pro-rated Applications.* Standard electric program applications, and telephone program applications will be prorated within their respective application categories to permit partial prepayments in the event that the aggregate amount of prepayment applications received during the application period exceeds the amount of prepayment authority allocated to that application category. In such circumstances, the amount of each borrower's permitted prepayment shall be determined within each respective application category, as follows:

(1) The principal amount of FFB advances under each individual application, which, if prepaid pursuant to this subpart, would result in an economic savings to the borrower, shall be divided by the aggregate principal amount of FFB advances, under all of the applications, which, if prepaid pursuant to this subpart, would result in an economic savings to the borrowers, in order to determine a percentage (hereinafter called a pro-rated percentage) for each borrower;

(2) Each borrower's share of the prepayment authority for its application category shall be equal to the product of (i) the prepayment authority times

(ii) the respective pro-rated percentage, and may be used to prepay a portion of any of the borrower's FFB loans listed pursuant to §1786.31(a)(2);

(3) If any approved prepayment transaction fails to be settled within 180 days of the date the borrower is notified by RUS of its prepayment allocation, RUS may rescind its approval. The unused prepayment authority represented by such a failed transaction is subject to being included in any subsequent notice of a new application period under this subpart; and

(4) In the event that applications from financially distressed borrowers exceed the amount prepayment authority remaining in the financially distressed borrowers' reserve, the Administrator at his discretion shall select one or more of such applications and allocate the reserve. In making such a selection and allocation, the Administrator may consider various factors, including without limitation, (i) the dollar amount of savings to be realized by the proposed prepayment; (ii) the interest rates on the FFB loans proposed to be prepaid; (iii) the magnitude of the default or potential default; and (iv) whether the borrower has previously completed a prepayment under section 306(A).

(c) *Notification of Borrowers' Allocations.* Promptly after allocating the prepayment authority to borrowers and completing any proration calculations that may be necessary, RUS will return to each borrower submitting a prepayment application pursuant to this subpart, a copy of their Notice of Intent to Prepay the Federal Financing Bank specifying the amount of the borrower's prepayment allocation.

[55 FR 1145, Jan. 11, 1990, as amended at 55 FR 49250, Nov. 27, 1990]

§1786.31 **Application procedure.**

Applications to make a prepayment pursuant to this subpart shall be submitted to RUS on such forms as RUS may prescribe in the following manner:

(a) *Application.* Each borrower desiring to make a prepayment pursuant to this subpart shall submit an application to RUS. No application from a borrower will be accepted by RUS prior to the commencement of the application period. An application shall not be deemed submitted to RUS until it is received by RUS, and the "Date Received" has been inscribed on the Notice of Intent to Prepay the Federal Financing Bank by an authorized official of RUS. Incomplete applications may be returned to the borrower at the discretion of RUS and thereafter must be resubmitted in order to be processed. To be considered complete, the application should include the following:

(1) "Notice of Intent to Prepay the Federal Financing Bank" in the form specified in §1786.33 hereof;

(2) A listing of each FFB loan advance to be prepaid by loan designation, RUS note number, RUS account number, advance date, maturity date, original amount, outstanding balance, and interest rate;

(3) Evidence that the borrower meets the qualification provisions of §1786.28(a) of these regulations;

(4) The certification set forth in part A of the Notice of Intent to Prepay the Federal Financing Bank executed by the chief executive officer of the borrower;

(5) In the event that a borrower submits a prepayment application which proposes to utilize a portion of the financially distressed borrowers' reserve, a certification signed by the chief executive officer of the system to the effect that the borrower is either (i) in default or near default on interest or principal payments due on loans made or guaranteed under the RE Act, and is making a good faith effort to increase rates and reduce costs to avoid or mitigate default; or (ii) participating in a work out or debt restructuring plan with RUS, either as the borrower being restructured or as a borrower providing assistance as part of the work out or restructuring and stating why the borrower is in default or near default.

(b) *Election of Method of Prepayment.* Prior to requesting RUS to schedule a settlement date, the borrower shall (1) elect whether it will use a private loan, internally generated funds, or a combination of a private loan and internally generated funds to make the prepayment, by completing part C of its Notice of Intent to Prepay the Federal Financing Bank; (2) specify in part C of the Notice of Intent to prepay the Federal Financing Bank a date after which a prepayment closing may be scheduled; (3) if appropriate, execute the certification set forth in part C of the Notice of Intent to Prepay the Federal Financing Bank; and (4) return a completed copy of the Notice of Intent to Prepay the Federal Financing Bank to the RUS area office.

(c) *Final Documentation.* All documentation in connection with a proposed prepayment made pursuant to this subpart shall have been submitted to RUS in final form, no later than 5 business days prior to the settlement date agreed to by the borrower and RUS. To be considered complete, the final documentation shall include the following material:

(1) A completed copy of the Notice of Intent to Prepay the Federal Financing Bank;

(2) In the event that a borrower proposes to utilize a private loan in connection with a prepayment or a portion of a prepayment,

(i) Evidence, in form and substance satisfactory to RUS, that the borrower has an irrevocable commitment from the lender to close the private loan on the settlement date at an interest rate that meets the requirements of §1786.28(c)(1);

(ii) Evidence that the lender meets the qualification provisions of §1786.28(b);

(iii) Evidence that the private loan meets the qualification provisions of §1786.28(c); and

(iv) The final documentation for the private loan;

(3) Estimate of fees, and expenses, including any taxes, in connection with the prepayment transaction;

(4) A certified copy of a resolution of the board of directors of the borrower approving the certification cited above and requesting RUS approval of the prepayment.

(5) In the case of financially distressed borrowers, evidence in form and substance satisfactory to the Administrator that the benefits of prepayment will not be used to reduce rates and that any Federal or state regulatory body having jurisdiction over the borrower's rates has acknowledged its awareness of this requirement;

(6) In the event that borrower is unable to deliver final documentation or

the evidence specified in accordance with, §1786.31(c), RUS may reschedule the settlement date at its discretion.

(Approved by the Office of Management and Budget under control number 0572–0088)

§ 1786.32 Settlement procedure.

(a) *General.* Settlements in connection with prepaying FFB loans pursuant to this subpart shall be conducted in accordance with the provisions of this section.

(b) *Settlement date.* The prepayment will be settled and if a private loan is utilized, the guarantee will be delivered, on a settlement date agreed upon by the borrower and RUS. Prior to scheduling a settlement date for a borrower's prepayment pursuant to this subpart, RUS shall have received the material specified in §1786.31(b).

(c) *Place of settlement.* All settlements will take place in Washington, DC, at a location of the borrower's choosing; provided however, if more than one settlement is proposed for the same settlement date, RUS reserves the right to coordinate the date and location of the settlements with borrowers involved.

(d) *Repayment of FFB.* Prior to 1:00 p.m. prevailing local time in New York, New York, on the settlement date, the borrower shall wire immediately available funds to RUS through the Department of the Treasury account at the Federal Reserve Bank of New York or shall provide for payment to RUS in another manner acceptable to RUS and FFB, in an amount sufficient to pay the outstanding principal of the FFB loan being prepaid plus accrued interest from the last payment date to and including the settlement date.

(e) *Documentation.* The borrower shall deliver, or cause to be delivered to RUS and FFB, not less than 3 business days prior to the settlement date, written notice of the settlement date and a complete listing of each FFB loan advance to be prepaid or partially prepaid, in the format required by §1786.31(a)(2). In the event that a private loan is used in connection with the prepayment, the following executed documents, opinions and material shall be delivered at the settlement:

(1) The guaranteed note evidencing the private loan.

(2) The guarantee.

(3) The loan guarantee agreement.

(4) Copy of the private loan agreement between the lender and the borrower.

(5) Evidence that the borrower has received all approvals which are required under Federal or state law, loan agreements, security agreements, existing financing arrangements, or any other agreement to which the borrower is a party.

(6) An amendment in recordable form revising the description of the obligations secured by the mortgage including the obligation of the borrower to reimburse RUS for any amounts that RUS may pay under the guarantee.

(7) An approving opinion of the borrower's legal counsel to the effect that the guaranteed note evidencing the private loan is a valid and legally binding obligation of the borrower which is secured under the mortgage, and the priority of the mortgage, as amended pursuant to paragraph (e)(6) of this section, remains undisturbed.

(8) An approving opinion of the lender's legal counsel to the effect that the loan guarantee agreement is a valid and legally binding obligation of the lender.

(9) Such other opinions of counsel as may be required by the Administrator.

(10) Copies of any other documentation required by the lender.

(11) Copies of any other documentation required by RUS to ensure that the obligations of the borrower to reimburse RUS for any amounts that RUS pays under the guarantee or may advance in connection with the private loan are adequately secured under the mortgage.

(Approved by the Office of Management and Budget under control number 0572–0088)

§ 1786.33 Forms.

Guarantees and loan guarantee agreements executed by RUS pursuant to this subpart will be on forms prescribed by RUS. Such forms will include, without limitation, additional details on servicing, procedures for notifying RUS of a default, the manner for requesting payment on a guarantee. The Notice of Intent to Prepay the Federal Financing Bank shall be substantially in the form specified by RUS. RUS may also prescribe standard forms of certifications to be used in connection with materials required to be furnished pursuant to § 1786.31 of this subpart.

§ 1786.34 Access to records of lenders, servicers, and trustees.

The lender, the servicer, or the trustee will permit representatives of RUS (or other agencies of the U.S. Department of Agriculture authorized by that Department) to inspect and make copies of any of their records pertaining to RUS guaranteed loans. Such inspection and copying may be made during regular office hours of the respective party or any other time the party and RUS find convenient.

§ 1786.35 Loss, theft, destruction, mutilation, or defacement of RUS guarantee.

(a) *Authorized representative.* Except where the evidence of debt was or is a bearer instrument, the RUS Administrator is authorized on behalf of RUS to issue a replacement guarantee(s) for one(s) which may have been lost, stolen, destroyed, mutilated, or defaced. Such replacement(s) shall be issued only to the lender or holder and only upon receipt of an acceptable certificate of loss and an indemnity bond.

(b) *Requirements.* When a guarantee(s) is lost, stolen, destroyed, mutilated, or defaced while in the custody of the lender, or holder, the lender will coordinate the activities of the party who seeks the replacement documents and will submit the required documents to RUS for processing. The requirements for replacement are as follows:

(1) A certificate of loss properly notarized which includes:

(i) Legal name and present address of the owner, requesting the replacement forms;

(ii) Legal name and address of lender of record;

(iii) Capacity of person certifying;

(iv) Full identification of the guarantee, including the name of the borrower, date of the guarantee, face amount of the evidence of debt purchased, date of evidence of debt and present balance of the loan. Any existing parts of the documents to be replaced should be attached to the certificate;

(v) A full statement of circumstances of the loss, theft, or destruction of the guarantee; and

(vi) The lender or holder, shall present evidence demonstrating current ownership of the guarantee and note. If the present holder is not the same as the original lender, a copy of the endorsement of each successive holder in the chain of transfer from the initial private lender to present holder shall be included. If copies of the endorsement cannot be obtained, best available records of transfer shall be presented to RUS (e.g., order confirmation, cancelled checks, etc).

(2) An indemnity bond acceptable to RUS shall accompany the request for replacement except when the holder is the United States, a Federal Reserve Bank, a Federal Government Corporation, a state or territory, or the District of Columbia. The bond may be with or without surety. The bond shall be with surety except when the outstanding principal balance and accrued interest due the present holder is less than $1,000,000 verified by the lender in writing in a letter of certification of balance due. The surety shall be a qualified surety company holding a certificate of authority from the Secretary of the Treasury and listed in Treasury Department Circular 580.

(3) All indemnity bonds shall be issued and/or payable to the United States of America acting through the Administrator of the Rural Utilities Service. The bond shall be in an amount not less than the unpaid principal and interest. The bond shall save RUS harmless against any claim or demand which might arise or against any damage, loss, costs, or expenses which might be sustained or incurred by reasons of the loss or replacement of the instruments.

§ 1786.36 Other prepayments.

Nothing contained in this subpart shall prohibit a borrower from making prepayments of FFB loans in accordance with the terms thereof.

§ 1786.37 Application of regulation to previous prepayments.

Nothing contained in this subpart shall affect the validity of prepayments made or guarantees issued pursuant to previous regulations. Those borrowers, however, that completed a prepayment pursuant to section 306(A) of the RE Act and closed loans prior to February 27, 1988, may, in their discretion request RUS approval and if required by prior regulations the concurrence of the Secretary of the Treasury, of any amendments necessary to make the terms and conditions of such loans consistent with, or to consolidate such loans with, loans guaranteed under these regulations.

§ 1786.38 Judicial review.

This subpart is intended to set forth RUS policies and procedures for the orderly administration of the provisions of section 306(A) of the RE Act, section 633 of the continuing resolution, and section 637 of the 1989 Appropriations Act and is not intended to create any right or benefit, substantive or procedural, enforceable at law by a party against the United States, its agencies, its officers or any person.

§§ 1786.39–1786.49 [Reserved]

Subpart C—Special Discounted Prepayments on RUS Direct/Insured Loans

AUTHORITY: 7 U.S.C. 901–950b; Title I, Subtitle B, Pub. L. 99–509; Pub. L. 103–354, 108 Stat. 3178 (7 U.S.C. 6941 et seq.).

SOURCE: 51 FR 46999, Dec. 29, 1986, unless otherwise noted. Redesignated at 55 FR 49250, Nov. 27, 1990.

§ 1786.50 Purpose.

This subpart sets forth the policies and procedures of RUS whereby electric and telephone borrowers may prepay outstanding RUS Notes at the Discounted Present Value of the RUS Notes with private financing.

§ 1786.51 Definitions.

As used in this subpart:

Act means the Rural Electrification Act of 1936, as amended (7 U.S.C. 901 *et seq.*).

Administrator means the Administrator of RUS.

Discounted Present Value shall have the meaning specified in §1786.53

Fund means the Rural Electrification and Telephone Revolving Fund established pursuant to the Act.

REA means the Rural Electrification Administration formerly an agency of the United States Department of Agriculture and predecessor agency to RUS with respect to administering certain electric and telephone loan programs.

RUS means the Rural Utilities Service, an agency of the Unites States Department of Agriculture, established pursuant to Section 232 of the Federal Crop Insurance Reform and Department of Agriculture Reorganization Act of 1994 (Pub. L. 103–354, 108 Stat. 3178), successor to REA with respect to administering certain electric and telephone programs. See 7 CFR 1700.1.

RUS Loan Agreement means the agreement between the borrower and RUS providing for loans pursuant to the Act.

RUS Notes means those notes, bonds or other obligations evidencing indebtedness created by loans made pursuant to Titles I, II or III of the Act (7 U.S.C. 901–940).

[51 FR 46999, Dec. 29, 1986. Redesignated at 55 FR 49250, Nov. 27, 1990, as amended at 59 FR 66441, Dec. 27, 1994]

§1786.52 Prepayment.

Through September 30, 1987, the Administrator may, pursuant to this subpart, permit eligible electric and telephone borrowers to prepay all outstanding RUS Notes issued or assumed by such borrowers and held in the Fund, upon paying the lesser of the outstanding balance or the Discounted Present Value.

§1786.53 Discounted present value.

The Discounted Present Value shall be calculated five business days before prepayment is made by summing the present values of all remaining payments by using the following formula:

$$\text{Present Value} = \sum_{k=1}^{n} \frac{P_k}{\prod_{i=1}^{k}\left[1.0 + \left\langle\frac{D1_i}{365} + \frac{D2_i}{366}\right\rangle \times I\right]}$$

Where:

P_k = Total payment including interest, due on the kth payment date following the prepayment date.

n = Total number of remaining payments dates.

I = The discount rate, in decimals, which shall be the average rate on utility bonds bearing a rating of "Aa" as set forth in that issue of Moody's Public Utility News Reports most recently published prior to the date on which Discounted Present Value is calculated.

$D1_i$ = Number of days in the ith payment period that are in a non-leap year (365 day year).

$D2_i$ = Number of days in the ith payment period that are in a leap year (366 day year).

§1786.54 Eligibility criteria.

To be eligible to prepay RUS Notes at the Discounted Present Value a borrower must comply with the following criteria:

(a) The borrower must be current on all payments due on its outstanding RUS Notes and all other payment obligations owed to RUS and the Rural Telephone Bank.

(b) The borrower must agree to prepay all of its outstanding RUS Notes.

(c) The borrower must identify the source of private financing that will be used to refinance its outstanding RUS Notes, which financing may not include obligations the income of which is exempt from taxation under the Internal Revenue Code of 1986.

(d) The borrower must have expended all funds advanced on account of the RUS Notes for the purposes for which such funds were advanced.

(e) The borrower must agree to a rescission of the unadvanced balance of the RUS Notes.

(f) The borrower must agree that the borrower, its successors or assigns, shall pay to the Government, as a condition of receiving additional loans or loan guarantees pursuant to Titles I, II and III of the Act, an amount equal to the aggregate of the difference with respect to each of the RUS Notes between the amount outstanding on the RUS Note and the Discounted Present Value of the RUS Note upon prepayment with interest accruing quarterly; the interest rates shall be the rates provided in the respective RUS Notes.

(g) If the borrower is a party to a wholesale power contract with a power supplier financed pursuant to the Act, the borrower must provide the Administrator with such assurances as the Administrator may request that it will meet its obligations to the power supplier.

§ 1786.55   Application procedure.

Any borrower seeking to prepay its RUS Notes under this subpart should apply to the appropriate RUS Area Director by submitting:

(a) A board resolution that:

(1) Requests approval of the prepayment of the borrower's outstanding RUS Notes, and

(2) States the intent of the borrower to comply with all eligibility criteria set forth in § 1786.54 of this subpart.

(b) A list of all RUS Notes together with the outstanding amount on such notes.

(c) Such additional information as the Administrator shall request.

§ 1786.56   Approval of applications.

The applications will ordinarily be reviewed and, if satisfactory, approved, and closing schedule based on the order in which executed prepayment agreements are received. The Administrator may limit the number of applications approved and closings scheduled from time to time taking into account, among other matters, the financial interests and administrative considerations of the Government.

§ 1786.57   Prepayment agreement.

Upon approving an application for prepayment under this subpart, the Administrator shall notify the borrower and deliver to the borrower for its execution a prepayment agreement which shall set forth and provide:

(a) The RUS Notes to be prepaid and when the Discounted Present Value will be calculated.

(b) The place and conditions for closing.

(c) Agreement that the unadvanced balance of RUS Notes shall be rescinded.

(d) Agreement that the borrower, or its successors or assigns, shall pay to the Government, as a condition of receiving additional loans or loan guarantees pursuant to Titles I, II and III of the Act, an amount equal to the aggregate of the difference with respect to each of the RUS Notes between the amount outstanding on the RUS Note and the Discounted Present Value of the RUS Note upon prepayment with interest accruing quarterly; the interest rates shall be the rates provided in the respective RUS Notes.

(e) Assurances that the borrower will meet its obligations to any power supplier financed pursuant to the Act.

(f) Such other terms and conditions as the Administrator deems appropriate.

§ 1786.58   Security.

If, after prepayment of RUS Notes, the Government should continue to hold liens on the borrower's property that secure loans made or guaranteed pursuant to the Act, the Administrator of RUS or the Governor of the Rural Telephone Bank, as the case may be, will consider request for the accommodation of such liens for the purpose of providing security for loans the proceeds of which were used to prepay RUS Notes. Such lien accommodations shall be limited in amount to the Discounted Present Value of the RUS Notes plus such costs, as the Administrator shall determine to be reasonable, incurred by the borrower in obtaining such loans.

§ 1786.59   Loan fund audit.

Within 6 months of closing RUS shall have the right to audit transactions involving the RUS construction fund established and maintained by the borrower pursuant to the terms of the RUS Loan Agreement and to inspect all books, records, accounts and other documents and papers of the borrower.

Should RUS determine that the borrower has made disbursements of funds advanced pursuant to RUS Notes which do not comply with the requirements of the RUS Loan Agreement, the borrower shall be required to pay to the Government an amount equal to the difference between the amount which the borrower prepaid on such RUS Notes evidencing RUS loan funds which were improperly disbursed and the amount which the borrower would otherwise have been required to return to the Government as a result of noncompliance if the borrower had not prepaid such RUS Notes. (See 7 CFR part 1721)

§ 1786.60 Closing.

(a) The borrower shall be responsible for obtaining all approvals necessary to consummate the transaction as required by the prepayment agreement including such approvals as may be required by regulatory bodies and other lenders.

(b) The RUS Notes shall be prepaid at a closing to be held in accordance with the prepayment agreement; *Provided, however,* That no closing may be scheduled for after September 30, 1987. At closing, a borrower shall prepay the RUS Notes by paying to the Government an amount equal to the Discounted Present Value of the RUS Notes. The closing shall otherwise be conducted as prescribed in the prepayment agreement.

§ 1786.61 Other prepayments.

RUS loan documentation generally permits borrowers to prepay RUS Notes by paying the outstanding balance due thereon. Nothing in this subpart shall prohibit any borrower from prepaying its outstanding RUS Notes in accordance with the terms thereof. The provisions of this subpart shall not be applicable to such prepayment.

§§ 1786.62–1786.74 [Reserved]

Subpart D [Reserved]

Subpart E—Discounted Prepayments on RUS Notes in the Event of a Merger of Certain RUS Electric Borrowers

Source: 56 FR 37268, Aug. 6, 1991, unless otherwise noted.

§ 1786.95 Purpose.

This subpart sets forth the policies and procedures of RUS whereby certain electric borrowers may prepay outstanding RUS Notes at the Discounted Present Value of the RUS Notes with private financing.

§ 1786.96 Definitions.

As used in this subpart:

Act means the Rural Electrification Act of 1936, as amended (7 U.S.C. 901 *et seq.*).

Administrator means the Administrator of RUS.

Consolidation means:

(1) The combination, pursuant to state law, of two or more borrower or nonborrower organizations into a new successor organization that takes over the assets and assumes the liabilities of those organizations; or

(2) Any other transaction including an acquisition which has substantially the same effect.

Discounted Present Value shall have the meaning specified in § 1786.98.

Fund means the Rural Electrification and Telephone Revolving Fund pursuant to the Act.

Merger means:

(1) The combination, pursuant to state law, of two or more borrower or nonborrower organizations into an existing survivor organization that takes over the assets and assumes the liabilities of the merged organizations; or

(2) Any other transaction including an acquisition which has substantially the same effect.

REA means the Rural Electrification Administration formerly an agency of the United States Department of Agriculture and predecessor agency to RUS with respect to administering certain electric and telephone loan programs.

RUS means the Rural Utilities Service, an agency of the United States Department of Agriculture established pursuant to Section 232 of the Federal

Crop Insurance Reform and Department of Agriculture Reorganization Act of 1994 (Pub. L. 103–354, 108 Stat. 3178), successor to REA with respect to administering certain electric and telephone programs. See 7 CFR 1700.1.

RUS Loan Agreement means the agreement between the borrower and RUS providing for loans pursuant to the Act.

RUS Notes means those notes, bonds or other obligations evidencing indebtedness created by loans made or guaranteed by RUS pursuant to titles I and III of the Act (7 U.S.C. 901–940).

[56 FR 37268, Aug. 6, 1991, as amended at 59 FR 66440, Dec. 27, 1994]

§ 1786.97 Prepayment.

There were 29 former RUS electric borrowers that prepaid their direct or insured loans under section 306B(a) of the Act prior to October 1, 1987. (See subpart C of this part.) These borrowers are listed in appendix A to subpart E of this part. Any RUS electric borrower which is the result of a merger or consolidation involving any of these 29 former borrowers and a borrower with outstanding Notes may, after meeting all requirements of this subpart, prepay all outstanding RUS Notes issued or assumed by the borrower upon paying the lesser of the outstanding balance or the Discounted Present Value. Such prepayment must be made not later than one year after the effective date of the merger or consolidation.

§ 1786.98 Discounted present value.

(a) The Discounted Present Value shall be calculated by RUS before prepayment is made by summing the present values of all remaining payments on all outstanding notes according to the following formula to compute the discounted present value of each note and adjusting as here and after provided for tax exempt financing.

$$\text{Present Value} = \sum_{k=1}^{n} \frac{P_k}{\prod_{i=1}^{k}\left[1.0 + \left\langle \frac{D1_i}{365} + \frac{D2_i}{366}\right\rangle \times I\right]}$$

Where:

P_k = Total payment, including interest, due on the kth payment date following the prepayment date. n = Total number of remaining payment dates. I = The discount rate applied to each transaction will be ascertained by using data specified in the "Federal Reserve Statistical Release" which is published each Monday. (See appendix B to subpart E of this part.) The specific discount rate will be the discount rate(s) specified in the "Treasury Constant Maturities" section of this publication eight working days prior to the closing. In applying the discount rate, the 1-year Treasury rate will be used for all notes with a remaining term of less than 2 years; the 2-year Treasury rate for notes with maturities between 2 and 3 years; the 3-year Treasury rate for all notes with maturities between 3 and 5 years; the 5-year Treasury rate for all notes with maturities between 5 and 7 years; the 7-year Treasury rate for all notes with maturities between 7 and 10 years; the 10-year Treasury rate for all notes with maturities between 10 and 30 years; and the 30-year Treasury rate for all notes with maturities longer than 30 years.

$D1_i$ = Number of days in the ith payment period that are in a non-leap year (365 day year).

$D2_i$ = Number of days in the ith payment period that are in a leap year (366 day year).

(b) Notwithstanding paragraph (a) of this section, in the event that the borrower shall elect to prepay using tax exempt financing, the calculation of the Discounted Present Value shall be adjusted to make the discount the equivalent of fully taxable financing.

§ 1786.99 Eligibility criteria.

To be eligible to prepay RUS Notes at the Discounted Present Value, a borrower must comply with the following criteria:

(a) The borrower must be current on all payments due on its outstanding RUS Notes and all other payment obligations owed to RUS;

(b) The borrower must agree to prepay all of its outstanding RUS Notes;

(c) The borrower must identify the source of financing that will be used directly or indirectly to refinance its outstanding RUS Notes. The borrower must certify in writing whether such financing will be tax exempt and, if so, shall furnish all information on the financing as RUS may request to enable RUS to adjust the discount to the equivalent to fully taxable financing;

(d) The borrower must have expended all funds advanced on account of the RUS Notes for the purposes for which such funds were advanced or repaid RUS for all unexpended funds;

(e) The borrower must agree to a rescission of the unadvanced balance of any RUS Notes outstanding as of the date of its application for prepayment;

(f) The borrower must agree that the borrower, its successors and assigns, shall pay to the Government, as a condition of receiving additional loans or loan guarantees pursuant to titles I and III of the Act, an amount equal to the aggregate of the difference with respect to each of the RUS Notes between the amount outstanding on the RUS Note and the Discounted Present Value of the RUS Note upon prepayment with interest accruing quarterly; the interest rates shall be the rates provided in the respective Notes; and

(g) If the borrower is a party to a wholesale power contract with a power supplier financed pursuant to the Act, the borrower must provide the Administrator with such assurances as the Administrator may request that it will meet its obligations to the power supplier. The borrower must also specifically agree to the following limitation: The borrower agrees that, for so long as the Wholesale Power Contract shall be in effect between the borrower and the power supplier, the borrower will not, without the approval in writing of the power supplier and the Administrator, take or suffer to be taken any steps for reorganization or to consolidate with or merge into any corporation or any other public power district, or to sell, lease or transfer (or make any agreement therefor) all or a substantial portion of its assets, whether now owned or hereafter acquired. Notwithstanding the foregoing, the borrower may take or suffer to be taken any steps for reorganization or to consolidate with or merge into any corporation or any other public power district, or to sell, lease or transfer (or make any agreement therefor) all or a substantial portion of its assets, whether now owned or hereafter acquired, so long as the borrower shall pay such portion of the outstanding indebtedness evidenced by the power supplier's Notes at the time outstanding as shall be determined by the power supplier with the prior written consent of the Administrator and shall otherwise comply with such reasonable terms and conditions as the Administrator and the Power Supplier shall require.

§1786.100 Application procedure.

Any borrower seeking to prepay its RUS Notes under this Subpart should apply to the appropriate RUS Area Director not less than 60 days prior to one year after the effective date of the merger or consolidation by submitting:

(a) A board resolution that:

(1) Requests approval of the prepayment of the borrower's outstanding RUS Notes;

(2) States the intent of the borrower to comply with all eligibility criteria set forth in §1786.99 of this subpart; and

(3) Identifies the source of financing.

(b) A list of all RUS Notes together with the outstanding amount on such notes.

(c) An opinion of counsel as to the effective date of the merger or consolidation.

(d) Such additional information as the Administrator will request.

§1786.101 Approval of application.

The applications will be reviewed and, if satisfactory, approved. Closing will be scheduled upon approval.

§1786.102 Prepayment agreement.

Upon approving an application for prepayment under this subpart, the Administrator shall notify the borrower and deliver to the borrower for its execution a prepayment agreement which shall set forth and provide:

(a) The RUS Notes to be prepaid and when the Discounted Present Value will be calculated.

(b) The place, date and conditions for closing.

(c) Agreement that the unadvanced balance of RUS Notes shall be rescinded.

(d) Agreement that the borrower, or its successors or assigns, shall pay to the Government, as a condition of receiving additional loans or loan guarantees pursuant to titles I and III of the Act, an amount equal to the aggregate of the difference with respect to each of the RUS Notes between the amount outstanding on the RUS Note and the Discounted Present Value of the prepaid RUS Note; with interest accruing quarterly. The interest rates shall be the rates provided in the respective RUS Notes.

(e) Assurances that the borrower will meet its obligations to any power supplier financed pursuant to the Act.

(f) Such other terms and conditions as the Administrator deems appropriate.

§ 1786.103   Security.

If, after prepayment of RUS Notes, the Government should continue to hold liens on the borrower's property, the Administrator of RUS will consider a request for the accommodation of such liens for the purpose of providing security for loans the proceeds of which were used to prepay RUS Notes. Such lien accommodations shall be limited in amount to the Discounted Present Value of the RUS Notes plus such costs, as the Administrator shall determine to be reasonable, incurred by the borrower in obtaining such loans.

§ 1786.104   Loan fund audit.

RUS shall have the right to audit within 6 months of closing, transactions involving the RUS construction fund established and maintained by the borrower pursuant to the terms of the RUS Loan Agreement and to inspect all books, records, accounts and other documents and papers of the bor-rower. Should RUS determine that the borrower has made disbursements of funds advanced pursuant to RUS Notes which do not comply with the requirements of the RUS Loan Agreement, the borrower shall be required to pay the Government an amount equal to the difference between the amount which the borrower prepaid on such RUS Notes evidencing RUS loans funds which were improperly disbursed and the amount which the borrower would otherwise have been required to return to the Government as a result of noncompliance if the borrower had not prepaid such RUS Notes. (See 7 CFR part 1721, Post-Loan Policies and Procedures for Insured Electric Loans.)

§ 1786.105   Closing.

(a) The borrower shall be responsible for obtaining all approvals necessary to consummate the transaction as required by the prepayment agreement, including such approvals as may be required by regulatory bodies and other lenders.

(b) The RUS Notes shall be prepaid at a closing to be held in accordance with the prepayment agreement. RUS shall designate the date of closing which in no event shall be later than one year after the effective date of the merger or consolidation. At closing, in addition to paying all current interest due on the date of prepayment, a borrower shall prepay the RUS Notes by paying to the Government an amount equal to the lesser of the outstanding balance or the Discounted Present Value of the RUS Notes. The closing shall otherwise be conducted as prescribed in the prepayment agreement.

§ 1786.106   Other prepayments.

RUS loan documentation generally permits borrowers to prepay RUS Notes by paying the outstanding balance due thereon. Nothing in this subpart shall prohibit any borrower from prepaying its outstanding RUS Notes in accordance with the terms thereof. The provisions of this subpart shall not be applicable to such prepayment.

APPENDIX A TO SUBPART E OF PART 1786—LISTING OF ELIGIBLE BORROWERS

State	Borrower name and address
Colorado	Colorado-Ute Electric Assn., Inc., Montrose.
Florida	Lee County Electric Coop. Inc., North Fort Myers.
Indiana	Clark County Rural Elec. Memb. Corp., Sellersburg.
Louisiana	Beauregard Electric Cooperative, Inc., Deridder.
Missouri	Culvre River Electric Cooperative, Inc., Troy.
Nebraska	Roosevelt Public Power District, Mitchell.
Nebraska	Howard Greely Rural Public Power Dist., St. Paul.
Nebraska	Cuming County Public Power District, West Point.
Nebraska	York County Rural Public Power District, York.
Nebraska ....:.......	Elkhorn Rural Public Power District, Battle Creek.
Nebraska	Southern Nebraska Rural P. P. D., Grand Island.
Nebraska	McCook Public Power District, McCook.
Nebraska	Niobrara Valley Electric Memb. Corp., O'Neill.
Nebraska	Cornhusker Public Power District, Columbus.
Nebraska	Custer Public Power District, Broken Bow.
Nebraska	Northwest Rural Public Power Dist., Hay Springs.
Nebraska	Southwest Public Power District, Palisade.
Nebraska	Loup Valleys Rural Public Power District, Ord.

State	Borrower name and address
Nebraska	South Central Public Power District, Nelson.
Oklahoma	Peoples' Electric Cooperative, Ada.
Texas	Deaf Smith County Electric Coop. Inc., Hereford.
Texas	Pedernales Electric Coop. Inc., Johnson City.
Texas	Bandera Electric Cooperative, Inc., Bandera.
Texas	Guadalupe Valley Electric Coop., Inc., Gonzales.
Texas	Bluebonnet Electric Cooperative, Inc., Giddings.
Texas	Cap Rock Electric Cooperative, Inc. Stanton.
Texas	San Bernard Electric Cooperative, Inc., Bellville.
Washington	Inland Power & Light Company, Spokane.
Washington	Pub. Util. Dist. No. 1 Grays Harbor Co., Aberdeen.

APPENDIX B TO SUBPART E OF PART 1786—FEDERAL RESERVE STATISTICAL RELEASE

FEDERAL RESERVE STATISTICAL RELEASE

These data are released each Monday. The availability of the release will be announced when the information is available, on (202) 452–3206.

H. 15 (519)

For immediate release February 4, 1991.

SELECTED INTEREST RATES

[Yields in percent per annum]

Instruments	1991 Jan. 28	1991 Jan. 29	1991 Jan. 30	1991 Jan. 31	1991 Feb. 1	This week	Last week	1991 Jan.
Federal Funds (effective) [1] [2] [3]	7.61	7.16	6.96	8.18	6.30	7.46	6.88	6.91
Commercial paper [3] [4] [5]								
1-Month	6.88	6.96	6.95	6.99	6.73	6.90	6.83	7.12
3-Month	6.92	6.96	6.94	6.95	6.67	6.89	6.92	7.10
6-Month	6.87	6.91	6.88	6.88	6.58	6.82	6.86	7.02
Finance paper placed directly [3] [4] [6]								
1-Month	6.76	6.85	6.83	6.83	6.55	6.76	6.68	6.95
3-Month	6.75	6.83	6.83	6.76	6.46	6.73	6.77	6.92
6-Month	6.53	6.53	6.59	6.53	6.19	6.47	6.55	6.59
Bankers acceptances (top rated) [3] [4] [7]								
3-Month	6.80	6.82	6.77	6.68	6.30	6.67	6.76	6.96
6-Month	6.67	6.70	6.65	6.55	6.15	6.54	6.63	6.84
CDS (secondary market) [3] [8]								
1-Month	6.78	6.85	6.87	6.82	6.52	6.77	6.77	7.10
3-Month	6.94	6.95	6.93	6.88	6.51	6.84	6.94	7.17
6-Month	6.95	6.98	6.95	6.88	6.51	6.85	6.97	7.17
Eurodollar deposits (London) [3] [9]								
1-Month	6.81	6.88	6.88	6.88	6.88	6.86	6.81	7.13
3-Month	6.94	7.06	7.00	6.94	6.94	6.98	7.01	7.23
6-Month	7.00	7.00	7.00	6.94	6.94	6.98	7.04	7.23

SELECTED INTEREST RATES—Continued

[Yields in percent per annum]

Instruments	1991 Jan. 28	1991 Jan. 29	1991 Jan. 30	1991 Jan. 31	1991 Feb. 1	This week	Last week	1991 Jan.
Bank prime loan [2][3][10]	9.50	9.50	9.50	9.50	9.50	9.50	9.50	9.52
Discount window borrowing [2][11]	6.50	6.50	6.50	6.50	6.00	6.50	6.50	6.50
U.S. Government securities								
Treasury bills								
Auction average [3][4][12]								
3-Month	6.22	....	....	....	....	6.22	6.14	6.30
6-Month	6.28	....	....	....	....	6.28	6.21	6.34
1-Year	....	....	....	....	....	....	....	6.22
Auction average (investment) [12]								
3-Month	6.41	....	....	....	....	6.41	6.32	6.49
6-Month	6.58	....	....	....	....	6.58	6.50	6.64
Secondary market [3][4]								
3-Month	6.25	6.22	6.20	6.19	6.00	6.17	6.12	6.22
6-Month	6.26	6.26	6.24	6.20	5.97	6.19	6.20	6.28
1-Year	6.24	6.20	6.17	6.13	5.91	6.13	6.19	6.25
Treasury Constant maturities [13]								
1-Year	6.64	6.59	6.56	6.51	6.27	6.51	6.58	6.64
2-Year	7.12	7.10	7.07	7.05	6.83	7.03	7.09	7.13
3-Year	7.38	7.35	7.34	7.30	7.10	7.29	7.35	7.38
5-Year	7.67	7.64	7.64	7.62	7.45	7.60	7.66	7.70
7-Year	7.93	7.90	7.90	7.89	7.75	7.87	7.92	7.97
10-Year	8.06	8.05	8.05	8.03	7.91	8.02	8.04	8.09
30-Year	8.23	8.20	8.23	8.21	8.09	8.19	8.22	8.27
Composite								
Over 10 years (long-term) [14]	8.29	8.26	8.29	8.27	8.15	8.25	8.28	8.33
Corporate bonds								
Moody's Seasoned								
AAA	9.03	9.01	9.00	8.99	8.96	9.00	9.05	9.04
BAA	10.43	10.37	10.35	10.33	10.24	10.34	10.44	10.45
A-Utility [15]	....	....	....	....	9.65	9.65	9.80	9.83
State and local bonds [16]	....	....	....	7.00	....	7.00	7.06	7.08
Conventional mortgages [17]	....	....	....	....	9.56	9.56	9.61	9.64

Footnotes:

[1] The daily effective federal funds rate is a weighted average of rates on trades through N.Y. brokers.

[2] Weekly figures are averages of 7 calendar days ending on Wednesday of the current week; monthly figures include each calendar day in the month.

[3] Annualized using a 360-day year or bank interest.

[4] Quoted on a discount basis.

[5] An average of offering rates on commercial paper placed by several leading dealers for firms whose bond rating is AA or the equivalent.

[6] An average of offering rates on paper directly placed by finance companies.

[7] Representative closing yields for acceptances of the highest rated money center banks.

[8] An average of dealer offering rates on nationally traded certificates of deposit.

[9] Bid rates for Eurodollar deposits at 11 a.m. London time.

[10] One of several base rates used by banks to price short-term business loans.

[11] Rate for the Federal Reserve Bank of New York.

[12] Auction date for daily data; weekly and monthly averages computed on an issue-date basis.

[13] Yields on actively traded issues adjusted to constant maturities. Source: U.S. Treasury.

[14] Unweighted average of rates on all outstanding bonds neither due nor callable in less than 10 years, including one very low yielding "flower" bond.

[15] Estimate of the yield on a recently offered, A-rated utility bond with a maturity of 30 years and call protection of 5 years; Friday quotations.

[16] Bond buyer Index, general obligation, 20 years to maturity, mixed quality; Thursday quotations.

[17] Contract interest rates on commitments for fixed-rate first mortgages. Source: FHLMC.

NOTE: Weekly and monthly figures are averages of business days unless otherwise noted.

Description of the Treasury Constant Maturity Series

Yields on Treasury securities at "constant maturity" are interpolated by the U.S. Treasury from the daily yield curve. This curve, which relates the yield on a security to its time to maturity, is based on the closing market bid yields on actively traded Treasury securities in the over-the-counter market. These market yields are calculated from composites of quotations reported by five leading U.S. Government securities dealers to the Federal Reserve Bank of New York. The constant maturity yield values are read from the yield curve at fixed maturities, currently 1, 2, 3, 5, 7, 10, and 30 years. This method provides a yield for a 10-year maturity, for example, even if no outstanding security has exactly 10 years remaining to maturity.

Subpart F—Discounted Prepayments on RUS Electric Loans

AUTHORITY: 7 U.S.C. 901 et seq.; Pub. L. 103–354, 108 Stat. 3178 (7 U.S.C. 6941 et seq.).

SOURCE: 59 FR 13620, Mar. 22, 1994, unless otherwise noted.

§ 1786.150 Purpose.

This subpart sets forth the policies and procedures of RUS whereby borrowers may prepay, with private financing or internally generated funds, outstanding RUS Notes evidencing electric loans at the Discounted present value of the RUS Notes, pursuant to the provisions of section 306(B) of the RE Act as amended by Public Law 102–428, 106 Stat. 2183, adopted October 21, 1992.

§ 1786.151 Definitions and rules of construction.

(a) *Definitions.* As used in this subpart:

Administrator means the Administrator of the Rural Utilities Service (RUS).

Borrower means any organization which has an outstanding note(s) evidencing electric loans made by RUS, or has previously prepaid such notes under subparts C and E of this part.

Business day means any day on which both the RUS and the Federal Reserve Bank of New York are open for business.

Construction Fund Account means the Cash—Construction Fund—Trustee Account, maintained by the borrower pursuant to the terms of the outstanding RUS Loan Contract.

Closing shall mean one of the several contemplated closings of the prepayment of the Qualified Notes prescribed by the Prepayment agreement.

Closing date shall mean any business day identified as such by the Government in its preclosing notice delivered to the Company pursuant to § 1786.158.

Closing request shall mean a request by the borrower of the Government to schedule a closing for certain Qualified Notes on the date requested therein.

Direct loan means a loan made pursuant to section 4 of the RE Act.

Discounted present value shall have the meaning set forth in § 1786.153.

Distribution borrower means a borrower that sells electric power and energy at retail in rural areas.

Electric loan means a Direct loan or an Insured loan made for the purpose of furnishing electric energy to persons in rural areas.

Final maturity means the final date on which all outstanding principal and accrued interest on an electric loan is due and payable.

Government means the United States of America, acting through the Administrator of the Rural Utilities Service.

Insured loan means a loan made pursuant to Section 305 of the RE Act.

Lien accommodation means the sharing of the Government's (RUS's) lien on property, usually all property, covered by the lien of the RUS Mortgage.

Loan guarantee means a loan guarantee under Section 306 of the RE Act.

Power supply borrower means a borrower that sells or intends to sell electric power at wholesale to distribution or power supply borrowers pursuant to RUS wholesale power contracts.

Preclosing notice shall mean a notice delivered by the Government to the borrower in response to a closing request, identifying the closing date, the Qualified Notes to be prepaid at such closing and documents to be delivered by the borrower to the Government prior to the closing date.

Prepayment agreement shall have the meaning set forth in § 1786.158.

Qualified Notes shall have the meaning set forth in § 1786.154.

RE Act means the Rural Electrification Act of 1936, as amended (7 U.S.C. 901 *et seq.*).

RUS means the Rural Utilities Service, an agency of the United States Department of Agriculture.

RUS Loan Contract means the agreement, as amended, supplemented, or restated from time to time, between a borrower and RUS providing for loans or loan guarantees pursuant to the RE Act.

RUS Mortgage means collectively those mortgages and security agreements made by and among the borrower, the Government, and third parties, if any, securing indebtedness evidencing electric loans or loan guarantees made pursuant to the RE Act.

Rural development loans means loans or grants made pursuant to Rural development programs.

Rural development programs means loan or grant programs under the authority of the Administrator pursuant to sections 313, 501, and 502 of the RE Act.

Supplemental lender means a private lender whose loan to the borrower is secured by the RUS mortgage.

Tax exempt financing means borrowing evidenced by bonds, notes and other evidence of indebtedness the income of which is excluded from gross income for the purposes of Chapter 1 of the Internal Revenue Code of 1986 (26 U.S.C. ch. 1).

(b) *Rules of construction.* Unless the context shall otherwise indicate, the terms defined in paragraph (a) of this section include the plural as well as the singular, and the singular as well as the plural.

§ 1786.152   **Prepayments of RUS loans.**

An electric loan made under the RE Act shall not be sold or prepaid at a value that is less than the outstanding principal balance, except that, on request of a borrower, an electric loan made under the RE Act, or a portion of such a loan, that was advanced before May 1, 1992, or has been advanced for not less than 2 years, shall be prepaid by the borrower at the lesser of the outstanding principal balance of the loan or the discounted present value thereof.

§ 1786.153   **Discounted present value.**

(a) The discounted present value shall be calculated by summing the present values of all remaining payments on all Qualified Notes to be prepaid according to the following formula and adjusted as provided in paragraph (b) of this section if tax exempt financing is used.

$$\text{Present Value} = \sum_{k=1}^{n} \frac{P_k}{\prod_{i=1}^{k}\left[1.0+\left(\frac{D1_i}{365}+\frac{D2_i}{366}\right)I\right]}$$

Where:

The Greek letter, Sigma (Σ) means the sum of the following terms.

The Greek letter, Pi (Π) means the product of the following terms.

P_k = Total payment, including interest due on the Kth payment date following the prepayment date.

n = Total number of remaining payment dates to final maturity.

$D1_i$ = Number of days in the ith payment period that are in a non-leap year (365-day year).

$D2_i$ = Number of days in the ith payment period that are in a leap year (366-day year).

I = The discount rate applied to each transaction ascertained by using data specified in the "Federal Reserve Statistical Release" (H.15 (519)), which is published each Monday. The availability of this Release will be announced when the information is available by telephone on (202) 452–3206. See adjustment for tax exempt refinancing at paragraph (b) of this section. The specific discount rate will be based on the discount rate(s) specified in the "Treasury Constant Maturities" section of this publication 8 business days prior to the closing and will be interpolated from that information as follows:

Remaining final maturity of RUS loan:		Treasury constant maturities
At least	But less than	
# years	# years	
0	2	1-year.
2	3	2-year.
3	4	3-year.
4	5	([1])
5	6	5-year.
6	7	([2])
7	8	7-year.
8	9	([3])
9	10	([3])
10	11	10-year.
11	20	([4])
20	21	20-year.
21	30	([5])
30	36	30-year.

NOTES:

[1] The arithmetic mean between the 3-year and 5-year. Treasury Constant Maturities; i.e., if 3-year. rate is 3.00% and the 5-year. rate is 4.00% then the rate used would be 3.5%.

[2] The arithmetic mean between the 5-year and 7-year Treasury Constant Maturities computed as above.

[3] A straight line interpolated rate between the 7-year rate and the 10-year rate. (See formula below)

[4] A straight line interpolated rate between the 10-year note and the 20-year Bond rate. (See formula below)

[5] A straight line interpolated rate between the 20-year bond and the 30-year bond using the following formula:

$$I = B + \frac{((C - E) \times (A - B))}{F - E}$$

Where:

I = The discount rate interpolated from the cost of money to the Treasury.

A = The Treasury interest rate for the most recently published maturity (in years) that is the shortest Treasury term (in years) which is greater than the borrower's remaining term (in years) to final maturity; i.e., (if the note to be prepaid has a final maturity of more than 10 years then this rate is the 20-year Treasury rate)

B = The Treasury interest rate for the most recently published maturity (in years) that is the longest Treasury term (in years) which is less than the borrower's remaining term (in years) to final maturity; i.e., (if the note to be prepaid has a final maturity of more than 10 years but less than 20 years then this term is the 10-year Treasury rate)

C = The remaining number of full years to the final maturity of the borrower's note. Drop all fractions of a year and use the remaining full years.

E = The published Treasury term (in years) to maturity which is the longest term to maturity for the published term that is less than the remaining term (in years) to final maturity of the borrower's note; i.e., (if the note to be prepaid has remaining years to maturity between 11 and 20 years then this term would be 10 or if the note to be prepaid has remaining years

to maturity between 21 years and 30 years then this term would be 20).

F = The published Treasury term (in years) to maturity which is the shortest term to maturity for the published term that is greater than the remaining term (in years) to maturity of the borrower's note; i.e., (if the note to be prepaid has remaining years to maturity between 11 and 20 years then this term would be 20 or if the remaining years to maturity is between 21 and 30 years then this term would be 30).

NOTE: The percentage terms used in the above formula will be truncated to two decimal places. For the purpose of the terms A, B, E, and F above the published Treasury rate and term shall mean the Treasury Constant Maturities from the Federal Reserve Statistical Release for 7 years, 10 years, 20 years, and 30 years.

(b)(1) In the event that the borrower prepays a loan under paragraph (a) of this section using, directly or indirectly, tax exempt financing, the discount shall be adjusted to ensure that the borrower receives a benefit that is no greater than the benefit the borrower would receive if the borrower used financing that was not tax exempt. The borrower shall certify in writing whether the financing will be tax exempt.

(2) The discount rate established in paragraph (a) of this section shall be adjusted for a tax exempt financing by substituting for the "I" term in the discount rate formula, a discount rate equal to the interest rate(s) published pursuant to 7 CFR 1714.5, determination of interest rates on municipal rate loans. This is the interest rate established for the new RUS loan program which is based on municipal interest rates for issues of comparable maturity. No interpolation or average will be used. If a note is to be prepaid under this subpart and is subject to this tax exempt adjustment, the discount rate will be determined from the published table in the FEDERAL REGISTER. For example, if the note to be discounted matures in the year 1999 then the discount rate will be the interest rate for the year 1999. RUS will publish a schedule of interest rates for municipal rate loans in the FEDERAL REGISTER at the beginning of each calendar quarter. The published rates in effect eight business days prior to closing will be used for the discount rates. All notes

to be prepaid that have remaining years to maturity of more than 20 years will be discounted at the interest rate in effect for new RUS municipal rate loans of comparable maturity at the time of closing.

§ 1786.154 Qualified Notes.

An eligible borrower may prepay Qualified Notes under this subpart at the discounted present value. A Qualified Note is a note evidencing an RUS electric loan, all advances of which were made prior to May 1, 1992, or not less than 2 years prior to the date of prepayment closing. See §§ 1786.155(a)(3) and 1786.158 (h) and (j).

§ 1786.155 Eligible borrower.

(a) To be eligible to prepay an electric loan under this subpart, the borrower must be in compliance with the following:

(1) The borrower shall be current on all payment obligations on outstanding loans made or guaranteed by RUS. For the purpose of determining eligibility for prepayment, a default by a power supply borrower from which a distribution borrower purchases wholesale power shall not be considered a default by the distribution borrower;

(2) There shall exist no material defaults under the borrower's RUS Loan Contract and Mortgage;

(3) The borrower shall have expended all funds advanced pursuant to the RUS Loan Contract for the purposes for which such funds were advanced. A borrower will not be eligible to prepay under this subpart if it has any funds advanced pursuant to the RUS Loan Contract in its Construction Fund Account; and

(4) The borrower shall be current on all obligations under any wholesale power contract with an RUS financed power supply borrower.

(b) The eligibility of borrowers that have had any indebtedness representing loans made or guaranteed by RUS restructured shall be determined on a case by case basis considering the terms and conditions of the restructuring agreement.

§ 1786.156 Application procedure.

Any borrower seeking to prepay Qualified Notes under this subpart

should apply to the appropriate RUS Regional Director or the Director of the Power Supply Division. The application shall provide the following:

(a) Borrower's RUS designation;

(b) Borrower's name and address;

(c) A certified copy of a resolution of the board of directors of the borrower that the borrower wishes to enter into a prepayment agreement providing for the prepayment of all or a portion of its Qualified Notes;

(d) Listing of each Qualified Note to be prepaid by loan designation, RUS account number, advance date, maturity date, original amount, and outstanding principal balance;

(e) Evidence that the borrower has the ability to obtain the financing necessary to prepay its Qualified Notes listed in paragraph (d) of this section and identification of the source of financing and the need if any of obtaining a lien accommodation from RUS; and

(f) Such additional information as the Administrator may request.

§ 1786.157 Approval of applications.

(a) Ordinarily, within 30 days of receipt, an application will be reviewed and the borrower will be notified as to whether the application has been approved. If the application has not been approved, the borrower will be informed as to the reasons. If the application is approved the borrower shall thereafter be provided with a prepayment agreement for execution.

(b) The Administrator may limit the number of applications approved and closings scheduled from time to time, taking into account, among other matters, administrative considerations of the RUS.

§ 1786.158 Terms and conditions of prepayment agreement.

Upon receipt of a satisfactory application, RUS shall provide to the borrower for its execution a prepayment agreement, in form and substance satisfactory to RUS, which may include the following:

(a) Provide for the prepayment of one or more Qualified Notes from time to time, but no more than two closings may be scheduled in any calendar year unless a third closing is for the prepayment of all outstanding electric loans of the borrower;

(b) Set forth procedures and forms through which the borrower will notify the Government of each election it makes to prepay certain Qualified Notes upon a requested closing date and the Government will notify the borrower of the established closing date and prepayment amount for the Qualified Notes for each closing;

(c) Reserve to the Administrator the right to reschedule closing dates to meet administrative considerations;

(d) Set forth closing requirements identifying the location and manner of payment, and all documentation and information to be delivered prior to or at closing, including opinions of counsel and certificates from the borrower;

(e) Provide for notice by either telephone or facsimile to be given by RUS to the borrower not more than 8 nor less than 3 business days before a scheduled closing date of the amount to be paid at closing which shall include all accrued interest and the discounted present value of the Qualified Notes to be prepaid;

(f) Provide for notice of the 120 month period during which the borrower's eligibility for direct or insured loans will be restricted;

(g) Set forth representations and warranties;

(h) Require the borrower to prepay each Qualified Note specified in full;

(i) Require the borrower to identify the source of the financing that will be used directly or indirectly to refinance the Qualified Notes. If the source is other than internally generated funds, the borrower must certify in writing whether such financing will be tax exempt, and if tax exempt financing will be used, furnish all information on the terms and conditions of the financing as RUS may require;

(j) Require the borrower to rescind the unadvanced balance of all outstanding electric loans as of the date of initial closing;

(k) Require the borrower, if it is a party to a wholesale power contract with a power supply borrower, to provide the Administrator with such assurances as the Administrator may require that it is in compliance with and

will continue to comply with its obligation to such power supply borrower;

(l) Provide RUS, if the Administrator determines it necessary, with security for all outstanding rural development loans and amendments to any outstanding rural development loan agreements in form and substance, and on terms and conditions, satisfactory to RUS;

(m) Prescribe remedies for violating the terms and conditions of the prepayment agreement;

(n) Provide for termination by RUS of the right for the borrower to prepay thereunder;

(o) Provide evidence that any approvals required from any supplemental lender have been obtained; and

(p) Set forth such other terms and conditions as the Administrator shall deem appropriate.

§1786.159 Initial closing.

(a) Upon receipt of the prepayment agreement, the borrower may submit, pursuant to the terms of the prepayment agreement, a closing request which shall request a closing date no less than 30 business days from the date of the request.

(b) The Government will respond to the borrower's closing request by delivering a preclosing notice to the borrower not less than 10 business days prior to the date which the Government, after reviewing the borrower's closing request, selects as a closing date.

§1786.160 Subsequent closings.

(a) Each subsequent prepayment after the initial closing shall be facilitated with the submission of an additional closing request by the borrower. Each closing request must request a closing date no less than 30 business days from the date of the request.

(b) The Government will respond to each subsequent closing request by delivering a preclosing notice to the borrower not less than 10 business days prior to the date which the Government, after reviewing the borrower's closing request, selects as a closing date in each case.

§ 1786.161 Return of Qualified Notes and release of lien.

Upon payment to RUS at closing of the full amount specified in the notice delivered by RUS to the borrower pursuant to the terms of the prepayment agreement (see § 1786.158(e)), RUS will deliver to the borrower at closing those Qualified Notes which have been paid in full at such closing, and upon payment and discharge of all outstanding RUS debt obligations by the borrower, RUS will deliver to the borrower at the final closing a release of lien prepared by the borrower pursuant to the terms of the prepayment agreement.

§ 1786.162 Outstanding loan documents.

(a) Except as expressly provided in this subpart, the borrower shall comply with all provisions of its RUS Loan Contract, its outstanding notes issued to RUS, and the RUS Mortgage.

(b) Nothing in this subpart shall affect any rights of supplemental lenders under the RUS Mortgage, or other creditors of the borrower.

(c) Nothing in this subpart shall prohibit a borrower from making prepayments of any loans pursuant to the RE Act in accordance with the terms of such loans.

§ 1786.163 Existing wholesale power contracts.

(a) If the borrower is a party to a wholesale power contract with a power supply borrower financed pursuant to the RE Act, the Administrator may require that the borrower and the power supply borrower enter into a supplement to the outstanding wholesale power contract providing substantially as follows:

SAMPLE CONTRACT TERMS

So long as any of the notes evidencing secured loans of the power supply borrower are outstanding, the borrower will not, without the approval in writing of the power supply borrower and the Administrator, take or suffer to be taken any steps for reorganization or dissolution, or to consolidate with or merge into any corporation, or to sell, lease or transfer (or make any agreement therefor) all or a substantial portion of its assets, whether now owned or hereafter acquired. The power supply borrower will not unreasonably withhold or condition its consent to

any such, reorganization, dissolution, consolidation, or merger, or to any such sale, lease or transfer (or any agreement therefor) of assets. The power supply borrower will not withhold or condition such consent except in cases where to do otherwise would result in rate increases for the other members of the power supply borrower or impair the ability of the power supply borrower to repay its secured loans in accordance with their terms, or adversely affect system performance in a material way. Notwithstanding the foregoing, the borrower may take or suffer to be taken any steps for reorganization or dissolution or to consolidate with or merge into any corporation or to sell, lease or transfer (or make any agreement therefor) all or a substantial portion of its assets, whether now owned or hereafter acquired without the power supply borrower's consent, so long as the borrower shall pay such portion of the outstanding indebtedness on the power supply borrower's notes or other obligations as shall be determined by the power supply borrower with the prior written consent of the Administrator and shall otherwise comply with such reasonable terms and conditions as the Administrator and power supply borrower may require either: (1) To eliminate any adverse effect that such action seems likely to have on the rates of the other members of the power supply borrower, or

(2) To assure that the power supply borrower's ability to repay the secured loans and other obligations of the power supply borrower in accordance with their terms is not impaired.

The Administrator may require, among other things, that any payment owed under (2) of the preceding sentence that represents a portion of the power supply borrower's indebtedness on Notes shall be paid by the borrower in the manner necessary to accomplish a defeasance of those obligations in accordance with the loan documents relating thereto, or be paid directly to the holders of the Notes for application by them as prepayments in accordance with the provisions of such documents, or be paid to the power supply borrower and held and invested in a manner satisfactory to the Administrator.

[End of sample contract terms]

(b) The Administrator may exempt a borrower from the requirement to enter into a supplement to its outstanding wholesale power contract if the Administrator determines that such requirement is burdensome and unnecessary in light of the provisions of the existing wholesale power contract, other security arrangements of the power supply borrower, and any other relevant facts and circumstances.

Normally such exemption will be granted only with the concurrence of the power supply borrower.

§1786.164 Loan fund audit.

In the event that a borrower shall prepay all its outstanding electric loans RUS shall have the right to audit within six (6) months of closing transactions involving the RUS Construction Fund Account established and maintained by the borrower pursuant to the terms of the RUS Loan Contract and to inspect all books, records, accounts, and other documents and papers of the borrower. Should RUS determine that the borrower has made disbursements of funds advanced pursuant to the RUS Loan Contracts which do not comply with the requirements thereof, the borrower shall be required to pay the RUS an amount equal to the difference between the amount which the borrower prepaid under this subpart with respect to such advances, and the amount which the borrower would otherwise have been required to return to the RUS as a result of noncompliance if the borrower had not prepaid such advances, plus interest. (See 7 CFR part 1721, Post-Loan Policies and Procedures for Insured Electric Loans.)

§1786.165 Reporting.

Borrowers that no longer have any loans made or guaranteed by RUS and are considering applying for other financial assistance pursuant to the RE Act are encouraged to file the end-of-year operating report, RUS Form 7.

§1786.166 Approvals.

The borrower shall be responsible for obtaining all approvals necessary to consummate the transaction as required by the prepayment agreement, including such approvals as may be required by regulatory bodies and other lenders.

§1786.167 Restrictions to additional RUS financing.

(a) No borrower that prepays an electric loan at a discount as provided under this subpart may apply for or receive direct or insured loans during the 120 months from the most recent closing date, except at the discretion of the Administrator. During the 120 month period the Administrator may consider providing an insured loan if, among other matters, it is necessary to assure repayment of, or protect the Government's security for any outstanding loans or loan guarantees, or the borrower's system has suffered severe physical plant related damage due to conditions beyond its control and the borrower is unable to obtain financing at reasonable terms to restore the system from non-RUS sources, including the Federal Emergency Management Agency, and from private sources. Upon expiration of the 120 months, such borrowers may apply for direct or insured loans in the same manner as other borrowers provided that such borrowers may not apply for direct or insured loans for facilities, construction of which commenced prior to the expiration of the 120 months. Special provisions for mergers involving a borrower that has prepaid pursuant to this subpart are in 7 CFR 1717.158.

(b) Borrowers that prepay their direct or insured RUS loans under this subpart remain eligible for certain types of financial assistance under the RE Act, including loan guarantees and rural development loans.

[59 FR 13620, Mar. 22, 1994, as amended at 61 FR 66874, Dec. 19, 1996]

§1786.168 Borrowers who prepaid under this part prior to October 21, 1992.

(a) A borrower that had prepaid, prior to the date of enactment of Public Law 102–428 (106 Stat. 2183) on October 21, 1992, at a discount rate as provided at 7 CFR part 1786, subpart C:

(1) Shall not be eligible except at the discretion of the Administrator as stated in paragraph §1786.167(a), to apply for or receive direct or insured loans during the 180-month period beginning on the date of the prepayment; and

(2) Shall not be eligible to apply for or receive direct or insured loans from RUS until the borrower has repaid to the RUS the sum of:

(i) The amount (if any) by which the discount the borrower received by reason of the prepayment exceeds the discount the borrower would have received had the discount been based on the cost of funds to the Department of the Treasury as calculated at § 1786.153 at the time of the prepayment; and

(ii) Interest on the amount described in paragraph (a)(2)(i) of this section for the period beginning on the date of the prepayment and ending on the date of the repayment, at a rate equal to the average annual cost of borrowing by the Department of the Treasury. This rate will be calculated first on the date of prepayment and at one year intervals from that date based on the same U.S. Treasury issues published in the Federal Reserve Statistical Release closest to that date. The Treasury rate of interest to be applied for each year will be the rate for the Treasury issue of comparable maturity to the number of years from the prepayment date to the repayment date and at one year intervals thereafter.

(b) If a borrower and the Administrator have entered into an agreement with respect to a prepayment occurring before October 21, 1992, this section shall supersede any provision in the agreement relating to the restoration of eligibility for loans under the RE Act.

(c) Borrowers who prepaid prior to October 1, 1987, are eligible for assistance under the RE Act in the same manner as other borrowers with respect to loan guarantees and the rural development loans.

(d) During the 180 month period described in paragraph (a)(1) of this section the Administrator may consider providing an insured loan, if the conditions described in § 1786.167(a) exist.

(e) Borrowers may not apply for direct or insured loans for facilities, construction of which commenced prior to the expiration of the 180 month period described in paragraph (a)(1) of this section.

§ 1786.169 Liability.

It is the intent of this subpart that any failure on the part of RUS to comply with any provisions of this subpart, including without limitation, those provisions setting forth specified timeframes for action by RUS on applications for prepayments or closing requests, shall not give rise to liability of any kind on the part of the Government or any employees of the Government including, without limitation, liability for damages, fees, expenses or costs incurred by or on behalf of a borrower, private lender or any other party.

§ 1786.170 Prepayment of loans approved after December 20, 1993. [Reserved]

§§ 1786.171–1786.199 [Reserved]

Subpart G—Refinancing and Prepayment of RUS Guaranteed FFB Loans Pursuant to Section 306(C) of the RE Act

AUTHORITY: 7 U.S.C. 901 et seq.; Pub. L. 103–354, 108 Stat. 3178 (7 U.S.C. 6941 et seq.); sec. 1201(b) of subtitle B of title 1 of Pub. L. 103–66, 107 Stat. 312.

SOURCE: 58 FR 51008, Sept. 30, 1993, unless otherwise noted.

§ 1786.200 Purpose.

This subpart sets forth the policies and procedures of RUS through the existing FFB program, whereby borrowers may prepay and refinance, outstanding FFB Notes evidencing electric or telephone loans with FFB, pursuant to the provisions of section 306(C) of the RE Act as added by Public Law 103–66, 107 Stat. 312, enacted August 10, 1993.

§ 1786.201 Definitions and rules of construction.

(a) *Definitions*. As used in this subpart:

Administrator means the Administrator of the Rural Utilities Service (RUS).

Borrower means any organization which has an outstanding note(s) evidencing electric or telephone loans guaranteed by RUS, from FFB.

Business day means any such day on which both the Federal Financing Bank and Federal Reserve Bank—New York are open for business.

Electric loan means a loan made by FFB and guaranteed by RUS under section 306 of the RE Act for electric service.

FFB means the Federal Financing Bank, an instrumentality and wholly owned corporation of the United States.

Government means the United States of America, acting through the Administrator of the Rural Utilities Service.

Loan guarantee means RUS's guarantee under section 306 of the RE Act of a loan from FFB.

Payment date means the date that payment is due and is the last day in a calendar quarter.

Prepayment penalty means the same as prepayment premium.

Prepayment premium shall have the meaning set forth at § 1786.207.

RE Act means the Rural Electrification Act of 1936, as amended (7 U.S.C. 901 *et seq.*).

REA means the Rural Electrification Administration formerly an agency of the United States Department of Agriculture and predecessor agency to RUS with respect to administering certain electric and telephone loan programs.

Refinancing note shall have the meaning set forth at § 1786.206.

RUS means the Rural Utilities Service, an agency of the United States Department of Agriculture established pursuant to Section 232 of the Federal Crop Insurance Reform and Department of Agriculture Reorganization Act of 1994 (Pub. L. 103–354, 108 Stat. 3178), successor to REA with respect to administering certain electric and telephone programs. See 7 CFR 1700.1.

RUS loan contract means the agreement, as amended, supplemented, or restated from time to time, between a borrower and RUS providing for loans or loan guarantees pursuant to the RE Act.

RUS mortgage means collectively those mortgages and security agreements made by and between the borrower and the Government, securing indebtedness evidencing electric and telephone loans or loan guarantees made pursuant to RE Act. The term includes such mortgages regardless whether third parties are mortgagees with RUS.

Supplemental lender means a private lender whose loan to the borrower is secured under an RUS mortgage.

Telephone loan means a loan made by FFB and guaranteed by RUS under section 306 of the RE Act for telephone service.

(b) *Rules of construction.* Unless the context shall otherwise indicate, the terms defined in paragraph (a) of this section include the plural as well as the singular, and the singular as well as the plural. The words "herein," "hereof" and "hereunder", and words of similar import, refer to this subpart as a whole.

[58 FR 51008, Sept. 30, 1993, as amended at 59 FR 66440, Dec. 27, 1994]

§ 1786.202 Prepayment and refinancing of RUS guaranteed FFB loans.

The borrower of an electric or telephone loan made by the FFB and guaranteed by RUS under section 306 of the RE Act may, at the option of the borrower, refinance or prepay a loan or an advance on the loan, or any portion of the loan or advance in accordance with section 306C of the RE Act, after meeting certain conditions using the procedures prescribed in the note. After refinancing existing notes under this section, additional prepayments or refinancings will be governed by the terms of the refinancing note(s).

§ 1786.203 Special considerations.

Generally all FFB borrowers with loans guaranteed by RUS whose FFB notes have not been accelerated are eligible to prepay or refinance under this part. All requests for prepayment or refinancing will be processed in accordance with this subpart except that some requests for refinancing and prepayments are more complicated and thus will involve special considerations. These requests will have to be handled on a case by case basis and include:

(a) Telephone borrowers who are required to meet certain terms of their indenture;

(b) Borrowers who have amended their old form note or have already repriced prior to September 30, 1993;

(c) Borrowers that have been involved in a merger or consolidation;

(d) Borrowers whose obligations to RUS, FFB notes, or security instruments differ from those normally used;

(e) A request to prepay or refinance an amount of less than $100,000 or an amount of less than the full amount of an advance outstanding; or

397

(f) A request to prepay or refinance a note that includes unadvanced loan funds.

§ 1786.204 Limitations.

(a) No more than three refinancing notes will be executed for any borrower per calendar year.

(b) The borrower may not select a term for the refinanced advance that ends after the maturity date set for that advance.

§ 1786.205 Application procedure.

(a) Any borrower seeking to prepay or refinance an advance from the FFB under this subpart should apply by letter to the appropriate RUS Regional Director or, in the case of power supply borrowers, to the Director of the Power Supply Division. The borrower will be required to submit applications and elections in a digital format to be supplied by RUS. The application letter shall provide the following:

(1) Borrower's RUS designation;

(2) Borrower's name and address;

(3) Listing of each note to be prepaid by loan designation, RUS note number, RUS account number, advance date, maturity date, original amount, outstanding balance, and date(s) of any substitute FFB note(s) amending the original FFB Note;

(4) A statement of the borrower's intention to finance the premium by an addition to principal balance or to pay the premium in cash or with unsecured debt;

(5) A statement of the maturity options that the borrower wishes to select;

(6) Such additional information as the Administrator may request.

(b) Requests for refinancing or prepayment will ordinarily be processed in the order that they are received. Borrower's may withdraw an application by notifying the appropriate RUS office in which they filed the application.

(c) When the request for prepayment or refinancing is approved for processing the borrower will be provided with appropriate instructions, documents and forms which may include but are not limited to the following:

(1) An FFB refinancing note;

(2) Resolution of Board of Directors;

(3) Legal Opinion;

(4) Certificate of Secretary;

(5) Waiver of Notice;

(6) Notice to borrower electing an effective date other than a scheduled quarterly payment date (if applicable);

(7) Documentation of obligations secured pursuant to section 1786.208 if any; and

(8) Security instrument.

(Approved by the Office of Management and Budget under control number 0572–0032)

§ 1786.206 Refinancing note.

(a) RUS will issue a replacement guaranty for refinancing notes delivered to FFB to replace and substitute for existing FFB notes in connection with any refinancing by FFB pursuant to section 306C of the RE Act.

(b) Generally, refinancing notes will, to the extent practicable, consolidate all of a borrower's existing FFB notes which have been guaranteed by RUS and containing terms and conditions as FFB may require and RUS and the borrower may accept.

(c) Notwithstanding any contrary provision contained in this subpart, RUS will give preference to processing refinancings that utilize a generic form of refinancing note in the event that FFB prescribes one.

[58 FR 51008, Sep. 30, 1993; 58 FR 58729, Nov. 3, 1993]

§ 1786.207 Prepayment premium.

(a) A premium shall be assessed against a borrower that refinances or prepays a loan or loan advance, or any portion of a loan or advance, under this section. RUS will collect the prepayment premium as calculated by FFB. FFB will calculate this premium as described in this section. Except as provided in paragraph (b) of this section, the premium shall be equal to the lesser of:

(1) The difference between the outstanding principal balance of the loan being refinanced and the present value of the loan discounted at a rate equal to the then current cost of funds to the Department of the Treasury for obligations of comparable maturity to the loan being refinanced or prepaid;

(2) 100 percent of the amount of interest for 1 year on the outstanding principal balance of the loan or loan advance, or any portion of the loan or advance, being refinanced, multiplied by the ratio that:

(i) The number of quarterly payment dates between the date of the refinancing or prepayment and the maturity date for the loan advance; bears to

(ii) The number of quarterly payment dates between the first quarterly payment date that occurs 12 years after the end of the year in which the amount being refinanced was advanced and the maturity date of the loan advance; and

(3)(i) The present value of 100 percent of the amount of interest for 1 year on the outstanding principal balance of the loan or loan advance, or any portion of the loan or advance, being refinanced or prepaid; plus

(ii) For the interval between the date of the refinancing or prepayment and the first quarterly payment date that occurs 12 years after the end of the year in which the amount being refinanced or prepaid was advanced, the present value of the difference between:

(A) Each payment scheduled for the interval on the loan amount being refinanced or prepaid; and

(B) The payment amounts that would be required during the interval on the amounts being refinanced or prepaid if the interest rate on the loan were equal to the then current cost of funds to the Department of the Treasury for obligations of comparable maturity to the loan being refinanced or prepaid.

(b)(1) Except as provided in paragraph (b)(2) of this section, the premium provided by paragraph (a)(1) of this section shall be required for refinancing or prepayment under this section.

(2) In the case of a loan advanced under an agreement that permits the refinancing or prepayment of the loan advance based on the payment of 1 year of interest on the outstanding principal balance of the loan advance, a borrower may, in lieu of the premium required by paragraph (a)(1) of this section, pay a premium as provided by:

(i) Paragraph (a)(2) of this section, if the loan advance has reached the 12-

year maturity required under the loan agreement for the refinancing or prepayment; or

(ii) Paragraph (a)(3) of this section, if the loan advance has not reached the 12-year maturity required under the loan agreement for the refinancing or prepayment.

§ 1786.208 Increased principal.

A borrower can meet the premium requirements by increasing the outstanding principal balance of the loan advance that is being refinanced. If it does so the borrower shall make a payment at the time of the refinancing equal to 2.5 percent of the amount of the premium that is added to the outstanding principal balance of the loan.

§ 1786.209 Outstanding loan documents.

(a) Except as expressly provided in this subpart, the borrower shall comply with all provisions of its RUS loan contract, its outstanding notes issued to RUS, and the RUS mortgage.

(b) Nothing in this subpart shall affect any rights of supplemental lenders under the RUS mortgage or the rights of any other creditors of the borrower.

(c) Nothing in this subpart shall prohibit a borrower from making prepayments on any loans pursuant to the RE Act in accordance with the terms thereof or as may be otherwise permitted by law.

§ 1786.210 Approvals.

The borrower shall be responsible for obtaining all approvals necessary to consummate the transaction as required by the refinancing note, including such approvals as may be required by regulatory bodies and other lenders.

PART 1787—THE "BUY AMERICAN" REQUIREMENT

AUTHORITY: 7 U.S.C. 903.

SOURCE: 83 FR 60730, Nov. 27, 2018, unless otherwise noted.

§ 1787.1   General.

(a) The "Buy American" provision of the Rural Electrification Act of 1936 (RE Act) requires, to the extent practicable and the cost of which is not unreasonable, that RUS Borrowers use loan funds only for such manufactured articles, materials, and supplies as have been manufactured in the United States or in any eligible country, substantially all from articles, materials, or supplies mined, produced or manufactured, as the case may be, in the United States or any eligible country.

(b) Each RUS Borrower is responsible for assuring that its use of loan funds complies with this requirement, and that the contracts it enters into for construction, materials and equipment, and purchases with vendors contain the Buy American requirement, along with certification as to compliance, made through RUS Form 213.

§ 1787.2   Definitions.

For purpose of this part, the following terms have the following meanings:

Administrator. The Administrator of the RUS, or his/her designee.

Buy American. A provision of the RE Act requiring that loan funds only be used to purchase products made in the U.S. or an eligible country.

Component. Any article, material, or supply, whether manufactured or unmanufactured, that is directly incorporated into the end product at the final assembly location.

Domestic product. A product or like product which both:

(1) Is manufactured in the United States or in any eligible country; and

(2) Contains components manufactured in the United States or in any eligible country consisting of more than 50 percent of the total cost of all components used in the product.

Eligible country. Any country that the United States Trade Representative determines as having corporations located therein, as eligible to enter into contract with an RUS Borrower, under which loan funds will be provided for unmanufactured and manufactured goods.

Loan funds. Funds provided under an RUS direct or guaranteed loan.

Manufactured. The application of processes to alter the form or function of materials or of elements of the product such that value is added or the materials or elements are transformed into a new end product functionally different from that which would result from mere assembly of the materials or elements.

Nondomestic bid. An offer to sell a nondomestic product to an RUS borrower.

Nondomestic product. Any product other than a domestic product or product from an eligible country.

Product. An item of manufactured material or assembled components, which is complete and capable of performing an intended practical purpose.

RE Act. Rural Electrification Act of 1936, as amended (7 U.S.C. 901 *et seq.*).

RUS. The Rural Utilities Service.

RUS Borrower. Any organization that has an outstanding RUS loan made or guaranteed by RUS pursuant to the RE Act.

Telecommunications. Any communication service for the transmission or reception of voice, data, sounds, signals, pictures, writings, or signs of all kinds, by wire, fiber, radio, light, or other visual or electromagnetic means, including all telephone lines, facilities, or systems used in the rendition of such service; but shall not be deemed to mean message telegram service or community antenna television system services or broadcasting facilities other than those intended exclusively for educational purposes, or radio broadcasting services or facilities within the meaning of section 3(o) of the Communications Act of 1934, as amended.

Unmanufactured. With respect to articles, materials, or supplies, refers to such goods that have not been manu-factured.

§1787.3 Products constituting a portion of a purchase order or contract.

Where a supplier or contractor offers or furnishes several products under a purchase order or contract, the provisions of this part apply to each product individually.

§1787.4 Unmanufactured articles, materials, and supplies.

The Buy American requirement also applies to unmanufactured articles, materials, and supplies to be financed with RUS loan funds, and will be considered domestic if mined or produced in the United States or in an eligible country.

§1787.5 Eligible countries.

The United State Trade Representative (USTR) determines what countries are eligible countries with respect to purchases made by electric borrowers or telecommunications borrowers. A particular country may be determined to be an eligible country for purchases made by telecommunications borrowers, for electric borrowers, or both. RUS maintains the latest FEDERAL REGISTER notice on its website which sets out the list of Eligible Countries for each RUS program at *https:// www.rd.usda.gov/files/ UEP_Engineering_EligibleCountries.pdf*.

§1787.6 Nondomestic products.

A product is considered to be non-domestic for the purpose of compliance with the "Buy American" requirement if:

(a) The product is manufactured outside the United States or any eligible country; or

(b) The product is manufactured in the United States or in any eligible country, but the cost of nondomestic components used therein constitutes 50 percent or more of the cost of all components. The cost of components shall be determined on a comparable basis, so that only the cost of domestic and nondomestic components, up to the point where they are combined and manufactured into a complete product shall be considered.

(1) The determination of the cost of the nondomestic components of a product shall include:

(i) The price paid to the nondomestic source;

(ii) The cost of shipment to the port of entry into the United States;

(iii) Applicable tariffs or duties;

(iv) The cost of transportation from the port of entry to the distributor's plant or warehouse; and

(v) Profit, overhead, and commissions of domestic and nondomestic suppliers and subcontractors of the components.

(2) The following items shall not be considered in determining the cost of components, although they are proper elements in the determination of the final selling price of the product:

(i) Fabrication or processing costs, if any, of nondomestic or domestic components at the assembly plant, or any other place of fabrication in the United States or any eligible country;

(ii) Testing costs at the assembly plant or at the installation site;

(iii) Direct profit, overhead, and commissions of the domestic distributor; and

(iv) Cost of transportation from the domestic assembly point to the installation site.

§1787.7 Components.

Where a component is manufactured only determines whether the component is classified as domestic or non-domestic even if all the materials and subcomponents comprising the component are manufactured in ineligible countries. A component manufactured in the United States or in an eligible country shall be considered domestic when determining whether a product is classified as domestic or nondomestic. A component manufactured in an ineligible country shall be considered non-domestic.

§1787.8 Purchase of nondomestic products.

An RUS Borrower may only use loan funds to purchase a nondomestic product if a waiver pursuant to §1787.10 has been received by the Administrator before entering into a contract with the vendor. Should the Administrator deny the waiver request, the RUS Borrower must use its own funds for the expenditure.

§ 1787.9 Waivers.

Under limited circumstances the Administrator may waive the Buy American requirement with respect to a specific contract entered into between an RUS Borrower and a third party which will be paid for with loan funds, subject to §§ 1787.10 through 1787.14.

§ 1787.10 Applications for specific waivers.

RUS borrowers may request a specific waiver of the Buy American requirement through a written, detailed explanation showing that:

(a) The cost between the nondomestic product and domestic product is unreasonable;

(b) There is a non-availability of domestic products; or

(c) It is not in the public interest or impractical for the RUS Borrower to purchase a domestic product.

§ 1787.11 Cost differential.

By application pursuant to § 1787.10, the Administrator may waive the Buy American requirement if the cost of the domestic product is unreasonable. Given that RUS loans terms normally range from 20 to 35 years, and that additional costs will be magnified with interest over these terms, the Administrator has determined that if the lowest bid or offered price is a non-domestic bid that is at least 6 percent lower than the next lowest bid or offered price, the RUS Borrower may request a cost differential waiver. With respect to contracts that are not required to be bid, prices of market-available, domestic products must be used for comparison in a request for waiver.

§ 1787.12 Non-availability or shortages.

By application pursuant to § 1787.10, the Administrator may waive the Buy American requirement upon a showing that there is no domestic product available in the market in sufficient and reasonable quantities and of satisfactory quality, and that such shortage of suitable domestic alternatives jeopardizes the project being completed on budget and/or according to scheduled planning. A lack of responsive and responsible bids to a well-publicized request for bids will be presumed to meet the conditions of a non-availability waiver. With respect to contracts that are not required to be bid, sufficient evidence must be presented to the Administrator in order to make a determination.

§ 1787.13 Public interest or impracticality.

(a) By application pursuant to § 1787.10, the Administrator may waive the Buy American requirement upon a showing that application of the requirement would be inconsistent with the public interest or impractical for the RUS Borrower. With respect to impracticality, an RUS Borrower may request a waiver upon a showing that the domestic product is incompatible or impractical to integrate with existing, significant capital infrastructure or existing, critical software already in use. Notwithstanding, the burden shall rest with the RUS Borrower to present how the use of the domestic product would create a hardship or negatively impact its project.

(b) With respect to contracts that were approved by RUS based on a bidder or offer that originally certified compliance with the Buy America requirements, but which can no longer comply with such certification, the Administrator may grant an impracticality waiver based on a showing that the original certification was made in good faith and that the product cannot now be obtained domestically due to commercial impossibility or impracticability, or without undue hardship or a negative impact to the project.

(c) In determining whether to issue any public interest waiver, the Administrator will consider all appropriate factors on a case-by-case basis, unless a general waiver has already been issued by the Administrator with respect to the product.

§ 1787.14 General waivers.

(a) The Administrator may issue a general waiver for all RUS Borrowers for a determinate period, if the Administrator finds that such manufactured or unmanufactured goods are in shortage regionally or nationally, so as to avoid the administrative burden of issuing individual, specific waivers.

(b) The Administrator has determined that it is in the best interest of RUS to issue a permanent general public interest waiver from the Buy America requirements for "small purchases," which shall be published in the FEDERAL REGISTER for each program under the RE Act and amended as needed from time to time. In carrying out this exception, however, the Administrator shall ensure that contracts are not artificially fragmented.

APPENDIX A TO PART 1787—PRODUCT PROCUREMENT

This appendix shows an example of how the 6 percent differential is applied to determine award of a bid. In response to a request for bids for a digital central office a borrower receives four responsive bids to the specification, three domestic bids and one nondomestic bid. The nondomestic bid is the apparent low bid. We will consider in our analysis the nondomestic bid and the lowest domestic bid as shown in the following table.

	Nondomestic bid	Domestic bid
Total materials	$895,000	$920,000
Installation	155,000	177,000
Freight	+1,000	+1,500
Total bid	$1,051,000	$1,098,500

Please note that once the product has been determined as nondomestic, the 6 percent cost differential shall be applied to all the material content in the nondomestic bid, even if the nondomestic product includes domestic components.

In this example, 6 percent of the total material content in the nondomestic bid ($895,000) equals $53,700. This cost differential is added to the total nondomestic bid as shown in the following table.

Total of the nondomestic bid	$1,051,000
6% of the all material cost	+53,000
Total evaluated bid	$1,104,700

This total evaluated bid, (that is the nondomestic bid plus the 6% of the cost of its material content), is compared with all the domestic bids for award of the bid. In our example the domestic bid ($1,098,500) is lower than the nondomestic evaluated bid ($1,104,700).

The domestic bid becomes the low bid and the domestic bidder gets award of the bid. This product is classified as domestic since the cost of the domestic components used in the product constitutes more than 50 percent of the cost of all the components used.

PART 1788—RUS FIDELITY AND INSURANCE REQUIREMENTS FOR ELECTRIC AND TELECOMMUNICATIONS BORROWERS

Subpart A—Borrower Insurance Requirements

AUTHORITY: 7 U.S.C. 901 *et seq.*; 7 U.S.C. 1921 *et seq.*; 7 U.S.C. 6941 *et seq.*

SOURCE: 64 FR 2, Jan. 4, 1999, unless otherwise noted.

Subpart A—Borrower Insurance Requirements

§ 1788.1 General and definitions.

(a) The standard forms of documents covering loans made or guaranteed by the Rural Utilities Service contain provisions regarding insurance and fidelity coverage to be maintained by each borrower. This part implements those provisions by setting forth the requirements to be met by all borrowers.

(b) As used in this part:

Borrower means any entity with any outstanding loan made or guaranteed by RUS.

Irregularity has the meaning found in § 1773.2.

Loan documents means the loan agreement, notes, and mortgage evidencing or used in conjunction with an RUS loan.

Mortgage means the mortgage, deed of trust, security agreement, or other security document securing an RUS loan.

Mortgaged property means any property subject to the lien of a mortgage.

RUS means the Rural Utilities Service and includes the Rural Telephone Bank.

RUS loan means a loan made or guaranteed by RUS.

(c) RUS may revise these requirements on a case by case basis for borrowers with unusual circumstances.

§ 1788.2 General insurance requirements.

(a) Borrowers will take out, as the respective risks are incurred, and maintain the classes and amounts of insurance in conformance with generally accepted utility industry standards for such classes and amounts of coverage for utilities of the size and character of the borrower and consistent with Prudent Utility Practice. Prudent Utility Practice shall mean any of the practices, methods, and acts which, in the exercise of reasonable judgment, in light of the facts, including but not limited to, the practices, methods, and acts engaged in or approved by a significant portion of the electric utility industry in the case of an electric borrower or of the telecommunications industry in the case of a telecommunications borrowers prior thereto, known at the time the decision was made, would have been expected to accomplish the desired result consistent with cost-effectiveness, reliability, safety, and expedition. It is recognized that Prudent Utility Practice is not intended to be limited to optimum practice, method, or act to the exclusion of all others, but rather is a spectrum of possible practices, methods, or act which could have been expected to accomplish the desired result at the lowest reasonable cost consistent with cost-effectiveness, reliability, safety, and expedition.

(b) The foregoing insurance coverage shall be obtained by means of bond and policy forms approved by regulatory authorities having jurisdiction, and, with respect to insurance upon any part of the mortgaged property securing an RUS loan, shall provide that the insurance shall be payable to the mortgagees as their interests may appear by means of the standard mortgagee clause without contribution. Each policy or other contract for such insurance shall contain an agreement by the insurer that, notwithstanding any right of cancellation reserved to such insurer, such policy or contract shall continue in force for at least 30 days after written notice to each mortgagee of suspension, cancellation, or termination.

(c) In the event of damage to or the destruction or loss of any portion of the mortgaged property which is used or useful in the borrower's business and which shall be covered by insurance, unless each mortgagee shall otherwise agree, the borrower shall replace or restore such damaged, destroyed, or lost portion so that such mortgaged property shall be in substantially the same condition as it was in prior to such damage, destruction, or loss and shall apply the proceeds of the insurance for that purpose. The borrower shall replace the lost portion of such mortgaged property or shall commence such restoration promptly after such damage, destruction, or loss shall have occurred and shall complete such replacement or restoration as expeditiously as practicable, and shall pay or cause to be paid out of the proceeds of such insurance form all costs and expenses in connection therewith.

(d) Sums recovered under any policy or fidelity bond by the borrower for a loss of funds advanced under a note secured by a mortgage or recovered by any mortgagee or holder of any note secured by the mortgage for any loss under such policy or bond shall, unless applied as provided in the preceding paragraph, be used as directed by the borrower's mortgage.

(e) Borrowers shall furnish evidence annually that the required insurance and fidelity coverage has been in force for the entire year, and that the borrower has taken all steps currently necessary and will continue to take all steps necessary to ensure that the coverage will remain in force until all loans made or guaranteed by RUS are paid in full. Such evidence shall be in a form satisfactory to RUS. Generally a certification included as part of the RUS Financial and Statistical Report filed by the borrower annually (RUS Form 7 or Form 12 for electric borrowers, RUS Form 479 for telecommunications borrowers, or the successors to these forms) is sufficient evidence of this coverage.

§ 1788.3 Flood insurance.

(a) Borrowers shall purchase and maintain flood insurance for buildings in flood hazard areas to the extent available and required under the National Flood Insurance Act of 1968, as amended (42 U.S.C. 4001, *et seq.*) The insurance should cover, in addition to the building, any machinery, equipment, fixtures, and furnishings contained in the building.

(b) The National Flood Insurance Program (see 44 CFR part 59 *et seq.*) provides for a standard flood insurance policy; however, other existing insurance policies which provide flood coverage may be used where flood insurance is available in lieu of the standard flood insurance policy. Such policies must be endorsed to provide:

(1) That the insurer give 30 days written notice of cancellation or nonrenewal to the insured with respect to the flood insurance coverage. To be effective, such notice must be mailed to both the insured and RUS and other mortgagees if any and must include information as to the availability of flood insurance coverage under the National Flood Insurance Program, and

(2) That the flood insurance coverage is at least as broad as the coverage offered by the Standard Flood Insurance Policy.

§ 1788.4 Disclosure of irregularities and illegal acts.

(a) Borrowers must immediately report, in writing, all irregularities and all indications or instances of illegal acts in its operations, whether material or not, to RUS and the Office of the Inspector General (OIG). See 7 CFR 1773.9(c)(3) for OIG addresses. The reporting requirements for borrowers are the same as those for CPA's set forth in § 1773.9

(b) Borrowers are required to make full disclosure to the bonding company of the dishonest or fraudulent acts.

§ 1788.5 RUS endorsement required.

In the case of a cooperative or mutual organization, RUS requires that the following:

Endorsement Waiving Immunity From Tort Liability'' be included as a part of each public liability, owned, non-owned, hired automobile, and aircraft liability, employers' liability policy, and boiler policy:

The Insurer agrees with the Rural Utilities Service that such insurance as is afforded by the policy applies subject to the following provisions:

1. The Insurer agrees that it will not use, either in the adjustment of claims or in the defense of suits against the Insured, the immunity of the Insured from tort liability, unless requested by the Insured to interpose such defense.

2. The Insured agrees that the waiver of the defense of immunity shall not subject the Insurer to liability of any portion of a claim, verdict or judgment in excess of the limits of liability stated in the policy.

3. The Insurer agrees that if the Insured is relieved of liability because of its immunity, either by interposition of such defense at the request of the Insured or by voluntary action of a court, the insurance applicable to the injuries on which such suit is based, to the extent to which it would otherwise have been available to the Insured, shall apply to officers and employees of the Insured in their capacity as such; provided that all defenses other than immunity from tort liability which would be available to the Insurer but for said immunity in suits against the Insured or against the Insurer under the policy shall be available to the Insurer with respect to such officers and employees in suits against such officers and employees or against the Insurer under the policy.

§ 1788.6 RUS right to place insurance.

If a borrower fails to purchase or maintain the required insurance and fidelity coverage, the mortgagees may place required insurance and fidelity coverage on behalf and in the name of the borrower. The borrower shall pay the cost of this coverage, as provided in the loan documents.

§§ 1788.7–1788.10　[Reserved]

Subpart B—Insurance for Contractors, Engineers, and Architects, Electric Borrowers

§ 1788.11　Minimum insurance requirements for contractors, engineers, and architects.

(a) Each electric borrower shall include the provisions in this paragraph in its agreements with contractors, engineers, and architects, said agreements that are wholly or partially financed by RUS loans or guarantees. The borrower should replace "Contractor" with "Engineer" or "Architect" as appropriate.

1. The Contractor shall take out and maintain throughout the period of this Agreement insurance of the following minimum types and amounts:

a. Worker's compensation and employer's liability insurance, as required by law, covering all their employees who perform any of the obligations of the contractor, engineer, and architect under the contract. If any employer or employee is not subject to workers' compensation laws of the governing State, then insurance shall be obtained voluntarily to extend to the employer and employee coverage to the same extent as though the employer or employee were subject to the workers' compensation laws.

b. Public liability insurance covering all operations under the contract shall have limits for bodily injury or death of not less than $1 million each occurrence, limits for property damage of not less than $1 million each occurrence, and $1 million aggregate for accidents during the policy period. A single limit of $1 million of bodily injury and property damage is acceptable. This required insurance may be in a policy or policies of insurance, primary and excess including the umbrella or catastrophe form.

c. Automobile liability insurance on all motor vehicles used in connection with the contract, whether owned, non-owned, or hired, shall have limits for bodily injury or death of not less than $1 million per person and $1 million each occurrence, and property damage limits of $1 million for each occurrence. This required insurance may be in a policy or policies of insurance, primary and excess including the umbrella or catastrophe form.

2. The Owner shall have the right at any time to require public liability insurance and property damage liability insurance greater than those required in paragraphs (a)(1)(b) and (a)(1)(c) of this section. In any such event, the additional premium or premiums payable solely as the result of such additional insurance shall be added to the Contract price.

3. The Owner shall be named as Additional Insured on all policies of insurance required in (a)(1)(b) and (a)(1)(c) of this section.

4. The policies of insurance shall be in such form and issued by such insurer as shall be satisfactory to the Owner. The Contractor shall furnish the Owner a certificate evidencing compliance with the foregoing requirements that shall provide not less than 30 days prior written notice to the Owner of any cancellation or material change in the insurance.

(b) Electric borrowers shall also ensure that all architects and engineers working under contract with the borrower have insurance coverage for Errors and Omissions (Professional Liability Insurance) in an amount at least as large as the amount of the architectural or engineering services contract but not less than $500,000.

(c) The borrower may increase the limits of insurance if desired.

(d) The minimum requirement of $1 million of public liability insurance does not apply to contractors performing maintenance work, janitorial-type services, meter reading services, rights-of-way mowing, and jobs of a similar nature. However, borrowers shall ensure that the contractor performing the work has public liability coverage at a level determined to be appropriate by the borrower.

(e) If requested by RUS, the borrower shall provide RUS with a certificate from the contractor, engineer, or architect evidencing compliance with the requirements of this section.

§ 1788.12　Contractors' bonds.

Electric borrowers shall require contractors to obtain contractors' bonds when required by part 1726, Electric System Construction Policies and Procedures, of this chapter. Surety companies providing contractors' bonds shall be listed as acceptable sureties in the U.S. Department of Treasury Circular No. 570. The circular is maintained through periodic publication in the FEDERAL REGISTER and is available on the Internet under ftp://ftp.fedworld.gov/pub/tel/sureties.txt, and on the Department of the Treasury's computer bulletin board at 202–874–6817.

Subpart C—Insurance for Contractors, Engineers, and Architects, Telecommunications Borrowers

§ 1788.46 General.

This subpart sets forth RUS policies for minimum insurance requirements for contractors, engineers, and architects performing work under contracts which are wholly or partially financed by RUS loans or guarantees with telecommunications borrowers.

§ 1788.47 Policy requirements.

(a) Contractors, engineers, and architects performing work for borrowers under construction, engineering, and architectural service contracts shall obtain insurance coverage, as required in §1788.48, and maintain it in effect until work under the contracts is completed.

(b) Contractors entering into construction contracts with borrowers shall furnish a contractors' bond, except as provided for in §1788.49, covering all of the contractors' undertaking under the contract.

(c) Borrowers shall make sure that their contractors, engineers, and architects comply with the insurance and bond requirements of their contracts.

§ 1788.48 Contract insurance requirements.

Contracts entered into between borrowers and contractors, engineers, and architects shall provide that they take out and maintain throughout the contract period insurance of the following types and minimum amounts:

(a) Workers' compensation and employers' liability insurance, as required by law, covering all their employees who perform any of the obligations of the contractor, engineer, and architect under the contract. If any employer or employee is not subject to the workers' compensation laws of the governing state, then insurance shall be obtained voluntarily to extend to the employer and employee coverage to the same extent as though the employer or employee were subject to the workers' compensation laws.

(b) Public liability insurance covering all operations under the contract shall have limits for bodily injury or death of not less than $1 million each occurrence, limits for property damage of not less than $1 million each occurrence, and $1 million aggregate for accidents during the policy period. A single limit of $1 million of bodily injury and property damage is acceptable. This required insurance may be in a policy or policies of insurance, primary and excess including the umbrella or catastrophe form.

(c) Automobile liability insurance on all motor vehicles used in connection with the contract, whether owned, non-owned, or hired, shall have limits for bodily injury or death of not less than $1 million per person and $1 million per occurrence, and property damage limits of $1 million for each occurrence. This required insurance may be in a policy or policies of insurance, primary and excess including the umbrella or catastrophe form.

(d) When a borrower contracts for the installation of major equipment by other than the supplier or for the moving of major equipment from one location to another, the contractor shall furnish the borrower with an installation floater policy. The policy shall cover all risks of damage to the equipment until completion of the installation contract.

§ 1788.49 Contractors' bond requirements.

Construction contracts in amounts in excess of $250,000 for facilities shall require contractors to secure a contractors' bond, on a form approved by RUS, attached to the contract in a penal sum of not less than the contract price, which is the sum of all labor and materials including owner-furnished materials installed in the project. RUS Form 168b is for use when the contract exceeds $250,000. RUS Form 168c is for use when the contractor's surety has accepted a Small Business Administration guarantee and the contract is for $1,000,000 or less. For minor construction contracts under which work will be done in sections and no section will exceed a total cost of $250,000, the borrower may waive the requirement for a contractors' bond.

§ 1788.50 Acceptable sureties.

Surety companies providing contractors' bonds shall be listed as acceptable sureties in the U.S. Department of Treasury Circular No. 570. The circular is maintained through periodic publication in the FEDERAL REGISTER and is available on the Internet under ftp://ftp.fedworld.gov/pub/tel/sureties.txt, and on the Department of the Treasury's computer bulletin board at 202-874-6817.

§§ 1788.51-1788.53 [Reserved]

§ 1788.54 Compliance with contracts.

It is the responsibility of the borrower to determine, before the commencement of work, that the engineer, architect, and the contractor have insurance that complies with their contract requirements.

§ 1788.55 Providing RUS evidence.

When RUS shall specifically so direct, the borrower shall also require the engineer, the architect, and the contractor, to forward to RUS evidence of compliance with their contract representative of the insurance company and include a provision that no change in or cancellation of any policy listed in the certificate will be made without the prior written notice to the borrower and to RUS.

PART 1789—USE OF CONSULTANTS FUNDED BY BORROWERS

Subpart A—Policy and Procedures With Respect to Consultant Services Funded by Borrowers—General

Subpart B—Escrow Account Funding and Payments

AUTHORITY: 7 U.S.C. 901-950b; Pub. L. 103-354, 108 Stat. 3178 (7 U.S.C. 6941 et seq.).

SOURCE: 61 FR 48606, Sept. 16, 1996, unless otherwise noted.

Subpart A—Policy and Procedures With Respect to Consultant Services Funded by Borrowers—General

§ 1789.150 Purpose.

This part sets forth policies and the procedures for implementing subsection (c) of section 18 of the Rural Electrification Act of 1936, as amended (7 U.S.C. 901 et seq.)(RE Act) which authorizes the Rural Utilities Service (RUS) to use the services of Consultants funded by the Borrowers to facilitate timely action on Applications by Borrowers for financial assistance and other approvals.

§ 1789.151 Definitions.

As used in this part:

Administrator means the Administrator of the Rural Utilities Service (RUS).

Application means a request for financial assistance under the RE Act or such other approvals as may be required of the RUS pursuant to the terms of outstanding loan or security instruments or otherwise.

Borrower means any organization which has an outstanding loan(s) made or guaranteed by RUS or its predecessor agency, the Rural Electrification Administration (REA) under the RE Act or any organization which has submitted or submits an Application before RUS.

Consultant means a person or firm which has been retained pursuant to this subpart under a contract to provide financial, legal, engineering, environmental, or other technical advice and services.

Consultant Contract means a contract for the performance of consulting services for RUS, to be paid using funds provided by a Borrower, which may be in the form of a Retainer Contract, purchase order, or other form as may be appropriate.

Escrow Account means an account established pursuant to § 1789.158.

Escrow Agreement means an agreement, between a Borrower, a Consultant and a Third-party Commercial Institution, meeting the requirements of § 1789.167.

Final Invoice means the closing Invoice prepared for a given Task Order.

Financial Consultant means a Consultant retained pursuant to this part to provide financial advisory services.

Funding Agreement means an agreement, between a Borrower and a Consultant, providing for the Borrower to fund the costs of a Task Order and otherwise meeting the requirements of § 1789.166.

Indemnification Agreement means an agreement by a Borrower meeting the requirements of § 1789.162.

Invoice means an invoice prepared by a Consultant pursuant to the terms of a Consultant Contract.

Legal Consultant means any Consultant retained pursuant to this part to provide legal services to RUS.

Notice of Proposal to Fund means a notice meeting the requirements of § 1789.156 provided to RUS by the Borrower.

Organizational conflict of interest means that because of other activities or relationships with other persons, a person is unable or potentially unable to render impartial assistance or advice to the Government, or the person's objectivity in performing the contract work is or might be otherwise impaired, or a person has an unfair competitive advantage.

Retainer Contract means a Consultant Contract providing for a minimum required payment to a Consultant irrespective of whether services are utilized by RUS thereunder.

Task Order means a written request for consultant services pursuant to the terms of a Consultant Contract.

Third-party Commercial Institution means a commercial financial institution mutually acceptable to the Borrower and the Consultant.

§ 1789.152 Policy.

(a) As provided in this subpart, RUS may, at its discretion, use the services of Consultants funded by a Borrower where such services will facilitate timely action on an Application by such Borrower for financial assistance or other approvals. Such Consultants may provide financial, legal, engineering, environmental or other technical advice and services in connection with the review of an Application.

(b) With the approval of RUS, a Borrower may fund the cost of consulting services in connection with the review by RUS of an Application by such Borrower. Such funding shall be provided pursuant to the terms of a Funding Agreement between the Borrower and the Consultant designated by RUS.

(c) RUS may not, without the consent of the Borrower, require, as a condition of processing any Application for approval, that the Borrower agree to pay the costs of a Consultant hired to provide services to RUS.

(d) The government shall retain sole discretion in the selection of Consultants to provide services to RUS and the form of contract utilized. RUS may either use the services of one or more Consultants retained under Retainer Contracts or the government may elect to retain a Consultant not otherwise on retainer. The government shall have sole discretion to prescribe terms and conditions of Consultant Contracts. The Borrower may bring considerations to the attention of the government which the Borrower deems pertinent to the selection process.

(e) RUS shall retain sole discretion as to whether to further pursue use of an outside consultant for the relevant application in the event the Borrower does not enter into the agreements referenced in § 1789.158(c)(3)(iii) within 60 days of the government providing to the Borrower the information set forth in § 1789.158(c)(3).

§ 1789.153 Borrower funding.

Borrowers shall use their general funds for the purposes of funding consultant services hereunder. Borrowers may not use the proceeds of loans made or guaranteed under the RE Act for costs incurred by Borrowers pursuant to the funding of consultant services for RUS.

§ 1789.154　Eligible borrowers.

All Borrowers are eligible to fund consultant services under this part.

§ 1789.155　Approval criteria.

RUS will consider approving the use of consultant services funded by a Borrower on a case by case basis taking into account, among other matters, the following:

(a) Whether such services are required to facilitate timely action on a Borrower's Application. RUS shall determine what represents timely action with respect to each Application considering, among other matters, the review period normally required for such projects by RUS and other lenders and the consequences to the Borrower of adjusting the review period.

(b) The availability of staff resources, the priorities of other projects then before RUS, and the efficiencies to be realized from the use of consultant services.

(c) Whether it is in the best interest of RUS to use Borrower-funded Consultants. Certain types of projects, such as those involving issues of program-wide significance, may not be well suited for the use of Borrower funded Consultants.

§ 1789.156　Proposal procedure.

(a) In the event RUS determines that consideration should be given to the use of a Borrower-funded consultant in connection with the review of an Application, the RUS Regional Director or the Director of the Power Supply Division, as appropriate, will discuss with the Borrower the nature of the Application and the projected review period required of RUS. If RUS concludes that the projected review period will not result in timely action on the Application, and after being so notified in writing by RUS the Borrower wishes to fund consultant services to facilitate RUS review, the Borrower shall submit to the same Director a funding proposal. The proposal shall set forth the following:

(1) Identification in the heading or caption as a Notice of Proposal to Fund Consulting Services;

(2) Borrower's REA/RUS designation;

(3) Borrower's legal name and address;

(4) A description of the Application, critical issues and concerns relating to the Application, time deadlines, and the consequences of any delays in RUS review;

(5) A description of the consulting service(s) that would facilitate timely RUS review of the Application; and

(6) Such additional documents and information as RUS may request.

(b) RUS will review the Notice of Proposal to Fund and any additional information RUS deems relevant in determining whether to proceed with procuring Borrower funded Consultants. If RUS proposes to utilize Legal Consultants, RUS must obtain the concurrence of the Office of General Counsel (OGC) of the Department of Agriculture. RUS will notify the Borrower in writing of its determination.

§ 1789.157　Consultant contract.

(a) The Federal Acquisition Regulation (FAR), 48 CFR Ch. 1, and the Agriculture Acquisition Regulation (AGAR), 48 CFR Ch. 4, shall apply to all Consultant Contracts entered into pursuant to this part except as provided in this section.

(1) Contracts for Legal Consultants shall provide for a technical representative from OGC.

(2) All Consultant Contracts shall provide for an escrow account funding mechanism pursuant to this part and for the government's sole discretion in determining whether payments are to be made from the Escrow Account to the Consultant.

(3) All Consultant Contracts shall provide that payment of all obligations for work performed thereunder must be satisfied by amounts available in the Escrow Account; with the exception of the annual retainer fee, if any, Consultants shall not be entitled to any payments from the government.

(b) The provisions of paragraph (a) of this section shall be given prominent emphasis in requests for proposals issued under this part.

§ 1789.158　Implementation.

(a) Upon making a determination to go forward with Borrower funding for consulting services, RUS shall initiate a procurement request for a Consultant to provide the services. The government may either contract with a Consultant on a case by case basis or elect to use a Consultant pursuant to an outstanding Retainer Contract. The Borrower will not be informed of the Consultant selected until such time as the government provides the information set forth in paragraph (c)(3) of this section.

(b) If the government determines to contract with a Consultant on a case by case basis, the government shall notify the Borrower of the applicable procedures.

(c) If the government determines to contract with a Consultant under an outstanding Retainer Contract, the following procedures will normally apply:

(1) Pursuant to the terms of the contract, the government will prepare a draft Task Order requesting consultant services in connection with the review of the Borrower's Application. The draft Task Order shall set forth for the Consultant's review and acceptance, a description of the services to be provided and applicable time frames for the provision of such services.

(2) The government will request that the Consultant:

(i) Notify the government as to the acceptability of the form and substance of the draft Task Order;

(ii) Notify the government as to its ability to provide a satisfactory conflict of interest certification consistent with the requirements of the FAR (48 CFR ch. 1); and

(iii) Provide a cost estimate for the draft Task Order.

(3) When the government is satisfied with the response(s) received pursuant to paragraph (c)(2) of this section, the government shall promptly provide to the Borrower:

(i) A copy of the draft Task Order identifying the Consultant;

(ii) The Consultant's cost estimate for the draft Task Order; and

(iii) Contract information required to enable the Borrower to develop a Funding Agreement, an Escrow Agreement and an Indemnification Agreement (the "agreements").

(4) The Borrower shall develop and submit to the government for approval executed originals of:

(i) The agreements; and

(ii) A certified copy of a resolution of the board of directors authorizing the Borrower to enter into the agreements and to take such other action as is necessary to effect the purposes of the agreements.

(5) Upon receiving written RUS approval of the agreements and the form and substance of the board resolution, the Borrower shall:

(i) Establish and fund the Escrow Account; and

(ii) Provide written notice to the government of the Escrow Account number, the funding thereof, and such other information as required pursuant to the agreements.

(6) After the Borrower has funded the Escrow Account, the government shall issue Task Order(s) for consultant services in accordance with the terms and conditions of the applicable Retainer Contract.

§1789.159 Contract administration.

The government shall be solely responsible for the administration of a Consulting Contract and shall have complete control over the scope of the Consultant's work, the timetable for performance, the standards to be applied in determining the acceptability of deliverables and the approval of payment of Invoices.

§1789.160 Access to information.

The Borrower shall not have rights in nor right of access to the work product of the Consultant. All analyses, studies, opinions, memoranda, and other documents and information provided by the Consultant pursuant to a Consulting Contract may be released and made available to the Borrower only with the approval of RUS. This section does not restrict release of information by RUS pursuant to the Freedom of Information Act (5 U.S.C. 552(a)(2)) or other legal process.

§1789.161 Conflicts of interest.

The standard for determining organizational conflicts of interest shall be as set forth in the FAR subpart 9.5 (48 CFR part 9, subpart 9.5); however, the identification of the existence of an organizational conflict of interest may be made by either the Administrator or the cognizant Contracting Officer. In the event an organizational conflict of interest is determined to exist, the cognizant Contracting Officer shall take the actions prescribed at FAR 9.504 (48 CFR 9.504) to attempt to avoid, neutralize or mitigate the conflict. Should these actions be deemed by the Administrator and the Contracting Officer to adequately resolve the conflict, the contracting action with the offeror/contractor may proceed. Should the Administrator or the Contracting Officer determine that an organizational conflict of interest still exists such that contract award or other contracting action cannot be taken (award of task/delivery order, etc.) the offeror/contractor shall be so informed by the Contracting Officer and be provided a reasonable opportunity to respond in accordance with FAR 9.504(e) (48 CFR 9.504(e)). After considering the contractor's response, if it is found by both the Administrator and Contracting Officer to remedy the conflict of interest, the contracting action may proceed. If the Administrator and Contracting Officer determine that the contractor's response does not resolve the conflict of interest, yet continuing with the contracting action with the offeror/contractor in question is considered in the best interest of the United States, a waiver in accordance with FAR 9.503 (48 CFR 9.503) may be executed. This waiver shall be submitted under the Contracting Officer's signature and approved by the Administrator. The Administrator has been delegated Head of Contracting Activity authority by the USDA Senior Procurement Executive solely for the purpose of waiver approval.

§ 1789.162 Indemnification agreement.

As a condition of approving Borrower funding, the government will require the Borrower to enter into an Indemnification Agreement, in form and substance satisfactory to RUS, providing that the Borrower will indemnify and hold harmless the government and any officers, agents or employees of the government from any and all liability, including costs, fees, and settlements arising out of, or in any way connected with the payment of the Consultant's fee pursuant to the Consultant Contract. The Indemnification Agreement may recognize, as a condition of liability thereunder, the rights of the borrower to prompt notice, to use of counsel of its own choosing, and to participation in any settlement of a claim against which indemnification is sought.

§ 1789.163 Waiver.

RUS may waive any requirement or procedure of this subpart by determining that its application in a particular situation would not be in the government's interest, except that certain provision that the subject contracts are subject to the provisions of the FAR (48 CFR ch. 1) and AGAR (48 CFR ch. 4).

§§ 1789.164–1789.165 [Reserved]

Subpart B—Escrow Account Funding and Payments

§ 1789.166 Terms and conditions of funding agreement.

Funding Agreements between the Borrower and a Consultant shall be in form and substance satisfactory to RUS and provide for, among other matters, the following:

(a) Specific reference by number to the applicable Consulting Contract entered into between the government and the Consultant;

(b) Specific reference by number to the applicable Task Order (where applicable);

(c) A brief description of the Application;

(d) A requirement that Invoices make specific reference to:

(1) The applicable contract and Task Order(s); and

(2) The Escrow Account from which payment is to be made;

(e) A requirement that the Final Invoice for a Task Order be clearly identified as such;

(f) A description of the services to be provided by the Consultant to RUS and the applicable time frames for the provision of such services;

(g) Agreement that the Borrower shall pay for the Consultant services provided to RUS under the applicable contract through an Escrow Account established pursuant to an Escrow Agreement, the Consultant shall not provide services to RUS under the applicable contract unless there are sufficient funds in the Escrow Account to pay for such services, the Consultant shall seek compensation for services provided under the applicable contract from, and only from, funds made available through the Escrow Account, and the Consultant must submit all Invoices to the government for approval.

(h) A form of Escrow Agreement satisfactory to the Borrower, Consultant and the designated Third-party Commercial Institution;

(i) A schedule setting forth when and in what amounts the Borrower shall fund the Escrow Account;

(j) Acknowledgment by the Consultant of the Indemnification Agreement provided by the Borrower to the government; and

(k) The Funding Agreement shall not be effective unless and until approved in writing by RUS.

§ 1789.167 Terms and conditions of escrow agreement.

Escrow Agreements between and among the Borrower, Consultant and Third-party Commercial Institution shall be in form and substance satisfactory to RUS and provide for, among other matters, the following:

(a) Specific reference by number to the applicable contract for services;

(b) Specific reference by number to the applicable Task Order;

(c) Specific reference by number to the Escrow Account into which funds are to be deposited;

(d) Invoices to specifically identify the applicable contract and Task Order(s);

(e) Funds to be held in the Escrow Account by the escrow agent until paid to the Consultant pursuant to the government's authorization;

(f) The Escrow Account to be closed and all remaining funds remitted to the Borrower after payment of the Final Invoice, unless otherwise directed by the government;

(g) The government, the Consultant and the Borrower to have the right to be informed, in a timely manner and in such form as they may reasonably request, as to the status of and activity in the Escrow Account; and

(h) The Escrow Agreement shall not be effective unless and until approved in writing by RUS.

§§ 1789.168–1789.175 [Reserved]

PART 1792—COMPLIANCE WITH OTHER FEDERAL STATUTES, REGULATIONS, AND EXECUTIVE ORDERS

Subparts A–B [Reserved]

Subpart C—Seismic Safety of Federally Assisted New Building Construction

Sec.
1792.101 General.
1792.102 Definitions.
1792.103 Seismic design and construction standards for new buildings.
1792.104 Seismic acknowledgments.

AUTHORITY: 7 U.S.C. 901 et seq., 1921 et seq., 6941 et seq.; 42 U.S.C. 7701 et seq.; E.O. 12699 (3 CFR, 1990 Comp., p. 269).

SOURCE: 58 FR 32437, June 10, 1993, unless otherwise noted.

Subparts A–B [Reserved]

Subpart C—Seismic Safety of Federally Assisted New Building Construction

§ 1792.101 General.

(a) The Earthquake Hazards Reduction Act of 1977 (42 U.S.C. 7701 et seq.) was enacted to reduce risks to life and property through the establishment and maintenance of an effective earthquake hazards reduction program (the National Earthquake Hazards Reduction Program or NEHRP). The Federal Emergency Management Agency (FEMA) is designated as the agency with the primary responsibilities to plan and coordinate the NEHRP. This program includes the development and implementation of feasible design and construction methods to make structures earthquake resistant. Executive Order 12699 of January 5, 1990, Seismic Safety of Federal and Federally Assisted or Regulated New Building Construction (3 CFR, 1990 Comp., p. 269), requires that measures to assure seismic safety be imposed on federally assisted new building construction.

(b) This subpart identifies acceptable seismic standards which must be employed in new building construction funded by loans, grants, or guarantees made by the Rural Utilities Service (RUS) or the Rural Telephone Bank (RTB) (or through lien accommodations or subordinations approved by RUS or RTB).

[58 FR 32437, June 10, 1993, as amended at 65 FR 76916, Dec. 8, 2000]

§ 1792.102 Definitions.

As used in this subpart, the following terms have the following meaning:

Administrator— Administrator of RUS or the Governor of the Rural Telephone Bank or his or her designee.

Borrower—An entity which borrows or seeks to borrow money from, or arranges financing with the assistance of RUS through guarantees, lien accommodations or lien subordinations.

Building—Any structure, fully or partially enclosed, used or intended for sheltering persons or property.

Federally assisted—The provision of financing assistance by RUS through loans, loan guarantees, grants, and lien accommodations and subordinations.

Grant recipient—Any entity which receives a grant from RUS.

Lien accommodation—The consensual sharing of the government's (RUS's) lien on property or the release of government's lien on property.

Lien subordination—Agreement that the government's (RUS) lien on property will rank below the lien of another entity.

Model Code—A building code developed for the adoption of local or state authorities or to be used as the basis of a local or state building code.

NEHRP—National Earthquake Hazards Reduction Program.

Registered—A person licensed by the State(s) or Authority(ies) to perform architectural or engineering services in the State(s) where construction occurs.

RUS—Rural Utilities Service, and for the purposes of this subpart, shall include the Rural Telephone Bank. For the purposes of RTB borrowers, as used in this subpart, RUS means RTB and Administrator means Governor.

State—Each of the 50 States of the United States, the District of Columbia, and territories and possessions of the United States which are authorized to receive loans, loan guarantees, or grants from RUS.

[58 FR 32437, June 10, 1993, as amended at 59 FR 66440, Dec. 27, 1994; 65 FR 76916, Dec. 8, 2000; 69 FR 23642, Apr. 30, 2004]

§ 1792.103 Seismic design and construction standards for new buildings.

(a) In the design and construction of federally assisted buildings, the borrowers and grant recipients must utilize the seismic provisions of the most recent edition of those standards and practices that are substantially equivalent to or exceed the seismic safety level in the 2000 edition of the NEHRP Recommended Provisions for the Development of Seismic Regulation for New Buildings.

(b) Each of the following model codes or standards provides a level of seismic safety substantially equivalent to that provided by the 2000 NEHRP Recommended Provisions and are appropriate for federally assisted new building construction:

(1) *2003 NFPA 5000 Building Construction and Safety Code.* Copies of the book are available from the NFPA (National Fire Protection Association), 1 Batterymarch Park, Quincy, MA 02269–7471. Telephone: (617) 770–3000. Fax: (617) 770–0700.

(2) 2002 American Society of Civil Engineers (ASCE) 7, *Minimum Design Loads for Buildings and Other Structures.* Copies are available from the American Society of Civil Engineers, Publications Marketing Department, 1801 Alexander Bell Drive, Reston, VA 20191–4400. E-mail: *marketing@asce.org.* Telephone: (800) 548–2723. Fax: (703) 295–6211.

(3) *2003 International Code Council (ICC) International Building Code (IBC).* Copies of the book or CD-ROM are available from the International Conference of Building Officials, 4051 West Flossmoor Rd., Country Club Hill, IL 60478. Telephone: (800) 786–4452. Fax: (800) 214–7167.

(c) The NEHRP Recommended Provisions for the Development of Seismic Regulations for New Buildings is available from the Office of Earthquakes and Natural Hazards, Federal Emergency Management Agency, 500 C Street, SW., Washington, DC 20472.

[69 FR 23642, Apr. 30, 2004, as amended at 71 FR 60658, Oct. 16, 2006]

§1792.104   Seismic acknowledgments.

For each applicable building, borrowers and grant recipients must provide RUS a written acknowledgment from a registered architect or engineer responsible for the building design stating that seismic provisions pursuant to §1792.103 of this subpart will be used in the design of the building.

(a) For projects in which plans and specifications are required to be submitted to RUS, this acknowledgement shall be on the title page of the drawings included with the final plans and specifications. This acknowledgement will include the identification and date of the model code or standard that is used in the seismic design of the building project. The plans and specifications must be dated, signed, and sealed by the registered architect or engineer.

(b) For projects in which plans and specifications are not submitted, this acknowledgement shall be in the form of a statement from the architect or engineer responsible for the building design. The statement shall identify the model code or standard identified that is used in the seismic design of the building or buildings and, shall be dated and signed.

[69 FR 23642, Apr. 30, 2004]

PARTS 1794–1799 [RESERVED]

CHAPTER XVIII—RURAL HOUSING SERVICE, RURAL BUSINESS-COOPERATIVE SERVICE, RURAL UTILITIES SERVICE, AND FARM SERVICE AGENCY, DEPARTMENT OF AGRICULTURE

EDITORIAL NOTE: Nomenclature changes to chapter XVIII appear at 61 FR 1109, Jan. 16, 1996, and 61 FR 2899, Jan. 30, 1996.

SUBCHAPTER A—GENERAL REGULATIONS

SUBCHAPTER B—LOANS AND GRANTS PRIMARILY FOR REAL ESTATE PURPOSES

SUBCHAPTERS C–D [RESERVED]

SUBCHAPTER E—ACCOUNT SERVICING

SUBCHAPTER F—SECURITY SERVICING AND LIQUIDATIONS

SUBCHAPTER G—MISCELLANEOUS REGULATIONS

SUBCHAPTER H—PROGRAM REGULATIONS

SUBCHAPTER A—GENERAL REGULATIONS

PARTS 1800–1805 [RESERVED]

PART 1806—INSURANCE

AUTHORITY: 5 U.S.C. 301; 7 U.S.C. 1989; 42 U.S.C. 1480.

Subpart A—Real Property Insurance

AUTHORITY: 7 U.S.C. 1989; 42 U.S.C. 1480; 42 U.S.C. 2942; 5 U.S.C. 301; delegation of authority by the Sec. of Agri., 7 CFR 2.23; delegation of authority by the Asst. Sec. for Rural Development, 7 CFR 2.70; delegations of authority by Dir., OEO 29 FR 14764, 33 FR 9850.

§ 1806.1   General.

(a) *Authority.* This subpart sets forth the policies and procedures regarding insurance requirements on real property which serves as security for a debt under the Multi-Family Housing Programs of the Rural Housing Service (RHS), herein referred to as the "Agency." This subpart is inapplicable to Farm Service Agency, Farm Loan Programs.

(b) *Borrower to furnish insurance.* The real estate mortgage executed by the borrower provides that he will furnish and continually maintain and pay for insurance on buildings situated or constructed on the property with companies, in amounts, and on terms and conditions satisfactory to RD until the loan is repaid.

(c) *Borrower's selection of company.* The borrower may select the insurance company provided that the company and insurance policy comply with all the requirements set forth in this Instruction.

(d) *Responsibility.* The County Supervisor is responsible for taking all actions in connection with insurance as may be necessary to protect the security interest of RD. Any unusual situation that may arise with respect to obtaining or servicing insurance should be referred to the State Director. The State Director will refer any questions of a legal nature to the Office of the General Counsel (OGC).

(e) *Use of Form RD 426–1, "Valuations of Buildings."* The minimum insurance required will be indicated in the appraisal report by the employee who makes the appraisal of property that includes insurable buildings. In the case where no real estate appraisal is required or the appraisal report does not indicate the minimum insurance coverage, Form RD 426–1 will be prepared by the County Supervisor. Reevaluation of the buildings will not be done on appraisal reports; however, when new buildings are constructed or values increase or decrease materially and reevaluation is necessary to properly reflect the buildings' security interest of RD, the County Supervisor will prepare or revise Form RD 426–1 as appropriate. Changes made on an existing Form RD 426–1 will be dated and initialed. The reason for any deletion will be noted on the Form.

[41 FR 34571, Aug. 16, 1976, as amended at 61 FR 59777, Nov. 22, 1996; 72 FR 64121, Nov. 15, 2007; 80 FR 9864, Feb. 24, 2015]

§ 1806.2   Companies and policies.

Property insurance policies or other evidence of insurance will be accepted from borrowers when the requirements outlined herein are complied with fully.

(a) *Companies.* It is desirable that companies be licensed to do business in the particular State or other jurisdiction where the property is located, or that they be otherwise authorized by law to transact business within such State or other jurisdiction (hereinafter called "State"). If the required insurance is not available locally at comparable rates from an insurance company licensed or otherwise authorized to do business in the State, insurance may be accepted from another company if (1) the OGC advises that policies issued by such company will not be rendered unenforceable by virtue of the company's failure to be licensed or otherwise authorized to transact business in the State and that the company is a legal entity which may be sued in the State where the insured property is located, and (2) the State Director determines that the company is reputable and financially sound. In making the above determinations, the State Director will consider all relevant available information such as that which may be obtained from financial statements, Best's Insurance Reports, State insurance authorities, and other lending institutions.

(b) *Insurance policies*—(1) *Standard policies.* If a standard fire insurance policy has been adopted for the State, it should be used unless State statutes exempt the company from the regulations requiring its use. The standard policy is one containing substantially the same standard provisions adopted or recommended by legislative action or by order of the supervisory insurance authorities of the State in which the security is located.

(2) *Other policies.* To be acceptable, any other insurance policies must conform to the requirements of this Instruction.

(i) "Homeowner's" policies, "All Physical Loss" policies, "Broad Form" policies, and other such all-inclusive policies are acceptable if they otherwise meet the requirements of this Instruction.

(ii) A builder's risk policy naming the borrower as the insured or a builder's risk endorsement for a policy

issued to the borrower may be accepted during the period a building is under construction if the policy otherwise meets the requirements of this Instruction. If such a policy or endorsement does not automatically convert to full coverage when the building is completed, acceptable insurance must be obtained simultaneously with the expiration of the builder's isk provisions of the policy.

(iii) A builder's risk insurance policy issued to a contractor only may not be substituted for the property insurance, the borrower is required to provide.

(iv) Borrowers eligible for insurance under the National Flood Insurance Act of 1968, as amended by the Flood Disaster Act of 1973, will be serviced in accordance with subpart B of this part.

(3) *State instructions.* If the State Director and the OGC consider it advisable, a State Instruction may be issued to help County Supervisors identify standard insurance policies adopted for the State. The Instruction should also furnish a guide to assist in identifying other acceptable insurance policy forms that are commonly used by insurance companies in the State, recognizing that such information is not all inclusive.

(4) *Binders.* Whenever there is a justifiable reason for not issuing a policy or endorsement, as required, a written binder will be acceptable for a period not to exceed 60 days from the effective date of the insurance. The written binder must have attached thereto the approved form of mortgage clause. Such a binder will be submitted to the County Supervisor in lieu of an insurance policy or endorsement and the insurance policy or endorsement will be submitted on or before the expiration date of the binder. The State Director, with the advice of the OGC and subject to prior approval of the National Office, may issue a State Instruction authorizing such binders to be accepted for periods longer than 60 days.

(5) *Submission of policies.* (i) For Farmer Program (FP) loans secured by a first lien, the original policy or declaration page must be delivered to the County Supervisor. The original policy or declaration page will be returned to the borrower after one year using Form

RD 426–4, "Notice of Expiration of Insurance."

(ii) For Single Family Housing (SFH) loans secured by a first lien, the original policy or declaration page must be delivered to the closing agent.

(iii) In cases where an FP or SFH loan is secured by other than a first lien and the mortgage clauses include the names of the prior mortgagees, a certificate of insurance, copy of the policy, or other evidence of insurance is acceptable.

(iv) The County Supervisor will process an advance to pay for insurance only in strict compliance with provisions of §1806.6 of this subpart.

(6) *Master sets.* If the master sets meet all of the requirements of this Instruction they may be accepted in lieu of an original policy for each Agency borrower.

(i) One complete master set of the different insurance forms for policies issued by the insurance company must be on file in each County Office where the company insures property of Agency borrowers.

(ii) The "Declaration Page" furnished by the insurance company for each borrower insured, in lieu of a complete policy, will be filed in the borrower's case folder. When a "Declaration Page" in the form of a computer printout is used by an insurance company an endorsement on every policy issued by that company or a letter from that company will be obtained and attached to the printout. However, a letter signed by an authorized official of the company and addressed to the State Director may cover all policies issued by that company in the State. Any such endorsements or letters should clearly state that the company considers the printout to be an original "Declaration Page". Such endorsements or letters are not necessary if the printout itself clearly states that it is an original "Declaration Page."

(7) *Name and location.* The policy should contain names of all the borrowers who are owners of the property being insured, and it will be returned for correction if it does not do so. The location of the property should be so described in the policy that the property can easily be identified. The complete legal description of the property by metes and bounds is not required. Any deviation from the requirements of this paragraph must first be cleared with the National Office.

(8) *Loss or damage covered.* Buildings must be insured against loss or damage by fire, lightning, windstorm, hail, explosion, riot, civil commotion, aircraft, vehicles, and smoke.

(9) *Effective date of insurance.* If there are insurable buildings located on the property, the borrower will arrange with his agent or company to have adequate insurance in force at the time the loan, assumption, or credit sale is closed so that the policy will properly insure the borrower and the mortgagees. When new buildings are erected or major improvements are made to existing buildings, such insurance will be made effective as of the date materials are delivered to the property. The County Supervisor will make no payments from loan funds for labor or materials until the borrower has furnished adequate insurance to protect the interest of the Agency in the buildings being erected or improved.

(10) *Term.* The borrower will be required to furnish insurance for a term of at least one year with evidence that a full year's premium is paid. The term "premium" as used herein includes any assessments which may be charged to the borrower. If the assessments are of the type imposed only after a loss occurs involving property insured by the insurance company, then the borrower must present evidence (such as a letter from the company) that he currently does not owe any such assessments. The borrower may receive a discount for insuring for a longer period such as three years or five years and with an annual premium. If the insurance contains an automatic renewal clause, its provision should be substantially the following to be acceptable to the Agency:

This policy will be automatically extended for successive terms at expiration of the original term and of each extension thereof, upon payment of renewal premiums. It is a condition of this policy that if the policy expires or is canceled for nonpayment of premium, or for any other reason, the mortgagee will be given 10 days notice.

(11) *Mortgage clause.* The standard mortgage clause adopted by the State

must be attached to or printed in the policy, or Form RD 426–2, "Property Insurance Mortgage Clause (Without Contribution)," must be attached to or the provisions thereof printed in the policy. A letter signed by an authorized official of an insurance company to the State Director, stating that all insurance policies the company issues in the State and in which the Agency has a mortgage interest incorporates all of the provisions of Form RD426–2 may be accepted in lieu of attaching Form RD 426–2 to each policy. If such a blanket letter is used, the Agency will be named in the loss payable clause and a State Instruction will be issued, after prior approval is obtained from the National Office, authorizing the use of such method.

(i) If the use of a mortgage clause, other than the standard mortgage clause (without contribution), has been made mandatory by State laws or insurance regulations, a State Instruction will be issued, after prior approval is obtained from the National Office, authorizing the use of such a form.

(ii) When an approved mortgage clause is printed in the policy a "Loss Payable Clause" is acceptable provided the Agency, as mortgagee, would receive payment in case of loss even though the company would not be liable to the borrower. A "Loss Payable Clause" which contains the statement that the mortgagee is "subject to all terms and conditions of the policy" is not acceptable.

(iii) Whenever a new mortgage clause including the interest of the Agency is issued after the policy has been in force, the new mortgage clause must be signed by an authorized agent or officer of the company that issued the policy. Form RD 426–6, "Transmittal of Property Insurance Mortgage Clause," may be used to transmit the mortgage clause to the insurance official.

(iv) The Agency and all other mortgagees whose interests are insured by the policy will be shown either in the mortgage clause or in the "Declaration Page" in the order of priority of their mortgages.

(A) "United States of America (Rural Development)" will be named in the mortgage clause for direct and insured loan mortgages naming the Agency as mortgagee, whether in its own right or as trustee under a 2(f) or other agreement with a State Rural Rehabilitation Corporation.

(B) "United States of America (Rural Development), as first mortgagee or as statutory agent and insurer of such mortgagee," will be named in the mortgage clause for insured FO mortgages naming the lender as mortgagee, whether the mortgage is held by the original or a subsequent lender or by the insurance fund or by the Agency under a trust agreement or declaration of trust.

(C) If the designation is not identical to that set forth in paragraphs (b)(11)(iv)(A) or (B) of this section, whichever is applicable, it will be sufficient if the mortgagee is readily identifiable as Rural Development.

(c) *Evidence of premium payment.* (1) When Form RD 426–2 is attached to or the provisions thereof are printed in the policy, or a blanket letter from an insurance company incorporating the provisions of Form RD 426–2 in all policies in which the Agency has a mortgagee interest in effect, in accordance with paragraph (b)(11) of this section, no evidence of premium or assessment payment is required except for the first year of the loan. When a subsequent FP or section 502 RH loan is made to build, buy or rehabilitate essential buildings an endorsement to the existing policy including coverage for the property improved will be sufficient.

(2) [Reserved]

(d) *Policy restrictions.* (1) Any insurance on essential buildings as defined in §1806.3 having restrictions which limit the amount of collectible insurance must meet the Agency requirements set forth below (except for the clause described in paragraph (d)(1)(iv) of this section which is never acceptable); otherwise, such restrictions must be eliminated or modified to afford the required protection.

(i) *Coinsurance clause.* This clause generally provides that in consideration of a reduced rate, the borrower agrees to maintain insurance on his buildings up to a specified percentage (usually 80 percent) of their value and that the company will not be liable for a greater proportion of any partial loss than the amount of insurance bears to the specified percentage of either the undepreciated replacement value or the depreciated replacement value or the depreciated replacement value (actual cash value) of the buildings at the time of the loss. When the buildings are insured for the specified percentage of their value, the company, in the event of a partial loss, will be liable for the full amount of the loss not to exceed the amount of insurance. A coinsurance clause can be accepted only where the amount of insurance is at least equal to the specified percentage of either the undepreciated replacement value or the depreciated replacement value (actual cash value). For example, an 80 percent coinsurance clause can be accepted only where the amount of insurance on each insured building is at least equal to 80 percent of the appropriate replacement value of the insured building.

(ii) *Three-fourths' value clause.* This clause provides that the liability of the company shall be limited to three-fourths of the depreciated replacement value of the buildings covered at the time of the loss, not to exceed the amount of insurance. This clause may be accepted if the unpaid balance of the loan is not greater than three-fourths of the depreciated replacement value of the building and the amount of insurance is at least equal to the unpaid balance of the loan and any prior liens and no building is insured for more than three-fourths of its depreciated replacement value.

(iii) *Loss deductible clause.* (A) For all loans other than RRH, RCH, and LH organizations this clause generally provides that loss to each building to the extent of the limitation is not recoverable. The company is liable only for loss to each building in excess of such limitation stated in the clause. This clause may be accepted where the limitation does not exceed $150, or one percent of the insurance coverage whichever is greater. In no case, however, may the limitation on any one building exceed $500.00.

(B) For RRH, RCH, and LH organization loans this clause generally provides that loss to each project to the extent of the limitation is not recoverable. The company is liable only for loss to each project in excess of such limitation stated in the clause. This clause may be accepted where the limitation does not exceed the option shown below that is chosen by the borrower and agreed to by the Loan Approving Official and properly annotated in the borrower file. The borrower and the Official should consider the economic impact to the project when selecting the appropriate option.

(1) Option 1—Up to one-fourth of one percent (0.0025) of the insurable value. Maximum deductible $5,000.

(2) Option 2—Up to a maximum deductible of $500 on any project with an insurable value not exceeding $200,000.

(3) Option 3—Option 1 may be chosen and increased above the maximum deductible by an amount equivalent to funds specifically escrowed in the project replacement reserve account as an offset to the increased deductible.

(4) Option 4—Option 2 may be chosen and increased above the maximum deductible by an amount equivalent to funds specifically escrowed in the project replacement reserve account as an offset to the increased deductible.

(5) The funds used to increase the deductible in Option 3 or Option 4 may be from project funds if it does not create an unsecure financial situation for the project. Also, non-project funds may be used for Option 3 or 4 and then repaid by withdrawal from the project at the rate of 75 percent of the annual insurance premium savings earned by the amount of escrow deposit, up to the amount deposited.

(6) The funds escrowed to increase the authorized deductible will be placed in the project reserve account as an increased amount in and above the amount required by the Loan Agreement/Resolution and so annotated in the borrower's accounting system.

(iv) *Three-fourths' loss clause.* This clause provides that the company will not pay more than three-fourths of any loss, nor more than three-fourths of the amount of insurance in force. This clause is never acceptable and must be eliminated.

(v) *Deferred loss payable clause.* This clause provides that, if the amount payable under the policy for any loss to any building insured shall be in excess of a specified portion, (usually 60 percent) of the amount of insurance on such building, the company will withhold from its initial loss payment any sum in excess of the specified portion of the amount of insurance on such building. If the building sustaining such loss is repaired or replaced within six months from the date of the fire and at or within 300 feet of the original location, as described in the policy, the company upon receipt of evidence to that effect from the insured will pay the full balance withheld from the initial payment, provided the amount expended in repairing or replacing the building damaged or destroyed will equal or exceed the amount of loss as determined under the terms of the policy. Failure to repair or replace any insured building within the time and manner provided will constitute acceptance of the initial payment as full and final settlement under the policy with respect to the loss. This clause may be accepted if the amount of insurance is for the full depreciated replacement value (actual cash value) of the building and the unpaid balance of the loan and any prior lien(s) is not greater than the initial loss payment made by the company.

(vi) *Construction specifications and use conditions.* If the insurance policy contains clauses which specify certain standards of construction or prescribes certain uses of the property for the insurance to be valid, the policy is acceptable only if the property meets such specifications or conditions at the time of acceptance. For example, if the policy provides that the chimney be constructed of a certain type of material, the County Supervisor should be assured that the required material has been used, or if the policy provides that farming operations are not carried out on the premises he should be assured that this condition is met.

(2) Policies generally will not be accepted if, under the terms of the policies or local laws, contributions or assessments may be made against the Agency. However, policies which impose assessments on the borrower may be accepted only if the Agency mortgage will be recorded prior to any failure of the borrower to pay any such assessments. Policies also will not be accepted if, by their terms or other conditions, loss payments are contingent

upon collective action by the Board of Directors, or the stockholders, or the members.

(e) *Buildings on leaseholds.* The policy will indicate that the insured is the lessee or tenant and not the owner of the buildings securing the Agency loan; or, if he is the owner of the building on the leased land, the policy will indicate that the insured is the owner of the building, but not of the land. State Directors, with the advice of the OGC will issue State Instructions to meet any other special requirements needed to conform with the insurance requirements of the State to enable leaseholders to obtain property insurance for buildings which are security for the Agency loans.

(7 U.S.C. 1989; 42 U.S.C. 1480; 40 U.S.C. 442; 42 U.S.C. 1480; 42 U.S.C. 2942; 5 U.S.C. 301; Sec. 10 Pub. L. 93–357, 88 Stat. 392; delegation of authority by the Secretary of Agriculture, 7 CFR 2.23; delegation of authority by the Assistant Secretary for Rural Development, 7 CFR 2.70; delegations of authority by Director, OEO, 29 FR 14764, 33 FR 9850)

[41 FR 34571, Aug. 16, 1976, as amended at 41 FR 49990, Nov. 12, 1976; 42 FR 33262, June 30, 1977; 43 FR 56013, Nov. 30, 1978; 44 FR 45115, Aug. 1, 1979; 51 FR 17921, May 16, 1986; 54 FR 35869, Aug. 30, 1989; 56 FR 6945, Feb. 21, 1991; 80 FR 9864, Feb. 24, 2015]

§ 1806.3 Coverage requirements.

The County Supervisor should encourage the borrower for his own protection to insure for their depreciated replacement value (actual cash value) all essential buildings. Essential buildings include the dwelling and any other buildings that are necessary for the operation of the property or that provide income to assure orderly repayment of the loan. If insurance is for less than the depreciated replacement value of all essential buildings, the County Supervisor will see that the coverage is obtained on one or more of the most essential buildings. The minimum amount of coverage will be furnished as prescribed below:

(a) *Loans secured by a first lien.* (1) When the unpaid balance of the Agency loan secured by a first lien is equal to or greater than the depreciated replacement value of the essential buildings, or the cost of adequate essential buildings which can be constructed for amounts less than the depreciated replacement value of the existing buildings, the essential buildings will be insured, to the nearest multiple of insurance that is available, for the lesser of (i) their depreciated replacement value, or (ii) the cost of constructing adequate essential buildings. For example, if insurance is available in only multiples of $1,000, the minimum insurance required on an essential building valued at $6,600 would be $7,000, and that required on an essential building valued at $6,400 would be $6,000.

(2) When the unpaid balance of the loan is less than the sum of the depreciated replacement value of the essential buildings to be insured, the total amount of insurance must be at least equal to the lesser of (i) the unpaid balance of the loan, or (ii) the cost of adequate essential buildings which can be constructed for amounts less than the depreciated replacement value of the existing buildings to be insured.

(3) When, by the use of loan funds or otherwise, buildings are erected or substantial improvements are made to essential buildings, the amount of insurance will be adjusted in accordance with paragraphs (a)(1) or (2) of this section, whichever is applicable.

(b) *Loans secured by other than first liens.* The amount of insurance on buildings in the case of Agency loans secured by other than a first lien will be the same as required in paragraph (a) of this section, with the understanding that the unpaid balance of the loan will be deemed for this purpose to be the amount of the total real estate mortgage indebtedness owed all prior mortgagees named in the mortgage clause, plus the debt to the Agency which is secured by real estate mortgage.

(c) *Exception of buildings from insurance.* (1) Insurance will not be required on a building:

(i) That is not essential.

(ii) In such a state of disrepair that the cost of insurance would be prohibitive.

(iii) Which has a depreciated replacement value of $2,500 or less.

(iv) Which is being or has been repaired with a section 504 loan of $7,500 or less. Families receiving section 504 loans should be encouraged but not re-quired to carry insurance on their home.

(v) On LH security property which was not built or repaired with Agency loan funds provided that the State Director determines that the land and other structures adequately secure the Agency loan and any prior liens.

(vi) On which the hazards are so slight because of the character and construction of the building, or the cost of the insurance is so high in comparison with the value of the building that, according to common standards of judgment, it should not be insured, including but not limited to windmills, silos, and fire-cured tobacco barns.

(vii) In cases where the unpaid balance of the Agency loans and any prior liens have been reduced to $2,500 or less, property insurance need not be required if the borrower wants to discontinue it, provided the County Supervisor determines that the value of the land security itself is sufficient to protect the Agency in its collection of the amount of the outstanding indebtedness.

(viii) If insurance for windstorm and hail to meet all Agency requirements is not available in a hurricane area, the County Supervisor may accept from the borrower or applicant the windstorm and hail insurance policy that most nearly conforms to Agency requirements. If such an exception is made, the situation should be fully documented in the borrower's case file. However, if the best insurance policy a borrower or applicant can obtain at the time he receives a loan contains a loss deductible clause for windstorm and hail damage exceeding $250 or 10 percent of the actual cash value of the buildings, whichever amount is greater, the insurance policy, with an explanation of the reasons why more adequate insurance is not available will be submitted to the State Office for prior approval.

(2) [Reserved]

[41 FR 34571, Aug. 16, 1976, as amended at 56 FR 6945, Feb. 21, 1991; 80 FR 9865, Feb. 24, 2015]

§1806.4 **Examining and general servicing of insurance.**

(a) *Examination by county office of policies, endorsements, binders, and other evidence of insurance.* Upon receipt in the County Office of a policy, endorsement, binder, or other evidence of insurance, submitted by a borrower, it will be examined promptly for compliance with the requirements of this Instruction. If the evidence of insurance is found to be acceptable, it will be placed in the borrower's case folder.

(1) *Unacceptable policies.* (i) When the borrower furnishes any policy or other evidence of insurance which does not meet the requirements of this Instruction such policy or other evidence of insurance will be returned to the borrower with the reasons why it is not acceptable.

(ii) If the borrower does not furnish acceptable insurance by the date the previous policy expired or was canceled, the County Supervisor will proceed as provided in § 1806.6.

(2) *Expiration records and notices.* (i) In cases other than those involving FP or section 502 RH borrowers, the County Supervisor will notify the borrower of the expiration of his insurance at least 30 days in advance of such expiration unless he has received written evidence that the insurance has been renewed.

(ii) FP and Section 502 RH borrowers will be informed during the tenth month after the date of loan closing of their responsibility to carry insurance. Form RD 426–4 will be sent to these borrowers, regardless of whether there is evidence that the insurance has been renewed. Thereafter, the County Supervisor will not be required to further determine whether the borrower has adequately maintained insurance; however, if a further notice of expiration is received in the County Office, the County Supervisor will again notify the borrower by using RD 426–4 of his responsibility.

(3) *Release of mortgage interest.* When the borrower's loan has been paid in full and the satisfaction or release of the mortgage has been executed, the County Supervisor or his delegate will execute the following Release of Mortgage Interest on the mortgage clause attached to the policy or other evidence of insurance and transmit it with the policy or other evidence of insurance, the paid-in-full note, and the satisfaction to the borrower:

It is understood and agreed that the interest of the United States of America in the property insured hereunder ceased as of (Date of Final Payment), and that the Government shall have no interest in any loss or damage to such property occurring thereafter.

(4) *Lost or misplaced policies.* When an unexpired insurance policy or other evidence of insurance is lost or misplaced, it will be necessary to obtain a replacement policy or other evidence of insurance. The County Supervisor is authorized to sign a Lost Policy Receipt on behalf of the Agency. For FP and section 502 RH loans, this paragraph applies only during the period the policy is retained in the County Office.

(5) *Disposition of expired and canceled policies.* An expired or canceled policy or other evidence of insurance will be returned to the borrower, unless there is a loss settlement pending.

(b) *Special servicing of insurance*—(1) *Vacancy or unoccupancy—tenant occupancy—increased hazard.* If the County Supervisor has knowledge that insured property is vacant or unoccupied or that the ownership or occupancy has changed from owner to tenant, or that the hazards otherwise are increased, he will examine the policy to determine whether the policy permits such conditions. Unless the insurance permits such conditions, the County Supervisor will immediately notify the company or agent in writing. In any case where there is an additional premium due because of vacancy, unoccupancy, tenant occupancy, or other increased hazard, upon demand to the Agency from the company or agent because the borrower cannot, or will not, pay the additional premium, it may be paid in accordance with RD Instruction 2024–A, to the company or agent. For FP and section 502 RH borrowers, property insurance will not be obtained except in cases where an unusual and severe hazard exists and insurance is necessary to protect the interests of the Government.

(2) *Transfer of property.* (i) When a borrower or transferee requests the consent of the Agency to a transfer of the security property which already has been made, or when the County Supervisor learns that any such transfer

has been made, he will immediately inform the transferee that the mortgage requires the owner to provide and maintain adequate insurance acceptable to, and with loss payable to, the Agency as mortgagee. The transferee may obtain a new insurance policy or the transferor may have the insurance company or agent issue an endorsement to the current insurance policy changing the name of the assured to that of the transferee. If a new insurance policy is obtained, the old policy or other evidence of insurance will be returned to the transferor unless there is an unsettled loss. If there is an unsettled loss, the policy or other evidence of insurance will not be returned until the claim has been settled. The County Supervisor, with the concurrence of the State Director and the OGC, will notify the borrower and transferee that acceptance of the new policy or endorsement will not constitute consent by the Government to the transfer even though the Government is protected by a loss payable clause in such an insurance policy.

(ii) In a transfer with assumption, insurance will be required in the same amount and according to the same provisions as for an initial loan of the same type.

(3) *Voluntary conveyance of property to the Government and foreclosure.* Insurance will not be carried on buildings which the Government has acquired. After a foreclosure sale has been held, or after a deed of conveyance to the Government in lieu of foreclosure has been filed for record, insurance will not be maintained by the Government (whether or not subject to redemption).

[41 FR 34571, Aug. 16, 1976, as amended at 42 FR 33262, June 30, 1977; 50 FR 39638, Sept. 30, 1985; 54 FR 35869, Aug. 30, 1989; 57 FR 36590, Aug. 14, 1992; 69 FR 69103, Nov. 26, 2004; 80 FR 9865, Feb. 24, 2015]

§ 1806.5 Losses.

(a) *Protecting property.* It is the responsibility of the borrower to immediately notify the County Supervisor and insurance company or agent of any loss or damage to insured property and collect the amount of the loss. When the County Supervisor learns of a loss to property which secures an Agency loan, he will:

(1) Check the borrower's casefile for an insurance policy or other evidence of insurance. When a policy or other evidence of insurance has not been retained by the Agency, such as for FP and section 502 RH borrowers, the County Supervisor will determine whether the property was insured and whether the Agency was named as mortgagee in the insurance policy.

(2) Determine that the borrower has taken such steps as are necessary to protect the interest of the Agency in the security property against further damage. When serious problems arise with respect to protecting the property from further damage, the borrower cannot or will not arrange adequate protection for the property, or when legal action appears to be necessary, the County Supervisor will arrange for emergency protection and immediately refer the case with complete information to the State Director.

(b) *Loss covered by insurance.* (1) If the Agency is listed as mortgagee in the insurance policy, the County Supervisor will collect the amount of the loss and may consent to the borrower using funds to repair or replace damaged or destroyed property or to apply loss proceeds to his loan account or to any prior liens that might exist in the order of their priority.

(2) If the Agency is not listed as mortgagee in the insurance policy, the County Supervisor will contact the borrower to determine whether he has received the loss proceeds. If the borrower has received the loss proceeds but not yet paid for improvements to repair or replace the property, or has not received the loss proceeds the County Supervisor will:

(i) Notify the insurance company in writing of the Agency's interest in the security property and request that the loss proceeds be made payable jointly to the Agency and the borrower.

(ii) Inform the borrower of his responsibility for repairing or replacing the damaged or destroyed property or for authorized disposition of the loss proceeds as outlined in paragraph (b)(1) of this section.

(c) *Loss drafts—when loan is secured by a first mortgage.* (1) A loss draft which in the opinion of the County Supervisor represents a satisfactory adjustment of the loss will be endorsed immediately without recourse and deposited in a supervised bank account to be used in repairing or replacing the damaged building, except:

(i) Where the amount of the loss is $1,000 or less and the borrower will use the funds for repairing or replacing an essential building, the loss draft may be endorsed without recourse and given to the borrower upon satisfactory proof that the repairs or replacements have been made, or upon satisfactory assurance that the work will be performed.

(ii) When (A) the essential buildings are not to be repaired or replaced and other suitable buildings are not to be erected, or (B) a balance remains after all repairs, replacements, and other authorized disbursements have been made, such insurance funds will be applied on prior liens or as an extra payment to the borrower's loan accounts secured by the real estate or disposed of in accordance with the general principles applicable to the use of proceeds from the sale of a part of the security contained in applicable security servicing regulations for the type loan involved.

(iii) An insurance payment for loss or damage to a nonessential building the borrower voluntarily insured will be (A) applied on prior liens, or to current delinquencies to the Agency or as an extra payment on the borrower's loan accounts secured by real estate, (B) disposed of as authorized by the State Director in accordance with the general principles applicable to the use of proceeds from the sale of a part of the security contained in applicable security servicing regulations for the type loan involved, or (C) used for other purposes as authorized by the State Director if the loan is adequately secured and the loan account is current.

(iv) When the indebtedness secured by the insured property has been paid in full or the draft is in payment for loss of property on which the Agency has no claim, a loss draft which includes the Agency as a joint payee may be endorsed without recourse and delivered to the borrower.

(2) [Reserved]

(d) *Loss drafts—When loan is secured by other than first mortgage.* (1) When the loss draft does not include the interest of a prior mortgagee, it will be processed as provided in paragraph (c) of this section.

(2) When the loss draft includes the interest of a prior mortgagee, the County Supervisor is authorized to endorse and process the draft as follows:

(i) When the prior mortgagee will permit the use of such loss funds to repair or replace the damaged building, the draft may be endorsed without recourse upon satisfactory proof that the repairs or replacements have been made or upon satisfactory assurance that the work will be performed.

(ii) When the amount of the draft does not exceed the amount of the indebtedness then secured by the prior mortgage as stated in writing by the holder of the prior mortgage, and the holder of the prior mortgage has agreed in a written statement to the County Supervisor that he will apply such funds as a payment on the borrower's prior mortgage indebtedness, the draft may be endorsed without recourse.

(iii) When the amount of the draft exceeds the amount of the indebtedness then secured by the prior mortgage, as stated in writing by the holder, and he has agreed in writing to pay such indebtedness from the loss funds, the draft will be endorsed without recourse only after all parties named as payees in the draft have signed an agreement to deliver the draft "in escrow" to a bank acceptable to the named parties. The agreement will specify the manner in which the funds will be disbursed by the bank, as escrow agent, to the several mortgagees named in the draft. After the loss funds have been collected by the bank, it will issue cashier's checks in the manner prescribed in the escrow agreement (see exhibit A for suggested form). If this procedure is found to be impractical in an individual instance, the State Director may authorize an alternative method for disbursing the loss funds to protect the Government's financial interest.

(iv) Drafts which have been endorsed by all other payees will be endorsed immediately without recourse. Such drafts or other loss funds will be processed in accordance with the methods described in paragraph (c) of this section.

(e) *Servicing insurance losses under special circumstances—*(1) *Foreclosures and voluntary conveyances.* Losses on properties in process of foreclosure or voluntary conveyance will be handled with the advice of the OGC. If the necessary cooperation of the borrower cannot be obtained, the State Director, with the advice of the OGC, will determine the proper action to be taken. To the extent feasible from a legal and practical standpoint, all loss payments should be received for a damaged or destroyed building and applied on the borrower's real estate indebtedness before title to the property is taken by the Government through foreclosure sale, voluntary conveyance, or otherwise, unless absolute assignment has been made by the borrower to the Government of all loss funds due from the insurance company.

(2) *Subrogation agreements.* When a company claims nonliability to the borrower and subrogation to the rights of the Agency, the County Supervisor will forward a full report of the facts in the case to the State Director. The State Director will upon advice from OGC, instruct the County Supervisor regarding further action to be taken.

(f) *Repairs and replacements.* When any loss payments have been deposited in a supervised bank account, all repairs and replacements done by or under the direction of the borrower, or by contract, will be planned, performed, inspected, and paid for in the same manner as improvements financed with loan funds.

(g) *Completing adjustment.* The borrower must complete the adjustment of the loss with the company or its authorized representatives. The County Supervisor, upon request of the borrower may consult with the borrower regarding the loss adjustment, but will not enter into negotiations with insurance adjusters or company representatives relative to the adjustment or settlement of losses on borrower property, or make any commitments, or sign any forms in connection with the adjustment of the loss. The Agency will not waive any rights which it may have against the company except when the borrower's account or the Agency claim has been paid-in-full.

(1) The County Supervisor will maintain a proper followup on all losses until satisfactory settlement has been made by the company.

(2) Where the County Supervisor has evidence that the adjustment agreed to by the borrower is significantly less than the amount of damage to which the borrower is entitled under the terms of the policy, the loss draft accompanied by a report will be sent to the State Director so that he may reopen the adjustment, if he considers it is in the interest of the Agency to do so.

(3) When it appears evident that the amount of the loss is $1,000 or less, the County Supervisor may rely on estimates of contractors, building supply firms, reliable carpenters, or other evidence rather than personal inspection in determining whether the adjustment is equitable and the Government's interest is protected.

(h) *Reinstatement after loss.* In cases where insurance in the amount of the loss is not reinstated automatically by the provisions of the policy, it will be the responsibility of the County Supervisor to have the borrower reinstate as much of the insurance as may be necessary to fulfill the requirements of the Agency.

(i) *Losses not covered by insurance.* When a loss occurs and insurance is not in force, the County Supervisor will:

(1) Inform the borrower that he has violated the security instrument by not providing insurance coverage and that it is his responsibility to make the needed replacements or repairs.

(2) If the borrower is unable or unwilling to make needed repairs or replacements from his own resources, the County Supervisor will submit complete information to the Agency official authorized to determine whether the Agency will or will not continue with the loan. The County Supervisor's report will include recommendations on the following items:

(i) The advisability and possibility of making a subsequent loan to pay for needed repairs.

(ii) Subordination of the Agency real estate lien to permit the borrower to obtain funds for needed repairs from another source.

(iii) The possibility of the borrower obtaining funds secured by a junior lien from another source.

(iv) Whether an advance is needed to protect the Government's interest in the property.

(3) If the loan will not be continued with the borrower, it must be serviced in accordance with the applicable Instructions.

(4) If the borrower has improperly disposed of loss proceeds, the County Supervisor will refer the case with complete information and recommendations to the State Director. The State Director will consult the Regional Attorney when necessary and advise the County Supervisor as to appropriate servicing actions.

[41 FR 34571, Aug. 16, 1976, as amended at 50 FR 39638, Sept. 30, 1985; 80 FR 9865, Feb. 24, 2015]

§ 1806.6 Failure of borrower to provide insurance.

When a borrower fails to provide and maintain property insurance which meets the requirements set forth in § 1806.2 of this subpart, every effort will be made to have the borrower provide coverage acceptable to the Agency. It will be emphasized that under the terms of the security instrument, it is the borrower's responsibility to provide and maintain proper insurance coverage. Existing borrowers required to escrow will be notified by letter at least 90 days prior to initiating escrowing for insurance. Failure to provide insurance is a nonmonetary default and will be a consideration in determining if the loan is to be continued. For FP or SFH borrowers *not* required to escrow, the County Supervisor will obtain insurance coverage and voucher for the insurance premium only in cases where: An unusual and severe hazard, such as recurring fires or unstable ground conditions, exists, or, an SFH borrower on a moratorium is unable to pay the insurance premium and the borrower requests that the Agency pay the premium. For SFH borrowers required to escrow, force placed insurance will be obtained if the borrower fails to provide acceptable insurance. Borrowers being phased into es-

crow will be given at least 30 days to obtain coverage, after which force placed insurance will be obtained. If the escrow account contains insufficient funds to pay the insurance when due, the County Supervisor will request the borrower to pay an amount equal to the difference between the premium due and the escrow balance in a lump sum within 30 days after notification. If the borrower fails to remit the amount requested, the amount will be advanced and charged to the borrower's account as a recoverable cost. The amortization period for an advance due to an escrow shortage will be one year. Insurance coverage shall be provided continuously unless the property is acquired by the Agency. The cost of obtaining such a policy shall be advanced and charged to the borrower's account as a recoverable cost. Amortization of the charge will be handled in accordance with 7 CFR part 3550. If a borrower indebted for other than an FP or SFH loan fails to provide acceptable insurance, the Servicing Official will take the following action:

(a) *Expired policies.* (1) The County Supervisor will request the insurance agency or broker who issued the expired policy to issue a new policy which is acceptable to the Agency.

(i) The new policy will be effective as of the date of the County Supervisor's contact with the insurance agency or broker or as soon thereafter as possible, and will be for a term of one year. If State insurance regulations require a longer term, the State Director will issue a State Instruction authorizing County Supervisors to obtain policies for the minimum period permitted by State insurance regulations.

(ii) The Agency will be shown in the loss payable clause and in the mortgage clause in the proper order of priority.

(iii) Insurance coverage on each building usually will be the same as shown on the expired policy if it meets or exceeds Agency requirements. If the coverage shown on the expired policy does not meet Agency requirements, proper coverage will be obtained.

(iv) The County Supervisor will, if possible, have an automatic renewal provision included in the policy.

(v) If the borrower refuses to pay the insurance premium with his own funds or arrange with the agent for subsequent payment by premium not or otherwise, the County Supervisor will pay the amount of the insurance premium in accordance with RD Instruction 2024–A. The amount of the premium payment will be charged to the borrower's Agency account with the highest lien priority as a recoverable cost item.

(vi) If the insurance agency or broker who issued the expired policy refuses to issue a new policy, the County Supervisor will have the borrower designate in writing another insurance agency or broker from whom the insurance can be obtained.

(vii) After the County Supervisor and the borrower exhaust all efforts to obtain acceptable insurance, the County Supervisor will request advice from the State Office as to companies issuing acceptable policies in the State and from which the borrower might be able to obtain an acceptable policy. If the borrower still cannot obtain an acceptable policy from any such company, and the determination has been made to continue with the borrower, the County Supervisor will temporarily accept from the borrower the available insurance policy the Agency determines most nearly conforms to the requirements of § 1806.2 of this subpart.

(A) In making this determination, the following deficiencies become more objectionable in the order from (1) to (5) paragraphs (a)(1)(vii)(A) of this section:

(1) A policy written for an initial term of less than one year.

(2) A policy which will insure the most essential buildings but will not cover all essential buildings.

(3) A policy which covers major risks such as fire and lightning, but does not include one or more of the other risks specified in § 1806.2(8).

(4) A policy for a lesser amount of insurance than is required by § 1806.3.

(5) A policy that is issued by a company which is not licensed to do business in the State or otherwise does not meet the requirements of § 1806.3.

(B) Whenever adequate insurance becomes available, the County Supervisor will require the borrower to deliver to the County Office an acceptable insurance policy. The temporary policy will be returned to the borrower for cancellation after all losses claimed under the policy have been settled.

(C) If the borrower is unable to furnish a property insurance policy of any kind, he is still responsible for the debt in the event of loss.

(D) If the County Supervisor accepts an inadequate insurance policy under these conditions or the borrower fails to furnish any insurance policy, the County Supervisor will include in his report to the State Director an explanation of the efforts he and the borrower made to obtain acceptable insurance and his justification for accepting an inadequate policy, or for not obtaining an insurance policy of any kind.

(b) *Insurance canceled for reasons other than nonpayment of insurance premium.* (1) The County Supervisor, immediately upon receipt of a 10-day notice of cancellation for a policy, will urge the borrower to provide acceptable insurance.

(2) If the borrower fails to provide acceptable insurance before the cancellation is effective, the County Supervisor will contact the insurance agency or broker who issued the insurance policy to determine the reasons for cancellation and, if possible, have the policy reinstated.

(3) If the insurance company will not reinstate the policy, the County Supervisor will attempt to obtain an acceptable insurance policy from another agency or broker in accordance with the provisions of paragraph (a) of this section.

(c) *Insurance canceled for nonpayment of premium.* (1) The County Supervisor, immediately upon receiving a 10-day cancellation notice for a policy, will, if possible, contact the borrower in an effort to have him pay the insurance premium from his own funds or arrange with the agent for subsequent payment by premium note, or otherwise.

(2) If the borrower does not pay or arrange to pay the premium before the policy cancellation is effective, the County Supervisor will, before the cancellation becomes effective, notify the insurance company or broker by certified mail (return receipt requested), that the Agency as mortgagee (or

trustee) will pay the premium for one year to continue the policy in effect for that period. The County Supervisor will, in accordance with RD Instruction 2024–A, pay the amount of the premium for a period of one year. The amount of the premium will be charged to the borrower's loan account as a recoverable cost item.

(3) If a property insurance mortgage clause other than Form RD 426–2 is used in connection with the policy and the insurance company or broker refuses to accept payment from the Agency in this manner to reinstate or continue the policy, the County Supervisor will attempt to obtain an acceptable insurance policy from another insurance company or broker in accordance with the provisions of paragraph (a) of this section.

(7 U.S.C. 1989; 42 U.S.C. 1480; 42 U.S.C. 2942; 5 U.S.C. 301; Sec. 10 Pub. L. 93–357, 88 Stat. 392; delegation of authority by the Secretary of Agriculture, 7 CFR 2.23; delegation of authority by the Assistant Secretary for Rural Development, 7 CFR 2.70; delegation of authority by Director OEO 29 FR 14764, 33 FR 9850)

[41 FR 34571, Aug. 16, 1976, as amended at 42 FR 33263, June 30, 1977; 43 FR 34430, Aug. 4, 1978; 50 FR 39638, Sept. 30, 1985; 56 FR 6945, Feb. 21, 1991; 57 FR 36590, Aug. 14, 1992; 67 FR 78326, Dec. 24, 2002; 80 FR 9865, Feb. 24, 2015]

EXHIBIT A TO SUBPART A OF PART 1806— ESCROW AGREEMENT REAL PROPERTY INSURANCE

Date _____

(Name of bank) _____

(City or town) _____

(State) _____

Gentlemen: Attached is Draft No. _____, for $_____, issued by the _____ Insurance Company in payment of _____ loss which damage the buildings on the farm of _____, of _____ County, State of _____.

This draft has been endorsed by the undersigned payees who request that you collect these funds and issue cashier's checks to the following payees for the following amounts:

_____, First Mortgage $_____

_____, Second Mortgage $_____

_____, Third Mortgage $_____

The balance only, if any, will be paid to _____, the owner of the property.

First Mortgagee _____

Second Mortgagee _____

Third Mortgagee _____

Owner _____

Subpart B—National Flood Insurance

AUTHORITY: 7 U.S.C. 1989; 42 U.S.C. 1480; 40 U.S.C. 442; 42 U.S.C. 2942; 5 U.S.C. 301; delegation of authority by the Sec. of Agri., 38 FR 14944 (7 CFR 2.23); delegation of authority by the Asst. Sec. for Rural Development, 38 FR 14944, 14952 (7 CFR 2.70).

SOURCE: 39 FR 17093, May 13, 1974, unless otherwise noted.

§ 1806.21 General.

(a) *Authority.* This subpart prescribes the policies and procedures to be followed in implementing the National Flood Insurance Act of 1968 as amended by the Flood Disaster Protection Act of 1973. The provisions of these Acts are applicable to Rural Development and Farm Service Agency, herein referred to as the "Agency" authorities permitting financing of buildings of any type now located in or to be located in special flood or mudslide prone areas as designated by the Federal Insurance Administration (FIA) of the Department of Housing and Urban Development (HUD), and any machinery, equipment, fixtures and furnishings contained or to be contained therein. This subpart does not apply to Farm Service Agency, Farm Loan Programs and to Rural Rental Housing, Rural Cooperative Housing, or Farm Labor Housing programs of the Rural Housing Service.

(b) *Background.* The Congress has found that annual losses throughout the nation caused by floods and mudslides are increasing at an alarming rate, largely as a result of the accelerated development and concentration of populations in areas subject to floods and mudslides. The availability of Federal funds in the form of loans, grants, guarantees, insurance and other forms of financial assistance are often determining factors in the utilization of land and the location and construction of industrial, commercial and residential facilities.

(c) *Scope.* The National Flood Insurance Program (the program) was authorized and created because the private insurance industry has been unable to provide insurance coverage at reasonable prices for such natural disasters as floods and mudslides. Subsidized and affordable insurance has been made available under the Act through an agreement between the Federal Insurance Administration and the National Flood Insurers Association.

[39 FR 17093, May 13, 1974, as amended at 69 FR 69103, Nov. 26, 2004; 72 FR 64121, Nov. 15, 2007; 80 FR 9865, Feb. 24, 2015]

§ 1806.22 Areas of responsibility.

(a) *Federal Insurance Administration (FIA).* (1) Identify and publish information with respect to all areas in the country which are subject to floods and mudslides and designate those areas on Flood Hazard Boundary maps.

(2) Notify affected communities of their designations and encourage them to adopt and enforce land use and other control measures and to adopt ordinances or laws which will regulate and control construction in areas designated as having special flood or mudslide hazards.

(3) Make flood insurance available at reasonable rates in sufficient amounts, within the statutory limits, to adequately protect owners against loss to their buildings and contents when those buildings are located in or will be located in designated special flood and mudslide prone areas in communities participating in the National Flood Insurance Program.

(b) *The Agency.* The State Director, after being notified by the National Office or FIA of designated flood or mudslide hazard areas and receiving flood hazard boundary maps identifying the hazard areas, FIA insurance rate charts, or other information concerning the program, will inform the appropriate County Supervisors and provide them the maps, rate charts, and other relevant information concerning the program in areas they serve. Permanent records indicating the date a community was notified as containing identified flood hazard areas, communities participating in the program, and communities eligible to participate but not participating in the program will be maintained in the State Office. County Supervisors will notify, in writing, those borrowers whose insurable buildings are located

in designated flood or mudslide hazard areas of the availability of national flood insurance and encourage them to obtain flood insurance to protect their and the Government's financial interest.

(c) *Community.* Communities are required to participate in the National Flood Insurance Program within 1 year after notification of its formal identification as a community containing one or more special flood and mudslide prone areas, or by July 1, 1975, whichever is later, or be denied Federal financial assistance or Federally-related financial assistance for acquisition or construction purposes in such areas. Communities wishing to qualify for the program may submit a completed application to: Administrator, Federal Insurance Administration, Department of Housing and Urban Development, 451 Seventh Street, SW., Washington, DC 20410.

(d) *Lender.* The lender must determine whether real property is located in an area identified as having special flood or mudslide hazards and cannot discharge the responsibility merely by obtaining a self-certification from the applicant that the property is not located in an area having special flood hazards.

[39 FR 17093, May 13, 1974, as amended at 80 FR 9865, Feb. 24, 2015]

§ 1806.23 Definitions.

For the purpose of this subpart, the following definitions apply:

(a) *Financial assistance* means any form of direct, insured or guaranteed loan, including reamortization and assumption on new terms of any loan, any form of grant, or other form of direct or indirect assistance extended by the Agency.

(b) *Financial assistance for acquisition or construction purposes* means any form of Federal financial assistance which is intended in whole or in part for the acquisition, construction, reconstruction, or substantial improvement of any building and for any machinery, equipment, fixtures and furnishings contained or to be contained in such buildings.

(c) *Community means* any state or political subdivision thereof, such as county, parish, township, city or other local government which has zoning and building code jurisdiction over a particular area having special flood hazards.

(d) *Eligible community* means a community in which the Administrator of FIA has authorized the sale of flood insurance under the program.

(e) *Designated special flood or mudslide prone* area means those areas in a community subject to flood or mudslide which have been identified by flood hazard boundary maps or those areas not identified by maps but where, due to emergency, the FIA Administrator has authorized the sale of flood insurance.

(f) *Flood* means a general and temporary condition of partial or complete inundation of normally dry land areas from the overflow of streams, rivers, or other inland water, the collapse or subsidence of land along the shore of a lake or other body of water as a result of erosion or undermining caused by waves or currents of water exceeding anticipated cyclical levels, or abnormally high tidal water or rising coastal waters resulting from severe storms, hurricanes, or tidal waves resulting from volcano eruptions or earthquakes.

(g) *Mudslide* or *mudflow* means a major occurrence involving the appearance of a large river or flow of "liquid mud" down a hillside, usually as a result of earlier brushfires followed by heavy rains over a widespread area.

(h) *Flood insurance* means insurance coverage for floods and/or mudslides under the program or otherwise acceptable to FIA.

(i) *Building* means any walled and roofed structure, other than a gas or liquid tank, that is principally above ground and affixed to a permanent site. Residential and most types of industrial, commercial, and agricultural buildings, such as lumber sheds, machinery storage sheds, grain storage bins, and silos, are included in this definition.

(j) *Substantial improvement* means any repair, reconstruction or improvement of a structure, the cost of which equals or exceeds 50 percent of the actual cash value of the structure either before the improvement is started or, if the structure has been damaged and is being restored, before the damage occurred.

[39 FR 17093, May 13, 1974, as amended at 80 FR 9865, Feb. 24, 2015]

§ 1806.24   Eligibility.

In addition to an applicant meeting the requirements for the type of financial assistance requested, the following requirements for eligibility of applicants for financial assistance for acquisition and construction purposes in designated special flood and mudslide prone areas must be met:

(a) If flood insurance is available, to be eligible after March 1, 1974, the applicant must have purchased a flood insurance policy at the time the loan or grant is closed.

(b) Applicants will not receive financial assistance in those communities that have been notified as having special flood and mudslide prone areas and where flood insurance is not available within 1 year after such notification or by July 1, 1975, whichever is later.

§ 1806.25   Conditions.

The Agency financial assistance may be extended to eligible applicants meeting the eligibility requirements of § 1806.24 of this subpart, provided the following conditions are also met:

(a) *Dwelling and multi-unit housing facilities.* (1) If the financial assistance is to buy a dwelling or multi-unit housing facility:

(i) The first floor elevation of the habitable space of the dwelling or housing unit must be above the 100-year flood level.

(ii) The housing must be served by public utilities and facilities, such as sewer, gas, electrical and water systems that are located and constructed to minimize or eliminate flood damage, or have an onsite water supply system and waste disposal system located so as to avoid impairment of such systems and contamination from the waste disposal system to the water supply system from flooding.

(2) If the financial assistance is to build or provide substantial improvement, the requirements of paragraph (a)(1) of this section must be met and all construction must meet requirements of the applicable development standards, and:

(i) A building permit must be issued by the appropriate governing officials having jurisdiction in the area and compliance must be had with the zoning code or other established legal requirements of the area for reducing or eliminating flood or mudslide damage.

(ii) The structure must be designed and anchored to prevent flotation, collapse or lateral movement of the structure.

(iii) Construction materials and utility equipment that are resistant to flood damage must be used.

(iv) Construction methods and practices that will minimize flood damage must be followed.

(3) If the financial assistance is to make minor repairs, the conditions of paragraphs (a)(1) (i) and (ii) and (2) (i), (ii) and (iii) of this section must be met or the building must have existed on the site prior to the date the area was identified as having special flood or mudslide hazards and the loan approval official must determine that the dwelling is suitable as a residence.

(4) When applications for financial assistance are received in areas identified as having special flood and mudslide hazards, the loan approval official will consider the expected severity and frequency of floods and mudslides in determining whether any housing loans should be made in the area. He should be sure, if loans are made, that the objectives of the loans can be accomplished and the Government's financial interest will be adequately protected.

(b) *Nonresidential buildings.* Construction plans and specifications for new buildings or improvements to existing buildings must comply with flood plain area management or control laws, regulations or ordinances.

(c) *Flood insurance coverage.* (1) Any property on which flood insurance is required must be covered by such insurance during its anticipated economic and useful life in an amount at least equal to its development or replacement cost (except estimated land cost), or to the maximum limit of coverage made available with respect to the particular type of property under the National Flood Insurance Program, whichever is less. However, if the financial assistance provided is in the form of a loan, the amount of flood insurance required need not exceed the outstanding principal balance of the loan and need not be required beyond the term of the loan.

(2) The contents of a building must be insured separately from a building but coverage cannot be written on the contents of a three-walled machinery shed or similar type open building.

(3) Flood insurance shall not be required on any state owned property that is covered under an adequate state policy of self-insurance satisfactory to the Secretary of HUD, who will publish a list of states with such policies.

(4) It will be emphasized that under the terms of the security instrument it is the borrower's responsibility to provide and maintain proper flood insurance coverage. If flood insurance is not provided on any property for which it is required, the flood insurance premium will be paid to protect the Government's security interest. For borrowers required to escrow for flood insurance, payment of the premium will be handled in accordance with §1806.28 of this subpart. Existing borrowers required to escrow will be notified by letter at least 90 days prior to initiating escrowing for flood insurance. If the Agency pays the flood insurance premium for borrowers not required to escrow, the cost will be charged to the borrower's account as a recoverable cost. Failure to provide flood insurance is a nonmonetary default and will be a consideration in determining if the loan is to be continued.

[39 FR 17093, May 13, 1974, as amended at 52 FR 8002, Mar. 13, 1987; 56 FR 6945, Feb. 21, 1991; 80 FR 9865, Feb. 24, 2015]

§1806.26 Coverage and premium rates.

Exhibit A sets forth limits of coverage and chargeable premium rates under the program. Insurance policies under the program can be obtained from any licensed property insurance agent or broker serving the eligible community or from the National Flood Insurers Association Serving Company (Serving Company) for the state. The Servicing Company for each state is shown in exhibit B.

435

§ 1806.27 Acceptable policies and servicing.

The general acceptance of policies and servicing of insurance will be performed in accordance with Subpart A of this part. Any unusual situations that may arise with respect to obtaining or servicing flood insurance should be referred to the State Director. The State Director will attempt to resolve any problems concerning the flood insurance program in the state with the Servicing Company. Flood hazard boundary maps, insurance rate tables, the insurability of specific structures, and other information concerning the program may be obtained from the Servicing Company. Difficulties in administering the program which the State Director is unable to resolve should be referred to the National Office for Assistance.

§ 1806.28 Borrowers required to escrow.

For borrowers required to use escrow accounts for the payment of real estate taxes and insurance, the flood insurance premium will be paid when due from funds contained in the escrow account. If the escrow account contains insufficient funds to pay the flood insurance premium when due, the County Supervisor will request the borrower to pay an amount equal to the difference between the premium due and the escrow balance in a lump sum within 30 days after notification. If the borrower fails to remit the amount requested, the amount will be advanced and charged to the borrower's account as a recoverable cost. The amortization period for an advance due to an escrow shortage will be one year. Amortization of the charge will be handled in accordance with 7 CFR part 3550. When a borrower has more than one loan secured by the real estate on which the flood insurance premium is being paid, the advance will be charged to the initial or lowest numbered loan.

[56 FR 6946, Feb. 21, 1991, as amended at 67 FR 78326, Dec. 24, 2002]

EXHIBIT A TO SUBPART B OF PART 1806—COVERAGE AND PREMIUM RATES

1. The following table sets forth the limits of coverage available under the program:

Type of structure	Structure coverage		Contents of coverage [4]	
	Subsidized	Total [3]	Subsidized	Total [3]
Single family, residential [1]	$35,000	$70,000	$10,000	$20,000
All other, residential [1]	100,000	200,000	10,000	20,000
All nonresidential [2]	100,000	200,000	100,000	200,000

[1] For Alaska, Hawaii, and the Virgin Islands, the following limits of coverage apply: Structure coverage for one family residential is $50,000 subsidized and $100,000 total coverage, and structure coverage for other residential is $150,000 subsidized and $300,000 total coverage.

[2] Includes hotels and motels with normal occupancy of less than 6 months.

[3] Coverage in amounts exceeding the subsidized limits is available only after an actuarial cost has been established and flood insurance rate may be issued.

[4] Contents of a building must be insured separately from the building. However, coverage is applicable to contents only while in an enclosed building. Therefore, coverage cannot be written on the contents of a three-walled machinery shed or a similar type open building.

2. The following table sets forth the applicable premium rates:

Type of structure	Rates per $100 of coverage (subsidized only)	
	Structures	Contents
All residential	[1] $0.25	$0.35
All nonresidential	[1] .40	.75

[1] Actuarial (nonsubsidized) rates are applicable to any structure, the construction or substantial improvement of which started after Dec. 31, 1974, or the date on which the initial rate map was issued, whichever is later, in identified areas having special flood or mudslide hazards.

EXHIBIT B TO SUBPART B OF PART 1806—SERVICING COMPANY

The servicing company office to be contacted for information relative to the availability of coverage under the national flood insurance program, flood hazard boundary maps, insurance rate tables, and related material.

E.D.S. Federal Corporation, National Flood Insurance, P.O. Box 34294, Bethesda, Md. 20034, phone toll-free 800–638–6620; commercial phone 301–898–5900.

(7 U.S.C. 1989; 42 U.S.C. 1480; 42 U.S.C. 2942; 5 U.S.C. 301; delegation of authority by the Secretary of Agriculture, 7 CFR 2.23; delegation of authority by the Assistant Secretary for Rural Development, 7 CFR 2.70)

[43 FR 18538, May 1, 1978]

PARTS 1807–1809 [RESERVED]

PART 1810—INTEREST RATES, TERMS, CONDITIONS, AND APPROVAL AUTHORITY

Subpart A—Interest Rates, Amortization, Guarantee Fee, Annual Charge, and Fixed Period

Sec.
1810.1 Information concerning interest rates, amortization, guarantee fee, annual charge, and fixed period.
1810.2 Adjustment of interest rates for certain loans involving use of or construction on prime or unique farmland.

Subpart B [Reserved]

AUTHORITY: 7 U.S.C. 1989; 14 U.S.C. 1480; 7 CFR 2.23; 7 CFR 2.70.

Subpart A—Interest Rates, Amortization, Guarantee Fee, Annual Charge, and Fixed Period

§ 1810.1 **Information concerning interest rates, amortization, guarantee fee, annual charge, and fixed period.**

(a) Tables for computing the interest rates (including the annual charge rates and length of fixed period for initial repurchase agreement for insured loans), tables for use in determining the amounts of interest on loans at different rates, tables providing factors in amortizing loans, and the guarantee fee for guaranteed loans, may be obtained from any County, District, or State Office of the Agency (Rural Business-Cooperative Service and Rural Housing Service of the U.S. Department of Agriculture) or from its National Office at 14th and Independence Avenue SW., Washington, DC 20250.

(b) In the event that the tables provided for in paragraph (a) of this section do not furnish adequate information, questions should be directed to the Assistant Administrator, Finance Office, U.S. Department of Agriculture, 1520 Market Street, St. Louis, Missouri 63103.

[56 FR 11503, Mar. 19, 1991, as amended at 80 FR 9865, Feb. 24, 2015]

§ 1810.2 **Adjustment of interest rates for certain loans involving use of or construction on prime or unique farmland.**

(a) For essential community facility loans, insured farm ownership loans for recreation or non-farm enterprises, insured farm operating loans for recreation enterprises, soil and water loans for recreation purposes, individual recreation loans, and insured business and industry loans, the interest rate will be increased by two per centum per annum if the project being financed will involve the use of, or construction on, prime or unique farmland. Prime or unique farmland is as defined in § 657.5 (a) and (b) of title 7, Code of Federal Regulations (1980).

(b) The two per centum interest rate increase will not apply if the applicant/borrower is a public body or Indian tribe and has demonstrated to the Agency that there are no suitable options for locating the proposed essential community facility project on land that is not prime or unique farmland.

(c) For each essential community facility loan and insured business and industry loan the District Director, after consultation with the Soil Conservation Service (SCS), will determine whether the proposed project will involve the use of, or construction on, prime or unique farmland. For each insured farm ownership loan for a recreation or non-farm enterprise, insured farm operating loan for a recreation enterprise, soil and water loan for a recreational purpose, or individual recreation loan, the County Supervisor, after consultation with SCS, will determine whether the proposed project will involve the use of, or construction on, prime or unique farmland. The determination will be documented by the Agency and made a part of the official case file.

[46 FR 47763, Sept. 30, 1981, as amended at 56 FR 11503, Mar. 19, 1991; 80 FR 9866, Feb. 24, 2015]

Subpart B [Reserved]

SUBCHAPTER B—LOANS AND GRANTS PRIMARILY FOR REAL ESTATE PURPOSES

PART 1822—RURAL HOUSING LOANS AND GRANTS

Subparts A–F [Reserved]

Subpart G—Rural Housing Site Loan Policies, Procedures, and Authorizations

AUTHORITY: 42 U.S.C. 1480; 5 U.S.C. 301; 7 CFR 2.23; 7 CFR 2.70.

Subparts A–F [Reserved]

Subpart G—Rural Housing Site Loan Policies, Procedures, and Authorizations

AUTHORITY: 5 U.S.C. 301; 42 U.S.C. 1480.

SOURCE: 35 FR 16087, July 1, 1970, unless otherwise noted.

§ 1822.261 General.

This subpart sets forth the policies and procedures and delegates authority for making Rural Housing Site (RHS) loans under sections 523 and 524 of the Housing Act of 1949. Any processing or servicing activity conducted pursuant to this subpart involving authorized Rural Development (RD) employees, members of their families, known close relatives, or business or close personal associates, is subject to the provisions of subpart D of part 1900 of this chapter. Applicants for this assistance are required to identify any known relationship or association with an RD employee. Section 523 loans are direct loans for the purchase and development of building sites for housing to be built by the self-help method; they have additional requirements which are contained in § 1822.278.

[35 FR 16087, July 1, 1970, as amended at 58 FR 224, Jan. 5, 1993; 80 FR 9866, Feb. 24, 2015]

§ 1822.262 Objective.

The basic objective of RHS loans is to assist public or private nonprofit organizations interested in providing sites for housing, to acquire and develop land in rural areas. This land will be subdivided into adequate building sites and sold on a nonprofit basis to (a) families eligible for low and moderate income section 502 Rural Housing (RH) loans, including self-help housing; (b) cooperative Rural Cooperative Housing (RCH) applicants and broadly based nonprofit Rural Rental Housing (RRH) applicants; and (c) applicants eligible for Housing and Urban Development (HUD) sections 235 and 236 insured mortgages.

§ 1822.263 Definitions.

As used in this subpart:

(a) A *private nonprofit organization* is a corporation which: is owned and controlled by private persons; is organized and operated for purposes other than making gains or profits for the corporation or members; and, is legally precluded from distributing to its members any gains or profits.

(b) A *public nonprofit organization* is a nonprofit corporation other than a private nonprofit corporation, including a municipal corporation or other corporate agency of a State or local government.

438

(c) *Rural area* is open country or rural places as defined in 7 CFR part 3550, subpart A.

(d) *Development cost* means the cost of purchasing and developing the sites including engineering and legal fees, streets, roads, utilities, minimum essential administrative costs, necessary equipment and estimated interest which the borrower cannot pay from other sources.

(e) *RHS section 523 loan* means a loan to an organization which will provide sites for housing to be built by the self-help method.

(f) *RHS section 524 loan* means a loan to an organization which will provide sites for housing to be built with no limitation as to the method of construction that will be used.

(g) *OGC* means the Office of the General Counsel, including the regional attorney or attorney in charge serving the State in which the RHS project is located.

[35 FR 10687, July 1, 1970, as amended at 40 FR 52837, Nov. 13, 1975; 46 FR 61989, Dec. 21, 1981; 67 FR 78326, Dec. 24, 2003]

§1822.264 Eligibility requirements.

(a) *Eligibility of applicant.* To be eligible for an RHS loan, the applicant must be a private or public nonprofit organization as defined in §1822.263 (a) or (b) which is authorized to provide housing sites on a nonprofit basis.

(1) If it is a private nonprofit organization as defined in §1822.263(a), it should also:

(i) Have a membership of at least 10 community leaders.

(ii) Plan to adopt, if it is being newly organized, articles of incorporation and bylaws that generally conform to model articles and bylaws provided by the State director which will be consistent with State law and with changes appropriate to the purposes and powers of an eligible applicant under this subpart.

(2) [Reserved]

(b) *Authorized representative of applicant.* RHS will deal only with the applicant or bona fide representative or the applicant or the representative's technical advisors. An authorized representative of the applicant must have no pecuniary interest in the award of the engineering, architectural or construction contracts, necessary equipment, or the purchase or development of the land.

[35 FR 16087, July 1, 1970, as amended at 80 FR 9866, Feb. 24, 2015]

§1822.265 Loan purposes.

RHS loans may be made to qualified applicants:

(a) For the purchase and development of adequate sites, including the construction of essential access roads, streets, utility lines, and necessary equipment which will become a permanent part of the development. If public water and waste disposal facilities are not available and cannot reasonably be provided on a community basis with other financing, including Water and Waste Disposal Association loans, funds may be included for this purpose.

(b) For the payment of necessary engineering fees, legal fees, and closing costs.

(c) For the payment of actual cash cost of incidental administrative expenses such as postage, telephone, advertising, and temporary secretarial help, if funds to pay these expenses are not otherwise available. The estimated cost of these items should be identified and shown in the budget.

(d) To provide for needed landscaping, planting, seeding, or sodding, or other necessary facilities related to buildings such as walks, parking areas, and driveways.

(e) When legally required by proper local, county, and State Governmental bodies as a condition for subdivision approval, RHS loan funds may be used to provide common areas playgrounds and tot lots, provided such facilities are dedicated to, and maintained by, a public body.

(42 U.S.C. 1480; delegation of authority by the Sec. of Agri., 7 CFR 2.23; delegation of authority by the Asst. Sec. for Rural Development, 7 CFR 2.70)

[35 FR 16087, July 1, 1970, as amended at 43 FR 24264, June 5, 1978; 80 FR 9866, Feb. 24, 2015]

§1822.266 Limitations.

(a) *Loan limits.* No RHS loan(s) will be made to any applicant which will result in the applicant's owning an unpaid principal balance of more than

$100,000 on such loan(s) unless prior authorization for a larger loan is obtained from the national office. No such loan will exceed the development cost as defined in § 1822.263(d), or the value of the property as improved with the loan. These limitations also apply to cases in which the same persons hold a majority of the membership interests or constitute a majority of the directors of two or more applicants.

(b) *Limitations of use of loan funds.* Loans will not be made for:

(1) The purchase of land in excess of the immediate and identified needs in the locality.

(2) The purchase of land from a member of an applicant-organization, or from another organization in which any member of the applicant-organization has an interest, without prior consent of the national office.

(3) Refinancing of debts, except in accordance with paragraph (e) of this section.

(4) Payment of any fee, charge, or commission to any broker, negotiator, or other person for the referral of a prospective applicant or solicitation of a loan.

(5) Payment of any fee, salary, commission, profit, or compensation to an applicant, or to any officer, director, trustee, stockholder, member or agent of an applicant, except as provided in § 1822.265(b). No contract or agreement for services to be paid for with loan funds should be executed by the applicant without prior approval by the State director.

(c) *Sale of developed sites.* The sites developed with a section 524 loan must be for housing low- and moderate-income families and may be sold to families, nonprofit organizations, public agencies, and cooperatives eligible for assistance under any section of title V of the Housing Act of 1949, or under any other law which provides financial assistance. For example, this may include:

(1) Individuals with low and moderate incomes eligible for HUD mortgages.

(2) Individuals with low and moderate incomes eligible for VA guaranteed loans.

(3) Individuals with low or moderate incomes eligible for a loan from any private lender which is authorized by law to provide financial assistance for housing.

(4) Nonprofit organizations funded by Federal, State, or local governments carrying out programs for low- and moderate-income families to obtain housing.

(5) State or local public agencies such as a housing authority or a housing finance development agency carrying out programs for low- and moderate-income families to obtain housing.

(d) *Suitability of sites.* Sites will meet the requirements of the planned use; for example, individual housing or multiple housing or any combination thereof. Building sites must be well located and designed to provide a desirable living environment. Generally a loan will not be made for the development of less than 10 units, but they need not be contiguous.

(e) *Obligations incurred before loan closing.* When an applicant files an application for a loan, the county supervisor will advise the applicant that development work must not be started and obligations for work, materials, or land purchase must not be incurred before the loan is closed. If, nevertheless, the applicant incurs obligations for work, materials, or land purchase before the loan is closed, the State director may authorize the use of loan funds to pay such obligations only when he finds that all the following conditions exist:

(1) The obligations were incurred after the applicant filed a written application for a loan.

(2) The applicant is unable to pay such obligations from its own resources or to obtain credit from other sources, and failure to authorize the use of loan funds to pay such debts would impair the applicant's financial position.

(3) The obligations were incurred for authorized loan purposes.

(4) Contracts, materials, development and any land purchase meet RD standards and requirements.

(5) Payment of the obligations will remove any liens which have attached, and any basis for liens that may attach, to the property on account of

such obligations or such work, materials, or land purchase.

[35 FR 10687, July 1, 1970, as amended at 40 FR 6951, Feb. 18, 1975; 80 FR 9866, Feb. 24, 2015]

§ 1822.267 Special conditions.

(a) *Evidence of need.* Loans will be made on the basis of the applicant providing firm information as to the number of sites to be developed and evidence of a need for the proposed building sites in the locality.

(b) *Nondiscrimination.* The borrower will be required to agree not to discriminate or permit discrimination, in accordance with section 3 of the loan resolution form "('Rural Housing Site' Loan to Nonprofit Corporation)," available at all RD offices.

(c) *Supervisory assistance.* Supervision will be provided borrowers to the extent necessary to achieve the objectives of the loan and to protect the interests of the Government. County supervisors will counsel with applicants in selecting locations that will provide essential services and facilities and will result in the development of desirable residential communities.

(d) *Loan resolution.* A Loan Resolution will be adopted by the applicant's Board of Directors or similar governing body. If any provisions are not appropriate to a particular case, proposed substitute language should be submitted to the national office with the recommendations of the State director.

(e) *Development policies.* Development will be planned and performed in accordance with subparts A and C of part 1924 of this chapter.

(f) *Water and waste disposal facilities.* If public water and waste disposal facilities are not available and these facilities will be provided on a community basis with funds included in the RHS loan or with other financing, provision should be made to form an organization with members who will provide continuing maintenance and management of facilities. The cost of the facilities should be considered as a cost of developing the sites and included in the price charged for the lots when they are sold.

(g) *Compliance with local codes and regulations.* Planning and development of sites will comply with all State,

county, and local planning and zoning requirements, and will be for housing that will conform with any applicable laws, ordinances, codes, and regulations governing such matters as construction, heating, plumbing, electrical installation, fire prevention, health, and sanitation.

(h) *Optioning of land.* If a loan includes funds to purchase real estate, the applicable provisions of subpart A of part 1943 regarding options will be followed. After the loan is approved, the county supervisor will have Form RD, "Form Letter—Acceptance of Option," or other appropriate form of acceptance, completed, signed by the applicant, and mailed to the seller.

(i) *Use of and accountability for loan funds.* Supervised bank accounts will not be used except when their requirement is made or authorized by the State director for cases where adequate bonding is not available. If a supervised bank account is used, collateral for deposits of funds will be pledged when the supervised bank account exceeds $100,000. All loan funds and funds from other sources to be used to pay the development costs of the site, as well as proceeds from the sale of any sites, will be deposited in accordance with part 1902, subpart A of this chapter. The county supervisor will see that funds for land purchase are paid to the seller simultaneously with loan closing. After the loan is closed, monthly reports will be provided to Rural Development of all disbursements made and income received by the borrower. Reports for each month will be submitted to the Rural Development county office during the first 10 days of the next month. No expenditures will be made without prior Rural Development consent for items which are not included in the Rural Development approved development cost estimate or for amounts greater than those set forth in such estimate.

(j) *Insurance.* The State director will determine the minimum amounts and types of insurance the applicant will carry.

(1) Suitable workman's compensation insurance will be carried by the applicant for all its employees.

(2) The applicant will be advised of the possibility of incurring liability

and encouraged, or required when appropriate, to obtain liability insurance.

(k) *Bonding.* (1) Approved corporate surety bonds will be required in all cases involving a development contract in excess of $20,000, unless an exception is made by the national office. In other cases, the county supervisor will determine whether a surety bond is required.

(2) The applicant will provide fidelity bond coverage for its officers and employees entrusted with the receipt, custody, and disbursement of its funds and the custody of any other negotiable or readily saleable personal property. The amount of the bond will be at least equal to the maximum amount of such funds including funds in bank accounts, and property that the applicant will have in its possession or control at any one time. If permitted by State law, the United States will be named coobligee in the bond. Form RD, "Position Fidelity Schedule Bond," may be used if permitted by State law.

(l) *Conditional commitments for construction of homes on developed sites.* Conditional commitments may be issued on sites developed with an RHS section 524 loan to permit homes to be constructed on sites prior to the sale of the site to an eligible purchaser in accordance with the following:

(1) The requirements of 7 CFR 3550.70 must be met and a conditional commitment issued prior to the start of construction of the home.

(2) The conditional commitment must be issued to an RHS borrower who can legally provide the proposed housing and has the experience and training in construction to the extent necessary to assure that the housing will be built or jointly to the RHS loan borrower and a builder who has the legal capacity, training and experience necessary to construct the housing. In all cases the following language will be added under "other conditions" on Form RD 1944-11, "Conditional Commitment":

(i) "Not withstanding the other provisions of this commitment the sale of completed homes on sites developed with section 524 Rural Housing Site loans will be limited to families eligible for assistance under any section of title V of the Housing Act of 1949 or under any other law which provides financial assistance for housing low- and moderate-income families. The approval of Rural Development will be obtained prior to the sale of each home. The request for approval shall be submitted to the local Rural Development office along with an application for an RH 502 loan or a financial statement from the purchaser and verification of the other credit that is available."

(ii) The benefits of the nonprofit development of the site(s) must be passed on to the purchaser. This will result in this site being sold for $_____ (price to be determined as provided for in (§ 1822.275(b))).

(3) In arriving at the commitment price for the site and the completed home, the value will be based on the present market value of the house only, plus the nonprofit selling price of the lot.

(4) If in order to obtain interim financing for the construction of the homes, the RHS loan borrower requests a subordination by Rural Development on individual lots, the State Director may approve the subordination by completing and executing a subordination in the format of exhibit C of this subpart.

(5) The government's lien on any lot will be released only at the time of sale to an eligible purchaser.

(6) The County Supervisor should provide the necessary supervision to assure that the RHS loan borrower takes the necessary action to assure that all qualified builders in the area are aware of the availability of rural housing sites and are given an equal opportunity to participate in this conditional commitment program. As a minimum, the borrower will be required to submit a signed statement indicating the actions taken including

names and dates of contacts with builders.

(7 U.S.C. 1989; 5 U.S.C. 301; sec. 10, Pub. L. 93–347, 88 Stat. 392; 42 U.S.C. 1480; delegation of authority by the Sec. of Agr., 7 CFR 2.23; delegation of authority by the Asst. Sec. for Rural Development, 7 CFR 2.70)

[35 FR 16087, July 1, 1970, as amended at 41 FR 47460, Oct. 29, 1976; 42 FR 44669, Sept. 6, 1977; 43 FR 24264, June 5, 1978; 44 FR 1702, Jan. 8, 1979; 45 FR 39793, June 12, 1980; 46 FR 36106, July 14, 1981; 46 FR 61989, Dec. 21, 1981; 52 FR 8002, Mar. 13, 1987; 52 FR 19283, May 22, 1987; 67 FR 78326, Dec. 24, 2002; 80 FR 9866, Feb. 24, 2015]

§ 1822.268 Rates, terms, and source of funds.

(a) *Interest rate.* Upon request of the applicant, the interest rate charged by Rural Development will be the lower of the interest rates in effect at the time of loan approval or loan closing. If an applicant does not indicate a choice, the loan will be closed at the interest rate in effect at the time of loan approval. Interest rates are specified in exhibit B of RD Instruction 440.1 (available in any RD office) for the type assistance involved.

(b) *Repayment period.* Final payment will be due 2 years after the date of the loan. When necessary to carry out the loan purposes, the national office may authorize extension of maturity dates. As lots are sold before the final due date of the note, the proceeds of the sales will be applied on the account or any prior lien, or, with the prior approval of the national office, used in a manner consistent with the purposes of the loan and the security interest of the Government.

(c) *Source of funds.* Loans under this subpart will be made as insured loans, except that loans under § 1822.278 to develop building sites for sale in connection with self-help projects will be made as direct loans.

[35 FR 10687, July 1, 1970, as amended at 51 FR 6733, Feb. 26, 1986; 80 FR 9866, Feb. 24, 2015]

§ 1822.269 Security.

Each loan will be secured by a mortgage on the property purchased or improved with the loan, and a security interest in the funds held by the corporation in trust for the Government, in ac-

cordance with the provisions of the required Loan Resolution.

§ 1822.270 Technical, legal, and other services.

(a) *Appraisals.* The property will be appraised by an RD employee authorized to make real estate appraisals. The appraisal will consist of a narrative statement prepared and signed by the authorized employee describing in detail the items considered in arriving at the value of the property. Two values will be established by the appraiser:

(1) The fair market value of the total property "as is".

(2) The aggregate fair market value of the building sites after development.

(i) In determining the value of the property, the appraiser will consider the value and selling prices of similar building sites in the area. The selling prices of similar sites must be fully documented.

(ii) [Reserved]

(b) *Title clearance and legal services.* For a loan to a public nonprofit organization, title clearance and legal services will be obtained in accordance with instructions from the OGC, observing the provisions of subpart B of part 1927 of this chapter to the extent feasible. For a loan to a private nonprofit organization, the provisions of subpart B of part 1927 of this chapter regarding title clearance and legal services will apply. The applicant will be encouraged to have the same approved closing agent, where practical, perform the title clearance work in connection with the purchase of the land and the sale of the individual sites.

(c) *Contracts for legal services.* On projects requiring more legal services than are customarily required for title clearance alone, the applicant will be required to have a written contract when loan funds will be used for legal services. All such contracts will be subject to review and approval by the State director and therefore should be submitted to the State Director before execution by the applicant. Contracts will provide for the types of service to be performed and the amount of fees to be paid either in lump sum on the completion of all services or in installments as services are performed.

(d) *Engineering services.* On projects requiring engineering services, a written contract will be required between the engineer and the borrower. All such contracts will be subject to review and approval by the State director and therefore should be submitted to the State Director before execution by the applicant. The form of contract must conform with standard professional practices and describe the types of services to be performed and fees to be paid.

[35 FR 16087, July 1, 1970, as amended at 51 FR 6733, Feb. 26, 1986; 56 FR 67471, Dec. 31, 1991; 80 FR 9866, Feb. 24, 2015]

§ 1822.271   Processing applications.

(a) *Application.* The application will be in the form of a letter to the county supervisor with the following information included in or attached to the letter:

(1) Name and address of applicant.

(2) A copy of, or an accurate citation to, the specific provisions of State law under which the applicant is organized; a copy of the applicant's articles of incorporation, bylaws, and other authorizing documents; the names and addresses of the applicant's members, directors, and officers; and if another organization is a member of the applicant organization its name, address, and principal business.

(3) A current, dated, and signed financial statement showing assets, and liabilities, together with information on the repayment schedule and status of each debt.

(4) Evidence of inability to obtain credit from other sources.

(5) General description of the project.

(i) Location and size of tract or tracts to be bought and/or developed.

(ii) Number and size of individual sites planned together with a detailed plot plan.

(iii) Preliminary engineering plans, if available.

(6) Estimated cost and amount of loan needed.

(7) Explanation of applicant's financial contribution to the project.

(8) A map showing the location of and other supporting information on neighborhood and existing facilities such as distance to shopping area, neighborhood churches, available transportation, drainage, sanitation facilities, water supply available or planned, and access to essential services such as doctors, dentists, and hospitals.

(9) If facilities such as water and sewage systems, paved streets, and utilities are not currently available, information on when and how they will be provided.

(10) Evidence of the need for the proposed sites in the locality by low- and moderate-income families and other qualified applicants that are likely to be able to obtain financing for a home.

(11) Written evidence of any State, county, or local planning, zoning, or other ordinances imposing additional restrictions or requirements upon the proposed sites.

(b) *County supervisor's review and evaluation of applications.* The county supervisor will:

(1) Determine that the applicant meets the eligibility requirements of § 1822.264.

(2) Verify that the information provided is accurate and complete.

(3) Determine that:

(i) The sites will be located in a good residential area and that essential facilities and services will be provided.

(ii) The lots will be reasonable in cost and of a type Rural Development can appropriately finance.

(iii) There is an immediate and ready market for the proposed sites in the planned location.

(iv) The total number of sites planned does not exceed the number of loans the county supervisor can reasonably expect to include in the rural housing program or for which other credit is reasonably assured when the sites are developed.

(v) Proposed subdivisions will comply with the local codes and ordinances and also meet the requirements of subpart C of part 1924 of this chapter.

(4) Evaluate the manner in which the applicant plans to conduct its business and financial affairs.

(5) Comment on the background of the members, directors and officials.

(6) If he has questions about the proposal, send the incomplete docket to the State office for advice.

(7) If for any reason the loan cannot be made, inform the applicant.

(c) *Completion of the docket.* If the county supervisor determines that the applicant is eligible and the loan will be sound and proper, he should request the applicant to make any needed revisions. In addition to the items required in the application the docket must include:

(1) A plot plan and detailed preliminary plans and specifications for development of the building sites.

(2) A detailed cost breakdown of the project for such items as land and rights-of-way, utility installations or connections, on-site improvements, engineering and legal services, and estimated interest.

(3) If water and sanitary facilities are not publicly owned, a complete statement as to how they will be provided and details about their ownership and operation.

(4) Satisfactory evidence of review and approval of the proposed development by applicable State and local officials whose approval is required by State or local laws, ordinances, or regulations.

(5) Satisfactory evidence that the appropriate public bodies will accept and maintain all public facilities, including common areas, playgrounds, and tot lots, when dedicated to such bodies.

(d) *Preparation of docket forms—*(1) *Request for obligation of funds and fund analysis.* Form RD 3560–51, "Multiple Family Housing Obligation Fund Analysis" will be completed in accordance with the Forms Manual Insert (FMI).

(2) *County committee certification or recommendation.* County committees will not be used to review RHS loan applications.

(e) *Assembly, review and distribution of complete loan docket items.* When all items required for the complete loan docket have been furnished, they will be examined thoroughly to make sure they are properly and accurately prepared and are complete in all respects, including dates and signatures. The loan docket items will be assembled in the following order and distributed as follows:

Form No.	Name of form or document	Total No. of copies	Signed by borrower	Number for loan docket	Copy for borrower
	Application Letter and Attachments	2		1–0	1–C
RD 1910–11	Applicant Certification, Federal Collection Policies for Consumer or Commercial Debts.	2	2–O&C	1–O	1–C
	Evidence of Legal Authority (copy or citation of specific provisions of State statutory authority).	2	1	1–0	1–C
	Proof of Organization (certified copy of Articles of Incorporation) ...	2	1	1–0	1–C
	Certified copy of Bylaws	2	1	1–0	1–C
	List of names and addresses of officers, directors and members ...	2	1	1–0	1–C
	Narrative plan and other supporting information .... Evidence of Need.	2	1	1–0	1–C
	Certified Copy of Loan Resolution	1	1	1–0	
RD 440–4	Assurance Agreement	2	1	1–0	1–C
RD 400–1	Equal Opportunity Agreement (when applicable)	2	1	1–0	1–C
RD 400–3	Notice to Contractors and Applicants	3		1–C	1–C
RD 400–6	Compliance Statement (when applicable)	3		1–C	1–C
	Survey of land given as security, plans specifications, cost estimates, and proposed manner of development.	3	1	1–0	1–C
	Operating budget (if administrative expenses are to be included in loan).	2	1	1–0	1–C
	Appraisal Report with Attachments	1		1–0	
	Preliminary Title Opinion and a Final Title Opinion or a title insurance binder and a mortgage title insurance policy.				
	Option or copy of deed, purchase contract, or other instruments of ownership.				
RD 3560–51	Multiple Family Housing Obligation-Fund Analysis	2	2	[1]1	1

[1] Data input to Finance Office through field office terminals.

(f) *Submission of complete docket.* The complete docket will be sent to the State office together with the District Director's comments and recommendations and a draft for a press release.

(g) *Loan approval authority and State Office action.* The State Director is authorized to approve loans in accordance with this subpart and subpart A of part 1901 of this chapter. As soon as it is evident that a loan will be approved, the State Director will complete exhibit A to subpart C of part 2015 of this chapter. The State Director may redelegate approval authority to qualified State Office employees. When a docket or preliminary application is received in the State Office, the State Director will:

(1) Utilize the services of technicians on his staff and from other agencies in evaluating the application.

(2) Review the applicant's articles of incorporation and bylaws. If they conform to approved forms for the State as provided in § 1822.264(a)(1)(ii), the State director need not obtain a preliminary opinion from the OGC. In all other cases the State director will, and in any case may, submit the docket with any comments or questions to the OGC for a preliminary opinion as to whether the applicant and the proposed loan meet or can meet the requirements of State law and this subpart.

(3) If additional information is needed to adequately evaluate the application, return the loan docket to the District Director with any comments and recommendations for further processing.

(4) If the docket is sufficiently complete to enable the State Director to determine that the applicant is eligible and the loan would be sound and proper, issue a proposed memorandum of approval listing any specific conditions that must be met before loan closing.

(5) If the applicant is not eligible or the loan would not be sound and proper and the deficiencies cannot be corrected, inform the District Director accordingly.

(42 U.S.C. 1480; delegation of authority by the Sec. of Agr., 7 CFR 2.23; delegation of authority by the Asst. Sec. for Rural Development, 7 CFR 2.70)

[35 FR 16087, July 1, 1970, as amended at 41 FR 7487, Feb. 19, 1976; 41 FR 20392, May 18, 1976; 43 FR 24264, June 5, 1978; 44 FR 4435, Jan. 22, 1979; 50 FR 8583, Mar. 4, 1985; 52 FR 19283, May 22, 1987; 54 FR 29330, July 12, 1989; 69103, Nov. 26, 2004; 80 FR 9866, Feb. 24, 2015]

§ 1822.272 Approval or disapproval of a loan.

The provisions of 7 CFR part 3560, subpart B will be followed.

[69 FR 69103, Nov. 26, 2004]

§ 1822.273 Actions subsequent to loan approval.

After the loan is approved, actions to be taken will be in accordance with 7 CFR part 3560, subpart B.

[69 FR 69103, Nov. 26, 2004]

§ 1822.274 Loan closing.

(a) *Applicable instructions.* The complete loan docket will be sent to the OGC for loan closing instructions. RHS loans will be closed in accordance with applicable provisions of subpart B of part 1927 of this chapter, and State Instructions which supplement this Instruction, and closing instructions of the OGC, and with the assistance of the approved attorney, representatives of the title insurance company, or local attorney, whichever is appropriate.

(b) *Mortgage.* Unless the OGC determines the Form to be inappropriate, real estate mortgage Form RD 3550-14, "Real Estate Mortgage or Deed of Trust for (state)," will be used for all RHS section 524 loans modified as prescribed by or with the advice of the OGC with respect to the name, address, and other identification of the borrower, the style of execution, and the acknowledgement. Additional paragraphs will be included in the mortgage to read as follows:

The borrower agrees not to discriminate in the sale of the dwelling financed under this mortgage due to a prospective purchaser's race, color, national origin, sex, religion, age, marital status, or handicap. The borrower further agrees to comply with all Federal, State, or local laws and ordinances prohibiting discrimination in the sale of housing. The borrower's failure or refusal to comply with this agreement will be a basis for Rural Development to deny future requests for participation in its rural housing programs and activities.

This instrument also secures the obligations and covenants of borrower set forth in Borrower's Loan Resolution of _____ (date), which is hereby incorporated herein by reference.

(c) *Promissory note.* Form RD 3560–52, "Promissory Note," will be used. Instructions for preparation will be in accordance with the FMI and the following:

(1) The total amount to be shown in the note will be the amount of the loan shown on Form RD 3560–51. The note will be dated the date of the loan closing.

(2) The note will be signed in accordance with subpart B of part 1927 of this chapter and the forms manual insert for Form RD 3560–52 (available in RD office).

(3) Payments shall not be deferred.

(d) *Recorded mortgage.* When the real estate mortgage is returned by the recording official, the county supervisor will retain the original in the borrower's case folder. If the original is retained by the recording official for the county records, a conformed copy including the recording data showing the date and place of recordation and book and page number will be prepared and filed in the borrower's case folder. A copy of the mortgage will be delivered to the borrower but will be conformed only if required by State law or if it is the custom of other lenders in the area.

(e) *Date of loan closing.* An RHS loan is considered closed when the mortgage is filed of record.

[35 FR 16087, July 1, 1970, as amended at 42 FR 4408, Jan. 25, 1977; 50 FR 8584, Mar. 4, 1985; 56 FR 67472, Dec. 31, 1991; 69 FR 69104, Nov. 26, 2004; 80 FR 9866, Feb. 24, 2015]

§ 1822.275 Actions after sites are developed.

The building sites will be sold on a nonprofit basis to eligible families or organizations as described in § 1822.266(c).

(a) An option, RD 440–34, "Option to Purchase Real Property," will be executed. The site will be clearly identified by a land survey.

(b) The sale price of each individual site will not be more than a sufficient amount to pay a proportionate part of the RHS loan and any other actual costs of buying, developing, and selling the building site.

(c) The proceeds from sale of the building sites will be applied on the RHS loan and any prior lien or, with the prior approval of the National Office, used in a manner consistent with the purpose of the loan and the security interest of the Government. The sites will be released from the mortgage in accordance with 7 CFR part 3550, subpart D or otherwise in accordance with prior approval of the National Office.

[35 FR 16087, July 1, 1970, as amended at 51 FR 4135, Feb. 3, 1986; 67 FR 78326, Dec. 24, 2002; 80 FR 9866, Feb. 24, 2015]

§ 1822.276 Subsequent RHS loans.

A subsequent RHS loan is an RHS loan to an applicant indebted for an initial RHS loan. Subsequent RHS loans will be made on the same basis as initial RHS loans.

§ 1822.277 Complaints regarding discrimination in opportunity to buy developed sites.

Any applicant wishing to purchase a site financed by an RHS loan who believes he or she has been discriminated against because of race, color, national origin, religion, sex, handicap, or age, may file a complaint with the County Supervisor or State Director. Any such complaint will be handled in accordance with 7 CFR 3560.2.

[56 FR 67472, Dec. 31, 1991, as amended at 69 FR 69104, Nov. 26, 2004]

§ 1822.278 Special requirements for RHS section 523 loans (loans to organizations providing sites for self-help housing).

Loans to organizations which will provide sites for self-help housing (RHS sec. 523 loans) will be made under the provisions of this subpart with the following exceptions:

(a) *Eligibility.* The applicant must be a nonprofit organization engaged in assisting self-help projects.

(b) *Interest.* The interest rate will be 3 percent per annum on the unpaid principal balance.

(c) *Source of funds.* These will be direct loans made from the self-help fund.

(d) *Evidence of need.* Loans to newly formed organizations will be made on the basis of the applicant's providing firm information as to the number of sites to be developed and the names of eligible bona fide prospective purchasers who are assured of available home financing. Loans to organizations currently involved in mutual self-help housing projects may be made without submitting a list of the names of prospective site purchasers. There must, however, be definite evidence that enough families are available who are eligible and who will buy the sites when they are developed.

(e) *Multiple advances.* These loans may be disbursed over a period not to exceed 18 months from the date of the first advance.

(f) *Note forms.* Form RD 3560-52, "Multiple Family Housing Promissory Note," will be used. See § 1822.274 (c).

(g) *Mortgage.* Unless the OGC determines the Form to be inappropriate, real estate mortgage Form RD 3550-14, "Real Estate Mortgage or Deed of Trust for (state)," will be used modified as prescribed by or with the advice of the OGC with respect to the name, address, and other identification of the borrower, the style of execution, and the acknowledgement. Additional paragraphs will be included in the mortgage to read as follows:

The borrower agrees not to discriminate in the sale of the dwelling financed under this mortgage due to a prospective purchaser's race, color, national origin, sex, religion, age, marital status, or handicap. The borrower further agrees to comply with all Federal, State, or local laws and ordinances prohibiting discrimination in the sale of housing. The borrower's failure or refusal to comply with this agreement will be a basis for Rural Development to deny future requests for participation in its rural housing programs and activities.

This instrument also secures the obligations and covenants of borrower set forth in Borrower's Loan Resolution of _____ (date), which is hereby incorporated herein by reference.

[35 FR 16087, July 1, 1970, as amended at 42 FR 4408, Jan. 25, 1977; 50 FR 8584, Mar. 4, 1985; 56 FR 67472, Dec. 31, 1991; 69 FR 69104, Nov. 26, 2004; 80 FR 9866, Feb. 24, 2015]

§ 1822.279 Loan supervision and servicing.

Loan supervision and loan servicing will be provided according to 7 CFR part 3560.

[69 FR 69104, Nov. 26, 2004]

EXHIBITS A–B TO SUBPART G TO PART 1822 [RESERVED]

EXHIBIT C TO SUBPART G OF PART 1822— SUBORDINATION BY THE GOVERNMENT FOR USE WITH RURAL HOUSING SITE LOANS

Whereas, The United States of America acting through the Farmers Home Administration or its successor agency under Public Law 103–354 (hereinafter called the "Government") is the holder of the following-described instrument(s) executed by

of _____
County, State of _____
(hereinafter called the "Borrower")

Title of instrument	Date of instrument	Date filed	Office filed	Book No.	Page No.

And whereas, _____ (hereinafter called the "Lender") has agreed to provide a loan to the borrower or to a builder designated by the borrower to construct a home on the property described in this instrument.

Now Therefore, in consideration of the Lender's agreement to make such loan to the borrower, the Government hereby consents to the Borrower obtaining said loan from the lender, and agrees to and hereby subordinates in favor of the Lender and his successors and assigns its liens or security interests created or evidenced by the above-described instrument(s) insomuch as they cover the following described property:

Except That, The Government shall retain a first lien or security interest in the above-described property in an amount of $_____. Such first lien will be released only when satisfactory evidence is provided indicating that the lot with completed home is being sold to a family eligible for assistance under any section of Title V of the Housing Act of 1949 or under any other law which provides financial assistance for housing low- and moderate-income families and that the benefits of the nonprofit development of the site are being passed on to the eligible purchaser and that the amount of that first lien is paid on the Borrower's Rural Housing Site Loan debt to the Government.

This subordination is limited to the amount actually loaned by the Lender to the Borrower for the foregoing purpose, but shall not exceed $_____.

Only the above described property is affected by this subordination. This subordination shall not otherwise affect or modify the obligations secured by the aforesaid lien instrument(s), and the said obligations shall continue in force and effect until fully paid, satisfied, and discharged.

No member of Congress shall be admitted to any share or part of this agreement or to any benefit that may arise thereupon.

In Witness Whereof, The United States of America has caused these presents to be signed on the _____ day of _____, 19—, pursuant to delegated authority published in 7 CFR, Part 1800.

Witness: UNITED STATES OF AMERICA
_____   By: _____
_____   Title: _____
                       Farmers Home Administration or its successor agency under Public Law 103–354, U.S. Department of Agriculture.

(42 U.S.C. 1480; delegation of authority by the Secretary of Agriculture, 7 CFR 2.23; delegation of authority by the Assistant Secretary for Rural Development, 7 CFR 2.70)

[41 FR 47460, Oct. 29, 1976]

PART 1823 [RESERVED]

SUBCHAPTERS C–D [RESERVED]
SUBCHAPTER E—ACCOUNT SERVICING

PARTS 1863–1866 [RESERVED]

SUBCHAPTER F—SECURITY SERVICING AND LIQUIDATIONS

PART 1872 [RESERVED]

SUBCHAPTER G—MISCELLANEOUS REGULATIONS

PARTS 1890–1899 [RESERVED]

SUBCHAPTER H—PROGRAM REGULATIONS

PART 1900—GENERAL

AUTHORITY: 5 U.S.C. 301; 7 U.S.C. 1989; 7 U.S.C. 6991, et. seq.; 42 U.S.C. 1480; Reorganization Plan No. 2 of 1953 (5 U.S.C. App.).

Subpart A—Delegations of Authority

§ 1900.1 General.

The authorities contained in this subpart apply to all assets, functions, and programs now or hereafter administered or serviced by the Rural Development, including but not limited to those relating to indebtedness, security, and other assets obtained or contracted through the Secretary of Agriculture, Resettlement Administration, Farm Security Administration, or Emergency Crop and Feed Loan Offices of the Farm Credit Administration, the Soil Conservation Service in connection with water conservation and utilization projects; the Puerto Rico Hurricane Relief Commission and successor agencies in connection with Puerto Rico Hurricane relief loans to individuals; State Rural Rehabilitation Corporations, the United States of America or its officials as trustees of the assets of State Rural Rehabilitation Corporations, Regional Agricultural Credit Corporations, Defense Relocation Corporations, land leasing and purchasing associations, corporations, and agencies, and whether the interest of the United States in the indebtedness, instrument of debt, security, security instrument, or other assets is that of obligee, owner, holder, insurer, assignee, mortgagee, beneficiary, trustee or other interest. This subpart is inapplicable to Farm Service Agency, Farm Loan Programs.

[44 FR 18162, Mar. 27, 1979, as amended at 80 FR 9866, Feb. 24, 2015]

§ 1900.2 National office staff and state directors.

The following officials of the Rural Development, in accordance with applicable laws, and the regulations implementing these laws, are severally authorized, for and on behalf of and in the name of the United States of America or the Rural Development, to do and perform all acts necessary in connection with making and insuring loans, making grants and advances, servicing loans and other indebtedness and obtaining, servicing and enforcing security and other instruments related thereto: The Deputy Administrator Program Operations, the Assistant Administrators for Farmer Programs, Housing, and Community and Business Programs, the Assistant Administrator Accounting and Director Finance Office; each Director and the Insured Loan Officer, Finance Office; the Directors for the Water and Waste Disposal Division, the Community Facilities Division, the Business and Industry Division, the Multi-Family Housing Processing Division, the Multi-Family Housing Servicing and Property Management Division, the Single Family Housing Processing Division, the Single Family Housing Servicing and Property Management Division, the Farm Real Estate and Production Division, the Emergency Division; and each State Director within the area of that State Director's jurisdiction; and in the absence or disability of any such official, the person acting in that official's position; and the delegates of any such official. The authority includes, but is not limited to, the authority to:

(a) Effect the assignment of, or the declaration of trust with respect to, insured security instruments to place them in trust with the United States of America as trustee for the benefit of any holder of the promissory note or bond secured by such security instrument.

(b) Acknowledge receipt of notice of sale or assignment of insured loans and security instruments.

(c) Appoint or request the appointment of substitute trustees in deeds of trust.

(d) Execute proofs of claim in bankruptcy, death, and other cases.

(e) Consent to sale or assignment of, or sell or assign, direct or insured loans and security instruments (except that in the case of Agency asset sales, District Directors and County Supervisors are delegated the authority to assign security instruments), endorsements, reinsurance agreements, or other instruments in connection therewith; and execute agreements to insure and reinsure, and to purchase and repurchase insured loans and security instruments.

(f) Compromise, adjust, cancel or charge off indebtedness (except that County Supervisors are delegated authority to approve all settlements of sections 502 and 504 single family housing debt(s)).

(g) Modify contracts and other instruments and compromise claims owed to the Rural Development and covered by the Federal Claims Collection Act of 1966 and the joint regulations issued under it by the Attorney General and the Comptroller General as provided for in applicable program regulations.

(h) Perform all actions pertaining to the sale (or other disposal) of real or chattel property or interests therein and to execute and deliver bills of sale or other instruments to effect such sale (or disposition), which includes but is not limited to offering property for sale; advertising; receiving and accepting offers or bids; and closing sale transactions, including the collection of sale proceeds, and delivery of quitclaim deeds, easements, and right-of-way conveyances after those documents have been executed. The authority to execute any deeds of conveyance of inventory real property, including quitclaim deeds, easements, rights-of-way, or sale of any use rights is reserved to the State Director, and this authority may not be redelegated.

(i) Approve and consent to transfers of security property to other parties with or without assumption of debts;

and approve and accept transfers of security property or interests therein to the United States of America, and execute release from liability after determination is made in accordance with applicable program regulations.

(j) Execute and deliver, or approve in writing, suspensions, releases or terminations of assignments, of income, renewals, extensions, partial and full releases and satisfactions of security, and personal or indemnity liability for indebtedness, waivers, subordination agreements, severance agreements, affidavits, acknowlegements, certificates of residence, evidence of consent, and other instruments or documents.

(k) Accelerate and declare entire real estate or chattel indebtedness due and payable, foreclose or request foreclosure of real estate security instruments by exercise of power of sale or otherwise, and bid for and purchase at any foreclosure or other sale or otherwise acquire real property pledged, mortgaged, conveyed, attached, or levied upon to collect indetedness, and accept title to any property so purchased or acquired.

(l) Require and accept further or additional security.

(m) Accelerate and declare entire non-real estate indebtedness due and payable, and foreclose or request foreclosure of chattel security instruments by exercise of power of sale or otherwise.

(n) Bid for and purchase at any foreclosure or other sale, or otherwise acquire personal property pledged, mortgaged, conveyed, attached, or levied upon to collect indebtedness, and accept title to any property so purchased or acquired.

(o) Take possession of, maintain, and operate security or acquired real or personal property or interests therein, sell or otherwise dispose of such personal property, and execute and deliver contracts, caretaker's agreements, leases, and other instruments in connection therewith, as appropriate.

(p) Execute proofs of loss on insurance contracts and endorse without recourse loss payment drafts and checks.

(q) Issue, publish and serve notices and other instruments.

(r) File or record instruments, whether separate instruments, or by making marginal entries, or by use of other methods permissible under State law.

[44 FR 18162, Mar. 27, 1979, as amended at 47 FR 5700, Feb. 8, 1982; 50 FR 23901, June 7, 1985; 52 FR 44375, Nov. 19, 1987; 56 FR 6946, Feb. 21, 1991; 80 FR 9867, Feb. 24, 2015]

§ 1900.3 State, district, and county office employees.

The following officials and employees of the Rural Development, in accordance with applicable laws, and the regulations implementing these laws, for and on behalf of, and in the name of the United States of America or the Rural Development, are also severally authorized within the area of their respective jurisdictions to perform the acts specified in paragraphs (g) through (r) of § 1900.2; and within their loan approval authority to sell or otherwise dispose of real or chattel property or interests therein and to execute and deliver bills of sale or other instruments to effect such sale or disposition: Chief, Farmer Programs/Specialist; Chief, Rural Housing/Specialist; Chief, Community Programs/Specialist; Chief, Business and Industry/Specialist; Chief, Community and Business Programs/Specialist; Chief, Appraisal Staff/Appraiser; Chief, Underwriting Staff/Underwriter; Chief, Underwriting and Appraisal Staff; Chief, Servicing and Inventory Staff/Credit Management Specialist/Realty Specialist; each District Director, Assistant District Director, Loan Specialist General, County (including Parish) Supervisor, Emergency Loan Supervisor, Assistant Emergency Loan Supervisor, or other supervisor or assistant supervisor, and in the absence or disability of any such official or employee, the person acting in the position.

[50 FR 23902, June 7, 1985, as amended at 55 FR 43325, Oct. 29, 1990; 80 FR 9867, Feb. 24, 2015]

§ 1900.4 Ratification.

All written instruments affecting title to real or personal property, including but not limited to deeds, releases, satisfactions, subordination agreements, severance agreements, consents, waivers, assignments, declarations of trust, and heretofore executed by officials or employees of the agencies or other entities referred to in

§1900.1 to carry out any purpose authorized by law, incident to the administration of programs under the jurisdiction of said agencies or other entities, are hereby approved, confirmed, and ratified.

[44 FR 18162, Mar. 27, 1979]

§1900.5 Assignment of cases.

The State Director may, in writing, assign responsibilities and functions to a different office or staff position within the Rural Development State organizational structure other than that referred to in regulations, provided no benefits, rights, or opportunities of the public are changed.

[55 FR 43325, Oct. 29, 1990, as amended at 80 FR 9867, Feb. 24, 2015]

§1900.6 Chair, Loan Resolution Task Force.

The Chair, Loan Resolution Task Force is delegated the following authorities, to be exercised until September 30, 1996:

(a) The responsibility for, under applicable Rural Development regulations, collecting and settling all delinquent direct Farmer Program loans as defined in the Consolidated Farm and Rural Development Act, as amended, that have received all primary servicing rights and pre-acceleration homestead and preservation loan servicing rights under 7 CFR part 1951, subpart S;

(b) The responsibility for making and directing the making of loan servicing decisions, under applicable Rural Development regulations, concerning delinquent direct Farmer Programs loans for which accrued principal and interest equals or exceeds one million dollars, to extend to borrowers their remaining primary servicing rights and pre-acceleration homestead and preservation loan servicing rights under 7 CFR part 1951, subpart S;

(c) Authority for approving the grant of exceptions pursuant to §§1951.916, 1955.21, 1956.99 and 1965.35 of this chapter, to the extent necessary to carry out the responsibilities described in paragraphs (a) and (b) of this section.

[59 FR 43441, Aug. 24, 1994, as amended at 80 FR 9867, Feb. 24, 2015]

§1900.7 Effect on other regulations.

This subpart does not revoke or modify any other delegation or redelegation, instruction, procedure, or regulation issued by, or under authority of, the Under Secretary for Rural Development.

[44 FR 18162, Mar. 27, 1979. Redesignated at 55 FR 43325, Oct. 29, 1990, and further redesignated at 59 FR 43441, Aug. 24, 1994, as amended at 80 FR 9867, Feb. 24, 2015]

Subpart B—Adverse Decisions and Administrative Appeals

Source: 60 FR 67318, Dec. 29, 1995, unless otherwise noted.

§1900.51 Definitions.

Act means the Federal Crop Insurance Reform and Department of Agriculture Reorganization Act of 1994, Public Law No. 103–354 (7 U.S.C. 6991 *et seq.*).

Agency means the Rural Utilities Service (RUS), the Rural Housing Service (RHS), and the Rural Business-Cooperative Development Service (RBS), or their successor agencies.

Refer to 7 CFR 11.1 for other definitions applicable to appeals of adverse decisions covered by this subpart.

§1900.52 General.

This subpart specifies procedures for use by USDA personnel and program participants to ensure that full and complete consideration is given to program participants who are affected by an agency adverse decision.

§1900.53 Applicability.

(a) Appeals of adverse decisions covered by this subpart will be governed by 7 CFR part 11.

(b) The provisions of this subpart apply to adverse decisions concerning direct loans, loan guarantees, and grants under the following programs: RUS Water and Waste Disposal Facility Loans and Grants Program; RHS Housing and Community Facilities Loan Programs; RBS Loan, Grant, and Guarantee Programs and the Intermediary Relending Program; and determinations of the Rural Housing Trust 1987–1 Master Servicer.

(c) This subpart does not apply to decisions made by parties outside an agency even when those decisions are used as a basis for decisions falling within paragraph (b) of this section, for example: decisions by state governmental construction standards-setting agencies (which may determine whether RHS will finance certain houses); Davis-Bacon wage rates; flood plain determinations; archaeological and historical areas preservation requirements; and designations of areas inhabited by endangered species.

§ 1900.54 Effect on assistance pending appeal.

(a) Assistance will not be discontinued pending the outcome of an appeal of a complete or partial adverse decision.

(b) Notwithstanding the provisions of paragraph (a) of this section, administrative offsets initiated under subpart C of part 1951 will not be stayed pending the outcome of an appeal and any further review of the decision to initiate the offset.

§ 1900.55 Adverse action procedures.

(a) If an applicant, guaranteed lender, a holder, borrower or grantee is adversely affected by a decision covered by this subpart, the decision maker will inform the participant of the adverse decision and whether the adverse decision is appealable. A participant has the right to request the Director of NAD to review the agency's finding of nonappealability in accordance with 7 CFR 11.6(a). In cases where the adverse decision is based on both appealable and nonappealable actions, the adverse action is not appealable.

(b) A participant affected by an adverse decision of an agency is entitled under section 275 of the Act to an opportunity for a separate informal meeting with the agency before commencing an appeal to NAD under 7 CFR part 11.

(c) Participants also have the right under section 275 of the Act to seek mediation involving any adverse decision appealable under this subpart if the mediation program of the State in which the participant's farming operation giving rise to the decision is located has been certified by the Sec-retary for the program involved in the decision. An agency shall cooperate in such mediation. Any time limitation for appeal will be stayed pending completion of the mediation process (7 CFR 11.5(c)).

§ 1900.56 Non-appealable decisions.

The following are examples of decisions which are not appealable:

(a) Decisions which do not fall within the scope of this subpart as set out in § 1900.53.

(b) Decisions that do not meet the definition of an "adverse decision" under 7 CFR part 11.

(c) Decisions involving parties who do not meet the definition of "participant" under 7 CFR part 11.

(d) Decisions with subject matters not covered by 7 CFR part 11.

(e) Interest rates as set forth in agency procedures, except for appeals alleging application of an incorrect interest rate.

(f) The State RECD Director's refusal to request an administrative waiver provided for in agency program regulations.

(g) Denials of assistance due to lack of funds or authority to guarantee.

§ 1900.57 [Reserved]

EXHIBIT A TO SUBPART B OF PART 1900 [RESERVED]

EXHIBIT B–1 TO SUBPART B OF PART 1900—LETTER FOR NOTIFYING APPLICANTS, LENDER, HOLDERS AND BORROWERS OF ADVERSE DECISIONS WHERE THE DECISION IS APPEALABLE

UNITED STATES DEPARTMENT OF AGRICULTURE

Farmers Home Administration or its successor agency under Public Law 103–354

(Insert Address)

Date _____

Dear _____:

After careful consideration, we [were unable to take favorable action on your application/request for Farmers Home Administration or its successor agency under Public Law 103–354 services] [are cancelling/reducing the assistance you are presently receiving]. The specific reasons for our decision are:

(Insert here the adverse decision and all of the specific reasons for the adverse action.)

If you have any questions concerning the decision or the facts used in making our decision and desire further explanation, you may call or write the County Office (insert phone number) to request a meeting with (this office) (The County Committee) within 15 calendar days of the date of this letter. You should present any new information or evidence along with possible alternatives for our consideration. You may also bring a representative [or legal counsel] with you. You also have the right to appeal this decision to a hearing officer in lieu of, or in addition to, a meeting with [this office] [the County Committee]. See attachment for your appeal rights. (Attach Form FmHA or its successor agency under Public Law 103–354 1900–1.) (For guaranteed loans, except loss claims, the applicant and lender must jointly request a meeting and/or an appeal.)

If you do not wish a meeting, as outlined above, a request for a hearing must be sent to the Area Supervisor, National Appeals Staff (address) _____, postmarked no later than (month) _____, (date) _____.

(insert date 30 days from date of letter.)

The Federal Equal Credit Opportunity Act prohibits creditors from discriminating against credit applicants on the basis of race, color, religion, national origin, sex, marital status, handicap, or age (provided that the applicant has the capacity to enter into a binding contract), because all or part of the applicant's income derives from any public assistance program, or because the applicant has in good faith exercised any right under the Consumer Credit Protection Act. The Federal Agency that administers compliance with the law concerning this creditor is the Federal Trade Commission, Equal Credit Opportunity, Washington, DC 20580.

Sincerely,

_____

(Decision Maker)
(County Supervisor may sign for County Committee)

_____

(Title)

[55 FR 9874, Mar. 16, 1990]

EXHIBIT B–2 TO SUBPART B OF PART 1900—LETTER FOR NOTIFYING APPLICANTS, LENDERS AND HOLDERS AND BORROWERS OF UNFAVORABLE DECISION REACHED AT THE MEETING

UNITED STATES DEPARTMENT OF AGRICULTURE

Farmers Home Administration or its successor agency under Public Law 103–354

(Insert Address)

_____

Date
  Dear _____:
We appreciated the opportunity to review the facts relative to [your application/request for FmHA or its successor agency under Public Law 103–354 services] [the assistance you are presently receiving]. We regret that our meeting with you did not result in a satisfactory conclusion.

(Insert here the adverse decision and all the specific reasons for the adverse action).

See attachment for your appeal rights. (Attach Form FmHA or its successor agency under Public Law 103–354 1900–1) (For guaranteed loans, except loss claims, the applicant and lender must jointly request an appeal.)

A request for a hearing must be sent to the Area Supervisor, National Appeals Staff _____, postmarked no later than (month) _____, (date) _____

(insert date 30 days from date of letter.)

The Federal Equal Credit Opportunity Act prohibits creditors from discriminating against credit applicants on the basis of race, color, religion, national origin, sex, marital status, handicap, or age (provided that the applicant has the capacity to enter into a binding contract), because all or part of the applicant's income derives from any public assistance program, or because the applicant has in good faith exercised any right under the Consumer Credit Protection Act. The Federal Agency that administers compliance with the law concerning this creditor is the Federal Trade Commission, Equal Credit Opportunity, Washington, DC 20580.

Sincerely,

_____

(Decision Maker)
(County Supervisor may sign for County Committee)

_____

(Title)

[55 FR 9874, Mar. 16, 1990]

EXHIBIT B-3 TO SUBPART B OF PART 1900—LETTER FOR NOTIFYING APPLICANTS, LENDER, HOLDERS AND BORROWERS OF ADVERSE DECISIONS WHERE THE DECISION INVOLVES AN APPRAISAL (NOT TO BE USED IN CASES INVOLVING FARMER PROGRAM PRIMARY LOAN SERVICING ACTIONS)

UNITED STATES DEPARTMENT OF AGRICULTURE

Farmers Home Administration or its successor agency under Public Law 103–354

(Insert Address)

Date

Dear_____:

After careful consideration, we [were unable to take favorable action on your application/request for Farmers Home Administration or its successor agency under Public Law 103–354 services] [are cancelling/reducing the assistance you are presently receiving]. The specific reasons for our decision are:

(Insert here the adverse decision and all of the specific reasons for the adverse action.)

If you have any questions concerning the decision or the facts used in making our decision and desire further explanation, you may call or write the County Office (insert phone number) to request a meeting with (this office) (The County Committee) within 15 calendar days of the date of this letter. You should present any new information or evidence along with possible alternatives for our consideration. You may also bring a representative or legal counsel with you.

If you do not wish to have a meeting as outlined above, you may contest the appraisal of the property value. In order to contest the appraisal you must first request review of the appraisal by the FmHA or its successor agency under Public Law 103–354 State Director. Your request for review by the State Director should be made through our office. You will be advised of the results of the State Director's review. If after the State Director's review you still disagree with the appraisal you may request a hearing. When you receive the results of the State Director's review you will be advised on how to ask for a hearing. Your request for review of the appraisal must be postmarked no later than (month)_____, (date) _____ (insert date 15 days from date of letter).

The Federal Equal Credit Opportunity Act prohibits creditors from discriminating against credit applicants on the basis of race, color, religion, national origin, sex,

marital status, handicap, or age (provided that the applicant has the capacity to enter

[55 FR 9874, Mar. 16, 1990]

EXHIBIT B-4 TO SUBPART B OF PART 1900—LETTER FOR NOTIFYING APPLICANTS, LENDERS AND HOLDERS AND BORROWERS OF UNFAVORABLE DECISION REACHED AFTER STATE DIRECTOR REVIEW OF AN APPRAISAL (NOT TO BE USED IN CASES INVOLVING FARMER PROGRAM PRIMARY LOAN SERVICING ACTIONS)

UNITED STATES DEPARTMENT OF AGRICULTURE

Farmers Home Administration or its successor agency under Public Law 103–354

(Insert Address)

Date

Dear _____:

At your request we have reviewed the appraisal of the property you wish to purchase. We have determined that the value estimate of the property is both supportable and defensible (as required by FmHA or its successor agency under Public Law 103–354 regulations and appraisal industry standards) and therefore acceptable.

You have the right to appeal this decision. You must show why the appraisal is in error. You may submit an independent appraisal, at your expense, from a qualified [appraiser who is a designated member of [the American Institute of Real Estate Appraisers, Society of Real Estate Appraisers, American Society of Farm Managers and Rural Appraisers, etc.,] or an equivalent organization requiring appraisal education, testing and experience. The appraisal must conform to Agency Appraisal regulations applicable to the loan program.

See attachment for your appeal rights.

A request for a hearing must be sent to the Area Supervisor, National Appeals Staff (address) _____, postmarked no later than (month) _____, (date) _____,

(insert date 30 days from date of letter)

The Federal Equal Credit Opportunity Act prohibits creditors from discriminating against credit applicants on the basis of race, color, religion, national origin, sex, marital status, handicap, or age (provided that the applicant has the capacity to enter into a binding contract), because all or part of the applicant's income derives from any public assistance program, or because the applicant has in good faith exercised any right under the Consumer Credit Protection Act. The Federal Agency that administers compliance with the law concerning this creditor

is the Federal Trade Commission, Equal Credit Opportunity, Washington, DC 20580.
Sincerely,

(State Director)

[55 FR 9875, Mar. 16, 1990]

EXHIBIT C TO SUBPART B OF PART 1900—
LETTER FOR NOTIFYING APPLICANTS, LENDERS, HOLDERS, AND BORROWERS OF ADVERSE DECISIONS WHEN PART OR ALL OF THE DECISION IS NOT APPEALABLE [NOT USED IN CONNECTION WITH DECISIONS RELATED TO NON-PROGRAM APPLICANTS, BORROWERS, OR PROPERTY]

UNITED STATES DEPARTMENT OF AGRICULTURE

Farmers Home Administration or its successor agency under Public Law 103–354

(Insert Address)

(Date)
    Dear _____ :
    After careful consideration we [were unable to take favorable action on your application/request for Farmers Home Administration or its successor agency under Public Law 103–354 services] [are cancelling/reducing the assistance you are presently receiving].

(Insert and number all of the *specific* reasons for the adverse action. Examples of non-appealable reasons are listed in § 1900.55(a)).

If you have any questions about this action, we would like the opportunity to explain in detail why your request has not been approved, explain any possible alternative, or provide any other information you would like. You may bring any additional information you may have and you may bring a representative or counsel if you wish. Please call (*telephone number*) for an appointment.

Applicants and borrowers generally have a right to appeal adverse decisions, but FmHA or its successor agency under Public Law 103–354 decisions based on certain reasons are not appealable. We have determined that the reason(s) numbered _____ for the decision in this case make(s) the decision not appealable under FmHA or its successor agency under Public Law 103–354 regulations. You may, however, write the Area Supervisor, National Appeals Staff (insert address) for a review of the accuracy of our finding that the decision is not appealable, postmarked no later than (month) _____, (date) _____ (insert date 30 days from date of letter).

The Federal Equal Credit Opportunity Act prohibits creditors from discriminating against credit applicants on the basis of race, color, religion, national origin, sex, marital status, handicap, or age (provided that the applicant has the capacity to enter into a binding contract), because all or part of the applicant's income derives from any public assistance program, or because the applicant has in good faith exercised any right under the Consumer Credit Protection Act. The Federal Agency that administers compliance with the law concerning this creditor is the Federal Trade Commission, Equal Credit Opportunity, Washington, DC 20580.
    Sincerely,

(Decision Maker)
(County Supervisor may sign for County Committee)

(Title)

[55 FR 9875, Mar. 16, 1990, as amended at 58 FR 52646, Oct. 12, 1993]

EDITORIAL NOTE: At 58 FR 52646, Oct. 12, 1993, the Farmers Home Administration attempted to amend exhibit C of subpart B of part 1900 by removing in the second paragraph the words "(month) _____,"; however, because "(month) _____" does not exist in the second paragraph, this amendment could not be incorporated.

EXHIBIT D TO SUBPART B OF PART 1900—HEARINGS/REVIEW OFFICER DESIGNATIONS

HEARING/REVIEW OFFICER DESIGNATIONS

Decisionmaker or decision	Hearing officer	Review officer
County Supervisor	National Appeals Staff Hearing Officer	State Director and/or Director, National Appeals Staff.
County Committee	National Appeals Staff Hearing Officer	State Director and/or Director, National Appeals Staff.
*District Director, *State Program Chief, *District Specialist.	National Appeals Staff Hearing Officer	**State Director and/or Director, National Appeals Staff.
*State Director, *Regional Director	As appointed by Director, National Appeals Staff.	Director, National Appeals Staff.
Division Director or Assistant Administrator.	As appointed by Director, National Appeals Staff.	Director, National Appeals Staff.
Assistant Administrator	As appointed by Director, National Appeals Staff.	Director, National Appeals Staff.

HEARING/REVIEW OFFICER DESIGNATIONS—Continued

Decisionmaker or decision	Hearing officer	Review officer
Deputy or Associate Administrator	As appointed by Director, National Appeals Staff.	Director, National Appeals Staff.

*Decisionmaker for Rural Development Administration or its successor agency under Public Law 103–354 (RDA or its successor agency under Public Law 103–354) cases for Regional Office Operations.
**Review officer will be the Regional Director and/or the Director, National Appeals Staff for RDA or its successor agency under Public Law 103–354 cases.

NOTES

1. District Director also means Assistant District Director or District Loan Specialist.

2. County Supervisor also means Assistant County Supervisor with loan approval authority.

3. The Director of the National Appeals Staff may designate a staff member to conduct a hearing or review. When the hearing/review is completed, the designee will send the complete case file, hearing notes, tape recordings, and a recommended decision to the Director for a final decision. The Director may, for individual cases, delegate final decision authority to a designee.

4. For decisions not directly covered above, advice should be sought from the Director of the National Appeals Staff.

5. An appellant may elect to have an appeal reviewed by the State Director, or the Director of the National Appeals Staff. The decision of the State Director will be subject to further review by the Director of the National Appeals Staff upon request of the appellant.

[58 FR 4065, Jan. 13, 1993]

Subpart C—Applicability of Federal Law

§ 1900.101 General.

This subpart provides Rural Development policy concerning:

(a) The applicability of Federal rather than State Law in the conduct of Rural Development operations, and

(b) The liability of an auctioneer for conversion of personal property mortgaged to Rural Development.

[44 FR 10979, Feb. 26, 1979, as amended at 45 FR 8934, Feb. 11, 1980; 80 FR 9867, Feb. 24, 2015]

§ 1900.102 Applicable law.

Loans made by Rural Development are authorized and executed pursuant to Federal programs adopted by Congress to achieve national purposes of the U.S. Government.

(a) Instruments evidencing or securing a loan payable to or held by the Rural Development, such as promissory notes, bonds, guaranty agreements, mortgages, deeds of trust, financing statements, security agreements, and other evidences of debt or security shall be construed and enforced in accordance with applicable Federal law.

(b) Instruments evidencing a guarantee, conditional commitment to guarantee, or a grant, such as contracts of guarantee, grant agreements or other evidences of an obligation to guarantee or make a grant, executed by the Rural Development, shall be construed and enforced in accordance with applicable Federal law.

(c) In order to implement and facilitate these Federal loan programs, the application of local procedures, especially for recordation and notification purposes, may be utilized to the fullest extent feasible and practicable. However, the use of local procedures shall not be deemed or construed to be any waiver by Rural Development of Federal immunity from any local control, penalty, or liability, or to subject Rural Development to any State required acts or actions subsequent to the delivery by Rural Development officials of the instrument to the appropriate local or State official.

(d) Any person, corporation, or organization that applies for and receives any benefit or assistance from Rural Development that offers any assurance or security upon which Rural Development relies for the granting of such benefit or assistance, shall not be entitled to claim or assert any local immunity, privilege, or exemption to defeat the obligation such party incurred in obtaining or assuring such Federal benefit or assistance.

(e) The liability of an auctioneer for conversion of personal property mortgaged to Rural Development shall be determined and enforced in acceptance with the applicable Federal law. "Auctioneer" for the purposes of this subpart includes a commission merchant, market agency, factor or agent. In all cases in which there has been a disposition without authorization by Rural Development of personal property mortgaged to that agency, any auctioneer involved in said disposition shall be liable to the Government for conversion—notwithstanding any State statute or decisional rule to the contrary.

[44 FR 10979, Feb. 26, 1979, as amended at 80 FR 9867, Feb. 24, 2015]

Subpart D—Processing and Servicing Rural Development Assistance to Employees, Relatives, and Associates

SOURCE: 58 FR 224, Jan. 5, 1993, unless otherwise noted.

§1900.151 General.

(a) The Standards of Ethical Conduct for Employees of the Executive Branch requires the maintenance of high standards of honesty, integrity, and impartiality by employees. To reduce the potential for employee conflict of interest, any processing, approval, servicing or review activity, including access through automated information systems, is conducted only by authorized Rural Development employees who:

(1) Are not themselves the recipient.

(2) Are not members of the family or known close relatives of the recipient.

(3) Do not have an immediate working relationship with the recipient, the employee related to the recipient, or the employee who would normally conduct the activity.

(4) Do not have a business or close personal association with the recipient.

(b) No provision of this subpart takes precedence over individual program requirements or restrictions relating to eligibility for Rural Development assistance to Rural Development employees, members of families of employees, close relatives, or business or close personal associates of employees.

(c) The determination of a case's need for special handling under the provisions of this subpart is not an adverse action and, therefore, is not subject to appeal.

(d) The provisions of this subpart do not apply to the Farm Service Agency. The relevant regulations applicable to the Farm Service Agency can be found at 5 CFR parts 2635 and 8301.

[58 FR 224, Jan. 5, 1993, as amended at 71 FR 38979, July 11, 2006; 80 FR 9867, Feb. 24, 2015]

§1900.152 Definitions.

Applicant or borrower. All persons or organizations, individually or collectively, applying for or receiving insured or guaranteed loan or grant assistance from or through Rural Development. Referred to as recipient.

Assistance. Loans or grants made, insured or guaranteed, or serviced by Rural Development.

Associates. All persons with whom an employee has a business or close personal association or immediate working relationship.

Business association. Business relationship between those with an identity of financial interest; including but not limited to a business partnership, being an officer, director, trustee, partner or employee of an organization, or other long-term contractual relationship.

Close personal association. Social relationship between unrelated residents of the same household.

Close relatives. The spouse, relatives and step-relatives of an employee or the employee's spouse, including Grandmother, Grandfather, Mother, Father, Aunt, Uncle, Sister, Brother, Daughter, Son, Niece, Nephew, Granddaughter, Grandson, and First Cousin.

Conflict of interest. A situation (or the appearance of one) in which one could reasonably conclude that a Rural Development employee's private interest conflicts with his or her Government duties and responsibilities, even though there may not actually be a conflict.

Employee. All Rural Development personnel, including gratuitous employees

459

and those negotiating for or having arrangements for prospective employment, except as otherwise specifically stated. For the purposes of this instruction only, the term also refers to county or area committee members, elected or appointed, and to closing agents who, although they are not employees, have a special relationship to Rural Development and therefore should be subject to these provisions.

Immediate working relationship. A relationship between a subordinate and a supervisor in a direct line, or between co-workers in the same office. For the purposes of this subpart, the relationships among a County Supervisor and members of the local County Committee are immediate working relationships.

Members of family. Blood and in-law relatives (such as by marriage or adoption) who are residents of the employee's household.

Recipient. One who has applied for or received Rural Development financial assistance in the form of a loan or grant. See definition of applicant or borrower.

[58 FR 224, Jan. 5, 1993, as amended at 80 FR 9867, Feb. 24, 2015]

§ 1900.153 Identifying and reporting an employee relationship.

(a) *Responsibility of applicant.* When an application for assistance is filed, the processing official asks if there is any known relationship or association with a Rural Development employee. The applicant is required to disclose the requested information under pertinent program regulations.

(b) *Responsibility of the Rural Development employee.* A Rural Development employee who knows he or she is related to or associated with an applicant or recipient, regardless of whether the relationship or association is known to others, is required to notify the Rural Development official who is processing or servicing the assistance, in writing. RD Guide Letter 1900–D–1 (available in any RD office) may be used as the notice. If the appropriate official is not known, the State Director should be notified. Regardless of whether the relationship or association is defined in § 1900.152, if the employee believes there may be a potential con-

flict of interest, the Rural Development official who is processing or servicing the assistance may be notified and special handling requested. An employee's request that the case receive special handling is usually honored.

(c) *Responsibility of the Rural Development official.* When any relationship or association is identified, the Rural Development official completes and submits RD Guide Letter 1900–D–2 to the State Director (or Administrator, under paragraph (e) of this section or § 1900.155(a)). When completed, RD Guide Letter 1900–D–3 is returned by the State Director, the processing official;

(d) *Relationship or association established after application for Rural Development assistance.* If a relationship or association is established after an application has been filed or assistance has been provided, both recipient and employee are required to notify the Rural Development official as described in paragraphs (a) and (b) of this section.

(e) *Relationship or association with a State Office, Finance Office or National Office employee.* If an identified relationship or association is with an employee at a State Office (other than a State Director), Finance Office or National Office, the processing/servicing official completes and submits RD Guide Letter 1900–D–2 to the State Director in the normal manner. The State Director reviews the information, determines the need for special handling, designates the processing/servicing official, completes and submits RD Guide Letter 1900–D–3 to the Administrator for written concurrence. When the Administrator's concurrence is received, the State Director returns completed RD Guide Letter 1900–D–3 to the original official who completes the action described in paragraph (c) of this section.

(f) *Relationship or association with a State Director.* If an identified relationship or association is with a State Director, the processing/servicing official completes and submits RD Guide Letter 1900–D–2 to the Administrator. The Administrator reviews, determines the need for special handling, designates the processing/servicing official, completes and returns RD Guide Letter

1900–D–3 to the original official who completes the action described in paragraph (c) of this section.

(g) *Change in relationship or association, status of Rural Development assistance, or employee's duty station.* If the relationship or association has changed, the application denied or the assistance otherwise terminated, or Rural Development employee's duty station changed, the designated processing/servicing official completes RD Guide Letter 1900–D–2 with the new information and submits it. The review process takes place as described in paragraphs (a) through (e) of this section to determine if processing/servicing activity may return to normal or requires another change. If the assistance is denied or otherwise terminated, the designated official notifies the original official.

[80 FR 9867, Feb. 24, 2015]

§ 1900.154 Determining the need for special handling.

The State Director (or Administrator, under § 1900.153(e) or § 1900.155(a) of this subpart):

(a) [Reserved]

(b) Determines whether the reported relationship or association is defined in § 1900.152 of this subpart and would violate the provisions of § 1900.151(a) of this subpart,

(c)–(f) [Reserved]

§ 1900.155 Designating the processing/servicing official.

(a) *Designating an official with equivalent authority.* The State Director (or Administrator, under § 1900.253(e) of this subpart or this paragraph) designates a nonrelated or nonassociated Rural Development official authorized to conduct the activity under program regulations, established delegation of authority and approval authority under subpart A of part 1901 of this chapter, and whose duty station is most convenient to the recipient and to the security property. A type and/or amount of assistance processed or serviced by a County Supervisor or at a County Office should be assigned only to another County Supervisor or County Office. A type and/or amount of assistance processed or serviced by a District Director or at a District Office

should be assigned only to another District Director or District Office.

(b) *County Committee.* For processing or servicing decisions to be made by a County Committee, if the recipient is a member, a different County Committee is designated. If the recipient is related to or associated with the member, notwithstanding the provisions of § 1900.151(a)(3) of this subpart, the State Director *may* permit the decision to be made by the local committee, if the related/associated member abstains.

(c) [Reserved]

[58 FR 224, Jan. 5, 1993, as amended at 80 FR 9868, Feb. 24, 2015]

§ 1900.156 Special handling—processing.

(a) [Reserved]

(b) *Eligibility determination.* The designated processing official reviews the application and develops additional data as necessary. Upon determination of whether the assistance will be provided, the designated processing official notifies the applicant of the decision in writing under program regulations, subpart A of part 1910 of this chapter, and subpart B of part 1900. If the determination is favorable, unless otherwise designated, the complete application is returned to the original processing official for docket preparation. If the determination is unfavorable, the designated processing official as decisionmaker participates in the appeal process to its conclusion.

(c)–(e) [Reserved]

(f) *Closing agent.* Unless there is a clear or apparent conflict of interest, closing will be at a location and by a closing agent chosen by the recipient.

(g) *Supervised bank account.* Unless there is a clear or apparent conflict of interest, any supervised bank account (or construction account) is established at a financial institution chosen by the recipient under subpart A of part 1902 of this chapter. Countersignature authority is delegated only to a nonrelated or nonassociated Rural Development official.

(h) *Construction inspection.* Construction inspections are delegated to a nonrelated or nonassociated employee authorized to conduct inspections, whose duty station is nearest the construction site. The designated processing/

servicing official notifies the builder (or architect/engineer) in writing of how and from whom to request inspections.

[58 FR 224, Jan. 5, 1993, as amended at 80 FR 9868, Feb. 24, 2015]

§§ 1900.157–1900.200 [Reserved]

PART 1901—PROGRAM-RELATED INSTRUCTIONS

Subparts A–D [Reserved]

Subpart E—Civil Rights Compliance Requirements

Sec.
1901.201 Purpose.
1901.202 Nondiscrimination in FmHA or its successor agency under Public Law 103–354 programs.
1901.203 Title VIII of the Civil Rights Act of 1968.
1901.204 Compliance reviews.
1901.205 Nondiscrimination in construction financed with FmHA or its successor agency under Public Law 103–354 loan or grant.

EXHIBIT A TO SUBPART E OF PART 1901—CIVIL RIGHTS COMPLIANCE REVIEWS
EXHIBIT B TO SUBPART E OF PART 1901—SUMMARY REPORT OF CIVIL RIGHTS COMPLIANCE REVIEWS
EXHIBIT C TO SUBPART E OF PART 1901—FmHA OR ITS SUCCESSOR AGENCY UNDER PUBLIC LAW 103–354 FINANCED CONTRACT
EXHIBIT D TO SUBPART E OF PART 1901—GOALS AND TIMETABLES FOR MINORITIES AND WOMEN
EXHIBIT E TO SUBPART E OF PART 1901—LIST OF REGIONAL OFFICES, OFFICE OF FEDERAL CONTRACT COMPLIANCE PROGRAMS (OFCCP), U.S. DEPARTMENT OF LABOR (USDL)

Subpart F—Procedures for the Protection of Historical and Archeological Properties

1901.251 Purpose.
1901.252 Policy.
1901.253 Definitions.
1901.254 Scope.
1901.255 Historical and archeological assessments.
1901.256–1901.258 [Reserved]
1901.259 Actions to be taken when archeological properties are discovered during construction.
1901.260 Coordination with other agencies.
1901.261 [Reserved]
1901.262 State supplement.

EXHIBIT A TO SUBPART F OF PART 1901—NATIONAL PARK SERVICE, U.S. DEPARTMENT OF THE INTERIOR REGIONAL OFFICES

Subparts G–J [Reserved]

Subpart K—Certificates of Beneficial Ownership and Insured Notes

1901.501 Purpose.
1901.502 Policy.
1901.503 Definitions.
1901.504 Authorities and responsibilities.
1901.505 Certificates of beneficial ownership in Rural Development loans.
1901.506 Book-entry procedure for Rural Development securities—issuance and redemption of certificate by Reserve bank.
1901.507 Certificates of beneficial ownership issued by the Rural Development Finance Office.
1901.508 Servicing of insured notes outstanding with investors.
1901.509 Loss, theft, destruction, mutilation, or defacement of insured notes, insurance contracts, and certificates of beneficial ownership.

Subparts L–N [Reserved]

AUTHORITY: 5 U.S.C. 301; 7 U.S.C. 1989; 40 U.S.C. 442; 42 U.S.C. 1480, 2942.

Subparts A–D [Reserved]

Subpart E—Civil Rights Compliance Requirements

AUTHORITY: 5 U.S.C. 301; 7 U.S.C. 1989; 40 U.S.C. 442; 42 U.S.C. 1480, 2942.

§ 1901.201 Purpose.

This subpart contains policies and procedures for implementing the regulations of the Department of Agriculture issued pursuant to Title VI of the Civil Rights Act of 1964, title VIII of the Civil Rights Act of 1968, Executive Order 11246 and the Equal Credit Opportunity Act of 1974, as they relate to the Farmers Home Administration (FmHA) or its successor agency under Public Law 103–354. Nothing herein shall be interpreted to prohibit preference to American Indians on Indian Reservations.

[41 FR 40112, Sept. 17, 1976]

§ 1901.202 Nondiscrimination in FmHA or its successor agency under Public Law 103–354 programs.

(a) *Nondiscrimination by recipients of FmHA or its successor agency under Public Law 103–354 assistance.* (1) No recipient of FmHA or its successor agency under Public Law 103–354 financial assistance will directly or through contractual or other arrangements subject any person or cause any person to be subjected to discrimination on the ground of race, color, or national origin, with respect to any program or facility. This prohibition applies but is not restricted to unequal treatment in priority, quality, quantity, methods, or charges for service, use, occupancy or benefit, participation in the service or benefit available, or in the use, occupancy or benefit of any structure, facility, or improvement provided with FmHA or its successor agency under Public Law 103–354 financial assistance.

(2) Specifically, and without limiting the general applicability of this subpart, such recipient will not on the grounds of race, color, or national origin:

(i) Deny any person the use, occupancy, or enjoyment of the whole or any part of real or personal property or service, financial aid, or other benefit under any program or facility.

(ii) Provide any person with any service, use, occupancy, or other benefit different from that provided others by the program or facility.

(iii) Subject any person to segregation or separate treatment in any matter related to his or her receipt of any service or other benefit.

(iv) Restrict in any way any person's enjoyment of any right, privilege, or advantage enjoyed by others through the facility or activity.

(v) Treat any person differently from others in determining whether he or she satisfies any requirements or conditions for any admission or membership in the recipient or in any other organization.

(vi) Deny any person an opportunity or restrict opportunity to participate in a program or facility by:

(A) Refusing or failing to provide notice or services provided others for the purpose of encouraging participation in the program or facility; or

(B) Providing any person with such notice or services different from the notice or services provided others.

(vii) Utilize criteria or methods of administration that have the effect of subjecting a person to discrimination with respect to any program or facility or defeating or substantially impairing the achievement of the objectives of a program or facility.

(viii) Select sites or locate facilities with the purpose or effect of:

(A) Excluding individuals from, denying them the benefits of, or subjecting them to discrimination under any programs to which the regulations in this subpart apply; or

(B) Defeating or substantially impairing the achievement of the objectives of the regulations in this subpart.

(ix) Continue any previous or existing discriminatory practices, but will take affirmative action to overcome the effects of such discrimination.

(x) Deny any person the opportunity to participate as a member of a planning or advisory body which is an integral part of the program.

(b) *Nondiscrimination by FmHA or its successor agency under Public Law 103–354 employees.* (1) No. FmHA or its successor agency under Public Law 103–354 employee will exclude from participation in, or deny the benefits of, any program or activity administered by FmHA or its successor agency under Public Law 103–354, or subject to discrimination any person in the United States on the ground of race, color, religion, sex, national origin, or marital status.

(2) No FmHA or its successor agency under Public Law 103–354 employee will:

(i) Be limited in the discharge of his or her responsibilities to working with applicants solely on the basis of race, color, religion, sex, national origin, or marital status.

(ii) Obstruct equal access to buildings, facilities, structures, or lands under the control of FmHA or its successor agency under Public Law 103–354.

(iii) Deny under any program or activity of FmHA or its successor agency under Public Law 103–354 equal opportunity for employment; for participation in meetings, demonstrations, training activities or programs; for receiving awards; for receipt of information disseminated by publication, news, radio, and other media; for obtaining contracts, grants, loans or other financial assistance, or for selection to assist in the administration of programs

463

or activities of FmHA or its successor agency under Public Law 103–354.

(3) No FmHA or its successor agency under Public Law 103–354 employee will, while conducting official business, participate in or attend any segregated meetings or meetings held in a segregated facility from which persons are excluded because of race, color, religion, sex, national origin, or marital status.

(c) *Intimidating or retaliatory acts.* No recipient or other person will intimidate, threaten, coerce, or discriminate against any person for the purpose of interfering with any right or privilege under this subpart, or because a person has made a complaint or has testified, assisted, or participated in any manner in an investigation, proceeding, or hearing related to a complaint. The identity of complainants will be kept confidential except to the extent necessary to carry out the purposes of this subpart.

(d) *Nondiscrimination Agreement.* The County Supervisor will, at the time FmHA or its successor agency under Public Law 103–354 assistance is requested, give all applicants for loans and grants listed in § 1901.204(a) a copy of Form FmHA or its successor agency under Public Law 103–354 400–4, "Nondiscrimination Agreement," and inform the applicant that assistance will be conditioned upon executing this form and complying with the requirements of this subpart.

(e) *Covenants.* Each instrument of conveyance for loans subject to title VI of the Civil Rights Act of 1964, as outlined in § 1901.204, must contain the following covenant: "The property described herein was obtained or improved through Federal financial assistance. This property is subject to the provisions of title VI of the Civil Rights Act of 1964 and the regulations issued pursuant thereto for so long as the property continues to be used for the same or similar purpose for which financial assistance was extended or for so long as the purchaser owns it, whichever is longer."

(f) *Posters.* The nondiscrimination poster, "And Justice For All," will be displayed at the facilities and/or office of any borrower or grantee if the facilities have been financed by an FmHA or its successor agency under Public Law 103–354 loan or grant and are subject to title VI of the Civil Rights Act of 1964. This poster also will be displayed in all FmHA or its successor agency under Public Law 103–354 State and County Offices.

(g) *Racial and ethnic data.* Recipients should maintain, for review by FmHA or its successor agency under Public Law 103–354 and other appropriate agencies, racial and ethnic data showing the extent to which members of minority groups are beneficiaries of FmHA or its successor agency under Public Law 103–354-assisted programs. The data should identify recipients as White, Negro or Black, American Indian, Spanish Surname, Oriental and Other.

(h) *Discrimination complaints.* (1) Any person or any specific class of persons, if they believe they have been subject to discrimination prohibited by this subpart, may file a written complaint with any FmHA or its successor agency under Public Law 103–354 office, or, if they prefer with the Secretary of Agriculture. Persons who complain of discrimination will be advised of their rights to file complaints. A complaint must be filed not later than 180 days after the date of the alleged discrimination, unless the time for filing is extended by the Secretary of Agriculture.

(2) A complaint filed with the County Supervisor or the State Director will be referred promptly to the Administrator, Attention: Equal Opportunity Officer. Attached to the complaint should be a statement by the County Supervisor or State Director identifying the recipient and type of assistance provided by FmHA or its successor agency under Public Law 103–354, indicating whether a nondiscrimination agreement has been signed, and giving any other available pertinent information about the complaint.

[41 FR 40112, Sept. 17, 1976]

§ **1901.203   Title VIII of the Civil Rights Act of 1968.**

FmHA or its successor agency under Public Law 103–354 employees, FmHA or its successor agency under Public Law 103–354 borrowers, contractors, packagers, and others who provide housing for sale or rent, are obligated under the provisions of title VIII of the Civil Rights Act of 1968 to provide fair housing to all persons regardless of race, color, religion, sex, or national origin.

(a) *Coverage.* The prohibitions against discrimination in the sale, rental, or financing of housing contained in title VIII apply to:

(1) All dwellings financed by loans made by the Federal Government and, therefore, to all RH borrowers.

(2) Any person in the business of selling or renting dwellings defined as:

(i) The owner of a dwelling intended for occupancy by five or more families.

(ii) Any person who has participated as principal in the sale or rental of three or more dwellings in the past year.

(iii) Any person who has served as sale or rental agent in two or more transactions in the past year.

(b) *Discrimination acts prohibited.* Title VIII prohibits FmHA or its successor agency under Public Law 103–354 employees, multiple housing borrowers, and those with whom FmHA or its successor agency under Public Law 103–354 does business (contractors, realtors, packagers) from:

(1) Refusing to sell or rent a particular dwelling because of a person's race, color, religion, or national origin. The following actions constitute violations of this provision:

(i) Refusing to package an RH loan application.

(ii) Refusing or failing to show a particular dwelling or home in a particular subdivision.

(iii) Directing persons only to areas populated by those of similar race, color, religion, or national origin when housing is available in other areas.

(iv) Representing unsold dwellings or sites as sold to prospective buyers.

(2) Requiring applicants for services to meet different terms or conditions because of their race, color, religion, or national origin; for example, requiring larger rents or downpayments from minority applicants.

(3) Including in any advertising either directly or through visual representation a preference for applicants of a particular race or ethnic origin.

(i) Words indicative of the race or ethnic background of the dwelling or landlord such as "White private home," or "all Black subdivision," may not be used in advertising housing financed or to be financed by FmHA or its successor agency under Public Law 103–354.

(ii) Selection of advertising media and the areas to be covered by any advertising must be made to reach potential applicants of all races or ethnic origins.

(c) *FmHA or its successor agency under Public Law 103–354 affirmative action.* (1) It is the policy of the Farmers Home Administration or its successor agency under Public Law 103–354 to administer its housing program affirmatively so individuals of similar income levels in the housing market area have housing choices available to them regardless of their race, color, religion, sex, or national origin. Each participant in FmHA or its successor agency under Public Law 103–354 housing program shall pursue affirmative fair housing marketing policies in soliciting buyers and tenants, in determining their eligibility and in concluding sales and rental transactions.

(2) Applicability. The affirmative fair housing marketing requirements shall apply as follows:

(i) Participants in FmHA or its successor agency under Public Law 103–354 housing programs who request approval for subdivision development involving five or more sites, multi-family projects with five or more units including self-help technical assistance grantees assisting five or more families or five or more conditional commitments for single family dwelling units during a 12-month period must submit an affirmative marketing plan.

(ii) An Affirmative Fair Housing Marketing Plan is required to be prepared and submitted to FmHA or its successor agency under Public Law 103–354 by the contractor when:

(A) A real estate broker is offering five or more single-family dwellings located in the same subdivision for sale under an exclusive listing contract with FmHA or its successor agency under Public Law 103–354.

(B) An auctioneer under contract with FmHA or its successor agency

465

under Public Law 103–354 is offering five or more single-family dwellings located in the same subdivision for sale by public auction.

(C) A contractor under a contract with FmHA or its successor agency under Public Law 103–354 is managing a multiple-family housing project of five or more units or five or more single-family dwellings located in the same subdivision.

(3) Affirmative fair housing marketing plans will be submitted on form HUD 935.2(3–76) or the participant must be a signatory to a voluntary affirmative marketing agreement approved by the Department of Housing and Urban Development. The plan, if submitted on form HUD 935.2(3–76) shall describe an affirmative program which will meet the following requirements:

(i) Reaching those prospective buyers or tenants, regardless of sex, of majority and minority groups in the marketing area who traditionally would not be expected to apply for such housing without special outreach efforts because of existing racial or socio-economic patterns.

(ii) Undertaking and/or maintaining a non-discriminatory hiring policy in recruiting from both majority and minority groups including both sexes, for staff engaged in the sale or rental of properties.

(iii) Training and instructing employees engaged in the sale or rental properties in the policy and application of nondiscrimination and fair housing.

(iv) Displaying in all sales and rental offices the "Fair Housing" poster.

(v) Posting in a conspicuous position on each property and FmHA or its successor agency under Public Law 103–354 construction site a sign displaying the equal opportunity logo or the following statement:

We are pledged to the letter and spirit of U.S. policy for the achievement of equal housing opportunity throughout the nation. We encourage and support an affirmative advertising and marketing program in which there are no barriers to obtaining housing because of race, color, religion, sex, or national origin.

(vi) Undertaking efforts to publicize the availability of housing opportunities to minority persons through the type of media customarily used by the applicant or participant, including minority publications and other minority outlets available in the housing market area. As part of these efforts all advertising must include either the equal housing opportunity logo or statement. When illustrations or persons are included they shall depict persons of both sexes and of majority and minority groups.

(4) The affirmative fair housing marketing plans or evidence that the participant is covered by an approved voluntary affirmative marketing agreement must be submitted as follows:

(i) For subdivisions with the preliminary submission of plans and specifications.

(ii) For multi-family projects, including rural rental housing, labor housing, cooperative housing, technical assistance grants and site development loans with SF 424.1, "Application for Federal Assistance (For Non-construction)", or SF 424.2, "Application for Federal Assistance (For Construction)", or with the letter of application. Subsequent loans or grants extended to the participant will necessitate a new or updated plan.

(iii) For conditional commitments for five or more individual dwelling units in a 12-month period with the application for the fifth conditional commitment.

(iv) For real estate brokers listing housing properties on an exclusive basis, at any time more than 5 properties are listed for sale by FmHA or its successor agency under Public Law 103–354 in the same subdivision.

(5) Affirmative fair housing marketing plans will cover the following time periods:

(i) For subdivision, from time of application until all lots are sold.

(ii) For multi-family projects from time of application until the loan is paid in full or for so long as the project is being used for the same or a similar purpose for which the funds were extended.

(iii) For conditional commitments involving individual dwelling units, one year or until all units built through conditional commitments issued within the one year period have been sold.

(iv) For real estate brokers who list acquired rural housing properties under an exclusive listing contract, one year or until all properties covered under the plan have been sold, whichever is later.

(6) Affirmative fair housing marketing plans will be reviewed and approved by the official authorized to approve the assistance requested. The County Supervisor will review and submit with comments to the official authorized to approve the assistance requested, those fair housing marketing plans where the assistance requested exceeds his approval authority. Any participant covered by this section must have an approved affirmative fair housing marketing plan for any assistance approved 90 or more days after the issuance of these regulations.

(7) Approved affirmative fair housing marketing plans will be made available by the participant for public inspection at the participant's place of business and at each sales or rental office. Participants who fulfill the requirements of this section by filing a Form HUD 9352(3–76) will maintain records to reflect their efforts in fulfilling the affirmative fair housing marketing plan. These records will be made available for review by FmHA or its successor agency under Public Law 103–354 personnel. Affirmative fair housing marketing plans will be reviewed by FmHA or its successor agency under Public Law 103–354 personnel in accordance with section 2006–M of this chapter.

(8) Applicants failing to comply with these requirements will be liable to sanctions authorized by regulations, rules or policies governing the program in which they are participating including but not limited to denial of further participation in FmHA or its successor agency under Public Law 103–354 programs and referral to the Department of Justice for suit by the United States for injunctive or other appropriate relief.

(d) *Discrimination complaints.* (1) Complaints against FmHA or its successor agency under Public Law 103–354 employees or borrowers under title VIII of the Civil Rights Act of 1968 received by the County Office will be sent to the State Director. The State Director will forward the complaints to the Adminis-

trator, Attention: Equal Opportunity Officer.

(2) Complaints of discrimination against packagers, contractors or others with whom FmHA or its successor agency under Public Law 103–354 deals should be filed with the Department of Housing and Urban Development. However, these complaints may be accepted by FmHA or its successor agency under Public Law 103–354 employees and routed through the State Director to the Administrator, Attention: Equal Opportunity Officer.

(e) *Relations to other regulations.* Nothing in this section in any way interferes with the administration of the nondiscrimination requirements of Title VI of the Civil Rights Act of 1964 or the "Equal Opportunity in Housing Certification," signed by all packagers.

[41 FR 40112, Sept. 17, 1976, as amended at 42 FR 45894, Sept. 13, 1977; 42 FR 58737, Nov. 11, 1977; 50 FR 23903, June 7, 1985; 53 FR 27825, July 25, 1988; 55 FR 13503, Apr. 11, 1990]

§ 1901.204 Compliance reviews.

(a) *Recipients subject to reviews.* Recipients of the following kinds of loans and/or grants who received their loans or advances of funds on or after January 3, 1965, will be reviewed for compliance in accordance with Title VI of the Civil Rights Act of 1964. Guaranteed loans are not covered by Title VI and, therefore, are not subject to compliance reviews.

(1) Economic Opportunity loans to individuals for nonagricultural enterprises.

(2) Loans for Water and Waste Disposal facilities, including Resource Conservation and Development loans for this purpose.

(3) Community Facility loans.

(4) Watershed loans and advances.

(5) Recreation Association loans including those made from Resource Conservation and Development funds.

(6) Economic Opportunity loans to incorporated cooperative associations (Compliance reviews on unincorporated Economic Opportunity cooperatives subject to title VI will be conducted only as the need arises or as directed by either the State Director or the Administrator).

(7) Loans to Timber Development organizations.

(8) Rural Renewal loans and advances.

(9) Rural Rental Housing (formerly Senior Citizen rental) and Rural Cooperative Housing loans.

(10) Labor Housing loans and/or grants.

(11) Rural Housing Site loans.

(12) Business and Industrial Insured loans or grants.

(13) Technical Assistance grants.

(14) Development grants for water and waste disposal.

(15) Technical Assistance and Training grants in accordance with Title XIII of Pub. L. 99–198.

(16) Rural Business Development Grants.

(17) Section 601 Energy Impacted Area Development Assistance grants.

(18) Nonprofit National Corporations grants.

(19) System for Delivery of Certain Rural Development Programs Panel Grants.

(20) Emergency Community Water Assistance grants.

(21) Section 306C WWD loans and grants.

(22) Housing Application Packaging Grants.

(23) Rural and Cooperative Development Grants in subpart F of part 4284 of this title.

(24) Community Facilities Grants in part 3570, subpart B, of this title.

(b) *Duration of obligation for conducting reviews.* Compliance reviews will be conducted on recipients of loans and grants listed in paragraph (a) of this section:

(1) Until the loan is paid in full or otherwise satisfied; or sold through the sale of FmHA or its successor agency under Public Law 103–354's assets; or

(2) Until the last advance of grant funds is made for the grants listed in paragraph (a) of this section.

(c) *Compliance reviews of loans and grants to individuals—*(1) *Compliance Review Officer.* The County Supervisor will conduct compliance reviews of loans made to individuals.

(2) *Type of review.* If the borrower is currently receiving loan supervision, the County Supervisor may complete the compliance review based on his knowledge of the borrower's operations from other visits. Otherwise the Coun-

ty Supervisor must visit the borrower's facilities. Before completing the compliance review, the County Supervisor should be aware of:

(i) The borrower's operating regulations, for example, the grounds for eviction from a Rural Rental Housing Project.

(ii) The borrower's method of advertising the facility to the public, if there is any advertising, including how well these methods reach the minority community.

(iii) Any records of request for use of the borrower's facility.

(3) *Recording results of review.* The County Supervisor's determination that the borrower is or is not in compliance with title VI, together with information such as that outlined in paragraph (b)(2) of this section, will be recorded in the running record. Review of individual Rural Rental Housing borrowers will be recorded on Form FmHA or its successor agency under Public Law 103–354 400–8, "Compliance Review (Nondiscrimination by Recipients of Financial Assistance Through FmHA or its successor agency under Public Law 103–354.)"

(4) *Reporting results of review.* If the borrower is in compliance, the County Supervisor will report his findings to the State Director. Exhibit A is a sample report. In the case of Rural Rental Housing borrowers, a copy of Form FmHA or its successor agency under Public Law 103–354 400–8 will be filed in the borrower's County Office loan docket, and the original will be sent to the State Director. If the borrower is not in compliance, the borrower's name, location, type of loan involved, and the reasons for the finding of noncompliance will be sent to the State Director.

(5) *Forwarding report of noncompliance.* The State Director will see that all compliance review reports are complete. If the recipient was found in noncompliance, the State Director will immediately send a copy of the compliance review report to the Administrator, Attention: Equal Opportunity Officer, with recommended action to take to bring the recipient into compliance.

(d) *Review of loans or grants to organizations (any borrower or grantee other* than an individual)—(1) *Designation of compliance review officer.* The State Director, except for Technical Assistance and Training grants (Pub. L. 99–198) and Nonprofit National Corporations grants, will designate the Compliance Review Officer for recipient organization. County Supervisors may be designated only if they have received approved compliance review training. Otherwise, the Compliance Review Officer must be a member of the State staff. For Technical Assistance and Training grants and Nonprofit National Corporations grants, the Assistant Administrator for Community and Business Programs will designate the Compliance Review Officer for recipient organizations.

(2) *Type of review.* Compliance reviews may be completed in connection with regular supervision visits to organizations and must include an inspection of the FmHA or its successor agency under Public Law 103–354-financed facility. Before determining that the recipient is or is not complying with the provisions in Form FmHA or its successor agency under Public Law 103–354 400–4, the Compliance Review Officer will:

(i) Observe the recipient's records, including records on the present membership by race, the handling of applications for use of the facility, the user rates and membership fees or dues, and the facility's operating regulations.

(ii) Determine if the recipient advertises for members or users. If so, observe the effectiveness of the recipient's methods of advertising the availability of the facility to the public, and especially the effectiveness of this advertising in reaching the minority community.

(iii) Interview organization officials, members, and employees. In reviews of recipients of Technical Assistance grants, members of the self-help housing groups should be interviewed to determine the way in which they were recruited.

(iv) Interview informed local community leaders, including minority leaders, if any to determine if the facility is operating without discrimination because of race, color, or national origin.

(3) *Recording results of reviews*—(i) *Association, Watershed, Resource Conserva-* tion and Development, and Rural Renewal loans involving recreation facilities.* Reviews will be recorded on Form FmHA or its successor agency under Public Law 103–354 400–7, "Compliance Reviews for Recreational Loans to Associations (FmHA or its successor agency under Public Law 103–354 Borrowers)." If the organization is found in compliance with title VI, the original of the form will be sent to the State Director, and a copy will be filed in the borrower's County Office loan docket. If the organization is found in noncompliance, any additional information which led to the finding will be sent with the form.

(ii) *Loans and/or grants for Water and Waste Disposal systems, incorporated Economic Opportunity cooperatives, Grazing associations, Rural Rental Housing, Labor Housing, and Rural Housing Sites.* Reviews will be completed on Form FmHA or its successor agency under Public Law 103–354 400–8. The original of the form will be sent to the State Director and a copy filed in the borrower's County Office loan docket. If the organization is found in noncompliance, any additional information which led to the finding will be sent with the form.

(iii) *Timber Development organizations, Rural Cooperative Housing loans, and Technical Assistance grants.* The information obtained during the compliance review as well as the Compliance Review Officer's determination of the borrower's compliance or noncompliance will be recorded in the running record. If the organization is found in compliance, a report (see exhibit A) will be sent to the State Director. If the organization is not in compliance, the organization's name, location, type of loan received, and all information which led to the finding will be sent to the State Director.

(iv) *Technical Assistance and Training grants (Pub. L. 99–198) and Nonprofit National Corporations grants.* The Compliance Review Officer will record in the running record information obtained during the compliance review and the determination of recipient's compliance or noncompliance. A report will be prepared and sent to the Assistant Administrator, Community and Business Programs, for each recipient.

(4) *Mandatory hook-up ordinance.* Compliance reviews of public entity borrowers or grantees for water and waste disposal facilities who are operating under the provisions of a mandatory hook-up ordinance will consist of a certification by the borrower or grantee that the ordinance is still in effect and is being enforced.

(5) *Forwarding noncompliance report.* The State Director will see that the reports are complete. If the recipient was found in noncompliance, the State Director will immediately send a copy of the report to the Administrator, Attention: Equal Opportunity Officer, with action proposed to bring the recipient into compliance. For Technical Assistance and Training grants and Nonprofit National Corporations grants, the Assistant Administrator, Community and Business Programs, will send a copy of the report to the Equal Opportunity Officer.

(e) *Timing of reviews*—(1) *Reporting year.* The State Director will schedule Civil Rights compliance reviews from November 1 to October 31 of each year. For example, compliance reviews scheduled during 1976 should be conducted after November 1, 1975, but before October 31, 1976.

(2) *Initial reviews*—(i) *Water and Waste Disposal loan and/or grant.* The initial compliance review will be conducted before loan or grant closing or before the construction begins, whichever occurs first.

(ii) *Technical Assistance grants,* Technical Assistance and Training grants (Pub. L. 99–198) and Nonprofit National Corporations grants. The initial compliance review will be conducted before the grant is closed.

(iii) *Rural Housing Site loan.* The initial compliance review will be conducted at the beginning of the sale of the sites developed with the FmHA or its successor agency under Public Law 103–354 loan.

(iv) *Watershed loans for future water supply.* The initial compliance review will be made when usage of the stored water begins.

(v) *All other loans and/or grants.* The initial compliance review of loans and/or grants listed in paragraph (a) of this section will be conducted within the first reporting year after the loan or grant is closed or after Form FmHA or its successor agency under Public Law 103–354 400–4 is signed.

(3) *Subsequent reviews.* The State Director is responsible for requiring subsequent compliance reviews at intervals not less than 90 days, or more than 3 years, after the previous compliance review.

(i) For Water and Waste Disposal organizations with loans that have had at least two compliance reviews after loan closing covering a six-year period, and where no discriminatory practices are indicated, the frequency of subsequent reviews may be reduced to six years.

(ii) If Water and Waste Disposal organizations have merged to form a new organization, two reviews will be conducted at 3-year intervals after the merger and one every 6 years thereafter, provided no discriminatory practices are noted.

(f) *State Office summary reports.* The State Director will keep a list of all compliance reviews conducted during the reporting year so as to schedule each year's reviews. The State Director will submit a copy of this list to the Administrator, Attention: Equal Opportunity Office, no later than July 31 of each year. Recipients found in noncompliance will also be listed on the summary report. Exhibit B is a sample report. For Technical Assistance and Training grants and Nonprofit National Corporations grants, the Assistant Administrator, Community and Business Programs, will submit a summary report, using exhibit B of this subpart as a guide, to the Equal Opportunity Officer by July 31 of each year.

[41 FR 40112, Sept. 17, 1976, as amended at 52 FR 41949, Nov. 2, 1987; 53 FR 3860, Feb. 10, 1988; 55 FR 5962, Feb. 21, 1990; 57 FR 11559, Apr. 6, 1992; 58 FR 5565, Jan. 22, 1993; 58 FR 58643, Nov. 3, 1993; 59 FR 41389, Aug. 12, 1994; 61 FR 3781, Feb. 2, 1996; 62 FR 16468, Apr. 7, 1997; 62 FR 33510, June 19, 1997; 62 FR 42387, Aug. 7, 1997; 68 FR 69952, Dec. 16, 2003; 80 FR 15667, Mar. 25, 2015]

§ 1901.205 Nondiscrimination in construction financed with FmHA or its successor agency under Public Law 103–354 loan or grant.

Executive Order 11246 provides for equal employment opportunity without regard to race, color, religion, sex, or national origin and the elimination of all facilities segregated on the basis of race, color, religion, or national origin on construction work financed by FmHA or its successor agency under Public Law 103–354 involving a construction contract of more than $10,000.

(a) *Compliance.* This section applies to Federal or federally assisted construction contracts or subcontracts in excess of $10,000 for on-site construction. It also applies to invitations for bids published for such construction. If construction work of over $10,000 is partially financed by another Federal Agency, the County Supervisor will try to reach an agreement as to which agency will administer the nondiscrimination requirements. If unable to reach an agreement, the County Supervisor will refer the case to the State Director.

(b) *Requirements of applicants, contractors, or subcontractors and responsible FmHA or its successor agency under Public Law 103–354 officials*—(1) *Applicant.* The applicant will be required to execute Form FmHA or its successor agency under Public Law 103–354 400–1, "Equal Opportunity Agreement," at the time the loan is closed or before construction is started, whichever occurs first. If the applicant is an incorporated association, a resolution of the governing body will authorize execution of the form. Municipalities or other public bodies will have to incorporate references to this form in the loan resolution before it is adopted. If the applicant wants to publish for bids, the applicant must obtain Form FmHA or its successor agency under Public Law 103–354 1924–5, "Invitation for Bid (Construction Contract)" which is in compliance with Executive Order 11246, from the local FmHA or its successor agency under Public Law 103–354 County Supervisor.

(2) *Contractor or Subcontractor.* (i) The prospective contractor or subcontractor must submit Form FmHA or its successor agency under Public Law 103–354 400–6, "Compliance Statement," to the County Supervisor before contract bid negotiations, and comply with the requirements of Executive Order 11246, which are included with Form FmHA or its successor agency under Public Law 103–354 1924–6, "Con-

struction Contract," during the performance of the contract. The contract will contain the required "Standard Federal Equal Employment Opportunity Construction Contract Specifications" goals and timetables as set forth in exhibit D.

(ii) The contractor or subcontractor will prepare and submit Form Contract Compliance (CC) 257, "Monthly Employment Utilization Report" to the appropriate regional office of the U.S. Department of Labor (USDL) (see exhibit E, "List of Regional Offices") by the fifth of each month through completion of the contract.

(3) *The County Supervisor or the responsible FmHA or its successor agency under Public Law 103–354 official will:* (i) Deliver to the contractor the following forms, as appropriate:

(A) Form FmHA or its successor agency under Public Law 103–354 400–3, "Notice to Contractors and Applicants," with an attached Equal Employment Opportunity Poster. Posters in Spanish will be provided when appropriate,

(B) Form FmHA or its successor agency under Public Law 103–354 400–6, and

(C) Form CC 257.

(ii) Deliver to the applicant Form FmHA or its successor agency under Public Law 103–354 1924–5 when contractors are to be invited to submit bids, and Form FmHA or its successor agency under Public Law 103–354 1924–6 to contract for construction.

(iii) Explain to applicant and contractor the requirements of Executive Order 11246, when needed. However, inquiries concerning compliance must be addressed to the appropriate regional office of USDL (see exhibit E).

(iv) Submit a report similar in form and content to exhibit C ("FmHA or its successor agency under Public Law 103–354 Financed Contract") of this Instruction to the appropriate regional office of USDL (Exhibit E) within 10 calendar days of the date a contract or subcontract in excess of $10,000 is awarded.

(c) *Contractors with 100 or more employees and contract over $10,000.* Contractors with 100 or more employees and contract over $10,000, will file the following with the Joint Reporting

Committee, 1800 G Street NW., Washington, DC 20006:

(1) SF-100 "Employer Information Report EEO-1," within 30 days of contract award unless the report has been submitted within the past 12 months, and

(2) An annual report by March 31, so long as the contractor holds any FmHA or its successor agency under Public Law 103-354 financed contract in excess of $10,000.

(d) *Contractor with at least 50 employees and contract of $50,000 or more.* Each contractor or subcontractor with at least 50 employees and contract of $50,000 or more, must develop a written affirmative action compliance program for each project. This must be on file in each contractor's or subcontractor's personnel file within 120 days after the beginning of the contract. Form AD-425 provides guidelines for developing compliance programs.

(e) *Compliance during construction.* The County Supervisor will:

(1) Check to see that:

(i) Required posters are displayed.

(ii) There is no evidence of discrimination in employment.

(2) Record findings on Form FmHA or its successor agency under Public Law 103-354 1924-12, "Inspection Report."

(3) If there is any evidence of noncompliance, the County Supervisor will report all the facts to the appropriate office of USDL (see exhibit E).

(f) *Hometown Plans.* All construction contracts and subcontracts in excess of $10,000, financed by FmHA or its successor agency under Public Law 103-354, in areas which have Hometown Plans regarding affirmative action and equal employment, are subject to the conditions set forth in the applicable plan. Each State Director should seek the advice of the OGC as to compliance with any such plans in the State Director's jurisdiction.

(g) *Discrimination complaints.* (1) Complaints alleging discriminatory acts may be filed directly with the appropriate regional office of USDL (see exhibit E) or with the County Supervisor or the State Director for subsequent

forwarding to the above address, by any employee or applicant for employment with a contractor or subcontractor.

(2) Each complaint must be in writing and signed by the complainant (The FmHA or its successor agency under Public Law 103-354 official receiving the complaint will assist complainant when necessary). The complaint will include:

(i) Name, address, and telephone number of complainant.

(ii) Name and address of the person allegedly discriminating.

(iii) Date and place of the discrimination.

(iv) Description of the discrimination.

(v) Any other information that will assist in investigating and resolving the complaint.

(3) Complaints must be filed not later than 180 days after the alleged act unless the State Director extends the time, for good cause shown by the complainant.

[43 FR 58356, Dec. 14, 1978, as amended at 44 FR 24852, Apr. 27, 1979; 52 FR 8002, Mar. 13, 1987]

EXHIBIT A TO SUBPART E OF PART 1901—
CIVIL RIGHTS COMPLIANCE REVIEWS

To: State Director, FmHA or its successor agency under Public Law 103-354.

Civil Rights compliance reviews have been conducted, and each recipient listed below was found in compliance with title VI of the Civil Rights Act of 1964. Information which led to this finding and my determination that the recipient is in compliance are in the running record of the recipient's file.

Recipient	Case No.	Type of assistance [1]	Date of review
Sam H. Smith	99-05-7031 (rec.).	OL	Jan. 3, 1975.
John A. Jones ...	99-05-8764 ...	RL	Feb. 17, 1975.
Medina Housing Association.	99-05-9176 grant.	TA	Mar. 5, 1975.

[1] Indicate only the loans or grants received which are subject to compliance reviews.

County Supervisor

EXHIBIT B TO SUBPART E OF PART 1901—SUMMARY REPORT OF CIVIL RIGHTS
COMPLIANCE REVIEWS

To: Administrator, FmHA or its successor agency under Public Law 103–354.
Attention: Director, Equal Opportunity Staff.

I. Civil Rights Compliance Reviews have been conducted, and the following recipients were
found in compliance with title VI of the Civil Rights Act of 1964.

Loan type	Loan number	Type of review	
		Pre-award* post-award**	
1.			
2.			
3.			

*A pre-award review is a compliance review conducted prior to loan or grant approval.
**A post-award review is a compliance review conducted after loan closing.

II. The following recipients were found in non-compliance:

Name of borrower	Loan type	Loan number	Type of review	Date report of noncompli-ance sent to nat. ofc.
			Pre-award post-award	
1.				
2.				
3.				

State Director.

(7 U.S.C. 1989; 42 U.S.C. 1480; 7 CFR 2.23; 7 CFR 2.70)

[47 FR 39127, Sept. 7, 1982]

EXHIBIT C TO SUBPART E OF PART 1901—
FMHA OR ITS SUCCESSOR AGENCY
UNDER PUBLIC LAW 103–354 FI-
NANCED CONTRACT

To: Area Director, Office of Federal Contract
Compliance Program, U.S. Department
of Labor (DOL) (Insert address for your
DOL area, from exhibit E, FmHA or its
successor agency under Public Law 103–
354 Instruction 1901–E)

We submit the following information relative to a construction contract in excess of
$10,000:

1. Contractor's name: _____
Address: _____
Telephone Number: _____
Employer's Identification Number: _____

2. Contract for: ____ $_____

Starting Date: _____
Completion Date: _____
Contract Number: _____
City: _____
DOL Region: _____

[52 FR 8002, Mar. 13, 1987]

EXHIBIT D TO SUBPART E OF PART 1901—
GOALS AND TIMETABLES FOR MINORI-
TIES AND WOMEN

The preamble to regulations establishing a
new part 60–4 to 41 CFR chapter 60 published

at 43 FR 14888–14894, April 7, 1978, states that
OFCCP contemplates proposing standards
and goals for minorities within the very near
future. Until that notice has been proposed
and final action taken, construction contractors and subcontractors will continue to be
subject to the goals and timetables for minority utilization on Federal and federally
assisted construction existing now under Executive order 11246. Such goals are published
in appendix B.

Now, therefore, based on the foregoing and
41 CFR part 60–4, each contracting agency,
each applicant, and each contractor shall include the appropriate goal set forth in appendix A and appendix B in all invitations for
bids or other solicitations for federally involved construction contracts in excess of
$10,000. The goals in appendix A hereby are
established on a nationwide basis as the
standards for female utilization for all
trades.

Appendix B established the goals for minority utilization which shall be applicable
for the respective areas set forth in appendix
B.

Appendix A and appendix B shall be effective with respect to transactions for which
the invitations for bids or other solicitations
or amendments thereto are sent, on or after
May 8, 1978.

Weldon J. Rougeau,
Director, OFCCP.

March 28, 1978.

473

APPENDIX A

The following goals and timetables for female utilization shall be included in all Federal and federally assisted construction contracts and subcontracts in excess of $10,000. The goals are applicable to the contractor's aggregate on-site construction workforce whether or not part of that workforce is performing work on a Federal or federally assisted construction contract or subcontract.

AREA COVERED

Goals for Women apply nationwide.

GOALS AND TIMETABLES

Timetable	Goals (percent)
From Apr. 1, 1978 until Mar. 31, 1979	3.1
From Apr. 1, 1979 until Mar. 31, 1980	5.1
From Apr. 1, 1980 until Mar. 31, 1981	6.9

APPENDIX B

Until further notice, the following goals and timetables for minority utilization shall be included in all Federal or federally assisted construction contracts and subcontracts in excess of $10,000 to be performed in the respective covered areas. The goals are applicable to the contractor's aggregate on-site construction workforce whether or not part of that workforce is performing work on a Federal or federally assisted construction contract or subcontract.

REGION [1]

BOSTON, MASS. AREA

Area covered—Arlington, Boston, Belmont, Brookline, Burlington, Cambridge, Canton, Chelsea, Dedham, Everett, Malden, Medford, Wakefield, Westwood, Winthrop, Winchester, Woburn, and the Islands of Boston Harbor, Mass.

GOALS AND TIMETABLES
[In percent]

Timetable	Trade	Goal
Until further notice	Asbestos workers	10.8–10.12
	Boilermakers	9.6–12.0
	Bricklayers	8.0–10.0
	Carpenters	11.6–14.5
	Cement masons	25.5–27.5
	Electricians	6.0–7.0
	Elevator constructors	9.5–11.4
	Glaziers	8.8–11.0

[1] Region refers to the 10 regions in which the U.S. Department of Labor has offices. These Regions are headquartered in Boston, New York, Philadelphia, Atlanta, Chicago, Dallas, Kansas City, Denver, San Francisco, and Seattle, which are numbered I through X respectively.

GOALS AND TIMETABLES—Continued
[In percent]

Timetable	Trade	Goal
	Ironworkers	5.9–6.9
	Lathers	6.9–8.9
	Operating engineers	14.1–15.0
	Painters	9.1–11.1
	Pipefitters	11.0–12.1
	Plasterers	20.5–22.5
	Plumbers	9.8–11.8
	Roofers	8.4–10.5
	Sheetmetal workers	10.1–12.1
	Sprinkler fitters	12.3–15.6
	All other trades	10.3–12.3

STATE OF RHODE ISLAND AREA

Area covered—Statewide.

GOALS AND TIMETABLES
[In percent]

Timetable	Trade	Goal
Until further notice	All	5.0

REGION II

BUFFALO, NY AREA

Area covered—Erie County and Buffalo, NY.

GOALS AND TIMETABLES
[In percent]

Timetable	Trade	Goal
Until further notice	All	10.6–13.2

CAMDEN, NJ AREA

Area covered—Camden, NJ, area of Camden, Salem, and Gloucester Counties.

GOALS AND TIMETABLES
[In percent]

Timetable	Trade	Goal
Until further notice	Asbestos workers	11.6–14.5
	Boilermakers	10.8–13.5
	Bricklayers	17.8–20.0
	Carpenters	11.2–13.0
	Cement masons	12.0–15.0
	Electricians	14.9–17.8
	Elevator constructors	10.8–13.5
	Glaziers	16.0–20.0
	Lathers	10.8–13.5
	Operating engineers	10.0–12.5
	Painters/decorators/paper-hangers.	8.8–12.8
	Plasterers	17.0–19.0
	Plumbers/pipefitters/steam-fitters.	8.4–10.5
	Roofers	8.4–10.5
	Sheetmetal workers	11.2–14.0
	Sprinkler fitters	10.8–13.5
	Structural metal workers ...	12.9–15.3
	Wharf 7 dock builders	10.8–13.5

ELMIRA, NY AREA

Area covered—Chemung, Steuben, Schuyler, Tioga, and Yates Counties, NY.

GOALS AND TIMETABLES
[In percent]

Timetable	Trade	Goal
Until further notice	All	4.0–5.0

LONG ISLAND, NY AREA

Area covered—Nassau and Suffolk Counties, NY.

GOALS AND TIMETABLES
[In percent]

Timetable	Trade	Goal
Until further notice	All	6.0–8.0

WESTCHESTER, NY AREA

Area covered—Westchester County, NY.

GOALS AND TIMETABLES
[In percent]

Timetable	Trade	Goal
Until further notice	All	11–13

REGION III

STATE OF DELAWARE AREA

Area covered—State of Delaware.

GOALS AND TIMETABLES
[In percent]

Timetable	Trade	Goal
Until further notice	All	11–13

PHILADELPHIA, PA, AREA

Area covered—Bucks, Chester, Delaware, Montgomery, and Philadelphia Counties, PA.

GOALS AND TIMETABLES
[In percent]

Timetable	Trade	Goal
Until further notice	Ironworkers	22–26
	Plumbers and pipefitters ...	20–24
	Steamfitters	20–24
	Sheetmetal workers	19–23
	Electrical workers	19–23
	Elevator construction workers.	19–23

PITTSBURGH, PA, AREA

Area covered—Allegheny County, PA.

GOALS AND TIMETABLES
[In percent]

Timetable	Trade	Goal
Until further notice	Asbestos workers	24.3–27.8
	Boilermakers	33.8–37.7
	Bricklayers	11.9–13.0
	Carpenters	11.8–12.9
	Cement masons	16.3–18.1
	Electricians	17.0–20.3
	Glaziers	26.9–30.4
	Ironworkers	25.5–29.9
	Lathers	12.7–13.8
	Operating engineers	44.2–48.3
	Painters	16.4–17.9
	Plasterers	34.3–38.0
	Plumbers	7.8–9.2
	Roofers	47.1–50.1
	Sheetmetal workers	26.0–26.9
	Steamfitters	10.1–12.9
	Tile setters	13.6–16.0
	All other	27.6–31.5

WASHINGTON, DC, AREA

Area covered—District of Columbia; the Virginia cities of Alexandria, Fairfax, and Falls Church; the Virginia counties of Arlington, Fairfax, Loudoun, and Prince William; and the Maryland counties of Montgomery and Prince Georges.

GOALS AND TIMETABLES
[In percent]

Timetable	Trade	Goal
Until further notice	Electricians	28.0–34.0
	Painters and paperhangers	35.0–42.0
	Plumbers, pipefitters and steamfitters.	25.0–30.0
	Iron workers	35.0–43.0
	Sheetmetal workers	25.0–31.0
	Elevator constructors	34.0–40.0
	Asbestos workers	26.0–32.0
	Lathers	34.0–40.0
	Boilermakers	24.0–30.0
	Tile and terrazzo workers	28.0–34.0
	Glaziers	28.0–34.0

REGION IV

ATLANTA, GA, AREA

Area covered—Atlanta, GA, Standard Metropolitan Statistical Area which includes Fulton, DeKalb, Cobb, Clayton, and Gwinnett Counties.

GOALS AND TIMETABLES
[In percent]

Timetable	Trade	Goal
Until further notice	Asbestos workers	8.6–10.3
	Bricklayers	16.3–18.2
	Carpenters	11.0–12.8
	Electricians	10.9–12.2
	Glaziers	10.2–12.2
	Ironworkers	14.0–16.0
	Metal lathers	10.0–12.0
	Painters	10.3–12.0

GOALS AND TIMETABLES—Continued

[In percent]

Timetable	Trade	Goal
	Plumbers	9.4–10.9
	Pipefitters	9.4–10.9
	Plasterers	24.4–25.8
	Roofers	18.0–20.0
	Sheetmetal workers	9.5–11.3
	Sprinkler fitters	8.3–9.9
	Operating engineers	24.0–27.7
	Elevator installers	9.6–11.5

BIRMINGHAM, AL, AREA

Area covered—Jefferson, Shelby, and Walker Counties, AL.

GOALS AND TIMETABLES

[In percent]

Timetable	Trade	Goal
Until further notice	All	20–24

CHARLOTTE, NC, AREA

Area covered—Mecklenburg and Union Counties, NC.

GOALS AND TIMETABLES

[In percent]

Timetable	Trade	Goal
Until further notice	All	24–30

JACKSONVILLE, FL, AREA

Area covered—Drival County, FL.

GOALS AND TIMETABLES

[In percent]

Timetable	Trade	Goal
Until further notice	All	20–23

LOUISVILLE, KY, AREA

Area covered—Adair, Barren, Bullitt, Carrol, Edmundson, Grayson, Green, Hardin, Hart, Henry, Jefferson, Larue, Meade, Nelson, Oldham, Shelby, Spencer, Taylor, Trimble, Warren, and Washington Counties, KY; and Clark, Floyd and Harrison Counties, IN.

GOALS AND TIMETABLES

[In percent]

Timetable	Trade	Goal
Until further notice	All	12.0–16.0

MIAMI, FL, AREA

Area covered—Dade County, FL.

GOALS AND TIMETABLES

[In percent]

Timetable	Trade	Goal
Until further notice	All	20.0–40.0

NASHVILLE, TN, AREA

Area covered—City of Nashville, TN.

GOALS AND TIMETABLES

[In percent]

Timetable	Trade	Goal
Until further notice	All	16.0–20.0

REGION V

AKRON, OH, AREA

Area covered—Summit, Portage, and Medina Counties, OH.

GOALS AND TIMETABLES

[In percent]

Timetable	Trade	Goal
Until further notice	All	10.0–12.5

CANTON, OH, AREA

Area covered—Carroll, Holmes, Stark, Tuscarawas, and Wayne Counties, OH.

GOALS AND TIMETABLES

[In percent]

Timetable	Trade	Goal
Until further notice	All	7.0–8.4

CHICAGO, IL, AREA

Area covered—Cook, DuPage, Kane, Lake, McHenry, and Will Counties.

GOALS AND TIMETABLES

[In percent]

Timetable	Trade	Goal
Until further notice	Asbestos workers	8.6–10.3
	Bricklayers	16.3–8.2
	Carpenters	11.0–12.8
	Electricians	10.9–12.2
	Elevator installers	9.6–11.5
	Glaziers	10.2–12.2
	Ironworkers	14.0–16.0
	Metal lathers	10.0–12.0
	Painters	10.3–12.1
	Plumbers	9.4–10.9
	Pipe fitters	9.4–10.9
	Plasterers	24.4–25.8
	Roofers	18.0–20.0
	Sheetmetal workers	9.5–11.3
	Sprinkler fitters	8.3–9.9
	Operating engineers	(¹)

¹ 15.7 and above.

CINCINNATI, OH, AREA

Area covered. Ohio counties of Clermont, Hamilton, and Warren and in the Kentucky counties of Boone, Campbell, and Kenton, and in the Indiana county of Dearborn.

GOALS AND TIMETABLES

[In percent]

Timetable	Trade	Goal
Until further notice	Asbestos workers	9.3–12.2
	Boilermakers	8.0–8.4
	Carpenters	9.0–10.7
	Elevator constructors	10.2–12.7
	Engineers (stationary)	26.9–28.4
	Floor layers	9.0–10.5
	Glaziers	9.1–11.1
	Lathers	9.3–10.6
	Marble, tile and terrazzo workers and helpers.	8.3–9.9
	Millwrights	9.1–10.3
	Painters	11.0–13.5
	Pipefitters	10.0–12.0
	Plasterers	8.7 to 9.6
	Plumbers	10.0–12.7
	Sheetmetal workers	10.1–11.3
	All other	11.0–11.8

CLEVELAND, OH, AREA

Area covered—Ashland, Ashtabula, Crawford, Cuyahoga, Erie, Geauga, Huron, Lake, Lorain, Sandusky, and Seneca Counties, OH.

GOALS AND TIMETABLES

[In percent]

Timetable	Trade	Goal
Until further notice	Art glass workers	25.4–28.6
	Asbestos workers	20.9–23.9
	Boilermakers	16.3–18.9
	Bricklayers	28.8–29.5
	Carpenters	8.0–8.6
	Cement masons	41.1–42.2
	Electricians	15.1–18.1
	Elevator constructors	28.9–32.5
	Glaziers	35.8–40.0
	Ironworkers	11.4–13.2
	Painters	17.7–18.4
	Pipefitters	15.7–17.9
	Plasterers	21.6–23.2
	Plumbers	20.8–23.4
	Roofers	28.9–31.8
	All other	17.0–18.8

DAYTON, OH, AREA

Area covered—Greene, Miami, Montgomery, and Preble Counties, OH.

GOALS AND TIMETABLES

[In percent]

Timetable	Trade	Goal
Until further notice	All	10.6–11.8

DETROIT, MI., AREA

Area covered—Wayne, Oakland, and Macomb Counties, MI.

GOALS AND TIMETABLES

[In percent]

Timetable	Trade	Goal
Until further notice	Electricians	17.0–19.0
	Operating engineers	16.9–18.0
	Lathers	18.6–19.6
	Painters	15.0–17.7
	Riggers	16.8–17.7
	Roofers	15.3–16.6
	Tile, terrazzo marble workers.	15.0–17.8
	Tile and marble helpers	16.0–18.5
	Terrazzo helpers	17.8–19.5
	All other	18.6–20.4

EVANSVILLE, IN, AREA

Area covered—Vanderburgh County, IN.

GOALS AND TIMETABLES

[In percent]

Timetable	Trade	Goal
Until further notice	All	6.3–7.6

FORT WAYNE, IN, AREA

Area covered—Adams, Allen, DeKalb, Huntington, LaGrange, Noble, Steuben, Wells, and Whitley Counties, IN.

GOALS AND TIMETABLES

[In percent]

Timetable	Trade	Goal
Until further notice	Plumbers	5.2–5.5
	Steamfitters	5.2–5.5
	Carpenters	5.7–5.2
	Bricklayers	9.3–10.4
	Electricians	5.2–5.9
	Sheetmetal workers	4.4–5.2
	Ironworkers	7.3–8.4
	Operating engineers	5.2–6.0
	Painters	11.0–12.0
	All other	7.1–8.0

INDIANAPOLIS, IN, AREA

Area covered—Marion County, IN.

GOALS AND TIMETABLES

[In percent]

Timetable	Trade	Goal
Until further notice	Asbestos workers	32.2–37.7
	Bricklayers	17.4–19.5
	Electricians	6.6–7.8
	Elevator constructors	15.5–18.0
	Glaziers	25.2–28.6
	Ironworkers	11.6–14.0
	Lathers	21.1–22.0
	Operating engineers	7.7–8.8
	Painters	22.4–25.0

GOALS AND TIMETABLES—Continued

[In percent]

Timetable	Trade	Goal
	Plasterers	27.5–30.4
	Plumbers	25.5–30.0
	Roofers	15.9–18.1
	Sheetmetal workers	9.3–10.9
	Steamfitters	14.9–17.1
	All other	14.1–16.2

PEORIA, IL, AREA

Area covered—Peoria, Fulton, Tazewell, Woodford, Knox, Stark, Marshall, Hancock, Mason, McLean, McDonough, Henderson, Warren, Livingston, Bureau, Henry, and Putnam Counties, IL.

GOALS AND TIMETABLES

[In percent]

Timetable	Trade	Goal
Until further notice	All	5.0–6.0

ROCKFORD, IL, AREA

Area covered—Boone, Winnebago, Stephenson, De Kalb, Ogle, Lee, and Jo Daviess Counties; Cherry Grove, Shannon, Rock Creek, Lima, Wysox, and Elkhorn Townships in Carroll County; Genesee, Jordan, Hopkins, Sterling, Hume, Montmorency, Tampico, and Hahnaman Townships in Whiteside County, IL.

GOALS AND TIMETABLES

[In percent]

Timetable	Trade	Goal
Until further notice	All	10.0–12.0

SOUTH BEND, IN, AREA

Area covered—St. Joseph, County, IN.

GOALS AND TIMETABLES

[In percent]

Timetable	Trade	Goal
Until further notice	All	8.0–10.0

TOLEDO, OH, AREA

Area covered—Defiance, Fulton, Hancock, Henry, Lusas, Ottawa, Williams, and Wood Counties, OH.

GOALS AND TIMETABLES

[In percent]

Timetable	Trade	Goal
Until further notice	All	10.7–12.3

YOUNGSTOWN, OH AREA

Area covered—Columbiana, Mahoning, and Trumbull Counties, OH; and Lawrence and Mercer Counties, PA.

GOALS AND TIMETABLES

[In percent]

Timetable	Trade	Goal
Until further notice	All	6.0–7.1

REGION VI

EL PASO, TX, AREA

Area covered—El Paso County, TX.

GOALS AND TIMETABLES

[In percent]

Timetable	Trade	Goal
Until further notice	All	55.1–66.2

LAWTON, OK, AREA

Area covered—Commanche County, OK.

GOALS AND TIMETABLES

[In percent]

Timetable	Trade	Goal
Until further notice	All	15.8–16.8

LITTLE ROCK, AR, AREA

Area covered—Pulaski County, AR.

GOALS AND TIMETABLES

[In percent]

Timetable	Trade	Goal
Until further notice	All	25.6–30.6

NEW ORLEANS, LA.

Area covered—Parishes of Orleans, Jefferson, St. Bernard, St. Tammany, St. Charles, St. John, Lafourche, Plaquemines, Washington, Terrebonne, Tangipahoa,[1] Livingston,[2] and St. James.[3]

---

[1] Area covered is east of the Illinois Central RR.

[2] Area covered is southeast of the line from a point off the Livingston and Tangipahoa Parish line adjacent from New Orleans and Baton Rouge.

[3] Area covered is southeast of a line drawn from the town of Gramercy to the point of intersection of St. James, Lafourche, and Assumption Parishes.

GOALS AND TIMETABLES

[In percent]

Timetable	Trade	Goal
Until further notice	All	20–23

TULSA, OK

Area covered—Tulsa, Creek, Mayes, Rogers, Okfuskee, Washington, Nowata, Craig, Ottawa, Delaware, Okmulgee (northern half), dividing line Highway 16; Osage (eastern half), dividing line Highway 18; Pawnee (eastern half), and Payne (eastern half) Counties, OK.

GOALS AND TIMETABLES

[In percent]

Timetable	Trade	Goal
Until further notice	Bricklayers	24.0–25.0
	Carpenters	17.0–18.0
	Cement masons	21.5–22.5
	Floor covers	12.0–14.0
	Glaziers, glass workers	14.7–17.3
	Operating engineers	22.0–24.0
	Painters	18.0–20.0
	Pipefitters	10.0–12.0
	Plumbers	11.6–13.2
	Roofers	12.0–14.0
	Sheetmetal workers	8.0–10.0
	All other trades	12.0–14.4

REGION VII

KANSAS CITY (KS) AND (MO)

Area covered—Clay, Platte, Jackson, Bates, Carroll, Lafayette, Ray, Johnson, Henry, and Cass Counties, Mo., and Wyandotte, Johnson, and Miami Counties, KS.

GOALS AND TIMETABLES

[In percent]

Timetable	Trade	Goal
Until further notice	Asbestos workers	10.3–11.7
	Boilermakers	5.9–6.4
	Bricklayers	19.4–20.7
	Carpenters	5.9–6.9
	Carpet, linoleum and resilient floor decorators.	5.5–6.4
	Cement masons	25.5–26.5
	Elevator constructors	9.2–10.7
	Electricians	8.0–9.4
	Glaziers	9.8 to 10.5
	Lathers	14.5–15.6
	Marble masons, tile layers and terrazzo workers.	7.5–9.0
	Marble and tile helpers	4.8–5.6
	Operating engineers	9.0–10.9
	Painters	14.3–15.0
	Pipefitters	6.9–7.7
	Plasterers	19.0–20.4
	Plumbers	8.3–9.3
	Roofers	14.0–15.0
	Sheetmetal workers	7.0–8.0
	Teamsters	25.0–26.0
	All other trades	11.4–12.5

OMAHA, NE

Area covered—Sharpy and Douglas Counties, NE, Council Bluffs, IA (city limits only).

GOALS AND TIMETABLES

[In percent]

Timetable	Trade	Goal
Until further notice	All	9.0–10.0

ST. LOUIS, MO

Area covered—City of St. Louis, Mo., and St. Louis, MO.

GOALS AND TIMETABLES

[In percent]

Timetable	Trade	Goal
Until further notice	Asbestos workers ....,	5.2–5.7
	Boilermakers	34.0–37.7
	Bricklayers	12.6–14.2
	Carpenters	8.2–8.9
	Cement and concrete finishers.	13.3–16.6
	Electricians	13.6–16.1
	Elevator constructors	8.7–9.3
	Glaziers	28.7–34.5
	Ironworkers	9.0–10.4
	Lathers and plasterers	24.2–29.7
	Operating engineers	13.2–15.7
	Painters and paperhangers	25.1–29.3
	Plumbers and pipefitters ...	13.2–15.4
	Roofers and slaters	17.1–19.6
	Sheetmetal workers	22.5–27.0
	Tilesetters and terrazzo workers.	8.8–10.4

TOPEKA, KS

Area covered—Shawnee County, KS.

GOALS AND TIMETABLES

[In percent]

Timetable	Trade	Goal
Until further notice	All	8.8–10.5

REGION VIII

COLORADO

Area covered—State of Colorado

GOALS AND TIMETABLES

[In percent]

Timetable	Trade	Goal
Until further notice	All	13–14

REGION IX

ALAMEDA COUNTY, CA, AREA

Area covered—Alameda County, CA.

GOALS AND TIMETABLES

[In percent]

Timetable	Trade	Goal
Until further notice	All	28.5-33.0

ARIZONA

Area covered—State of Arizona.

GOALS AND TIMETABLES

[In percent]

Timetable	Trade	Goal
Until further notice	All	25.0-30.0

CONTRA COSTA COUNTY, CA

Area covered: Contra Costa County, CA.

GOALS AND TIMETABLES

[In percent]

Timetable	Trade	Goal
Until further notice	All	17.0-19.5

FRESNO COUNTY, CA

Area covered. Fresno, Madera, Kings, and Tulare Counties, CA.

GOALS AND TIMETABLES

[In percent]

Timetable	Trade	Goal
Until further notice	All	20.0-27.0

LAS VEGAS, NV

Area covered. Area of jurisdiction of the Building & Construction Trades Council of Clark, Lincoln, Nye and Esmeralda Counties, NV.

GOALS AND TIMETABLES

[In percent]

Timetable	Trade	Goal
Until further notice	Asbestos workers	17.7-20.2
	Bricklayers	18.8-21.3
	Carpenters	16.2-17.5
	Glaziers, floorcoverers, painters, tapers and wallcoverers.	16.3-17.7
	Plasterers	24.6-27.2
	Plumbers and pipefitters ...	15.2-16.2
	Sheet metal workers	16.2-17.7
	Wood, wire and metal lathers.	18.1-19.3
	All other trades	18.0-19.5

LOS ANGELES COUNTY, CA

Area covered. Area of jurisdiction of the Los Angeles Building & Construction Trades Council.

GOALS AND TIMETABLES

[In percent]

Timetable	Trade	Goal
Until further notice	All	21.7-25.1

MONTEREY, CA

Area covered. Monterey County, CA, and within the jurisdiction of the Monterey County Building & Construction Trades Council, AFL-CIO.

GOALS AND TIMETABLES

[In percent]

Timetable	Trade	Goal
Until further notice	All	27.0-29.8

NORTH BAY, CA

Area covered. Solano, Napa, Lake, Marin, Mendocino, and Sonoma Counties.

GOALS AND TIMETABLES

[In percent]

Timetable	Trade	Goal
Until further notice	All	10.5-12.6

SACRAMENTO, CA

Area covered. Sacramento, Yolo, Amador, Placer, El Dorado, Nevada, and Sierra Counties, CA.

GOALS AND TIMETABLES

[In percent]

Timetable	Trade	Goal
Until further notice	All	17.5-20.0

SAN DIEGO COUNTY, CA

Area covered. San Diego County, CA.

GOALS AND TIMETABLES

[In percent]

Timetable	Trade	Goal
Until further notice	All	24.0-30.0

SAN FRANCISCO CITY AND COUNTY, CA

Area covered. City and County of San Francisco, CA.

GOALS AND TIMETABLES

[In percent]

Timetable	Trade	Goal
Until further notice	Electricians	17.0
	Plumbers, pipefitters and steamfitters.	14.0

GOALS AND TIMETABLES—Continued

[In percent]

Timetable	Trade	Goal
	Structural metal workers ...	20.0
	Sheet metal workers	19.0
	Asbestos workers	40.0

SAN MATEO COUNTY, CA

Area covered. San Mateo County, CA.

GOALS AND TIMETABLES

[In percent]

Timetable	Trade	Goal
Until further notice	All	12.0–14.0

SANTA CLARA COUNTY, CA

Area covered. Santa Clara County, CA.

GOALS AND TIMETABLES

[In percent]

Timetable	Trade	Goal
Until further notice	All	18.0–21.7

SANTA CRUZ COUNTY, CA

Area covered. Santa Cruz County, CA.

GOALS AND TIMETABLES

[In percent]

Timetable	Trade	Goal
Until further notice	All	17.0–20.4

REGION X

ALASKA

Area covered. State of Alaska.

GOALS AND TIMETABLES

[In percent]

Timetable	Trade	Goal
Until further notice	Asbestos workers	26.4–28.0
	Carpenters	25.7–28.0
	Electricians	25.7–28.0
	Ironworkers	25.7–28.0
	Operating engineers	26.1–28.0
	Painters	25.8–28.0
	Pile drivers	25.1–28.0
	Plumbers and steamfitters	25.4–28.0
	Roofers	27.6–28.0
	Sheetmetal workers	25.6–28.0
	Teamsters	25.6–28.0
	All other	26.1–28.1

PASCO, WA

Area covered. The area of jurisdiction of the Southeastern Washington Building & Construction Trades Council as follows: all of Benton, Franklin, and Walla Walla Counties, Grant County to Highway 2 and the southwest corner of Adams County, WA.

GOALS AND TIMETABLES

[In percent]

Timetable	Trade	Goal
Until further notice	Boilermakers	12.5–15.0
	Bricklayers	11.0–13.5
	Carpenters	9.8–12.3
	Cement finishers	11.5–14.0
	Electricians	10.0–12.5
	Ironworkers	10.0–12.5
	Operating engineers	10.2–12.7
	Painters	10.0–12.5
	Plumbers and fitters	.9–12.4
	Sheetmetal workers	10.8–13.3
	Laborers	9.5–13.3
	All other	10.0–12.5

PORTLAND, OR

Area covered—Multnomah, Clackamas, and Washington Counties, OR.

GOALS AND TIMETABLES

[In percent]

Timetable	Trade	Goal
Until further notice	All	5.5–6.5

SEATTLE, WA

Area covered—King County, WA.

GOALS AND TIMETABLES

[In percent]

Timetable	Trade	Goal
Until further notice	All	8.8–11.5

SPOKANE, WA

Area covered—Washington Counties: Spokane, Whitman, Lincoln, Adams, Stevens, Pend Oreille, Columbia, Garfield, Asotin, Ferry, Okanogan, Chelan, Douglas and Grant (north of Highway 2), and in connection with Indian employment, parts of any other counties included in reservations incorporating portions of the above area; Idaho: Boundary, Bonner, Kootenai, Shoshone, Benewah, Latah, Clearwater, Nez Perce, Lewis, and Idaho, and in connection with Indian employment, any other territory included in reservations, part of which are in the above counties.

GOALS AND TIMETABLES

[In percent]

Timetable	Trade	Goal
Until further notice	All	(¹)

¹ 2.0 and above.

TACOMA, WA

Area covered—Pierce, Thurston, Mason, Lewis, Grays Harbor, and Pacific Counties, WA.

GOALS AND TIMETABLES

[In percent]

Timetable	Trade	Goal
Until further notice	All	12.2–15.0

[43 FR 58357, Dec. 14, 1978]

EXHIBIT E TO SUBPART E OF PART 1901— LIST OF REGIONAL OFFICES, OFFICE OF FEDERAL CONTRACT COMPLIANCE PROGRAMS (OFCCP), U.S. DEPARTMENT OF LABOR (USDL)

Region I (ME, NH, VT, MA, RI, CT)
Associate Regional Administrator, USDL/ OFCCP, JFK Building, Room 1612–C, Government Center, Boston, MA 12203, (617) 223–4232.

Region II (NY, NJ, PR, VI)
Associate Regional Administrator, USDL/ OFCCP, 1515 Broadway, Room 3306, New York, NY 10036, (212) 662–5563.

Region III (PA, MD, DE, VA, WV, DC)
Associate Regional Administrator, USDL/ OFCCP, Gateway Building, Room 15434, 3535 Market Street, Philadelphia, PA 19104, (215) 596–1213.

Region IV (NC, SC, KY, TN, MS, AL, GA, FL)
Associate Regional Administrator, USDL/ OFCCP, 1371 Peachtree Street, NE, Room 729, Atlanta, GA 30309, (402) 881–4211).

Region V (OH, IN, MI, IL, WI, MN)
Associate Regional Administrator, USDL/ OFCCP, New Federal Building, 16th Floor, 2340 South Dearborn Street, Chicago, IL 60604, (312) 353–8887.

Region VI (LA, AR, OK, TX, NM)
Associate Regional Administrator, USDL/ OFCCP), 555 Griffin Square Building, Room 506, Dallas, TX 75202, (214) 767–4771.

Region VII (MO, IA, NE, KS)
Associate Regional Administrator, USDL/ OFCCP Regional Administrator, Federal Office Building, Room 2000, 911 Walnut Street, Kansas City, MO 64106, (816) 374– 5384.

Region VIII (ND, SD, MT, WY, CO, UT)
Associate Regional Administrator, USDL/ OFCCP, 14431 Federal Office Building, 1961 Stout Street, Denver, CO 80202, (303) 837–5011.

Region IX (CA, NV, AZ, HI, GU)
Associate Regional Administrator, USDL/ OFCCP, Federal Office Building, Room 10341, 450 Golden Gate, San Francisco, CA 94102, (415) 556–3597.

Region X (WA, OR, ID)

Associate Regional Administrator, USDL/ OFCCP, Federal Office Building, 909 First Avenue, Room 4095, Seattle, WA 98174, (206) 442–4508.

[44 FR 24852, Apr. 27, 1979]

Subpart F—Procedures for the Protection of Historical and Archeological Properties

AUTHORITY: 16 U.S.C. 470; 7 U.S.C. 1989; 42 U.S.C. 1480; 42 U.S.C. 2942; 5 U.S.C. 301; sec. 10, Pub. L. 93–357, 88 Stat. 392; delegation of authority by Sec. of Agri., 7 CFR 2.23; delegation of authority by the Asst. Sec. for Rural Development, 7 CFR 2.70; delegations of authority by Dir., OEO, 29 FR 14764, 33 FR 9850.

SOURCE: 42 FR 62141, Dec. 9, 1977, unless otherwise noted.

§ 1901.251 Purpose.

This subpart prescribes Farmers Home Administration (FmHA) or its successor agency under Public Law 103– 354 policies, procedures, and guidelines for compliance with section 106 of the National Historic Preservation Act of 1966 (Pub. L. 89–665), the Reservoir Salvage Act of 1960 (Pub. L. 86–523), as amended May 24, 1974, by the Archeologic and Historic Preservation Act (Pub. L. 93–291), and section 1(3) of Executive Order 11593. This subpart is inapplicable to Farm Service Agency, Farm Loan Programs.

[42 FR 62141, Dec. 9, 1977, as amended at 72 FR 64121, Nov. 15, 2007]

§ 1901.252 Policy.

(a) The FmHA or its successor agency under Public Law 103–354 recognizes that significant scientific, prehistorical, historical and archeological (HA) resources are an important part of our National Heritage.

(b) The FmHA or its successor agency under Public Law 103–354 will consult with appropriate Federal, State, and local Agencies; other organizations; the State Historic Preservation Officer (SHPO) and individuals to assess the impact of any proposed FmHA or its successor agency under Public Law 103–354 undertaking on properties having historical or archeological significance in order to avoid or mitigate any adverse effects on the properties.

(c) The procedures in this subpart have been developed in accordance with section 1(3) of Executive Order 111593.

§1901.253 Definitions.

(a) *Undertaking* means any new or continuing projects or program activities supported in whole or in part through FmHA or its successor agency under Public Law 103–354 contracts, grants, subsidies, loans, or other forms of funding assistance. This does not include any actual construction by FmHA or its successor agency under Public Law 103–354.

(b) *National Historic Preservation Act. The National Register* means the National Register of Historic Places, which is a register of districts, sites, buildings, structures, and objects, significant in American history, architecture, archeology, and culture maintained by the Secretary of the Interior under the authority of section 2(b) of the Historic Sites Act of 1935 and section 101(a)(1) of the National Preservation Act. The National Register is published in its entirety in the FEDERAL REGISTER each year in February. Addenda are published on the first Tuesday of each month.

(c) *National Register Property* means a district, site, building, structure, or object included in the National Register.

(d) *Property eligible for inclusion in the National Register* means any district, site, building, structure, or object which the Secretary of the Interior determines is likely to meet the National Register criteria.

(e) *State Historic Preservation Officer* (SHPO) means the official within each State, designated by the Governor at the request of the Secretary of the Interior, to administer the National Register and historic preservation grants program and to coordinate preservation planning within the State.

(f) *Criteria of effect* means when any condition of an undertaking causes or may cause any change, beneficial or adverse, in the scientific, historical, architectural, archeological, or cultural character of a National Register property that qualifies the property under the National Register criteria.

(g) *Historical and archeological assessment* means a determination by the FmHA or its successor agency under Public Law 103–354 State Director using the criteria of effect as a guide, as to whether a proposed undertaking may have an effect upon any properties located within the project area which are included or eligible for inclusion in the National Register.

(h) *National Register criteria* means the following criteria established by the Secretary of the Interior for use in evaluating and determining the eligibility of properties for listing in the National Register: The quality of significance in American History, Architecture, Archeology, and the culture is present in districts, sites, buildings, structures, and objects of State and local importance, that possess integrity of location, design, setting, materials, workmanship, feeling, and association; and

(1) That are associated with events that have made a significant contribution to the broad patterns of our history; or

(2) That are associated with the lives of persons significant in our past; or

(3) That embody the distinctive characteristics of a type, period, or method of construction, or that represent the work of a master, or that possess high artistic values, or that represent a significant and distinguishable entity whose components may lack individual distinction; or

(4) That have yielded, or may be likely to yield, information important in prehistory or history.

(i) *FmHA or its successor agency under Public Law 103–354 official* means the FmHA or its successor agency under Public Law 103–354 County Supervisor, the FmHA or its successor agency under Public Law 103–354 State Director or his designated representative.

(j) *Project area* means those geographical or legally defined areas directly under or to be under the applicants control that are affected by the undertaking such as building sites, easements, rights-of-way, leasehold interests and those areas which are directly and significantly impacted by the undertaking.

(k) *Advisory council* means the Advisory Council on Historic Preservation, Suite 430, 1522 K Street NW., Washington, DC 20005, created by title II of

Pub. L. 89-665 and charged with the responsibility of advising the President, Congress, and others on matters relating to historic preservation.

(1) *HA* as used in this regulation is an abbreviation of the term "scientific, prehistorical, historical, and archeological."

§ 1901.254 Scope.

FmHA or its successor agency under Public Law 103-354 will evaluate all undertakings for possible HA significance. This subpart covers the following types of undertakings:

(a) *Undertakings requiring a historical and archeological assessment.* Although the following undertakings are presumed to involve nonfederally owned lands, they may have an effect on properties having HA significance and, therefore, will require a historical and archeological assessment:

(1) Loans and grants for the development of business and industry including guaranteed loans.

(2) Loans and grants for multiple family housing projects of 25 or more dwelling units.

(3) Subdivision plans submitted for approval having 25 or more building sites.

(4) Loans and grants in rural areas to construct, enlarge, extend, or otherwise improve:

(i) Community water, sanitary sewage, solid waste disposal, and storm waste water disposal systems.

(ii) Other essential community facilities such as fire and rescue, health, safety, public buildings, schools, transportation, traffic, and law enforcement.

(5) Loans to develop community irrigation, drainage, and other soil and water conservation and use facilities.

(6) Loans to acquire and develop grazing land for livestock of an association of members.

(7) Loans in areas designated by the Soil Conservation Service (SCS), U.S. Department of Agriculture (USDA), to conserve and develop natural resources and to contribute to economic improvement of the area.

(8) Loans to protect and develop land and water resources in small watersheds.

(9) Loans to permit Indian tribes to buy land within their reservations.

(b) *Undertakings presumed not to require a historical and archeological assessment.* The following undertakings are generally presumed to involve nonfederally owned lands and not to have an effect on properties of historical and archeological value and will therefore not usually require a historical and archeological assessment. However, when the State Director or County Supervisor finds or has had communication or obtains information from a recognized historical and archeological authority that a specific undertaking may have an effect on a property included or eligible for inclusion in the National Register, a historical and archeological assessment will be made.

(1) Loans to farmers and ranchers in rural areas for the purchase, development, and operation of farms and ranches.

(2) Loans to individual families in rural areas for the purchase, construction, or improvement of single family residences.

(3) Loans and grants for multiple family housing projects of not more than 24 family dwelling units.

(4) Subdivision plans submitted for approval having 24 or less building sites.

(5) Loans to farmers, ranchers, and other rural residents to develop land, water, and other related resources for increased production of food and other crops, improved pastures, feed crops, water facilities for livestock, and improved habitats for fish and wildlife.

(6) Emergency and disaster loans to farmers, ranchers and other rural residents in declared or designated areas as a result of a major or national disaster.

§ 1901.255 Historical and archeological assessments.

(a) The FmHA or its successor agency under Public Law 103-354 official, normally the FmHA or its successor agency under Public Law 103-354 County Supervisor, who receives a preapplication or application for loan or grant assistance on an undertaking that may have an effect on HA properties will, as part of the process, take the following actions:

(1) Carefully review the State supplements issued by the State Director pursuant to § 1901.262(a) to determine whether there are any properties within the project area that appear in the National Register.

(2) Document the following:

(i) A brief narrative report of the findings and conclusions of an on-site reconnaissance of the project area.

(ii) Any "in-house" knowledge of known or suspected HA sites in the project area.

(3) Submit the information outlined in paragraph (a)(2) of this section to the FmHA or its successor agency under Public Law 103–354 State Director as part of the preapplication or application.

(b) Upon receipt of the preapplication/application the FmHA or its successor agency under Public Law 103–354 State Director will, as a concurrent part of the preapplication/application review, prepare a historical and archeological assessment of the undertaking. In making the assessment the State Director will consider information from the following sources:

(1) State and Regional Clearinghouse comments.

(2) Information submitted by the County Supervisor pursuant to paragraph (a)(2) of this section.

(3) Factual comments or recommendations of the SHPO or other responsible Federal, State, or local officials.

(4) Any other reliable information concerning properties in the project area having HA significance.

(c) Upon completion of the preapplication or application review, the State Director will take the following actions:

(1) When his assessment indicates that no properties of HA significance will be effected by the proposed undertaking, he will proceed with processing of the preapplication or application.

(2) When his assessment indicates that there are properties included in the National Register that may be effected by the proposed undertaking, he will in consultation with the SHPO, the applicant and its representatives, and other appropriate historical and archeological authorities plan appropriate measures to avoid or mitigate any adverse effects. He will also notify the Advisory Council and Secretary of the Interior of the proposed undertaking, and of its possible effect on the National Register properties and provide them with a copy of the proposed plan in order to afford them a reasonable opportunity for comment. Comments that are received with 45 calendar days of notification in accordance with the requirements for comment as outlined in section 106 of the National Historic Preservation Act of 1966, will be considered in further development of the undertaking.

(3) When his assessment indicates that there are properties thay may be eligible for inclusion in the National Register, based on his application of the National Register criteria, he will request the Regional Director of the National Park Service, U.S. Department of the Interior, Attention: Interagency Archeological Services, in writing, to cause a survey of the project area to be made to determine the significance of the properties in accordance with section 3(b) of Pub. L. 93–291. The State Director's letter to the Regional Director should request a response within 45 calendar days as to whether the National Park Service intends to cause a survey to be made, declines to undertake a survey, or that a survey is not warranted based on available data. The addresses of the Regional Offices of the National Park Service are listed in exhibit A of this subpart. If no response is received within the 45-day period, the State Director will proceed as outlined in paragraph (c)(7) of this section.

(4) The State Director will cooperate fully with the National Park Service in the conduct of a survey should one be undertaken to assure that:

(i) The professional archeologist/historian conducting the survey provides his written opinion as to the eligibility of any identified properties for inclusion in the National Register.

(ii) When the professional archeologist/historian recommends recovery, protection, or preservation of identified properties, the National Park Service is requested to undertake this project.

(5) When the survey made in paragraph (c)(3) of this section does not

identify any historical and archeological properties that may be eligible for inclusion in the National Register, or the National Park Service is not going to undertake activity pursuant to paragraph (c)(4)(ii) of this section, the State Director, after consultation with the SHPO and the National Park Service, will document the findings and proceed with processing of the application.

(6) When the survey identifies properties that may be eligible for inclusion in the National Register, the State Director will request the SHPO to proceed with the nomination of such properties. The State Director will then proceed as outlined in paragraph (c)(2) of this section for any properties accepted for inclusion in the National Register.

(7) When the National Park Service declines to cause a survey to be made or determines that one is not warranted, the State Director will document such facts and proceed with processing of the application.

§§ 1901.256–1901.258 [Reserved]

§ 1901.259 Actions to be taken when archeological properties are discovered during construction.

(a) When properties of significant HA value are discovered during construction, the State Director will immediately consult with the applicant, the SHPO and the Regional Director of the National Park Service to determine whether there is sufficient factual evidence to warrant a decision to stop construction and undertake detailed survey and recovery.

(b) When the consultations in paragraph (a) of this section result in a determination by the National Park Service to request the applicant to stop construction, such stop action should be taken so that the Park Service can initiate measures for immediate recovery within 60 days after notification of a discovery.

(c) When the consultations in paragraph (a) of this section do not result in a determination by the National Park Service to stop construction and to undertake a survey and recovery, construction should be permitted to proceed with caution. In the event that the National Park Service determines

that recovery is necessary, the FmHA or its successor agency under Public Law 103–354 applicant/borrower and the Park Service should determine that the consent of all persons, associations, or public entities having legal interests in the property involved has been secured. Also, the applicant should be informed that the Secretary of the Interior is authorized to compensate any person, association, or public entity damaged as a result of delay in construction or as a result of the temporary loss of the use of public or any nonfederally owned land.

(d) No survey or recovery work will be required which in the determination of the State Director would seriously impede FmHA or its successor agency under Public Law 103–354 actions in providing assistance where the State Director determines that immediate action is required to avoid loss or damage of life or property. Nevertheless, appropriate measures will be taken to the extent practical to preserve, protect, or mitigate any damage to properties having HA significance.

§ 1901.260 Coordination with other agencies.

(a) When other Agencies are directly involved in any undertaking that requires a historical and archeological assessment, the State Director will contact the Agencies concerned to determine if a joint assessment will be prepared and whether a single lead Agency will assume primary responsibility for preparing the assessment.

(b) When a lead Agency is agreed upon other than FmHA or its successor agency under Public Law 103–354, FmHA or its successor agency under Public Law 103–354 will provide that Agency with information about its respective areas of responsibility. Assessments will indicate Agency participation and concurrence.

(c) When FmHA or its successor agency under Public Law 103–354 program activities are planned that primarily supplement those of the SCS, USDA, such as watershed projects, resource conservation and development measures, and irrigation and drainage projects, the SCS will be designated as the lead Agency.

§1901.261 [Reserved]

§1901.262 State supplement.

(a) The State Director shall be responsible for preparing a list of all properties included in the National Register in his area of jurisdiction and issuing such list as a part of a State supplement. Such a list will be updated as needed to reflect changes in the National Register.

(b) State Directors may also supplement this subpart and its exhibit as appropriate to meet State and local laws and regulations.

EXHIBIT A TO SUBPART F OF PART 1901—
NATIONAL PARK SERVICE, U.S. DE-
PARTMENT OF THE INTERIOR RE-
GIONAL OFFICES

Contact should be made to: Chief, Interagency Archeological Services Division, Office of Archeological and Historic Preservation, National Park Service.

The three Regional Offices are:

San Francisco Office: Old Post Office Building, Mission and 7th Streets, Post Office Box 5700, San Francisco, Calif. 94104.
States covered: Arizona, Utah, Idaho, and West, including Hawaii and Alaska. Attention: Mr. Garland Gordon. Telephone: 415–556–7711.
Denver Office: 1978 South Garrison Street, Denver, Colo. 80225.
States covered: Wisconsin, Iowa, Missouri, Oklahoma, Texas and West to San Francisco area. Attention: Mr. Jack R. Rudy. Telephone: 303–234–2560.
Atlanta Office: 730 Peachtree Street, Atlanta, Ga. 30308.
States covered: All others East of Denver area. Attention: Mr. Wilford Susted. Telephone: 404–526–2611.

Subparts G–J [Reserved]

Subpart K—Certificates of Beneficial Ownership and Insured Notes

AUTHORITY: 7 U.S.C. 1989; 42 U.S.C. 1480; delegation of authority by the Secretary of Agriculture, 7 CFR 2.23; delegation of authority by the Assistant Secretary for Rural Development, 7 CFR 2.70.

SOURCE: 41 FR 51799, Nov. 24, 1976, unless otherwise noted.

§1901.501 Purpose.

This subpart prescribes policies and procedures for Rural Development certificates of beneficial ownership and insured notes. This subpart is inapplicable to Farm Service Agency, Farm Loan Programs.

[41 FR 51799, Nov. 24, 1976, as amended at 72 FR 64121, Nov. 15, 2007; 80 FR 9868, Feb. 24, 2015]

§1901.502 Policy.

It is the current policy to sell all certificates of beneficial ownership to the Federal Financing Bank for financing activities from the Agricultural Credit Insurance Fund and the Rural Development Insurance Fund. Sales from the Rural Housing Insurance Fund will be made to the Federal Financing Bank to the extent necessary to service certificates of beneficial ownership held by the Federal Financing Bank. Sales in excess of those needed for servicing requirements will be made to the public. In addition to sales, this subpart provides policy for the servicing of outstanding certificates of beneficial ownership, insurance contracts, and insured notes held by investors.

[51 FR 24301, July 3, 1986]

§1901.503 Definitions.

(a) As used in §§1901.505, 1901.507, 1901.508 and 1901.509 the following definitions will apply:

(1) *Announcement of sale.* Any notice of terms and conditions respecting a sale of certificates.

(2) *Certificate.* A certificate of beneficial ownership issued by Rural Development under this subpart.

(3) *Director, Finance Office.* The Director or the Insured Loan Officer of the Finance Office of Rural Development.

(4) *Rural Development.* The United States acting through the Rural Housing Service, the Rural Utilities Service, or the Rural Business-Cooperative Service or their successor agencies.

(5) *Finance Office.* The office which maintains the Rural Development finance records. It is located at 1520 Market Street, St. Louis, Missouri 63103. (Phone: 314–425–4400)

(6) *Fixed period.* Any time interval (preceding an option period) during which the insured holder is not entitled

to require Rural Development to purchase the insured note, as specified in the insurance agreement.

(7) *Insurance agreement.* The entire contract evidencing and setting forth the terms and conditions of Rural Development insurance of the payment for the insured note. The insurance agreement with respect to any particular loan may be evidenced by Form RD 440–5, "Insurance Endorsement (Insured Loan)," RD 440–30, "Insurance Endorsement (Insure Loans)," or any other form or forms prescribed by the National Office and executed by an authorized official of Rural Development. It may include such provisions as, for example, an agreement of Rural Development to purchase or repurchase the loan, or to make supplementary payments from the insurance fund.

(8) *Insurance fund.* The Agricultural Credit Insurance Fund authorized by section 309 of the Consolidated Farm and Rural Development Act, the Rural Development Insurance Fund authorized by section 309A of the Consolidated Farm and Rural Development Act, or the Rural Housing Insurance Fund authorized by section 517 of title V of the Housing Act of 1949.

(9) *Insured holder.* The current owner of an insured note other than Rural Development, according to the records of Rural Development as insurer of the note.

(10) *Insured note.* Any promissory note or bond evidencing an insured loan regardless of whether it is held by Rural Development in the insurance fund, by a private holder, or by Rural Development as trustee.

(11) *Loan.* Loans made and held in the Agricultural Credit Insurance Fund, Rural Development Insurance Fund, or the Rural Housing Insurance Fund.

(12) *National Office.* The Administrator or other authorized officer of Rural Development in Washington, DC.

(13) *Option period.* Any period during which the insured holder has the optional right to require Rural Development to purchase the insured note, as specified in the insurance agreement.

(14) *Par value.* The total amount to which the insured holder is entitled under the terms of the insurance agreement.

(15) *Private buyer.* A buyer of an insured note other than Rural Development.

(16) *Private holder.* An insured holder other than Rural Development.

(17) *Repurchase agreement.* A provision in the insurance agreement obligating Rural Development to buy the insured note at the option of the holders.

(18) *Sale, or seller, and buyer.* The transfer of ownership (including possession or the right of possession), the transferor, and the transferee respectively.

(19) *State Director.* The State Director of Rural Development for the State in which is located the real estate improved, purchased, or refinanced with the loan evidenced by the insured note.

(b) As used in § 1901.506 the following definitions will apply:

(1) *Reserve bank.* The Federal Reserve Bank of New York (and any other Federal Reserve Bank which agrees to issue securities in book-entry form) as fiscal agent of the United States acting on behalf of Rural Development and, when indicated, acting in its individual capacity.

(2) *Rural Development security.* A certificate representing beneficial ownership of notes, bonds, debentures, or other similar obligations held by Rural Development under the Consolidated Farm and Rural Development Act and title V of the Housing Act of 1949, issued in the form of a definitive Rural Development security or a book-entry Rural Development security.

(3) *Definitive Rural Development security.* A Rural Development security in engraved on printed form.

(4) *Book-entry Rural Development security.* A Rural Development security in the form of an entry made as prescribed in this subpart on the records of a Reserve bank.

(5) *Pledge.* A pledge of, or any other security interest in, Rural Development securities as collateral for loans or advances, or to secure deposits of public moneys or the performance of an obligation.

(6) *Date of call.* The date fixed in the official notice of call published in the FEDERAL REGISTER on which Rural Development will make payment of the security before maturity in accordance with its terms.

(7) *Member bank.* Any national bank, state bank, or bank or trust company which is a member of a Reserve bank.

[41 FR 51799, Nov. 24, 1976, as amended at 80 FR 9868, Feb. 24, 2015]

§1901.504 Authorities and responsibilities.

The Administrator will approve all methods of Rural Development financing and major changes in existing methods. The Director, Finance Office, is responsible for servicing of all certificates of beneficial ownership and insured notes issued by the Finance Office, the Federal Reserve Bank of New York for the servicing of insurance contracts, and the Federal Reserve banks for certificates of beneficial ownership for which the Reserve banks are Rural Development's fiscal agents.

[41 FR 51799, Nov. 24, 1976, as amended at 80 FR 9868, Feb. 24, 2015]

§1901.505 Certificates of beneficial ownership in Rural Development loans.

(a) *Special trust of loans*—(1) *Establishment of special trusts.* From time to Rural Development will place in special trusts unmature loans evidenced by notes or other instruments. Loans may be placed into or removed from a special trust, but there will always be maintained in such trusts loans on which the unpaid amount is at least equal to the face value of the outstanding unmature certificates evidencing beneficial ownership in such special trust as provided in paragraph (a)(2) of this section.

(2) *Beneficial ownership of special trusts.* To permit interested persons to acquire a beneficial ownership of loans comprising a special trust established under paragraph (a)(1) of this section, Rural Development will sell certificates which will evidence beneficial ownership of an interest in the special trust to the extent of the face value of such certificates. Rural Development will own an interest in special trusts equal to the amount by which the unpaid principal amount of loans comprising the trusts exceeds the face value of all outstanding certificates evidencing beneficial ownership in such trusts.

(b) *Sale of certificates.* Rural Development will offer certificates for sale from time to time on such terms and conditions it may deem appropriate. Sales made by the Finance Office shall be made by its Director. No sale in excess of $1 million will be made to any one investor without prior approval of the Associate Administrator or his designee. The terms and limitations of sales are subject to change from time to time, and may be obtained from the Finance Office.

(1) *Form of certificates.* The certificates may be interest-bearing or non-interest-bearing. The certificates may be made payable to the bearer or registered holder thereof, and will be negotiable. The certificates will be issued in denominations specified in the invitations for bid or other announcement of sale.

(2) *Issue date and maturity date of certificates.* The certificates will be issued on such dates and mature on such dates as specified in the invitation for bids or other announcement of sale. Such dates will appear on the face of the certificates.

[41 FR 51799, Nov. 24, 1976, as amended at 80 FR 9868, Feb. 24, 2015]

§1901.506 Book-entry procedure for Rural Development securities— issuance and redemption of certificate by Reserve bank.

(a) *Authority of Reserve bank.* Each Reserve bank is hereby authorized in accordance with the provisions of this subpart to:

(1) Issue book-entry Rural Development securities by means of entries on its records which shall include the name of the depositor, the amount, the securities title (or series) and maturity date.

(2) Effect conversions between book-entry Rural Development securities and definitive Rural Development securities.

(3) Otherwise service and maintain book-entry Rural Development securities.

(4) Issue a confirmation of transaction in the form of a written advice (serially numbered or otherwise) which specifies the amount and description of any securities (that is, the securities title (or series) and the maturity date)

sold or transferred and the date of the transaction.

(b) *Scope and effect of book-entry procedure.* (1) A Reserve bank as fiscal agent of the United States acting on behalf of Rural Development may apply the book-entry procedure provided for in this subpart to any Rural Development securities which have been or are hereafter deposited for any purpose in accounts with it in its individual capacity under terms and conditions which indicate that the Reserve bank will continue to maintain such deposit accounts in its individual capacity, notwithstanding application of the book-entry procedure to such securities. This paragraph shall be applicable but not limited to Rural Development securities deposited:

(i) As collateral pledged to a Reserve bank (in its individual capacity) for advances by it.

(ii) By a member bank for its sole account.

(iii) By a member bank held for the account of its customers.

(iv) In connection with deposits in a member bank of funds of States, Municipalities, or other political subdivisions.

(v) In connection with the performance of an obligation or duty under Federal, State, Municipal, or local law, or judgments or decrees of courts.

(2) The application of the book-entry procedure under paragraph (b)(1) of this section shall not detract from or adversely affect the relationships that would otherwise exist between a Reserve bank in its individual capacity and its depositors concerning any deposit under this paragraph. Whenever the book-entry procedure is applied to such Rural Development securities, the Reserve bank is authorized to take all action necessary in respect of the book-entry procedure to enable such Reserve bank in its individual capacity to perform its obligation as depositary with respect to such Rural Development securities.

(3) A Reserve bank as fiscal agent of the United States acting on behalf of Rural Development may apply the book-entry procedure to Rural Development securities deposited as collateral pledged to the United States under Treasury Department Circular Nos. 92

and 176, both as revised and amended, and may apply the book-entry procedure, with the approval of the Secretary of the Treasury, to any other Rural Development securities deposited with a Reserve bank as fiscal agent of the United States.

(4) Any person having an interest in Rural Development securities which are deposited with a Reserve bank (in either its individual capacity or as fiscal agent of the United States) for any purpose shall be deemed to have consented to their conversion to book-entry Rural Development securities pursuant to the provisions of this subpart and in the manner and under the procedure prescribed by the Reserve bank.

(5) No deposits shall be accepted under this section on or after the date of maturity or call of Rural Development securities.

(c) *Transfer or pledge.* (1) A transfer or pledge of book-entry Rural Development securities to a Reserve bank (in its individual capacity or as fiscal agent of the United States), or to the United States, or to any transferee or pledgee eligible to maintain an appropriate book-entry account in its name with a Reserve bank under this subpart is effected and perfected, notwithstanding any provision of law to the contrary, by a Reserve bank making an appropriate entry in its records of the securities transferred or pledged. The making of such an entry in the records of a Reserve bank shall:

(i) Have the effect of a delivery in bearer form of definitive Rural Development securities.

(ii) Have the effect of a taking of delivery by the transferee or pledgee.

(iii) Constitute the transferee or pledgee a holder.

(iv) If a pledge, effect a perfected security interest therein in favor of the pledgee. A transfer or pledge of book-entry Rural Development securities effected under this paragraph shall have priority over any transfer, pledge, or other interest, theretofore or thereafter effected or perfected under paragraph (c)(2) of this section or any other manner.

(2) A transfer or pledge of transferable Rural Development securities, or

any interest therein, which is maintained by a Reserve bank (in its individual capacity or as fiscal agent of the United States) in a book-entry account under this subpart, including securities in book-entry form under §1901.506(b)(1)(iii) is effected, and a pledge is perfected by any means that would be effective under applicable law to effect a transfer or to effect and perfect a pledge of Rural Development securities, or any interest therein, if the securities were maintained by the Reserve bank in bearer definitive form. For purposes of transfer or pledge hereunder, book-entry Rural Development securities maintained by a Reserve bank shall, notwithstanding any provision of law to the contrary, be deemed to be maintained in bearer definitive form. A Reserve bank maintaining book-entry Rural Development securities, either in its individual capacity or as fiscal agent of the United States, is not a bailee for the purposes of notification of pledges of these securities under this paragraph, or a third person in possession for the purposes of acknowledgment of transfers thereof under this paragraph. Where transferable Rural Development securities are recorded on the books of a depositary (a bank, banking institution, financial firm, or similar party, which regularly accepts in the course of its business Rural Development securities as a custodial service for customers, and maintains accounts in the names of such customers reflecting ownership of or interest in such securities) for account of the pledgor or transferor thereof and such securities are on deposit with a Reserve bank in a book-entry account hereunder, such depositary shall, for purposes of perfecting a pledge of such securities or effecting delivery of such securities to a purchaser under applicable provisions of law, be the bailee to which notification of the pledge of the securities may be given or the third person in possession from which acknowledgment of the holding of the securities for the purchaser may be obtained. A Reserve bank will not accept notice or advice of a transfer or pledge effected or perfected under this paragraph and any such notice or advice shall have no effect. A Reserve bank may continue to deal with its depositor

in accordance with the provisions of this subpart, notwithstanding any transfer or pledge effected or perfected under this paragraph.

(3) No filing or recording with a public recording office or officer shall be necessary or effective with respect to any transfer or pledge of book-entry Rural Development securities or any interest therein.

(4) A Reserve bank shall, upon receipt of appropriate instructions, convert book-entry Rural Development securities into definitive Rural Development securities and deliver them in accordance with such instructions. No such conversion shall affect existing interest in such Rural Development securities.

(5) A transfer of book-entry Rural Development securities within a Federal Reserve Bank shall be made in accordance with procedures established by the Reserve bank not inconsistent with this subpart. The transfer of book-entry Rural Development securities by a Reserve bank may be made through a telegraphic transfer procedure.

(6) All requests for transfer or withdrawal must be made prior to the maturity or date of call of the securities.

(d) *Withdrawal of Rural Development securities.* (1) A depositor of book-entry Rural Development securities may withdraw them from a Reserve bank by requesting delivery of like definitive Rural Development securities to itself or on its order to a transferee.

(2) Rural Developmentsecurities which are actually to be delivered upon withdrawal may be issued in bearer or registered form.

(e) *Delivery of Rural Development securities.* A Reserve bank which has received Rural Development securities and effected pledges, made entries regarding them, or transferred or delivered them according to the instructions of its depositor is not liable for conversion or for participation in breach of fiduciary duty even though the depositor had no right to dispose of or take other action in respect of the securities. A Reserve bank shall be fully discharged of its obligations under this subpart by the delivery of Rural Development securities in definitive form to its depositor or upon the order of such depositor. Customers of a member bank or other depositary (other than a Reserve bank) may obtain Rural Development securities in definitive form only by causing the depositor of the Reserve bank to order the withdrawal thereof from the Reserve bank.

(f) *Registered securities.* (1) No formal assignment shall be required for the conversion to book-entry Rural Development securities of registered Rural Development securities held by a Reserve bank (in either its individual capacity or as fiscal agent of the United States) on the effective date of this subpart for any purpose specified in § 1901.506(b)(1). Registered Rural Development securities deposited thereafter with a Reserve bank for any purpose specified in § 1901.506(b) shall be assigned for conversion to book-entry Rural Development securities.

(2) The assignment which shall be executed in accordance with the provisions of subpart F of 31 CFR part 306, so far as applicable, shall be to Federal Reserve Bank of _____, as fiscal agent of the United States acting on behalf of the Rural Development, United States Department of Agriculture, for conversion to book-entry Rural Development securities.

(g) *Servicing book-entry Rural Development securities, payment of interest, payment at maturity or upon call.* Interest becoming due on book-entry Rural Development securities shall be charged to the general account of the Treasurer of the United States on the interest due date and remitted or credited in accordance with the depositor's instructions. Such securities shall be redeemed and charged to the same account on the date of maturity or call, and the redemption proceeds, principal, and interest shall be disposed of in accordance with the depositor's instructions.

(h) *Issuance and redemption.* (1) In those instances where the Reserve bank is acting as fiscal agent of the United States acting on behalf of Rural Development, the following subparts of Treasury Department Circular No. 300 (31 CFR part 306), so far as applicable, shall apply to such certificates.

(i) Subpart B, Registration.

(ii) Subpart C, Transfers, Exchanges and Reissues.

(iii) Subpart D, Redemption or Payment.

(iv) Subpart E, Interest.

(v) Subpart G, Assignments of Registered Securities—General.

(vi) Subpart F, Assignments by or in Behalf of Individuals.

(vii) Subpart H, Assignments in Behalf of Estates of Deceased Owners.

(viii) Subpart I, Assignments by or in Behalf of Trustees and Similar Fiduciaries.

(ix) Subpart J, Assignments in Behalf of Private or Public Organizations.

(x) Subpart K, Attorneys in Fact.

(xi) Subpart L, Transfer Through Judicial Proceedings.

(xii) Subpart M, Requests for Suspension of Transactions.

(xiii) Subpart N, Relief for Loss, Theft, Destruction, Mutilation, or Defacement of Securities.

[41 FR 51799, Nov. 24, 1976, as amended at 80 FR 9868, Feb. 24, 2015]

§ 1901.507 Certificates of beneficial ownership by the Rural Development Finance Office.

(a) *Orders and payment.* Orders for investment in certificates may be placed with the Finance Office by mail, telephone, or in person. Payment for purchase of certificates may be made by a wire transfer to the Federal Reserve Bank of St. Louis for credit to the Farmers Home Administration or its successor agency under Public Law 103–354, by a certified check or bank draft payable to the Farmers Home Administration or its successor agency under Public Law 103–354. The rate of interest paid on the certificate will be the rate in effect on the date the Finance Office receives the payment.

(b) *Registration.* (1) The registration used must express the actual ownership of a certificate and may not restrict the authority of the owner to dispose of it in any manner. Rural Development reserves the right to treat the registration as conclusive ownership. Request for registration must be clear, accurate, and complete, and include the appropriate taxpayer identifying number or social security number.

(2) The registration of all certificates owned by the same person, organization, or fiduciary should be uniform with respect to the name of the owner and, in case of fiduciary, the description of the fiduciary capacity. Individual owners should be designated by the names by which they are ordinarily known or under which they do business, preferably including at least one full given name. The name of an individual may be preceded by an applicable title, as, for example "Mrs.", "Mr.", "Miss", "Ms.", "Dr.", or "Rev.", or followed by a designation such as "M.D.", "D.D.", "Sr.", or "Jr.", Any other similar suffix should be included when ordinarily used or when necessary to distinguish the owner from another member of his family. The address should include, where appropriate, the name and street, route, or any other location feature, and zip code.

(3) If an erroneously inscribed certificate is received, it should not be altered in any respect. Rural Development should be given full particulars about the error and asked to furnish instructions.

(c) *Transfers and exchanges—closed periods—*(1) *General.* Transfer of registered certificates should be made by assignment in accordance with this section. Registered securities are eligible for denominational exchange. Specific instructions for issuance and delivery of new certificates signed by the owner or the owner's authorized representative must accompany the certificates presented. Certificates presented for transfer must be received by Rural Development not less than 1 full month before the date on which they mature. Any certificates so presented which are received too late to comply with this provision will be accepted for payment only.

(2) *Closing of transfer books.* The transfer books are closed for 1 full month preceding interest payment dates. If the date set for closing falls on Saturday, Sunday, or a legal holiday, the books will be closed as of the close of business on the last business day preceding that date. The books are reopened on the first business day following the date on which interest falls due. Registered certificates which have not matured, or have been submitted for transfer and are received when the books are closed for that certificate, will be processed on or after the date such books are reopened. If certificates are received for transfer when the books are closed for payment of final interest at maturity, the following action will be taken in the absence of different instructions:

(i) Payment of final interest will be made to the registered owner of record on the date the books were closed.

(ii) Payment of principal will be made to the assignee under a proper assignment of the certificate.

(d) *Redemption or payment*—(1) *General.* Certificates are payable in regular course of business at maturity. Rural Development may provide for the exchange of maturing certificates. The registered certificates should be presented and surrendered for redemption at the Rural Development Finance Office. No assignments or evidence in support of them will be required by or on behalf of the registered owner or assignee for redemption for his or its account, or for redemption-exchange if the new certificates are to be registered in exactly the same names and forms as in the registrations or assignments of the certificates surrendered.

(2) *Redemption at maturity.* Registered certificates presented and surrendered for redemption at maturity need not be assigned unless the owner desires that payment be made to some other person. Should the owner so desire assignments should be made to the " Rural Development for redemption for the account of (inserting name and address of person to whom payment is to be made)." Specific instructions for the issuance and delivery of the redemption check signed by the owner or the owner's authorized representative must accompany the certificates unless included in the assignment. Payment of the principal and interest will be made by a check drawn on the Treasurer of the United States to the order of the person entitled and mailed in accordance with the instructions received. If instructions are not received concerning interest, interest will be paid to the registered owner.

(3) *Interest.* The interest on Rural Development certificates accrues and is

payable annually. A full interest period does not include the day on which the last preceding interest became due, but does include the day on which the next succeeding interest payment is due. Certificates will cease to bear interest on the date of their maturity. The interest on registered certificates is payable by checks drawn on the Treasurer of the United States to the order of the registered owners, except as otherwise provided in this section. Rural Development prepares the interest checks in advance of the interest payment date and ordinarily mails them in time to reach the addressees on that date. Interest on a registered certificate which has not matured and which is presented for any transaction when the books for that certificate are closed will be paid by check drawn to the order of the registered owner of record. On receipt of notice of the death or incompetency of an individual named as registered owner, a change in the name or in the status of a partnership, corporation, or unincorporated association, the removal, resignation, succession, or death of a fiduciary or trustee, delivery of interest checks will be withheld pending receipt and approval of evidence showing who is entitled to receive the interest checks. If the inscriptions on certificates do not clearly identify the owners, delivery of interest checks may be withheld pending reissue of the certificates in the correct registration, except as provided in this section. The final installment of interest will be paid by check drawn to the order of the registered owner of record on presentation and surrender of the certificate for redemption. To assure timely delivery of interest checks, owners should promptly notify Rural Development of any change of address.

(e) *Assignments.* Assignments of certificates should be executed by the owner or the owner's authorized representative in the presence of an officer authorized to certify assignments. Registered certificates may be assigned to a specified transferee or to Rural Development for redemption or for exchange for other certificates offered at maturity. Assignments to "United States, Rural Development," "Farmers Home Administration for Transfer," or "Rural Development for Exchange"

will not be accepted unless supplemented by specific instructions by or in behalf of the owner. If an alteration or erasure has been made in an assignment, a new assignment from the assignor should be obtained. Otherwise, an affidavit or explanation by the person responsible for the alteration or erasure should be submitted for consideration.

(f) *Death of certificate holder.* The Finance Office should be notified of the death of the registered owner of a certificate. The following documents should be forwarded with the notice if available.

(1) A certified copy of the death certificate.

(2) A certified copy of the court order appointing the Administrator or Executor (include the mailing address of the Administrator or Executor). The Finance Office will notify the person submitting such notice and/or documentation if any other records or documents are needed. Legal opinions and advice will be obtained by the Finance Office as needed from the Regional Attorney. After all legal requirements are met, the certificate should be reissued in the name of the current owner.

(g) *Replacement.* Lost, stolen, destroyed, or mutilated certificates will be replaced by the Finance Office on the registered owner's compliance with the requirements of § 1901.509.

[41 FR 51799, Nov. 24, 1976, as amended at 80 FR 9869, Feb. 24, 2015]

§ 1901.508 Servicing of insured notes outstanding with investors.

The Director, or the insured loan officer of the Finance Office, is authorized in connection with the sale of any insured note to execute required documents on behalf of Rural Development and to take other appropriate action, including, but not limited to, acknowledging notice of sale of an insured note, or requiring an insured holder to sell an insured note to Rural Development in connection with any voluntary conveyance or foreclosure, or transfer related to liquidation of the borrower's account or any other servicing action so related. Upon recommendation by the State Director that purchase of an insured note is necessary for any servicing action not related to liquidation of the borrower's account, authorization may be given by the National Office to request the Director, Finance Office, to require a holder to sell an insured note to Rural Development.

(a) *Assignments*—(1) *Effective date of assignment.* When an insured note is sold by a private holder to a private buyer, notice of such sale executed by the seller must be given to and acknowledged by Rural Development in order for the sale to be binding on Rural Development, as to Rural Development, the effective date of the sale will be the acknowledgment date specified in the acknowledgement of notice executed by Rural Development.

(2) *Assignment to Rural Development at request of Rural Development.* At any time Rural Development considers it necessary for proper servicing of the loan, Rural Development may require, in writing, a private holder to sell an insured note to Rural Development.

(3) *Assignment to Rural Development at option of holder.* A private holder at any time during the option period may require, in writing, Rural Development to purchase an insured note.

(4) *Price.* If Rural Development is the buyer of an insured note, the price will be the par value as of the effective date of the sale. In other cases, the price will be determined by an agreement between the parties.

(b) *Sale of insured notes by private holders to private buyers.* (1) On receipt of notice from a private holder of intention to assign an insured note, the Director, Finance Office, will send the holder:

(i) Form RD 471–7 "Notice and Acknowledgment of Sale of Insured or Guaranteed Loan."

(ii) A statement of the unpaid principal. If requested the Director, Finance Office, will furnish a statement of account instead of or in addition to a statement of the unpaid principal.

(iii) Appropriate information on how to complete the assignment.

(2) If the Director, Finance Office, is informed that an insured note has been assigned and Rural Development is requested to recognize the assignment, the Director, Finance Office, will send the assignor Form RD 471–7, with directions for its execution.

(3) On receipt of Form RD 471–7 properly executed by the assignor, the Director, Finance Office, will complete and execute the acknowledgment section of the form. The Director, Finance Office, will retain the original of the form, have two facsimile copies made and send one to the assignor, and one to the assignee. For any correction or other change to be made in the record of the name or address of a private holder, or of a designated agent of a private holder, a request will be made to Rural Development in writing.

(4) As of the date of the acknowledgment, executed by the Director, Finance Office, on Form RD 471–7 the Director, Finance Office, will transfer the insured note from the assignor to the assignee as the insured holder on the records of Rural Development. The name and address of the assignee will be recorded by Rural Development exactly as they appear on Form RD 471–7.

(5) Payments transmitted by Rural Development on or after the acknowledgment date shown on Form RD 471–7 will be transmitted to the assignee. The Director, Finance Office, will give notice to the assignor and the assignee of any payments transmitted by Rural Development to the assignor before the acknowledgment date and after either the date of sale, or the date of the statement of account, whichever is earlier. However, Rural Development will not be liable for any failure to give such notice.

(c) *Assignment of insured notes to Rural Development*—(1) *Assignment at the request of the holder.* For assignment of an insured note to Rural Development during the option period at the request of the holder, the following procedure will apply:

(i) The holder will endorse the insured note as follows: "Pay to the order of the United States of America. Without recourse." The holder will then deliver the endorsed note, together with the insurance agreement, to the Director, Finance Office.

(ii) On receipt of the endorsed note with the accompanying insurance agreement, the Director, Finance Office, will acknowledge receipt of the note and process payment to the assignor of the par value of the note as of the date of the Treasury check.

(2) *Assignment at the request of Rural Development.* The procedure for assigning an insured note at the request of Rural Development will be the same as that prescribed in paragraph (c)(1) of this section, except that the Director, Finance Office, will send a written request to the holder requiring that the insured note be assigned to Rural Development and delivered to the Director, Finance Office, with the accompanying insurance agreement. The Director, Finance Office, will explain that the assignment is necessary to enable Rural Development to service the account properly and will give the holder all necessary information as to the manner of making the assignment and the amount to be paid by Rural Development.

(d) *Replacement of called or fully paid notes.* Certain insurance endorsements contain a clause or rider providing for a replacement note when the original note is paid in full, or is called by Rural Development. This provision applies to loans sold for a fixed period of 10 years or longer for loans sold on or after December 1, 1969, and a fixed period of 15 years or longer for loans sold before December 1, 1969. If a note is paid in full or called by the Government and the lender is entitled to a replacement note, the lender may obtain a certificate of beneficial ownership in lieu of the replacement note. The certificate will carry the rates and terms applicable to the replacement note.

(e) *Death of a noteholder.* The Finance Office should be notified of the death of a holder of an insured note. The following documents should be forwarded with the notice if available:

(1) A certified copy of the death certificate.

(2) A certified copy of the court order appointing the Administrator or Executor (include the mailing address of the Administrator or Executor). The Finance Office will notify the person submitting the notice and/or documentation if any other records or documents are needed, and will provide any additional instructions that are needed. Legal opinions and advice will be obtained by the Finance Office as needed from the Regional Attorney.

[41 FR 51799, Nov. 24, 1976, as amended at 80 FR 9869, Feb. 24, 2015]

§ 1901.509 Loss, theft, destruction, mutilation, or defacement of insured notes, insurance contracts, and certificates of beneficial ownership.

(a) *Block sale insurance contracts.* The Associate Administrator is authorized in connection with block sale insurance contracts to authorize the Rural Development's fiscal agent to establish requirements for issuance of a replacement insurance contract when the original issued by the Federal Reserve Bank of New York (Rural Development's fiscal agent) is lost, stolen, destroyed, mutilated, or defaced. When a block sale insurance contract is lost, stolen, or destroyed, a duplicate may be issued to the registered holder upon receipt of an acceptable certificate of loss and an indemnity bond without surety. The certificate of loss should include the legal name and present address of the owner and address when issued, if different from the present address; the capacity of person certifying, if other than owner; the identity of the insurance contract, including series number, contract number, denomination, issue date, and form of inscription of registry, and the full statement of circumstances of loss. All available portions of an insurance contract that is mutilated, defaced, or partially destroyed should be submitted to the Federal Reserve Bank of New York (Rural Development's fiscal agent) for determination as to whether a duplicate insurance contract can be issued without a certificate of loss and posting of an indemnity bond. In the event the holder of a block sales insurance contract obtains possession of the underlying notes, the requirements of paragraph (b) of this section apply.

(b) *Notes and certificates of beneficial ownership sold by County Office and Finance Office.* The Director, or the insured loan officer of the Finance Office, is authorized on behalf of the Government, in connection with insured notes or certificates of beneficial ownership sold through the Rural Development Finance Office to require indemnity bonds from a noteholder when a note or certificate is lost, stolen, destroyed, mutilated, or defaced while in the custody of the holder or his designee. When a note or certificate of beneficial ownership is lost, stolen, or destroyed while in the custody of the holder or his designee, the following will apply:

(1) A certificate of loss should be filed with Rural Development Finance Office. The certificate should include:

(i) Legal name and present address of owner when issued, if different from present address.

(ii) Capacity of person certifying, if other than the owner.

(iii) Identity of the note or certificate of beneficial ownership, including the name and Rural Development case number of the maker thereof, issue date, interest rate of obligation, face amount of note or certificate of beneficial ownership, and a full description of any assignment, endorsement, or any other writing.

(iv) A full statement of circumstances of the loss, theft, or destruction of the note.

(2) An indemnity bond in the amount of the unpaid principal and interest will be required except in the following instances:

(i) Substantially the entire note or certificate of beneficial ownership is presented and surrendered by the owner or holder, and the Director, Finance Office, is satisfied as to the identity of the instruments and that any missing portions are not sufficient to form the basis of a valid claim against the United States or the borrower; or

(ii) The owner or holder is the United States, a Federal Reserve Bank, a Federal Government Corporation, a State or territory, or the District of Columbia.

(3) An indemnity bond without surety will be provided in the following cases:

(i) Cases involving registered unassigned obligations held by banks, trust companies, savings and loan associations, or companies holding certificates of authority from Secretary of the Treasury as acceptable sureties on Federal Bonds (companies listed on Treasury Department Circular 570) where the financial responsibilities of such claimants are well known or readily ascertainable.

(ii) Cases involving registered unassigned obligations where the evidence reasonably justifies a conclusion that the obligations were destroyed and the unpaid principal and interest amount does not exceed $1,000.

(4) An indemnity posted with a qualified surety is required in all cases involving registered unassigned obligations other than those cited in paragraphs (b)(2)(i), (b)(2)(ii), (b)(3)(i) and (b)(3)(ii) of this section. A qualified surety is a company holding a certificate of authority from the Secretary of the Treasury as acceptable sureties on Federal Bonds, and listed in Treasury Department Circular 570.

(5) All indemnity bonds for notes must be payable to both the borrower and Rural Development. All indemnity bonds for certificates of beneficial ownership must be payable to Rural Development. The bond may be posted at the time the note or certificate of beneficial ownership becomes eligible for repurchase by Rural Development. If the holder desires to continue to hold the note for the life of the note, an indemnity bond will not be required.

(6) An assignment of the note or certificate of beneficial ownership shall be made to the United States of America, acting through Rural Development, United States Department of Agriculture. An acceptable form of assignment is available from the Director, Finance Office.

(c) *Other cases.* Cases involving bearer obligations and other cases not discussed in this section will be forwarded to the Director, Finance Office, for requirements.

(d) *Replacement of notes.* Rural Development will not attempt to obtain replacement notes from borrowers.

[41 FR 51799, Nov. 24, 1976, as amended at 80 FR 9869, Feb. 24, 2015]

Subparts L–N [Reserved]

PART 1902—SUPERVISED BANK ACCOUNTS

Subpart A—Supervised Bank Accounts of Loan, Grant, and Other Funds

Sec.
1902.1 General.

Subparts B–C [Reserved]

AUTHORITY: 5 U.S.C. 301; 7 U.S.C. 1989; 7 U.S.C. 6991, *et seq.;* 42 U.S.C. 1480; Reorganization Plan No. 2 of 1953 (5 U.S.C. App.).

Subpart A—Supervised Bank Accounts of Loan, Grant, and Other Funds

SOURCE: 46 FR 36106, July 14, 1981, unless otherwise noted.

§ 1902.1 General.

This subpart prescribes the policies and procedures in establishing and using supervised bank accounts, and in placing Multi-Family Housing (MFH) reserve accounts in supervised bank accounts. 7 CFR part 2018, subpart D, provides the procedures Servicing Officials should follow in ordering loan and grant disbursements. This subpart is inapplicable to Farm Service Agency, Farm Loan Programs.

(a) Borrowers referred to in this subpart include both loan and grant recipients. They are referred to as "depositors" in the deposit agreements hereinafter described. References herein and in deposit agreements to "other lenders" include lenders and grantors other than Rural Development.

(b) Banks and savings associations referred to in this subpart are those in which deposits are insured by the FDIC.

(c) Credit unions referred to in this subpart are those in which deposits are insured by the NCUA.

(d) Financial institutions as referred to in this subpart include banks, savings associations, and credit unions which are covered by the proper insurance coverage cited in paragraphs (b) and (c) of this section.

(e) Supervised bank accounts referred to in this subpart are bank, savings association, or credit union accounts established through deposit agreements entered into between the borrower, the United States of America acting through Rural Development, and the Financial Institution on Form RD 402–1, "Deposit Agreement".

(f) Form RD 402–1 provides for the deposit of funds in a supervised bank account to ensure the performance of the borrower's obligation to Rural Development in connection with a loan and/or grant.

(g) "Interest-Bearing Deposit Agreement" (Exhibit B of this subpart), provides for the deposit of loan or grant funds that are not required for immediate disbursement in specified interest-bearing deposits, and it is executed in conjunction with Form RD 402–1.

(h) Servicing officials referred to in this instruction include county supervisors, district directors, local supervisors, area supervisors, and National Office grant program managers.

(i) Automated systems referred to in this instruction refers to the loan accounting systems; e.g., Program Loan Accounting System, Automated Multi-Housing Accounting System, and Dedicated Loan Origination System, from which loan and grant disbursements are ordered.

(j) This subpart includes the National Office directly servicing a grant recipient or recipient of cooperative agreement funds.

[70 FR 59225, Oct. 12, 2005, as amended at 80 FR 9869, Feb. 24, 2015]

§ 1902.2 Policies concerning disbursement of funds.

(a) Generally, loan and grant disbursements may be requested on an as needed basis, thereby reducing the need for supervised bank accounts. For all construction loans and those loans using multiple advances, only the actual amount to be disbursed at loan closing will be requested through the automated systems. Subsequent disbursements will be ordered as needed. However, supervised bank accounts may be used in certain circumstances. For example:

(1) When a construction loan is made and the construction is substantially completed, but a small amount is being withheld pending completion of landscaping or some similar item. In this case, funds not disbursed may be placed in a supervised bank account for future disbursement as appropriate.

(2) When a large number of checks will be issued in the construction of a dwelling or other development. In such cases, loan and grant disbursements will be requested in accordance with 7 CFR part 2018, subpart D as necessary, deposited in a supervised bank account, and disbursed as necessary to suppliers, sub-contractors, etc.

(3) Association loan and grant funds made on a multiple advance basis may be deposited in a supervised bank account when required by State statutes or when determined necessary by the loan approval official.

(4) Supervised bank accounts may be used when needed as defined in paragraph (a)(5) of this section to ensure the correct expenditures of all or a part of loan and grant funds, borrower contributions, and borrower income. Such accounts will be limited in amount and duration to the extent feasible through the prudent disbursement of funds and the prompt termination of the interests of Rural Development and other lenders when the accounts are no longer required.

(5) When it is determined by the Servicing Official that special supervision is needed in the management of the borrower's finances, funds may be deposited in a supervised bank account. This supervisory technique will be used for a temporary period to help the borrower learn to properly manage his/her finances. Such a period will not exceed 1 year unless extended by the Servicing Official.

(b) Program instructions provide information as to the type of note to be utilized and the method of handling advances and the interest accrued.

(c) The debt instruments executed at the time of loan closing constitute an obligation on the part of the Government to disburse all funds at one time or in multiple advances, provided the funds are for purposes authorized by the Government at the time of loan closing. This obligatory commitment takes priority over any intervening liens or advances by other creditors, regardless of the provisions of the State laws involved.

[70 FR 59225, Oct. 12, 2005]

§ 1902.3 Procedures to follow in fund disbursement.

(a) The Servicing Official will determine during loan approval the amount(s) of loan or grant disbursement(s)—full or partial—and will process the request to the appropriate automated system in accordance with 7 CFR part 2018, subpart D.

(b) When Treasury check(s) are delivered to the Servicing Official, the Servicing Official will make sure that the name of the borrower and the amount(s) of check(s) coincide with the request on file. The Servicing Official should be sure that the check is properly endorsed to ensure payment to the intended recipient. Examples of such restrictive endorsements are:

(1) "For Deposit only to Account No. (Number of Construction Account) of (Name of Borrower) in (Name of Financial Institution)."

(2) "Pay to the order of (3rd party payee)"—(Contractor, Developer, Sub-Contractor, Building Supply House, etc.) for the purpose of _____.

(c) When necessary, and only under the circumstances listed in § 1902.2, the Servicing Official will establish, or cause to be established, a supervised bank account. Funds deposited in a supervised bank account are to be recorded and accounted for on Form RD

402–2, "Statement of Deposits and Withdrawals".

[46 FR 36106, July 14, 1981, as amended at 53 FR 26588, July 14, 1988; 53 FR 35670, Sept. 14, 1988; 54 FR 39727, Sept. 28, 1989; 70 FR 59226, Oct. 12, 2005]

§ 1902.4 Establishing MFH reserve accounts in a supervised bank account.

(a) *General requirements.* All MFH borrowers required to maintain reserve accounts must place the reserve accounts in a supervised bank account(s) which meets the following requirements:

(1) *Countersignature requirements.* The reserve account must require that any funds withdrawn be countersigned by an authorized government official.

(2) *Restrictions on collateral.* The financial institution holding the reserve account must ensure that the funds are not pledged or taken as security without the Agency's prior consent.

(3) *Interest bearing.* The reserve account funds are encouraged to be maintained in an interest-bearing account. The "Interest-Bearing Deposit Agreement" set out in Exhibit B of this subpart is not required to be used for reserve accounts.

(4) *Restricted investments.* Reserve funds must be placed in investments authorized in 7 CFR part 3560, subpart G. The authorized investments are deemed to be of acceptable risk such that the potential for any loss is minimal.

(5) *Financial institutions.* The reserve account must be maintained in authorized financial institutions set out in 7 CFR part 3560, subpart G; e.g., banks, savings associations, credit unions, brokerage firms, mutual funds. Generally, any financial institution may be used provided invested or deposited funds are insured to protect against theft and dishonesty. The reserve account funds need not be Federally insured, but must be otherwise covered by non-Federal insurance against theft and dishonesty.

(6) *Rules where multiple projects are involved.* A reserve account(s) must be maintained for each borrower. When a borrower owns multiple projects, reserve accounts may be established for each project. A single reserve account may also be established by a borrower owning multiple projects, provided the conditions set out in 7 CFR part 3560, subpart G are met.

(7) *Term.* Reserve accounts are expected to be kept for the full term of the loan.

(b) *Deposits and account activity statements*—(1) *Deposits.* Generally, Rural Development will not require the review or approval of deposits or the use of Form RD 402–1 or 402–2.

(2) *Account activity statements.* Generally, Rural Development will not monitor or reconcile the reserve account activity statements issued periodically by the financial institutions holding the funds. Rural Development will monitor reserve account levels through budget reports, audits, and Agency reserve tracking systems. If disputes arise or the borrower is in violation of Agency regulations, the Agency may require account activity statements. When account activity statements are sought, it will normally be sufficient to obtain the statement which reflects balances as of the last activity statement ending period. Form RD 402–2 is not required to be used.

[59 FR 3778, Jan. 27, 1994, as amended at 69 FR 69104, Nov. 26, 2004; 70 FR 59226, Oct. 12, 2005; 70 FR 73347, Dec. 12, 2005; 80 FR 9869, Feb. 24, 2015]

§ 1902.5 [Reserved]

§ 1902.6 Establishing supervised bank accounts.

(a) Each borrower will be given an opportunity to choose the financial institution in which the supervised bank account will be established, provided the financial institution is a member of the FDIC or NCUA, as applicable.

(b) When accounts are established, it should be determined that:

(1) The financial institution is fully informed concerning the provisions of the applicable deposit agreement,

(2) Agreements are reached with respect to the services to be provided by the financial institution including the frequency and method of transmittal of checking account statements, and

(3) An agreement is reached with the financial institution regarding the place where the counter-signature will be on the checks.

(c) When possible, Servicing Officials will make arrangements with financial institutions to waive service charges in connection with supervised bank accounts. However, there is no objection to the payment by the borrower of a reasonable charge for such service.

(d) For each borrower, if the amount of any loan and grant funds, plus any borrower contributions and funds from other sources to be deposited in the supervised bank account will exceed the maximum amount insurable by the Federal government, the financial institution will be required to pledge collateral for the excess over that limit before the deposit is made (see § 1902.7 of this subpart). If the supervised bank account is a joint account, any amount over the maximum amount insurable by the federal government must be collateralized.

(e) Only one supervised bank account will be established for any borrower regardless of the amount or source of funds, except for *Rural Rental Housing* loans where separate accounts will be established for each project.

(f) When a supervised bank account is established, an original and two copies of the applicable Deposit Agreement and the Interest-Bearing Deposit Agreement (Exhibit B of this subpart), when applicable, will be executed by the borrower, the financial institution, and a Servicing Office employee. The original will be retained in the borrower's case file, one executed copy will be delivered to the financial institution and one executed copy to the borrower. An extra copy of the Interest-Bearing Deposit Agreement, when applicable, will be prepared and attached to the certificate, passbook, or other evidence of deposit representing the interest-bearing deposit.

[46 FR 36106, July 14, 1981, as amended at 53 FR 231, Jan. 6, 1988; 70 FR 59227, Oct. 12, 2005; 77 FR 41258, July 13, 2012]

§ 1902.7 Pledging collateral for deposit of funds in supervised bank accounts.

(a) Funds in excess of the maximum amount insurable by the Federal government, per financial institution, deposited for borrowers in supervised bank accounts, must be secured by pledging acceptable collateral with the Federal Reserve Bank (FRB) in an amount not less than the excess. If the supervised bank account is a joint account, any amount over the maximum amount insurable by the federal government must be collateralized.

(b) As soon as it is determined that the loan will be approved and the applicant has selected or tentatively selected a financial institution for the supervised bank account, the Servicing Official will contact the financial institution to determine:

(1) That the financial institution selected is insured by the FDIC (banks and savings associations) or NCUA (credit unions).

(2) Whether the financial institution is willing to pledge collateral with the FRB under 31 CFR part 202 (Treasury Circular 176) to the extent necessary to secure the amount of funds being deposited in excess of the FDIC or NCUA insurance limit.

(3) If the financial institution is not a member of the Federal Reserve System, it will be necessary for the financial institution to pledge the securities with a correspondent bank who is a member of the System. The correspondent bank should contact the FRB informing them they are holding securities pledged for the supervised bank account under 31 CFR part 202 (Treasury Circular 176).

(c) If the financial institution agrees to pledge collateral, the Servicing Official should complete RD Form Letter 1902–A–2, "Designated Financial Institution—Collateral Pledge", in an original and two copies: The original for the National Office, Policy and Analysis Division; the first copy for the State Office; and the second copy for the Servicing Official. The Rural Development Form Letter 1902–A–2 should be forwarded to the National Office, Policy and Analysis Division, at least 30 days before the date of loan closing.

(d) The National Office, Policy and Analysis Division, will arrange for the financial institution under its designation as a depository and financial agent of the U.S. Government to pledge the requested collateral.

(e) If, two days before loan closing, the local Rural Development office which requested the collateral has not received notification from the National

Office, Policy and Analysis Division, that collateral has been pledged, contact should be made with the financial institution to ascertain whether they have pledged collateral with their local FRB under 31 CFR part 202 (Treasury Circular 176). If the financial institution has pledged collateral, the local Rural Development office should contact the National Office, Policy and Analysis Division, who will follow-up with the local FRB concerning the collateral.

(f) When the amount of deposit in the supervised bank account has been reduced to a point where the financial institution desires part or all of the collateral released, it should contact the National Office, Policy and Analysis Division. The local Rural Development office will be contacted for release authorization. The authorization release will be made through the local FRB, with notification to the financial institution. The local Rural Development office may also request release through the National Office, Policy and Analysis Division.

[46 FR 36106, July 14, 1981, as amended at 53 FR 231, Jan. 6, 1988; 53 FR 24437, June 29, 1988; 56 FR 50648, Oct. 8, 1991; 70 FR 59227, Oct. 12, 2005; 77 FR 41258, July 13, 2012]

§ 1902.8 Authority to establish and administer supervised bank accounts.

Servicing Officials are authorized to establish supervised bank accounts, deposit loan checks and other funds, countersign checks, close accounts, and execute all forms in connection with supervised bank account transactions and redelegate this authority to a person under their supervision who is considered capable of exercising such authority. State Directors will make written demand upon the bank for withdrawals outlined in § 1902.16.

[70 FR 59227, Oct. 12, 2005]

§ 1902.9 Deposits.

(a) *Deposit by Rural Development personnel.* (1) Checks made payable solely to the Federal Government or any Agency thereof, and a joint check when the Treasurer of the United States is a joint payee, may not be deposited in a supervised bank account.

(2) Rural Development personnel will accept funds for deposit in a borrower's supervised bank account ONLY in the form of: A check or money order endorsed by the borrower "For Deposit Only;" a check drawn to the order of the financial institution in which the funds are to be deposited; a loan check drawn on the U.S. Treasury; or a Rural Development electronic funds transfer disbursement.

(i) A joint check that is payable to the borrower and Rural Development will be endorsed by the Servicing Official as provided in 7 CFR part 1951, subpart B, Exhibit B, section 4.

(ii) Ordinarily, when deposits are made from funds which are received as the result of consent or subordination agreements or assignments of income, the check should be drawn to the order of the financial institution in which the supervised bank account is established or jointly to the order of the borrower and Rural Development. All such checks should be delivered or mailed to the Servicing Office.

(3) If direct or insured loan funds or borrower contributions are to be deposited in a supervised bank account, such funds will be deposited on the date of loan closing after it has been determined that the loan can be closed. However, if it is impossible to deposit the funds on the day the loan is closed due to reasons such as distance from the financial institution or banking hours, the funds will be deposited on the first banking day following the date of loan closing.

(4) Grant funds will be deposited when such funds are delivered.

(5) When funds from any source in the form of cash, check, or money order are deposited by Rural Development personnel in a supervised bank account, a deposit slip will be prepared in an original and two copies with distribution as follows: Original to the financial institution, one copy to the borrower, and one copy for the borrower's case folder. The name of the borrower, the sources of funds, "Subject to Rural Development Countersignature" and, if applicable, the account number, will be entered on each deposit slip.

(6) A loan or grant check drawn on the U.S. Treasury may be deposited in a supervised bank account without endorsement by the borrower when it will

facilitate delivery of the check and is acceptable to the financial institution. The borrower will be notified immediately of any deposit made and will be furnished a copy of the deposit slip. When a deposit of this nature is made, the following endorsement will be used:

For deposit only in the supervised bank account of (*name of borrower*) in the (*name of financial institution and address when necessary for identification*) pursuant to Deposit Agreement dated _____.

(7) Accounts established through the use of Interest-Bearing Deposit Agreement will be in the name of the depositor and the Government.

(b) *Deposits by borrowers.* Funds in the form of cash, check, or money order may be deposited in the supervised bank account by the borrower if authorized by Rural Development, provided the financial institution has agreed that when a deposit is made to the account by other than Rural Development personnel, the financial institution will promptly deliver or mail a copy of the deposit slip to the Rural Development Servicing Office.

(1) A loan or grant check drawn on the U.S. Treasury may be deposited in a supervised bank account by a borrower, provided the following endorsement is used and is inserted thereon prior to delivery to the borrower for signature:

For deposit only in my supervised bank account in the (*name of financial institution and address when necessary for identification*) pursuant to Deposit Agreement dated _____.

(2) Funds other than loan or grant funds may be deposited by the borrower in those exceptional instances where an agreement is reached between the Servicing Official and the borrower, whereby the borrower will make deposits of income from any source directly into the supervised bank account. In such instances the borrower will be instructed to prepare the deposit slip in the manner described in §1902.9 (a)(5) of this subpart.

[46 FR 36106, July 14, 1981, as amended at 70 FR 59227, Oct. 12, 2005; 80 FR 9869, Feb. 24, 2015]

§1902.10 Withdrawals.

(a) The Servicing Official will not countersign checks on the supervised bank account for the use of funds unless the funds deposited by the borrower from other sources were cash deposits, checks which the Servicing Official knows to be good, or deposited checks which have cleared.

(b) Withdrawals of funds deposited under the applicable deposit agreement are permitted only by order of the borrower and countersignature of authorized Rural Development personnel, or upon written demand on the financial institution by the State Director.

(c) Upon withdrawal or maturity of interest-bearing accounts established through the use of an Interest-Bearing Deposit Agreement, such funds will be credited to the supervised bank account established through the use of Form RD 402-1.

(d) The issuance of checks on the supervised bank account will be kept to the minimum possible without defeating the purpose of such accounts. When major items of capital goods are being purchased, or a limited number of relatively costly items of operating expenses are being paid, or when debts are being refinanced, the checks will be drawn to the vendors or creditors. If minor capital items are being purchased or numerous items of operating and family living expenses are involved as in connection with a monthly budget, a check may be drawn to the borrower to provide the funds to meet such costs.

(1) A check will be issued payable to the appropriate payee but will never be issued to "cash." The purpose of the expenditure will be clearly shown on Form FmHA or its successor agency under Public Law 103–354 402–2 and indicated on the fact of the check. When checks are drawn in favor of the borrower to cover items too numerous to identify, the expenditure will be identified on the check, as "miscellaneous."

(2) Ordinarily, a check will be countersigned before it is delivered to the payee. However, in justifiable circumstances, such as when excessive travel on the part of the borrower or Servicing Official would be involved, and purchase would be prevented, and the borrower can be relied upon to select goods and services in accordance with the plans, a check may be delivered to

the payee by the borrower before being countersigned.

(i) When a check is to be delivered to the payee before being countersigned, the Servicing Official must make it clear to the borrower and to the payee, if possible, that the check will be countersigned only if the quantity and quality of items purchased are in accordance with approved plans.

(ii) Checks delivered to the payee before counter-signature will bear the following legend in addition to the legend for countersignature: Valid only upon countersignature of Rural Development.''

(iii) The check must be presented by the payee or a representative to the Rural Development Servicing Office for the required countersignature.

(iv) Such check must be accompanied by a bill of sale, invoice, or receipt signed by the borrower identifying the nature and cost of goods or services purchased, or similar information must be indicated on the check.

(3) For real estate loans or grants, whether the check is delivered to the payee before or after countersignature, the number and date of the check will be inserted on all bills of sale, invoices, receipts, and itemized statements for materials, equipment, and services.

(4) Bills of sale, invoices, receipts, or itemized statements may be returned to the borrower with the canceled check for the payment of the bill.

(5) Checks to be drawn on a supervised bank account will bear the legend:

• "Countersigned," not as co-maker or endorser.

(Title)

Rural Development

[46 FR 36106, July 14, 1981, as amended at 54 FR 47959, Nov. 20, 1989; 70 FR 59227, Oct. 12, 2005; 80 FR 9870, Feb. 24, 2015]

§ 1902.11 Servicing Office records.

A record of funds deposited in a supervised bank account will be maintained on Form RD 402–2 in accordance with the Forms Manual Insert. The record of funds provided for operating purposes by another creditor or grant-

or will be on a separate Form RD 402–2 so that they can be clearly identified.

[70 FR 59228, Oct. 12, 2005]

§§ 1902.12–1902.13 [Reserved]

§ 1902.14 Reconciliation of accounts.

(a) A checking account statement will be obtained periodically in accordance with established practices in the area. If the checking statement does not include sufficient information to reconcile the account (the name of the payee or the check number and the amount of each check; i.e., a negotiable demand draft drawn on a financial institution), the original cancelled check or either a copy or other reasonable facsimile of the cancelled check must be provided to the Servicing Office with the statement. Checking account statements will be reconciled promptly with Servicing Office records. The person making the reconciliation will initial the record and indicate the date of the action.

(b) All checking account statements and, if necessary, original cancelled checks or either a copy or other reasonable facsimile of the cancelled checks will be forwarded immediately to the borrower when bank statements and Servicing Office records are in agreement. If a transmittal is used, Form RD 140–4, "Transmittal of Documents", is prescribed for that purpose.

(c) If the financial institution did not return the original cancelled check(s) to the Agency with the statements, and Rural Development has a need for the original cancelled check(s), the financial institution, upon request by the Agency, will furnish to the Agency the requested original cancelled check(s) or a certified copy or other reasonable certified facsimile of the cancelled check(s) and will provide this service to Rural Development with no fees being assessed the Agency or the Depositor's account for the service.

[70 FR 59228, Oct. 12, 2005]

§ 1902.15 Closing accounts.

When Rural Development loan or grant funds and those of any other lender or grantor have all been properly expended or withdrawn, Form RD

402–6 may be used to give Rural Development's consent (and of another lender or grantor, if involved) to close the supervised bank account in the following situations:

(a) When Rural Development loan funds in the supervised bank account of a borrower have been reduced to $100 or less, and a check for the unexpended balance has been issued to the borrower to be used for authorized purposes.

(b) For all loan accounts, after completion of authorized loan funds expenditures, and after promptly refunding any remaining unexpended loan funds on the borrower's loan account with Rural Development or another lender, as appropriate.

(c) Promptly upon death of a borrower, except when the loan is being continued with a joint debtor, when a borrower is in default and it is determined that no further assistance will be given, or when a borrower is no longer classified as "active."

(1) *Deceased borrowers.* (i) Ordinarily, upon notice of the death of a borrower, the District Director or the County Supervisor will request the State Director to make demand upon the bank for the balance on deposit and apply all the balance after payment of any bank charges to the borrower's Rural Development indebtedness. When the State Director approves continuation with a survivor, the supervised bank account of deceased borrower may be continued with a remaining joint debtor who is liable for the loan and agrees to use the unexpended funds as planned, provided:

(i) Ordinarily, upon notice of the death of a borrower, the Servicing Official will request the State Director to make demand upon the bank for the balance on deposit and apply all the balance after payment of any bank charges to the borrower's Rural Development indebtedness. When the State Director approves continuation with a survivor, the supervised bank account of a deceased borrower may be continued with a remaining joint debtor who is liable for the loan and agrees to use the unexpended funds as planned, provided:

(A) The account is a joint survivorship supervised bank account, or

(B) If not a joint survivorship account, the financial institution will agree to permit the addition of the surviving joint debtor's name to the existing signature card and the appropriate Deposit Agreement and continue to disburse checks out of the existing account upon Rural Development's countersignature and the joint debtor's signature in place of the deceased borrower, or

(C) The financial institution will permit the State Director to withdraw the balance from the existing supervised bank account with a check jointly payable to the Rural Development and the surviving joint debtor and deposit the money in a new supervised bank account with a surviving joint debtor, and will disburse checks from this new account upon the signature of such survivor and the countersignature of an authorized Rural Development official.

(ii) The State Director, before applying the balance remaining in the supervised bank account to the Rural Development indebtedness, is authorized upon approval by the Office of the General Counsel (OGC) to refund any unobligated balances of funds from other lenders to the Rural Development borrower for specific operating purposes in accordance with subordination agreements or other arrangements between Rural Development, the lender and the borrower.

(iii) The State Director, upon the recommendation of an authorized representative of the estate of the deceased borrower and the approval of the OGC, is authorized to approve the use of deposited funds for the payment of commitments for goods delivered or services performed in accordance with the deceased borrower's plans approved by Rural Development.

(2) *Borrowers in default.* Whenever it is impossible or impractical to obtain a signed check from a borrower whose supervised bank account is to be closed, the Servicing Official will request the State Director to make demand upon the financial institution for the balance on deposit in the borrower's supervised bank account for application as appropriate:

(i) To the borrower's Rural Development indebtedness, or

(ii) As refunds of any unobligated advance provided by other lenders which were deposited in the account, or

(iii) For the return of Rural Development grant funds in accordance with 7 CFR part 1951, subpart B or

(iv) For the return of grant funds to other grantors.

(3) *Inactive borrowers.* An inactive borrower is one whose loan has not been paid in full, but is no longer classified as "active."

(4) *Paid up borrowers.* A paid-up borrower is one who has a balance remaining in the supervised bank account and has repaid the entire indebtedness to Rural Development and has properly expended all funds advanced by other lenders. In such cases the Servicing Official will:

(i) Notify the borrower in writing that the interests in the account of Rural Development have been terminated, and

(ii) Inform the borrower of the balance remaining in the supervised bank account.

[46 FR 36106, July 14, 1981, as amended at 53 FR 231, Jan. 6, 1988; 54 FR 47196, Nov. 13, 1989; 66 FR 1569, Jan. 9, 2001; 70 FR 59228, Oct. 12, 2005; 80 FR 9870, Feb. 24, 2015]

§ 1902.16 Request for withdrawals by State Director.

When the State Director is requested to make written demand upon the financial institution for the balance on deposit in the supervised bank account, or any part thereof, the request will be accompanied by the following information.

(a) Name of borrower as it appears on the applicable Deposit Agreement.

(b) Name and location of financial institution.

(c) Amount to be withdrawn for refund to another lender of any balance that may remain of funds received by the borrower from such lender as a loan or grant, or under a subordination agreement or other arrangement between the Rural Development, the other lender, and the borrower.

(d) Amount to be withdrawn, excluding any service charges, for a refund due to Rural Development.

(e) Other pertinent information including reasons for the withdrawal.

[46 FR 36106, July 14, 1981, as amended at 80 FR 9870, Feb. 24, 2015]

§§ 1902.17–1902.49 [Reserved]

§ 1902.50 OMB control number.

The reporting and recordkeeping requirements contained in this regulation have been approved by the OMB and have been assigned OMB Control Number 0575–0158.

[70 FR 59228, Oct. 12, 2005]

EXHIBIT A TO SUBPART A OF PART 1902 [RESERVED]

EXHIBIT B TO SUBPART A OF PART 1902— UNITED STATES DEPARTMENT OF AGRICULTURE, FARMERS HOME ADMINISTRATION OR ITS SUCCESSOR AGENCY UNDER PUBLIC LAW 103–354—INTEREST-BEARING DEPOSIT AGREEMENT

BECAUSE certain funds of _____ referred to as the "Depositor," are now on deposit with the _____, referred to as the "Financial Institution," under a Deposit Agreement, dated _____, 20___, providing for supervision by the United States of America, acting through the Farmers Home Administration or its successor agency under Public Law 103–354, referred to as the "Government," which Deposit Agreement grants to the Government security and/or other interest in the funds covered by that Deposit Agreement, and

BECAUSE certain of these funds are not now required for immediate disbursement and it is the desire of the Depositor to place these funds in interest-bearing deposits with the Financial Institution:

THEREFORE, the Depositor and the Government authorize and direct the Financial Institution to place _____ Dollars ($_____) of the funds subject to that Deposit Agreement in interest-bearing deposits as follows:

_____ for a period of _____ months at _____ % interest.

_____ for a period of _____ months at _____ % interest.

_____ for a period of _____ months at _____ % interest.

These interest-bearing deposits and the income earned on them at all times shall be considered a part of the account covered by said Deposit Agreement except that the right of the Depositor and the Government to jointly withdraw all or a portion of the funds in the account covered by the Deposit

Agreement by an order of the Depositor countersigned by a representative of the Government, and the right of the Government to make written demand for the balance or any portion of the balance, is modified by the above time deposit maturity schedule. The evidence of such time deposits shall be issued in the names of the Depositor and the Farmers Home Administration or its successor agency under Public Law 103–354.

A copy of this Agreement shall be attached to and become a part of each certificate, passbook, or other evidence of deposit that may be issued to represent such interest-bearing deposits.

Executed this _____ day of _____, 20___.

UNITED STATES OF AMERICA

By: _____

County Supervisor
Farmers Home Administration or its successor agency under Public Law 103–354
U.S. Department of Agriculture

_____

(Depositor)

By: _____

Title:

Accepted on the above terms and conditions this _____ day of _____, 20___.

_____

(Financial Institution)

_____

(Office or Branch)

By: _____

Title: _____

[53 FR 35671, Sept. 14, 1988, as amended at 55 FR 21524, May 25, 1990; 70 FR 59228, Oct. 12, 2005]

Subparts B–C [Reserved]

PART 1904—LOAN AND GRANT PROGRAMS (INDIVIDUAL) [RESERVED]

PART 1910—GENERAL

Subpart A [Reserved]

Subpart B—Credit Reports (Individual)

AUTHORITY: 5 U.S.C. 301; 7 U.S.C. 1989; 42 U.S.C. 1480.

SOURCE: 43 FR 56643, Dec. 4, 1978, unless otherwise noted.

Subpart A [Reserved]

Subpart B—Credit Reports (Individual)

SOURCE: 49 FR 40790, Oct. 18, 1984, unless otherwise noted.

§ 1910.51   Purpose.

This subpart prescribes the policies and procedures of Rural Development for individual and joint type credit reports. Credit reports will be ordered to determine the eligibility of applicants requesting Rural Development loans. A nonrefundable fee will be charged the applicant. This subpart is inapplicable to Farm Service Agency, Farm Loan Programs.

[80 FR 9870, Feb. 24, 2015]

§ 1910.52   [Reserved]

§ 1910.53   Policy.

The County Supervisor will be responsible for ordering individual credit reports. These will be obtained on initial and rescheduled Farmer Program loans and on all initial Single Family Housing applications, except for those situations outlined in paragraph (c) of this section, to help determine the eligibility of the loan applicant, and when it appears the credit report will not have to be updated before loan closing.

[55 FR 46188, Nov. 2, 1990]

§§ 1910.54–1910.100   [Reserved]

Subpart C—Commercial Credit Reports

SOURCE: 52 FR 6498, Mar. 4, 1987, unless otherwise noted.

§ 1910.101   Preface.

This subpart (§§ 1910.101 through 1910.150) describes the procedure to be used by Rural Development in obtaining commercial credit reports. A nonrefundable fee, set forth in § 1910.106(d) of this Instruction will be collected from the applicant, general contractor or dealer contractor who is the subject of the report. This subpart is inapplicable to Farm Service Agency, Farm Loan Programs.

[52 FR 6498, Mar. 4, 1987, as amended at 72 FR 64122, Nov. 15, 2007; 80 FR 9870, Feb. 24, 2015]

§§ 1910.102–1910.150 [Reserved]

PART 1922 [RESERVED]

PART 1924—CONSTRUCTION AND REPAIR

Subpart A—Planning and Performing Construction and Other Development

Subpart B [Reserved]

Subpart C—Planning and Performing Site Development Work

Subparts D–E [Reserved]

Subpart F—Complaints and Compensation for Construction Defects

1924.272 [Reserved]
1924.273 Approval or disapproval.
1924.274 Final inspection.
1924.275 [Reserved]
1924.276 Action against contractor.
1924.277–1924.299 [Reserved]
1924.300 OMB control number.

AUTHORITY: 5 U.S.C. 301; 7 U.S.C 1989; 42 U.S.C 1480.

EDITORIAL NOTE: Nomenclature changes to part 1924 appear at 80 FR 9870, Feb. 24, 2015.

Subpart A—Planning and Performing Construction and Other Development

SOURCE: 52 FR 8002, Mar. 13, 1987, unless otherwise noted.

§1924.1 Purpose.

This subpart prescribes the basic Agency policies, methods, and responsibilities in the planning and performing of construction and other development work for insured Rural Housing (RH), single unit Labor Housing (LH). It also provides supplemental requirements for Rural Rental Housing (RRH) loans, Rural Cooperative Housing (RCH) loans, multi-unit (LH) loans and grants, and Rural Housing Site (RHS) loans. This subpart is inapplicable to Farm Service Agency, Farm Loan Programs.

[53 FR 35679, Sept. 14, 1988, as amended at 72 FR 64122, Nov. 15, 2007]

§1924.2 [Reserved]

§1924.3 Authorities and responsibilities.

The County Supervisor and District Director are authorized to redelegate, in writing, any authority delegated to them in this subpart to the Assistant County Supervisor and Assistant District Director, respectively, when determined to be qualified. Agency Construction Inspectors, District Loan Assistants, and County Office Assistants are authorized to perform duties under this subpart as authorized in their job descriptions.

§1924.4 Definitions.

(a) *Construction.* Such work as erecting, repairing, remodeling, relocating, adding to or salvaging any building or structure, and the installation or repair of, or addition to, heating and electrical systems, water systems, sewage disposal systems, walks, steps, driveways, and landscaping.

(b) *Contract documents.* The borrower-contractor agreement, the conditions of the contract (general, supplementary, and other), the drawings, specifications, warranty information, all addenda issued before executing the contract, all approved modifications thereto, and any other items stipulated as being included in the contract documents.

(c) *Contractor.* The individual or organization with whom the borrower enters into a contract for construction or land development, or both.

(d) *County Supervisor and District Director.* In Alaska, for the purpose of this subpart, "County Supervisor" and "District Director" also mean "Assistant Area Loan Specialist" and "Area Loan Specialist," respectively. The terms also include other qualified staff who may be delegated responsibilities under this subpart in accordance with the provisions of subpart F of part 2006 (available in any Agency office).

(e) *Date of commencement of work.* The date established in a "Notice to Proceed" or, in the absence of such notice, the date of the contract or other date as may be established in it or by the parties to it.

(f) *Date of substantial completion.* The date certified by the Project Architect/Engineer or County Supervisor when it is possible, in accordance with any contract documents and applicable State or local codes and ordinances, and the Agency approved drawings and specifications, to permit safe and convenient occupancy and/or use of the buildings or other development.

(g) *Development.* Construction and land development.

(h) *Development standards.* Any of the following codes and standards:

(1) A standard adopted by the Agency for each state in accordance with §1924.5(d)(1)(i)(E) of this subpart.

(2) *Voluntary national model building codes (model codes).* Comprehensive documents created, referenced or published by nationally recognized associations of building officials that regulate the construction, alteration and

repair of building, plumbing, mechanical and electrical systems. These codes are listed in exhibit E of this subpart.

(3) *Minimum Property Standards (MPS).* The Department of Housing and Urban Development (HUD) Minimum Property Standards for Housing, Handbook 4910.1, 1984 Edition with Changes. (For One and Two Family Dwellings and Multi-Family Housing).

(i) *Identity of interest.* Identity of interest will be construed as existing between the applicant (the party of the first part) and general contractors, architects, engineers, attorneys, subcontractors, material suppliers, or equipment lessors (parties of the second part) under any of the following conditions:

(1) When there is any financial interest of the party of the first part in the party of the second part. The providing of normal professional services by architects, engineers, attorneys or accountants with a client-professional relationship shall not constitute an identity of interest.

(2) When one or more of the officers, directors, stockholders or partners of the party of the first part is also an officer, director, stockholder, or partner of the party of the second part.

(3) When any officer, director, stockholder or partner of the party of the first part has any financial interest whatsoever in the party of the second part.

(4) Between the spouse, significant other, relatives, and step-relatives of the principal owners of the party of the first part and its management, such as Grandmother, Aunt, Daughter, Granddaughter, Grandfather, Uncle, Son, Grandson, Mother, Sister, Niece, Cousin, Father, Brother, Nephew;

(5) When the party of the second part advances any funds to the party of the first part.

(6) When the party of the second part provides and pays on behalf of the party of the first part the cost of any legal services, architectural services or engineering services other than those of a surveyor, general superintendent, or engineer employed by a general contractor in connection with obligations under the construction contract.

(7) When the party of the second part takes stock or any interest in the party of the first part as part of the consideration to be paid them.

(8) When there exist or come into being any side deals, agreements, contracts or undertakings entered into thereby altering, amending, or cancelling any of the required closing documents except as approved by the Agency.

(9) An identity of interest will also exist when another party can significantly influence the management or operating policies of the transacting parties or if it has an ownership interest in one of the transacting parties and can significantly influence the other to an extent that one or more of the transacting parties might be prevented from fully pursuing its own separate interests.

(j) *Land development.* Includes items such as terracing, clearing, leveling, fencing, drainage and irrigation systems, ponds, forestation, permanent pastures, perennial hay crops, basic soil amendments, pollution abatement and control measures, and other items of land improvement which conserve or permanently enhance productivity. Also, land development for structures includes the applicable items above, and items such as rough and finish grading, retaining walls, water supply and waste disposal facilities, streets, curbs and gutters, sidewalks, entrancewalks, driveways, parking areas, landscaping and other related structures.

(k) *Manufactured housing.* Housing, constructed of one or more factory-built sections, which includes the plumbing, heating and electrical systems contained therein, which is built to comply with the Federal Manufactured Home Construction and Safety Standards (FMHCSS), and which is designed to be used with or without a permanent foundation. Specific requirements for manufactured homes sites, rental projects and subdivisions are in exhibit J of this subpart.

(1) *Mechanic's and materialmen's liens.* A lien on real property in favor of persons supplying labor and/or materials for the construction for the value of labor and/or materials supplied by

them. In some jurisdictions, a mechanic's lien also exists for the value of professional services.

(m) *Modular/panelized housing.* Housing, constructed of one or more factory-built sections, which, when completed, meets or exceeds the requirements of one or more of the recognized development standards for site-built housing, and which is designed to be permanently connected to a site-built foundation.

(n) *Project representative.* The architect's or owner's representative at the construction site who assists in the administration of the construction contract. When required by the Agency, a full-time project representative shall be employed.

(o) *Technical services.* Applicants are responsible for obtaining the services necessary to plan projects including analysis of project design requirements, creation and development of the project design, preparation of drawings, specifications and bidding requirements, and general administration of the construction contract.

(1) Architectural services. The services of a professionally qualified person or organization, duly licensed and qualified in accordance with state law to perform architectural services.

(2) Engineering services. The services of a professionally qualified person or organization, duly licensed and qualified in accordance with State law to perform engineering services.

(p) *Warranty.* A legally enforceable assurance provided by the builder (warrantor) to the owner and the Agency indicating that the work done and materials supplied conform to those specified in the contract documents and applicable regulations. For the period of the warranty, the warrantor agrees to repair defective workmanship and repair or replace any defective materials at the expense of the warrantor.

[52 FR 8002, Mar. 13, 1987, as amended at 59 FR 6882, Feb. 14, 1994]

§1924.5 Planning development work.

(a) *Extent of development.* For an FO loan, the plans for development will include the items necessary to put the farm in a livable and operable condition consistent with the planned farm and home operations. For other types of loans, the plans will include those items essential to achieve the objectives of the loan or grant as specified in the applicable regulation.

(b) *Funds for development work.* The total cash cost of all planned development will be shown on Form RD 1924–1, "Development Plan," except Form RD 1924–1 may be omitted when: (1) All development is to be done by the contract method, (2) adequate cost estimates are included in the docket, and (3) the work, including all landscaping, repairs, and site development work, is completely described on the drawings, in the specifications, or in the contract documents. Sufficient funds to pay for the total cash cost of all planned development must be provided at or before loan closing. Funds to be provided may include loan proceeds, any cash to be furnished by the borrower, proceeds from cost sharing programs such as Agricultural Stabilization and Conservation Service (ASCS) and Great Plains programs or proceeds from the sale of property in accordance with paragraph (g) of this section.

(c) *Scheduling of development work.* (1) All construction work included in the development plan for RH loans will be scheduled for completion as quickly as practicable and no later than 9 months from the date of loan closing, except for mutual self-help housing where work may be scheduled for completion within a period of 15 months.

(2) Development for farm program loans will be scheduled for completion as quickly as practicable and no later than 15 months from the date of loan closing unless more time is needed to establish land development practices in the area.

(d) *Construction.* (1) All new buildings to be constructed and all alterations and repairs to buildings will be planned to conform with good construction practices. The Agency Manual of Acceptable Practices (MAP) Vol. 4930.1 (available in any Agency office), provides suggestions and illustrative clarifications of design and construction methods which are generally satisfactory in most areas. All improvements to the property will conform to applicable laws, ordinances, codes, and regulations related to the safety and sanitation of buildings; standards referenced in Appendices C through F of HUD Handbook 4910.1, Minimum Property Standards for Housing; Thermal Performance Construction Standards contained in exhibit D of this subpart and, when required, to certain other development standards described below.

(i) The development standard applicable to a proposal will be selected by the loan applicant or recipient of an RH Conditional Commitment in accordance with the following. The standard selected must:

(A) Relate to the type(s) of building proposed.

(B) Meet or exceed any applicable local or state laws, ordinances, codes and regulations.

(C) Include all referenced codes and standards.

(D) Exclude inapplicable administrative requirements.

(E) Be the current edition(s) of either paragraph (d)(1)(i)(E)(*1*) or (*2*) of this section:

(*1*) The development standard, consisting of building, plumbing, mechanical and electrical codes, adopted by the Agency for use in the state (identified in a State Supplement to this section) in which the development is proposed, in accordance with the following:

(*i*) The adopted development standard shall include any building, plumbing, mechanical or electrical code adopted by the State, if determined by the State Director to be based on one of the model codes listed in exhibit E to this subpart, or, if not available,

(*ii*) The adopted development standard shall include any building, plumbing, mechanical or electrical code adopted by the state, if determined by the Administrator to be acceptable, or, if not available,

(*iii*) The adopted development standard shall include the model building, plumbing, mechanical or electrical code listed in exhibit E to this subpart that is determined by the State Director to be most prevalent and appropriate for the state.

(*2*) Any of the model building, plumbing, mechanical and electrical codes listed in exhibit E to this subpart or the standards defined in § 1924.4(h)(3) of this subpart.

(ii) Guide 2, " Rural Development Design Guide," of this subpart (available in any Rural Development office), includes guidelines for the evaluation of the design features which are not fully addressed in the development standards.

(iii) In new housing, all design, materials and construction will meet or exceed the applicable development standard as provided in paragraph (d)(1)(i) of this section.

(iv) For multi-family residential rehabilitation, as defined in exhibit K of this subpart, all substantial rehabilitation work on existing buildings will meet or exceed the applicable development standard. All moderate rehabilitation work should comply with Guide 3, "Quality and Performance Criteria for Moderate Rehabilitation," of this subpart (available in any Agency office).

(v) The design and construction of housing repairs made with Agency loan or grant funds will, as near as possible, comply with the applicable development standard.

(vi) Farm LH design and construction will comply with the following:

(A) Family projects, where the length of occupancy will be:

(*1*) Year-round, will meet or exceed the applicable development standard.

(*2*) Less than 12 months, but more than 6 months, will be in substantial conformance with the applicable development standard and constructed to facilitate conversion to year-round occupancy standards.

(*3*) Six months or less, may be less than the applicable development standard but should be constructed in accordance with exhibit I of this subpart.

(B) Dormitory and other nonfamily type projects, where the length of occupancy will be:

(*1*) More than 6 months, will be in substantial conformance with the applicable development standard and will at least meet or exceed the requirements of the Department of Labor, Bureau of Employment Security (29 CFR 1910.140).

(2) Six months or less, will comply with §1924.5(d)(1)(vi)(A)(3).

(vii) Farm service buildings should be designed and constructed for adaptation to the local area. In designing and locating farm service buildings, consideration will be given to practices recommended by agriculture colleges, the Extension Service (ES), Soil Conservation Service (SCS) and other reliable sources.

(2) Drawings, specifications, and estimates will fully describe the work. Technical data, tests, or engineering evaluations may be required to support the design of the development. The "Guide for Drawings and Specifications," exhibit C of this subpart, describes the drawings and specifications that are to be included in the application for building construction, and subpart C of part 1924 of this chapter describes the drawings that should be included for development of building sites. The specific development standard being used, if required under paragraph (d)(1) of this section will be identified on all drawings and specifications.

(3) Materials acceptance shall be the same as described in paragraph X of exhibit B to this subpart.

(4) Except as provided in paragraphs (d)(4)(i) through (iii) of this section, new building construction and additions shall be designed and constructed in accordance with the earthquake (seismic) requirements of the applicable Agency's development standard (building code). The analysis and design of structural systems and components shall be in accordance with applicable requirements of an acceptable model building code.

(i) Agricultural buildings that are not intended for human habitation are exempt from these earthquake (seismic) requirements.

(ii) Single family conventional light wood frame dwellings of two stories or 35 feet in height maximum shall be designed and constructed in accordance with the 1992 Council of American Building Officials (CABO) One and Two Family Dwelling Code or the latest edition.

(iii) Single family housing of masonry design and townhouses of wood frame construction and additions fi-

nanced (either directly or through a guarantee) under title V of the Housing Act of 1949 are recommended to be designed and constructed in accordance with the earthquake (seismic) requirements of one of the building codes that provides an equivalent level of safety to that contained in the latest edition of the National Earthquake Hazard Reduction Program's (NEHRP) Recommended Provisions for the Development of Seismic Regulations for New Building (NEHRP Provisions).

(iv) Acknowledgment of compliance with the applicable seismic safety requirements for new construction will be contained in the certification of final plans and specification on the appropriate Agency Form.

(e) *Land development.* (1) In planning land development, consideration will be given to practices, including energy conservation measures, recommended by agricultural colleges, ES, SCS or other reliable sources. All land and water development will conform to applicable laws, ordinances, zoning and other applicable regulations including those related to soil and water conservation and pollution abatement. The County Supervisor or District Director also will encourage the applicant to use any cost-sharing and planning assistance that may be available through agricultural conservation programs.

(2) Site and subdivision planning and development must meet the requirements of subpart C of part 1924 of this chapter.

(3) Plans and descriptive material will fully describe the work.

(4) The site planning design, development, installation and set-up of manufactured home sites, rental projects and subdivisions shall meet the requirements of exhibit J of this subpart and subpart C of part 1924 of this chapter.

(i) Plans for land leveling, irrigation, or drainage should include a map of the area to be improved showing the existing conditions with respect to soil, topography, elevations, depth of topsoil, kind of subsoil, and natural drainage, together with the proposed land development.

(ii) When land development consists of, or includes, the conservation and

513

use of water for irrigation or domestic purposes, the information submitted to the County Supervisor will include a statement as to the source of the water supply, right to the use of the water, and the adequacy and quality of the supply.

(f) *Responsibilities for planning development.* Planning construction and land development and obtaining technical services in connection with drawings, specifications and cost estimates are the sole responsibility of the applicant, with such assistance from the County Supervisor or District Director (whichever is the appropriate loan processing and servicing officer for the type of loan involved), as may be necessary to be sure that the development is properly planned in order to protect the Agency's security.

(1) *Responsibility of the applicant.* (i) The applicant will arrange for obtaining any required technical services from qualified technicians, tradespeople, and recognized plan services, and the applicant will furnish the Agency sufficient information to describe fully the planned development and the manner in which it will be accomplished.

(ii) When items of construction or land development require drawings and specifications, they will be sufficiently complete to avoid any misunderstanding as to extent, kind, and quality of work to be performed. The applicant will provide the Agency with one copy of the drawings and specifications. Approval will be indicated by the applicant and acceptance for the purposes of the loan indicated by the County Supervisor or District Director on all sheets of the drawings and at the end of the specifications, and both instruments will be a part of the loan docket. After the loan is closed, the borrower will retain a conformed copy of the approved drawings and specifications, and provide another conformed copy to the contractor. Items not requiring drawings and specifications may be described in narrative form.

(iii) The Agency will accept final drawings and specifications and any modifications thereof only after the documents have been certified in writing as being in conformance with the applicable development standard if required under paragraph (d)(1) of this section. Certification is required for all Single Family Housing (SFH) thermal designs (plans, specifications, and calculations).

(A) Certifications may be accepted from individuals or organizations who are trained and experienced in the compliance, interpretation or enforcement of the applicable development standards for drawings and specifications. Plan certifiers may be any of the following:

(1) Licensed architects,

(2) Professional engineers,

(3) Plan reviewers certified by a national model code organization listed in exhibit E to this subpart,

(4) Local building officials authorized to review and approve building plans and specifications, or

(5) National codes organizations listed in exhibit E to this subpart.

(B) The license or authorization of the individual must be current at the time of the certification statement. A building permit (except as noted in paragraph (f)(1)(iii)(C)(2) of this section) or professional's stamp is not an acceptable substitute for the certification statement. However, a code compliance review conducted by one of the National recognized code organizations indicating no deficiencies or the noted deficiencies have been corrected is an acceptable substitute for the certification statement.

(C) For Single Family Housing (one to four family dwelling units) the Agency may also accept drawings and specifications that have been certified by:

(1) Registered Professional Building Designers certified by the American Institute of Building Design.

(2) A local community, if that community has adopted, by reference, one of the model building codes and has trained official(s) who review(s) plans as well as inspect(s) construction for compliance as a requisite for issuing a building permit. The building permit, issued by the community, may serve as evidence of acceptance. The State Director will determine eligible communities and publish, as a State supplement to this section, a list of those communities that qualify.

(3) A plan service that provides drawings and specifications that are certified by individuals or organizations as listed in paragraph (f)(1)(iii)(A) or (f)(1)(iii)(C) (1) and (2) of this section as meeting the appropriate state adopted development standard.

(4) Builders/Contractors who provide 10-year warranty plans for the specific Agency finance dwelling unit that meet the requirements of exhibit L of this subpart.

(5) Builders/Contractors that are approved by the United States Department of Housing and Urban Development (HUD) for self-certification.

(D) The modifications of certified drawings or specifications must be certified by the same individual or organization that certified the original drawings and specifications. If such individual or organization is not available, the entire set of modified drawings and specifications must be recertified.

(E) The certification of modifications for single family housing (SFH) construction may be waived if the builder or original author of the drawings and specifications provides a written statement that the modifications are not regulated by the applicable development standard. The County Supervisor may consult with the State Office Architect/Engineer as to acceptance of the statement and granting a waiver.

(F) All certifications of final drawings, specifications, and calculations shall be on Form RD 1924–25, "Plan Certification."

(2) *Responsibility of the County Supervisor or District Director.* In accordance with program regulations for loans and grants they are required to process, the County Supervisor or District Director, for the sole benefit of the Agency, will:

(i) Visit each farm or site on which the development is proposed. For an FO loan, the County Supervisor and the applicant will determine the items of development necessary to put the farm in a livable and operable condition at the outset. Prepare Form RD 1924–1, when applicable in accordance with the Forms Manual Insert (FMI) for the form, after a complete understanding has been reached between the applicant and the County Supervisor regarding the development to be accomplished, including the dates each item of development will be started and completed.

(ii) Notify the loan or grant applicant in writing immediately if, after reviewing the preliminary proposal and inspecting the site, the proposal is not acceptable. If the proposal is acceptable, an understanding will be reached with the applicant concerning the starting date for each item of development.

(iii) Discuss with the applicant the Agency requirements with respect to good construction and land development practices.

(iv) Advise the applicant regarding drawings, specifications, cost estimates, and other related material which the applicant must submit to the Agency for review before the loan can be developed. Advise the applicant of the information necessary in the drawings, how the cost estimates should be prepared, the number of sets of drawings, specifications, and cost estimates required, and the necessity for furnishing such information promptly. Advise the applicant that the Agency will provide appropriate specification forms, Form RD 1924–2, "Description of Materials," and Form RD 1924–3, "Service Building Specifications." The applicant may, however, use other properly prepared specifications.

(v) Advise the applicant regarding publications, plans, planning aids, engineering data, and other technical advice and assistance available through local, state, and Federal agencies, and private individuals and organizations.

(vi) Review the information furnished by the applicant to determine the completeness of the plans, adequacy of the cost estimates, suitability and soundness of the proposed development.

(vii) When appropriate, offer suggestions as to how drawings and specifications might be altered to improve the facility and better serve the needs of the applicant. The County Supervisor or District Director may assist the applicant in making revisions to the drawings. When appropriate, the contract documents will be forwarded to the State architect/engineer for review. For revisions requiring technical determinations that the Agency is not able to make, the applicant will be requested to obtain additional technical assistance.

(viii) Provide the applicant with a written list of changes required in the contract documents. The applicant will submit two complete revised (as requested) sets of contract documents, for approval. On one set, the County Supervisor or District Director will indicate acceptance on each sheet of the drawings, and on the cover of the specifications and all other contract documents. At least the date and the initials of the approval official must be shown. On projects where a consulting architect or engineer has been retained, this acceptance will be indicated only after the State Director has given written authorization. The marked set of documents shall be available at the job site at all times for review by the Agency. The second set will become part of the loan docket.

(ix) Review the proposed method of doing the work and determine whether the work can be performed satisfactorily under the proposed method.

(x) Instruct the applicant not to incur any debts prior to loan closing for materials or labor or make any expenditures for such purposes with the expectation of being reimbursed from loan funds.

(xi) Instruct the applicant not to commence any construction nor cause any supplies or materials to be delivered to the construction site prior to loan closing.

(xii) Under certain conditions prescribed in exhibit H of this subpart, provide the applicant with a copy of the leaflet, "Warning—Lead-Based Paint Hazards," which is attachment 1 of exhibit H (available in any Agency office), and the warning sheet, "Caution Note on Lead-Based Paint Hazard," which is attachment 2 of exhibit H (available in any Agency office).

(g) *Surplus structures and use or sale of timber, sand, or stone.* In planning the development, the applicant and the County Supervisor or District Director should, when practicable, plan to use salvage from old buildings, timber, sand, gravel, or stone from the property. The borrower may sell surplus buildings, timber, sand, gravel, or stone that is not to be used in per-

forming planned development and use net proceeds to pay costs of performing planned development work. In such a case:

(1) An agreement will be recorded in the narrative of Form RD 1924-1 which as a minimum will:

(i) Identify the property to be sold, the estimated net proceeds to be received, and the approximate date by which the property will be sold.

(ii) Provide that the borrower will deposit the net proceeds in the supervised bank account and apply any funds remaining after the development is complete as an extra payment on the loan, or in accordance with § 1965.13(f) of subpart A of part 1965 of this chapter for farm program loans.

(2) The agreement will be considered by the Government as modifying the mortgage contract to the extent of authorizing and requiring the Government to release the identified property subject to the conditions stated in the agreement without payment or other consideration at the time of release, regardless of whether or not the mortgage specifically refers to Form RD 1924-1 or the agreement to release.

(3) If the Agency loan will be secured by a junior lien, all prior lienholders must give written consent to the proposed sale and the use of the net proceeds before the loan is approved.

(4) Releases requested by the borrower or the buyer will be processed in accordance with applicable release procedures in 7 CFR part 3550, as appropriate.

(h) *Review prior to performing development work.* For the sole benefit of the Agency, prior to beginning development work, the County Supervisor or District Director will review planned development with the borrower. Adequacy of the drawings and specifications as well as the estimates will be checked to make sure the work can be completed within the time limits previously agreed upon and with available funds. Items and quantities of any materials the borrower has agreed to furnish will be checked and dates by which each item of development should be started will be checked in order that the work may be completed on schedule. If any changes in the plans and specifications are proposed, they should be within the general scope of the work as originally planned. Changes must be approved and processed in accordance with § 1924.10 of this subpart. The appropriate procedure for performing development should be explained to the borrower. Copies of RD forms that will be used during the period of construction should be given to the borrower. The borrower should be advised as to the purpose of each form and at what period during construction each form will be used.

(i) *Time of starting development work.* Development work will be started as soon as feasible after the loan is closed. Except in cases in which advance commitments are made in accordance with 7 CFR part 3550 or according to § 1924.13(e)(1)(vi)(A) or § 1924.13(e)(2)(ix)(A) of this subpart, no commitments with respect to performing planned development will be made by the Agency or the applicant before the loan is closed. The applicant will be instructed that before the loan is closed, debts should not be incurred for labor or materials, or expenditures made for such purposes, with the expectation of being reimbursed from funds except as provided in subpart A of part 1943 of this chapter, 7 CFR part 3550, and subpart E of part 1944 of this chapter. However, with the prior approval of the National Office, a State Supplement may be issued authorizing County Supervisors to permit applicants to commence welldrilling operations prior to loan closing, provided:

(1) It is necessary in the area to provide the water supply prior to loan closing,

(2) The applicant agrees in writing to pay with personal funds all costs incurred if a satisfactory water supply is not obtained,

(3) Any contractors and suppliers understand and agree that loan funds may not be available to make the payment,

(4) Such action will not result under applicable State law in the giving of priority to mechanics and materialmen's liens over the later recorded Agency mortgage, and

(5) The Agency does not guarantee that the cost will be paid.

[52 FR 8002, Mar. 13, 1987, as amended at 52 FR 19283, May 22, 1987; 52 FR 48391, Dec. 22, 1987; 52 FR 48799, Dec. 28, 1987; 53 FR 43676, Oct. 28, 1988; 59 FR 43723, Aug. 25, 1994; 61 FR 65156, Dec. 11, 1996; 67 FR 78326, Dec. 14, 2002]

§ 1924.6 Performing development work.

All construction work will be performed by one, or a combination, of the following methods: Contract, borrower, mutual self-help, or owner-builder. All development work must be performed by a person, firm or organization qualified to provide the service. The mutual self-help method is performance of work by a group of families by mutual labor under the direction of a construction supervisor, as described in 7 CFR part 3550.

(a) *Contract method.* This method of development will be used for all major construction except in cases where it is clearly not possible to obtain a contract at a reasonable or competitive cost. Work under this method is performed in accordance with a written contract.

(1) *Forms used.* Form RD 1924–6, "Construction Contract," will be used for SFH construction. Other contract documents for more complex construction, acceptable to the loan approval official and containing the requirements of subpart E of part 1901 of this chapter, may be used provided they are customarily used in the area and protect the interest of the borrower and the Government with respect to compliance with items such as the drawings, specifications, payments for work, inspections, completion, nondiscrimination in construction work and acceptance of the work. If needed, the Office of the General Counsel (OGC) will be consulted. The United States (including the Agency) will not become a party to a construction contract or incur any liability under it.

(2) *Contract provisions.* Contracts will have a listing of attachments and the provisions of the contract will include:

(i) The contract sum.

(ii) The dates for starting and completing the work.

(iii) The amount of liquidated damages to be charged.

(iv) The amount, method, and frequency of payment.

(v) Whether or not surety bonds will be provided.

(vi) The requirement that changes or additions must have prior written approval of the Agency.

(3) *Surety requirements.* (i) Unless an exception is granted in accordance with paragraph (a)(3)(iii) of this section or when interim financing will be used, surety that guarantees both payment and performance in the amount of the contract will be furnished when one or more of the following conditions exist:

(A) The contract exceeds the applicable Rural Development Single Family Housing area loan limit as per 7 CFR 3550.63. (Loan limits are available at the local Rural Development field office.)

(B) The loan approval official determines that a surety bond appears advisable to protect the borrower against default of the contractor.

(C) The applicant requests a surety bond.

(D) The contract provides for partial payments in excess of the amount of 60 percent of the value of the work in place.

(E) The contract provides for partial payments for materials suitably stored on the site.

(ii) If surety bonds are required the construction contract must indicate that the contractor will furnish properly executed surety bonds prior to the start of any work. Exhibits F and G of this subpart as revised by OGC if necessary to comply with local or state statutory requirements will be used as the forms of payment bond and performance bond to be provided. Unless noncorporate surety is provided, the surety bonds may only be obtained from a corporate bonding company listed on the current Department of the Treasury Circular 570 (published annually in the FEDERAL REGISTER), as holding a certificate of authority as an acceptable surety on Federal bonds and as legally doing business in the State where the land is located. Noncorporate sureties are not recommended and the State Director will be responsible for determining the acceptability of the individual or individuals proposed as sureties on the bonds. The

State Director must determine that an individual or individuals proposed as sureties must have cash or other liquid assets easily convertible to cash in an amount at least equal to 25 percent more than the contract amount in order to be acceptable. The individual(s) will pledge such liquid assets in an amount equal to the contract amount. Fees charged for noncorporate sureties may not exceed fees charged by corporate sureties on bonds of equal amount and, in no case, may surety be provided by the applicant or any person or organization with an identity of interest in the applicant's operation. The United States (including the Agency) will incur no liability related in any way to a performance or payment bond provided in connection with a construction contract. The Agency will be named as co-obligee in the performance and payment bonds unless prohibited by state law.

(iii) When an experienced and reliable contractor cannot obtain payment and performance bonds meeting the surety requirements of paragraph (a)(3)(ii) of this section, the State Director may entertain a request from the applicant for an exception to the surety requirements. The applicant's request must specifically state why the proposed contractor is unable to obtain payment and performance bonds meeting the surety requirements, and why it is financially advantageous for the applicant to award the contract to the proposed contractor without the required bonds.

If the applicant's request is reasonable and justified, and if the proposed contractor is reliable and experienced in the construction of projects of similar size, design, scope, and complexity, the State Director may grant an exception to the surety requirements for loans or grants within the State Director's approval authority and accept one or a combination of the following:

(A) An unconditional and irrevocable letter of credit issued by a lending institution which has been reviewed and approved by OGC. In such cases, the construction contract must indicate that the contractor will furnish a properly executed letter of credit from a lending institution acceptable to the Agency prior to the start of any work.

The letter of credit must remain in effect until the date of final acceptance of work by the owner and the Agency. In addition, the letter of credit must stipulate that the lending institution, upon written notification by the Agency of the contractor's failure to perform under the terms of the contract, will advance funds up to the amount of the contract (including all Agency approved contract change orders) to satisfy all prior debts incurred by the contractor in performing the contract and all funds necessary to complete the work. Payments may be made to the contractor in accordance with paragraph (a)(12)(i)(C) of this section as if full surety bonds were being provided.

(B) If a letter of credit satisfying the conditions of paragraph (a)(3)(iii)(A) of this section cannot be obtained, the State Director may accept a deposit in the amount of the contract, into an interest or non-interest bearing supervised bank account. In such cases, the construction contract must indicate that the contractor will furnish the required deposit prior to the start of any work and that the funds shall remain on deposit until final acceptance of work by the owner and the Agency. Payments may be made to the contractor in accordance with paragraph (a)(12)(i)(C) of this section as if full surety bonds were being provided.

(C) When the provisions of paragraph (a)(3)(iii) (A) or (B) of this section can be met except that a surety bond, a letter of credit, and/or deposits are not obtainable in full amount of the contract, the State Director may accept an amount less than the full amount of the contract provided all of the following conditions are met:

(1) The contractor provides a surety bond, a letter of credit, or deposits in the greatest amount possible, and provides documentation indicating the reasons why amounts exceeding the proposed amount cannot be provided.

(2) The applicant agrees to the amount of the surety bond, letter of credit, or deposits proposed, and the State Director determines that the applicant has the financial capability to withstand any financial loss due to default of the contractor.

(3) In the opinion of the State Director, the proposed amount and the method of payment will provide adequate protection for the borrower and the Government against default of the contractor.

(4) The contract provides for partial payments not to exceed 90 percent of the value of the work in place for that portion of the total contract which is guaranteed by an acceptable surety bond, letter of credit, or deposits, and partial payments not to exceed 60 percent of the value of the work in place for that portion of the total contract which is not guaranteed by surety, letter of credit, or deposits.

Example:

Contractor has a surety bond which guarantees payment and performance in an amount of $150,000 which represents 75 percent of the total contract amount of $200,000. The contractor's first request for payment appears thus:

—Value of work in place is $10,000.
—Payment for work guaranteed by surety is 75 percent times $10,000 times 90 percent is $6,750.
—Payment for work not guaranteed by surety is 25 percent times $10,000 times 60 percent is $1,500.
—Authorized payment is $8,250.

(Each partial payment shall reflect values for work guaranteed by surety, letter of credit, or deposits, and work not so guaranteed).

(iv) In cases where the contractor does not obtain payment and performance bonds in accordance with the surety requirements of paragraph (a)(3)(ii) of this section, or where an exception to the surety requirements is granted by the State Director, the following steps will be taken to protect the borrower and the government against latent obligations or defects in connection with the construction:

(A) The contractor will furnish a properly executed corporate latent defects bond or a maintenance bond in the amount of 10 percent of the construction contract; or

(B) An unconditional and irrevocable letter of credit in the amount of 10 percent of the construction contract issued by a lending institution which has been reviewed and approved by OGC; or

(C) A cash deposit into an interest or non-interest bearing supervised bank account in the amount of 10 percent of the construction contract;

(D) The period of protection against latent obligations and/or defects shall be one year from the date of final acceptance of work by the owner and the Agency;

(E) Final payment shall not be rendered to the contractor until the provisions of paragraph (a)(3)(iv) (A), (B) or (C) of this section have been met;

(F) The contract will contain a clause indicating that the contractor agrees to provide surety or guarantee acceptable to the owner and the Agency against latent obligations and/or defects in connection with the construction.

(4) *Equal opportunity.* Section 1901.205 of subpart E of part 1901 of this chapter applies to all loans or grants involving construction contracts and subcontracts in excess of $10,000.

(5) *Labor standards provisions.* The provisions of the Davis-Bacon and related acts, which are published by the Department of Labor (29 CFR parts 1, 3 and 5), will apply when the contract involves either LH grant assistance, or 9 or more units in a project being assisted under the HUD section 8 housing assistance payment program for new construction.

(6) *Historical and archaeological preservation.* The provisions of subpart F of part 1901 of this chapter concerning the protection of historical and archaeological properties will apply to all construction financed, in whole or in part, by Agency loans and grants. These provisions have special applicability to development in areas designated by NRCS as Resource Conservation and Development (RC&D) areas. (See part 1942, subpart I of this chapter.)

(7) *Air and water acts.* Under Executive Order 11738, all loans or grants involving construction contracts for more than $100,000 must meet all the requirements of section 114 of the Clean Air Act (42 U.S.C. 7414) and section 308 of the Water Pollution Control Act (33 U.S.C., section 1813). The contract should contain provisions obligating the contractor as a condition for the award of the contract as follows:

(i) To notify the owner of the receipt of any communication from Environmental Protection Agency (EPA) indicating that a facility to be utilized in the performance of the contract is under consideration to be listed on the EPA list of Violating Facilities. Prompt notification is required prior to contract award.

(ii) To certify that any facility to be utilized in the performance of any nonexempt contractor subcontract is not listed on the EPA list of Violating Facilities as of the date of contract award.

(iii) To include or cause to be included the above criteria and requirements of paragraphs (a)(7) (i) and (ii) of this section in every nonexempt subcontract, and that the contractor will take such action as the Government may direct as a means of enforcing such provisions.

(8) *Architectural barriers.* In accordance with the Architectural Barriers Act of 1968 (Pub. L. 90–480), as implemented by the General Services Administration regulations (41 CFR 101–19.6) and section 504 of the Rehabilitation Act of 1973 (Pub. L. 93–112) as implemented by 7 CFR, parts 15 and 15b, all facilities financed with Agency loans and grants and which are accessible to the public or in which people with disabilities may be employed or reside must be developed in compliance with this Act. Copies of the Act and Federal accessibility design standards may be obtained from the Executive Director, Architectural and Transportation Barriers Compliance Board, Washington, DC 20201.

(9) *National Environmental Policy Act.* Loans and grants, including those being assisted under the HUD section 8 housing assistance payment program for new construction, must comply with the environmental review requirements in accordance with 7 CFR part 1970.

(10) *Obtaining bids and selecting a contractor.* (i) The applicant may select a contractor and negotiate a contract or contact several contractors and request each to submit a bid. For complex construction projects, refer also to § 1924.13(e) of this subpart.

(ii) When a price has already been negotiated by an applicant and a contractor, the County Supervisor, District Director or other appropriate

Agency official will review the proposed contract. If the contractor is qualified to perform the development and provide a warranty of the work and the price compares favorably with the cost of similar construction in the area, further negotiation is unnecessary. If the Agency official determines the price is too high or otherwise unreasonable, the applicant will be requested to negotiate further with the contractor. If a reasonable price cannot be negotiated or if the contractor is not qualified, the applicant will be requested to obtain competitive bids.

(iii) When an applicant has a proposed development plan and no contractor in mind, competitive bidding will be encouraged. The applicant should obtain bids from as many qualified contractors, dealers or tradespeople as feasible depending on the method and type of construction.

(iv) If the award of the contract is by competitive bidding, Form RD 1924–5, "Invitation for Bid (Construction Contract)," or another similar invitation bid form containing the requirements of subpart E of part 1901 of this chapter, may be used. All contractors from whom bids are requested should be informed of all conditions of the contract including the time and place of opening bids. Conditions shall not be established which would give preference to a specific bidder or type of bidder. When applicable, copies of Forms RD 1924–6 and RD 400–6, "Compliance Statement," also should be provided to the prospective bidders.

(11) *Awarding the contract.* The borrower, with the assistance of the County Supervisor or District Director, will consider the amount of the bids or proposals, and all conditions which were listed in the "Invitation for Bid." On the basis of these considerations, the borrower will select and notify the lowest responsible bidder.

(i) Before work commences, the County Supervisor, District Director or other Agency employee having knowledge of contracts and construction practices will hold a preconstruction conference with the borrower(s), contractor and architect/engineer (if applicable). The purpose of the conference is to reach a mutual understanding of each party's responsibilities under the terms and conditions of the contract documents and the loan agreement during the construction and warranty periods. Form RD 1924–16, "Record of Preconstruction Conference," may be used as a guide for an agenda.

(ii) A summary of the items covered will be entered in the running case record.

(iii) The contract will then be prepared, signed and copies distributed in accordance with the FMI for Form RD 1924–6.

(iv) After a borrower/contractor's contract or subcontract in excess of $10,000 is received in the Agency County or District Office, the responsible Agency official will send within 10 calendar days of the date of the contract or subcontract, a report similar in form and content to exhibit C of subpart E of part 1901 of this chapter to the Area Director, Office of Federal Contract Compliance Programs, U.S. Department of Labor, at the applicable address listed in exhibit E, subpart E of part 1901 of this chapter. The report must contain, at least, the following information: contractor's name, address and telephone number; employer's identification number; amount, starting date and planned completion date of the contract; contract number; and city and DOL region of the contract site. The information for this report should be obtained from the contractor when the contract is awarded.

(12) *Payments for work done by the contract method.* (i) Payments will be made in accordance with one of the following methods unless prohibited by state statute, in which case the State Director shall issue a State Supplement to this section:

(A) The "One-Lump-Sum" payment method will be used when the payment will be made in one lump-sum for the whole contract.

(B) The "Partial payments not to exceed 60 percent of the value of the work in place" payment method will be used when the contractor does not provide surety bond, a letter of credit, or deposits.

(C) The "Partial payments in the amount of 90 percent of the value of the work in place and of the value of the materials suitably stored at the site" payment method will be used when the contractor provides a surety bond equal to the total contract amount.

(D) The "Partial payments which reflect the portions of the contract amount which is guaranteed" method will be used when the contractor provides surety bonds, a letter of credit, or deposits less than the total amount of the contract in accordance with the provisions of paragraph (a)(3)(iii)(C) of this section.

(ii) When Form RD 1924-6 is used, the appropriate payment clause will be checked and the other payment clauses not used will be effectively crossed out.

(iii) When a contract form other than Form RD 1924-6 is used, the payment clause must conform with paragraph (a)(12)(i) of this section and the appropriate clause as set forth in Form RD 1924-6.

(iv) The borrower and FmHA or its successor agency under Public Law 103-354 must take precautionary measures to see that all payments made to the contractor are properly applied against bills for materials and labor procured under the contract. Prior to making any partial payment on any contract where a surety bond is not used, the contractor will be required to furnish the borrower and the FmHA or its successor agency under Public Law 103-354 with a statement showing the total amount owed to date for materials and labor procured under the contract. The contractor also may be required to submit evidence showing that previous partial payments were applied properly. When the borrower and the County Supervisor or District Director have reason to believe that partial payments may not be applied properly, checks may be made jointly to the contractor and persons who furnished materials and labor in connection with the contract.

(v) When partial payments are requested by the contractor and approved by the owner, the amount of the partial payment will be determined by one of the following methods:

(A) Based upon the percentage completed as shown on a recently completed and properly executed Form RD 1924-12, "Inspection Report."

(B) When the structure will be covered by an insured 10-year warranty, the insurer's construction inspector must provide the Agency with any available copies of inspection reports showing percentage of completion immediately after the inspections are completed. To make partial payments when copies of inspection reports are not available, the responsible Agency official will make the inspections or will be guided by the provisions of § 1924.6(a)(12)(v)(C) of this subpart. If further assurance is deemed necessary to justify partial payments, the Agency official may make onsite inspections or require additional information.

(C) Based upon an application for payment containing an estimate of the value of work in place which has been prepared by the contractor and accepted by the borrower and the Agency. When the contract provides for partial payments for materials satisfactorily stored at the site, the application for payment may include these items. Prior to receiving the first partial payment, the contractor should be required to submit a list of major subcontractors and suppliers and a schedule of prices or values of the various phases of the work aggregating the total sum of the contract such as excavation, foundations, framing, roofing, siding, mill work, painting, plumbing, heating, electric wiring, etc., made out in such form as agreed upon by the borrower, the Agency, and the contractor. In applying for payments, the contractor should submit a statement based upon this schedule. See exhibit A of this subpart for guidance in reviewing the contractor's schedule of prices and estimating the value of the work in place.

(vi) *Final payment.* (A) When the structure will be covered by an insured 10-year warranty, the insurer must provide an insured 10-year warranty policy (or a binder if the policy is not available) before final payment is made to the builder.

(B) Final payment of the amount due on the contract or disbursal of the Agency loan funds where an interim loan was used will be made only upon completion of the entire contract, final inspection by the Agency, acceptance of the work by the Agency and the borrower, issuance of any and all final permits and approvals for the use and occupancy of the structure by any applicable state and local governmental authorities, and compliance by the contractor with all terms and conditions of the contract. In the event the work of construction is delayed or interrupted by reason of fire, flood unusually stormy weather, war, riot, strike, an order, requisition or regulation of any governmental body (excluding delays related to possible defects in the contractor's performance and excluding delays caused by the necessity of securing building permits or any required inspection procedures connected therewith) or other contingencies reasonably unforeseeable and beyond the reasonable control of the contractor, then with the written consent of the Agency, the date of completion of the work may be extended by the owner by the period of such delay, provided that the contractor shall give the owner and the Agency written notice within 72 hours of the occurrence of the event causing the delay or interruption.

(C) Prior to making final payment on the contract when a surety bond is not used or disbursing Agency loan funds when an interim loan was used, the Agency will be provided with a Form RD 1924–9, "Certificate of Contractor's Release," and Form RD 1924–10, "Release by Claimants." executed by all persons who furnished materials or labor in connection with the contract. The borrower should furnish the contractor with a copy of the "Release by Claimants" form at the beginning of the work in order that the contractor may obtain these releases as the work progresses.

(1) If such releases cannot be obtained, the funds may be disbursed provided all the following can be met:

(i) Release statements to the extent possible are obtained;

(ii) The interests of the Agency can be adequately protected and its security position is not impaired; and

(iii) Adequate provisions are made for handling the unpaid account by withholding or escrowing sufficient funds to pay any such claims or obtaining a release bond.

(2) The State Director may issue a State Supplement which will:

(i) Not require the use of Form RD 1924–10, if, under existing state statutes, the furnishing of labor and materials gives no right to a lien against the property, or

(ii) Provide an alternative method to protect against mechanic's and materialmen's liens. In this case, the use of Form RD 1924–10 is optional.

(b) *Borrower method.* The borrower method means performance of work by or under the direction of the borrower, using one or more of the ways specified in this paragraph. Development work may be performed by the borrower method only when it is not practicable to do the work by the contract method; the borrower possesses or arranges through an approved self-help plan for the necessary skill and managerial ability to complete the work satisfactorily; such work not interfere seriously with the borrower's farming operation or work schedule, and the County Office caseload will permit a County Supervisor to properly advise the borrower and inspect the work.

(1) *Ways of performing the work.* The borrower will:

(i) Purchase the material and equipment and do the work.

(ii) Utilize lump-sum agreements for (A) minor items or minor portions of items of development, the total cost of which does not exceed $5,000 per agreement, such as labor, material, or labor and material for small service buildings, repair jobs, or land development; or (B) material and equipment which involve a single trade and will be installed by the seller, such as the purchase and installation of heating facilities, electric wiring, wells, painting, liming, or sodding. All agreements will be in writing, however, the County Supervisor may make an exception to this requirement when the agreement involves a relatively small amount.

(2) *Acceptance and storage of material on site.* The County Supervisor will advise the borrower that the acceptance of material as delivered to the site and the proper storage of material will be the borrower's responsibility.

(3) *Payment for work done by the borrower method*—(i) *Payments for labor.* Before the County Supervisor

countersigns checks for labor, the borrower must submit a completed Form RD 1924–11, "Statement of Labor Performed," for each hired worker performing labor during the pay period. Ordinarily, checks for labor will be made payable to the workers involved. However, under justifiable circumstances, when the borrower has paid for labor with personal funds and has obtained signatures of the workers on Form RD 1924–11 as having received payment, the County Supervisor may countersign a check made payable to the borrower for reimbursement of these expenditures. Under no circumstances will the County Supervisor permit loan funds or funds withdrawn from the supervised bank account to be used to pay the borrower for the borrower's own labor or labor performed by any member of the borrower's household.

(ii) *Payment for equipment, materials or lump-sum agreements.* (A) Before countersigning checks for equipment or materials, the County Supervisor must normally have an invoice from the seller covering the equipment or materials to be purchased. When an invoice is not available at the time the check is issued, an itemized statement of the equipment or materials to be purchased may be substituted until a paid invoice from the seller is submitted, at which time the prepurchase statement may be destroyed.

(B) When an invoice is available at the time the check is drawn, the check will include a reference to the invoice number, the invoice date if unnumbered and, if necessary, the purpose of the expenditure.

(C) The check number and date of payment will be indicated on the appropriate Form RD 1924–11, invoice, itemized statement of equipment or materials and/or lump-sum agreement.

(D) Ordinarily, checks for equipment or materials will be made payable to the seller. Under justifiable circumstances, when the borrower has paid for equipment or materials with personal funds and furnished a paid invoice, the County Supervisor may countersign a check made payable to the borrower for reimbursement of these expenses.

(E) When an invoice includes equipment or materials for more than one item of development, the appropriate part of the cost to be charged against each item of development will be indicated on the invoice by the borrower, with the assistance of the County Supervisor.

(F) Payment made under lump-sum agreements will be made only when all items of equipment and materials have been furnished, labor has been performed as agreed upon, and the work has been accepted by the borrower and the Agency.

(G) Each paid Form RD 1924–11, invoice, itemized statement for equipment or material and/or lump-sum agreement will be given to the borrower in accordance with the FMI.

(c) *Mutual self-help method.* The mutual self-help method is performance of work by a group of families by mutual labor under the direction of a construction supervisor, as described in 7 CFR part 3550. The ways of doing the work, buying materials, and contracting for special services are like those used for the borrower method. Materials can be bought jointly by the group of families, but payments will be made individually by each family. In the case of RH loans to families being assisted by Self-Help Technical Assistance (TA) grants in accordance with subpart I of part 1944 of this chapter, the County Supervisor may countersign checks for materials and necessary contract work made payable directly to the TA grantee, provided the District Director determines that:

(1) The grantee acts in the same capacity as a construction manager in the group purchase of material and services.

(2) The grantee has an adequate bookkeeping system approved by the District Director to assure that funds in each RH account are properly distributed and maintained.

(3) The grantee receives no compensation in the way of profit or overhead for this service and all discounts and rebates received in connection with the purchase of materials or services are passed on to the participating families.

(4) The grantee has a record-keeping system which shows that the costs of

the materials and services were pro-rated to each borrower's account in relation to the actual material and service used by each borrower.

(d) *Owner-builder method.* This method of construction applies only to RRH loans made under subpart E of part 1944 of this chapter. Regulations governing this method are found at §1924.13(e)(2) of this subpart.

[52 FR 8002, Mar. 13, 1987, as amended at 55 FR 41833, Oct. 16, 1990; 60 FR 55122, Oct. 27, 1995; 61 FR 56116, Oct. 31, 1996; 71 FR 25740, May 2, 2006; 81 FR 11029, Mar. 2, 2016]

§1924.7 [Reserved]

§1924.8 Development work for modular/panelized housing units.

(a) Exhibit B of this subpart applies to all loans involving modular/panelized housing units.

(b) Complete drawings and specifications will be required as prescribed in exhibit C of this subpart. Each set of drawings will contain the design of the foundation system required for the soil and slope conditions of the particular site on which the modular/panelized house is to be placed.

(c) The manufacturer will provide a certification (exhibit B, attachment 5 of this subpart), stating that the building has been built substantially in accordance with the drawings and specifications. The builder will also provide a certification that the onsite work complies with drawings, specifications, and the applicable development standard (eExhibit B, attachment 5 of this subpart).

(d) Responsibility for field inspections will be in accordance with §1924.9(a) of this subpart. Frequency and timing of inspections will be in accordance with §1924.9(b) of this subpart, except that the Stage 2 inspection should be made during the time and in no case later than two working days after the crews commence work on the site and the house is being erected or placed on the foundation, to determine compliance with the accepted drawings and specifications.

(e) Periodic plant inspections will be performed in accordance with paragraphs II and III of exhibit B of this subpart. Agency employees responsible for inspections in the area in which the manufacturing plant or material supply yard is located will perform such inspections as deemed necessary under paragraph III of exhibit B of this subpart.

(1) Plant inspections will be made if the type construction method used could restrict adequate inspections on the building site.

(2) Plant inspections will be made as often as necessary; however, after initial inspection and acceptance of the unit, only when it appears advisable to ascertain the performance and continuing stability of accepted materials and construction.

(f) Only one contract will be accepted for the completed house on the site owned or to be bought by the borrower. The manufacturer of the house or the manufacturer's agent may be the prime contractor for delivery and erection of the house on the site or a builder may contract with the borrower for the complete house in place on the site. Such contracts should provide that payments will be made only for work in place on the borrower's site.

(g) Payments for modular/panelized units will be made in accordance with the terms of the contract and in compliance with §1924.6(a)(12) of this subpart.

§1924.9 Inspection of development work.

The following policies will govern the inspection of all development work.

(a) *Responsibility for inspection.* The County Supervisor or District Director, accompanied by the borrower when practicable, will make final inspection of all development work and periodic inspections as appropriate to protect the security interest of the government. In this respect, inspections other than final inspections, may be conducted by other qualified persons as authorized in paragraph (d) of this section, in 7 CFR part 3550, in RD Instruction 2024-A (available in any Agency office), and as authorized under other agreements executed by, or authorized by, the National Office.

The borrower will be responsible for making inspections necessary to protect the borrower's interest. Agency inspectionsare not to assure the borrower that the house is built in accordance with the plans and specifications. The inspections create or imply no duty or obligation to the particular borrower. Agency inspections are for the dual purpose of determining that the Agency has adequate security for its loan and is achieving the statutory goal of providing adequate housing. If difficult technical problems are encountered, the County Supervisor or District Director should request the assistance of the State Office or a qualified technician from SCS or the State University Cooperative Extension Service.

(b) *Frequency of inspections.* The County Supervisor or District Director will inspect development work as frequently as necessary to assure that construction and land development conforms to the drawings and specifications. The final inspection will be made at the earliest possible date after completion of the planned development. When several major items of development are involved, final inspection will be made upon completion of each item.

(1) For new buildings and additions to existing buildings, inspections will be made at the following stages of construction and at such other stages of construction as determined by the County Supervisor or District Director except as modified by paragraph (b)(3) of this section.

(i) *Stage 1.* Customarily, the initial inspection in construction cases is made just prior to or during the placement of concrete footings or monolithic footings and floor slabs. At this point, foundation excavations are complete, forms or trenches and steel are ready for concrete placement and the subsurface installation is roughed in. However, when it is not practicable to make the initial inspection prior to or during the placement of concrete, the County Supervisor or District Director will make the initial inspection as soon as possible after the placement of concrete and before any backfill is in place.

(ii) *Stage 2.* The Stage 2 inspection will be made when the building is enclosed, structural members are still exposed, roughing in for heating, plumbing, and electrical work is in place and

visible, and wall insulation and vapor barriers are installed. Customarily, this is prior to installation of brick veneer or any interior finish which would include lath, wallboard and finish flooring.

(iii) *Stage 3.* The final inspection will be made when all on-site and off-site development has been completed and the structure is ready for occupancy or its intended use.

(2) For rehabilitation of existing buildings, inspections will be made in accordance with paragraphs (b)(1) (ii) and (iii) of this section, and at such other stages of construction to assure that construction is being performed in a professional manner and in accordance with Agency approved drawings and specifications.

(3) For new construction when the structure will be covered by an insured 10-year warranty plan as described in exhibit L of this subpart, only the final inspection is required, except in cases when partial payments are required when the provisions of § 1924.6(a)(12)(v) of this subpart will be followed.

(4) Arrangements should be made to have the borrower join the County Supervisor or the District Director in making periodic inspections as often as necessary to provide a mutual understanding with regard to the progress and performance of the work.

(5) The Borrower should make enough periodic visits to the site to be familiar with the progress and performance of the work, in order to protect the borrower's interest. If the borrower observes or otherwise becomes aware of any fault or defect in the work or nonconformance with the contract documents, the borrower should give prompt written notice thereof to the contractor with a copy to the County Supervisor or District Director responsible for servicing the type of loan or grant involved.

(6) The borrower should, when practicable, join the County Supervisor or District Director in making all final inspections.

(7) When irrigation equipment and materials are to be purchased and installed, a performance test under actual operating conditions by the person or firm making the installation should be required before final acceptance is

made. The test should be conducted in the presence of the borrower, a qualified technician, and, when practicable, the County Supervisor or District Director. If the Agency official is not present at the performance test, he or she should request the technician to furnish a report as to whether or not the installation meets the requirements of the plans and specifications.

(8) For irrigation and drainage construction or any dwelling construction where part or all of the work will be buried or backfilled, interim inspections should be made at such stages of construction that compliance with plans and specifications can be determined.

(c) *Recording inspections and correction of deficiencies.* All periodic and final inspections made by the County Supervisor or District Director will be recorded on Form RD 1924–12 in accordance with the FMI. The County Supervisor or District Director will be responsible for following up on the correction of deficiencies reported on Form RD 1924–12. When an architect/ engineer is providing services on a project, the District Director should notify the architect/engineer immediately of any fault or defect observed in the work or of any nonconformance with the contract document. If the borrower or the contractor refuses to correct the deficiencies, the District Director will report the facts to the State Director who will determine the action to be taken. No inspection will be recorded as a final inspection until all deficiencies or nonconforming conditions have been corrected.

(d) *Acceptance by responsible public authority.* When local (city) county, state, or other public authority) codes and ordinances require inspections, final acceptance by the local authority having jurisdiction will be required prior to final inspection or acceptance by the Agency.

(e) *Acceptance by project architect.* If architectural services pursuant to §1924.13(a) of this subpart have been obtained, final acceptance by the project architect pursuant to §1924.13(a)(5)(v)

of this subpart will be required prior to acceptance by the Agency.

[52 FR 8002, Mar. 13, 1987, as amended at 60 FR 55122, Oct. 27, 1995; 61 FR 2899, Jan. 30, 1996; 67 FR 78327, Dec. 24, 2002]

§ 1924.10 Making changes in the planned development.

The borrower may request changes in the planned development in accordance with this section.

(a) *Authority of the County Supervisor.* The County Supervisor is authorized to approve changes in the planned development involving loans and grants within the County Supervisor's approval authority provided:

(1) The change is for an authorized purpose and within the scope of the original proposal.

(2) Sufficient funds are deposited in the borrower's supervised bank account or with the interim lender, as appropriate, to cover the contemplated changes when the change involves additional funds to be furnished by the borrower.

(3) The change will not adversely affect the soundness of the operation or the Agency's security. If uncertain as to the probable effect the change would have on the soundness of the operation or Agency security, the County Supervisor will obtain advice from the District Director on whether to approve the change.

(4) If a surety bond has been provided on the full amount of the construction contract, the aggregate amount of all contract change orders on Form RD 1924–7, "Contract Change Order," or other acceptable form will not exceed 20 percent of the original contract amount. Change orders for contracts on which a surety bond has been provided which increases the original contract amount by more than 20 percent may only be approved if additional surety is provided in the full revised amount of the contract. For purposes of this paragraph, letters of credit and deposits are not considered surety.

(5) Change orders for contracts on which letters of credit or deposits have been provided on the full amount of the contract which will increase the original contract amount are approved only

if additional letters of credit or deposits are provided in the full revised amount of the contract.

(6) Modifications have been certified in accordance with § 1924.5(f)(1)(iii) or certification has been waived in accordance with § 1924.5(f)(1)(iii)(C) of this subpart.

(b) *Authority of the District Director.* The District Director is authorized to approve changes in the development planned with RRH, RCH, and RHS loans and LH loans and grants within the District Director's approval authority, provided the conditions in § 1924.10(a) have been met. For such loans in excess of the District Director's approval authority, the borrower's request with the District Director's recommendation will be forwarded to the State Director for consideration.

(c) *Recording changes in the planned development.* (1) Changes should be accomplished only after Agency written approval. Changes will not be included in payment requests until approved by the borrower; the contractor, if applicable; the architect/engineer, if applicable; and the Agency loan approval official. Examples of changes requiring documentation are:

(i) Any changes in labor and materials and their respective costs.

(ii) Changes in facility design.

(iii) Any decrease or increase in unit-price on final measurements that are different from those shown in the bidding schedule.

(iv) Any increase or decrease in the time to complete the project.

(2) All changes shall be recorded in chronological order as follows:

(i) Contract method. Changes shall be numbered in sequence as they occur using Form RD 1924-7 with necessary attachments.

(ii) Borrower method. An increase or decrease in the cash cost, extension of time, transfer of funds between items, or an addition or deletion of items of development, will be summarized on the front of Form RD 1924-1 by striking through the original figures on items and writing in the changes. Changes made in the "Development Plan" in the working drawings, or in the plans and specifications will be dated and initialed by all parties.

(iii) Mutual self-help method. [See paragraph (c)(2)(ii) of this section.]

(iv) Owner-builder method. [See paragraph (c)(2)(i) of this section.]

(3) All changes in facility design and/or materials must be certified in accordance with § 1924.5(f)(1)(iii) of this subpart.

§ 1924.11 **District Director's review of incomplete development.**

During monthly District Office work organization meetings and during regular visits to the County Office, the District Director will review the progress that is being made in completing development financed with loans within the District Director's and County Supervisor's responsibility.

(a) Once each year the District Director will make a comprehensive review of all development work not completed within the time scheduled. For incomplete development financed with loan or grant funds within the responsibility of the District Director, the District Director will take the necessary actions to assure that the borrower or grantee completes the planned development. For incomplete development financed with loan or grant funds within the responsibility of the County Supervisor, the District Director will give the necessary direction to the County Supervisor to assure completion of the work. In connection with these responsibilities, the District Director will consider:

(1) The current farm and home operations with respect to the need for the development as originally planned.

(2) Revisions to the development plan.

(3) Funds remaining in the supervised bank account.

(4) Need for additional funds.

(5) Personal funds that could be furnished by the borrower.

(6) Estimated completion dates.

(7) The borrower's attitude with respect to completing the development.

(b) After a complete review of the status of development in both the District and County Offices has been made, the District Director will make a written report to the State Director which will include observations and recommendations regarding incomplete development. The report may be included in the District Director's regular report, and will include:

(1) The number of cases in which borrowers have not completed their development within 9, 15 or 24 months when authorized, and also the number of cases in which funds have been exhausted and the work is incomplete.

(2) The number of borrowers who have not completed their development within 3 years from the loan closing, and indicate the action that was taken in each such case.

(c) If the borrower has not completed development work within 3 years after the date of loan closing and the District Director has determined that the borrower cannot or will not complete the development, the District Director will so indicate on Form RD 1924-1 and request the State Director to withdraw, for application on the loan, any unused development funds remaining in the borrower's supervised bank account, if the borrower will not sign a check for a refund to the loan account.

§ 1924.12 Warranty of development work.

(a) Form RD 1924-19, "Builder's Warranty," or an insured 10-year home warranty as described in exhibit L of this subpart, and normal trade warranties on items of equipment will be issued to the borrower at the completion of new building construction, dwelling rehabilitation by the contract method, all cases of newly completed and previously unoccupied dwellings or construction under conditional commitments issued to builders and sellers.

(b) If the warranty is not an insured 10-year warranty, a completed Form RD 1924-19, with warranty protection for 1 year, must be provided by the builder upon final acceptance of the work by the owner and the Agency. If an insured 10-year warranty is provided, the requirements of exhibit L of this subpart apply, and a copy of the warranty insurance policy or a binder must have been received by the Agency prior to disbursement of the final payment to the builder.

(c) If, for some reason, the warranty insurance policy cannot be issued, the contractor will be required to execute Form RD 1924-19 and the case will be forwarded to the State Director for consideration of debarment under the provisions of subpart M of part 1940 (available in any Agency office). The County Supervisor will assist the borrower to the extent necessary under the provisions of the warranty and subpart F of part 1924 of this chapter.

(d) The County Supervisor will take the following action prior to the expiration of the first year of the warranty period:

(1) As soon as the warranty has been executed, the follow-up date for sending Form RD 1924-21, "Notice of Expiration of First Year of Warranty," which will be used for the 1 year warranty or the first year of the insured 10-year warranty, will be posted to the "Servicing and Supervision" section of the Management System card.

(2) Form RD 1924-21 is provided for use in notifying the borrower of the expiration date of the first year of the warranty. This letter will be mailed to the borrower early in the second month preceding the expiration date of the first year of the warranty period.

(3) If the County Supervisor or District Director does not hear from the borrower within 30 days, it can reasonably be assumed that no complaint exists or that any complaint has been satisfied unless information to the contrary has been received.

(4) If the borrower notifies the Agency that any complaint has not been satisfied, an onsite inspection shall be made as early as possible, but not later than 1 month preceding the expiration date of the first year of the warranty. The results of the inspection will be recorded on Form RD 1924-12. If the borrower has complaints, the case should be handled in accordance with the provisions of subpart F of part 1924 of this chapter, or as otherwise provided in this subpart.

[52 FR 8002, Mar. 13, 1987, as amended at 54 FR 14334, Apr. 11, 1989]

§ 1924.13 Supplemental requirements for more complex construction.

This section includes additional provisions that apply to planning and conduct of construction work on all multiple family housing projects and other projects that are more extensive in scope and more complex in nature than individual housing units or farm buildings. This section will apply in addition to all other requirements contained elsewhere in this subpart.

(a) *Architectural services.* Complete architectural services, as defined in §1924.4(o)(1) of this subpart are recommended on all projects. They are required for projects involving an LH grant and for all loans for RRH, RCH, and LH projects consisting of more than 4 units unless prior consent to making an exception to the requirements for complete architectural services is obtained from the National Office. If the applicant or contractor is an architect or organization with architectural capability, the applicant must, nevertheless, hire an independent qualified architect or architectural firm to inspect the construction work and perform other needed services during the construction and warranty phases. See Guide 4, attachment 1, "Attachment to AIA Document— Standard Form of Agreement Between Owner and Architect," for further information (available in any Agency office).

(1) *Exception.* Any request for National Office consent to an exception being made for complete architectural services should include the proposed drawings and specifications, method of providing specific services, the comments and recommendations of the Agency State Architect, and any other pertinent information. The State Director must determine that any services for which an exception is requested can be performed by qualified State or District Office staff members.

(2) *Selecting the architect.* The applicant is responsible for selecting the architect. The District Director with the advice of the State architect/engineer should discuss with the applicant the selection of the architect for the job as early as possible to assist in the site selection and participate in early consultations regarding project scope and design.

(3) *Architectural fees.* Fees for architectural services shall not exceed the fee ordinarily charged by the profession for similar work when Agency financing is not involved. The fee should cover only the architectural services rendered by the architect. The reduc-

tion or elimination of any services described in paragraph (a)(5) of this section shall be directly reflected in the fee. Fees for special services rendered by the architects, such as the packaging of the loan application or additional nonarchitectural services, will not be authorized to be paid with loan funds.

(4) *Agreement between borrower and architect.* The borrower and the architect will execute a written agreement. The agreement must provide:

(i) The services listed in paragraph (a)(5) of this section.

(ii) The amount of the fee and how it will be determined and paid.

(iii) That the agreement and any amendments to the agreement shall not be in full force and effect until concurred with in writing by the State Director or the State Director's delegate, and it will contain the following provision:

The Agency, as potential lender or insurer of funds to defray the costs of this agreement and without liability for any payments thereunder, hereby concurs in the form, content and the execution of this agreement.

Date _____

Agency Approval Official _____

Title _____

(5) *Specific services.* Architectural services will include six consecutive phases as follows:

(i) *Schematic design phase.* The architect will:

(A) Consult with the applicant to obtain available information pertinent to the project requirements.

(B) Consult with Agency State architect/engineer about Agency requirements and procedures.

(C) Assist in preparing the project design after analyzing engineering and survey data on the site selected by applicant.

(D) Prepare schematic design studies consisting of drawings and other documents illustrating the scale and relationship of project components for the applicant's approval.

(E) Submit estimates of current development costs based on current area, volume, or other unit costs.

(F) When the applicant and the Agency have accepted the schematic design studies and estimated development costs, the project architect may be authorized to proceed with the next phase.

(ii) *Design development phase.* The architect will:

(A) Prepare the design development exhibits from the accepted schematic design studies for approval by the applicant. These exhibits should consist of drawings and other documents to fix and describe the size and character of the entire project as to structural, mechanical, and electrical systems, materials, and other essentials as appropriate.

(B) Submit a further statement of probable construction cost.

(C) Obtain applicant and Agency approval of drawings, specifications, and authorization to proceed with next phase.

(iii) *Construction documents phase.* The architect will:

(A) Prepare the working drawings and specifications from the approved design development drawings and set forth in detail the requirements for the construction of the entire project in accordance with applicable regulations and codes; for example, necessary bidding information, assistance in preparing bidding forms, conditions of the construction contract, and the form of agreement between applicant/owner and contractor.

(B) Submit a final and more comprehensive statement of probable development cost. It should show a breakdown of the estimated total development cost of the project and the various trades in enough detail for an adequate review.

(C) Obtain the acceptance of the applicant and the Agency for contract documents, including approval of the final drawings and specifications and authorization to proceed.

(D) Discuss with the applicant various items as they develop.

(iv) *Bidding or negotiation phase.* The architect will, as appropriate, for a bidded or negotiated contract:

(A) Assist in review and selection of bidders and submission of contract documents to selected bidders.

(B) Assist in the interpretation of drawings and specifications, and other contract documents.

(C) Receive and tabulate all bids.

(D) Review the bids and the negotiated proposals and assist in the award and preparation of construction contracts.

(v) *Construction phase.* This phase includes the administration of the construction contract. It will commence with the award of the construction contract and end when the borrower makes final payment to the contractor. The architect will:

(A) Attend the preconstruction conference. Advise and consult with the borrower (or the borrower's representative) and issue the borrower's instructions to the contractor.

(B) Prepare change orders.

(C) Keep construction accounts and work as the general administrator of the project during construction.

(D) Interpret the contract documents and have the authority to reject all work and materials which do not comply.

(E) Review and approve shop drawings, samples, and other submissions of the contractor for conformance with the design concept and for compliance with the contract documents.

(F) Conduct periodic inspections of all phases of construction to determine compliance with the contract documents and certify as to the amount is in place and materials suitably stored on site for partial payment estimates. These inspections will be augmented, when necessary, by inspections performed by structural, mechanical, and electrical representatives. Periodic inspections should be made as frequently as is necessary to verify that the work conforms with the intent of the contract documents and that a high quality of workmanship is maintained. The State Director may require a full-time project representative on projects with a total development cost of $750,000 or more, when in the opinion of the State Director there is a need for such representative, and the State Director states the reasons for such need to the borrower.

(G) Determine, based on the inspections, the dates of substantial completion and final completion; receive on the borrower's behalf all written guarantees and related documents assembled by the contractor; and issue a final certificate for payment.

(vi) *Warranty phase.* The architect will advise and consult with the borrower, as the borrower's representative, about items to be corrected within the warranty period. The architect will accompany the Agency representative during the inspection required one month prior to expiration of the warranty period.

(b) *Other professional services.* The State Director, on the recommendation of the State architect/engineer, may request that additional professional services be provided.

(1) Professional services typically include soils engineering, structural engineering, civil engineering, surveying, land planning, or professional cost estimation or certification. Fees for these services may be paid directly by the borrower or by the architect as reimbursable expenses.

(2) When a project representative is utilized, unless otherwise agreed, the representative will be provided by the consulting architect/engineer. Prior to the preconstruction conference, the architect/engineer will submit a resume of qualifications of the project representative to the applicant and to the Agency for acceptance in writing. If the applicant provided the project representative, the applicant must submit a resume of the representative's qualifications to the project architect/engineer and the Agency for acceptance in writing, prior to the preconstruction conference. The project representative will attend the preconstruction conference where duties and responsibilities will be fully discussed. The project representative will work under the general supervision of the architect/engineer. The project representative will maintain a daily diary in accordance with the following:

(i) The diary shall be maintained in a hard-bound book.

(ii) The diary shall have all pages numbered and all entries in ink.

(iii) All entries shall be on daily basis, beginning with the date and weather conditions.

(iv) Daily entries shall include daily work performed, number of men and equipment used in the performance of the work, and all significant happenings during the day.

(v) The diary shall be made available to Agency personnel and will be reviewed during project inspections.

(vi) The project representative's diary will become the property of the owner after the project is accepted and final payments are made.

(c) *Drawings.* The type and kinds of drawings should be in accordance with exhibit C of this subpart and subpart D of part 1944 of this chapter.

(1) The drawings must be clear, accurate, with adequate dimensions and of sufficient scale for estimating purposes.

(2) Construction sections and large-scale details sufficient for accurate bidding and for the purpose of correlating all parts of the work should be part of the general drawings. This is particularly important where the size of a project makes necessary the preparation of the general drawings at a scale of ⅛ inch equals 1 foot or less.

(3) Mechanical and electrical work should be shown on separate plans.

(4) Schedules should be provided for doors, windows, finishes, electrical fixtures, finish hardware, and any other specialty items necessary to clarify drawings.

(d) *Specifications.* Trade-type specifications (specifications divided into sections for various trades) should be used. The specifications should be complete, clear, and concise, with adequate description of the various classes of work shown under the proper sections and headings.

(e) *Methods of administering construction.* Projects involving a total development cost of less than $100,000 which do not include an LH grant may, with the approval of the State Director, follow the contract procedure in § 1924.6(a) of this subpart without modification. Construction of all other projects, however, will be administered by the contract method or owner-builder method as set forth in this section.

(1) *Contract method.* This method of development will be used for all complex construction except in cases where owner-builder method is authorized. Development under this method is done in accordance with § 1924.6(a) of this subpart except as modified by this paragraph. All construction work will

be completed under one written construction contract. Guide 1, "Contract Documents," of this subpart (available in any Agency office) is provided to assist Agency personnel and applicants in assembling and reviewing contract documents for more complex construction such as that administered under this section.

(i) *Competitive bidding methods.* (A) All construction contracts must be awarded on the basis of competitive bidding unless an exception is granted in accordance with paragraph (e)(1)(vii) of this section thereby permitting contract negotiation. The applicant's architect should prepare the bidding documents. Public notice must be given inviting all interested bidders to submit a bid. Prospective bidders may be contacted asking for their bids; however, public notice is necessary so that all local contractors have the opportunity to submit bids.

(B) A bid bond is required from each bidder in the amount of 5 percent of the bid price as assurance that the bidder will, upon acceptance of the bid, execute the required contract documents within the time specified.

(C) The construction contract will be awarded based on the contract cost, and all conditions listed in the "Invitation to Bid."

(D) If advertising does not provide a satisfactory bid in the opinion of the applicant and the Agency, the applicant shall reject all bids and will then be free to negotiate with bidders on anyone else to obtain a satisfactory contract. The following conditions must be met:

(*1*) The State Director determines that the original competitive bid process was handled in a satisfactory manner and that there is no advantage to advertising for competitive bid again.

(*2*) The requirements of paragraph (e)(1)(vii) of this section are met.

(E) If there is no agreement by the Agency and the applicant as to the construction cost, the State Director will cease any further action on the preapplication and inform the applicant of the right to appeal in accordance with subpart B of part 1900 of this chapter.

(ii) *Contract documents.* Contract documents will conform with recognized professional practices as prescribed in this paragraph. Such contract documents will contain substantially the following:

Item I Invitation for Bids (Form RD 1924–5)
Item II Information for Bidders
Item III Bid
Item IV Bid Bond
Item V Agreement (Construction Contract)
Item VI Compliance Statement (Form RD 400–6)
Item VII General Conditions
Item VIII Supplemental General Conditions
Item IX Payment Bond (exhibit F of this subpart)
Item X Performance Bond (exhibit G of this subpart)
Item XI Notice of Award
Item XII Notice of Proceed
Item XIII Drawings and Specifications
Item XIV Addenda
Item XV Contract Change Order (Form RD 1924–7)
Item XVI Labor Standards Provisions [Where applicable]
Item XVII Monthly Employment Utilization Report (Form CC–257)
Item XVIII Partial Payment Estimate (Form RD 1924–18)
Item XIX Builder's Warranty (Form RD 1924–19)

(A) Substitution of term "architect" for "engineer" may be necessary on some of the forms. Other modifications may be necessary in some cases to conform to the nature and extent of the project. All such contract documents and related items will be concurred with by the State Director, with the assistance of OGC prior to the release of invitations to bid.

(B) Items listed as I through IV and item XI of paragraph (e)(1)(ii) of this section may be omitted when an exception to the competitive bidding requirement is granted in accordance with paragraph (e)(1)(vii) of this section, thereby permitting a negotiated contract.

(C) All negotiated contracts shall include a provision to the effect that the borrower, USDA, the Comptroller General of the United States, or any of their duly authorized representatives, shall have access to any books, documents, papers, and records of the contractor which are directly pertinent to a specific Federal loan program for the purpose of making audit, examination, excerpts, and transcriptions.

(D) A provision of liquidated damages *will* be included in all contracts. The liquidated damage amount must be reasonable and represent the best estimate possible of how much interest or other costs will accrue on the loan, and also represent any loss of rent or other income which would result from a delay in the completion of the project beyond the estimated completion date.

(E) All contracts shall include a provision for compliance with the Copeland "Anti-Kickback" Act (18 U.S.C. 874) as supplemented in Department of Labor regulations (29 CFR part 3). This Act prohibits anyone from inducing any person in connection with the construction to give up any part of the compensation to which the person is otherwise entitled.

(F) All contracts will contain a certification by the applicant indicating that there is not now nor will there be an identity of interest between the applicant and any of the following: Contractor, architect, engineer, attorney, subcontractors, material suppliers, equipment lessors, or any of their members, directors, officers, stockholders, partners, or beneficiaries unless specifically identified to the Agency in writing prior to the award of the contract. All contracts must also indicate that when any identity of interest exists or comes into being, the contractor agrees to have construction costs as reported to the Agency on Form 1924-13, "Estimate and Certificate of Actual Cost," audited by a Certified Public Accountant (CPA) or Licensed Public Accountant (LPA) licensed prior to December 31, 1970, who will provide an opinion as to whether the Form RD 1924-13 presents fairly the costs of construction in conformity with eligible construction costs as prescribed in Agency regulations.

(G) All contracts on any form other than Form RD 1924-6, must contain the language of clause (D) of Form RD 1924-6, which is available in all Agency offices. The language of clause (D) of Form RD 1924-6 sets forth the Notice of Requirement for Affirmative Action to Ensure Equal Employment Opportunity required by Executive Order 11246, the Equal Opportunity clause published at 41 CFR 60-1.4 (a) and (b), and the Standard Federal Equal Employment Opportunity Construction Contract Specifications required by Executive Order 11246. For contract forms other than Form RD 1924-6, Form AD 767, "Equal Employment Opportunity Contract Compliance Notices," which can be obtained from the Finance Office, should be attached and made a part of the contract.

(H) All contracts will contain a provision that they are not in full force and effect until concurred with by the State Director or the State Director's delegate, in writing. Therefore, before loan closing or before the start of construction, whichever occurs first, the State Director or the State Director's delegate will concur in the contract form, content, and execution if acceptable, by including the following paragraph at the end of the contract:

The Agency, as potential lender or insurer of funds to defray to costs of this contract, and without liability for any payments thereunder, hereby concurs in the form, content, and execution of this contract.

Date _____

Agency Official _____

Title _____

(I) The requirements of § 1924.6 (a)(11)(iv) of this subpart apply to all contracts or subcontracts in excess of $10,000.

(iii) *Surety.* When multiple advances of loan or grant funds are utilized, surety that guarantees both payment and performance in the full amount of the contract will be provided in accordance with § 1924.6(a)(3)(ii) of this subpart. Exceptions to the surety requirements shall be governed by the following:

(A) In accordance with the guidance and recommendations of OMB Circulars A-102 and A-110, exceptions to the surety requirements of § 1924.6(a)(3)(ii) of this subpart will not be granted for nonprofit organization or public body applicants.

(B) For loans or grants to applicants other than non-profit organizations or public bodies that are within the State Director's approval authority, the State Director may, upon request of the borrower or grantee, grant exceptions to the surety requirements in accordance with the provisions of § 1924.6(a)(3)(iii) of this subpart. Before granting such an exception, however, the State Director should be provided the following information from the proposed contractor in order to fully evaluate the experience and capabilities of the contractor:

(1) A resume indicating the contractor's history, ability and experience.

(2) A current, dated and signed financial statement of the contractor's operations indicating the payment status of accounts and any contingent liabilities that may exist. Agency personnel will be responsible for analyzing the financial statement as to the sufficiency of the contractor's financial capability to carry out construction. The financial strength must demonstrate the ability of the contractor to pay all bills prior to receiving periodic draws of funds from the lender.

(3) A credit report (obtained at no expense to the Agency) attesting to the contractor's credit standing.

(4) A listing of trade references that could be contacted to substantiate the contractor's experience and good standing.

(5) Statements from owners for whom the contractor has done similar work, indicating the scope of the work and the owner's evaluation of the contractor's performance.

(C) For loans or grants to applicants other than non-profit organization or public bodies that are in excess of the State Director's approval authority, the State Director may request National Office authorization to grant one of the exceptions to the surety requirements as indicated in §1924.6(a)(3)(iii) of this subpart. The following information must be submitted with the request to the National Office:

(1) An explanation of why interim financing is not available.

(2) An explanation of why the proposed contractor cannot obtain surety bonds meeting the requirements of §1924.6(a)(3)(ii) of this subpart.

(3) The information listed in paragraph (e)(1)(iii)(B) of this section.

(4) The drawings and specifications for the proposed project, together with the comments of the State architect/engineer.

(5) The applicant's written request for an exception.

(6) An explanation of why the requirements of §1924.6(a)(3)(iii) (A) or (B) of this subpart cannot be met in those cases where the State Director requests authorization to grant an exception as indicated in §1924.6(a)(3)(iii)(C) of this subpart. When such a request is made, the documentation required of the contractor under the provision must also be forwarded.

(7) The State Director's recommendation.

(D) Adequate steps will be taken to protect the interests of the borrower and the government in accordance with the payment provisions of §1924.6(a)(12)(i) of this subpart and any alternative as outlined in §1924.6(a)(3)(iii)(c) of this subpart.

(iv) *Contract cost breakdown.* In any case where the loan approval official feels it appropriate, and prior to the award or approval of any contract in which there is an identity of interest as defined in §1924.4 (i) of this subpart, the contractor and any subcontractor, material supplier or equipment lessor sharing an identity of interest must provide the applicant and the Agency with a trade-item cost breakdown of the proposed contract amount for evaluation. The cost of any surety as required by §1944.222 (h) and (i) of subpart E of part 1944 of this chapter and §1924.6(a)(3) of this subpart, or cost certification as required by paragraph (e)(1)(v) of this section, will be included in the proposed contract amount and shown under General Requirements on Form RD 1924–13, which is available in all Agency offices. Agency personnel will be responsible for reviewing the estimates on Form RD 1924–13 to determine if the dollar amounts total correctly, to assure that costs are categorized under their appropriate columns, and to confirm that the estimated costs for all line items are reasonable and customary for the State.

(v) *Cost certification.* Whenever the State Director determines it appropriate, and in all situations where there is an identity of interest as defined in §1924.4(i) of this subpart, the borrower, contractor and any subcontractor, material supplier, or equipment lessor having an identity of interest must each provide certification using Form RD 1924-13 as to the actual cost of the work performed in connection with the construction contract. The construction costs, as reported on Form RD 1924-13, must also be audited, in accordance with Government Auditing Standards, by a CPA, or LPA licensed on or before December 31, 1970. In addition, certain agreed upon procedures (available in any Agency office) will be performed in accordance with Attestation Standards. In some cases, the Agency will contract directly with a CPA or LPA for the cost certification. In that event, documentation necessary to have the costs of construction certified by an Agency contractor that they were the actual costs of the work performed, as reported on Form RD 1924-13, will be provided. Funds which were included in the loan for cost certification and which are ultimately not needed because Agency contracts for the cost certification will be returned on the loan. Agency personnel will utilize exhibit M of this subpart (available in any Agency office) and Form RD 1924-26, "Cost Certification Worksheet," to assist in the evaluation of the cost certification process.

(A) Prior to the start of construction, the borrower, contractor and any subcontractor, material supplier, or equipment lessor sharing an identity of interest must submit, to the CPA or LPA, the accounting system that the borrower, contractor, subcontractor, material supplier or equipment lessor and/or the CPA or LPA proposes to set up and use in maintaining a running record of the actual cost. In order to be acceptable, the borrower must provide a written assertion that it has an accounting system that is suitably designed to provide for a trade-item basis comparison of the actual cost as compared to the estimated cost submitted on Form RD 1924-13. Costs pertaining to a specific line item will be set up in the accounting system for that particular account. For instance, only costs of materials, supplies, equipment, and labor associated with concrete will be shown in the concrete account. The accounting system must also restrict costs to those pertaining to a specific project so that costs from multiple projects will not be co-mingled. The

independent CPA or LPA shall report on the borrower's assertion in accordance with the Standards for Attestation Engagements of the American Institute of Certified Public Accountants (AICPA). The borrower's and the CPA or LPA's reports on the accounting system shall be provided to the Agency by the borrower.

(B) Prior to final payment to anyone required to cost certify, a trade-item breakdown showing the actual cost compared to the estimated cost must be provided to the owner and the Agency. Form RD 1924-13 is the form of comparative breakdown that must be used, and contains the certifications required of the applicant and contractor prior to final payment. The amounts for builder's general overhead, builder's profit, and general requirements, respectively, shall not exceed the amounts represented on the estimate of cost breakdown provided in accordance with paragraph (e)(1)(iv) of this section for any contractor, subcontractor, material supplier, or equipment lessor having or sharing an identity of interest with the borrower. The amounts for general overhead, builder's profit, and general requirements must be established prior to the Agency approving the construction contract and will not be changed during the course of construction. This applies to all contractors, subcontractors, material suppliers, or equipment lessors having or sharing an identity of interest with the applicant. Contract change orders will be processed to adjust the contract amount downward prior to the final payment to the contractor, if necessary, to assure that the amounts shown in the certificate of actual costs do not exceed the amounts represented in the contract cost breakdown. Reduction in the builder's profit, and general overhead if needed, will counterbalance any increase reflected in the contract costs. Any funds remaining as a result of hard cost savings will be applied to the account as an extra payment or used for eligible loan purposes approved by the Agency as long as the improvements are genuinely needed and will enhance marketability of the project. All increases or decreases of 15 percent or more in line item costs will require documentation as to the reason for the increases and/or decreases. The State Director may require documentation for increases and/or decreases of less than 15 percent, if he/she determines it necessary. This information will be required with the cost certification.

(C) The CPA or LPA audit, performed in accordance with Government Auditing Standards, will include such tests of the accounting records and such other auditing procedures of the borrower and the contractor (and any subcontractor, material supplier or equipment lessor sharing an identity of interest) concerning the work performed, services rendered, and materials supplied in accordance with the construction contract he/she considers necessary to express an opinion on the construction costs as reported on Form RD 1924–13. The CPA or LPA shall also perform the additional agreed upon procedures specified by the Agency (available in any Agency office), performed in accordance with Attestation Standards, for the applicant and the contractor (and any subcontractor, material supplier, or equipment lessor sharing an identity of interest) concerning the work performed, services rendered, and materials supplied in accordance with the construction contract.

(D) Upon completion of construction and prior to final payment, the CPA or LPA will provide an opinion concerning whether the construction costs, as reported on Form RD 1924–13, present fairly the costs of construction in conformity with eligible construction costs as prescribed in Agency regulations.

(E) In some cases, cost certification will be obtained by the Agency through direct contract with the CPA or LPA. The borrower and his/her CPA or LPA will cooperate fully with the contract CPA or LPA by providing all documentation necessary to conduct the certification. The Agency reserves the right to determine, upon receipt of the certified Form RD 1924–13 and the auditor's report, whether they are satisfactory to the Agency. If not satisfactory to the Agency, the borrower will be responsible for providing additional information.

(F) There will exist no business relationship between the CPA or LPA and the borrower except for the performance of the examination of the cost certification, accounting systems work, and tax preparation. Any CPA or LPA who acts as the borrower's accountant (performing manual or automated bookkeeping services or maintains the official accounting records) will not be the same CPA or LPA who cost certifies the project.

(G) Forms RD 1944–30, "Identity of Interest (IOI) Disclosure Certificate" and RD 1944–31, "Identity of Interest (IOI) Qualification Form," provide written notification to the borrower that willful and intentional falsification of cost certification documents will result in debarment of all violators in accordance with the provisions of RD Instruction 1940–M (available in any Agency office). These forms require the disclosure of all identities of interest associated with project construction, certify the entity's ability to provide the contracted service, and cite the penalties for failure to disclose or falsify such certification. Each applicant/borrower will be required to complete and sign the forms (available in any Agency office).

(H) (*Subcontracting development work.* (*1*) Contractors will not be allowed to obtain a profit and overhead unless they are performing actual construction. "Actual construction" means "work" as defined in American Institute of Architects (AIA) documents: "* * * labor, materials, equipment, and services provided by the contractor to fulfill the contractor's obligations." Under this definition, contractors who choose to subcontract out construction of the project to another contractor will not obtain a builder's fee (general overhead and profit) when:

(*i*) More than 50 percent of the contract sum in the construction contract is subcontracted to one subcontractor, material supplier, or equipment lessor, and/or

(*ii*) Seventy-five percent or more with three or fewer subcontractors, material suppliers and/or equipment lessors.

(*2*) NOTE: If two or more subcontractors have common ownership, they are considered as one subcontractor.

(*3*) How to apply rule:

(*i*) The 50 percent rule will apply when division of the amount of the largest subcontract by the contract sum of the construction contract results in more than 50 percent.

(*ii*) The 75 percent rule will apply when division of the sum of the amounts of the three largest subcontracts by the contract sum of the construction contract results in 75 percent or more.

(I) (*Qualified contracting entities.* Contractors, subcontractors, material suppliers, and any other individual or organization sharing an identity of interest and providing materials or services for the project must certify that it is a viable, ongoing trade or business qualified and properly licensed to undertake the work for which it intends to contract. Form RD 1944–31 will be prepared and executed by the contracting entities. The form provides notification to the entities of the penalty, under law, for erroneously certifying to the statements contained therein. Debarment actions will be instituted against entities who fail to disclose an identity of interest in accordance with the provisions of RD Instruction 1940–M (available in any Agency office).

(vi) *Method of payments.* Partial payments may be requested in accordance with the terms of the construction contract on Form RD 1924–18, "Partial Payment Estimate," or other professionally recognized form that contains the architect's certification, approval of the owner, and conditional acceptance of the Agency as shown in Form RD 1924–18.

(A) If interim financing is available at reasonable rates and terms for the construction period, such financing shall be obtained. exhibit B of subpart E of part 1944 of this chapter shall be used to inform the interim lender that the Agency will not close its loan until the project is substantially complete, ready for occupancy, evidence is furnished indicating that all bills have been paid or will be paid at loan closing for work completed on the project, all inspections have been completed and all required approvals have been obtained from municipal and governmental authorities having jurisdiction over the project.

Upon presentation of proper partial payment estimates approved by the applicant and accepted by the Agency, the interim lender may advance construction funds in accordance with the payment terms of the contract. It is suggested that partial payments not exceed 90 percent of the value of work in place and materials suitably stored on site.

(B) When interim financing is not available, payments will be made in accordance with § 1924.6(a)(12) of this subpart.

(vii) *Exception to competitive bidding—* (A) *For all applicants.* An applicant may negotiate a construction contract provided the State Director grants an exception and documentation shows that:

(*1*) The contract price is competitive with other projects similar in construction and design being built in the area.

(*2*) The proposed contractor is experienced in construction of projects of similar size, scope, and complexity, and is recognized as a reliable builder.

(*3*) The proposed development work meets all requirements of this subpart.

(*4*) If appropriate for nonprofit organizations and public bodies, the applicant provides a copy of a duly authorized resolution by its governing body requesting the Agency to permit awarding the construction contract without formal bidding.

(*5*) The applicant is permitted by state law, local law and/or organizational by-laws to negotiate a construction contract.

(*6*) The requirements of paragraphs (e)(1) (ii), (iii), (iv) and (v) of this section are met.

(B) In considering an exception to competitive bidding, the following additional steps will be taken in all cases.

(*1*) If, after a full review of the case documents by the appropriate members of the State Office staff, the State Director determines that the requirements have been met and the costs are reasonable, an exception to competitive bidding may be granted. Written documentation of the State Office review results will be placed in the application file.

(*2*) If after the full review by the State Office staff, the State Director determines that the negotiated contract price is not competitive with other similar projects in construction and design being built in the area, the applicant will be requested to competitively bid the construction of the project in accordance with paragraph (e)(1)(i) of this section.

(3) If there is no agreement by the Agency and the applicant as to the construction cost, the State Director will cease any further action on the preapplication and inform the applicant of the right to appeal in accordance with subpart B of part 1900 of this chapter.

(C) Any requests for exceptions to competitive bidding that are not covered in this section may be submitted to the National Office for consideration.

(viii) *Exception to contract method—public body.* With the approval of the National Office, the State Director may grant to a public body an exception to the requirement for using contract method construction under the following circumstances:

(A) The loan or grant is for repair or rehabilitation of existing facilities and it is not practicable to perform all work by the contract method.

(B) The applicant has the managerial ability and qualified employees necessary to complete the work successfully.

(C) That applicant submits a written request to the District Director indicating:

(1) The scope of work and construction timetable;

(2) What phases of work can be contracted and what cannot;

(3) Why is it not practicable to contract all phases;

(4) Management ability and employee qualifications for performing the work;

(5) Proposed method of fund control and frequency of payments;

(6) How changes in scope of work and construction timetable will be approved; and,

(7) Proposed method of certifying progress and requesting payments.

(D) The request, recommendations of the District Director, appropriate members of the State Office staff and the State Director and the application file will be sent to the National Office.

(2) *Owner-builder method.* This method of development is used only when re-quested by profit or limited profit RRH applicants when the applicant or any of its controlling principals (such as stockholders, members, partners other than limited partners, directors, or officers), are general contractors by profession, and will serve as the builder of the project without a written construction contract. The State Director may make an exception to the contract method of construction and authorize proceeding by the owner-builder method of construction in accordance with the provisions of this section if the amount of the loan(s) does not exceed the State Director's approval authority. For projects over the State Director's authority, prior written consent of the National Office is required. In such cases, the drawings, specifications, cost estimates, copy of the State Architect/Engineer's review and detailed information on the applicant's qualifications will be submitted to the National Office along with the State Director's recommendations.

(i) The applicant's request to construct a project by the owner-builder method of construction shall be in the form of a letter giving specific and detailed information concerning the owner-builder's proposal, and the qualifications and past experience of the owner-builder. The following information must be included with the request:

(A) A resume indicating the owner-builder's history, ability, and experience.

(B) Dated and signed financial statements on the owner-builder's operation (including balance sheets and statements of income and expense) from current and prior years indicating the payment status of the owner-builder's accounts and any contingent liabilities that may exist. Agency personnel will be responsible for analyzing the financial statement as to the sufficiency of the owner-builder's financial capability to carry out construction. The financial strength must demonstrate the ability of the owner-builder to pay all bills prior to receiving periodic draws of funds from the lender.

(C) A written, dated, and signed statement agreement to provide any funds necessary in excess of the applicant's contribution and the loan amount to complete the project.

(D) A credit report (obtained at no expense to the Agency) attesting to the owner-builder's credit standing.

(E) A listing of trade references that could be contacted to substantiate the owner-builder's experience and good standing.

(F) Statements from other persons for whom the owner-builder has done similar work, indicating the scope of the work and that person's evaluation of the owner-builder's performance.

(G) A current, dated, and signed trade-item cost breakdown of the estimated total development cost of the project which has been prepared by the applicant/owner-builder. Form RD 1924–13 will be used for this purpose. If cost certification services are required by the Agency, the cost of such services may be included in the total development cost of the project. Any subcontractor, material supplier, or equipment lessor sharing an identity of interest with the applicant/owner-builder as defined in § 1924.4(i) of this subpart must also provide a trade-item cost breakdown of the proposed amount.

(H) Prior to the start of construction, the owner-builder and any subcontractor, material supplier, or equipment lessor sharing an identity of interest must submit, to the CPA or LPA, the accounting system that the owner-builder, subcontractor, material supplier or equipment lessor and/or the CPA or LPA proposes to set up and use in maintaining a running record of the actual cost. In order to be acceptable, the owner-builder must provide a written assertion that it has an accounting system that is suitably designed to provide for a trade-item basis comparison of the actual cost as compared to the estimated cost submitted on Form RD 1924–13. Costs pertaining to a specific line item will be set up in the accounting system for that particular account. For instance, only costs of materials, supplies, equipment, and labor associated with concrete will be shown in the concrete account. The accounting system must also restrict costs to those pertaining to a specific project so that costs from multiple projects will not be co-mingled. The independent CPA or LPA shall report on the owner-builder's assertion in accordance with the Standards for Attestation Engage-

ments of the AICPA. The owner-builder's and the CPA or LPA's reports on the accounting system shall be provided to the Agency by the owner-builder.

(I) A written, dated, and signed statement agreeing to permit U.S. Department of Agriculture, the Comptroller General of the United States, or any of their duly authorized representatives, to have access to any books, documents, papers, and records which are directly pertinent to the specific Federal program for the purpose of making audit, examination, excerpts and transcriptions.

(ii) In order to grant an exception to the contract method of construction and proceed with the owner-builder method of construction, the State Director must determine that the following conditions exist:

(A) The applicant or at least one of its principals is a fully qualified and licensed (if necessary under applicable local law) builder by profession, has adequate experience in constructing the type of units proposed as well as projects of similar size, scope, and complexity and will be able to complete the work in accordance with the Agency approved drawings and specifications.

(B) Based upon the information presented in the applicant's financial statements, the applicant is presently able and is likely to continue to be able to provide any funds necessary in excess of the applicant's contribution and the loan amount to complete the project.

(C) The total development cost of the project does not exceed that which is typical for similar type projects in the area. The total development cost recognized by the Agency for each individual case will be determined by the MFH Coordinator with the advice of the State Architect.

(D) The owner-builder has provided sufficient information on all contracts or subcontracts in excess of $10,000 to permit compliance with § 1924.6(a)(11)(iv) of this subpart.

(iii) In addition to the requirements for the State Director to authorize the owner-builder method of construction as indicated in § 1924.13(e)(2) (i) and (ii) of this subpart, the following additional steps will be taken by the State Director.

(A) If, after a full review of the case documents by the appropriate members of the State Office staff, the State Director determines that the requirements have been met and the construction cost is reasonable, an exception to competitive bidding may be granted. Written documentation of the State Office review results will be placed in the application file.

(B) If, after the full review by the State Office staff, the State Director determines that the construction cost is not competitive with other similar projects in construction and design being built in the area, the applicant will be requested to competitively bid the construction of the project in accordance with paragraph (e)(1)(i) of this section.

(C) If there is no agreement by the Agency and the applicant as to construction cost and the applicant is not agreeable to any of the aforementioned alternatives, the State Director will cease any further action on the preapplication and inform the applicant of the right to appeal, in accordance with subpart B of part 1900 of this chapter.

(iv) The development cost of the project may include a typical allowance for general overhead, general requirements and a builder's profit. These amounts may be determined by local investigation and also from HUD data for the area. The applicant/owner-builder and any subcontractors, material suppliers and equipment lessors having or sharing an identity of interest with the applicant/owner-builder may not be permitted a builder's profit, general overhead, and general requirements which exceed the amounts represented on their cost breakdown.

(v) Under no circumstances will loan funds be used to pay the owner/builder or its stockholders, members, directors or officers, directly or indirectly, any profits from the construction of the project except a typical builder's fee for performing the services that would normally be performed by a general contractor under the contract method of construction. Discounts and rebates given the owner-builder in advance must be deducted before the invoices are paid. If discounts or rebates are given after the invoices are paid, the funds must be returned to the supervised bank account or applied on the interim construction loan, as appropriate. Under no circumstances will the dollar amount be placed in the reserve account.

(vi) The plan and specifications must be specific and complete so that there is a clear understanding as to how the facility will be constructed and the materials that will be used.

(vii) When architectural services are required by §1924.13(a) during the construction and warranty phases they must be provided by an architect who has no identity of interest with the applicant/owner-builder. The services to be rendered during the construction and warranty phases include, but are not limited to inspections, changes in the scope of project or work to be done, administration of construction accounts, rejection of work and materials not conforming to the Agency approved drawings and specifications, and other appropriate service listed in §1924.13(a)(5) (v) and (vi) of this subpart.

(viii) The applicant/owner-builder and any subcontractor, material supplier, or equipment lessor sharing an identity of interest as defined in §1924.4(i) of this subpart must each provide certification as to the actual cost of the work performed in connection with the construction of the project on Form RD 1924–13 prior to final payment. The construction costs, as reported on Form RD 1924–13, must be audited by a CPA, or LPA licensed on or before December 31, 1970, in accordance with Government Auditing Standards, and certain agreed upon procedures (available in any Agency office) performed in accordance with Attestation Standards. In some cases, FmHA or its successor agency under Public Law 103–354 will contract directly with a CPA or LPA for the cost certification. In that event, documentation necessary to have the costs of construction certified by an Agency contractor that they were the actual costs of the work performed, as reported on Form RD 1924–13, will be provided. Funds which were included in the loan for cost certification and which are ultimately not needed because Agency contracts for the cost certification will be returned on the loan.

(A) The CPA or LPA's audit, performed in accordance with Government Auditing Standards, will include such tests of the accounting records and such other auditing procedures of the applicant/owner-builder (and any subcontractor, material supplier, or equipment lessor sharing an identity of interest) concerning the work performed, services rendered, and materials supplied in connection with the construction of the project he/she considers necessary to express an opinion on the construction costs as reported on Form RD 1924-13. Upon completion of construction and prior to final payment, the CPA or LPA will provide an opinion as to whether the construction costs as reported on Form RD 1924-13 present fairly the costs of construction in conformity with eligible construction costs as prescribed in Agency regulations. The Agency reserves the right to determine, upon receipt of the certified Form RD 1924-13 and the auditor's report, whether they are satisfactory to the Agency. At a minimum, the CPA or LPA shall also perform any additional agreed upon procedures (available in any Agency office) specified by the Agency, performed in accordance with Attestation Standards, of the owner-builder (and any subcontractor, material supplier, or equipment lessor sharing an identity of interest) concerning the work performed, services rendered, and materials supplied in connection with the construction. There will exist no business relationship between the CPA or LPA and the borrower except for the performance of the examination of the cost certification, accounting systems work, and tax preparation. Any CPA or LPA who acts as the borrower's accountant (performing manual or automated bookkeeping services or maintains the official accounting records) will not be the same CPA or LPA who cost certifies the project.

(B) Prior to final payment to anyone required to cost certify, the Agency must be provided with a certification and a trade-item breakdown showing the actual cost compared to the estimated cost furnished in accordance with paragraph (e)(2)(i)(G) of this section. Form RD 1924-13 is the form of comparative breakdown that must be used, and contains the certification required of the applicant/owner-builder prior to final payment. The amounts for builder's general overhead, general requirements, and builder's profit shall not exceed the amounts represented on the estimate of cost breakdown provided in accordance with paragraph (e)(2)(i)(G) of this section for the owner-builder or any subcontractor, material supplier, or equipment lessor having or sharing an identity of interest with the applicant/owner-builder. Final payment to the owner-builder will be adjusted, if necessary, to assure that the amounts shown on the certificate of actual cost do not exceed the amounts represented on the cost breakdown. Any funds remaining as a result of hard cost savings will be applied to the account as an extra payment or used for eligible loan purposes approved by the Agency as long as the improvements are genuinely needed and will enhance marketability of the project. All increases or decreases of 15 percent or more in line item costs will require documentation as to the reason for the increases or decreases. The State Director may require documentation for increases or decreases of less than 15 percent, if he/she determines it necessary. This information will be required with the cost certification.

(C) Subcontracting development work.

(1) Owner-builders will not be allowed to obtain a profit and overhead unless they are performing actual construction. "Actual construction" means "work" as defined in AIA documents: "* * * labor, materials, equipment, and services provided by the contractor to fulfill the contractor's obligations." Under this definition, owner-builders who choose to subcontract out construction of the project to another contractor will not obtain a builder's fee (general overhead and profit) when:

(i) More than 50 percent of the total cost of the building construction is subcontracted to one subcontractor, material supplier, or equipment lessor, and/or (*ii*) Seventy-five percent or more with three or fewer subcontractors, material suppliers, and/or equipment lessors.

(*2*) NOTE: If two or more subcontractors have common ownership, they are considered as one subcontractor.

(*3*) How to apply rule:

(*i*) The 50 percent rule will apply when division of the amount of the largest subcontract by the total amount of the building cost results in more than 50 percent.

(*ii*) The 75 percent rule will apply when division of the sum of the amounts of the three largest subcontracts by the total building cost results in 75 percent or more.

(D) Qualified contracting entities. Contractors, subcontractors, material suppliers, and any other individual or organization sharing an identity of interest and providing materials or services for the project must certify that it is a viable, ongoing trade or business qualified and properly licensed to undertake the work for which it intends to contract. Form RD 1944–31 will be prepared and executed by the contracting entities. The form provides notification to the entities of the penalty, under law, for erroneously certifying to the statements contained therein. Debarment actions will be instituted against entities who fail to disclose an identity of interest in accordance with the provisions of RD Instruction 1940–M (available in any Agency office).

(ix) Requests for payment for work performed by the owner-builder method, shall be permitted to the Agency District Director for review and approval prior to each advance of funds in order to insure that funds are used for authorized purposes. Requests for payment shall be made on Form RD 1924–18 or other professionally recognized form containing the following certification to the Agency:

The undersigned certifies that the work has been carefully inspected and to the best of their knowledge and belief, the quantities shown in this estimate are correct and the work has been performed in accordance with the contract documents.

(Name of Architect)

By: _____

(Title   (Date)

Approved by Owner's Representative: By: __

(Title)

Accepted by Agency Representative: By: ___

(Title)

The review and acceptance of partial payment estimates by the Agency does not attest to the correctness of the quantities shown or that the work has been performed in accordance with the plans and specifications.

(A) If interim financing is available at reasonable rates and terms for the construction period, such financing shall be obtained. Exhibit B of subpart E of part 1944 of this chapter shall be used to inform the interim lender that the Agency will not close its loan until the project is complete, ready for occupancy, evidence is furnished indicating that all bills have been paid for work completed on the project, all inspections have been completed and all required approvals have been obtained from any governmental authorities having jurisdiction over the project. Upon presentation of proper partial payment estimates containing an estimate of the value of work in place which has been prepared and executed by the owner-builder, certified by the applicant's architect, and accepted by the Agency, the interim lender may advance construction funds in accordance with the provisions of this section. It is suggested that the partial payment not exceed 90 percent of the value of work in place and material suitably stored on site.

(B) If interim financing is not available, partial payments not to exceed 90 percent of the value of work in place and materials suitably stored on site may be made to the owner-builder for that portion of the estimated cost of development guaranteed by a letter of credit or deposits meeting the requirements of § 1924.6(a)(3)(iii) (A), (B) or (C) of this subpart. Partial payments may not exceed 60 percent of the value of work in place in all other cases. The determination of the value of work in place will be based upon an application for payment containing an estimate of the value of work in place which has been prepared and executed by the owner-builder, certified by the borrower's architect, and accepted by the

543

Agency. Prior to receiving the first partial payment, the owner-builder must submit a schedule of prices or values of the various trades or phases of the work aggregating the total development cost of the project as required in § 1924.13(e)(2)(i) (G) and (H) of this subpart. Each application for payment must be based upon this schedule, and show the total amount owed and paid to date for materials and labor procured in connection with the project. With each application for payment, the owner-builder must also submit evidence showing how the requested partial payment is to be applied, evidence showing that previous partial payments were properly applied, and a signed statement from the applicant's attorney, title insurance company, or local official in charge of recording documents certifying that the public records have been searched and that there are no liens of record. When the District Director has reason to believe that partial payments may not be applied properly, checks will be made payable to persons who furnish materials and labor for eligible purposes in connection with the project.

(x) Under no circumstances shall funds be released for final payment or to pay any items of the builder's profit until the project is 100 percent complete, ready for occupancy, and the owner-builder has completed and properly executed Form RD 1924–13 or complied with the cost certification procedures of § 1924.13(e)(2)(viii) of this subpart.

[52 FR 8002, Mar. 13, 1987; 52 FR 26139, July 13, 1987, as amended at 53 FR 2155, Jan. 26, 1988; 59 FR 6882, Feb. 14, 1994; 61 FR 56116, Oct. 31, 1996]

§§ 1924.14–1924.48   [Reserved]

§ 1924.49   State supplements.

State Supplements or policies will not be issued or adopted to either supplement or set requirements different from those of this subpart, unless specifically authorized in this subpart, without prior written approval of the National Office.

§ 1924.50   OMB control number.

The reporting and recordkeeping requirements contained in this regulation have been approved by the Office of Management and Budget (OMB) and have been assigned OMB control number 0575–0042. Public reporting burden for this collection of information is estimated to vary from 5 minutes to 4 hours per response, with an average of 37 minutes per response, including time for reviewing instructions, searching existing data sources, gathering and maintaining the data needed, and completing and reviewing the collection of information. Send comments regarding this burden estimate or any other aspect of this collection of information, including suggestions for reducing this burden, to U.S. Department of Agriculture, Clearance Officer, OIRM, AG Box 7630, Washington, DC 20250; and to the Office of Management and Budget, Paperwork Reduction Project (OMB #0575–0042), Washington, DC 20503.

[59 FR 6885, Feb. 14, 1994]

EXHIBIT A TO SUBPART A OF PART 1924— ESTIMATED BREAKDOWN OF DWELLING COSTS FOR ESTIMATING PARTIAL PAYMENTS

[In percent]

	With slab on grade	With crawl space	With base-ment
1. Excavation	3	5	6
2. Footings, foundations columns	8	8	11
3. Floor slab or framing	6	4	4
4. Subflooring	0	1	1
5. Wall framing, sheathing	7	7	6
6. Roof and ceiling framing, sheathing	6	6	5
7. Roofing	5	5	4
8. Siding, exterior trim, porches	7	7	6
9. Windows and exterior doors	9	9	8
10. Plumbing—roughed in	3	2	3
11. Sewage disposal	1	1	1
12. Heating—roughed in	1	1	1
13. Electrical—roughed in	2	2	2
14. Insulation	2	2	2
15. Dry wall or plaster	8	8	7
16. Basement or porch floor, steps	1	1	6
17. Heating—finished	3	3	3
18. Flooring	6	6	5
19. Interior carpentry, trim, doors	6	6	5
20. Cabinets and counter tops	1	1	1
21. Interior painting	4	4	3
22. Exterior painting	1	1	1
23. Plumbing—complete fixtures	4	4	3
24. Electrical—complete fixtures	1	1	1
25. Finish hardware	1	1	1

[In percent]

	With slab on grade	With crawl space	With base-ment
26. Gutters and downspouts	1	1	1
27. Grading, paving, land-scaping	3	3	3
Total	100	100	100

EXHIBIT B TO SUBPART A OF PART 1924— REQUIREMENTS FOR MODULAR/ PANELIZED HOUSING UNITS

For the benefit of the Agency this exhibit prescribes evaluation, acceptance, inspection and certification procedures formodular/ panelized housing units proposed for use in Agency Rural Housing programs. It applies to proposed development packages provided either under a contract between an Agency borrower and a single contractor or under a conditional commitment. This exhibit also describes the use of background information available through the Department of Housing and Urban Development (HUD) for analysis of manufactured products. This exhibit also applies to the evaluation of manufactured farm service buildings in paragraph XI, below. For the purpose of this exhibit, County Supervisor and County Office also mean District Director and District Office, respectively.

I. Applicable Standards and Manuals.

A. The HUD Handbook 4950.1, Technical Suitability of Products Program Technical and Processing Procedures, must be followed by housing manufacturers to obtain acceptance of their products. Acceptance documents issued by HUD include: Structural Engineering Bulletins (SEB) on a national basis, Area Letters of Acceptance (ALA) which when accepted by all Area HUD Offices in a HUD region will, in essence, become Regional Letters of Acceptance (RLA), Truss Connector Bulletins (TCB): and, Mechanical Engineering Bulletins (MEB). These documents as well as the Use of Material Bulletins (UM) and Materials Release Bulletins (MR) are addendums to the HUD Minimum Property Standards (MPS), Under handbook guidelines, HUD also examines state agency regulations concerning design, construction and labeling of modular/ panelized housing units and designates those states having procedures acceptable for use under HUD programs. Modular/panelized housing produced in these states is called *Category III* and is considered technically suitable for use without further structural analysis.

B. All Agency Offices should maintain a close working relationship with each HUD office in their jurisdiction to assure coordination. Any deviations in structure, materials or design from HUD acceptance documents must comply with one of the other applicable development standards.

II. Modular Housing Units that Require Factory Inspections.

Only those types which cannot be completely inspected on site are required to obtain acceptance from HUD. Those that receive acceptance will be periodically factory inspected by HUD or HUD's designated agency, usually about every 6 months.

III. Panelized Housing Units that Do Not Require Factory Inspections.

A. Housing completely assembled on the building site does not require HUD acceptance. This includes housing that is manufactured but is assembled on the site such as: Precut pieces, log wall houses, trussed roof rafters or floor trusses; open panel walls, and other types that can be completely inspected on site.

B. Housing that is assembled in local materials dealers' yards for moving to local sites and to be purchased by an Agency applicant, will be inspected during construction in the yard by the local Agency County representative. These units must be constructed according to the applicable development standard and not transported out of the local Agency County Office jurisdiction. The inspection must be recorded on Form RD 1924–12, "Inspection Report."

IV. Manufacturer's Actions Required for Submissions to the Agency are listed in attachment 1 to this exhibit B.

V. State Agency Office Actions when Manufacturing Facilities are in its Jurisdiction. The State Office, upon receipt of manufacturer's submission, must:

A. Determine that the unit structural system has been accepted by HUD as appropriate under HUD Handbook 4950.1 requirements.

B. Review the thermal characteristics and approach of the calculations to determine actions to be taken in compliance with paragraph IV C of exhibit D of this subpart.

C. Review the proposal for compliance with §1924.5(d)(1) of this subpart.

D. Determine that the prerequisites for consideration of acceptance by the Agency are met. The prerequisites include all of the following:

1. A current acceptance document from HUD (SEB, RLA, ALA), except for Category III housing (modular/panelized housing that does not have to have a Structural Engineering Bulletin as designated by HUD). In Category III states, the state government requirements for manufactured housing must be followed.

2. A current HUD Factory Inspection Report, Form No. 2051m, or in the case of Category III housing, a copy of the inspection report from the state government or accepted third party performing the factory inspection. Each report must be made by HUD or a

545

HUD authorized agency, and must be no older than 6 months.

3. A letter from the manufacturer requesting a review for acceptance. Enclosed with the letter shall be all the information listed in attachment 1 to this exhibit B.

E. Issue acceptance letters to the manufacturer stating the conditions of acceptance in the format of attachment 2 to this exhibit B. The letter shall have an attachment listing all models accepted in the format of attachment 3 to this exhibit B. A copy of the acceptance letter and list of models shall be sent to each County Office in the state and, when requested by the manufacturer, to each other Agency State Office in which the product is to be marketed.

F. After initial review of a submission, maintain a master file of accepted manufacturers and models and review the file twice yearly to determine the currency of the factory inspection reports and HUD or state government acceptance documents.

G. Notify manufacturers of overdue factory inspection reports, for acceptance of documents review and updating, using the format of attachment 4 to this exhibit B. Accompanying the notification will be a temporary acceptance sheet (Attachment 3 to this exhibit B) indicating to the manufacturer that the company models have temporary acceptance for 60 days. If the manufacturer provides evidence that a review is being processed by HUD, a maximum of an additional 90 days may be granted. Otherwise, the acceptance shall terminate on the last extension date and it will be necessary for the manufacturer to resubmit as if for initial acceptance.

H. Distribute a list of added models, deleted models, or notice of deletion of any manufacturer's product to the County Offices and other State Agency Offices as necessary.

I. Issue an initial supply of Manufacturer's and Builder's Certification forms (Attachment 5 to this exhibit B) to each existing and newly accepted manufacturer. Manufacturers are to duplicate this form as necessary in their market areas.

J. Resolve any problems with the manufacturer, as reported by the County Office. Action may include coordination, Agency plant inspections or cancellation of acceptance letters when problems persist.

VI. County Office Actions:

A. When an application is received involving any of the manufacturer's products on the accepted list, the County Office Agency authorized personnel will:

1. Review the drawings and description of materials described in paragraphs A and B of attachment 1 to this exhibit B. The floor plans and elevations must be identifiable with the model listed in the accepted list issued by the State Office.

2. Require the builder/dealer or manufacturer to provide any drawings necessary to adapt the house to the site conditions where the house will be located.

3. Require site plan drawing such as those illustrated in attachments 1 and 2 to exhibit C of this subpart (available in any Agency office).

4. Inspect and identify the model delivered against the manufacturer's certification and the accepted drawings and description of materials before the unit has been set on the foundation.

5. Require the builder/dealer to certify that the work for which the builder/dealer is responsible has been erected in compliance with the applicable development standard. This certification will be completed on a copy of attachment 5 to this exhibit B, and filed in County Office case file.

6. Observe any noncompliance with the applicable development standard or with paragraphs IV and V of this exhibit B. In this respect:

a. Minor noncompliance will be resolved by the manufacturer through the builder/dealer. In cases where there is no builder/dealer, the County Office may resolve such issues with the manufacturer directly.

b. Noncompliance that cannot be resolved at the County Office level will be reported to the State Office.

7. Inspect manufactured housing according to § 1924.8(d) of this subpart.

8. Be aware that the accepted list may include many models from which loan applicants may choose. No changes from accepted model designs are permitted. The model selected by an applicant should be appropriate to the needs of that particular family in accordance with 7 CFR part 3550.

VII. Noncompliance Issues.

A. When minor issues are noted, the County Office will attempt to resolve them as described above. If they cannot be resolved locally, they will be referred to the State Office. When any issues cannot be resolved at State Office level, the National Office Program Support Staff (PSS) will be contacted for guidance.

B. The National Office PSS coordinating with HUD, will take the appropriate actions to resolve the issues reported.

C. Manufacturers and builder/dealers must be aware that if the Agency inspector finds any of the following conditions, the inspector may refuse to accept the construction until corrections have been made:

1. Evidence of noncompliance with any option of the method described in the HUD—SEB, RLA, or ALA.

2. Faulty shop fabrication, including surface defects.

3. Damage to shop fabricated items or materials due to transportation, improper storage, handling or assembly operation.

4. Unsatisfactory field or site workmanship.

VIII. Actions by Other State Offices. When a State Office receives a copy of the accepted list from the State Office in which a manufacturing plant is located, it will:

A. Maintain a file, by manufacturer, of each accepted list of models.

B. Provide copies of the accepted list of models to each County Office in the State.

C. Request a copy of the drawings, description of materials, and thermal calculations to determine compliance with the thermal requirements for the county in which the house is to be located according to exhibit D of this subpart.

D. Check to see that County Offices within the state will act as prescribed in paragraph VI of this exhibit B.

E. When two or more State Offices have different interpretations of the acceptability of a particular model, there must be an agreement between the states so that they will have the same requirements. If the states cannot agree, the National Office PSS will be consulted for guidance.

IX. Subsequent Review.

The Agency will make periodic reviews of houses, both site-built and houses manufactured offsite, to determine acceptability of the finished product. If, in the judgment of the Agency, the product has failed to perform satisfactorily, acceptance may be withdrawn. The State Director will notify the manufacturer and/or the builder/dealer of the reasons for the withdrawal no later than the time of withdrawal. Negotiations for corrections will be carried out by the County Office with the assistance of the State Office or National Office, as necessary.

X. Materials and Products Acceptance—Material Release Bulletins, Use of Materials Bulletins, Manufacturer's Instructions.

A. The Materials Release (MR) and Use of Materials Bulletins (UM) provide for the national acceptance of specific nonstandard materials and products not covered in the current HUD MPS.

B. When contractors or builders intend to use products or materials not listed as approved in the MPS, the Agency personnel reviewing or concerned with the approval of construction in which the product is to be used, will require the contractor or builder to furnish a Materials Release Bulletin or Use of Materials Bulletin on the materials or products. If the product has been accepted, the supplier should be able to obtain the bulletin for the contractor or builder from the manufacturer. These bulletins describe the products or materials limitations to use, method of installing or applying, approved type of fasteners, if used, etc. and will provide the contractor with instructions as to proper installation or application.

C. When Agency personnel are unfamiliar with any materials or products which have

been accepted in the MPS, they will request the contractor or builder to furnish the manufacturer's instructions to assure that the materials or products are properly installed or applied. Any questions on any product that cannot be resolved in the County Office should be referred to the State Office. When the question cannot be resolved at the State Office level, the National Office PSS should be consulted for guidance.

XI. Manufactured Farm Service Buildings.

A. When a loan application is received that involves a manufactured building or special equipment that cannot be completely inspected on the site, the local State Land Grant University recommendations should be requested.

B. When the County Office questions the advisability of making a loan on a manufactured building, the State Office should also be consulted.

C. The State Office should review and make recommendations to the County Office. If doubt still exists, the National Office PSS should be consulted for guidance.

ATTACHMENT 1—REQUIRED INFORMATION FOR ACCEPTANCE OF MODULAR/PANELIZED HOUSING UNITS

The manufacturer or sponsor of modular/panelized housing units wishing to participate in the Agency Rural Housing programs shall submit to the Agency State Director having jurisdiction over the state in which the proposed housing is to be manufactured, two complete sets of the information listed below for evaluation. Submissions not including all the information requested will be returned.

A. *Statements:*

1. Name and location of organization, including titles and names of its principal officers.

2. A brief description of plant facilities.

3. Extent of intended market distribution, including a list of any other states in which units will be marketed.

4. The method of quality control during site installation.

5. A copy of the applicable current HUD Structural Engineering Bulletin (SEB), Regional Letter of Acceptance (RLA), or Area Letter of Acceptance (ALA).

6. A current factory inspection report made within 6 months by HUD or HUD authorized agency.

7. Name and address of any third party inspection agency.

8. Location of nearest assembled product for inspection.

9. Field manuals for site installation and/or set-up procedures.

10. Specifications or descriptions of materials using either Form RD 1924–2, (HUD-FHA Form 2005), "Description of Materials," including sizes, species and grade of all building and finishing materials. All blanks

should be filled and additional sheets may be attached as well as equipment manufacturer's brochures. Use an asterisk (*) to denote all items of onsite construction that will be provided by the builder-dealer. The builder-dealer must complete a form for the builder-dealer's portion of the work. Use N/A in any blank which is not applicable.

11. Names and addresses of other public and private agencies which have rendered or been asked to render a technical suitability or acceptance determination with respect to the products or structural methods employed.

12. Written certification that construction drawings and specifications conform with the applicable development standard.

13. Any other pertinent information.

14. An index of all documents submitted.

B. *Working Drawings.* For emphasis as to the details required for modular/panelized housing proposals, the following items are listed in addition to and in more detail than the requirements in exhibit C of this subpart. In some cases, the drawing presentation sheets may be required to be reduced to 200 mm by 266 mm (8 × 10½ inches) sheet size:

1. Foundation and/or Basement Plan. This plan shall include anchorage details, exterior and interior dimensions, typical footings, wall thickness, pilaster sizes and locations, column or pier sizes and locations and girders required to support the structures. Show location of all equipment (furnace, water heater, laundry tubs, sump, etc.) floor drains, electrical outlets, electrical entrance panels, and all doors and windows or crawl space vents with all sizes indicated.

2. Floor Plans of all levels. Show square footage of each habitable room with square footage of each area of natural light and ventilation. In addition, a design sketch scaled properly to illustrate a typical furniture arrangement for all habitable levels is required to indicate intended occupancy functions of the design. A window and door schedule should also be provided indicating glazed size, sash size, and thermal conductance of each type.

3. All exterior elevations including opening and sizes; wall finish materials, flashing, finish grades intended, depth of footings when known, finish floor, ceiling heights, roof slope, location of downspouts, gutters, vents for both structural spaces and for equipment. Indicate construction joint locations and details of connections between sections, modules or components.

4. Building cross sections showing size and spaces of all framing members from lowest member (bottom of footing) to highest point of roof (ridge) plus;

(a) Type of material and method of application of all covering materials, such as subflooring, combination subflooring and

underlayment, sheathing, interior and exterior finishes;

(b) Complete details including computations of trussed rafter systems with the architect/engineer's stamp of those responsible for the design.

(c) Details of insulation and vapor barrier installation and attic ventilation. If the thermal characteristics to be provided are determined according to optional method for overall structure performance allowed in exhibit D of this subpart, the submission and complete engineering calculations with all details of construction shall be sent to Administrator, Attn. PSS, U.S. Department of Agriculture, Washington, DC 20250, for analysis as prescribed in paragraph IV C of exhibit D of this subpart.

(d) Special details as necessary to show any special features of construction, including method of fabricating, erection, joining, and finishing of all elements; and

(e) Details and sections of stairways including all critical dimensions, such as, riser, run and headroom.

5. Interior elevations of kitchen cabinets and bathroom elevations with schedule of all shelf, counter-top and drawer footage. Indicate whether kitchen cabinets are to be custom made for each model or made for any model by a cabinet manufacturing company.

6. Plumbing schematics, including pipe materials, sizes and plumbing code compliance.

7. Heating plan, including heat loss of each room, is needed for heating systems, sizings and capacities, forced air, electric baseboard, or electric space heaters and, if applicable, heat gain. For forced air systems, include supply and return duct layout and location of appropriate diffusers.

8. Electrical plan, including circuit chart or diagram.

9. Any other pertinent facts or drawings that will better explain why and how certain unusual materials or structural methods are employed.

ATTACHMENT 2

John Dough Manufacturing Company,
3444 Residence Avenue,
Elkton, Indiana 00051.

Dear Sirs: Although the documents submitted to this office have only received a cursory review, they appear to be in substantial compliance to qualify your firm for the type of acceptance indicated on the attached list.

The acceptance being issued is subject to this letter of conditions, compliance with HUD Handbook 4950.1 Technical Suitability of Products Program Technical and Processing Procedures, compliance with Agency) Thermal Performance Construction Standards, and compliance with the conditions set forth in the HUD acceptance document, if applicable, whose number appears on the acceptance.

The manufacturer and the authorized builder-dealer bear the responsibility of complying with the above, the exhibits submitted and the applicable development standards.

The manufacturer and/or builder-dealer also shall:

1. Provide positive identification of the modular unit by model, date of manufacture and factory in which the unit was manufactured.

2. Furnish with each home to be financed by the Agency in _____(State)_____, a written certificate (Attachment 5 to this exhibit B) endorsed by the builder-dealer certifying that all requirements have been satisfied.

3. Furnish the local Agency County Supervisor with a complete set of drawings including site plans, description of materials, structural engineering bulletins when applicable in the state, and documentation relating to the manufacture, transportation, erection, and installation for each model of modular/panelized housing to be financed in the county. Electrical, plumbing and heating plans must be furnished for each model in addition to the basic drawings. Floor plans and elevation drawings may vary from those listed in attachment 1 of exhibit B to RD Instruction 1924–A to reflect each of the manufacturer's models provided they are in compliance with the applicable development standard and the Agency Thermal Performance Construction Standards and provided they have been accepted and listed in this state's approval of manufactured structures. No field alterations to the accepted models will be allowed.

4. Furnish, when required by the County Supervisor, foundation drawings (including special foundation design considerations when the unit is to be erected in seismic zones 1, 2 or 3) adapting the modular home to any unusual site conditions needing information additional to that furnished by the standard drawings.

5. Furnish the County Office with a copy of inspection reports of the manufacturing facilities immediately after the inspection reports have been completed.

6. Allow RD personnel to inspect the manufacturing facilities at any time and furnish all Agency State Offices, where acceptance has been obtained, with a copy of any Agency inspection reports immediately after the inspection reports have been completed.

7. In the event there are major changes to the submitted drawings, obtain approval under the HUD Technical Suitability of Products Program and submit verification of this approval to the County Office for listing on the state's accepted list. Any modular home shipped with major changes incorporated, without such changes on file at the County Office may be rejected.

(Add state and local requirements appropriate to this letter of conditions.)

This acceptance may be subject to corrective action when deficiencies are noted in the product, field inspections, manufacturing facilities, or when there is noncompliance with the provisions of the HUD Technical Suitability of Products Program.

The inclusion of these models on the accepted list is based only on the material and structural aspects of the manufactured units. Final determination of acceptability rests with RD personnel. Other factors relating to the property in its entirety such as appraisal, location, sustained market acceptance, architectural planning and appeal, thermal qualities, mechanical and electrical equipment, etc., must be considered in the final determination.

Your cooperation in this acceptance program is appreciated.

Sincerely,

State Director

ATTACHMENT 3

Date _____ File No. _____

ACCEPTANCE OF MODULAR/PANELIZED HOUSING UNITS

(BASED ON HUD HANDBOOK 4950.1)

Manufacturer:

_____ Acceptance Document _____
_____ Type of Acceptance:
_____ ___ Regular
_____ ___ Temporary, Expires _____
Plant Locations: _____
Date of Latest Plans
Reviewed _____
Date of Latest Factory
Inspection _____
Acceptance Document Review
Date _____

FmHA or its successor agency under Public Law 103–354 Instruction 1924–A, exhibit D

THERMAL PERFORMANCE CONSTRUCTION STANDARDS

State Office Review

(Exh. D, IV, C, 1, a or b)
National Office Review

(Exh. D, IV, C, 2)

Maximum Winter Degree Days for
State _____ Walls R _____
Glazing/Gross Wall Area Ratio _____%
Ceilings R _____
Glazing _____ Pane(s)
Floor R _____
Glazing _____ Pane(s)
Insulated Door _____
Wood and Storm _____
Insulated Door _____
Wood and Storm _____

Models Accepted:

ATTACHMENT 4

John Dough Manufacturing Company,
3444 Residence Avenue,
Elktown, Indiana 00051.

Dear Sirs: As set forth in acceptance letters issued by this office, acceptance of modular/panelized homes in this state is based on HUD's Technical Suitability of Products Program and the conditions stated in the acceptance letter. Your file has been reviewed and the following has been noted.

____An inspection report of your manufacturing facilities is overdue. Inspections are required twice yearly. The last inspection report on file at this office is dated ____.

____Your Structural Engineering Bulletin No. ____ dated ____ has not been reviewed by HUD. Reviews are generally required every three years. Temporary acceptance will be considered when you provide evidence that the review documents have been submitted to HUD.

____The drawings being used for the construction of your homes are not listed in your Structural Engineering Bulletins. Drawings used in the field should be those upon which the Structural Engineering Bulletin was issued.

____There have been ____ revisions to the development standards since ____, the date of the last drawings we have on file for your homes. It is recommended that you review the revisions to ascertain whether your drawings need to be updated.

Please submit a written response and appropriate documents for the above items within ____ days, or your product will be removed from the accepted list until your firm can again qualify. If you have any problems furnishing the above within the time stated, please contact this office.

We look forward to receiving the materials indicated so that your firm's listing may be continued.

Sincerely,
State Director

ATTACHMENT 5

CERTIFICATION BY MANUFACTURER

Delivery location of structure
for component _____
_____

This is to certify that
Model: _____,
Serial # _____,
manufactured ____
(date) ____, 19 __ in
____ (location) ____
and being sold to ____
(name of _____
builder-dealer or borrower) has been manufactured in accordance with drawings and

specifications on file in the Agency State Office and that the construction complies with applicable development standards, except as modified by HUD Acceptance Document (SEB, RLA, ALA,) NO. _____,
dated _____,
and in compliance with the Agency Thermal Performance Construction Standards.

Date

Signature of Authorized Official

Title

CERTIFICATION BY BUILDER-DEALER

_____ (Name of builder-dealer) _____ certifies that the foundation and other on-site work has been constructed in accordance with the drawings and specifications and the above structure or component has been erected, installed or applied in compliance with the applicable development standards.

It is understood that the manufacturer's certification does not relieve the builder/dealer of responsibility under the terms of the builder's warranty required by the National Housing Act.

Date

Signature of Authorized Official

Title

[52 FR 8002, Mar. 13, 1987, as amended at 67 FR 78327, Dec. 24, 2002; 80 FR 9872, Feb. 24, 2015]

EXHIBIT C TO SUBPART A OF PART 1924— GUIDE FOR DRAWINGS AND SPECIFICATIONS

This exhibit applies to all new buildings to be constructed, including all single family housing and related facilities and, as applicable, farm housing and farm service buildings.

I. General

The documents recommended in this exhibit correspond with the list of exhibits in Chapter 3 of the Department of Housing and Urban Development (HUD) "Architectural Handbook for Building Single-Family Dwellings" No. 4145.2. This exhibit may be used as a public handout and shall be used as a guide for drawings and specifications to be submitted in support of any type of application involving construction of major new buildings or extensive rehabilitation, alterations or additions to existing buildings. Descriptions of work for minor alterations or repairs need pertain only to work to be done and may be in narrative form when acceptable to the County Supervisor. Complete and accurate drawings and specifications are necessary:

A. To determine the acceptability of the proposed development,

B. To determine compliance with the applicable standards and codes,

C. To prepare a cost estimate, and

D. To provide a basis for inspections and the builder's warranty.

II. Drawings for a Specific Structure

Drawings for individual single dwellings shall provide at least the following:

A. *Plot Plan.* Refer to Example Plot Plan No. 1, attachment 1 to this exhibit C (available in any Agency office). Ratio: 1:240 (1″ = 20′) (at scale, 1″ = 20′ or ⅟₁₆″ = 1′ 0″ minimum):

1. Lot and block number.

2. Dimensions of plot and north point.

3. Dimensions of front, rear and side yards.

4. Location and dimensions of garage, carport and other accessory buildings.

5. Location and sizes of walks, driveways and approaches.

6. Location and sizes of steps, terraces, porches, fences and retaining walls.

7. Location and dimensions of easements and established setback requirements, if any.

8. Elevations at the following points: (a) first floor of dwelling and floor of garage, carport and other accessory building; (b) finish curb or crown of street at points of extension of lot lines; (c) finish grade elevation at each principal corner of structure; (d) finish grade at bottom of drainage swales at extension of each side of structure as feasible.

9. The following additional elevations, as applicable, if the topography of the site or the design of the structure is such that special grading, drainage or foundations may be necessary. Examples are irregular or steeply sloping sites, filled areas on sites, or multilevel structure designs; (a) finish and existing grade elevations at each corner of the plot; (b) existing and finish grade at each principal corner of dwelling; (c) finish grade at both sides of abrupt changes of grade such as retaining walls, slopes, etc.; (d) other elevations that may be necessary to show grading and drainage.

10. Indication of type and approximate location of drainage swales.

11. When an individual water supply and/or sewage system is proposed, drawings, specifications and other items prescribed in paragraph V of this exhibit.

B. *Floor Plans.*

1. Scale, 1:50 (¼″ = 1′ 0″).

2. Floor plan of each floor and basement, if any. Show typical furniture locations to suggest intended use of each habitable space.

3. Plan of all attached terraces and porches, and of garage or carport.

4. If dwelling is of crawl-space type, a separate foundation plan. Slab-type foundation may be shown on sections.

5. Direction, size and spacing of all floor and ceiling framing members, girders, columns or piers.

6. Location of all partitions and indication of door sizes, and direction of door swing.

7. Location and size of all permanently installed construction and equipment such as kitchen cabinets, closets, storage shelving, plumbing fixtures, water heaters, etc. Details of kitchen cabinets may be on separate drawing.

8. Location and symbols of all electrical equipment, including switches, outlets, fixtures, etc.

9. Heating system on separate drawing, or when it may be shown clearly it may be part of the floor or basement plan showing: (a) layout of system; (b) location and size of ducts, piping, registers, radiators, etc.; (c) location of heating unit and room thermostat; (d) total calculated heat loss of dwelling including heat loss through all vertical surfaces, ceiling and floor. When a duct or piped distribution system is used, calculated heat loss of each heated space is required.

10. Cooling system, on separate drawings or, as part of heating plan, floor or basement plan showing: (a) layout of system; (b) location and size of ducts, registers, compressors, coils, etc.; (c) heat gain calculations, including estimated heat gain for each space conditioned; (d) model number and Btu capacity of equipment or units in accordance with applicable Air Conditioning and Refrigeration Institute (ARI) or American Society of Refrigerating Engineers (ASRE) Standard; (e) Btu capacity and total kilowatt (KW) input at stated local design conditions; (f) if room or zone conditioners are used, provide location, size and installation details.

C. *Exterior Elevations.*

1. Scale, 1:50 (¼″ = 1′ 0″). Elevations, other than main elevation, which contain no special details may be drawn at 1:100 (⅛″ = 1′ 0″).

2. Front, rear and both side elevations, and elevations of any interior courts.

3. Windows and doors—indicate size unless separately scheduled or shown on floor plan.

4. Wall finish materials where more than one type is used.

5. Depth of wall footings, foundations, or piers, if stepped or at more than one level.

6. Finish floor lines.

7. Finish grade lines at buildings.

D. *Details and Sections.*

1. Section through exterior wall showing all details of construction from footings to highest point of road. Where more than one type of wall material is used, show each type. Scale 1:25 (⅜″ = 1′ 0″) minimum.

2. Section through any portion of dwelling where rooms are situated at various levels or where finished attic is proposed, Scale, 1:50 (¼″ = 1′ 0″) minimum.

3. Section through stair wells, landings, and stairs, including headroom clearances and surrounding framing. Scale, 1:50 (¼″ = 1′ 0″) minimum.

4. Details of roof trusses, if proposed, including connections and stress or test data with seal of architect or engineer responsible. Scale of connections, 1:25 (⅜″ = 1′ 0″) minimum.

5. Elevation and section through fireplace. Scale, 1:25 (⅜″ = 1′ 0″) minimum.

6. Elevations and section through kitchen cabinets, indicating shelving. Scale, 1:50 (¼″ = 1′ 0″) minimum.

7. Sections and details of all critical construction points, fastening systems, anchorage methods, special structural items or special millwork. Scale as necesaary to provide information, 1:25 (⅜″ = 1′ 0″) minimum.

III. Master Drawings for Group Structures

Drawings for a group of structures (such as for several conditional commitments) may be submitted in lieu of drawings for each individual property when a number of applications are simultaneously submitted involving repetition of the same type structure.

A. *Master plot plan* shall include the following:

1. Scale which will provide the following information in a clear and legible manner.

2. North point.

3. Location and width of streets and rights-of-way.

4. Location and dimensions of all easements.

5. Dimensions of each lot.

6. Location of each dwelling on lot with basic dimensions.

7. Dimensions of front, rear and side yards.

8. Location and dimensions of garage, carports and other accessory buildings.

9. Identification of each lot by number and indication of basic plan and elevation type.

10. Location of walks, driveways and other permanent improvements.

B. *Typical plot plan* for each basic type dwelling may be submitted in lieu of fully detailing each lot on master plot plan, when topography and lot arrangements present no individual planning or construction problems.

1. Information not shown on the typical plot plan shall be included on the master plot plan.

2. Typical plot plans shall not be used for corner lots, lots with irregular boundaries, lots involving pronounced topographic variations or other lots where individual detailing is necessary.

3. Location of dwelling on typical lot and full dimensions.

4. Location and dimensions of all typical improvements, such as garage, carport, accessory buildings, walks, drives, steps, porches, terraces, trees, shrubs, retaining walls, fences, etc.

C. *Grading* may be shown on separate grading plan or on the master plot plan. Scale shall be sufficiently large to provide the following information in clear and legible manner:

1. Contours of existing grade at intervals of not more than 1.524 m (5 feet). Intervals less the 1.524 m (5 feet) may be required when indicated by the character of the topography.

2. Location of house and accessory buildings on each lot.

3. Identification of each lot by number.

4. Elevations in accordance with individual plot plan including bench mark and datum or, in lieu of finish grade elevations, contours of proposed finish grading may be submitted. Contour intervals selected shall be appropriate to the topography of the site.

5. Lot grading shall be shown by indicating protective slopes and approximate location of drainage swales.

6. Location of drainage outfall, if any drainage is not to a street.

D. *Floor plans, elevations, sections and details* shall be submitted for each basic plan. Alternate elevations to basic plan may be shown at scale, 1:100 (⅛″ = 1′ 0″).

IV. Specifications

Form RD 1924-2, "Description of Materials," or other acceptable and comparable descriptions of all materials forms shall be submitted with the drawings. The forms shall be completed in accordance with the instructions on Form RD 1924-2 to describe the materials to be used in the construction.

A. Form RD 1924-2 may be reproduced if size, format and printed text are identical to the current official form. When it is reproduced, the following deletions must be made:

1. All lines indicating RD form numbers or other Government agency initials and/or numbers, and

2. The United States Government Printing Office (GPO) imprint and reference number.

B. The material identification shall be in sufficient detail to fully describe the material, size, grade and when applicable, manufacturer's model or identification numbers. When necessary, additional sheets must be attached as well as manufacturer's specification sheets for equipment and/or special materials, such as aluminum siding or carpeting.

V. Individual Water Supply and Sewage Disposal Systems

When an individual water and/or sewage disposal system is proposed, the following additional information must be submitted:

A. *Approval and recommendations of other authorities.*

1. A written opinion by the health authority having jurisdiction that the site is suitable and acceptable for the proposed systems(s) and,

2. If available, a soils report from the local USDA-Soil Conservation Service and any recommendations they may have.

3. Approval of appropriate environmental control authority.

4. A signature of the health authority on the plot plan indicating approval of the design of the proposed system.

B. *Plot Plan.* Refer to Example Plot Plan No. 2, attachment 2 to this exhibit C (available in any Agency office).

1. Location and size of septic tank, distribution box, absorption field or bed, seepage pits and other essential parts of the sewage disposal system and distance to all individual wells, open streams or drainageways.

2. Location of well, service line and other essential parts of the water supply system and distance to other wells and/or sewage disposal systems.

3. Exact location of individual systems (water or sewage) on adjacent properties and description of system, if available.

C. Construction details of all component parts of individual water supply and sewage disposal systems shall clearly indicate material, equipment and construction. Extra sheets and drawings should be added as necessary to fully explain the proposed installation.

EXHIBIT D TO SUBPART A OF PART 1924—THERMAL PERFORMANCE CONSTRUCTION STANDARDS

I. Purpose

This exhibit prescribes thermal performance construction standards to be used in all housing loan and grant programs. These requirements shall supersede the thermal performance requirements in any of the development standards in §1924.4(h) of this subpart.

II. Policy

All loan or grant applications involving new construction (except for new Single Family Housing (SFH)) and all applications for conditional commitments (except for new SFH) shall have drawings and specifications prepared to comply with paragraphs IV A or C and IV D of this exhibit. All new SFH construction shall have drawing and specifications prepared to comply with paragraph IV F of this exhibit.

III. Definitions

A. *British thermal unit* (Btu) means the quantity of heat required to raise the temperature of one pound (.4535 Kg.) of water by one degree Fahrenheit (F). For example, one Btu is the amount of heat needed to raise the temperature of one pound of water from 59 degrees F to 60 degrees F.

B. *Glazing* is the material set into a sash or door when used as a natural light source and/ or for occupant's views of the outdoors.

C. *"R" value*, thermal resistence, is a unit of measure of the ability to resist heat flow. The higher the R value, the higher the insulating ability.

D. *"U" value* is the overall coefficient of heat transmission and is the combined thermal value of all the materials in a building section. U is the reciprocal of R. Thus U = 1/R or R = 1/U or 1/C where C is the thermal conductance and is the unit of measure of the rate of heat flow for the actual thickness of a material one square foot in area at a temperature of one degree Fahrenheit. The lower the U value, the higher the insulating ability.

E. *Winter degree-day* is a unit based on temperature difference and time. For any one day, when the mean temperature is less than 65 degrees F (18.3 degrees Celsius), there are as many degree-days as the number of degrees difference between the mean temperature for the day and 65 degrees F. The daily mean temperature is computed as half the total of the daily maximum and daily minimum temperatures.

F. *CABO Model Energy Code, 1992 Edition (MEC-92)*—This code sets forth the minimum energy/thermal requirements for the design of new buildings and structures or portions thereof and additions to existing buildings. The MEC is maintained by the Council of American Building Officials (CABO).

IV. Minimum Requirements

A. All multifamily dwellings to be constructed with Agency loan and/or grant funds and all repair, remodeling, or renovation work performed on single family and multifamily dwellings with Agency loan and/or grant funds shall be in conformance with the following, except as provided in paragraphs IV C 3 and IV D of this exhibit:

NEW CONSTRUCTION—MAXIMUM U VALUES FOR CEILING, WALL AND FLOOR SECTION OF VARIOUS CONSTRUCTION

Winter degree days [1]	Ceilings [2]	Walls	Floors [3]	Glazing [4]	Doors [5]
1000 or less	0.05	0.08	0.08	1.13	
1001 to 2500	.04	.07	.07	.69	
2501 to 4500	.03	.05	.05	.69	Storm door if hollow core door or if over 25% glass.
4501 to 6000	.03	.05	.05	.47	Storm Door.

NEW CONSTRUCTION—MAXIMUM U VALUES FOR CEILING, WALL AND FLOOR SECTION OF VARIOUS CONSTRUCTION—Continued

Winter degree days [1]	Ceilings [2]	Walls	Floors [3]	Glazing [4]	Doors [5]
6001 or more	.026	.05	.05	.47	Storm Door.

NOTE. U values are not adjusted for framing. Values calculated for components may be rounded. For example, a total R Value of 18.88 converts to a U value of .0529 rounded to .05.

[1] Winter degree-days may be obtained from the ASHRAE Handbook; the "NAHB Insulation Manual for Homes/Apartments"; local utilities; and the National Climatic Center, Federal Building, Asheville, NC. Manuals are available from NAHB RF, Rockville, MD 20850, or NMWIA, 382 Springfield Avenue, Summit, NJ 07901. Other sources of degree day data may be used if available from a recognized authority.

[2] Insulation must be continuous (i.e. no gaps) above all ceiling joists. In pitched roof construction, compression of insulation at the outside building walls is permitted to allow a 1″ ventilation space under the roof sheathing. For any loose fill insulation, a baffle must be provided. Raised trusses are not required.

[3] For floors of heated spaces over unheated basements, unheated garages or unheated crawl spaces, the U value of floor section shall not exceed the value shown. A basement, crawl space, or garage shall be considered unheated unless it is provided with a positive heat supply to maintain a minimum temperature of 50 degrees F. Positive heat supply is defined by ASHRAE as "heat supplied to a space by design or by heat losses occurring from energy-consuming systems or components associated with that space."

Where the walls of an unheated basement or crawl space are insulated in lieu of floor insulation, the total heat loss attributed to the floor from the heated area shall not exceed the heat loss calculated for floors with required insulation.

Insulation may be omitted from floors over heated basement areas or heated crawl spaces if foundation walls are insulated. The U value of foundation wall sections shall not exceed the value shown. This requirement shall include all foundation wall area, including header joist (band joist), to a point 50 percent of the distance from a finish grade to the basement floor level. Equivalent Uo configurations are acceptable.

MAXIMUM U VALUES OF THE FOUNDATION WALL SECTIONS OF HEATED BASEMENT NOT CONTAINING HABITABLE LIVING AREA OR HEATED CRAWL SPACE

Winter degree-days (65 F base)	Maximum U value	Glazing*
2500 or less	No requirement	1.13
2501 to 4500	0.17	1.13
4501 or more	0.10	.69

* Glazing in heated basement shall be limited to 5 percent of floor area unless alternative Uo combination is documented.

[4] Sliding glass doors are considered as glazing. The glazing value is for glass only. Glazing shall be limited to 15 percent of the gross area of all exterior walls enclosing heated space, except when demonstrated that the winter daily solar heat gain exceeds the heat loss and the glass area is properly screened from summer solar heat gain.

[5] 1¾ inch metal-faced door systems with rigid insulation core and durable weatherstripping providing a "U" value equivalent to a wood door with storm door and an infiltration rate no greater than .50 cfm per foot of crack length tested according to ASTM E–283 at 1.567 psf of air pressure, may be substituted for a conventional door and storm door. All doors shall be weatherstripped. Any glazed areas must be double-glazed.

MINIMUM R VALUES OF PERIMETER INSULATION FOR SLABS-ON-GRADE

Winter degree-days (65 F base)	Minimum R values*	
	Heated slab	Unheated slab
500 or less	2.8	
1000	3.5	
2000	4.0	2.5
3000	4.8	2.8
4000	5.5	3.5
5000	6.3	4.2
6000	7.0	4.8
7000	7.8	5.5
8000	8.5	6.2
9000	9.2	6.8
10000 or greater	10.0	7.5

* For increments between degree days shown, R values may be interpolated.

B. [Reserved]

C. *Optional Standards*

Housing design not in compliance with the requirements of paragraph IV A of this exhibit may be approved in accordance with the provisions of this paragraph. Requests for acceptance proposed under paragraph C 1 of this exhibit, must be approved by the State Director. Requests for acceptance of site-built housing proposed under paragraph C 2 of this exhibit must be approved by the Administrator. Requests for acceptance of manufactured housing proposed under paragraph C 2 of this exhibit may be approved by the State Director. All submissions of proposed options to the State Director or Administrator shall contain complete descriptions of materials, engineering data, test data (when U values claimed are lower than the ASHRAE Handbook of Fundamentals), and calculations to document the validity of the proposal. All data and calculations will be based upon the current edition of the ASHRAE Handbook of Fundamentals or other universally accepted data sources.

1. *Overall "U" values for enveloped components.* The following requirements shall be used in determining acceptable options to the requirements of paragraph IV A of this exhibit.

a. Uo (gross wall)—Total exterior wall area (opaque wall and window and door) shall have a combined thermal transmittance value (Uo value) not to exceed the values shown in attachment 1 to this exhibit D (available in any Agency office). Equation 1 in attachment 1 shall be used to determine acceptable combinations to meet the requirements.

b. Uo (gross ceiling)—Total ceiling area (opaque ceiling and skylights) shall have a combined thermal transmittance value (Uo value) not to exceed the values shown in attachment 2 to this exhibit D (available in any Agency office). Equation 2 in attachment 2 shall be used to determine acceptable combinations to meet the requirements.

2. *Overall structure performance.* The following requirements shall be used in determining acceptable options to the requirements of paragraph IV A of this exhibit.

a. The methodology must be cost effective to the energy user, and must not adversely affect the structural capacity, durability or safety aspects of the structure.

b. All data and calculations must show valid performance comparisons between the proposed option and a structure comparable in size, configuration, orientation and occupant usage designed in accordance with paragraph IV A. Structures may be considered for Agency loan consideration which can be shown by accepted engineering practice to have energy consumption equal to or less than those which would be attained in a representative structure utilizing the requirements of paragraph IV A.

3. *Special consideration for seasonally occupied farm labor housing.* The following sets forth the minimum acceptable options to the requirements of paragraph IV A of this exhibit for seasonally occupied housing serving as security for farm labor housing loans and grants.

a. When the period of occupancy does not encounter 500 or more heating degree-days (HDD) as determined by an average of the previous 10 years based upon local climatological data published by the National Oceanic and Atmospheric Administration, Environmental Data Service, the standards of paragraph IV A will not apply.

b. When the period of use exceeds 500 HDD, the 10-year average value for the period of occupancy shall be used to determine the degree to which the thermal insulation requirements of paragraph IV A shall apply.

c. If mechanical cooling is provided and the period of occupancy encounters more

than 700 cooling degree-days (CDD), as determined by an average of the previous 8 years based upon local climatological data published by the same source cited in paragraph IV C3a above, the thermal insulation requirements for 1,000 and less degree-days as stated in paragraph IV A shall apply.

D. *Energy efficient construction practices.* This section prescribes those items of design and quality control which are necessary to guarantee the energy efficiency of homes built according to the standards of this exhibit. Also included are recommendations for extra energy efficiency in dwellings. This section does not apply to new SFH construction.

1. *Infiltration.* a. Requirements: All construction shall be performed in such a manner as to provide a building envelope free of excessive infiltration.

(i) Caulking and sealants. Exterior joints around windows and door frames, between wall cavities and window or door frames, between wall and foundation, between wall and roof, between wall panels, at penetrations of utility services through walls, floors and roofs, and all other openings in the exterior envelope shall be caulked, gasketed, weatherstripped, or otherwise sealed. Caulking shall be silicone rubber base or butyl rubber base, conforming to Federal Specifications TT-S-1543 and TT-S-1657 respectively, or materials demonstrating equivalent performance in resilience and durability.

(ii) Windows shall comply with ANSI 134.1, NWMA 15–2; the air infiltration rate shall not exceed 0.5 ft 3/min per ft. of sash crack.

(iii) Sliding glass doors shall comply with ANSI 134.2, NWM 15–3; the air infiltration rate shall not exceed .5 ft 3/min per square ft. of door area.

(iv) All insulation placed in open cavity walls shall be installed so that all space behind electrical switches and receptacles, plumbing, ductwork and other obstructions in the cavity are insulated as completely as possible. Insulation shall be omitted on the side facing the conditioned area; however, the vapor barrier in walls must not be cut or destroyed.

b. Recommendations: (i) Wrap outside corners of wall sheathing with 15 lb. asphalt impregnated building felt before siding application.

(ii) Utilize vestibules for entry doors, especially those facing into the direction of winter wind.

(iii) Install plumbing, mechanical and electrical components in interior partitions as much as possible. All water piping should be insulated from freezing temperatures.

2. *Heating and/or Cooling Equipment.* a. Requirements: All mechanical equipment for heating and/or cooling habitable space shall be designed to provide economy of operations.

(i) All space heating equipment (including fireplaces) requiring combustion air shall be sealed combustion types, or be located in a nonconditioned area (such as unheated basements) or adequate combustion air must be provided from outside the conditioned space.

(ii) All ductwork shall be designed and installed to minimize leakage. All metal to metal connections shall be mechanically joined and taped.

b. Recommendations: (i) Whenever possible, locate ductwork inside of conditioned areas in dropped ceilings, interior partitions or other similar areas.

(ii) Locate outside cooling units in areas not subject to direct sunlight or heat build-up.

3. *Vapor Barrier.* a. Requirements: Adequate vapor barriers must be provided adjacent to the interior finish material of the wall or other closed envelope components which do not have ventilation space on the non-conditioned side of the insulation.

(i) A vapor barrier at the inside of the wall or other closed envelope component must have a permeability (perm) rating less than that of any other material in the component and in no case have a perm rating greater than one. All vapor barriers must be sealed around all openings in the interior surface. Vapor barriers are not required in ceilings and floors. Continuous vapor barriers on ceilings, walls, and floors require adequate moisture vapor control in the conditioned space.

(ii) All vapor producing or exhaust equipment shall be ducted to the outside and equipped with dampers. This equipment includes rangehoods, bathroom exhaust fans and clothes dryers. If a dwelling design proposes the use of windows to satisfy the kitchen and/or bathroom ventilation requirements of the development standards, the incorporation of dehumidification equipment should be considered in accordance with paragraph IV D 3 b. Exhaust of any equipment shall not terminate in an attic or crawl space.

b. Recommendation: Forced air heating/cooling systems should include humidification/dehumidification systems where conditions indicate.

E. [Reserved]

F. *New SFH construction.* New SFH construction shall meet the requirements of CABO Model Energy Code, 1992 Edition (MEC–92).

G. *New manufactured housing.*

The Uo Value Zone indicated on the "Heating Certificate" for comfort heating shall be equal to or greater than the HUD Zone listed in the following table:

RHS climate zones (winter degree days)	FMHCSS (HUD code) Uo value zones
0–1000	1

RHS climate zones (winter degree days)	FMHCSS (HUD code) Uo value zones
1001–2500	2
2501–4500	2
4501–6000	3
>6000	3

Example: If a manufactured home is to be located in a geographic area having between 2501 and 4500 RHS winter degree days, the Agency will accept a Uo value Zone 2 unit or Zone 3 unit constructed to the HUD FMHCSS.

If a central air conditioning system is provided by the home manufacturer, a "Comfort Cooling Certificate" must be permanently affixed to an interior surface of the unit that is readily visible. This certificate may be combined with the heating certificate on the data plate.

V. *General Design Recommendations:*

A. Orient homes with greatest glass area facing south with adequate overhangs to control solar gain during non-heating periods. Examples of proper roof overhangs are given in attachment 3 to this exhibit D (available in any Agency office).

B. Arrange plantings with evergreen wind buffers on north side and deciduous trees on south.

C. Whenever possible, orient entry door away from winter winds.

D. Design house with simple shape to minimize exterior wall area.

E. Minimize glass areas within constraints of required light and ventilation, applicable safety codes and other appropriate consideration.

F. Minimize the amount of paved surface adjacent to the structure where heat gain is not desirable.

VI. *State Supplements:* State supplements or policies will not be issued or adopted to either supplement or set requirements different from those of this exhibit without the prior written approval of the National Office.

[52 FR 8002, Mar. 13, 1987, as amended at 54 FR 6874, Feb. 15, 1989; 59 FR 43723, Aug. 25, 1994; 64 FR 48085, Sept. 2, 1999; 72 FR 70221, Dec. 11, 2007]

EXHIBIT E TO SUBPART A OF PART 1924—
VOLUNTARY NATIONAL MODEL
BUILDING CODES

The following documents address the health and safety aspects of buildings and related structures and are voluntary national model building codes as defined in § 1924.4(h)(2) of this subpart. Copies of these documents may be obtained as indicated below:

Building code	Plumbing code	Mechanical code	Electrical code
BOCA Basic/National Building Code [1]. Standard Building Code [2] .... Uniform Building Code [3] .... CABO One and Two Family Dwelling Code [4].	BOCA Basic/National Plumbing Code [1]. Standard Plumbing Code [2] .... Uniform Plumbing Code [3]	BOCA Basic/National Mechanical Code [1]. Standard Mechanical Code [2]. Uniform Mechanical Code [3].	National Electrical Code [5]

[1] Building Officials and Code Administrators International, Inc., 4051 West Flossmoor Road, Country Club Hills, Illinois 60477.
[2] Southern Building Code Congress International, Inc., 900 Montclair Road, Birmingham, Alabama 35213–1206.
[3] International Conference of Building Officials, 5360 South Workman Mill Road, Whittier, California 90601.
[4] Council of American Building Officials, 5203 Leesburg Pike, Falls Church, Virginia 22041.
[5] National Fire Protection Association, Batterymarch Park, Quincy, Massachusetts 02269.

EXHIBIT F TO SUBPART A OF PART 1924—
PAYMENT BOND

KNOW ALL PERSONS BY
THESE PRESENTS: that

_____
(Name of Contractor)

_____
(Address of Contractor)
a _____,
(Corporation, Partnership or Individual)
hereinafter called
PRINCIPAL and

_____
(Name of Surety)
hereinafter called SURETY, are held and
firm
bound unto _____

_____
(Name of Owner)

_____
(Address of Owner)
hereinafter called OWNER and the United
States of America acting through the Farmers Home Administration or its successor
agency under Public Law 103–354 hereinafter
referred to as GOVERNMENT, and unto all
persons, firms, and corporations who or
which may furnish labor, or who furnish materials to perform as described under the
contract and to their successors and assigns
in the total aggregate penal sum of _____,
_____ Dollars ($_____) in lawful money of
the United States, for the payment of which
sum well and truly to be made, we bind ourselves, our heirs, executors, administrators,
successors, and assigns, jointly and severally, firmly by these presents.
THE CONDITION OF THIS OBLIGATION is
such that whereas, the PRINCIPAL entered
into a certain contract with the OWNER,
dated the _____ day of
_____19___, a copy of which is
hereto attached and made a part hereof for
the construction of:

_____
_____

_____
NOW, THEREFORE, if the PRINCIPAL shall
promptly make payment to all persons,
firms, and corporations furnishing materials
for or performing labor in the prosecution of
the WORK provided for in such contract, and
any authorized extension or modification
thereof, including all amounts due for materials, lubricants, oil, gasoline, coal and coke,
repairs on machinery; equipment and tools,
consumed or used in connection with the
construction of such WORK, and for all labor
cost incurred in such WORK including that
by a SUBCONTRACTOR, and to any mechanic or materialman lienholder whether it
acquires its lien by operation of State or
Federal law; then this obligation shall be
void, otherwise to remain in full force and effect.

PROVIDED, that beneficiaries or claimants hereunder shall be limited to the SUBCONTRACTORS, and persons, firms, and corporations having a direct contract with the
PRINCIPAL or its SUBCONTRACTORS.

PROVIDED, FURTHER, that the said
SURETY for value received hereby stipulates
and agrees that no change, extension of
time, alteration or addition to the terms of
the contract or to the WORK to be performed
thereunder or the SPECIFICATIONS accompanying the same shall in any way affect its
obligation on this BOND, and it does hereby
waive notice of any such change, extension
of time, alteration or addition to the terms
of this contract or to the WORK or to the
SPECIFICATIONS.

PROVIDED, FURTHER, that no suit or action shall be commenced hereunder by any
claimant: (a) Unless claimant, other than
one having a direct contract with the PRINCIPAL (or with the GOVERNMENT in the
event the GOVERNMENT is performing the
obligations of the OWNER), shall have given
written notice to any two of the following:
The PRINCIPAL, the OWNER, or the SURETY above named within ninety (90) days
after such claimant did or performed the last
of the work or labor, or furnished the last of
the materials for which said claim is made,
stating with substantial accuracy the
amount claimed and the name of the party
to whom the materials were furnished, or for
whom the work or labor was done or performed. Such notice shall be served by mailing the same by register mail or certified

557

mail, postage prepaid, in an envelope addressed to the PRINCIPAL, OWNER, or SURETY, at any place where an office is regularly maintained for the transaction of business, or served in any manner in which legal process may be served in the state in which the aforesaid project is located, save that such service need not be made by a public officer. (b) After the expiration of one (1) year following the date of which PRINCIPAL ceased work on said CONTRACT, it being understood, however, that if any limitation embodied in the BOND is prohibited by any law controlling the construction hereof, such limitation shall be deemed to be amended so as to be equal to the minimum period of limitation permitted by such law.

PROVIDED, FURTHER, that it is expressly agreed that the BOND shall be deemed amended automatically and immediately, without formal and separate amendments hereto, upon amendment to the Contract not increasing the contract price more than 20 percent, so as to bind the PRINCIPAL and the SURETY to the full and faithful performance of the Contract as so amended. The term "Amendment", wherever used in this BOND and whether referring to this BOND, the contract or the loan Documents shall include any alteration, addition, extension or modification of any character whatsoever.

PROVIDED, FURTHER, that no final settlement between the OWNER or GOVERNMENT and the CONTRACTOR shall abridge the right of any benficiary hereunder, whose claim may be unsatisfied.

IN WITNESS WHEREOF, this instrument is executed in [number] counterparts, each one of which shall be deemed an original, this the ____ day of _____.

ATTEST:

_____

Principal

_____

(Principal) Secretary

(SEAL)

By _____(s)

_____

(Address)

_____

Witness as to Principal

_____

(Address)

_____

Surety

ATTEST:

_____

Witness as to Surety

_____

(Address)

By _____

    Attorney-in-Fact

_____

(Address)

NOTE. Date of BOND must not be prior to date of Contract.

If CONTRACTOR is partnership, all partners should execute BOND.

Important: Surety companies executing BONDS must appear on the Treasury Department's most current list (Circular 570 as amended) and be authorized to transact business in the state where the project is located.

EXHIBIT G TO SUBPART A OF PART 1924—
PERFORMANCE BOND

KNOW ALL PERSONS BY THESE PRESENTS: that _____

_____

(Name of Contractor)

_____

(Address of Contractor)

_____

(Corportion, Partnership, or Individual) hereinafter called PRINCIPAL, and

_____

(Name of Surety)

_____

(Address of Surety) hereinafter called SURETY, are held and firmly bound unto

_____

_____

(Name of Owner)

_____

(Address of Owner) hereinafter called OWNER, and the United States of America acting through the Farmers Home Administration or its successor agency under Public Law 103–354 hereinafter referred to as the GOVERNMENT in the total aggregate penal sum of

_____

Dollars ($_____) in lawful money of the United States, for the payment of which sum well and truly to be made, we bind ourselves, our heirs, executors, administrators, successors, and assigns, jointly and severally, firmly by these presents.

THE CONDITION OF THIS OBLIGATION is such that whereas, the PRINCIPAL entered into a certain contract with the OWNER, dated the ____ day of _____ 19 __, a copy of which is hereto attached and made a part hereof for the construction of:

_____

NOW, THEREFORE, if the PRINCIPAL shall well, truly and faithfully perform its duties, all the undertakings, covenants, terms, conditions, and agreements of said contract during the original term thereof, and any extensions thereof which may be granted by the OWNER, or GOVERNMENT, with or without notice to the SURETY and during the guaranty period and if the PRINCIPAL shall satisfy all claims and demands incurred under such contract, and shall fully indemnify and save harmless the OWNER and GOVERNMENT from all costs and damages which it may suffer by reason of failure to do so, and shall reimburse and repay the OWNER and GOVERNMENT all outlay and expense which the OWNER and GOVERNMENT may incur in making good any default, then this obligation shall be void, otherwise to remain in full force and effect.

PROVIDED, FURTHER, that the liability of the PRINCIPAL AND SURETY hereunder to the GOVERNMENT shall be subject to the same limitations and defenses as may be available to them against a claim hereunder by the OWNER, provided, however, that the GOVERNMENT may, at its option, perform any obligations of the OWNER required by the contract.

PROVIDED, FURTHER, that the said SURETY, for value received hereby stipulates and agrees that no change, extension of time, alteration or addition to the terms of the contract or to WORK to be performed thereunder or the SPECIFICATIONS accompanying same shall in any way affect its obligation on this BOND, and it does hereby waive notice of any such change, extension of time, alteration or addition to the terms of the contract or to the WORK or to the SPECIFICATIONS.

PROVIDED, FURTHER, that it is expressly agreed that the BOND shall be deemed amended automatically and immediately, without formal and separate amendments hereto, upon amendment to the Contract not increasing the contract price more than 20 percent, so as to bind the PRINCIPAL and the SURETY to the full and faithful performance of the CONTRACT as so amended. The term "Amendment", wherever used in this BOND, and whether referring to this BOND, the Contract or the Loan Documents shall include any alteration, addition, extension, or modification of any character whatsoever.

PROVIDED, FURTHER, that no final settlement between the OWNER or GOVERNMENT and the PRINCIPAL shall abridge the right of the other beneficiary hereunder, whose claim may be unsatisfied. The OWNER and GOVERNMENT are the only beneficiaries hereunder.

IN WITNESS WHEREOF, this instrument is executed in [Number] counterparts, each one of which shall be deemed an original, this the ___ day of _____.

ATTEST:

Principal

_____
(Principal) Secretary

(SEAL)

_____

Witness as to Principal

_____
(Address)

By _____(s)

_____
(Address)

_____

Surety

ATTEST:

_____

Witness as to Surety

_____
(Address)

By _____
    Attorney-in Fact

_____
(Address)

EXHIBIT H TO SUBPART A OF PART 1924—
PROHIBITION OF LEAD-BASED PAINTS

I. Purpose

This exhibit prescribes the methods to be used to comply with the requirements of the Lead-Based Paint Poisoning Prevention Act, Public Law 91–695, as amended, (42 U.S.C. 4801 *et seq.*) and the amendment to section 501 (3) of Public Law 91–695 (42 U.S.C. 4841 (3)) as amended by the National Consumer Health Information and Health Promotion Act of 1976, Public Law 94–317.

II. Policy

The Agency shall not permit the use of lead-based paint on applicable surfaces of any housing or buildings purchased, repaired, or rehabilitated for human habitation with financial assistance provided by this agency. Paints used on applicable surfaces shall not contain more than 0.06 percent lead by weight calculated as lead metal in the total nonvolatile content of liquid paints or in the dried film of paint already applied.

III. Definitions

A. Housing and buildings mean any house, apartment, or structure intended for human habitation. This includes any institutional structure where persons reside, such as an orphanage, boarding school, dormitory, day care center or extended care facility, college housing, domestic or migratory labor housing, hospitals, group practice facilities, community facilities, and business or industrial facilities.

B. Applicable surfaces means all interior surfaces, whether accessible or not, and those exterior surfaces which are readily accessible to children under 7 years of age, such as stairs, decks, porches, railings, windows, and doors.

C. Lead-based paint means any paint containing more than .5 of 1 percentum lead by weight, or with respect to paint manufactured after June 22, 1977, lead-based paint containing more than six one-hundredths of 1 percentum lead by weight.

IV. Requirements

A. All new housing and buildings shall comply with paragraph II of this exhibit H.

B. For all existing housing and buildings built *after* 1950, on which a loan is closed after July 19, 1978, the Agency requires that the applicant, borrower or tenant be notified of the potential hazard of lead-based paints, of the symptoms and treatment of lead poisoning, and of the importance and availability of maintenance and removal techniques for eliminating such hazards. This will be accomplished by providing each applicant, borrower and/or tenant with a copy of attachment 1 to this exhibit H, "Lead-based Paint Hazards, Symptoms, Treatment and Techniques for Eliminating Hazards," available in any Agency County Office. Copies of attachment 1 may be obtained by the County Supervisor from the Finance Office, 1520 Market Street, St. Louis, MO 63103.

C. For all existing housing or buildings built *before* 1950 on which a loan is closed after July 19, 1978, the Agency requires that the applicant, borrower and/or tenant be notified as in paragraph IV B and a copy of attachment 2 to this exhibit H, "Caution Note on Lead-Based Paint Hazard," available in any Agency County Office, shall be delivered to the hands of the applicant, borrowers and/or tenant.

D. For all property transfers and inventory property sales, attachments 1 and 2 to this Exhibit H (available in any Agency office) shall be handed to the purchaser by the Agency representative.

E. All inventory housing or buildings built before 1950 to be repaired, renovated, or rehabilitated shall have tests for lead content, and where found to be hazardous, shall have any interior lead-based paint removed entirely. Loose or cracked surfaces shall be cleaned down to the base surface before repainting with a paint containing not more than six one-hundredths of 1 percentum lead by weight in the total nonvolatile content of the paint or the equivalent measure of lead in the dried film of paint already applied or both. Contracting officers shall include the following provision prohibiting the use of lead-based paint in all contracts and subcontracts for construction or rehabilitation of housing or buildings:

Lead-Based Paint Prohibition

No lead-based paint containing more than .5 of 1 percentum lead by weight (calculated as lead metal) in the total nonvolatile content of the paint, or the equivalent measure of lead in the dried film of paint already applied, or both, or with respect to paint manufactured after June 22, 1977, no lead-based paint containing more than .06 of 1 percentum lead by weight (calculated as lead metal) in the total nonvolatile content of the paint, or the equivalent measure of lead in the dried film of paint already applied, or both, shall be used in the construction or rehabilitation of residential structures under this contract or any subsequent subcontractors.

Authority: This amendment is made under provisions of 5 U.S.C. 301, 40 U.S.C. 486 (c).

Done at _____, _____ this _____ day of _____, 19___.

_____
Agency Representative

V. Summary

Section 401 of the Lead-Based Paint Poisoning Prevention Act as amended by the National Consumer Health Information and Health Promotion Act of 1976, Pub. L. 94–317, provides a requirement that each federal agency issue regulations and to take such other steps necessary to prohibit the use of lead-based paint on all applicable surfaces in Federal and Federally-assisted construction or rehabilitation of residential structures. The Lead-Based Paint Poisoning Prevention Act, Pub. L. 91–695, January 13, 1971, provides for grants to units of general local government in any state for the purpose of detecting and treating incidents of lead-based paint poisoning. Title II of this Act also provides for grants to the same units to identify those areas of risk including testing to detect the presence of lead-based paint on surfaces of residential housing.

EXHIBIT I TO SUBPART A OF PART 1924—
GUIDELINES FOR SEASONAL FARM
LABOR HOUSING

Section 100

General—This exhibit sets forth the guidelines and minimum standards for planning and construction of new Labor Housing (LH) that will be occupied on a seasonal basis. Rehabilitation LH projects will be in substantial conformance with these guidelines and standards. A "seasonal basis" is defined as 6 months or less per year. Seasonal housing for the farmworker need not be convertible to year-round occupancy; however, the living units shall be designed for the intended type of tenant, the time of occupancy, the location, the specific site, and the planned method of operation. It is important that the design of the LH site and buildings will help to create a pleasing lifestyle which will promote human dignity and pride among its tenants.

Section 200

Codes and Regulations—Compliance is required with National, state and local codes or regulations affecting design, construction, mechanical, electrical, fire prevention, sanitation, and site improvement.

Section 300

Planning

300–1 Complete architectural/engineering services in accordance with this subpart will be required if an LH grant is involved or the LH loan will involve more than four individual family units, or any number of group living units, or dormitory units accommodating 20 or more persons.

300–2 Buildings and site design shall provide for a safe, secure, economical, healthful, and attractive living facility and environment suited to the needs of the domestic farm laborer and his/her family.

300–3 At least 5 percent of the individual family units in a project, or one unit, whichever is greater, and all common use facilities will be accessible to or adaptable for physically handicapped persons. This requirement may be modified if a recipient/borrower shows, through a market survey acceptable to the Agency, that a different percentage of accessible or adapatable units is more appropriate for a particular project and its service area.

Site Design

301–1 General—The site design shall be arranged to utilize and preserve the favorable features and characteristics of the property and to avoid or minimize the potential harmful effect of unfavorable features. Particular attention is directed to § 1944.164 (1), (m) and (n) of subpart D of part 1944 of this chapter with reference to compliance with 7 CFR part 1970. Some of the features which must be considered are the topography, drainage, access, building orientation to sun and breezes; and advantageous features, such as vegetation, trees, good views, etc. or disadvantageous features, such as offensive odors, noxious plants, noise, dust, health hazards, etc.

301–2 Drainage—Surface and subsurface drainage systems shall be provided in accordance with the applicable development standard and subpart C of part 1924 of this chapter.

301–3 Water and Sewage Disposal—Water supply and sewage disposal installations shall comply with subpart C of part 1924 of this chapter, the applicable development standard and all governing state and local department of health requirements. Where environmentally and economically feasible, the LH facility shall connect to public water and waste disposal systems.

301–4 Electrical—Adequate electrical service shall be provided for exterior and interior lighting and for the operation of equipment.

301–5 Vehicular Access and Parking.

301–5.1 Safe and convenient all-weather roads shall be provided to connect the site and its improvements to the off-site public road.

301–5.2 All-weather drives and parking shall be provided for tenants, and for trucks and buses as needed within the site. Driveways, parking areas and walkway locations shall be in substantial conformance with the applicable development standard.

301–6 Walks:

301–6.1 Walks shall be provided for safe convenient access to all dwellings and for safe pedestrian circulation throughout the development between locations and facilities where major need for pedstrian access can be anticipated, such as laundry, parking to dwelling units, common dining rooms, etc.

301–6.2 Walkways shall be hard surface, such a concrete, asphalt, or stablized gravel, and shall be adequately drained.

301–7 Building Location:

301–7.1 Side and rear yards and distances between buildings shall conform to the applicable development standard.

301–8. Garbage and Refuse:

301–8.1 Garbage and refuse containers for individual units are required and shall be stored on durable functional racks or shall be located in a central screened area with easily cleaned surfaces. Single containers for multiple units shall be screened and in locations designed to accommodate collection vehicle functions.

301–9 Fencing:

301–9.1 Fencing used in the site design for project privacy or building security shall be harmonious in appearance with other fences and surrounding facilities which fall within the same view.

301–10 Outdoor living:

301–10.1 All public areas where pedestrian use can be anticipated after sunset shall be adequately lighted for security purposes, such as walkways to common use facilities—laundry, dining halls, building entrances, parking areas, etc.

301–11 Planting and Landscaping:

301–11.1 Planting and lawns or ground covers shall be provided as required to protect the site from erosion, control dust, for active and passive recreation areas, and provide a pleasant environment.

Building Design

302–1.1 Living Units Design:

302–1.1 *Individual Family Unit*—One family or extended family to a unit which shall contain adequate space for living, dining, kitchen, bath and bedrooms. Multifamily type units are required whenever possible for economy of site and building construction.

a. The minimum total net living unit size shall be 400 square feet. This size assumes occupancy of four persons. Units planned for additional occupants shall include an additional 60 square feet of living area per person.

b. A living/dining area shall be provided to accommodate a table and chairs with adequate dining and circulation space for the intended number of occupants. The living/dining area should be combined with the kitchen area.

c. The kitchen shall contain a sink, cooking range and refrigerator. A minimum free countertop area of six square feet is required. A minimum of 40 square feet of shelf area is required.

d. Each bathroom shall contain adequate space and circulation for a bathtub and/or shower, water closet and lavatory. Access to the bathroom shall not be through another bedroom in dwelling units containing more than one bedroom.

e. Bedroom areas separate from living areas are required. The design of the unit shall provide a minimum of 50 square feet of sleeping area per intended occupant including storage. Housing for families with children shall have a separate bedroom or sleeping area for the adult couples. A two foot by two foot shelf with a two foot long clothes hanging rod is required for each occupant.

302–1.2 *Group Living Unit*—A living unit designed for the occupancy of more than one family or for separate occupancy of male and/or female groups. Common bath spaces shall be contained in the same building. Group living units for families shall have separate bedrooms for each adult couple.

a. The design of the unit shall provide for a minimum of 620 square feet of total net living area for eight persons and an additional 60 square feet for each additional occupant. Additional area shall be planned for a second bathroom when anticipated occupancy will exceed eight persons, or if it will be occupied by persons of both sexes.

b. The kitchen shall contain an adequate sink, cooking range, refrigerator, and space the size of which is commensurate with the needs of the group living unit. A minimum of free countertop area of eight square feet is required. A minimum of 50 square feet of shelf area is required.

c. Refer to paragraph 302–1.1 b for living/dining requirements.

d. Each bathroom shall contain adequate space and circulation for comfortable access to, and use of, fixtures which will include a bathtub and/or shower, water closet and lavatory. In no case shall minimum fixtures be less than that required per paragraph 302–1.3 c below.

e. Refer to paragraph 301–1.1 e for bedroom requirements.

302.1.3 *Dormitory Living Unit*—A building which provides common sleeping quarters for persons of the same sex and may or may not contain kitchen and/or dining facilities in the same building as the sleeping quarters.

a. The design of areas for sleeping purposes, using single beds, shall provide for not less than 72 square feet per occupant including storage.

b. The design of areas for sleeping purposes, using double bunk beds, shall provide for not less than 40 square feet per occupant. Triple bunk beds will not be allowed.

c. The design of each dormitory building must include a water closet and a bathtub or shower for each 12 occupants, and a lavatory for each 8 persons. Urinals may be substituted for men's water closets on the basis of one urinal for one water closet, up to maximum of one-third of the required water closets.

d. Adequate kitchen and dining facilities must be provided which may be in the dormitory building or detached at a distance of not more than 200 feet from the sleeping quarters. In either case, the space must contain adequate cooking ranges, refrigerators, sinks, countertop, food storage shelves, tables and chairs, and circulation space. These facilities will comply with the requirements of the "Food Service Sanitation Ordinance and Code," part V of the "Food Service Sanitation Manual," U.S. Public Health Service Publication 934 (1965).

302–2 Other Facilities:

302–2.1 *General*—Other facilities, authorized by subpart D of part 1944 of this chapter, needed by farm workers may be provided in several ways: part of a living unit, located in the project, or, with the exception of laundry facilities, available nearby.

302–2.2 *Laundry Facilities*—Laundry facilities shall be required on-site. Drying yards shall be provided if dryer units are not provided. The design of washing facilities shall plan for a minimum rate of one washer for each 20 occupants. One drying unit may be provided for every two washers, if automatic dryers are customarily provided for rental housing in the community. Laundry facilities shall have adequate space for loading the units, circulation, and clothes folding.

302–2.3 *Office and Maintenance*—An office and maintenance space shall be provided or available, commensurate with the number of living units served, and shall meet the criteria of the Agency Manual of Acceptable Practices. If necessary, the maintenance space shall have sufficient area to accommodate furniture storage.

302–2.4 *Child Care Center*—Where feasible, a child care center may be included to provide supervised activity and safety for children while the parents work. Supervisors and workers for such centers are sometimes enlisted on a volunteer basis and the cost borne by nonprofit associations or community organizations. Grants are sometimes available through Federal or state programs. Consequently, the design of the child care center should meet the requirements of those sources providing organizational personnel and/or financing.

302–2.5 *Manager's Dwelling*—If a manager's dwelling unit is to be provided as a part of the Agency loan or grant, it will meet these guidelines. However, if it is necessary to provide a year-round caretaker/manager dwelling unit with the Agency loan or grant funds, it will meet the applicable development standard.

302–2.6 *Recreation*—Outdoor recreation space is required and shall be commensurate with the needs of the occupants. Active and passive recreation areas will be provided which may consist of outdoor sitting areas, playfields, tot lots and play equipment.

General Requirements

303–1 Materials and Construction—All materials and their installation in a LH facility shall meet the applicable development standard. Any exceptions to these requirements for materials and their installation must be obtained with the approval of the Agency National Office. Material should be selected that is durable and easily cleaned and maintained.

303–2 Fire Protection—Fire protection and egress shall be provided to comply with the applicable development standard.

303–3 Light, Ventilation, Screening—Natural light and ventilation requirements as specified in the applicable development standard shall be followed. Screening of all exterior openings is required.

303–4 Ceiling Heights—Ceiling heights of habitable rooms shall be a minimum of seven feet six inches clear, and seven feet in halls or baths in dwelling units. Public rooms shall have a minimum of eight feet clear ceiling height. Sloping ceilings shall have at least seven feet six inches for ½ the room with no portion less than five feet in height.

303–5 Heating and Cooling—Heating and cooling and/or air circulation equipment shall be installed as needed for the comfort of the tenants, considering the climate and time of year the facility will be in operation. Maximum feasible use of passive solar heating and cooling techniques shall be required. All equipment installed will be in accordance with the applicable development standard to protect the health and safety of occupants.

303–6 Plumbing—Plumbing materials and their installation shall meet the applicable development standard. Hot water will be required to all living units, baths, kitchens and laundry facilities.

303–7 Insulation, Thermal Standards, Winterization—Insulation will be required where either heating or cooling is provided as per paragraph 303–5 above or when climatic conditions dictate a need for insulation. Insulation Standards will comply with exhibit D,

paragraph IV C 3, of this subpart, or the state insulation standards, whichever are the more stringent.

303–8 Electrical—Electrical design, equipment and installation shall comply with the requirements of the latest edition of the National Electrical Code, and the applicable development standard for materials and their installation. Individual family units may be separately metered; other types of dwelling units may be separately metered as required.

303–9 Security and Winterization—Adequate management and physical measures will be provided as necessary to protect the facility during off-season periods, including adequate heating and insulation as required.

[52 FR 8002, Mar. 13, 1987, as amended at 52 FR 19283, May 22, 1987; 58 FR 38922, July 21, 1993; 81 FR 11029, Mar. 2, 2016; 81 FR 26667, May 4, 2016]

EXHIBIT J TO SUBPART A OF PART 1924—MANUFACTURED HOME SITES, RENTAL PROJECTS AND SUBDIVISIONS: DEVELOPMENT, INSTALLATION AND SET-UP

Part A—Introduction
Part B—Construction and Land Development
Part C—Drawings, Specifications, Contract Documents and Other Documentation
Part D—Inspection of Development Work

Part A—Introduction

I. *Purpose and Scope.* This exhibit describes and identifies acceptable site development, installation and set-up practices and concepts for manufactured homes. It is intended for Agency field personnel, builders, developers, sponsors, and others participating in Agency housing programs.

This exhibit applies to all manufactured homes (except those referenced in exhibit B of this subpart) on scattered sites or in rental projects and subdivisions and covers the requirements for design and construction of manufactured home communities. The Agency may approve alternatives or substitutes if it finds the proposed design satisfactory for the proposed use, and if the materials, installation, device, arrangement, or method of work is at least equivalent to that prescribed in this exhibit considering quality, strength, effectiveness, durability, safety and protection of life and health.

The Agency will require satisfactory evidence to be submitted to substantiate claims made regarding the use of any proposed alternative.

II. *Background.* The Agency has authority to make (1) section 502 Rural Housing (RH) loans with respect to manufactured homes and lots, and (2) section 515 Rural Rental Housing (RRH) loans with respect to manufactured home rental projects.

563

The manufactured home must be constructed in conformance with the Federal Manufactured Home Construction and Safety Standard (FMHCSS) and be permanently attached to a site-built permanent foundation which meets or exceeds the Minimum Property Standards (MPS) for One- and Two-Family Dwellings or Model Building Codes acceptable to the Agency. The manufactured home must be permanently attached to that foundation by anchoring devices adequate to resist all loads identified in the MPS. This includes resistance to ground movements, seismic shaking, potential shearing, overturning and uplift loads caused by wind. Note that anchoring straps or cables affixed to ground anchors other than footings will *not* meet these requirements.

7 CFR part 1970 applies on scattered sites, in subdivisions and rental projects with regard to the ·development, installation and set-up of manufactured homes. To determine the level of environmental analysis required for a particular application, each manufactured home or lot involved will be considered as equivalent to one housing unit or lot. Because the development, installation and set-up of manufactured home communities, including scattered sites, rental projects, and subdivisions, differ in some requirements from conventional site and subdivision development, two of the purposes of this exhibit are to:

A. Encourage economical and orderly development of such communities and nearby areas, and

B. Promote the safety and health of residents of such communities.

Therefore, this exhibit identifies those required standards and regulations and suggested guidelines for eliminating and preventing health and safety hazards and promoting the economical and orderly development and utilization of land for planning and development of manufactured home communities. The exhibit also provides the requirements for meeting the following:

A. *Resistance to Wind.* Foundations and anchorages shall be designed to resist wind forces specified in American National Standards Institute (ANSI) A–58.1–1982 for the geographic area in which the manufactured home will be sited;

B. *Proper Installation.* The manufacturer's installation instructions provided with each manufactured home shall contain instructions for at least one site-built foundation with interior and/or perimeter supports. Agency field office personnel shall review to determine its adequacy as security for an Agency loan only, the foundation design concept for compliance with this exhibit, the Agency/MPS and any Model Building Code acceptable to the Agency in that particular geographic area; and

C. *Proper Foundation Design.* Manufactured homes shall be installed on a foundation sys-tem which is designed and constructed to sustain, within allowable stress and settlement limitations, all applicable loads. Any foundation and anchorage system or method of construction to be used should be analyzed in accordance with well-established principles of mechanics and structural engineering.

III. *Definitions.* For the purpose of this exhibit the following definitions apply:

Accessory Building or Structure.

A subordinate building or structure which is an addition to or supplements the facilities provided by a manufactured home.

Anchoring Systems. An approved system for securing the manufactured home to the ground or foundation system that will, when properly designed and installed, resist overturning and lateral movement of the home from wind forces.

Contiguous. Sharing a boundary, adjoining or adjacent. A lot or subdivision is considered to be contiguous to other lots or subdivisions if it is adjoining, touching or adjacent.

Federal manufactured Home Construction and Safety Standards (FMHCSS). A 1976 federal standard, commonly known as the HUD Standard, for the construction, design and performance of a manufactured home which meets the needs of the public including the need for quality, durability and safety. Units conforming to the FMHCSS are certified by an affixed label that reads as follows:

AS EVIDENCED BY THIS LABEL NO. _____ THE MANUFACTURER CERTIFIES TO THE BEST OF THE MANUFACTURER'S KNOWLEDGE AND BELIEF THAT THIS MANUFACTURED HOME HAS BEEN INSPECTED IN ACCORDANCE WITH THE REQUIREMENTS OF THE DEPARTMENT OF HOUSING AND URBAN DEVELOPMENT AND IS CONSTRUCTED IN CONFORMANCE WITH THE FEDERAL MANUFACTURED HOME CONSTRUCTION AND SAFETY STANDARDS IN EFFECT ON THE DATE OF MANUFACTURE. SEE DATA PLATE.

Manufactured Home. A structure which is built to the Federal Manufactured Home Construction and Safety Standards and Agency's thermal requirements. It is transportable in one or more sections, which in the traveling mode is ten body feet or more in width, and when erected on site is four hundred or more square feet, and which is built on a permanent foundation when connected to the required utilities. It is designed and constructed for permanent occupancy by a single family and contains permanent eating, cooking, sleeping and sanitary facilities. The plumbing, heating, and electrical systems are contained in the structure.

Manufactured Home Community. A parcel or contiguous parcels of land which contains two or more manufactured home sites available to the general public for occupancy. Sites and units may be for rent, or sites may be sold for residential occupancy (as in a subdivision).

Manufactured Home Rental Project. A parcel or multiple parcels of land which have been so designated and improved to contain manufactured homes with sites available for rent.

Manufactured Home Site. A designated parcel of land in a manufactured home rental project, subdivision or scattered site designed for the accommodation of a unit and its accessory structures for the exclusive use of the occupants.

Manufactured Home Subdivisions. Five or more contiguous (developed or undeveloped) lots, or building sites that meet the requirements of subpart C of part 1924 of this chapter.

Permanent Perimeter Enclosure. A permanent perimeter structural system completely enclosing the space between the floor joist of the manufactured home and the ground. If separate from the foundation system, the permanent perimeter enclosure shall be secured to the perimeter of the manufactured home, properly ventilated and accessible and constructed of materials that conform to the Agency adopted MPS requirements for foundations.

Pier Support System. Consists of footings, piers, caps, leveling spacers, or approved prefabricated load bearing devices.

Related Facilities. Any nonresidential structure or building used for rental housing related purposes.

Site-Built Permanent Foundation System. A foundation system (consisting of a combination of footings, piers, caps and shims and anchoring devices or required structural connections) which is designed and constructed to support the unit and sustain, within allowable stress and settlement limitations, all applicable loads specified in ANSI A58.1–1982. All loads shall be transferred from the manufactured home to the earth at a depth below the established frost line without exceeding the safe bearing capacity of the supporting soil.

Set-Up. The work performed and operations involved in the placement of a manufactured home on a foundation system, to include installation of accessories or appurtenances and anchoring devices, and when local regulations permit, connection of utilities, but excluding preparation of the site.

IV. *Compliance with Local Regulations.* These requirements do not replace site development standards established by local law, ordinances, or regulations. Whenever such local standards contain more stringent provisions than any of the site development, installation and set-up minimums of the Agency, the more stringent standards shall govern.

V. *Applicable Standards, Regulations and Manuals.* A. Manufactured housing to be financed by the Agency must comply with the following standards:

1. Federal Manufactured Home Construction and Safety Standards, 24 CFR part 3280, mandated by Congress under title VI of the Federal Housing and Community Development Act of 1974, except for § 3280.506, "Heat Loss," of subpart F, "Thermal Protection," to part 3280.

2. Foundation requirements of the Minimum Property Standards as adopted by the Agency or a Model Building Code acceptable to the Agency.

3. [Reserved]

4. Uniform Federal Accessibility Standard (UFAS).

5. ANSI A58.1–1982, Minimum Design Loads for Buildings and Other Structures.

B. Manufactured housing to be financed by the Agency shall comply with all applicable Agency regulations, including but not limited to the following:

1. Subpart C of part 1924 of this chapter, "Planning and Performing Development Work."

2. Subpart A of part 1924, exhibit D, "Thermal Performance Construction Standards."

3. 7 CFR part 1970.

4. 7 CFR part 3550, "Direct Single Family Housing Loans and Grants."

5. Subpart E of part 1944, "Rural Rental Housing Loan Policies, Procedures, and Authorizations."

The requirements of the above references have not been repeated in this exhibit. Those requirements contained above are either mandatory or minimums and every effort should be made by the applicant, builder-developer or dealer-contractor to utilize higher standards, when appropriate.

Part B—Construction and Land Development

I. *General Acceptability Criteria.* The following criteria apply to development on scattered sites, in subdivisions and in rental project communities.

A. A manufactured home development including a site, rental project or subdivision shall be located on property designated for that use, where designations exist, by the local jurisdiction.

B. Conditions of soil, ground water level, drainage, flooding and topography shall not create hazards to the property and health or safety of the residents.

C. The finished grade elevation beneath the manufactured home or the first floor elevation of the habitable space, whichever is lower, must be above the 100-year flood elevation. This requirement applies wherever manufactured homes may be installed, not just in locations designated by the National Flood Insurance Program as areas of special flood hazards. The use of fill to accomplish this is a last resort. As is stated in EO 11988 and 7 CFR part 1970, it is the Agency's policy not to approve or fund any proposal in a 100-year floodplain area unless there is no practicable alternative to such a floodplain location.

D. Essential services such as employment centers, shopping, schools, recreation areas, police and fire protection, and garbage and trash removal shall be convenient to the development and any site, community, or subdivision must comply with the environmental review requirements in accordance with 7 CFR part 1970.

E. Manufactured home sites, rental projects and subdivisions shall not be subject to any adverse influences of adjacent land uses. An adverse influence is considered as one that is out of the acceptable level or range of a recognizable standard or where no standard exists is considered a nuisance irrespective of a site being zoned for manufactured home use. Health, safety and aesthetic consequences of location shall be carefully assessed by inspection of the site prior to selection of development. Undesirable land uses sush as deteriorated residential or commercial areas and noxious industrial properties shall be avoided to ensure compatibility. Other undesirable elements such as heavily traveled highways, airport runways, railroad, or fire hazards and other areas subject to recognizably intolerable noise levels shall be avoided.

F. The requirements for streets shall be those found in subpart C of part 1924 of this chapter.

G. The site design and development shall be in accordance with sound engineering and architectural practices and shall provide for all utilities in a manner which allows adequate, economic, safe, energy efficient and dependable systems with sufficient easements for their required installation and maintenance.

H. Utilities for each manufactured home site, rental housing project or subdivision shall be designed and installed in accordance with subpart C of part 1924 of this chapter; and the State health authority having jurisdiction, and all local laws and regulations requiring approval prior to construction.

I. Exhibit C, section V of this subpart shall be complied with by the applicant, dealer-contractor or builder-developer for manufactured home projects with individual water supply and sewage disposal systems. This exhibit shall be used by the Agency County Supervisors, District Directors, and State Directors in reviewing submissions.

J. During the planning, design, and construction of the foundation system and/or perimeter enclosure, provisions shall be made for the installation and connection of on-site water, gas, electrical and sewer systems, which are necessary for the normal operation of the manufactured home. Water and sewer system hookups shall be adequately protected from freezing.

II. *Development on Scattered Sites and in Subdivisions.*—A. General. Scattered sites and subdivision developments will be planned and constructed in accordance with specific requirements of this subpart, subpart C of part 1924, and 7 CFR part 1970, and the applicable Agency/MPS or Model Building Codes acceptable to the Agency. Manufactured homes for development in a manufactured home community shall:

1. Be erected with or without a basement on a site-built permanent foundation that meets or exceeds applicable requirements of the Agency/MPS for One- and Two-Family Dwellings or Model Building Codes acceptable to the Agency;

2. Be permanently attached to that foundation by anchoring devices adequate to resist all loads identified in the Agency adopted MPS (this includes resistance to ground movements, seismic shaking, potential shearing, overturning and uplift loads caused by wind, etc.);

3. Have had the towing hitch or running gear, which includes tongues, axles, brakes, wheels, lights and other parts of the chassis that operate only during transportation removed;

4. Have any crawl space beneath the manufactured home properly ventilated and enclosed by a continuous permanent perimeter enclosure. If it is not the supporting foundation, designed to resist all forces to which it may be subject without transmitting to the building superstructure movements or any effects caused by frost heave, soil settlement (consolidation), or shrinking or swelling of expansive soils; and be constructed of materials that conform to Agency adopted MPS requirements for foundations;

5. Have the manufactured home insulated to meet the energy conserving requirements contained in exhibit D of this subpart;

6. Have a manufactured home site, site improvements, and all other features of the mortgaged property not addressed by the Federal Manufactured Home Construction and Safety Standards, meet or exceed applicable requirements of this subpart and subpart C of part 1924 of this chapter, the Agency adopted MPS except paragraph 31-2.2 or a Model Building Code acceptable to the Agency;

7. Have had the manufactured unit itself braced and stiffened where necessary before it leaves the factory to eliminate racking and potential damage during transportation; and

8. Be eligible for financing in accordance with the requirements of either section 502, or section 515 of the Agency's Housing Program, for which purpose the beginning of construction will be the commencement of on-site work even though the manufactured home itself may have been produced and

temporarily stored prior to the date of application for financing.

B. *Site Planning and Development*. The site planning and development of manufactured home scattered sites and subdivisions shall also comply with the following:

1. *Arrangement of Structures and Facilities*. The site, including the manufactured home, accessory structures, and all site improvements shall be harmoniously and efficiently organized in relation to topography, the shape of the plot, and the shape, size and position of the unit. Particular attention shall be paid to use, appearance and livability.

2. *Adaptation to Site Assets*. The manufactured home shall be fitted to the terrain with a minimum disturbance of the land. Existing trees, rock formations, and other natural site features shall be preserved to the extent practical. Favorable views or outlooks shall be emphasized by the plan.

3. *Site Plan*. The site plan shall provide for a desirable residential environment which is an asset to the community in which it is located.

4. *Lot Size*. The size of manufactured home lots (scattered sites and subdivision) shall be determined by 7 CFR part 3550 and subpart C of part 1924 of this chapter.

C. *Foundation Systems, Anchoring and Set-up.*

1. The foundation system shall be constructed in accordance with this subpart and one of the following: (a) The foundation system included in the manufacturer's installation instructions meeting Agency/MPS requirements, (b) the Agency/MPS 4900.1, which specifies performance requirements for foundations in section 600 "General" and paragraph 601–16 "Foundations," or (c) an FmHA or its successor agency under Public Law 103–354 recognized model building code.

2. The manufactured home permanent foundation system shall constitute a permanent load bearing support system for the manufactured home. The manufacturer or applicant shall be permitted to design or specify the installation of a foundation system which meets Agency/MPS design requirements for foundations and the general requirements above.

3. The applicant's responsibility for proper design and installation of the permanent foundation system, anchoring and set-up shall be in accordance with § 1924.5(f)(1), of this subpart.

4. The builder/developer of the manufactured home property, for proposed construction, shall submit with the application for financing by the applicant or for a conditional commitment design calculations, details and drawings for the installation, anchorage and construction of permanent foundation and perimeter enclosure to be used.

III. *Rental Housing Project Development*. A. General. Manufactured housing rental developments shall be planned and constructed in accordance with requirements of subpart C of part 1924; this subpart; 7 CFR part 1970, the Agency/MPS; and the requirements of subpart E of part 1944 of this chapter.

B. *Site Planning and Development*. Site planning and development shall adapt to individual site conditions and the type of market to be served, reflect advances in site planning and development techniques, and be adaptable to the trends in design of the manufactured home. Site planning and development shall utilize existing terrain, trees, shrubs and rocks formations to the extent practicable. A regimental style site plan design should be avoided.

C. *Foundation Systems, Anchoring and Set-up*. Foundation systems, anchoring and set/ups for manufactured home rental projects (site and home) developed under Agency section 515 Rural Rental Housing program shall comply with the requirements of paragraphs II A and II C above.

IV. *Accessory Structures and Related Facilities*. A. *General*. Accessory structures and related facilities are dependent upon the manufactured home and its environment.

1. Accessory structures and related facilities shall be planned, designed and constructed in accordance with the applicable provisions of this subpart; the Agency/MPS; and local criteria of the authority having jurisdiction.

2. Accessory structures and related facilities shall be designed in a manner that will eliminate and prevent health and safety hazards and enhance the appearance of the manufactured home and its environment.

3. Accessory structures and related facilities shall not obstruct required openings for light and ventilation of the manufactured home and shall not hamper installation and utility connections of the unit.

B. *Accessory Structures*. 1. Accessory structures shall not include spaces for pantries, bath, toilet, laundries, closets or utility rooms.

2. Accessory structures shall be carefully designed and constructed for the convenience and comfort of the manufactured home occupant. These features significantly affect the visual appearance of the community and influence livability.

C. *Related Facilities (Rental Housing Projects)*. 1. This includes those facilities as defined in § 1944.212(e) of subpart E of part 1944 of this chapter.

2. Related facilities built on-site must meet the Agency/MPS and subpart A of part 1924 of this chapter or other building codes approved by by the Agency.

3. Workmanship shall be of a quality equal to good standard practice. Material shall be of such kind and quality as to assure reasonable durability and economy of maintenance, all commensurate with the class of building under consideration.

4. All members and parts of the construction shall be properly designed to carry all loads imposed without detrimental effect on finish or covering materials.

5. The structure shall be adequately braced against lateral stresses and each member shall be correctly fitted and connected.

6. Adequate precautions shall be taken to protect against fire and accidents.

7. All related facilities which require accessibility to the handicapped must comply with the Uniform Federal Accessibility Standard (UFAS).

V. *Fire Protection and Safety.* A. The design of the site plan for each manufactured community and scattered site shall meet the fire protection and safety requirements of the local authority responsible for providing the necessary fire protection services.

B. All fire detection and alarm systems, and water supply requirements for fire protection for manufactured communities shall be in accordance with the local authority responsible for providing the necessary fire protection services.

C. Any portion of a manufactured home shall not be closer than the local separation requirements of the development standard for side to side, end to end, and end to side siting. If the exposed composite wall and roof of two or more manufactured homes are proposed to be joined they shall be without openings and constructed of materials which will provide a minimum one-hour fire rating each, or the manufactured homes are separated by a one-hour fire rated barrier designed and approved for such installation and permitted by the authority having jurisdiction.

D. Manufactured homes shall not be positioned vertically (stacked) with one over the other in whole or in part without the specific approval of the authority having jurisdiction.

Part C—Drawings, Specifications, Contract Documents and Other Documentation

I. *General.* Adequate site development and foundation installation drawings and specifications shall be provided by the applicant or dealer-contractor to the Agency to fully describe the construction and other development work. These documents shall be provided according to the requirements of §1924.5(f)(1) of this subpart. Contract documents will be prepared in accordance with §1924.6 and, in the case of multiple family housing construction and development, §1924.13 of this subpart.

A. The documents recommended shall be used as a guide for drawings and specifications to be submitted in support of all types of loan and/or grant applications involving manufactured homes. Adequate and accurate drawings and specifications are necessary to:

1. Determine the acceptability of the physical environment and improvements,

2. Determine compliance with the applicable standards and codes,

3. Review cost estimates, and

4. Provide a basis for financing, inspections, and the warranty.

B. Detailed floor plans, drawings and specifications are not required for any manufactured home to be installed on a scattered site, in a subdivision or rental housing project. However, a schematic floor plan should be submitted by the applicant when applying for Agency financing. The unit must have an affixed label as specified in exhibit D of this subpart indicating that the unit is constructed to the Agency thermal requirements for the appropriate winter degree days. This will indicate that the manufacturer certifies that the unit has been properly inspected and it meets the Agency Thermal Performance Construction Standard.

C. For proposed construction, the builder or dealer-contractor shall submit with the loan or grant application design calculations, details and drawings for the installation, anchorage and construction of the permanent foundations and perimeter enclosure to be used. Drawings and specifications for foundation systems will be reviewed and examined by either the Agency County Supervisor, District Director, or State Architect/Engineer for foundation support locations, loads and connection requirements specified by the manufacturer as a basis for evaluating foundation compliance with the Agency/MPS or Model Building Code, and for determining design suitability for soil conditions. Drawings and specifications will also be examined by the Agency to determine compliance with all other on-site features not covered by the FMHCSS.

D. Foundation design sections and details of all critical construction points systems, anchorage methods, and structural items shall be scaled as necessary to provide all appropriate information 1:30 (3/8″ = 1′-0″) minimum.

II. *Scattered Sites.* Drawings for single family manufactured housing shall be submitted by the applicant in addition to the requirements of paragraph I above and the requirements of paragraphs II A and D-7 of exhibit C of this subpart.

III. *Subdivisions.* Subpart C of part 1924 of this chapter will be used in preparing and providing supporting documents.

IV. *Rental Housing Projects.* Subpart C of part 1924 of this chapter will be used in preparing and providing supporting documents.

V. *Specifications.* A. Form RD 424-2, "Description of Materials," or other acceptable and comparable descriptions of all materials used for site development, foundation installation and the permanent perimeter enclosure shall be submitted with the drawings by the applicant.

B. The material identification information shall be in sufficient detail to fully describe the material, size and grade. Where necessary, additional sheets shall be attached as well as manufacturer's specification sheets for equipment and/or special materials.

Part D—Inspection of Development Work

I. *General.* The following policies will govern the inspection of all manufactured housing development work. This includes scattered sites, subdivisions, rental housing projects and all accessory structures and related facilities unless otherwise indicated.

II. *Inspections.* A. The responsibility for frequency and propose of inspections shall be in accordance with § 1924.9(b) (1), (2) and (3) of this subpart. The inspection requirements of § 1924.13 apply to the planning and conduct of construction work on all 515 housing developments that are more extensive in scope and more complex in nature than those involving an individual manufactured housing unit. The Stage 2 inspection customary for site-built housing when the building is enclosed is not required for manufactured homes.

The Stage 2 inspection for manufactured homes will be made within two working days after erection or placement on the foundation to determine compliance with accepted installation drawings and specifications for installation and set-up and to verify that the correct unit is on the site.

Stages 2 and 3 inspections for manufactured homes may be combined when authorized by the State Director.

B. The borrower will join the County Supervisor or the District Director in making periodic inspections as often as possible and always for the final inspection.

C. The borrower should be encouraged to make enough periodic visits to the site to be familiar with the progress and performance of the work in order to protect the borrower's interest. If the borrower observes or otherwise becomes aware of any fault or defect in the work or nonconformance with the contract documents, the borrower should give prompt written notice thereof to the dealer-contractor and a copy of the notice to the appropriate County Supervisor or District Director.

D. During inspection, it will generally be infeasible to determine whether a manufactured unit erected on a site was properly braced and stiffened during transportation. Inspectors should examine these units to determine that there is no obvious damage or loosening of fastenings that may have occurred during transportation. The dealer-contractor must warrant these units against such damage, which should protect the Agency's interest.

III. *Warranty Plan Coverage.* The warranty requirements for all development work shall

be in accordance with § 1924.9(d) of this subpart and 7 CFR part 3550, subpart B.

[51 FR 41603, Nov. 18, 1986, as amended at 52 FR 19283, May 22, 1987; 53 FR 2156, Jan. 26, 1988; 67 FR 78327, Dec. 24, 2002; 81 FR 11029, Mar. 2, 2016]

EXHIBIT K TO SUBPART A OF PART 1924— CLASSIFICATIONS FOR MULTI-FAMILY RESIDENTIAL REHABILITATION WORK

I. General

This exhibit distinguishes between what the Agency considers maintenance and repair work, moderate rehabilitation and substantial rehabilitation. In all cases, the building or project to be rehabilitated shall be structurally sound. The applicant shall have a structural analysis of the existing building made to determine the adequacy of all structural systems for the proposed rehabilitation.

II. Definitions

Maintenance and Repair—Work involved in the selective replacement and general maintenance and repair of certain materials, appliances or components of an existing residential building.

Moderate Rehabilitation—All work directly involved in the rearrangement of interior space, the replacement of finish materials or components of the electrical, plumbing, heating or conveyance systems of an existing multi-family residential building. Work and improvements are considered to be more than routine maintenance and repair.

Substantial Rehabilitation—All work directly involved in the rearrangement of interior space that involves alteration of load bearing partitions and columns; the replacement of the electrical, plumbing, heating or conveyance systems; and the addition to and/ or major conversion of existing multi-family residential buildings or other building structures.

Moderate rehabilitation and repair shall not be limited to building changes for cosmetic or convenience purposes. In all cases moderate rehabilitation shall involve a minimum of three (3) components of building rehabilitation listed as moderate. Unless combined with other improvements in a project that are considered to be moderate or substantial rehabilitation the items identified as maintenance and repair are considered to be cosmetic and convenience changes.

When a rehabilitation project consists of both moderate and substantial rehabilitation components, those substantial rehabilitation components shall be in accordance with the Agency's development standards and local codes and regulation requirements. Where the majority of project components of building rehabilitation are considered substantial the project shall be considered in the substantial rehabilitation category.

Those site components of rehabilitation such as landscaping, grading, drainage, fencing, parking areas, recreation areas, water and waste disposal systems, etc., whether considered either maintenance and repair, moderate rehabilitation or substantial rehabilitation shall be in accordance with the Agency's development standards for site development work; all local codes and regulation requirements; and sound engineering and architectural practices.

Any alteration of a structure listed or eligible for listing on the National Register of Historic Places may be considered either moderate or substantial rehabilitation; however, it shall conform first to the Secretary of the Interior's Standards for Rehabilitation and Guidelines for Rehabilitating Historic Buildings and then to the Agency's requirements. In cases where the Secretary of the Interior's standards cannot be met, rehabilitation will conform to the agreed upon approaches, treatments and techniques resulting from the consultation process between the Agency, the borrower, the State Historic Preservation Officer and the Advisory Council of Historic Preservation.

III. Components of Multi-Family Building Rehabilitation

The components of multi-family building rehabilitation necessary and generally considered by the Agency to be either maintenance and repair, moderate rehabilitation or substantial rehabilitation include but are not limited to those listed in the following chart.

COMPONENTS OF MULTI-FAMILY BUILDING REHABILITATION

Components	Maintenance and repair	Moderate rehabilitation	Substantial rehabilitation
Air conditioning	o		
Appliance replacement or repair	o		
Cabinet replacement or repair	o		
Carpeting	o		
Caulking	o		
Ceiling framing	o		
Clothes closets or shelving improvements	o		
Door repair	o		
Drywall repair	o		
Gutters and downspouts	o		
Hardware replacement or repair	o		
Kitchen cabinet improvement	o		
Lighting fixture replacement or repair	o		
Mail boxes	o		
Painting	o		
Paneling	o		

COMPONENTS OF MULTI-FAMILY BUILDING REHABILITATION—Continued

Components	Maintenance and repair	Moderate rehabilitation	Substantial rehabilitation
Partition repair	o		
Roof repair	o		
Signage	o		
Stair repair	o		
Tile work	o		
Wallpapering	o		
Window shades and curtains	o		
Door replacement		o	
Drywall replacement		o	
Elevator components replacement		o	
Exterior entrance redesign, relocation		o	
Finish flooring materials		o	
Flashing		o	
Furnace replacement		o	
Gas pipes		o	
Insulation		o	
Lath and plaster replacement		o	
New shingles or roof replacement		o	
Partition (nonbearing) replacement, or relocation		o	
Plumbing fixture replacement		o	
Pointing		o	
Porch and steps alterator or replacement		o	
Stair replacement, or relocation		o	
Storm windows and weatherstripping		o	
Subfloor material replacement		o	
Trim—exterior and interior		o	
Window replacement		o	
New or alteration to the:			
Mechanical system			o
Soil pipes			o
Vent pipes			o
Waste pipes			o
Alteration or replacement of structural components:			
Beams.			
Chimneys and vents			o
Columns and post			o
Electrical service—replacement or new			o
Elevator replacement			o
Exterior walls			o
Floor construction			o
Footing			o
Foundation wall			o
Foundation waterproofing			o
Interior walls			o

Moderate repair and rehabilitation shall not be limited to building changes for cosmetic purposes. In all cases moderate rehabilitation shall involve a minimum of three (3) components of building rehabilitation listed as moderate. Unless combined with other improvements in a project that are considered to be moderate or substantial rehabilitation the items identified as maintenance and repair are considered to be cosmetic and convenience changes.

EXHIBIT L TO SUBPART A OF PART 1924—
INSURED 10-YEAR HOME WARRANTY
PLAN REQUIREMENTS

I. Purpose

In recent years, numerous third-party home warranty plans have been developed offering new homeowners varying degrees of protection against builder default and/or major structural defects in their homes. This exhibit establishes the criteria and procedures by which a warranty plan is found acceptable for new construction of single family homes financed by the Agency. An acceptable warranty plan will:

A. Assure that the Agency borrowers receive adequate warranty coverage,

B. In certain circumstances, eliminate the requirement for the Agency personnel to make the first two construction inspections, and

C. Permit a loan up to the market value of the security (less the unpaid principal balance and past due interest of any other liens against the security), even though the Agency personnel may not have performed period inspections during construction.

II. Types of Warranty Companies

A. An insured warranty company is underwritten by an insurance carrier, licensed to operate as an insurer by the states where the warranty company plans to operate, and has an acceptable rating from a nationally recognized rating company such as A.M. Best Company.

B. A risk retention group is an insurer which is licensed in one state and is authorized, under the Products Liability Risk Retention Act of 1981, to issue its policies in all states. This authority is not challenged by the Agency; however, there remains some question as to the legal propriety of a 10-year insured warranty insurer to be a risk-retention group. If at some future time any state insurance commission or regulatory agency challenges the legal authority of such group, the Agency will reconsider its acceptance of the group.

C. Individual state warranty plans, such as that offered by the State of New Jersey, are backed by the full faith and credit of the state government.

III. Plan Requirements

To be considered acceptable, a warranty plan must include the following features:

A. The entire cost (fee, premium, etc.) of the coverage is prepaid and coverage automatically transfers to subsequent owners without additional cost.

B. The coverage is not cancellable by the warrantor (builder), warranty company or insurer.

C. The coverage age includes at least the following:

(1) For one year from the effective date, any defects caused by faulty workmanship of defective materials.

(2) During the second year after the effective date, the warranty continues to cover the wiring, piping and duct work of the electrical, plumbing, heating and cooling systems, plus the items in (3).

(3) During the third through the tenth years, the warranty continues to cover major structural defects. A major structural defect is actual damage to the load-bearing portion of the home including damage due to subsidence, expansion or lateral movement of the soil (excluding movement caused by flood or earthquake) which affects its load-bearing function and which vitally effects or is imminently likely to affect use of the home for residential purposes.

D. A system is provided for complaint (claims) handling which includes a conciliation and, if necessary to resolve matters in dispute, arbitration arranged by the American Arbitration Association or similar organization.

E. A construction inspection plan is required if the Agency is to eliminate the first two Agency inspections or permit a full market value loan when Agency inspections are not conducted.

IV. Information for Review

A. Companies submitting warranty plans for a determination of acceptability must support requests with the following information.

(1) Evidence that the insured warranty company has met the applicable state licensing and/or regulatory requirements in the state in which the company plans to operate.

(2) Evidence that the insurance carrier underwriting the warranty plan is licensed to operate as an insurer in the states in which the company plans to operate and has an acceptable rating from a nationally recognized company such as A.M. Best Company.

(3) State warrenty plan agencies will provide evidence that the plan is backed by the full faith and credit of the state.

(4) A full description of the warranty plan including information on the fees, builder and home registration procedures, required

construction standards, construction inspection procedures, coverage provided and claims procedures.

(5) A sample copy of the warranty information and/or policy which is provided to the homeowner.

(6) Suggested means by which Agency field offices can readily assure that the builder is a member in good standing prior to loan approval and that a warrant will be issued upon the completion of construction prior to the final release of funds.

B. Submission and Acceptance:

(1) Insured warranty companies, except those operating as risk retention groups, and state warranty plan agencies will submit their requests and supporting information to the Agency State Director in the state in which they plan to operate. State Directors will determine the acceptability of insured warranty plans and state warranty plans in their jurisdictions, notify the company or agency of the decision in writing and notify field offices by issuance of a State Supplement including the names and addresses of acceptable warranty companies and any other pertinent information.

(2) Warranty companies claiming authority as risk retention groups will submit their requests and supporting information including certification that it has complied with all requirements of the Products Liability Risk Retention Act of 1981 (Pub. L. 97–45) and information indicating the state in which it is licensed, information to the Agency National Office, Single Family Housing Processing Division. The National Office will determine the acceptability of the warranty of a risk retention group, notify the company of the decision in writing and notify field offices by issuance of an attachment to this exhibit.

V. Warranty Performance

A. County Supervisors will report inadequate warranty performance through their District Director to the State Director. State Directors will review the situation, assist in resolving any problems and, if necessary, initiate action under subpart F of part 1942 of this chapter. State Directors will inform, by memorandum, the Director, Single Family Housing Processing Division, National Office, of any problems with warranty performance and if any debarment action is initiated.

B. State Directors will annually monitor each warranty company and/or its insurer to assure continued compliance with state licensing and/or regulatory requirements.

Attachment 1—Acceptable Warranty Companies

The warranty companies listed below claim authority to act as a risk retention group under the Products Liability Risk Retention Act of 1981 and as such, to operate in

all States to provide 10-year home warranties. This authority remains subject to future challenges by any State insurance commissioner or regulatory agency; however, until such challenge is made, the Agency accepts their warranty.

Name and address	Area of operation
Home Owners Warranty Corporation/ HOW Insurance Company, 11 North Glebe Road, Arlington, Virginia 22201, (703) 516–4100.	All States.
Home Buyers Warranty, 89 Liberty Street, Asheville, North Carolina 22801, Telephone: (704) 254–4478.	All States.
Residential Warranty Corporation, P.O. Box 641, Harrisburg, Pennsylvania 17108–0641, Telephone: 1–800–247–1812.	All States.
Manufactured Housing Warranty Corporation, P.O. Box 641, Harrisburg, Pennsylvania 17108–0641, Telephone: 1–800–247–1812.	All States.

[52 FR 8002, Mar. 13, 1987, as amended at 56 FR 29167, June 26, 1991]

Subpart B [Reserved]

Subpart C—Planning and Performing Site Development Work

SOURCE: 60 FR 24543, May 9, 1995, unless otherwise noted.

§ 1924.101 Purpose.

This subpart establishes the basic Rural Housing Service (RHS) policies for planning and performing site development work. It also provides the procedures and guidelines for preparing site development plans consistent with Federal laws, regulations, and Executive Orders.

§ 1924.102 General policy.

(a) *Rural development.* This subpart provides for the development of building sites and related facilities in rural areas. It is designed to:

(1) Recognize community needs and desires in local planning, control, and development.

(2) Recognize standards for building-site design which encourage and lead to the development of economically stable communities, and the creation of attractive, healthy, and permanent living environments.

(3) Encourage improvements planned for the site to be the most cost-effective of the practicable alternatives. Encourage utilities and services utilized to be reliable, efficient, and available at reasonable costs.

(4) Provide for a planning process that will consider impacts on the environment and existing development in order to formulate actions that protect, enhance, and restore environmental quality.

(5) No site will be approved unless it meets the requirements of this part and all state and local permits and approvals in connection with the proposed development have been obtained.

(b) *Subdivisions.* RHS does not review or approve subdivisions. Each site approved by RHS must meet the requirements of § 1924.115, on a site by site basis.

(c) *Development related costs*—(1) *Applicant.* The applicant is responsible for all costs incurred before loan or grant closing associated with planning, technical services, and actual construction. These costs may be included in the loan or grant as authorized by RHS regulations.

(2) *Developer.* The developer is responsible for payment of all costs associated with development.

§ 1924.103 Scope.

This subpart provides supplemental requirements for Rural Rental Housing (RRH) loans, Rural Cooperative Housing (RCH) loans, Farm Labor Housing (LH) loans and grants, and Rural Housing Site (RHS) loans. It also provides a site development standard, as indicated in exhibit B of RD Instruction 1924–C, which supplements this subpart to provide the minimum for the acceptability of development. All of this subpart applies to Single Family Housing unless otherwise noted. All of this subpart also applies to Multiple Family Housing except §§ 1924.115 and 1924.120, and any paragraph specifically designated for Single Family Housing only. In addition, RHS will consult with appropriate Federal, state, and local agencies, other organizations, and individuals to implement the provisions of this subpart.

§ 1924.104 Definitions.

As used in this subpart:

Applicant. Any person, partnership, limited partnership, trust, consumer cooperative, corporation, public body, or association that has filed a preapplication, or in the case of RHS programs that do not require a preapplication, an official application, with RHS in anticipation of receiving or utilizing RHS financial assistance.

Community. A community includes cities, towns, boroughs, villages, and unincorporated places which have the characteristics of incorporated areas with support services such as shopping, post office, schools, central sewer and water facilities, police and fire protection, hospitals, medical and pharmaceutical facilities, etc., and are easily identifiable as established concentrations of inhabited dwellings and private and public buildings.

Developer. Any person, partnership, public body, or corporation who is involved with the development of a site which will be financed by RHS.

Development. The act of building structures and installing site improvements on an individual dwelling site, a subdivision, or a multiple family tract.

Multiple Family Housing. RHS RRH loans, RCH loans, LH loans and grants, and RHS loans.

Single Family Housing. RHS Rural Housing loans for individuals for construction of, repair of, or purchase of a dwelling to be occupied by one household.

Site. A parcel of land proposed as a dwelling site, with or without development.

Site approval official. The RHS making the determination that a site meets the requirements in this subpart to be acceptable for site loans. (See § 1924.120.)

Street surfaces. Streets may be hard or all-weather surfaced.

(1) *Hard surface*—a street with a portland cement concrete, asphaltic concrete, or bituminous wearing surface or other hard surfaces which are acceptable and suitable to the local public body for use with local climate, soil, gradient, and volume and character of traffic.

(2) *All-weather*—a street that can be used year-round with a minimum of

573

maintenance, such as the use of a grader and minor application of surface material, and is acceptable and suitable to the local public body for use with local climate, soil, gradient, and volume and character of traffic.

Subdivision. Five or more contiguous (developed or undeveloped) lots or building sites. Subdivisions may be new or existing.

§ 1924.105 Planning/performing development.

(a) *General.* Planning is an evaluation of specific development for a specific site. Planning must take into consideration topography, soils, climate, adjacent land use, environmental impacts, energy efficiency, local economy, aesthetic and cultural values, public and private services, housing and social conditions, and a degree of flexibility to accommodate changing demands. All planning and performing development work is the responsibility of the applicant or developer. All development will be arranged and completed according to applicable local, state, or Federal regulations including applicable health and safety standards, environmental requirements, and requirements of this subpart. When a public authority requires inspections prior to final acceptance, written assurance by the responsible public authority of compliance with local, city, county, state or other public codes, regulations, and ordinances is required prior to final acceptance by RHS.

(1) [Reserved]

(2) *Technical Services.* [Reserved]

(i) [Reserved]

(ii) An applicant or developer for a Multiple Family Housing project or a Single Family Housing site which requires technical services under § 1924.13(a), must contract for the technical services of an architect, engineer, land surveyor, landscape architect, or site planner, as appropriate, to provide complete planning, drawings, and specifications. Such services may be provided by the applicant's or developer's "in house" staff subject to RHS concurrence. Technical services must be performed by professionals who are qualified and authorized to provide such services in the state in which the project would be developed. All tech-

nical services must be provided in accordance with the requirements of professional registration or licensing boards. At completion of all construction or completion of a phase or phases of the total project, the persons providing technical services under this section must notify the RHS field office in writing that all work has been completed in substantial conformance with the approved plans and specifications.

(iii) For developments not specifically required to have technical services under paragraph (a)(2)(ii) of this section, such services may be required by the state director when construction of streets or installation of utilities is involved.

(3) *Drawings, specifications, contract documents, and other documentations.* Adequate drawings and specifications must be provided by the applicant or developer to RHS in sufficient detail to fully and accurately describe the proposed development. Contract documents must be prepared in accordance with § 1924.6 or, in the case of more complex construction, § 1924.13.

(b) *Single Family Housing.* Proposals for development of individual dwelling sites must meet the following requirements:

(1) *Site development design requirements.* Exhibit B (available in any RHS field office) will be used as a minimum by applicants or developers in preparing proposals and supporting documents for Single Family Housing loans, in addition to specific requirements made in this subpart.

(2) [Reserved]

(c) *Multiple Family Housing.* Exhibit C (available in any RHS office) should be used as a guide by the applicant or developer in preparing a proposal and supporting documents for multiple family housing projects.

§ 1924.106 Location.

(a) *General.* It is RHS's policy to promote compact community development and to finance projects that avoid or minimize conversion of wetlands or important farmlands, avoid unwarranted alterations or encroachment on floodplains, and avoid unwarranted adverse effects to historic properties (including those listed or eligible for listing on the National Register of Historic Places), when practicable alternatives exist to meet development needs; RHS is prohibited from financing development within the Coastal Barrier Resource System, or on a barrier island. A complete listing of the environmental review requirements is found in 7 CFR part 1970. In order to be eligible for RHS participation:

(1) The site must be located in an eligible area as defined in the program regulations under which the development is being funded or approved.

(2) The site must comply with the environmental review requirements in accordance with 7 CFR part 1970.

(b) *Single Family Housing.* In addition to the general requirements in paragraph (a) of this section, sites must provide a desirable, safe, functional, convenient, and attractive living environment for the residents.

(c) *Multiple Family Housing.* Multiple family housing projects shall be located in accordance with the requirements in paragraph (r) of § 1944.215. Locating sites in less than desirable locations of the community because they are in close proximity to undesirable influences such as high activity railroad tracks; adjacent to or behind industrial sites; bordering sites or structures which are not decent, safe, or sanitary; or bordering sites which have potential environmental concerns such as processing plants, etc., is not acceptable. Screening such sites does not make them acceptable. Sites which are not an integral part of a residential community and do not have a reasonable access, either by location or terrain, to essential community facilities such as water, sewerage, schools, shopping, employment opportunities, medical facilities, etc., are not acceptable.

[60 FR 24543, May 9, 1995, as amended at 81 FR 11030, Mar. 2, 2016]

§ 1924.107 Utilities.

All development under this subpart must have adequate, economic, safe, energy efficient, dependable utilities with sufficient easements for installation and maintenance.

(a) *Water and wastewater disposal systems—*(1) *Single Family Housing.* If sites are served by central water or sewer systems, the systems must meet the requirements of paragraphs (a)(2) (i) and (ii) of this section. If sites have individual water or sewer systems, they must meet the requirements of the state department of health or other comparable reviewing and regulatory authority and the minimum requirements of exhibit B (available in any RHS field office), paragraphs V and VI. Sites in subdivisions of more than 25 dwelling units on individual systems, or sites that do not meet the requirements of exhibit B, paragraphs V and VI, must have state director concurrence.

(2) *Multiple Family Housing.* Proposals processed under this paragraph shall be served by centrally owned and operated water and wastewater disposal systems unless this is determined by RHS to be economically or environmentally not feasible. All central systems, whether they are public, community, or private, shall meet the design requirements of the state department of health or other comparable reviewing and regulatory authority. The regulatory authority will verify in writing that the water and wastewater systems are in compliance with the current provisions of the Safe Drinking Water Act and the Clean Water Act, respectively.

(i) Sites which are not presently served by a central system, but are scheduled for tie-in to the central system within 2 years, should have all lines installed during the initial construction. Such sites must have an approved interim water supply or wastewater disposal system installed capable of satisfactory service until the scheduled tie-in occurs.

(ii) In addition to written assurance of compliance with state and local requirements, there must be assurance of continuous service at reasonable rates for central water and wastewater disposal systems. Public ownership is preferred whenever possible. In cases where interim facilities are installed pending extension or construction of permanent public services, the developer must assume responsibility for the operation and maintenance of the interim facility or establish an entity for its operation and maintenance which is acceptable to the local governing body. If a system is not or will not be publicly owned and operated, it must comply with one of the following:

(A) Be an organization that meets the ownership and operating requirements for a water or wastewater disposal system that RHS could finance under 7 CFR part 1942, subpart A or be dedicated to and accepted by such an organization.

(B) Be an organization or individual that meets other acceptable methods of ownership and operation as outlined in HUD Handbook 4075.12, "Ownership and Organization of Central Water and Sewerage Systems." RHS should be assured that the organization has the right, in its sole discretion, to enforce the obligation of the operator of the water and sewerage systems to provide satisfactory continuous service at reasonable rates.

(C) Be adequately controlled as to rates and services by a public body (unit of Government or public services commission).

(iii) Multiple family developments of more than 25 units with individual system must have national office concurrence.

(A) [Reserved]

(B) Supporting information for the proposed individual water systems, covering the following points:

(1) In areas where difficulty is anticipated in developing an acceptable water supply, the availability of a water supply will be determined before closing the loan.

(2) Documentation must be provided that the quality of the supply meets the chemical, physical, and bacteriological standards of the regulatory authority having jurisdiction. The maximum contaminant levels of U.S. EPA shall apply. Individual water systems must be tested for quantity and bacteriological quality. Where problems are anticipated with chemical quality, chemical tests may be required. Chemical tests would be limited to analysis for the defects common to the area such as iron and manganese, hardness, nitrates, pH, turbidity, color, or other undesirable elements. Polluted or contaminated water supplies are unacceptable. In all cases, assurance of a pota-

ble water supply before loan closing is required.

(C) Supporting information for individual wastewater disposal systems with subsurface discharge provided by a soil scientist, geologist, soils engineer, or other person recognized by the local regulatory authority. This data must include the following:

(1) Assurance of nonpollution of ground water. The local regulatory authority having jurisdiction must be consulted to ensure that installation of individual wastewater systems will not pollute ground water sources or create other health hazards or otherwise violate State water quality standards.

(2) Records of percolation tests. Guidance for performing these tests is included in the EPA design manual, "Onsite Wastewater Treatment and Disposal Systems" and the minimum RHS requirements are in exhibit B, paragraph VI. (These may be waived by the state director when the state has established other acceptable means for allowing onsite disposal.)

(3) Determination of soil types and description. The assistance of the SCS or other qualified persons should be obtained for soil type determination and a copy of its recommendations included in the documentation.

(4) Description of ground water elevations, showing seasonal variations.

(5) Confirmation of space allowances. An accurate drawing to indicate that there is adequate space available to satisfactorily locate the individual water and wastewater disposal systems; likewise, documented assurance of compliance with all local requirements. Structures served by wastewater disposal systems with subsurface discharge require larger sites than those structures served by another type system.

(6) Description of exploratory pit observations, if available.

(D) Supporting information for individual wastewater disposal systems with surface discharge covering the following points:

(1) Effluent standards issued by the appropriate regulatory agency that controls the discharge of the proposed individual systems. Assurance from this regulatory agency that the effluent standards will not be exceeded by

the individual systems being proposed must be included.

(2) Program of maintenance, parts, and service available to the system-owner for upkeep of the system.

(3) A plan for local inspection of the system by a responsible agency with the authority to ensure compliance with health and safety standards.

(b) *Electric service.* The power supplier will be consulted by the applicant to assure that there is adequate service available to meet the needs of the proposed site. Underground service is preferred.

(c) *Gas service.* Gas distribution facilities, if provided, will be installed according to local requirements where adequate and dependable gas service is available.

(d) *Other utilities.* Other utilities, if available, will be installed according to local requirements.

§1924.108 Grading and drainage.

(a) *General.* Soil and geologic conditions must be suitable for the type of construction proposed. In questionable or unsurveyed areas, the applicant or developer will provide an engineering report with supporting data sufficient to identify all pertinent subsurface conditions which could adversely affect the structure and show proposed solutions. Grading will promote drainage of surface water away from buildings and foundations, minimize earth settlement and erosion, and assure that drainage from adjacent properties onto the development or from the development to adjacent properties does not create a health hazard or other undesirable conditions. Grading and drainage will comply with exhibit B, paragraphs III and IV, of this subpart.

(b) *Cuts and fills.* Development requiring extensive earthwork, cuts and fills of 4 feet or more shall be designed by a professional engineer. Where topography requires fills or extensive earthwork that must support structures and building foundations, these must be controlled fills designed, supervised, and tested by a qualified soils engineer.

(c) *Slope protection.* All slopes must be protected from erosion by planting or other means. Slopes may require temporary cover if exposed for long periods during construction.

(d) *Storm water systems.* The design of storm water systems must consider convenience and property protection both at the individual site level and the drainage basin level. Storm water systems should be compatible with the natural features of the site. In areas with inadequate drainage systems, permanent or temporary storm water storage shall be an integral part of the overall development plan. Design of these facilities shall consider safety, appearance, and economical maintenance operations.

§§1924.109–1924.114 [Reserved]

§1924.115 Single Family Housing site evaluation.

(a) *Site review.* The site approval official will evaluate each site (developed or undeveloped) to determine acceptance for the program. Information on the site will be provided by the appraiser or site approval official on a form provided by RHS and available in any RHS field office.

(b) *Site access.* Each site must be contiguous to and have direct access from:

(1) A hard surfaced or all weather road which is developed in full compliance with public body requirements, is dedicated for public use, and is being maintained by a public body or a home owners association that has demonstrated its ability or can clearly demonstrate its ability to maintain the street; or

(2) An all weather extended driveway which can serve no more than two sites connecting to a hard surface or all weather street or road that meets the requirements of paragraph (b)(1); or

(3) A hard surfaced street in a condominium or townhouse complex which:

(i) Is owned in common by the members or a member association and is maintained by a member association that has demonstrated its ability or can clearly demonstrate its ability to maintain the street; and

(ii) Connects to a publicly owned and dedicated street or road.

(c) *Exceptions to street requirements.* A site not meeting the conditions in paragraph (b) of this section will be acceptable if:

(1) The applicant is a builder for a conditional commitment (a loan will not be approved until the site meets the conditions in paragraph (b) of this section), or the builder posts an irrevocable performance and payment bond (or similar acceptable assurance) that assures the site approval official that the site will be developed to meet the conditions in paragraph (b) of this section; or

(2) The site is recommended by the site approval official and approved by the state director. A request for state director approval must justify that it is in the best interest of both the government and the applicant to approve the site.

(d) *Site layout.* (1) Sites shall be surveyed and platted. Permanent markers shall be placed at all corners.

(2) Sites shall meet all requirements of state and local entities and RHS.

(e) *Covenants, conditions and restrictions.* Sites in subdivisions shall be protected by covenants, conditions, and restrictions (CC&Rs) to preserve the character, value, and amenities of the residential community and to avoid or mitigate potential environmental impacts unless, an exception is granted by RHS after considering the suitability of local ordinances, zoning, and other land use controls.

(1) CC&Rs shall be recorded in the public land records and specifically referenced in each deed.

(2) The intent of the CC&Rs is to assure the developers that the purchasers will use the land in conformance with the planned objectives for the community. In addition, the CC&Rs should assure the purchasers that the land covered by the CC&Rs will be used as planned and that other purchasers will use and maintain the land as planned to prevent changes in the character of the neighborhood that would adversely impact values or create a nuisance.

§§ 1924.116–1924.118   [Reserved]

§ 1924.119   Site Loans.

Subdivisions approved under subpart G of part 1822 (RD Instruction 444.8) or exhibit F of subpart I of part 1944, will meet the general requirements of this subpart to insure lots in the subdivi-

sion will meet the requirements of § 1924.115.

§§ 1924.120–1924.121   [Reserved]

§ 1924.122   Exception authority.

The Administrator of RHS may in individual cases, make an exception to any requirement or provision of this subpart or address any omission of this subpart which is not inconsistent with the authorizing statute or other applicable law if the Administrator determines that application of the requirement or provision would adversely affect the Government's interest. The Administrator will exercise this authority upon the written request of the state director or the appropriate program assistant administrator. Requests for exceptions must be supported with documentation to explain the adverse effect on the Government, proposed alternative courses of action, and show how the adverse effect will be eliminated or minimized if the exception is granted.

§§ 1924.123–1924.149   [Reserved]

§ 1924.150   OMB Control Number.

The reporting requirements contained in this subpart have been approved by the Office of Management and Budget (OMB) and have been assigned OMB control number 0575–0164. Public reporting burden for this collection of information is estimated to vary from 5 minutes to 10 minutes per response, with an average of .13 hours per response, including time for reviewing instructions, searching existing data sources, gathering and maintaining the data needed, and completing and reviewing the collection of information. Send comments regarding this burden estimate or any other aspect of this collection of information, including suggestions for reducing this burden to the Department of Agriculture, Clearance Officer, OIRM, Ag Box 7630, Washington, DC 20250; and to the Office of Management and Budget, Paperwork Reduction Project (OMB #0575–0164), Washington, DC 20503.

EXHIBIT A TO SUBPART C OF PART 1924
[RESERVED]

EXHIBIT B TO SUBPART C OF PART 1924—
SITE DEVELOPMENT DESIGN RE-
QUIREMENTS

This exhibit prescribes site development requirements to be used in developing residential sites in all housing programs. These requirements cover only those areas which involve health and safety concerns. They are not intended to cover all aspects of site development. Applicants and developers are expected to follow local practice, as a minimum, in all areas of site development not addressed in this exhibit. When State, local, or other requirements are applicable in addition to FmHA or its successor agency under Public Law 103–354's requirements, the most stringent requirement shall apply.

Proper integration of the natural features of a site with the manmade improvements is one of the most critical aspects of residential development. Poor site planning in large scale subdivisions, rental projects and individual sites, has resulted in a loss of valuable private and public natural resources and caused economic burdens and conditions unsuitable for healthy and pleasant living. Proper site design can preserve desirable natural features of the site, minimize expenses for streets and utilities, and provide a safe and pleasant living environment.

TABLE OF CONTENTS

I. Streets

A. *Types*—1. *Collector streets.* Collector streets are feeder streets which carry traffic from local streets to the major system of arterial streets and highways. They include the principal entrance streets of residential developments and streets for circulation within such developments.

2. *Local streets.* Local streets are minor streets used primarily for access to abutting properties. These include drives serving multi-family housing units.

B. *Design Features*—1. *Emergency Access.* Access for fire equipment and other emergency vehicles shall be within 100 feet of main building entrances.

2. *Cul-de-sacs.* Cul-de-sac streets shall have a turn-around with an outside roadway diameter of at least 80 feet, and a right-of-way diameter of at least 100 feet.

3. *Intersection Angle.* Streets shall be laid out to intersect as nearly as possible at right angles and no street shall intersect any other street at an angle less than 75 degrees.

Curb radii shall be a minimum of 20 feet for street intersections.

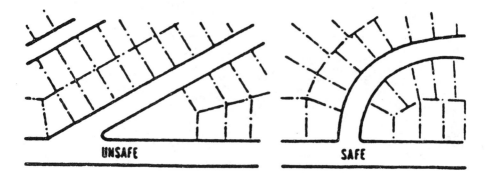

4. *Intersection Sight Distance.* Adequate distances must be maintained at intersections. Vehicles must be visible when within 75 feet of the centerlines of uncontrolled intersecting streets.

C. *Street Geometry*—1. *Definitions.* The definitions in Sections I.C.1.a and I.C.1.b. apply to the requirements in Section I.C.2.

a. *Terrain Classifications.*

(1) Ordinary—Slope less than 8%.

(2) Rolling—Slope range of 8% to 15%.

(3) Hilly—Slope greater than 15%.

b. *Development Density* (Number of Lots). (Land Area minus Undeveloped Areas greater than Average Lot Size)

(1) Low—Less than 2 lots per acre.

(2) Medium—2 to 6.0 lots per acre.

(3) High-More than six lots per acre.

2. *Design Requirements.* Collector streets and local streets shall comply with the requirements in tables 1 and 2 unless an exception is granted by the State Director. These requirements may need modification in localities having winter icing conditions.

TABLE 1—PAVEMENT WIDTHS (FEET)

Street type	On-street parallel parking	Development density		
		Low	Medium	High
Collector	Prohibited	26	32	36
Collector	No Restrictions	36	36	40
Local	Prohibited	18	18	20
Local	Partial, One Side [1]	18	20	26
Local	Partial, One Side [1]	22	26	32
Local	Total, One Side [2]	22	26	26
Local	Total, Both Sides [2]	26	32	36

([1]) At least one parking space per dwelling is provided off-street.
([2]) No parking spaces are provided off-street.

TABLE 2—STREET DESIGN (FEET)

	Terrain		
	Ordinary	Rolling	Hilly
(1) *Collector street:*			
(a) Minimum centerline radius of curvature	300	225	150
(b) Minimum sight distance	250	200	150
(c) Minimum right-of-way width	60	60	60
(2) *Local Street:*			
(a) Minimum centerline radius of curvature	200	150	100
(b) Minimum sight distance	200	150	100
(c) Minimum right-of-way width [1]	50	50	50

([1]) For cul-de-sac streets, the minimum right-of-way width is 40 feet.

D. *Construction.* Street configuration and wearing surfaces must provide safe and economical access to all building sites. The design and construction of the street shall be appropriate for all anticipated traffic, climatic and soil conditions. Streets shall meet or exceed all local, county, and State requirements.

II. Walks and Steps

A. *Walks.* Where walks are provided, they shall be located to assure a minimum vertical clearance of 7 feet from all permanent or temporary obstructions. Walks shall have a slip resistant surface.

B. *Exterior Steps Not Contiguous to Dwelling or Building*—1. *Flight.* a. Single steps or flights of steps exceeding a vertical height of 12 feet shall not be accepted.

b. Steps shall be set back from an intersecting walk or drive a minimum of 1 foot at a retaining wall and 2 feet at slopes.

2. *Risers and Treads.* a. Risers shall be a maximum of 6 inches, a minimum of 3 inches and uniform throughout the flight.

b. Treads shall be a minimum of 12 inches and uniform throughout the flight.

c. Treads shall have a slip resistant surface.

d. Treads shall be pitched appropriately to ensure drainage.

3. *Landings.* a. Minimum length shall equal 3 feet or walk width whichever is greater.

b. A change in direction in a flight of stairs shall be accomplished only at a landing or by a winder which has a tread width at a point 18 inches from the converging end, equal to the full straight stair tread width.

4. *Handrails.* Stairways having a flight rise exceeding 30 inches shall have a 36 inch high handrail located on one side for stairs 5 feet or less in width and on both sides of stairways over 5 feet wide.

III. Grading

A. *Compaction*—All fill for street or home construction shall have compaction of not less than 95 percent maximum density, as determined by proctor or other accepted testing methods. Maximum thickness of compaction layers shall be 6 inches except where compaction equipment of demonstrated capability is used under the direction of a qualified soils engineer. Earth fill used to support a building foundation shall be a controlled fill which is designed, supervised, and tested by a qualified soils engineer in accordance with good practice.

B. *Gradients.* Grading design shall be arranged to assure safe and convenient all-weather pedestrian and vehicular access to residential buildings and to all other necessary site facilities. Site grading shall be designed to establish building floor elevations and ground surface grades which allow drainage of surface water away from buildings and adjacent sites. Grading design shall conform with tables 3 and 4.

TABLE 3—ACCESS AND PARKING GRADIENTS [1]

[In percent]

	Minimum		Maximum	
	Center line	Crown or cross slope	Center line	Crown or cross slope
Streets	0.5	1.0	14.0	5.0
Street Intersections	0.5	1.0	[2]5.0	5.0
Driveways ([3])	.05	1.0	14.0	5.0
Sidewalks ([4]):				
Concrete		0.5		
Bituminous		1.0		
Building Entrances & Short Walks	1.0		12.0	5.0
Main Walks	0.5		10.0	5.0
Adjoining Steps			2.0	
Landings		1.0		
Stepped Ramp Treads	1.0		2.0	5.0
Parking		0.5	5.0	5.0

[1] Approximate Equivalents .5% =$\frac{1}{16}$″ ft., 1.0 =$\frac{1}{8}$″ ft., 2.0% =$\frac{1}{4}$″ ft., 5.0% =$\frac{5}{8}$″ ft., 10.0% = 1$\frac{1}{4}$″ ft., 12.0% = 1$\frac{1}{2}$″ ft., 21% = 2$\frac{5}{8}$″ ft.

[2] Grades approaching intersections shall not exceed 5 percent for a distance of not less than 100 feet from the centerline of the intersection.

[3] Vertical transitions shall percent contact of car undercarriage of bumper with driveway surface.

[4] Five percent maximum for major use by elderly tenants.

TABLE 4—SLOPE GRADIENTS [1]

[In percent]

	Minimum	Maximum
Slope Away From Foundations:		
Pervious Surfaces	[2]5.0	[3]21.0
Impervious Surfaces	[2]1.0	21.0

TABLE 4—SLOPE GRADIENTS [1]—Continued

[In percent]

	Minimum	Maximum
Pervious Surfaces:		
Ground Frost Area	2.0	
Non-Ground Frost Areas	[4]1.0	
Impervious Surfaces	0.5	
Slopes to be maintained by Machine		[3]33.0

[1] See table 3, footnote (1).
[2] Minimum length of 10 feet or as limited by property lines.
[3] Minimum length of 4 feet.
[4] The minimum is 2.0% if the annual precipitation is more than 50 inches.

IV. Drainage

A. *General*—1. *Collection and Disposal.* Surface and subsurface drainage systems shall be provided, as appropriate, for collection and disposal of storm drainage and subsurface water. These systems shall provide for the safety and convenience of occupants. They shall protect dwellings, other improvements and useable lot areas from water damage, flooding, and erosion.

2. *Concentrated Flow.* Where storm drainage flow is concentrated, permanently maintained facilities shall be provided to prevent significant erosion and other damage or flooding on site or on adjacent properties.

B. *Drainage Design and Flood Hazard Exposure*—1. *Storm Frequency.* Drainage facilities shall be designed for a 10 year storm frequency of 24-hour duration. Full potential development of all contributing areas shall be used as a basis for this determination.

2. *Street Drainage.* Streets shall be useable during runoff equivalent to a 10-year return frequency. Where drainage outfall is inadequate to prevent runoff equivalent to a 10-year return frequency from ponding over 6 inches deep, streets shall be made passable for local commonly used emergency vehicles during runoff equivalent to a 25-year return frequency except where an alternative access street not subject to such ponding is available.

3. *Foundation Drainage.* Appropriate crawl space and foundation drainage shall be provided for the removal of subsurface moisture.

C. *Primary Storm Sewer*—1. *Pipe Size.* Pipe size for the primary storm sewer (any storm sewer or inlet lateral located in a street or other public right-of-way) shall have an inside diameter based on design analysis but not less than 15 inches. Where anticipated runoff from the five-year return frequency rainfall will not fill a 15 inch pipe, a primary storm sewer system usually is unnecessary.

2. *Minimum Gradient.* Minimum gradient shall be selected to provide for self-scouring of the conduit under low-flow conditions and for removal of sediments foreseeable from the drainage area.

3. *Easements.* Easements for storm sewers shall be a minimum of 10 feet in width.

D. *Drainage Swals and Gutters*—1. *Design.* Paved gutters shall have a minimum grade of 0.5 percent. Paved gutters and unpaved drainage swales shall have adequate depth and width to accommodate the maximum foreseeable runoff without overflow. Swales and gutters shall be seeded, sodded, sprigged or paved as appropriate to minimize potential erosion. Side slopes shall be no steeper than 2:1.

2. *Easements.* Surface channels shall have an easement which is at least the width of the channel plus 10 feet.

E. *Downspouts*—1. *Outfall.* Where downspouts are provided, they shall either be connected to an available storm sewer, provided with suitable splash blocks, or empty at acceptable locations onto paved areas so that water drains away from buildings. Downspouts shall not connect to sanitary sewers.

2. *Piped Drainage.* Piped roof drainage from buildings shall be connected to available storm sewers or empty at locations where no erosion or other damage will be caused.

F. *Storm Inlets and Catch Basins*—1. *Openings.* Where inlets are accessible to small children, openings shall have one dimension limited to 6 inch access. Inlet openings in paved areas shall be designed to avoid entrapment or impedence of bicycles, baby carriages, etc.

2. *Access.* Access for cleaning shall be provided to all inlet boxes and catch basins.

G. *Drywells*—Drywells for the disposal of water from foundation drains, crawl spaces, and other small quantity sources shall be permissible where the bottom of drywells project into strata of undistributed porous soil at a level where the bottom of the drywell will be above the ground water table at its highest seasonal elevation.

V. Water Supply Systems

A. *Individual Water Systems*—1. *General.* a. In this subpart, an individual water system is a system which serves fewer customers or connections than the lower threshold for community systems stated in the Safe Drinking Water Act.

b. The system for an individual household should be capable of delivering a sustained flow of 5 gpm. A system supplying water to multiple household shall be designed by a Professional Engineer and have sufficient capacity to serve estimated demand. A test of at least 4 hours duration shall be conducted to determine the yield and maximum drawdown for all wells developed as part of an individual water system. This test may be waived by the State Office based on the hydrologic and geologic conditions in the area.

c. Water that requires continual or repetitive treatment to be safe bacterially is not acceptable.

d. After installation, the system should be disinfected in accordance with the recommendations of the health authority. In the absence of a health authority, system cleaning and disinfection should conform with the current EPA Manual of Individual Water Supply Systems.

e. Any method for individual water supply contained herein which is not permitted by the local health authority having jurisdiction shall not be used.

2. *Well Location*—a. A well located within the foundation walls of a dwelling is not acceptable except in arctic and sub-arctic regions.

b. Water which comes from soil formation which may be polluted or contaminated or is fissured or creviced or which is less than 20 feet below the natural ground surface (subject to the requirements of the local health authority) is not acceptable.

c. Individual water supply systems are not acceptable for individual lots in areas where chemical soil poisoning is practiced if the overburden of soil between the ground surface and the water bearing strata is coarsegrained sand, gravel, or porous rock, or is creviced in a manner which will permit the recharge water to carry the toxicants into the zone of saturation.

d. Table 5 shall be used in establishing the minimum acceptable distances between wells and sources of pollution located on either the same or adjoining lots. These distances may be increased by either the health authority having jurisdiction or the FmHA or its successor agency under Public Law 103-354 State Director.

TABLE 5—DISTANCE FROM SOURCE OF POLLUTION

Source of pollution	Minimum horizontal distance (feet)
Property Line	10
Septic Tank	50
Absorption field	[1]100
Seepage pit	[1]100
Absorption Bed	[1]100
Sewer Lines w/Permanent Watertight Joints	10
Other Sewer Lines	50
Chemically Poisoned Soil	[1]100
Dry Well	50

TABLE 5—DISTANCE FROM SOURCE OF POLLUTION—Continued

Source of pollution	Minimum horizontal distance (feet)
Other	[2]—

NOTES:
[1] The horizontal distance between the sewage absorption system and the well, or the chemically poisoned soil and the well, may be reduced to 50 feet only where the ground surface is effectively separated from the water bearing formation by an extensive, continuous impervious strata of clay, hardpan, or rock. The well shall be constructed so as to prevent the entrance of surface water and contaminants.
[2] Other sources of pollution could be fuel oil or gasoline storage tanks, farm yards or chemical storage tanks, etc. The well should be separated from these sources of pollution a distance recommended by the local health authority.

3. *Well Construction*—a. The well shall be constructed to allow the pump to be easily placed and to function properly.

b. All drilled wells shall be provided with a sound, durable and watertight casing capable of sustaining the loads imposed. The casing shall extend from a point several feet below the water level at drawdown or from an impervious strata above the water level, to 12 inches above either the ground surface or the pump room floor. The casing shall be sealed at the upper opening.

c. Bored wells shall be lined with concrete, vitrified clay, or equivalent materials.

d. The space between the casing or liner and the wall of the well hole shall be sealed with cement grout.

e. The well casing shall not be used to convey water except under positive pressure. A separate drop pipe shall be used for suction line.

f. When sand or silt is encountered in the water-bearing formation, the well shall either be gravel packed, or a removable strainer or screen shall be installed.

g. The surface of the ground above and around the well shall be graded to drain surface water away from the well.

h. Openings in the casing, cap, or concrete cover for the entrance of pipes, pump or manholes, shall be made watertight.

i. If a breather is provided, it shall extend above the highest level to which surface water may rise. The breather shall be watertight, and the open end shall be screened and positioned to prevent entry of dust, insects and foregin objects.

4. *Pumps and Equipment*—a. Pumps shall be capable of delivering the volume of water required herein under normal operating pressures within the living unit. Well pump capacity shall not exceed the output of the well.

b. Pumps and equipment shall be mounted to be free of objectionable noises, vibrations, flooding, pollution, and freezing.

c. Suction lines shall terminate below maximum drawdown of the water level in the well.

d. Horizontal segments of suction line shall be placed below the frost line in a sealed casing pipe or in at least 4 inches of concrete. The distance from suction line to sources of pollution shall be not less than shown in table 5.

5. *Storage Tanks*—a. A system for an individual household shall include a pressure tank having a minimum capacity of 42–gallons. However, prepressured tanks and other pressurizing devices are acceptable provided that delivery between pump cycles equals or exceeds that of a 42 gallon tank. Storage capacity on a system for multiple households must be sufficient to meet estimated peak demands.

b. Tanks shall be equipped with a clean-out plug at the lowest point, and if pressurized, a suitable pressure relief valve.

c. When additional storage is necessary because the well yield will not meet the system peak demands, all nonpressurized intermediate tanks shall be designed and installed in a manner that will prevent the pollution or degradation of the water supply.

B. *Community Water Systems*—1. *Definition.* In this subpart, a community water system is a system which meets the definition in the Safe Drinking Water Act.

2. *Design.* A community water system shall be designed by a qualified, professional engineer licensed in the state in which the water system will be located. Community water systems shall comply with all Federal and State laws.

VI. Wastewater Disposal Systems

Each dwelling shall be provided with a water-carried system adequate to dispose of domestic wastes in a manner which will not create a nuisance, contaminate any existing or prospective water source or water supply, or in any way endanger the public health.

A. *Individual Wastewater Disposal Systems*—1. *General.* a. In this subpart, an individual wastewater disposal system is a sewage disposal system which serves only 1 dwelling unit.

b. When service from an acceptable public or community system is not available or feasible, and ground water and soil conditions are acceptable, an individual system may be used.

c. Each individual wastewater disposal system shall consist of a house sewer, a pretreatment unit (e.g., septic tank, individual package treatment plant), and acceptable absorption system (subsurface absorption field, seepage pit(s), or subsurface absorption bed). The system shall be designed to receive all sanitary sewage (bathrooms, kitchen and laundry) from the dwelling, but not footing or roof drainage. It shall be designed so that gases generated anywhere in the system can easily flow back to the building sewer stack.

2. *Percolation Tests*—a. Percolation tests are required unless a waiver is granted by the National Office. Waivers may be granted on a statewide or local basis in cases where an onsite evaluation of soils would be performed by a qualified soil technician, soil scientist, or engineer. Requests for waivers must describe the qualifications of the person evaluating the soils and discuss the criteria to be used in designing the absorption system.

b. In uniform soils one percolation test shall be made within each area proposed for an absorption system. If significant soil variations are encountered or expected, additional tests shall be made for each variation.

c. Percolation tests shall be conducted in accordance with good practice. Guidance for performing these tests is included in the EPA design manual, "Onsite Wastewater Treatment and Disposal Systems."

3. *Subsurface Absorption System*—a. Where percolation rates, soil characteristics and site conditions are acceptable, an absorption system may be installed in an area which is well drained, has an acceptable slope, and is acceptable for excavation.

b. Soils with percolation rates less than 1 minute per inch may be used if the soil is replaced with a layer of loamy or fine sand at least 2 feet thick. (Refer to the EPA Design Manual, "Onsite Wastewater Treatment and Disposal System".)

c. Soils with percolation rates greater than 60 minutes per inch are not acceptable for subsurface wastewater disposal systems.

B. *Community Wastewater Disposal Systems*—1. *Definition.* In this subpart, a community wastewater disposal system is any wastewater disposal system which serves more than 1 dwelling unit.

2. *Design.* A community wastewater disposal system shall be designed by a qualified, professional engineer licensed in the state in which the system will be located.

[52 FR 19284, May 22, 1987]

EXHIBIT C TO SUBPART C OF PART 1924—
CHECKLIST OF VISUAL EXHIBITS AND
DOCUMENTATION FOR RRH, RCH,
AND LH PROPOSALS

U.S. Department of Agriculture

Farmers Home Administration or its
successor agency under Public Law 103–354

This exhibit lists visual exhibits and documentation necessary for FmHA or its successor agency under Public Law 103–354 to properly evaluate proposed development. Intermediate consultation by the applicant, builder-developer and others hereafter referred to as the sponsor with the FmHA or its successor agency under Public Law 103–

354 District or State Offices should be as frequent as necessary to reduce chances of misunderstandings and limit the amount of nonproductive time and expense for all parties concerned.

I. *Preapplication Submission Documents:* The sponsor will submit the following information to the District Director to determine feasibility of the project and general conformance with FmHA or its successor agency under Public Law 103–354 policy:

A. *Environmental review requirements.* As requested by the Agency, the applicant is responsible for providing details of the project's potential impact on the human environment and historic properties, in accordance with 7 CFR part 1970. Guidance concerning the environmental review requirements is available at any Agency office or on the Agency's Web site.

B. *Location Map.* A general site location map of the area indicating the adjacent land zoning and uses, the present and future access roads to the site as well as the proximity to shopping, schools, churches, and major transportation facilities with note of traffic volumes. If a satisfactory map of the locality is not available, a clear and preferably scaled rough sketch map that provides the required information will be sufficient.

C. *Property Survey Map.* A current survey map of the project site showing the boundaries as well as all existing known features specifically including utilities, easements, access roads, floodplains, drainageways, rock outcroppings and wooded areas or specimen trees. If a current survey does not exist, the most accurate document which is available will be submitted.

D. *Soils Map and Report.* A complete soils map and report, including "site specific" interpretations and recommendations, from the local or county representative of the U.S. Department of Agriculture, Soil Conservation Service (SCS) Office will be included with the location and feasibility submission. A determination of whether or not any lands described in USDA Regulation 9500–3 are impacted by the proposed development should also be included. The local SCS office may provide recommendations for the development of suitable drainage and landscaping plans later in the planning process.

E. *Market survey.* A market survey will be submitted in accordance with the requirements of the respective loan program as indicated in part 1944, subparts D and E of this chapter.

F. *Request for Exceptions.* Any need for State or National Office exception(s) should be identified at this stage in the processing. Appropriate exception(s) should be requested and obtained before proceeding to the preliminary submission.

G. *Other.* The applicant will need to submit any additional information that may be needed as indicated in subpart D or exhibit A–7 of subpart E of part 1944 of this chapter. This may include but is not limited to:

1. Schematic design drawings showing the proposed plot plan, typical unit plans, and elevations. If available, the proposed preliminary drawings and specifications may be submitted. This would be of assistance if it is determined that the loan must receive National Office authorization.

2. Type of construction.

3. The total number of living units and the number of each type of living unit proposed.

4. Type of utilities such as water, sewer, gas, and electricity and whether each is public, community, or individually owned.

II. *Application Submission Documents:* After it is determined by FmHA or its successor agency under Public Law 103–354 that the project is feasible and the location conforms with the intent of the funding program, the sponsor will submit the following information to the District Director in addition to those materials submitted previously.

A. *Property Survey.* A survey (where 1 inch represents no more than 100 feet) of the property lot showing the exact boundaries and corners of the property accompanied by a written description of said boundaries. Also, locations of predominant features such as utilities, easements, access points, floodplains, drainageways, rock outcroppings and wooded areas or specimen trees affecting the proposed development must be included. This document shall bear the seal of a professional licensed to provide surveying services in the State in which the project will be located. This survey could be a part of item D below.

B. *Topographic Map.* An accurate topographic map showing existing and proposed contours with a scale compatible with the size of the project. The site shall be shown at a reasonable scale with 5-foot contour intervals. Where the site is unusually level or steep, the contour intervals may be varied accordingly.

C. *Preliminary Site Plan.* A line drawing, to scale, showing proposed street locations with profiles and widths, lot layouts, major drainageways, and other development planned. Preliminary sections and details shall be provided for the street construction, curbs and gutters, drainageways, and other physical improvements.

D. *Preliminary Dwelling Drawings and Specifications.* Drawings of the dwelling units, preliminary floor plans and specifications, elevations and sample site plans showing the placement of the individual buildings should be submitted.

E. *Statement of Planning and Zoning Compliance.* Local, county and State approvals as applicable. If change of zoning or variance is required, the status of the variance or change of zoning shall be documented.

F. *Technical Service Contracts.* Executed contracts for the professional services of an architect, engineer, land surveyor, landscape architect, site planner and/or soil engineer will be submitted as appropriate for the planning of the proposed development.

G. *Utility Approvals.* Statements of approval and feasibility for utility systems as follows:

1. Verification of adequate capacity and approval to tie-in with local existing water, wastewater disposal, electric, telephone, and other utility systems, as appropriate.

2. Tentative approval of local or State health authority for individual water and/or wastewater disposal systems when it is clear that central systems are unfeasible at this time. Use § 1924.108(a)(5) of this subpart when preparing information required.

H. *Facility Acceptance.* Evidence that the appropriate public body is willing to accept and maintain streets, common areas, lighting, fire hydrants, sidewalks, drainageways, and utilities, as appropriate, when dedicated to said body.

I. *Preliminary Specifications.* Outline specifications describing all the proposed materials to be used and how they are to be applied. These are only the materials used in the land development and construction of the streets, drainage, and utility work.

J. *Incremental Slopes Plan.* If areas of common slope are not identified elsewhere in adequate detail, this information should be provided in a separate plan.

K. *Preliminary Grading Plan.* This plan will indicate degree of work required to provide positive drainage of all building sites and control measures to be taken to eliminate soil erosion. Dwelling locations may be shown if they can be predetermined.

L. *Other.* The applicant will need to submit any additional information that may be needed as indicated in the respective loan program regulations as indicated in part 1944, subparts D and E and part 1822, subpart F of this chapter (FmHA or its successor agency under Public Law 103–354 Instruction 444.7). This may include but not be limited to:

1. A detailed trade-item cost breakdown of the project for such items as land and right-of-way, building construction, equipment, utility connections, architectural/engineering and legal fees, and both on- and off-site improvements. The cost breakdown also should show separately the items not included in the loan, such as furnishings and equipment. This trade-item cost breakdown should be updated just prior to loan approval.

2. Information on the method of construction, on the proposed contractor if a construction contract is to be negotiated and on the architectural, engineering, and legal services to be provided.

3. For all projects containing over four units the applicant will submit an Affirmative Fair Housing Marketing Plan for approval by FmHA or its successor agency under Public Law 103–354 in accordance with § 1901.203 of subpart E to part 1901 of this chapter. The Affirmative Fair Housing Marketing Plan must be prepared in a complete, meaningful, responsive and detailed manner.

4. A description and justification of any related facilities (including but not limited to workshops, community buildings, recreation center, central cooking and dining facilities, or other similar facilities to meet essential needs) to be financed wholly or in part with loan funds.

III. *Technical Documents Necessary for the Obligation of Funds.* All decisions regarding the conceptual design of the proposed project should be made prior to this submission. This effort is mainly to demonstrate that those agreed upon concepts have been transformed into construction documents and the necessary approvals have been granted. All items requiring revision or more detailed information as determined by the review of the preliminary submission will be resolved before the sponsor prepares the final submission. All documents shall be executed in a professional manner and shall carry the appropriate designation attesting to the professional qualifications of the architect, engineer, land surveyor or site planner. All documents will be accurately drawn at an appropriate scale.

[52 FR 19284, May 22, 1987, as amended at 56 FR 2202, Jan. 22, 1991; 81 FR 11030, Mar. 2, 2016]

Subparts D–E [Reserved]

Subpart F—Complaints and Compensation for Construction Defects

Source: 56 FR 40241, Aug. 14, 1991, unless otherwise noted.

§ 1924.251 Purpose.

This subpart contains policies and procedures for receiving and resolving complaints concerning the construction of dwellings and construction, installation and set-up of manufactured homes (herein called "units"), financed by the Rural Development, and for compensating borrowers for structural defects under section 509(c) of the Housing Act of 1949, as amended. Provisions of this subpart do not apply to dwellings financed with guaranteed section 502 loans.

§ 1924.252 Policy.

Rural Development is responsible for receiving and resolving all complaints concerning the construction of dwellings and the construction, installation and set-up of units financed by Rural Development. Rural Development must determine whether defects are structural or non-structural. If the defect is structural and is covered by the builder's/dealer-contractor's (the "contractor") warranty, the contractor is expected to correct the defect. If the contractor cannot or will not correct the defect, the costs of correcting the defect may be paid by the Government, or the borrower may be compensated for correcting the defect, under the provisions of this subpart. If the defect is non-structural but is covered under the provisions of the contractor's warranty or independent home warranty, the contractor is still expected to correct the defect. Rural Development will assist the borrower in obtaining assistance through the independent home warranty company's and/or manufacturer's complaint resolution process. However, if the contractor cannot or will not correct a non-structural defect covered under the provisions of the contractor's warranty, the Government will not pay the costs for correcting the defect, nor will the borrower be compensated for doing so.

§ 1924.253 Definitions.

As used in this subpart, the following definitions apply:

(a) *Newly constructed dwelling.* One which:

(1) Is financed with a section 502 insured loan;

(2) Was constructed substantially or wholly under the contract method, or under a conditional commitment, or, as to only work performed by a contractor or covered by a manufacturer's warranty, under the mutual self-help program;

(3) Was not more than one year old and not previously occupied as a residence at the time financial assistance was granted unless Rural Development has extended the conditional commitment issued on a newly constructed dwelling in accordance with 7 CFR part 3550; and

(4) Had the required construction inspections performed by Rural Development, the Department of Housing and Urban Development (HUD), or the Veterans Administration (VA).

(b) *Newly constructed manufactured home (unit).* One which:

(1) Is financed with a section 502 insured loan;

(2) Was not more than one year old and not previously occupied as a residence at the time financial assistance was granted; and

(3) Is built to the Federal Manufactured Home Construction and Safety Standards (FMHCSS) and is certified by an affixed label as shown in exhibit J of subpart A of part 1924 of this chapter.

(c) *Non-structural defect.* A construction defect which does not affect the overall useful life, habitability, or structural integrity of the dwelling or unit. Some non-structural defects may be covered under the contractor's warranty. Examples of non-structural defects include, but are not limited to:

(1) Cracks attributed to normal curing or settlement.

(2) Cosmetic defects in cabinets, woodwork, floorcovering, wallcovering, ornamental trim, etc.

(3) Improper or incomplete seeding or sodding of yard, or failure of trees, shrubs, grass and other landscaping items to thrive.

(4) Improper grading of yard, unless the grade is causing damage which may lead to a structural defect.

(d) *Structural defect.* A defect in the dwelling or unit, installation or set-up of a unit, or a related facility or a deficiency in the site or site development which directly and significantly reduces the useful life, habitability, or integrity of the dwelling or unit. The defect may be due to faulty material, poor workmanship, or latent causes that existed when the dwelling or unit was constructed. The term includes, but is not limited to:

(1) Structural failures which directly and significantly affect the basic integrity of the dwelling or unit such as in the foundation, footings, basement walls, slabs, floors, framing, walls, ceiling, or roof.

(2) Major deficiencies in the utility components of the dwelling or unit or

site such as faulty wiring, or failure of sewage disposal or water supply systems located on the property securing the loan caused by faulty materials or improper installation.

(3) Serious defects in or improper installation of heating systems or central air conditioning.

(4) Defects in or improper installation of safety and security devices, such as windows, external doors, locks, smoke detectors, railings, etc., as well as failure to provide or properly install devices to aid occupancy of dwellings by handicapped individuals, where required.

(5) Defects in or improper installation of protective materials, such as insulation, siding, roofing material, exterior paint, etc.

[56 FR 40241, Aug. 14, 1991, as amended at 67 FR 78327, Dec. 24, 2002]

§§ 1924.254–1924.257　[Reserved]

§ 1924.258　Notification of borrowers.

Rural Development will notify by letter all borrowers who receive Section 502 RH financial assistance for a newly constructed dwelling or unit of the provisions of this subpart. Subsequent owners of eligible dwellings will also be notified in accordance with this section. Borrowers will be notified within 30 days after the loan is closed, or within 30 days after final inspection, whichever is later. This notification will contain information concerning time frames for filing claims under this subpart. Rural Development will also notify and advise borrowers of the construction defects procedure at any time construction defects are apparent within the statutory time frame and favorable results cannot be obtained from the contractor. This notification will be documented in the borrower's case file.

§ 1924.259　Handling dwelling construction complaints.

This section describes the procedure for handling construction defect complaints.

(a) Each borrower who complains about construction defects will be requested to make a written complaint using a Rural Development approved format. All known defects will be listed. An oral complaint may be accepted if making a written complaint will impose a hardship on the borrower. If an oral complaint is made, Rural Development will notify the contractor on behalf of the borrower.

(b) The borrower will be informed that if, after 30 calendar days, the defects have not been corrected or other satisfactory arrangements made by the contractor, the borrower should notify Rural Development using a Rural Development approved format.

(c) Rural Development will advise the contractor in writing of the borrower's complaint, the time and date of planned inspection by Rural Development personnel, and request that the contractor accompany the inspector and borrower on a joint inspection of the property in an attempt to resolve the complaint.

(d) If, prior to the planned inspection, the contractor informs Rural Development that the alleged defect(s) has been or will be corrected within 30 calendar days, Rural Development will notify the borrower.

(e) If the case is not resolved as outlined in paragraph (d) of this section, Rural Development will:

(1) [Reserved]

(2) Notify the borrower, contractor and manufacturer, if applicable, in writing of the findings and who has been determined responsible for correcting the defect(s).

(i) If the defects are determined to be covered under the contractor's warranty, Rural Development will advise the contractor that the repairs must be completed within 30 calendar days or other time period agreed to by the borrower, the contractor, and Rural Development.

(ii) Rural Development will further advise the contractor and/or manufacturer that if the defect(s) are not corrected, the Government will consider compensating the borrower for the costs of correcting the defect(s). In such a case, the contractor and/or manufacturer may be liable for costs paid by the Government and may be subject to suspension and/or debarment pursuant to subpart M of part 1940 of this chapter (available in any Rural Development office). Even if the manufacturer is determined to be solely responsible for the defect, the contractor will still be held liable for correction of the defect.

(3) Should a contractor refuse to correct a defect after being officially requested in writing to do so, Rural Development will promptly institute formal suspension and debarment proceedings against the contractor (as a company and as individual(s)) in accordance with subpart M of part 1940 of this chapter (available in any Rural Development office). The contractor's failure to reply to official correspondence or inability to correct a defect constitutes noncompliance.

(4) If the contractor is willing to correct legitimate defects but the borrower refuses to permit this, Rural Development will document the facts in the borrower's case file. If the borrower chooses to file a claim for compensation for these defects, the circumstances of the borrower's refusal will be reviewed and may be sufficient grounds for disapproval of the claim.

(f)–(h) [Reserved]

§1924.260 Handling manufactured housing (unit) construction complaints.

When a borrower who has purchased a manufactured home (or "unit") complains about construction defects, the borrower will be instructed to first contact the dealer-contractor from whom the unit was purchased. Rural Development will assist the borrower in obtaining assistance through the dealer-contractor's and/or HUD's complaint resolution process. If the dealer-contractor cannot resolve the complaint, the borrower should contact the appropriate State Administrative Agency (SAA) or HUD. If the complaint resolution process does not result in the correction of the defect, the borrower's complaint will be handled in accordance with §1924.259 of this subpart.

§1924.261 Handling complaints involving dwellings covered by an independent or insured home warranty plan.

Borrowers with complaints about dwellings covered by an independent or insured home warranty plan will be instructed to first contact the warranty company and follow the complaint resolution process for that company, with the assistance of Rural Development, if needed. If the complaint is not resolved in this manner, it will be handled under §1924.259 of this subpart.

§1924.262 Handling complaints involving dwellings constructed by the self-help method.

When a borrower whose dwelling was constructed by the self-help method complains about construction defects, Rural Development will determine whether the defect is the result of work performed by a contractor or work performed by the borrower under the guidance of the self-help group. Defects which are determined to be the responsibility of a contractor will be handled in accordance with §1924.259 of this subpart. Defects determined to be the result of work performed by the borrower are not eligible for compensation under this subpart.

§§1924.263–1924.264 [Reserved]

§1924.265 Eligibility for compensation for construction defects.

(a) To be eligible for assistance under this subpart, the following criteria must be met:

(1) The approval official, in consultation with the State Architect/Engineer and/or Construction Inspector, must determine that:

(i) The construction is defective in workmanship, material or equipment, or

(ii) The dwelling or unit has not been built in substantial compliance with the approved drawings and specifications, or

(iii) The dwelling or unit does not comply with the Rural Development construction standards in effect at the time the loan was approved or the conditional commitment was issued, or

(iv) The property does not meet code requirements.

(2) The claim must be for one or more of the following:

(i) To pay for repairs;

(ii) To compensate the owner for repairs;

589

(iii) To pay emergency living or other expenses resulting from the defect; or

(iv) To acquire title to property.

(3) The dwelling or unit must be newly constructed as defined in § 1924.253 of this subpart and financed with an insured Section 502 RH loan.

(4) The claim seeking compensation from Rural Development must be filed with Rural Development within 18 months after the date financial assistance is granted. Defects for which claims are filed beyond the 18-month period must have been documented by Rural Development in the borrower's case file or on the form designated by Rural Development (available in any Rural Development office), prior to expiration of the 18-month period. For loans made to construct a new dwelling or erect a new manufactured housing unit, financial assistance is granted on the date of final construction inspection and acceptance by the borrower and Rural Development. Claims must be submitted by completing the designated form (available in any Rural Development office).

(5) Any obligation of the contractor to correct the defect(s) under a contractor's warranty must have expired, or the contractor is responsible for making corrections under the contractor's warranty but is unable or unwilling to do so.

(b) Subsequent owners of eligible dwellings or units who are also Section 502 borrowers may be eligible to receive compensation for construction defects. These owners will be notified in accordance with § 1924.258 of this subpart. However, the claim for compensation must be filed in accordance with paragraph (a)(4) of this section within the 18-month period established for the original rural housing (RH) borrower.

§ 1924.266 Purposes for which claims may be approved.

(a) *Eligible purposes.* A claim may be approved to:

(1) Pay, or reimburse the borrower for costs already paid, to repair major structural defects which are completed in accordance with plans and specifications approved by Rural Development. Repairs must be made by a reputable licensed contractor and a warranty covering the repairs will be issued by the contractor when the repairs are completed, as prescribed in subpart A of this part. Payment will be based on actual cost of the development and the borrower must provide evidence to reasonably establish the development cost. Workmanship and materials used in repairs must be consistent with the level of quality specified in the original dwelling or unit specifications and/or comparable to the items being replaced. Payment may be made:

(i) To cover damages which are a direct result of the defect to permanent enhancements made, such as landscaping, completion of unfinished living spaces, etc., of the dwelling or unit, installation or set-up of the unit, or related facilities, and

(ii) For costs approved by Rural Development for professional reports by engineers, architects or others needed to determine cause of or means to repair the defect.

(2) Reimburse the borrower for funds expended for emergency repairs. Emergency repairs are those repairs necessary to preserve the integrity of the structure, to prevent damage or further damage to personal property or fixtures in the dwelling or unit and related facilities, or to prevent or eliminate immediate health hazards. Receipts or other evidence of borrower's expenditures must be provided.

(3) Acquire title to the property by the Government and, when appropriate, compensate the claimant for any loss of borrower contribution at the time the loan was closed. Conveyance of properties under this section will be handled in accordance with 7 CFR part 3550.

(i) Before Rural Development accepts a conveyance, the borrower must attempt to sell the dwelling or unit in accordance with 7 CFR part 3550, if the dwelling or unit is considered decent, safe and sanitary as prescribed in 7 CFR part 3550. If the property is sold, Rural Development will:

(A) Pay the borrower's relocation expenses, including temporary living expenses as prescribed in paragraph (a)(4) of this section, until another suitable property can be located;

(B) Pay related sales expenses, as prescribed in 7 CFR part 3550, if the property is sold for less than the debt against it;

(C) Release the borrower from personal liability for the remaining Rural Development debt; and

(D) Process an application for a new RH loan if the borrower so desires and is still eligible for Rural Development assistance.

(ii) If the dwelling or unit is not considered decent, safe and sanitary as prescribed in 7 CFR part 3550, Rural Development should accept a voluntary conveyance of the property under the provisions of 7 CFR part 3550. Compensation for properties taken into inventory under this paragraph may not exceed the difference between the present market value of the security as established by the appraisal when the loan was made and the amount of the Rural Development loan and any prior liens.

(iii) A borrower contribution which may be compensated for under this paragraph may be such things as:

(A) A borrower's land or cash contribution,

(B) Development work done by the borrower under the self-help program or borrower method of construction, the cost of which was not included in the loan funds,

(C) Attorney fees, abstract costs or title insurance costs actually paid by the claimant in connection with closing the loan.

(4) Pay or reimburse the borrower for temporary living expenses, miscellaneous expenses, storage of household goods and moving expenses incurred as a result of the defect.

(i) Payment under this paragraph may be made under either of the following circumstances:

(A) The property is acquired by the Government in accordance with 7 CFR part 3550 and Rural Development determines that the dwelling is not habitable and the severity of the defect(s) prevents the property from being repaired and made suitable as a permanent residence for the borrower.

(B) The property is not acquired by the Government but Rural Development determines that the dwelling is not habitable or must be vacated in order to repair the defects.

(ii) Claims for compensation under paragraph (a)(4) of this section are limited as follows:

(A) Compensation may be granted for temporary living expenses for not more than 45 calendar days per claim unless a longer period is authorized by Rural Development. Compensation will be paid for actual cost to the claimant not to exceed the Government per diem rate for the area where the borrower's dwelling or unit is located. Reimbursement may be claimed for expenses such as food, lodging, laundering, etc., which would not have been incurred had the claimant remained in the house.

(B) Compensation may be granted for actual miscellaneous expenses not to exceed $500 to cover such items as utility connect and disconnect fees.

(C) Compensation may be granted for moving and storage expenses not to exceed $5,000 unless authorized by Rural Development and not to exceed the actual cost of moving the claimant household with personal belongings a distance of not more than 50 miles from the original residence. Compensation for storage expenses may not exceed that amount paid to store household furnishings for 45 days.

(D) A strict accounting of the use of such funds must be maintained by the borrower and will be verified by Rural Development.

(5) Compensate the claimant for reasonable interest paid on loans obtained for the sole purpose of correcting structural defects or other approved purposes under this section.

(b) *Ineligible purposes.* Compensation will not be granted for:

(1) Completion of a dwelling or unit or installation of materials/items required under the construction contract and/or specifications.

(2) Defective items which were not completed under the contract method or under the conditional commitment and supported by a builder's warranty. Work performed under the borrower method or self-help program without a warranty by a responsible party is not eligible for compensation.

(3) Damage caused by defective design, workmanship, or material in

making enhancements to or remodeling the dwelling or unit or related facilities which were not financed or approved by Rural Development.

(4) The loss of past, present or future wages or salary directly or indirectly resulting from the defect.

(5) Treatment for physical or psychological damages including medical and dental claims.

(6) Death benefits or funeral expenses.

(7) Damages encountered as a result of war, civil disorder, flood, tornado, lightning, earthquake or acts of nature which the structure was not designed to withstand.

(8) Damages resulting from the homeowner's negligence or failure to properly maintain the property.

(9) Damage to personal property.

[56 FR 40241, Aug. 14, 1991, as amended at 67 FR 78327, Dec. 24, 2002]

§§ 1924.267–1924.270 [Reserved]

§ 1924.271 Processing applications.

An application for compensation for construction defects shall be submitted by the claimant to Rural Development on the designated form. The application shall be completed in its entirety. All structural defects and claims for which compensation is sought will be listed. Borrowers will be told not to incur any expenses for repairs or temporary living expenses, except for emergency situations, until funds have been allocated and the request has been approved under § 1924.273 of this subpart.

§ 1924.272 [Reserved]

§ 1924.273 Approval or disapproval.

(a) Claimants will be notified in writing of the decision on the claim within 60 days of the date the designated form is signed by the borrower. If the claim or any part of the claim is denied at any level, the claimant will be informed in writing of the reason(s) for the denial and advised of appeal rights in accordance with 7 CFR part 11.

(b) [Reserved]

[56 FR 40241, Aug. 14, 1991, as amended at 67 FR 78327, Dec. 24, 2002]

§ 1924.274 Final inspection.

Except for emergency repairs, all repair work must be performed in accordance with subpart A of this part. In all cases, Rural Development will make a final inspection of the repair work performed before final payment is made for the work.

§ 1924.275 [Reserved]

§ 1924.276 Action against contractor.

If Rural Development pays for correction of construction defects which are the responsibility of the contractor, debarment proceedings will be initiated against the contractor in accordance with subpart M of part 1940 of this chapter (available in any Rural Development office), even if the contractor has gone out of business, declared bankruptcy, cannot be located, etc. The debarment will be pursued in both the contractor's company name and the principal parties as individuals, and any successor entities, if known. If the manufacturer of the defective product is determined to be solely responsible, no action will be taken against the contractor. In such a case, debarment will be initiated against the manufacturer. An assignment of the borrower's claim against the contractor or other party will be obtained if it appears to the approval officials, with any necessary advice from the Office of the General Counsel, that recovery is reasonably possible.

§§ 1924.277–1924.299 [Reserved]

§ 1924.300 OMB control number.

The reporting and recordkeeping requirements contained in this regulation have been approved by the Office of Management and Budget (OMB) and have been assigned OMB control number 0575–0082. Public reporting burden for this collection of information is estimated to vary from 15 minutes to 2 hours per response, with an average of .28 hours per response including time for reviewing instructions, searching existing data sources, gathering and maintaining the data needed, and completing and reviewing the collection of information. Send comments regarding this burden estimate or any other aspect of this collection of information, including suggestions for reducing this burden, to Department of Agriculture, Clearance Officer, OIRM, room 404–W, Washington, DC 20250; and to the Office of Management and Budget, Paperwork Reduction Project (OMB #575–0082), Washington, DC 20503.

PART 1925—TAXES

Subpart A—Real Estate Tax Servicing

Sec.
1925.1 General.
1925.2 Definition of tax.
1925.3 Servicing taxes.
1925.4 Servicing delinquent taxes.
1925.5–1925.50 [Reserved]

AUTHORITY: 5 U.S.C. 301; 7 U.S.C. 1989; 42 U.S.C. 1480.

SOURCE: 57 FR 36590, Aug. 14, 1992, unless otherwise noted.

EDITORIAL NOTE: Nomenclature changes to part 1925 appear at 80 FR 9876, Feb. 24, 2015.

Subpart A—Real Estate Tax Servicing

§1925.1 General.

This Instruction applies to borrowers with Rural Rental Housing (RRH), Rural Cooperative Housing (RCH), Labor Housing (LH),and Non-Program (NP) loans secured by real estate. It also applies to section 502 and section 504 Rural Housing borrowers (Single Family Housing (SFH)) who also have a Farmer Program loan. It does not apply to borrowers who have a SFH loan only; those will be serviced under 7 CFR part 3550. Borrowers are responsible for paying taxes on the real estate security to the proper taxing authorities before taxes become delinquent. This obligation is set forth in the security instrument securing the loan. This subpart is inapplicable to Farm Service Agency, Farm Loan Programs.

[57 FR 36590, Aug. 14, 1992, as amended at 67 FR 78327, Dec. 24, 2002; 72 FR 64122, Nov. 15, 2007]

§1925.2 Definition of tax.

For the purpose of this instruction, the word "tax" means all taxes, assessments, levies, irrigation and water charges or other similar obligations which are or will, on nonpayment, become a lien upon the real estate prior to the mortgage securing the Agency loan.

§1925.3 Servicing taxes.

(a) The County Supervisor will be responsible for ascertaining that all mortgaged real estate is listed properly for tax purposes.

(b) The County Supervisor will be responsible for taking all actions in connection with taxes as may be necessary to protect the Government's security interests. Any unusual situations that may arise with respect to tax servicing should be referred to the State Office for consideration.

(c) The County Supervisor will encourage each borrower to pay taxes promptly in order to avoid any penalties. Normally, this can be accomplished through routine servicing of loans by emphasizing the advantages of setting aside sufficient income to meet tax obligations when they become due. Taxes will be adequately budgeted for those borrowers with whom Form RD 431–2, "Farm and Home Plan," is developed. Each borrower will be encouraged to notify the County Supervisor when he has paid his taxes. After the delinquent date, it will be necessary for the County Supervisor to determine the borrowers whose taxes are delinquent. The Multi-Family Housing Information System (MFIS) will be used in posting servicing actions on delinquent taxes.

[57 FR 36590, Aug. 14, 1992, as amended at 69 FR 69104, Nov. 26, 2004]

§1925.4 Servicing delinquent taxes.

(a) The County Supervisor will contact each borrower with a delinquent tax and make every practical effort to have him pay the tax with his own funds. He will use the Management System Card for follow-up of delinquent taxes. If the delinquent tax is not paid and the borrower comes to the office with proceeds for application on the Agency account secured by the real estate, the County Office personnel will endeavor to get the borrower to use the proceeds to pay the delinquent tax. If the amount of the delinquent tax is less than the amount of the proposed payment, the difference will be applied on the Agency account in accordance with the policy outlined in subpart A of part 1951 of this chapter.

(b) Prior (usually about 90 days) to the time it is legally possible for action to be taken that will cause the borrower to lose title or right of possessions of the security property or the use of essential water, the County Supervisor will contact the borrower and definitely determine if he will pay the delinquent tax immediately. If the borrower is unable or unwilling to pay the delinquent tax with his own funds after every appropriate effort has been made to have him do so, the County Supervisor will refer to RD Instruction 2024–A and utilize the Type 60 Purchase Order System to pay the amount of the delinquent taxes plus the amount of any accrued penalty to bring taxes current.

(1) In an exceptional case where reasons for delinquent taxes have been removed and planned income during the next year covers payment of current obligations plus delinquent taxes not vouchered, only the delinquent taxes will be paid that could cause the borrower to lose title or right of possession of security property.

(2) If the Government is holding a mortgage other than a first mortgage on the property, do not initiate payment request until the County Supervisor has determined that (i) the prior lien holder will not pay the delinquent tax, (ii) the Government's security will be jeopardized if the delinquent tax is not paid, and (iii) the value of the security is sufficient to justify the advance.

[57 FR 36590, Aug. 14, 1992, as amended at 67 FR 78327, Dec. 24, 2002]

§§ 1925.5–1925.50 [Reserved]

PART 1927—TITLE CLEARANCE AND LOAN CLOSING

Subpart A [Reserved]

Subpart B—Real Estate Title Clearance and Loan Closing

AUTHORITY: 5 U.S.C. 301; 7 U.S.C. 1989; 42 U.S.C. 1480.

SOURCE: 61 FR 11711, Mar. 22, 1996, unless otherwise noted.

Subpart A [Reserved]

Subpart B—Real Estate Title Clearance and Loan Closing

§ 1927.51 General.

(a) *Types of loans covered by this subpart.* This subpart sets forth the authorities, policies, and procedures for real estate title clearance and closing of loans, assumptions, voluntary conveyances and credit sales in connection with the following types of Rural Housing Service (RHS), Rural Housing (RH), Farm Labor Housing (LH), Rural Rental Housing (RRH), Rural Cooperative Housing (RCH), and NonProgram (NP) loans. This subpart does not apply to guaranteed loans. This subpart is inapplicable to Farm Service Agency, Farm Loan Programs.

(b) *Programs not covered by this subpart.* Title clearance and closing for all other types of agency loans and assumptions will be handled as provided in the applicable program instructions or as provided in special authorizations from the National Office.

(c) [Reserved]

(d) Copies of all agency forms referenced in this regulation and the agency's internal administrative procedures for title clearance and loan closing are available upon request from the agency's State Office. Forms and title clearance and loan closing requirements which are specific for any individual state must be obtained from the agency State Office for that state.

[61 FR 11711, Mar. 22, 1996, as amended at 72 FR 64122, Nov. 15, 2007]

§ 1927.52 Definitions.

Agency. The Rural Housing Service (RHS) or its successor agencies.

Approval official. The agency employee who has been delegated the authority to approve, close, and service the particular kind of loan, will approve an attorney or title company as closing agent for the loans. If a loan must be approved at a higher level, the initiating office may approve the closing agent.

Approved attorney. A duly licensed attorney, approved by the agency, who provides title opinions directly to the agency and the borrower or upon whose certification of title an approved title insurance company issues a policy of title insurance. Approved attorneys also close loans, assumptions, credit sales, and voluntary conveyances and disburse funds in connection with agency loans. Approved attorney is further defined in §1927.54(c).

Approved title insurance company. A title insurance company, approved by the agency, (including its local representatives, employees, agents, and attorneys) that issues a policy of title insurance. Depending on the local practice, an approved title insurance company may also close loans, assumptions, credit sales, and voluntary conveyances and disburse funds in connection with agency loans. If the approved title insurance company does not close the loan itself, the loan closing functions may be performed by approved attorneys or closing agents authorized by the approved title insurance company.

Borrower. The party indebted to the agency after the loan, assumption, or credit sale is closed.

Certificate of title. A certified statement as to land ownership, based upon examination of record title.

Closed loan. A loan is considered to be closed when the mortgage is filed for record and the appropriate lien has been obtained.

Closing agent. The approved attorney or title company selected by the applicant and approved by the agency to provide closing services for the proposed loan. Unless a title insurance company also provides loan closing services, the term "title company" does not include "title insurance company."

Closing protection letter. An agreement issued by an approved title insurance company which is an American Land Title Association (ALTA) form closing protection letter or which is otherwise acceptable to the agency and which protects the agency against damage, loss, fraud, theft, or injury as a result of negligence by the issuing agent, approved attorney, or title company when title clearance is done by means of a policy of title insurance. Depending on the area, closing protection letters may also be known as "Insured Closing Letters," "Indemnification Agreements," "Insured Closing Service Agreements," or "Statements of Settlement Service Responsibilities."

Cosigner. A party who joins in the execution of a promissory note or assumption agreement to guarantee repayment of the debt.

Credit sale. A sale in which the agency provides credit to the purchasers of agency inventory property. Title clearance and closing of a credit sale are the same as for an initial loan except the property is conveyed by quitclaim deed.

Deed of trust. See trust deed.

Exceptions. Exceptions include, but are not limited to, recorded covenants; conditions; restrictions; reservations; liens; encumbrances; easements; taxes and assessments; rights-of-way; leases; mineral, oil, gas, and geothermal rights (with or without the right of surface entry); timber and water rights; judgments; pending court proceedings in Federal and State courts (including bankruptcy); probate proceedings; and agreements which limit or affect the title to the property.

Fee simple. An estate in land of which the owner has unqualified ownership and power of disposition.

General warranty deed. A deed containing express covenants by the grantor or seller as to good title and right to possession.

Indemnification agreement. An agreement that protects the agency against damage, loss, fraud, theft, or injury as a result of useful conduct or negligence on behalf of the issuing agent, approved attorney, or title company. This agreement may also be entitled closing protection letter, insured closing letter, insured closing service agreement, statement of settlement service responsibilities, or letters which provide similar protection.

Issuing agent. An individual or entity who is authorized to issue title insurance for an approved title insurance company.

Land purchase contract (contract for deed). An agreement between the buyer and seller of land in which the buyer has the right to possession and use of the land over a period of time (usually in excess of 1 year) and makes periodic payments of a portion of the purchase price to the seller. The seller retains legal title to the property until the final payment is made, at which time the buyer will receive a deed to the land vesting fee title in the buyer.

Mortgage. Real estate security instrument which pledges land as security for the performance of an obligation such as repayment of a loan. For the purpose of this regulation the term "mortgage" includes deed of trust and deed to secure debt. A real estate mortgage or deed of trust form for the state in which the land to be taken as security is available in any agency office, and will be used to secure a mortgage to the agency.

National Office. The National Headquarters Office of RHS.

OGC. The Office of the General Counsel, United States Department of Agriculture.

Program regulations. The agency regulations for the particular loan program involved (e.g., 7 CFR part 3550 for single family housing (SFH) loans).

Quitclaim deed. A transfer of the seller's interest in the title, without warranties or covenants. This type of deed is used by the agency to convey title to purchasers of inventory property.

RHS. The Rural Housing Service, an agency of the United States Department of Agriculture, or its successor agency.

Seller. Individual or other entity which convey ownership in real property to an applicant for an agency loan or to the agency itself.

Special warranty deed. A deed containing a covenant whereby the grantor agrees to protect the grantee against any claims arising during the grantor's period of ownership.

State Office. This term refers to the Rural Development State Director.

Title clearance. Examination of a title and its exceptions to assure the agency

that the loan is legally secured and has the required priority.

Title company. A company that may abstract title, act as an issuing agent of title insurance for a title insurance company, act as a loan closing agent, and perform other duties associated with real estate title clearance and loan closing.

Title defects. Any exception or legal claim of ownership (through deed, lien, judgment, or other recorded document), on behalf of a third party, which would prevent the seller from conveying a marketable title to the entire property.

Trust deed. A three party security instrument conveying title to land as security for the performance of an obligation, such as the repayment of a loan. For the purpose of this regulation a trust deed is covered by the term "mortgage." A trust deed is the same as a deed of trust.

Voluntary conveyance. A method of liquidation by which title to agency security is transferred by a borrower to the agency by deed in lieu of foreclosure.

Warranty deed. A deed in which the grantor warrants that he or she has the right to convey the property, the title is free from encumbrances, and the grantor shall take further action necessary to perfect or defend the title.

[61 FR 11711, Mar. 22, 1996, as amended at 67 FR 78327, Dec. 24, 2002; 80 FR 9876, Feb. 24, 2015]

§ 1927.53 Costs of title clearance and closing of transactions.

The borrower or the seller, or both, in compliance with the terms of the sales contract or option will be responsible for payment of all costs of title clearance and closing of the transaction and will arrange for payment before the transaction is closed. These costs will include any costs of abstracts of title, land surveys, attorney's fees, owner's and lender's policies of title insurance, obtaining curative material, notary fees, documentary stamps, recording costs, tax monitoring service, and other expenses necessary to complete the transaction.

§1927.54 Requirements for closing agents.

(a) *Form of title certification.* State Offices are directed to require title insurance for all loan closings unless the agency determines that the use of title insurance is not available or is economically not feasible for the type of loan involved or the area of the state where the loan will be closed. If title insurance is used, State Offices are authorized to require a closing protection letter issued by an approved title insurance company to cover the closing agent, if available. A closing protection letter need not be furnished when the closing is conducted by the title insurance company.

(b) *Approval of closing agent.* An attorney or title company may act as a closing agent and close agency real estate loans, provide necessary title clearance, and perform such other duties as required in this subpart. A closing agent will be responsible for closing agency loans and disbursing both agency loan funds and funds provided by the borrower in connection with the agency loan so as to obtain title and security position as required by the agency. The closing agent must be covered by a fidelity bond which will protect the agency unless a closing protection letter is provided to the agency. The borrower will select the approved closing agent. If title clearance is by an attorney's opinion, the agency will approve the attorney who will perform the closing in accordance with paragraph (c) of this section. The attorney will be approved after submitting a certification acceptable to the agency. If title certification is by means of a policy of title insurance, the title company which will issue the policy must have been approved in accordance with paragraph (d) of this section. A closing agent's delay in providing services without justification in connection with agency loans may be a basis for not approving the closing agent in future cases.

(c) *Approval of attorneys.* Any attorney selected by an applicant, who will be providing title clearance where the certificate of title will be an attorney's opinion, must submit an agency form certifying to professional liability insurance coverage. If the attorney is also the closing agent, fidelity coverage for the attorney and any employee having access to the funds must be provided. The agency will determine the appropriate level of such insurance. Required insurance will, as a minimum, cover the amount of the loan to be closed. The agency will approve the form stipulating the bond coverage. The agency will approve any attorney who is duly licensed to practice law in the state where the real estate security is located and who complies with the bonding and insurance requirements in this section. If the certification of title will be by means of title insurance, any attorney or closing agent designated as an approved attorney or closing agent by the approved title insurance company which will issue the policy of title insurance will be acceptable, and when covered by a closing protection letter, will not be required to obtain professional liability insurance or a fidelity bond. Each approved title insurance company may provide a master list of their approved attorneys that are covered by its closing protection letters to the State Office and, in such cases the attorneys are approved for closings for that title insurance company. Delay in providing closing services without justification may be a basis for not approving the attorney in future cases.

(d) *Approval of title companies.* A title company acting as a closing agent, or as an issuing agent for a title insurance company, must be covered by a title insurance company closing protection letter or submit an agency form certifying to fidelity coverage to cover all employees having access to the loan funds. The agency will determine the appropriate level of such coverage and will approve the form stipulating the bond coverage. Delay in providing closing services without justification may be a basis for not approving the company in future cases. Each approved title insurance company may provide a master list of their approved title companies that are covered by its closing protection letter to the State Office and, in such cases the title companies on the list are approved for closings for that title insurance company.

(e) *Approval of title insurance companies.* The agency will approve any title insurance company which issues policies of title insurance in the State where the security property is located if:

(1) The form of the owner's and lender's policies of title insurance (including required endorsements) to be used in closing agency loans are acceptable to the agency, and will contain only standard types of exceptions and exclusions approved in advance by the agency;

(2) The title insurance company is licensed to do business in the state (if a license is required); and

(3) The title insurance company is regulated by a State Insurance Commission, or similar regulator, or if not, the title insurance company submits copies of audited financial statements, or other approved financial statements satisfactory to the agency, which show that the company has the financial ability to cover losses arising out of its activities as a title insurance company and under any closing protection letters issued by the title insurance company.

(4) Delay in providing services without justification may be a basis for not approving the company.

(f) [Reserved]

(g) *Conflict of interest.* A closing agent who has, or whose spouse, children, or business associates have, a financial interest in the real estate which will secure the agency debt shall not be involved in the title clearance or loan closing process. Financial interest includes having either an equity, creditor, or debtor interest in any corporation, trust, or partnership with a financial interest in the real estate which will secure the agency debt.

(h) *Debarment or suspension.* No attorney, title company, title insurance company, or closing agent, currently debarred or suspended from participating in Federal programs, may participate in any aspect of the agency loan closing and title clearance process.

(i) *Special provisions.* Closing agents are responsible for having current knowledge of the requirements of State law in connection with loan closing and title clearance and should advise the agency of any changes in State law which necessitate changes in the agen-

cy's State mortgage forms and State Supplements.

(j) [Reserved]

§ 1927.55 Title clearance services.

(a) *Responsibilities of closing agents.* Services to be provided to the agency and the borrower by a closing agent in connection with the transaction vary depending on whether a title insurance policy or title opinion is being furnished. The closing agent is expected to perform these services without unnecessary delay.

(b) [Reserved]

(c) *Ordering title services.* Application for title examination or insurance will be made by the borrower to a title company or attorney. The lender's policy will be for at least the amount of the loan. The United States of America will be named as the insured lender.

(d) *Use of title opinion.* If a title opinion will be issued, a title examination will include searches of all relevant land title and other records, so as to express an opinion as to the title of the property and the steps necessary to obtain the appropriate title and security position to issue a title opinion as required by this subpart. The closing agent or approved attorney will determine:

(1) The legal description and all owners of the real property;

(2) Whether there are any exceptions affecting the property and advise the approval official and borrower of the nature and effect of outstanding interests or exceptions, prior sales of part of the property, judgments, or interests to assist in determining which exceptions must be corrected in order for the borrowers to obtain good and marketable title of record in accordance with prevailing title examination standards, and for the agency to obtain a valid lien of the required priority;

(3) Whether there are outstanding Federal, State, or local tax claims (including taxes which under State law may become a lien superior to a previously attaching mortgage lien) or homeowner's association assessment liens;

(4) Whether outstanding judgments of record, bankruptcy, insolvency, divorce, or probate proceedings involving

any part of the property, whether already owned by the borrower, or to be acquired by assumption or with loan funds, or involving the borrower or the seller exist;

(5) If a water right is to be included in the security for the loan, and if so, the full legal description of the water right;

(6) In addition to paragraph (d)(2) of this section, if wetlands easements or other conservation easements have been placed on the property;

(7) What measures are required for preparing, obtaining, or approving curative material, conveyances, and security instruments, and

(8) That sufficient copies of these interests and exceptions are provided as requested by the approval official.

(e) *Use of title insurance.* When title insurance is to be obtained, the approval official will be furnished with a title insurance binder disclosing any defects in, exceptions to, and encumbrances against, the title, the conditions to be met to make the title insurable and in the condition required by the agency, and the curative or other actions to be taken before closing of the transaction. The binder must include a commitment to issue a lender policy in an amount at least equal the amount of the loan, except in instances where there may be an outstanding owner's policy in favor of the borrower. Not withstanding the provisions of this section, the instance of an assumption without a subsequent loan, the existing policy may be continued if the coverage meets or exceeds the assumption balance and the title company agrees in writing to extend coverage in full force and effect.

(f) [Reserved]

§ 1927.56 Scheduling loan closing.

The agency, in coordination with the closing agent, will arrange a loan closing and send loan closing instructions, on an agency form to the closing agent when the agency determines that the exceptions shown on the preliminary title opinion or title insurance binder will not adversely affect the suitability, security value, or successful operation of the property and all other agency conditions to closing have been satisfied.

§ 1927.57 Preparation of closing documents.

(a) *Preparation of deeds.* The closing agent, unless prohibited by law, will prepare, complete, or approve documents, including deeds, necessary for title clearance and closing of the transaction and provide the agency with the policy of title insurance or title opinion providing the lien priority required by the agency and subject only to exceptions approved by the agency. Agency forms will be used when required by this part.

(1)–(2) [Reserved]

(b) *Preparation of mortgages.* The closing agent will insure that all mortgages are properly prepared, completed, executed, and filed for record. Where applicable, the mortgages should recite that it is a purchase money mortgage. The following requirements will be observed in preparing agency morgages:

(1)–(8) [Reserved]

(9) *Alteration of mortgage form.* An agency mortgage form may be altered pursuant to a State Supplement having prior approval of the National Office, or in a special case, to comply with the terms of loan approval prescribed in accordance with program instructions. No other alterations in the printed mortgage forms will be made without prior approval of the National Office. Any changes made by deletion, substitution, or addition (excluding filling in blanks) will be initialed in the margin by all persons signing the mortgage.

(10) [Reserved]

(11) *Mortgages on leasehold estates.* When the agency security interest is a leasehold estate, unless State law or State Supplement otherwise provides, the real estate mortgage or deed of trust form, available in any agency office, will be modified as follows:

(i) In the space provided on the mortgage for the description of the real property security, the leasehold estate and the land covered by the lease must be described. The following language must be used unless modified by a State Supplement:

All of borrower's right, title, and interest in and to a leasehold estate for an original term of ____ years, commencing on ____, 19 ____, created and established by and between ____ as lessor and owner and ____ as lessee,

599

including any extensions and renewals thereof, a copy of which lease was recorded or filed in book ___, page ___, as instrument number ___, in the Office of the (e.g., County Clerk), for the aforesaid county and State and covering the following real property: ___.

(ii) Immediately preceding the covenant starting with the words "should default," the following covenant will be added:

( ) Borrower covenants and agrees to pay when due all rents and any and all other charges required by said lease, to comply with all other requirements of said lease, and not to surrender or relinquish, without the Government's prior written consent, any of borrower's right, title, or interest in or to said leasehold estate or under said lease while this mortgage remains of record.

(12) *Mortgages on land purchase contract.* When the agency security interest is on a borrower's interest in a land purchase contract, OGC will provide language used to modify agency forms.

(13) [Reserved]

(c) [Reserved]

(d) *Preparation of protective instruments.* The closing agent will properly prepare, complete, and approve releases and curative documents necessary for title clearance and closing, in recordable form and record them if required.

(1) *Prior lienholder's agreement.* If any liens (other than agency liens or tax liens to local governmental authorities) or security agreements (hereafter called "liens"), with priority over the agency mortgage will remain against the real property securing the loan, the lienholders must execute, in recordable form, agreements containing all of the following provisions unless prior approval for different provisions has been obtained from the National Office:

(i) The prior lienholder shall agree not to declare the lien in default or accelerate the indebtedness secured by the prior lien for a specific period of time after notice to the agency. The agreement must:

(A) Provide that the specified period of time will not commence until the lienholder gives written notice of the borrower's default and the prior lienholder's intention to accelerate the indebtedness to the agency office servicing the loan,

(B) Include the address of the agency servicing office,

(C) Give the agency the option to cure any monetary default by paying the amount of the borrower's delinquent payments to the prior lienholder, or pay the obligation in full and have the lien assigned to the agency, and

(D) Provide that the prior lienholder will not declare the lien in default for any nonmonetary reason if the agency commences liquidation proceedings against the property and thereafter acquires the property.

(ii) When the prior lien secures future advances, including the lienholder's costs for borrower liquidation or bankruptcy, which under State law have priority over the mortgage being taken (or an agency mortgage already held), the prior lienholder shall agree not to make advances for purposes other than taxes, insurance or payments on other prior liens without written consent of the agency.

(iii) The prior lienholder shall consent to the agency making (or transferring) the loan and taking (or retaining) the related mortgage if the prior lien instrument prohibits a loan or mortgage (or transfer) without the prior lienholder's consent.

(iv) The prior lienholder shall consent to the agency transferring the property subject to the prior lien after the agency has obtained title to the property either by foreclosure or voluntary conveyance if the prior lien instrument prohibits such transfer without the prior lienholder's consent.

(2)–(3) [Reserved]

(4) *Agreement by holder of seller's interest under land purchase contract.* If the buyer's interest in the security property is that of a buyer under a land purchase contract, it will be necessary for the seller to execute, in recordable form, an agreement containing all of the following provisions:

(i) The seller shall agree not to sell or voluntarily transfer the seller's interest under the land purchase contract without the prior written consent of the State Office.

(ii) The seller shall agree not to encumber or cause any liens to be levied against the property.

(iii) The seller shall agree not to commence or take any action to accelerate, forfeit, or foreclose the buyer's interest in the security property until a specified period of time after notifying the State Office of intent to do so. This period of time will be 90 days unless a State Supplement provides otherwise. The agreement shall give the agency the option to cure any monetary default by paying the amount of the buyer's delinquent payments to the seller, or paying the seller in full and having the contract assigned to the agency.

(iv) The seller shall consent to the agency making the loan and taking a security interest in the borrower's interest under the land purchase contract as security for the agency loan.

(v) The seller shall agree not to take any actions to foreclose or forfeit the interest of the buyer under the land purchase contract because the agency has acquired the buyer's interest under the land purchase contract by foreclosure or voluntary conveyance, or because the agency has subsequently sold or assigned the buyer's interest to a third party who will assume the buyer's obligations under the land purchase contract.

(vi) When the agency acquires a buyer's interest under a land purchase contract by foreclosure or deed in lieu of foreclosure, the agency will not be deemed to have assumed any of the buyer's obligations under the contract, provided that the failure of the agency to perform any such obligations while it holds the buyer's interest is a ground to commence an action to terminate the land purchase contract.

(5)–(6) [Reserved]

(e) [Reserved]

§ 1927.58 Closing the transaction.

The closing agent will cooperate with the approval official, borrower, seller, and other necessary parties to arrange the time and place of closing. The transaction may be closed when the agency determines that the agency requirements for the loan have been satisfied and the closing agent or approved attorney can issue or cause to be issued a policy of title insurance or final title opinion as of the date of closing showing title vested as required by the agency, the lien of the agency's mortgage in the priority required by the agency, and title to the mortgaged property subject only to those exceptions approved in writing by the agency. The loan will be considered closed when the mortgage is filed for record and the required lien is obtained.

(a) *Disbursement of loan funds.* When the closing agent indicates that the conditions necessary to close the loan have been met, loan funds will be forwarded to the closing agent. Loan funds will not be disbursed prior to filing of the mortgage for record; however, when necessary, loan funds may be placed in escrow before the mortgage is filed for record and disbursed after it is filed. No development funds will be kept in escrow by the closing agent after loan closing, unless approved by the agency. Loan funds for the payment of a lien may be disbursed only upon the recording of a discharge, satisfaction, or release of prior lien interests (or assignment where necessary to protect the interests of the agency).

(b) *Title examination and liens or claims against borrowers.* If there are exceptions or recorded items which have arisen since the preliminary title opinion, the transaction will not be closed until these entries have been cleared of record or approved by the agency. The closing agent will advise the approval official of the nature of such intervening instruments and the effect they may have on obtaining a valid mortgage of the priority required or the title insurance policy to be issued.

(c) *Taxes and assessments.* The closing agent will determine if all taxes and assessments against the property which are due and payable are paid at or before the time of loan closing. If the seller and the borrower have agreed to prorate any taxes or assessments which are not yet due and payable for the year in which the closing of the transaction takes place, the seller's proportionate share of the taxes and assessments will be deducted from the proceeds to be paid to seller at closing and will be added to the amount required to be paid by borrower at closing. Appropriate prorations as agreed upon between the borrower and seller may also be made for taxes paid by the seller which are applicable to a period after the closing date, and for common area maintenance fees, prepaid rentals, insurance (unless the borrower is to obtain a new policy of insurance), and growing crops.

(d) *Affidavit regarding work of improvement*—(1) *Execution by borrower.* If required by State Supplement, the closing agent will require that an affidavit regarding work of improvement, provided by the agency, be completed and executed when a loan is being made to a borrower who already owns the real estate to be mortgaged. This affidavit will be executed by the borrower at closing.

(2) *Execution by seller.* If required by State Supplement, the closing agent will require that an affidavit regarding work of improvement, provided by the agency, be completed and executed (including acknowledgment) by the seller when the agency is making a loan to a borrower to enable the borrower to acquire the property (including transfers). This affidavit will be executed by the seller at closing.

(3) *Legal insufficiency of affidavit form.* If the agency affidavit regarding work of improvement is not legally sufficient in a particular State, a State form approved by OGC will be used. A similar form that may be required by a title insurance company may be substituted for the agency form.

(4) *Recording.* The affidavit will not be recorded unless the closing agent deems it necessary and State law permits.

(5) *Delay in closing.* The loan will not be closed if, at the loan closing, the seller (in a sale transaction) or the borrower (in a nonpurchase money loan situation) indicates that construction, repair, or remodeling has been commenced or completed on the property, or related materials or services have been delivered to or performed on the property within the time limit specified in the affidavit, unless a State Supplement provides otherwise. The closing agent will notify the approval official, who will determine if the work of improvement could result in a lien prior to the agency lien. The State Office will, with the advice and concurrence of OGC, provide in a State Supplement the period of time to be used in completing the affidavit.

(e)–(f) [Reserved]

(g) *Return of loan documents to approval official after loan closing.* Within 1 day after loan closing, the closing agent will return completed and executed copies of the loan closing instructions, the executed original promissory note, and all other documents required for loan closing (except the mortgage), to the approval official. If the recorded mortgage is customarily returned to the borrower or closing agent after recording, then it must be forwarded to the approval official immediately.

(h) *Final title opinion or title insurance policy.* As soon as possible after the transaction has been closed.

(1) *Final title opinion.* The attorney will issue a final title opinion to the agency and the borrower on a form provided by the agency. Issuance of the final title opinion should not be held up pending the return of recorded instruments. If it is not possible for the final title opinion to show the book and page of recording of the agency security instrument, the words "and is recorded" in the final title opinion form provided by the agency office, may be deleted and the blank space completed to show the filing office and the filing instrument number, if available. Attached to the final title opinion will be required documents then available, including any which the approval official has furnished to the attorney which were not previously returned. The attorney will ensure that all recorded instruments are forwarded or delivered to the proper parties after recording. The certification of title will be forwarded for a voluntary conveyance.

(2) *Title insurance policy.* The closing agent will send or deliver the title insurance policy, with the United States listed as mortgage holder, to the approval official. The policy will be subject only to standard exceptions and those outstanding encumbrances, and exceptions, approved by the approval official. If an owner's policy of title insurance is requested, the closing agent will send or deliver it to the borrower. The closing agent will ensure that all recorded instruments are delivered or sent to the proper parties after recording.

(3) [Reserved]

(i) *Other services of the closing agent.* (1) The closing agent will assist the approval official in preparing, completing, obtaining execution and acknowledgment, and recording the required documents when necessary. The closing agent will keep the approval official advised as to the progress of title clearance and preparation of material for closing the transaction.

(2) The closing agent will provide services for deeds in lieu of foreclosure as set forth in §1927.62 of this subpart, and §1955.10 of subpart A of part 1955 of this chapter.

§1927.59 Subsequent loans and transfers with assumptions.

Title services and closing for subsequent loans to an existing borrower will be done in accordance with previous instructions in this subpart, except that:

(a) *Loans closed using title insurance or title opinions.* (1) Title insurance or title opinions will be obtained unless:

(i) The cost of title services is excessive in relationship to the size of the loan,

(ii) The agency currently has a first mortgage security interest,

(iii) The applicant has sufficient income to service the additional loan,

(iv) The borrower is current on the existing agency loan, and

(v) The best mortgage obtainable adequately protects the agency security interests.

(2) Title insurance or a final title opinion will not be obtained for a subsequent Section 504 loan where the previous Section 504 loan was unsecured or secured for less than $7,500 and the outstanding debt amount plus the new loan is less than $7,500.

(3) Loans closed using a new lender title insurance policy:

(i) Will cover the entire real property which is to secure the loan, including the real property already owned and any additional real property being acquired by the borrower with the loan proceeds.

(ii) Will cover the entire amount of any subsequent loan plus the amount of any existing loan being refinanced (if the existing loan is not being refinanced, the new lender policy will insure only the amount of the subsequent loan).

(b) *Title services required in connection with assumptions.* These regulations are contained in subparts A and B of part 1965 of this chapter and 7 CFR part 3550 as appropriate for the loan type.

[61 FR 11711, Mar. 22, 1996, as amended at 67 FR 78327, Dec. 24, 2002]

§§1927.60–1927.99 [Reserved]

§1927.100 OMB control number.

The reporting requirements contained in this regulation have been approved by the Office of Management and Budget and have been assigned OMB control number 0575–0147. Public reporting burden for this collection of information is estimated to vary from 5 minutes to 1.5 hours per response, with an average of .38 hours per response, including time for reviewing instructions, searching existing data sources, gathering and maintaining the data needed, and completing and reviewing the collection of information. Send comments regarding this burden estimate or any other aspect of this collection of information, including suggestions for reducing this burden, to Department of Agriculture, Clearance Officer, OIRM, Ag Box 7630, Washington, DC 20250; and to the Office of Management and Budget, Paperwork Reduction Project (OMB #0575–0147), Washington, DC 20503. You are not required to respond to the collection of information unless it displays a currently valid OMB control number.

PART 1930—GENERAL

AUTHORITY: 5 U.S.C. 301; 7 U.S.C. 1989; 16 U.S.C. 1005.

Subparts A–C [Reserved]

PARTS 1931–1939 [RESERVED]

FINDING AIDS

A list of CFR titles, subtitles, chapters, subchapters and parts and an alphabetical list of agencies publishing in the CFR are included in the CFR Index and Finding Aids volume to the Code of Federal Regulations which is published separately and revised annually.

Table of CFR Titles and Chapters
Alphabetical List of Agencies Appearing in the CFR
List of CFR Sections Affected

Table of CFR Titles and Chapters

(Revised as of January 1, 2020)

Title 1—General Provisions

Title 2—Grants and Agreements

Title 2—Grants and Agreements—Continued

Title 3—The President

Title 4—Accounts

Title 5—Administrative Personnel

Title 6—Domestic Security

Title 7—Agriculture

Title 7—Agriculture—Continued

611

612

Title 12—Banks and Banking—Continued

Title 13—Business Credit and Assistance

Title 14—Aeronautics and Space

Title 15—Commerce and Foreign Trade

Title 15—Commerce and Foreign Trade—Continued

Title 16—Commercial Practices

Title 17—Commodity and Securities Exchanges

Title 18—Conservation of Power and Water Resources

Title 19—Customs Duties

Title 20—Employees' Benefits

615

Title 23—Highways—Continued

Title 24—Housing and Urban Development

Title 25—Indians

Title 26—Internal Revenue

Title 27—Alcohol, Tobacco Products and Firearms

Title 28—Judicial Administration

Title 29—Labor

Title 29—Labor—Continued

Title 30—Mineral Resources

Title 31—Money and Finance: Treasury

Title 31—Money and Finance: Treasury—Continued

Title 32—National Defense

Title 33—Navigation and Navigable Waters

Title 34—Education

Title 34—Education—Continued

Title 35 [Reserved]

Title 36—Parks, Forests, and Public Property

Title 37—Patents, Trademarks, and Copyrights

Title 38—Pensions, Bonuses, and Veterans' Relief

Title 39—Postal Service

Title 40—Protection of Environment

Title 41—Public Contracts and Property Management

Title 42—Public Health

Title 43—Public Lands: Interior

Title 44—Emergency Management and Assistance

Title 45—Public Welfare

Title 47—Telecommunication—Continued

Title 48—Federal Acquisition Regulations System

Title 49—Transportation

Title 50—Wildlife and Fisheries

Title 50—Wildlife and Fisheries—Continued

VI   Fishery Conservation and Management, National Oceanic and Atmospheric Administration, Department of Commerce (Parts 600—699)

Alphabetical List of Agencies Appearing in the CFR

(Revised as of January 1, 2020)

Agency	CFR Title, Subtitle or Chapter
Administrative Conference of the United States	1, III
Advisory Council on Historic Preservation	36, VIII
Advocacy and Outreach, Office of	7, XXV
Afghanistan Reconstruction, Special Inspector General for	5, LXXXIII
African Development Foundation	22, XV
Federal Acquisition Regulation	48, 57
Agency for International Development	2, VII; 22, II
Federal Acquisition Regulation	48, 7
Agricultural Marketing Service	7, I, VIII, IX, X, XI; 9, II
Agricultural Research Service	7, V
Agriculture, Department of	2, IV; 5, LXXIII
Advocacy and Outreach, Office of	7, XXV
Agricultural Marketing Service	7, I, VIII, IX, X, XI; 9, II
Agricultural Research Service	7, V
Animal and Plant Health Inspection Service	7, III; 9, I
Chief Financial Officer, Office of	7, XXX
Commodity Credit Corporation	7, XIV
Economic Research Service	7, XXXVII
Energy Policy and New Uses, Office of	2, IX; 7, XXIX
Environmental Quality, Office of	7, XXXI
Farm Service Agency	7, VII, XVIII
Federal Acquisition Regulation	48, 4
Federal Crop Insurance Corporation	7, IV
Food and Nutrition Service	7, II
Food Safety and Inspection Service	9, III
Foreign Agricultural Service	7, XV
Forest Service	36, II
Information Resources Management, Office of	7, XXVII
Inspector General, Office of	7, XXVI
National Agricultural Library	7, XLI
National Agricultural Statistics Service	7, XXXVI
National Institute of Food and Agriculture	7, XXXIV
Natural Resources Conservation Service	7, VI
Operations, Office of	7, XXVIII
Procurement and Property Management, Office of	7, XXXII
Rural Business-Cooperative Service	7, XVIII, XLII
Rural Development Administration	7, XLII
Rural Housing Service	7, XVIII, XXXV
Rural Utilities Service	7, XVII, XVIII, XLII
Secretary of Agriculture, Office of	7, Subtitle A
Transportation, Office of	7, XXXIII
World Agricultural Outlook Board	7, XXXVIII
Air Force, Department of	32, VII
Federal Acquisition Regulation Supplement	48, 53
Air Transportation Stabilization Board	14, VI
Alcohol and Tobacco Tax and Trade Bureau	27, I
Alcohol, Tobacco, Firearms, and Explosives, Bureau of	27, II
AMTRAK	49, VII
American Battle Monuments Commission	36, IV
American Indians, Office of the Special Trustee	25, VII
Animal and Plant Health Inspection Service	7, III; 9, I
Appalachian Regional Commission	5, IX
Architectural and Transportation Barriers Compliance Board	36, XI

Agency	CFR Title, Subtitle or Chapter
National Technical Information Service	15, XI
National Telecommunications and Information Administration	15, XXIII; 47, III, IV, V
National Transportation Safety Board	49, VIII
Natural Resources Conservation Service	7, VI
Natural Resource Revenue, Office of	30, XII
Navajo and Hopi Indian Relocation, Office of	25, IV
Navy, Department of	32, VI
Federal Acquisition Regulation	48, 52
Neighborhood Reinvestment Corporation	24, XXV
Northeast Interstate Low-Level Radioactive Waste Commission	10, XVIII
Nuclear Regulatory Commission	2, XX; 5, XLVIII; 10, I
Federal Acquisition Regulation	48, 20
Occupational Safety and Health Administration	29, XVII
Occupational Safety and Health Review Commission	29, XX
Ocean Energy Management, Bureau of	30, V
Oklahoma City National Memorial Trust	36, XV
Operations Office	7, XXVIII
Patent and Trademark Office, United States	37, I
Payment From a Non-Federal Source for Travel Expenses	41, 304
Payment of Expenses Connected With the Death of Certain Employees	41, 303
Peace Corps	2, XXXVII; 22, III
Pennsylvania Avenue Development Corporation	36, IX
Pension Benefit Guaranty Corporation	29, XL
Personnel Management, Office of	5, I, XXXV; 5, IV; 45, VIII
Human Resources Management and Labor Relations Systems, Department of Homeland Security	5, XCVII
Federal Acquisition Regulation	48, 17
Federal Employees Group Life Insurance Federal Acquisition Regulation	48, 21
Federal Employees Health Benefits Acquisition Regulation	48, 16
Pipeline and Hazardous Materials Safety Administration	49, I
Postal Regulatory Commission	5, XLVI; 39, III
Postal Service, United States	5, LX; 39, I
Postsecondary Education, Office of	34, VI
President's Commission on White House Fellowships	1, IV
Presidential Documents	3
Presidio Trust	36, X
Prisons, Bureau of	28, V
Privacy and Civil Liberties Oversight Board	6, X
Procurement and Property Management, Office of	7, XXXII
Public Contracts, Department of Labor	41, 50
Public and Indian Housing, Office of Assistant Secretary for	24, IX
Public Health Service	42, I
Railroad Retirement Board	20, II
Reclamation, Bureau of	43, I
Refugee Resettlement, Office of	45, IV
Relocation Allowances	41, 302
Research and Innovative Technology Administration	49, XI
Rural Business-Cooperative Service	7, XVIII, XLII
Rural Development Administration	7, XLII
Rural Housing Service	7, XVIII, XXXV
Rural Utilities Service	7, XVII, XVIII, XLII
Safety and Environmental Enforcement, Bureau of	30, II
Saint Lawrence Seaway Development Corporation	33, IV
Science and Technology Policy, Office of	32, XXIV
Science and Technology Policy, Office of, and National Security Council	47, II
Secret Service	31, IV
Securities and Exchange Commission	5, XXXIV; 17, II
Selective Service System	32, XVI
Small Business Administration	2, XXVII; 13, I
Smithsonian Institution	36, V
Social Security Administration	2, XXIII; 20, III; 48, 23

Agency	CFR Title, Subtitle or Chapter
Soldiers' and Airmen's Home, United States	5, XI
Special Counsel, Office of	5, VIII
Special Education and Rehabilitative Services, Office of	34, III
State, Department of	2, VI; 22, I; 28, XI
Federal Acquisition Regulation	48, 6
Surface Mining Reclamation and Enforcement, Office of	30, VII
Surface Transportation Board	49, X
Susquehanna River Basin Commission	18, VIII
Tennessee Valley Authority	5, LXIX; 18, XIII
Trade Representative, United States, Office of	15, XX
Transportation, Department of	2, XII; 5, L
Commercial Space Transportation	14, III
Emergency Management and Assistance	44, IV
Federal Acquisition Regulation	48, 12
Federal Aviation Administration	14, I
Federal Highway Administration	23, I, II
Federal Motor Carrier Safety Administration	49, III
Federal Railroad Administration	49, II
Federal Transit Administration	49, VI
Maritime Administration	46, II
National Highway Traffic Safety Administration	23, II, III; 47, IV; 49, V
Pipeline and Hazardous Materials Safety Administration	49, I
Saint Lawrence Seaway Development Corporation	33, IV
Secretary of Transportation, Office of	14, II; 49, Subtitle A
Transportation Statistics Bureau	49, XI
Transportation, Office of	7, XXXIII
Transportation Security Administration	49, XII
Transportation Statistics Bureau	49, XI
Travel Allowances, Temporary Duty (TDY)	41, 301
Treasury, Department of the	2, X;5, XXI; 12, XV; 17, IV; 31, IX
Alcohol and Tobacco Tax and Trade Bureau	27, I
Community Development Financial Institutions Fund	12, XVIII
Comptroller of the Currency	12, I
Customs and Border Protection	19, I
Engraving and Printing, Bureau of	31, VI
Federal Acquisition Regulation	48, 10
Federal Claims Collection Standards	31, IX
Federal Law Enforcement Training Center	31, VII
Financial Crimes Enforcement Network	31, X
Fiscal Service	31, II
Foreign Assets Control, Office of	31, V
Internal Revenue Service	26, I
Investment Security, Office of	31, VIII
Monetary Offices	31, I
Secret Service	31, IV
Secretary of the Treasury, Office of	31, Subtitle A
Truman, Harry S. Scholarship Foundation	45, XVIII
United States and Canada, International Joint Commission	22, IV
United States and Mexico, International Boundary and Water Commission, United States Section	22, XI
U.S. Copyright Office	37, II
Utah Reclamation Mitigation and Conservation Commission	43, III
Veterans Affairs, Department of	2, VIII; 38, I
Federal Acquisition Regulation	48, 8
Veterans' Employment and Training Service, Office of the Assistant Secretary for	41, 61; 20, IX
Vice President of the United States, Office of	32, XXVIII
Wage and Hour Division	29, V
Water Resources Council	18, VI
Workers' Compensation Programs, Office of	20, I, VII
World Agricultural Outlook Board	7, XXXVIII

List of CFR Sections Affected

All changes in this volume of the Code of Federal Regulations (CFR) that were made by documents published in the FEDERAL REGISTER since January 1, 2015 are enumerated in the following list. Entries indicate the nature of the changes effected. Page numbers refer to FEDERAL REGISTER pages. The user should consult the entries for chapters, parts and subparts as well as sections for revisions.

For changes to this volume of the CFR prior to this listing, consult the annual edition of the monthly List of CFR Sections Affected (LSA). The LSA is available at *www.govinfo.gov*. For changes to this volume of the CFR prior to 2001, see the "List of CFR Sections Affected, 1949–1963, 1964–1972, 1973–1985, and 1986–2000" published in 11 separate volumes. The "List of CFR Sections Affected 1986–2000" is available at *www.govinfo.gov*.